Elements of ENGINEERING MATHEMATICS–I

(For Engineering Students)

Dr. V. P. Mishra

Formerly, Head, Department of Applied Mathematics,
Delhi College of Engineering, Delhi.
Presently with M.A.I.T, (G.G.S. Indraprastha University, Delhi)

Dr. Jyoti Sinha

Associate Professor
Department of Applied Mathematics,
Indraprastha College of Engineering (M.M.T.U.)
Ghaziabad, (U.P.)

S.K. KATARIA & SONS®

Publisher of Engineering Books
4885/109, Prakash Mahal, Dr. Subhash Bhargav Lane,
Opposite Delhi Medical Association, Daryaganj, New Delhi–110002 (INDIA)
Phone: +91-11-43551243 Mobile: +91-9871775858, 7678425209
e-mail: katariabook@gmail.com, katariabooks@yahoo.com
Website: www.skkatariaandsons.com
(e-books, subscriptions and bulk buying also available on our website)

Published by:

S.K. Kataria & Sons®
4885/109, Prakash Mahal, Dr. Subhash Bhargav Lane,
Opposite Delhi Medical Association, Daryaganj, New Delhi-110002 (INDIA)
Phone: +91-11-43551243, Mobile: +91-9871775858, 7678425209
e-mail: katariabook@gmail.com, katariabooks@yahoo.com
Website: www.skkatariaandsons.com

First Edition: July 2009; Third Edition: 2012; Reprint: 2026

Price : **₹ 695/-**

Laser Type Setting by:
SARA ASSIGNMENTS, Delhi

Printed at:
A.S. OFFSET PRINTERS, Delhi

Preface to the Third Edition

This book has been written to serve the mathematical needs of the students engaged in a first course in engineering at B.Tech. level. I caters to the need of the students of every engineering disciplines. Actually mathematics is the language of engineering these days. The book precisely conforms to the Mathematics–I B.Tech. syllabi of GBTU, Lucknow – MTU, Noida and UTU, Dehradun.

Our endeavour throughout has been to make clear even the less easy mathematical concepts and for quite a large number of problems of various nature has been solved. In essence the book encourages the integration of engineering and mathematics. Important results are highlighted for easy reference. Exercises and solutions are provided at the end of most sections. It very important to attempt these as the only way to develop competence and understanding is through practice. Efforts have been made to adopt an informal approach and describe mathematical processes using every day language.

Units I and II cover the Higher Level Differential Calculus with Applications. Unit III covers the Theory and Applications of Double and Triple Integration alongwith Special Functions—Beta and Gamma. Unit IV develops the Concepts of Vector Calculus in detail. Unit V builds the topic on Matrices alongwith their Applications.

We are indebted to all those who supported and gave suggestions in completing this assignment. Valued readers are requested to point out errors in the form of misprint or otherwise to us which will be gratefully acknowledged. We extend our sincere thanks to the publishers and DTP agency for their full help and cooperation in bringing out this book in time.

Delhi July, 2012 **Authors**

Contents

SYMBOLS AND BASIC FORMULAE

1. Greek Letters

α alpha	ϕ phi	β beta	Φ capital phi
Γ capital gamma	Ψ capital psi	γ gamma	ψ psi
Δ capital delta	η eta	δ delta	ξ xi
ε epsilon	ζ zeta	ι iota	χ chi
λ lambda	σ sigma	θ theta	π pi
μ mu	Σ capital sigma	ν nu	τ tau
ω omega	ρ rho	Ω capital omega	κ kapha

2. Algebraic Formulae

(*i*) Arithmetic Progression $a, a + d, a + 2d, ...,$

n^{th} term $T_n = a + (n - 1)d$;

Sum of n terms $= \frac{n}{2}[2a + (n - 1)d]$.

(*ii*) Geometrical Progression: $a, ar, ar^2, ...,$

n^{th} term $T_n = ar^{n-1}$;

Sum of n terms $= \frac{a(1-r^n)}{1-r}$.

(*iii*) Arithmetic Mean of two number a and b is $\frac{1}{2}(a+b)$.

(*iv*) Geometric Mean of two numbers a and b is $\sqrt{ab}$.

(*v*) Harmonic Mean of two numbers a and b is $\frac{2ab}{a+b}$.

(*vi*) If $ax^2 + bx + c = 0$ is quadratic, then

(*a*) its roots are given by $\frac{-b \pm \sqrt{b^2 - 4ac}}{2a}$

(*b*) the sum of the roots is equal to $-\frac{b}{a}$,

(*c*) product of the roots is equal to $\frac{c}{a}$,

(*d*) $b^2 - 4ac = 0 \Rightarrow$ the roots are equal,

(*e*) $b^2 - 4ac > 0 \Rightarrow$ the roots are real and distinct

(*f*) $b^2 - 4ac < 0$ the roots are complex,

(*g*) if $b^2 - 4ac$ is a perfect square, the roots are rational.

3. Properties of Logarithm

(*i*) $\log_a 1 = 0$, $\log_a 0 = -\infty$ for $a > 1$, $\log_a a = 1$.

$\log_e 2 = 0.6931$, $\log_e 10 = 2.3026$ $\log_{10} e = 0.4343$

(*ii*) $\log_a p + \log_a q = \log_a pq$.

(*iii*) $\log_a p - \log_a q = \log_a \frac{p}{q}$.

(*iv*) $\log_a p^q = q \log_a p$.

(*v*) $\log_a n = \log_a b \cdot \log_b n = \frac{\log_b n}{\log_b a}$

4. Angles Relations

(*i*) 1 radian $= \frac{180^\circ}{\pi}$

(*ii*) $1^\circ = 0.0174$ radian

5. Algebraic Signs of Trigonometrical Ratios:

(*a*) First quadrant: All trig. Ratios are positive.

(*b*) Second quadrant: sin θ and cosec θ *are positive, all others negative.*

(*c*) Third quadrant: tan θ and cot θ are positive, all others negative.

(*d*) Fourth quadrant: cos θ and sec θ are positive, all others negative.

6. Commonly Used Values of Triganometrical Ratios:

$\sin\frac{\pi}{2}=1, \cos\frac{\pi}{2}=0, \tan\frac{\pi}{2}=\infty,$

$\sin\frac{\pi}{6}=\frac{1}{2}, \cos\frac{\pi}{6}=\frac{\sqrt{3}}{2}, \tan\frac{\pi}{6}=\frac{1}{\sqrt{3}},$

$\sin\frac{\pi}{3}=\frac{\sqrt{3}}{2}, \cos\frac{\pi}{3}=\frac{1}{2}, \tan\frac{\pi}{3}=\sqrt{3},$

$\sin\frac{\pi}{4}=\frac{1}{\sqrt{2}}, \cos\frac{\pi}{4}=\frac{1}{\sqrt{2}}, \tan\frac{\pi}{4}=1,$

$\text{cosec}\frac{\pi}{2}=1, \sec\frac{\pi}{2}=\infty, \cos\frac{\pi}{2}=0,$

$\text{cosec}\frac{\pi}{6}=2, \sec\frac{\pi}{6}=\frac{2}{\sqrt{3}}, \cot\frac{\pi}{6}=\sqrt{3},$

$\text{cosec}\frac{\pi}{3}=\frac{2}{\sqrt{3}}, \sec\frac{\pi}{3}=2, \cot\frac{\pi}{3}=\frac{1}{\sqrt{3}},$

$\text{cosec}\frac{\pi}{4}=\frac{1}{\sqrt{2}}, \sec\frac{\pi}{4}=\sqrt{2}, \cot\frac{\pi}{4}=1.$

7. Trig. Ratios of Allied Angles

(*a*) $\sin(-\theta) = -\sin\theta$, $\cos(-\theta) = \cos\theta$, $\tan(-\theta) = -\tan\theta$,
$\text{cosec}(-\theta) = -\text{cosec}\,\theta$, $\sec(-\theta) = \sec\theta$, $\cot(-\theta) = -\cot\theta$.

(*b*) Any trig ratio of $(n.90 \pm \theta) = \begin{cases}\pm \text{ same trig ratio of } \theta \text{ when } n \text{ is even} \\ \pm \text{ co-ratio of } \theta \text{ when } n \text{ is odd}\end{cases}$

For example: $\sin(4620) = \sin[90°(52) - 60°] = \sin(-60°) = -\sin 60° = -\frac{\sqrt{3}}{2}$.

Similarly, $\text{cosec}(270° - \theta) = \text{cosec}(90°(3) - \theta) = -\sec\theta$.

8. Transformations of Products and Sums

(*a*) $\sin(A + B) = \sin A \cos B + \cos A \sin B$

(*b*) $\sin(A - B) = \sin A \cos B - \cos A \sin B$

(*c*) $\cos(A + B) = \cos A \cos B - \sin A \sin B$

(*d*) $\cos(A - B) = \cos A \cos B + \sin A \sin B$

(*e*) $\tan(A + B) = \frac{\tan A + \tan B}{1 - \tan A \tan B}$

(*f*) $\tan(A - B) = \frac{\tan A - \tan B}{1 + \tan A \tan B}$

(*g*) $\sin 2A = 2\sin A \cos A = \frac{2\tan A}{1 + \tan^2 A}$

(*h*) $\cos 2A = \cos^2 A - \sin^2 A = 1 - 2\sin^2 A$

(*i*) $\tan 2A = \frac{\sin 2A}{\cos 2A} = \frac{2\tan A}{1 - \tan^2 A}$

(*j*) $\sin 3A = 3\sin A - 4\sin^3 A$

(*k*) $\cos 3A = 4\cos^3 A - 3\cos A$

(*l*) $\tan 3A = \frac{3\tan A - \tan^3 A}{1 - 3\tan^2 A}$

(*m*) $\sin A + \sin B = 2\sin\frac{A+B}{2}\cos\frac{A-B}{2}$

(*n*) $\sin A - \sin B = 2\cos\frac{A+B}{2}\sin\frac{A-B}{2}$

(*o*) $\cos A + \cos B = 2\cos\frac{A+B}{2}\cos\frac{A-B}{2}$

(*p*) $\cos A - \cos B = 2\sin\frac{A+B}{2}\sin\frac{B-A}{2}$

(*q*) $\sin A \cos B = \frac{1}{2}[\sin(A + B) + \sin(A - B)]$

(*r*) $\cos A \sin B = \frac{1}{2}[\sin(A + B) - \sin(A - B)]$

(*s*) $\cos A \cos B = \frac{1}{2}[\cos(A + B) + \cos(A - B)]$

(*t*) $\sin A \sin B = \frac{1}{2}[\cos(A - B) - \cos(A + B)]$

9. Expressions for $\sin\frac{A}{2}, \cos\frac{A}{2}$ and $\tan\frac{A}{2}$

(a) $\sin\frac{A}{2} = \pm\sqrt{\frac{1-\cos A}{2}}$

(b) $\cos\frac{A}{2} = \pm\sqrt{\frac{1+\cos A}{2}}$

(c) $\tan\frac{A}{2} = \pm\sqrt{\frac{1-\cos A}{1+\cos A}}$

(d) $\sin\frac{A}{2} + \cos\frac{A}{2} = \pm\sqrt{1+\sin A}$

(e) $\sin\frac{A}{2} - \cos\frac{A}{2} = \pm\sqrt{1-\sin A}$

10. Relations Between Sides and Angles of a Triangle

(a) $\frac{a}{\sin A} = \frac{b}{\sin B} = \frac{c}{\sin C}$ (sine formulae)

(b) $\left.\begin{aligned} \cos A &= \frac{b^2+c^2-a^2}{2bc} \\ \cos B &= \frac{c^2+a^2-b^2}{2ca} \\ \cos C &= \frac{a^2+b^2-c^2}{2ab} \end{aligned}\right\}$ cosine formulae

(c) $\left.\begin{aligned} a &= b\cos C + c\cos B \\ b &= c\cos A + a\cos C \\ c &= a\cos B + b\cos A \end{aligned}\right\}$ Projection formulae

11. Permutations and Combinations Formulae

(a) ${}^nP_r = \frac{n!}{(n-r)!}$, ${}^nC_r = \frac{n!}{r!(n-r)!} = {}^nC_{n-r}$, ${}^nC_0 = {}^nC_n = 1$

12. Differentiation Formulae

(a) $\frac{d}{dx}(\sin x) = \cos x$ (b) $\frac{d}{dx}(\cos x) = -\sin x$

(c) $\frac{d}{dx}(\tan x) = \sec^2 x$ (d) $\frac{d}{dx}(\cot x) = -\text{cosec}^2 x$

(e) $\frac{d}{dx}(\sec x) = \sec x \tan x$

(f) $\frac{d}{dx}(\text{cosec}\, x) = -\text{cosec}\, x \cot x$

(g) $\frac{d}{dx}(e^x) = e^x$ (h) $\frac{d}{dx}(a^x) = a^x \log_e a$

(i) $\frac{d}{dx}(\log_a x) = \frac{1}{x\log a}$ (j) $\frac{d}{dx}(\log_e x) = \frac{1}{x}$

(k) $\frac{d}{dx}(ax+b)^n = na(ax+b)^{n-1}$

(m) $\frac{d}{dx}(\sin^{-1} x) = \frac{1}{\sqrt{1-x^2}}$ (n) $\frac{d}{dx}(\cos^{-1} x) = -\frac{1}{\sqrt{1-x^2}}$

(o) $\frac{d}{dx}(\tan^{-1} x) = \frac{1}{1+x^2}$ (p) $\frac{d}{dx}(\cot^{-1} x) = -\frac{1}{1+x^2}$

(q) $\frac{d}{dx}(\sec^{-1} x) = \frac{1}{x\sqrt{x^2-1}}$

(r) $\frac{d}{dx}(\text{cosec}^{-1} x) = -\frac{1}{x\sqrt{x^2-1}}$

(s) $\frac{d}{dx}(\sinh x) = \cosh x$ (t) $\frac{d}{dx}(\cosh x) = \sinh x$

(u) $D^n(uv) = D^n u + {}^nC_1 D^{n-1}uDv + {}^nC_2 D^{n-2}uD^2v + \ldots + {}^nC_r D^{n-r}uD^rv + \ldots + {}^nC_n uD^nv$

(Leibnitz's Formula)

(l) $\frac{d^n}{dx^n}(ax+b)^m = m(m-1)(m-2)\ldots(m-n+1)(ax+b)^{m-n}$

13. Integration Formulae

(*a*) $\int \sin x dx = -\cos x$ (*b*) $\int \cos x dx = \sin x$

(*c*) $\int \tan x dx = -\log \cos x$ (*d*) $\int \cot x dx = \log \sin x$

(*e*) $\int \sec x dx = \log(\sec x + \tan x)$

(*f*) $\int \operatorname{cosec} x dx = \log(\operatorname{cosec} x - \cot x)$

(*g*) $\int \sec^2 x dx = \tan x$ (*h*) $\int \operatorname{cosec}^2 x dx = -\cot x$

(*i*) $\int e^x dx = e^x$ (*j*) $\int a^x dx = \dfrac{a^x}{\log_e a}$

(*k*) $\int \dfrac{1}{x} dx = \log_e x$

(*l*) $\int x^n dx = \dfrac{x^{n+1}}{n+1}, n \neq -1$

(*m*) $\int \dfrac{dx}{a^2+x^2} = \dfrac{1}{a}\tan^{-1}\dfrac{x}{a}$ (*n*) $\dfrac{dx}{a^2-x^2} = \dfrac{1}{2a}\dfrac{a+x}{a-x}$

(*o*) $\int \dfrac{dx}{x^2-a^2} = \dfrac{1}{2a}\dfrac{x-a}{x+a}$ (*p*) $\dfrac{dx}{\sqrt{a^2-x^2}} = \sin^{-1}\dfrac{x}{a}$

(*q*) $\int \dfrac{dx}{\sqrt{a^2+x^2}} = \sinh^{-1}\dfrac{x}{a}$ (*r*) $\int \dfrac{dx}{\sqrt{x^2-a^2}} = \cosh^{-1}\dfrac{x}{a}$

(*s*) $\int \sqrt{a^2+x^2}\,dx = \dfrac{x\sqrt{a^2+x^2}}{2} + \dfrac{a^2}{2}\sinh^{-1}\dfrac{x}{a}$

(*t*) $\int \sqrt{x^2-a^2}\,dx = \dfrac{x\sqrt{x^2-a^2}}{2} - \dfrac{a^2}{2}\cosh^{-1}\dfrac{x}{a}$

(*u*) $\int \sqrt{a^2-x^2}\,dx = \dfrac{x\sqrt{a^2-x^2}}{2} + \dfrac{a^2}{2}\sin^{-1}\dfrac{x}{a}$

(*v*) $\int e^{ax}\sin bx dx = \dfrac{e^{nx}}{a^2+b^2}(a\sin bx - b\cos bx)$

(*w*) $\int e^{ax}\cos bx dx = \dfrac{e^{ax}}{a^2+b^2}(a\cos bx + b\sin bx)$

(*x*) $$\int_0^{\pi/2} \sin^n x dx = \int_0^{\pi/2} \cos^n x dx = \begin{cases} \dfrac{(n-1)(n-3)(n-5)\ldots}{n(n-2)(n-4)\ldots} \ldots \text{if } n \text{ is odd} \\[2ex] \dfrac{(n-1)(n-3)(n-5)\ldots}{n(n-2)(n-4)\ldots}\dfrac{\pi}{2} \text{ if } n \text{ is even} \end{cases}$$

(*y*) $$\int_0^{\pi/2} \sin^m x \cos^n x dx = \begin{cases} \dfrac{(n-1)(n-3)\ldots(n-1)(n-3)\ldots}{(m+n)(m+n-2)(m+n-4)\ldots} & \text{if } m \text{ and } n \text{ are not simultaneously even} \\[2ex] \dfrac{(m-1)(m-3)\ldots(n-1)(n-3)\ldots}{(m+n)(m+n-2)(m+n-4)\ldots}\dfrac{\pi}{2} & \text{if both } m \text{ and } n \text{ are even} \end{cases}$$

14. Beta and Gamma Functions

(*a*) $\beta(m, n) = \int_0^1 x^{m-1}(1-x)^{n-1}dx$ converges for $m, n > 0$.

(*b*) $\Gamma(n) = \int_0^\infty e^{-x}x^{n-1}\,dx$ converges for $n > 0$

(*c*) $\Gamma(n+1) = n\,\Gamma(n)$ and $\Gamma(n+1) = n!$ if n is positives integer

(*d*) $\Gamma(1) = 1 = \Gamma(2)$ and $\Gamma\left(\dfrac{1}{2}\right) = \sqrt{\pi}$

(*e*) $\beta(m, n) = \dfrac{\Gamma(m)\,\Gamma(n)}{\Gamma(m+n)}$

(*f*) $$\int_0^{\pi/2} \sin^p x \cos^q x dx = \frac{1}{2}\beta\left(\frac{p+1}{2}, \frac{q+1}{2}\right) = \frac{\Gamma\left(\frac{p+1}{2}\right)\Gamma\left(\frac{q+1}{2}\right)}{2\Gamma\left(\frac{p+q+2}{2}\right)}$$

(*g*) $$\int_0^{\pi/2} \sqrt{\tan\theta}\,d\theta = \int_0^{\pi/2} \sin^{1/2}\cos^{-1/2}\theta\,d\theta = \frac{\Gamma\left(\frac{3}{4}\right)\Gamma\left(\frac{1}{4}\right)}{2\Gamma(1)} = \frac{1}{2}\Gamma\left(\frac{1}{4}\right)\Gamma\left(\frac{3}{4}\right)$$

UNIT-I
DIFFERENTIAL CALCULUS-I

Successive Differentiation & Leibnitz's Theorem,
Taylor's Maclaurin's Series for one Variable
Asymptotes for Cartesian Coordinates,
Curvature–Cartesian formula for radius of
Curvature and Centre of Curvature
Curve Tracing (Cartesian and Polar coordinates)

CHAPTER

1 Successive Differentiation, Expansion of Functions of one Variable

Successive Differentiation and Leibnitz's Theorem. Taylor's and Maclaurin's Series for Functions of one Variable. Approximate Errors.

INTRODUCTION

If a function $f(x)$ is differentiable we can obtain its **derivative** and denote it by $f'(x)$. If $f'(x)$ is also differentiable we can differentiate it and get its derivative which is called **second order derivative** of $f(x)$ and is denoted by $f''(x)$. Continuing this process we get third and higher order derivatives and denote them by $f'''(x)$, $f^{\text{iv}}(x)$, ..., $f^{(n)}(x)$, ... This notation is somewhat cumbersome in practice and a better notation is given by

$$\frac{dy}{dx}, \frac{d^2y}{dx^2}, \frac{d^3y}{dx^3}, \ldots, \frac{d^ny}{dx^n}, \ldots \quad \text{or} \quad y^{(1)}, y^{(2)}, y^{(3)}, \ldots, y^{(n)}, \ldots$$

or, alternately, the suffix notation $y_1, y_2, y_3, \ldots\ldots, y_n, \ldots\ldots.$

where $y = f(x)$ and the suffix n is a positive integer giving the order of derivative.

Yet another notation is the operator notation as

$$\mathrm{D}y, \mathrm{D}^2y, \mathrm{D}^3y, \ldots, \mathrm{D}^ny, \ldots$$

where D symbolizes the operation of differentiation $\dfrac{d}{dx}$.

n^{th} Order Derivatives of some Standard Functions

1. Let $y = e^{ax}$, then,

$$y_1 = a\,e^{ax},\ y_2 = a^2 e^{ax},\ \ldots\ldots,\ y_n = a^n e^{ax}, \quad \text{that is,} \quad \boxed{\frac{d^n}{dx^n} e^{ax} = a^n e^{ax}}$$

2. Let $y = (ax+b)^m$ then,

$$y_1 = m\,a\,(ax+b)^{m-1},\ y_2 = m\,(m-1)\,a^2\,(ax+b)^{m-2}, \ldots$$

In general, $\boldsymbol{y_n = m\,(m-1)\,(m-2)\,\ldots\,(m-n+1)\,a^n\,(ax+b)^{m-n}}$.

In particular, if m is a positive integer greater than n, then we can write,

$$\boxed{y_n = \frac{d^n}{dx^n}(ax+b)^m = \frac{m!}{(m-n)!}a^n\,(ax+b)^{m-n}}$$

and if m is a positive integer less than n then $y_n = 0$.
Further, if $m = n$, then,

$$\boxed{y_n = \frac{d^n}{dx^n}(ax+b)^n = n\,!\,a^n}$$

and if $m = -1$,

$$y_n = \frac{d^n}{dx^n}(ax+b)^{-1} = (-1)(-2)(-3)\ldots(-n)\,a^n\,(ax+b)^{-1-n}$$

or we can write $$\boxed{y_n = \frac{d^n}{dx^n}\left(\frac{1}{ax+b}\right) = \frac{(-1)^n n!\,a^n}{(ax+b)^{n+1}}}$$

3. Let $y = \log(ax+b)$, then,

$$y_1 = \frac{a}{ax+b} = a\,(ax+b)^{-1}$$

and $$y_n = \frac{d^n}{dx^n}\log(ax+b) = \frac{d^{n-1}}{dx^{n-1}}\left(\frac{a}{ax+b}\right) = a.(-1)^{n-1}(n-1)!\,a^{n-1}(ax+b)^{-1-(n-1)}$$

or $$\boxed{y_n = \frac{d^n}{dx^n}(\log(ax+b)) = (-1)^{n-1}\frac{(n-1)!\,a^n}{(ax+b)^n}}$$

4. Let $y = a^x$ then,

$$y_1 = a^x \log a, \quad y_2 = a^x(\log a)^2, \ldots, \quad \boxed{y_n = \frac{d^n}{dx^n}(a^x) = a^x(\log a)^n}$$

5. Let $y = \sin(ax+b)$ then,

$$y_1 = a\cos(ax+b) = a\sin(ax+b+\pi/2)$$
$$y_2 = a^2\cos(ax+b+\pi/2) = a^2\sin(ax+b+2.\pi/2)$$

....

....

$$\boxed{y_n = \frac{d^n}{dx^n}(\sin(ax+b)) = a^n\sin(ax+b+n\,\pi/2)}$$

Similarly, if $y = \cos(ax+b)$ then,

$$y_1 = -a\sin(ax+b) = a\cos(ax+b+\pi/2)$$
$$y_2 = -a^2\sin(ax+b+\pi/2) = a^2\cos(ax+b+2\,\pi/2)$$

....

....

$$y_n = \frac{d^n}{dx^n}(\cos(ax+b)) = a^n \cos(ax+b+n\pi/2)$$

The above two formulae, in brief, can be written as

$$\frac{d^n}{dx^n} \begin{matrix}\sin\\\cos\end{matrix} (ax+b) = a^n \begin{matrix}\sin\\\cos\end{matrix} (ax+b+n\pi/2).$$

6. Let $\quad y = e^{ax}\sin(bx+c)$ then,

$$y_1 = a\,e^{ax}\sin(bx+c) + e^{ax}\,b\cos(bx+c) = e^{ax}[a\sin(bx+c) + b\cos(bx+c)]$$

If we put $a = r\cos\alpha,\; b = r\sin\alpha$ then

$$y_1 = e^{ax}[\,r\sin(bx+c)\cos\alpha + r\cos(bx+c)\sin\alpha] = r\,e^{ax}\sin(bx+c+\alpha)$$

where $\quad r = \sqrt{a^2+b^2}$ and $\alpha = \tan^{-1}(b/a)$.

Again differentiating and making the above substitution, gives

$$y_2 = r^2 e^{ax}\sin(bx+c+2\alpha)$$

and continuing the process, in general, we get

$$y_n = r^n e^{ax}\sin(bx+c+n\alpha).$$

Similarly, if $y = e^{ax}\cos(bx+c)$, and proceeding as above, we get

$$y_n = r^n e^{ax}\cos(bx+c+n\alpha).$$

The above two results can be together written as

$$\frac{d^n}{dx^n}\left(e^{ax} \begin{matrix}\sin\\\cos\end{matrix} (bx+c)\right) = r^n e^{ax} \begin{matrix}\sin\\\cos\end{matrix} (bx+c+n\alpha)$$

where $\quad r = \sqrt{a^2+b^2}$ and $\alpha = \tan^{-1} b/a$.

RULES OF PARTIAL FRACTIONS

A fraction is said to be **proper fraction** if degree of the numerator is strictly less than the degree of the denominator, otherwise it is known as **improper fraction.**

To find the nth derivative of any fraction first we have to convert it into proper fraction and then the fraction be resolved into partial fractions by applying the following rules:

(*i*) $\dfrac{1}{(x-a)(x-b)(x-c)}$ into $\dfrac{A}{x-a} + \dfrac{B}{x-c} + \dfrac{C}{x-c}$

(*ii*) $\dfrac{1}{(x-a)(x-b)^2}$ into $\dfrac{A}{x-a} + \dfrac{B}{x-b} + \dfrac{B}{(x-b)^2}$

(*iii*) $\dfrac{1}{(x-a)(x-b)^3}$ into $\dfrac{A}{x-a} + \dfrac{B}{x-b} + \dfrac{C}{(x-b)^2} + \dfrac{D}{(x-b)^3}$

(*iv*) $\dfrac{1}{(x-a)(x-b)(\alpha x^2+\beta x+\gamma)}$ into $\dfrac{A}{x-a} + \dfrac{B}{x-b} + \dfrac{Cx+D}{\alpha x^2+\beta x+\gamma}$

where A, B, C, D etc. are the constants to be evaluated.

EXAMPLE 1.1. **(a)** Find the nth derivative of $\dfrac{x^3}{(x-1)(2x+3)}$, $(n > 1)$.

(b) Find y_n if $y = \dfrac{x^n - 1}{x - 1}$. **(UPTU 2011-12)**

SOLUTION: (a) First of all we have to resolve the given rational function into partial fractions for which the given function which is not a proper fraction, must be converted into a proper fraction. Therefore, we divide the numerator by the denominator so as to get

$$f(x) = \frac{1}{2}x - \frac{1}{4} + \frac{\frac{7}{4}x - \frac{3}{4}}{(x-1)(2x+3)} = \frac{x}{2} - \frac{1}{4} + \frac{1}{5(x-1)} + \frac{27}{20(2x+3)}$$

Now, differentiation n times $(n > 1)$, gives

$$\frac{d^n}{dx^n} f(x) = 0 + 0 + \frac{1}{5}\frac{(-1)^n n!}{(x-1)^{n+1}} + \frac{27}{20}\frac{(-1)^n n!\, 2^n}{(2x+3)^{n+1}} = \frac{(-1)^n n!}{10}\left[\frac{2}{(x-1)^{n+1}} + \frac{27.\, 2^{n-1}}{(2x+3)^{n+1}}\right]. \quad \textbf{Ans.}$$

(b)
$$y = \frac{x^n - 1}{x - 1} = x^{n-1} + x^{n-2} + x^{n-3} + \ldots + 1$$

$$\therefore \quad y_n = \frac{d^n y}{dx^n} = \frac{d^n}{dx^n}\left[x^{n-1} + x^{n-2} + x^{n-3} + \ldots + 1\right] = 0 \quad \textbf{Ans.}$$

EXAMPLE 1.2. Find the nth derivative of $\sin^2 x \cos^3 x$.

SOLUTION: Let $y = \sin^2 x \cos^3 x$, we can write

$$y = \left(\frac{1 - \cos 2x}{2}\right)\cdot\left(\frac{\cos 3x + 3\cos x}{4}\right)$$

$$= \frac{1}{8}\left[\cos 3x + 3\cos x - \cos 2x \cos 3x - 3\cos 2x \cos x\right]$$

$$= \frac{1}{8}\left[\cos 3x + 3\cos x - \frac{1}{2}(\cos 5x + \cos x) - \frac{3}{2}(\cos 3x + \cos x)\right]$$

$$= \frac{1}{16}\left[2\cos x - \cos 3x - \cos 5x\right]$$

$$\therefore \quad y_n = \frac{1}{16}\left[2\cos\left(x + \frac{n\pi}{2}\right) - 3^n \cos\left(3x + \frac{n\pi}{2}\right) - 5^n \cos\left(5x + \frac{n\pi}{2}\right)\right]. \quad \textbf{Ans.}$$

EXMAPLE 1.3. If $y = e^{5x}\sin(2x+3)\sin(x-1)\cos(x+1)$, find y_n.

SOLUTION: We have $y = \frac{1}{2} e^{5x}\sin(2x+3)(\sin 2x - \sin 2)$

$$= \frac{1}{2} e^{5x}\{\sin(2x+3)\sin 2x - \sin(2x+3)\cdot\sin 2\}$$

$$= \frac{1}{4} e^{5x}\{\cos 3 - \cos(4x+3)\} - \frac{1}{2}\sin 2\cdot e^{5x}\sin(2x+3)$$

$$= \frac{1}{4}\cos 3\cdot e^{5x} - \frac{1}{4} e^{5x}\cos(4x+3) - \frac{1}{2}\sin 2\cdot e^{5x}\sin(2x+3)$$

Therefore,
$$y_n = \frac{1}{4}\cos 3\cdot 5^n e^{5x} - \frac{e^{5x}}{4} r_1^n \cos(4x+3+n\alpha) - \frac{1}{2}\sin 2\cdot r_2^n e^{5x}\sin(2x+3+n\beta)$$

where
$$r_1 = \sqrt{5^2+4^2} = \sqrt{41},\quad r_2 = \sqrt{5^2+2^2} = \sqrt{29},$$

$$\alpha = \tan^{-1}\frac{4}{5} \quad\text{and}\quad \beta = \tan^{-1}\frac{2}{5}.$$ **Ans.**

EXAMPLE 1.4. Find the nth derivative of $\tan^{-1}(x/a)$.

SOLUTION: Let $y = \tan^{-1}(x/a)$ then $y_1 = \dfrac{1}{a\left(1+\dfrac{x^2}{a^2}\right)} = \dfrac{a}{x^2+a^2}$

or
$$y_1 = \frac{a}{(x+ai)(x-ai)} = \frac{a}{2ai}\left\{\frac{1}{x-ai} - \frac{1}{x+ai}\right\} \quad \text{where } i = \sqrt{-1}.$$

Differentiating the above relation $(n-1)$ times w.r.t. x, gives

$$y_n = \frac{1}{2i}\left\{\frac{d^{n-1}}{dx^{n-1}}\frac{1}{x-ai} - \frac{d^{n-1}}{dx^{n-1}}\frac{1}{x+ai}\right\} = \frac{1}{2i}\left\{\frac{(-1)^{n-1}(n-1)!}{(x-ai)^n} - \frac{(-1)^{n-1}(n-1)!}{(x+ai)^n}\right\}$$

$$= \frac{(-1)^{n-1}(n-1)!}{2i}\left\{\frac{1}{(x-ai)^n} - \frac{1}{(x+ai)^n}\right\}.$$

The above expression can be simplified by using De Moivre's theorem, for which we put $x = r\cos\theta$, $a = r\sin\theta$, then

$$y_n = \frac{(-1)^{n-1}(n-1)!}{2i}\left[\frac{1}{r^n(\cos\theta - i\sin\theta)^n} - \frac{1}{r^n(\cos\theta + i\sin\theta)^n}\right]$$

$$= \frac{(-1)^{n-1}(n-1)!}{2i\,r^n}\left[(\cos\theta - i\sin\theta)^{-n} - (\cos\theta + i\sin\theta)^{-n}\right]$$

$$= \frac{(-1)^{n-1}(n-1)!}{2i\,r^n}[\cos n\theta + i\sin n\theta - \cos n\theta + i\sin n\theta)] = \frac{(-1)^{n-1}(n-1)!}{r^n}\sin n\theta$$

$$= \frac{(-1)^{n-1}(n-1)!}{(a/\sin\theta)^n}\sin n\theta = \frac{(-1)^{n-1}(n-1)!}{a^n}\sin n\theta\sin^n\theta \qquad (\text{using } a = r\sin\theta)$$

where $\tan\theta = \dfrac{a}{x}$ or $\theta = \tan^{-1}(a/x) = \cot^{-1}(x/a)$. **Ans.**

EXAMPLE 1.5. If $y=\dfrac{x}{x^2+a^2}$, find y_n, the nth derivative of y.

SOLUTION: Here $y=\dfrac{x}{(x+ai)(x-ai)} = \dfrac{1}{2}\left\{\dfrac{1}{x-ai}+\dfrac{1}{x+ai}\right\}$

$$\therefore \qquad y_n = \frac{1}{2}\frac{(-1)^n n!}{(x-ai)^{n+1}}+\frac{1}{2}\frac{(-1)^n n!}{(x+ai)^{n+1}}$$

To simplify the above expression we apply DeMoivre's theorem for which, as in previous example, we put $x = r\cos\theta$, $a = r\sin\theta$, to get

$$y_n = \frac{(-1)^n n!}{2}\left[\frac{1}{r^{n+1}(\cos\theta - i\sin\theta)^{n+1}}+\frac{1}{r^{n+1}(\cos\theta+i\sin\theta)^{n+1}}\right]$$

$$= \frac{(-1)^n n!}{2\,r^{n+1}}\,[(\cos\theta - i\sin\theta)^{-n-1}+(\cos\theta+i\sin\theta)^{-n-1}]$$

$$= \frac{(-1)^n n!}{2\,r^{n+1}}\,[\{\cos(n+1)\theta + i\sin(n+1)\theta\}+\{\cos(n+1)\theta - i\sin(n+1)\theta\}]$$

$$= \frac{(-1)^n n!}{r^{n+1}}\cos(n+1)\theta = \frac{(-1)^n n!}{a^{n+1}}\cos(n+1)\theta\sin^{n+1}\theta \quad (\text{using } a = r\sin\theta)$$

where $\cot\theta = x/a$ or $\theta = \cot^{-1}(x/a) = \tan^{-1}(a/x)$. **Ans.**

EXAMPLE 1.6. If $p^2 = a^2\cos^2\theta + b^2\sin^2\theta$, show that

$$p+\frac{d^2p}{d\theta^2}=\frac{a^2b^2}{p^3}.$$

SOLUTION: Differentiating $p^2 = a^2\cos^2\theta + b^2\sin^2\theta$, we get ...(1)

$$2p\frac{dp}{d\theta} = -2a^2\cos\theta\sin\theta + 2b^2\sin\theta\cos\theta$$

or $$p\frac{dp}{d\theta} = (b^2-a^2)\sin\theta\cos\theta. \qquad ...(2)$$

Differentiating w.r.t. θ again, we get

$$p\frac{d^2p}{d\theta^2}+\left(\frac{dp}{d\theta}\right)^2 = (b^2-a^2)(\cos^2\theta-\sin^2\theta).$$

Multiplying throughout by p^2 and using (2), gives

$$p^3\frac{d^2p}{d\theta^2}+(b^2-a^2)^2\sin^2\theta\cos^2\theta = (b^2-a^2)p^2(\cos^2\theta-\sin^2\theta)$$

$$= p^2\{(b^2\cos^2\theta + a^2\sin^2\theta)-(a^2\cos^2\theta+b^2\sin^2\theta)\}$$

$$= p^2(b^2\cos^2\theta + a^2\sin^2\theta) - p^4$$

or $$p^3\frac{d^2p}{d\theta^2}+p^4= (a^2\cos^2\theta + b^2\sin^2\theta)(b^2\cos^2\theta + a^2\sin^2\theta) - (b^2-a^2)^2\sin^2\theta\cos^2\theta$$

$$= a^2b^2(\cos^4\theta + \sin^4\theta) + (a^4+b^4)\sin^2\theta\cos^2\theta - (a^4+b^4-2a^2b^2)\sin^2\theta\cos^2\theta$$

$$= a^2b^2(\cos^4\theta+\sin^4\theta+2\sin^2\theta\cos^2\theta) = a^2b^2(\cos^2\theta+\sin^2\theta)^2 = a^2b^2$$

or $$\frac{d^2p}{d\theta^2}+p = \frac{a^2b^2}{p^3}.$$ **Hence Proved.**

EXAMPLE 1.7. If $y = x\log\left(\frac{x-1}{x+1}\right)$, show that

$$y_n = (-1)^{n-2}(n-2)!\left[\frac{x-n}{(x-1)^n}-\frac{x+n}{(x+1)^n}\right].$$ **(UPTU 2003)**

SOLUTION: Here $y = x\log\left(\frac{x-1}{x+1}\right) = x\log(x-1) - x\log(x+1)$

$$\therefore \quad y_1 = \frac{x}{x-1}+\log(x-1)-\frac{x}{x+1}-\log(x+1)$$

$$= 1+\frac{1}{x-1}+\log(x-1)-1+\frac{1}{x+1}-\log(x+1)$$

$$= \frac{1}{x-1}+\frac{1}{x+1}+\log(x-1)-\log(x+1)$$

Differentiating both sides $(n-1)$ times with respect to x, we get

$$y_n = \frac{(-1)^{n-1}(n-1)!}{(x-1)^n}+\frac{(-1)^{n-1}(n-1)!}{(x+1)^n}+\frac{(-1)^{n-2}(n-2)!}{(x-1)^{n-1}}-\frac{(-1)^{n-2}(n-2)!}{(x+1)^{n-1}}$$

$$= (-1)^{n-2}(n-2)!\left\{\frac{-(n-1)+x-1}{(x-1)^n}\right\}+(-1)^{n-2}(n-2)!\left\{\frac{-(n-1)-(x+1)}{(x+1)^n}\right\}$$

$$= (-1)^{n-2}(n-2)!\left[\frac{x-n}{(x-1)^n}-\frac{x+n}{(x+1)^n}\right].$$ **Hence the result.**

EXAMPLE 1.8. Find the nth derivative of $\frac{1}{1+x+x^2}$.

SOLUTION: $$y = \frac{1}{1+x+x^2} = \frac{1}{(x-w)(x-w^2)}$$ where $w = \frac{-1+i\sqrt{3}}{2}$, $w^2 = \frac{-1-i\sqrt{3}}{2}$

$$= \left(\frac{1}{x-w}-\frac{1}{x-w^2}\right)\frac{1}{w-w^2} = \frac{-i}{\sqrt{3}}\left(\frac{1}{x-w}-\frac{1}{x-w^2}\right)$$

$$\therefore \quad y_n = \frac{d^n}{dx^n}\left[\frac{1}{1+x+x^2}\right] = \frac{-i}{\sqrt{3}}\left[\frac{d^n}{dx^n}\frac{1}{x-w}-\frac{d^n}{dx^n}\frac{1}{x-w^2}\right]$$

$$= \frac{-i}{\sqrt{3}}\left[\frac{(-1)^n n!}{(x-w)^{n+1}} - \frac{(-1)^n n!}{(x-w^2)^{n+1}}\right]$$

$$= i(-1)^{n+1}\frac{n!}{\sqrt{3}}\left[\frac{1}{(x-w)^{n+1}} - \frac{1}{(x-w^2)^{n+1}}\right]$$

$$= \frac{(-1)^{n+1}}{\sqrt{3}} n!\, i\left[\frac{1}{\left(x+\frac{1}{2}-\frac{i\sqrt{3}}{2}\right)^{n+1}} - \frac{1}{\left(x+\frac{1}{2}+\frac{i\sqrt{3}}{2}\right)^{n+1}}\right]$$

$$= (-1)^{n+1}\frac{2^{n+1}}{\sqrt{3}} n!\, i\left[\frac{1}{(2x+1-i\sqrt{3})^{n+1}} - \frac{1}{(2x+1+i\sqrt{3})^{n+1}}\right]$$

Putting, $2x + 1 = r\cos\theta$, $\sqrt{3} = r\sin\theta$, we get

$$y_n = \frac{(-1)^{n+1}2^{n+1}n!\,i}{\sqrt{3}\cdot r^{n+1}}\,[(\cos\theta - i\sin\theta)^{-(n+1)} - (\cos\theta + i\sin\theta)^{-(n+1)}]$$

$$= \frac{(-1)^{n+1}2^{n+1}n!\,i}{\sqrt{3}\left(\frac{\sqrt{3}}{\sin\theta}\right)^{n+1}}\,[\cos(n+1)\theta + i\sin(n+1)\theta - \cos(n+1)\theta + i\sin(n+1)\theta]$$

$$= \frac{(-1)^{n+1}\,2^{n+1}n!\cdot i\sin^{n+1}\theta}{(\sqrt{3})^{n+2}}\,.\,2i\sin(n+1)\theta$$

$$= (-1)^{n+2}\frac{2^{n+2}\,n!}{(\sqrt{3})^{n+2}}\sin^{n+1}\theta\sin(n+1)\theta = \frac{(-1)^n\,n!}{a^{n+2}}\sin^{n+1}\theta\sin(n+1)\theta$$

where $\tan\theta = \frac{\sqrt{3}}{2x+1}$ and $a = \frac{\sqrt{3}}{2}$. **Ans.**

EXERCISE 1A

1. Find the nth derivative of the following functions

(i) $\dfrac{x^2}{(x+2)(2x+3)}$ (ii) $\dfrac{x^4}{(x-1)(x-2)}$.

2. Obtain the nth derivative of

(i) $\sin^4 x$ (ii) $\dfrac{4x}{(x-1)^2(x+1)}$.

3. Determine $\dfrac{d^2y}{dx^2}$ if $x = a(\theta + \sin\theta)$, $y = a(1 - \cos\theta)$.

4. If $y = e^{-kt}\cos(bt + c)$ then prove that

$$\frac{d^2y}{dt^2} + 2k\frac{dy}{dt} + n^2 y = 0 \quad \text{where } n^2 = k^2 + b^2.$$

5. If $y = \sin ax + \cos ax$, show that

$$y_n = a^n\sqrt{1+(-1)^n \sin 2ax} \quad \text{and hence show that } y_8(\pi) = \left(\frac{1}{2}\right)^{3/2} \text{ when } a = \frac{1}{4}.$$

(UPTU 2011-12)

6. If $y = e^{-ax}\cos(bx + c)$, show that

$$y_n = (-1)^n (a^2 + b^2)^{n/2} e^{-ax}\cos(bx + c + n\tan^{-1} b/a).$$

7. Find the nth derivative of $\tan^{-1}\left[\dfrac{1-x}{1+x}\right]$.

8. If $y = \tan^{-1}x$, prove that

$$y_n = (n-1)!\cos\{ny + (n-1)\pi/2\}\cos^n y.$$

9. If $(1 - x^2)\tan y = 2x$, show that

$$y_n = 2(-1)^{n-1}(n-1)!\sin^n\theta \sin n\theta, \quad \text{where } \cot\theta = x.$$ **(UPTU 2001)**

10. Find the nth derivative of $e^x \sin 4x \cos 6x$.

11. If $y = \dfrac{1}{1 - 5x + 6x^2}$ find y_n. **(UPTU 2005)**

❑❑❑

LEIBNITZ'S THEOREM

(UPTU 2000-01)

This theorem provides the nth derivative of the product of two functions. It states as follows:

If $y = u.v$, where u and v are functions of x and their nth derivatives are separately known, then $y_n = (uv)_n = u_n v + {}^nC_1\, u_{n-1}\, v_1 + {}^nC_2\, u_{n-2}\, v_2 + \ldots + {}^nC_r\, u_{n-r}\, v_r + \ldots + u.\, v_n$.

Proof : The theorem will be proved by the principle of mathematical induction.

Here $\quad y_1 = \dfrac{du}{dx}v + u\dfrac{dv}{dx} = u_1 v + u\, v_1$

which shows that the theorem is true for $n = 1$.

Now, let us suppose that the statement of the theorem is true for $n = \text{K}$, *i.e.* suppose

$$(uv)_K = u_K\, v + {}^KC_1\, u_{K-1}\, v_1 + {}^KC_2\, u_{K-2}\, v_2 + \ldots + u\, v_K \quad \text{is true.}$$

Then differentiating both sides w.r.t. x, gives

$$(uv)_{K+1} = (u_{K+1}\, v + u_K\, v_1) + {}^KC_1\, (u_K\, v_1 + u_{K-1}\, v_2) + {}^KC_2\, (u_{K-1}\, v_2 + u_{K-2}\, v_3) + \ldots + (u_1\, v_K + u\, v_{K+1})$$

$$= u_{K+1}\, v + (1 + {}^KC_1)\, u_K\, v_1 + ({}^KC_1 + {}^KC_2)\, u_{K-1}\, v_2 + \ldots + ({}^KC_{K-1} + {}^KC_K)\, u_1\, v_K + \ldots + u\, v_{K+1}$$

$$= u_{K+1}\, v + {}^{K+1}C_1\, u_K\, v_1 + {}^{K+1}C_2\, u_{K-1}\, v_2 + \ldots + {}^{K+1}C_K\, u_1\, v_K + u\, v_{K+1}$$

$$\left(\text{on using the result } {}^KC_r + {}^KC_{r+1} = {}^{K+1}C_{r+1} \text{ and } 1 = {}^KC_0\right)$$

This shows that the statement of the theorem is true for $n = \text{K} + 1$ under the supposition that it is true for $n = \text{K}$. Therefore, by the principle of mathematical induction the theorem is true for every positive integral value of n.

EXAMPLE 1.9. Find the nth derivative of $x^2 e^{3x} \sin 4x$.

SOLUTION: Let $y = u\, v$ where $u = e^{3x} \sin 4x$ and $v = x^2$

Hence $\quad u_n = e^{3x} r^n \sin(4x + n\alpha)$

where $\quad r = \sqrt{3^2 + 4^2} = 5 \quad$ and $\quad \tan\alpha = 4/3$,

Here $\quad v_1 = 2x, \quad v_2 = 2, \quad v_3 = v_4 = \ldots = v_n = 0.$

Therefore, by Leibnitz's theorem

$$y_n = u_n\, v + {}^nC_1\, u_{n-1}\, v_1 + {}^nC_2\, u_{n-2}\, v_2$$

$$= 5^n e^{3x} \sin(4x + n\alpha)\, x^2 + {}^nC_1\, 5^{n-1} e^{3x} \sin(4x + (n-1)\alpha)\,(2x)$$

$$+ {}^nC_2\, 5^{n-2} e^{3x} \sin(4x + (n-2)\alpha) \cdot (2)$$

$$= 5^n e^{3x}\left[x^2 \sin(4x + n\alpha) + \frac{2\,nx}{5} \sin(4x + (n-1)\alpha) + \frac{n\,(n-1)\cdot(2)}{2.25} \sin(4x + (n-2)\alpha)\right]$$

or $\quad y_n = 5^n e^{3x}\left[x^2 \sin(4x + n\alpha) + \dfrac{2\,nx}{5} \sin(4x + n\alpha - \alpha) + \dfrac{n\,(n-1)}{25} \sin(4x + n\alpha - 2\alpha)\right].$

where $\alpha = \tan^{-1}(4/3)$. **Ans.**

EXAMPLE 1.10. If $y = x^2e^x$ show that $y_n = \dfrac{n(n-1)}{2} y_2 - n(n-2)\, y_1 + \dfrac{(n-1)\,(n-2)}{2} y.$

SOLUTION: $y = x^2e^x$ so $y_1 = 2xe^x + x^2e^x = 2xe^x + y$

Differentiating again, we get $y_2 = 2xe^x + 2e^x + y_1$

or $y_2 = y_1 - y + 2e^x + y_1$ or $2e^x = y_2 - 2y_1 + y.$

Now applying Lebneitz theorem on x^2e^x, we get

$$y_n = x^2e^x + n\,.2xe^x + \frac{n(n-1)}{2}.2.e^x$$

$$= y + n\,(y_1 - y) + \frac{n(n-1)}{2}(y_2 - 2y_1 + y)$$

$$= \frac{n(n-1)}{2} y_2 + y_1 [n - n(n-1)] + y\left[1 - n + \frac{n(n-1)}{2}\right]$$

$$= \frac{n(n-1)}{2} y_2 - n\,(n-2) y_1 + \frac{1}{2}(n-1)\,(n-2)\,y.$$

Hence the result.

EXAMPLE 1.11. Prove that the value of nth derivative of $\dfrac{x^3}{x^2-1}$ for $x = 0$, is 0 if n is even and $-n!$ if n is odd and greater than 1.

SOLUTION: Let $y = \dfrac{x^3}{x^2-1} = x + \dfrac{x}{x^2-1} = x + \dfrac{1}{2}\left(\dfrac{1}{x-1}\right) + \dfrac{1}{2}\left(\dfrac{1}{x+1}\right)$

Then $y_n = 0 + \dfrac{1}{2}\dfrac{d^n}{dx^n}\dfrac{1}{x-1} + \dfrac{1}{2}\dfrac{d^n}{d^{x^n}}\dfrac{1}{x+1}$ when $n > 1$.

$$= \frac{1}{2}\frac{(-1)^n\, n!}{(x-1)^{n+1}} + \frac{1}{2}\frac{(-1)^n n!}{(x+1)^{n+1}}$$

At $x = 0$ $y_n(0) = \dfrac{(-1)^n n!}{2}\left[\dfrac{1}{(-1)^{n+1}} + 1\right] = \dfrac{(-1)^n n!}{2(-1)^{n+1}}[1 + (-1)^{n+1}]$

$$= \frac{-n!}{2}[1 + (-1)^{n+1}] = \begin{cases} 0 & \text{if } n \text{ is even} \\ -n! & \text{if } n \text{ is odd} \end{cases}.$$

Hence Proved.

EXAMPLE 1.12. If $\log y = \tan^{-1} x$, show that

$(1 + x^2)\, y_{n+2} + \{2\,(n+1)x - 1\}\, y_{n+1} + n\,(n+1)\, y_n = 0$

and hence find y_3, y_4 and y_5 at $x = 0$.

SOLUTION: We have $\log y = \tan^{-1} x$ or $y = e^{\tan^{-1} x}$ ...(1)

hence $y_1 = e^{\tan^{-1} x} \cdot \dfrac{1}{1+x^2}$ or $(1 + x^2)\, y_1 = y$ ...(2)

Differentiating (2), $(n + 1)$ times using Leibnitz's theorem, gives

$$(1 + x^2)\, y_{n+2} + {}^{n+1}C_1.\, 2\, x\, y_{n+1} + {}^{n+1}C_2.\, 2\, y_n = y_{n+1}$$

or $(1 + x^2)\, y_{n+2} + \{2\,(n+1)\,x - 1\}\, y_{n+1} + n\,(n+1)\, y_n = 0$ ...(3)

At $x=0$, relation (3) becomes

$$(y_{n+2})_0-(y_{n+1})_0+n(n+1)(y_n)_0=0 \qquad ...(4)$$

Taking $n=0$ in (4), we get

$$(y_2)_0-(y_1)_0+0(y)_0=0 \quad \text{or} \quad (y_2)_0=(y_1)_0$$

But from (2), $(y_1)_0=y(0)=1$ hence $(y_2)_0=1$

Taking $n=1$ in (4), we get

$$(y_3)_0-(y_2)_0+2(y_1)_0=0 \quad \Rightarrow \quad (y_3)_0=-1$$

Taking $n=2$ in (4), we get

$$(y_4)_0-(y_3)_0+2.3(y_2)_0=0 \quad \Rightarrow \quad (y_4)_0=-7$$

and lastly, taking $n=3$ in (4), we get

$$(y_5)_0-(y_4)_0+3.4(y_3)_0=0 \quad \Rightarrow \quad (y_5)_0=5.$$ **Ans.**

EXMAPLE 1.13. If $y=\dfrac{\sin^{-1}x}{\sqrt{1-x^2}}$, prove that

$(1-x^2)\,y_{n+1}-(2n+1)\,xy_n-n^2\,y_{n-1}=0$ and hence find $(y_n)_0$.

SOLUTION: The given relation can be written as

$$y\sqrt{1-x^2}=\sin^{-1}x \qquad ...(1)$$

On differentiating w.r.t. x on both sides of (1), gives

$$\sqrt{1-x^2}\,y_1-\frac{2x}{2\sqrt{1-x^2}}\,y=\frac{1}{\sqrt{1-x^2}}$$

or $$(1-x^2)\,y_1-xy=1 \qquad ...(2)$$

Differentiating n times using Leibnitz's theorem, we get

$$(1-x^2)\,y_{n+1}+{}^nC_1(-2x)\,y_n+{}^nC_2(-2)\,y_{n-1}-x\,y_n-{}^nC_1.\,1.\,y_{n-1}=0$$

or $$(1-x^2)\,y_{n+1}-(2n+1)\,x\,y_n-n^2\,y_{n-1}=0,$$ **Hence the result**

At $x=0$ the above relation becomes $(y_{n+1})_0=n^2(y_{n-1})_0$

or $$(y_n)_0=(n-1)^2\,(y_{n-2})_0$$

$$=(n-1)^2\,(n-3)^2\,(y_{n-4})_0=(n-1)^2\,(n-3)^2\,(n-5)^2\,(y_{n-6})_0=.....$$

$$\Rightarrow \quad (y_n)_0=\begin{cases}(n-1)^2\,(n-3)^2\,(n-5)^2\,......3^2.1^2\,(y)_0 & \text{when } n \text{ is even}\\(n-1)^2\,(n-3)^2\,(n-5)^2\,......4^2.2^2\,(y_1)_0 & \text{when } n \text{ is odd}\end{cases}$$

Since $y(0)=0$ we have $(y_n)_0=0$ when n is even,

and since, from (2), $(y_1)_0=1$, we have

$$(y_n)_0=(n-1)^2(n-3)^2(n-5)^2\,......\,2^2.\,1 \text{ when } n \text{ is odd.}$$ **Ans.**

EXMAPLE 1.14. If $x+y=1$, prove that

$$D^n(x^n y^n) = n!\left[y^n - \left({}^nC_1\right)^2 y^{n-1}x + \left({}^nC_2\right)^2 y^{n-2}x^2 - \dots + (-1)^n x^n\right]$$

where $D \equiv \dfrac{d}{nx}$.

SOLUTION: Here $x + y = 1$ therefore, $\dfrac{dy}{dx} = Dy = -1$.

Applying Leibnitz's theorem, to $D^n(x^n y^n)$, we get

$$D^n(x^n y^n) = y^n D^n(x^n) + {}^nC_1 D(y^n) D^{n-1}(x^n) + {}^nC_2 D^2(y^n) D^{n-2}(x^n) + \dots + D^n(y^n).x^n.$$

$$= y^n n! + {}^nC_1 . n y^{n-1}(-1). n!\, x + {}^nC_2\, n(n-1)(-1)^2 y^{n-2} n! \frac{x^2}{2!} + \dots + n!(-1)^n x^n$$

$$= n!\left[y^n - {}^nC_1 \cdot n y^{n-1} x + {}^nC_2 \cdot \frac{n(n-1)}{2!} y^{n-2} x^2 - \dots + (-1)^n x^n\right]$$

or $$D^n(x^n y^n) = n!\left[y^n - \left({}^nC_1\right)^2 \cdot y^{n-1}x + \left({}^nC_2\right)^2 y^{n-2}x^2 - \left({}^nC_3\right)^2 y^{n-3}x^3 + \dots + (-1)^n x^n\right].$$

Hence the result.

EXAMPLE 1.15. If $V_n = \dfrac{d^n}{dx^n}\left(x^n \log x\right)$, establish the relation

$V_n = n V_{n-1} + (n-1)!$ and hence show that

$V_n = n!\left(\log x + 1 + \dfrac{1}{2} + \dfrac{1}{3} + \dots + \dfrac{1}{n}\right)$. **(UPTU 2001)**

SOLUTION: $$V_n = \frac{d^n}{dx^n}\left(x^n \log x\right) = \frac{d^{n-1}}{dx^{n-1}}\left\{\frac{d}{dx}\left(x^n \log x\right)\right\} = \frac{d^{n-1}}{dx^{n-1}}\left\{x^n \frac{1}{x} + n x^{n-1} \log x\right\}$$

$$= \frac{d^{n-1}}{dx^{n-1}} x^{n-1} + n \frac{d^{n-1}}{dx^{n-1}}\left(x^{n-1} \log x\right) = (n-1)! + n V_{n-1} \qquad \text{...(1)}$$

Dividing throughout by $n!$, we get

$$\frac{V_n}{n!} = \frac{(n-1)!}{n!} + \frac{n}{n!} V_{n-1} = \frac{1}{n} + \frac{V_{n-1}}{(n-1)!} \qquad \text{...(2)}$$

Replacing n by $n-1$ in the above relation, we get

$$\frac{V_{n-1}}{(n-1)!} = \frac{1}{n-1} + \frac{V_{n-2}}{(n-2)!}$$

Therefore, (2) becomes

$$\frac{V_n}{n!} = \frac{1}{n} + \frac{1}{n-1} + \frac{V_{n-2}}{(n-2)!}$$ (again using (2) replacing n by $n-2$, we have)

$$= \frac{1}{n} + \frac{1}{n-1} + \frac{1}{n-2} + \frac{V_{n-3}}{(n-3)!}$$

$$= \ldots\ldots\ldots\ldots\ldots\ldots\ldots$$

$$= \ldots\ldots\ldots\ldots\ldots\ldots\ldots$$

$$= \frac{1}{n} + \frac{1}{n-1} + \frac{1}{n-2} + \frac{1}{n-3} + \ldots + \frac{1}{2} + \frac{1}{1} + \frac{V_0}{0!}$$

But $\quad V_0 = x^0 \log x = \log x$

Therefore, $\quad V_n = n!\left[\log x + 1 + \frac{1}{2} + \frac{1}{3} + \ldots + \frac{1}{n}\right].$ **Hence Proved.**

EXAMPLE 1.16. If $y = e^{x^2/2} \cos x$, show that

$$y_{n+2}(0) - 2n\, y_n(0) + n(n-1)\, y_{n-2}(0) = 0.$$

SOLUTION: We have $y = e^{x^2/2} \cos x$ which on differentiation, gives

$$y_1 = e^{x^2/2}\, x\cos x - e^{x^2/2} \sin x = xy - e^{x^2/2} \sin x$$

and $\quad y_2 = xy_1 + y - e^{x^2/2}\, x \sin x - e^{x^2/2} \cos x$

$$= xy_1 + y + x(y_1 - xy) - y = 2xy_1 - x^2 y$$

Now, differentiating the above relation n times and using Leibnitz's theorem, gives

$$y_{n+2} = 2xy_{n+1} + {}^nC_1 \cdot 2 \cdot y_n - (x^2 y_n + {}^nC_1\, 2xy_{n-1} + {}^nC_2\, 2\, y_{n-2})$$

or $\quad y_{n+2} - 2xy_{n+1} + (x^2 - 2n)\, y_n + 2nxy_{n-1} + n(n-1)\, y_{n-2} = 0$

Putting $x = 0$ throughout, we get

$$y_{n+2}(0) - 2n\, y_n(0) + n(n-1)\, y_{n-2}(0) = 0.$$ **Hence the result.**

EXAMPLE 1.17. If $y = \sqrt{\dfrac{1+x}{1-x}}$ prove that $y = (1 - x^2)\, y_1$ and hence show that

$$(1 - x^2)\, y_n - \{2(n-1)x + 1\}\, y_{n-1} - (n-1)(n-2)\, y_{n-2} = 0.$$ **(UPTU 2011-12)**

SOLUTION: The given relation can be written as

$$(1 - x)\, y^2 = 1 + x \qquad \ldots(1)$$

This, on differentiation w.r.t. x, gives $(1 - x)\, 2yy_1 - y^2 = 1$

or $\quad (1 - x)\, 2yy_1 = 1 + y^2 = 1 + \frac{1+x}{1-x} = \frac{2}{1-x}$

Multiplying throughout by y, gives

$$(1 - x)\, 2\, y^2\, y_1 = \frac{2y}{1-x}.$$ Using (1) here, we get $(1 - x)^2 \left(\frac{1+x}{1-x}\right) y_1 = y$

or $\quad (1 - x^2)\, y_1 = y$ which is the desired relation.

Now, differentiating this relation n times, with respect to x and using Lebneitz's theorem, we get

$$(1 - x^2)\, y_{n+1} + {}^nC_1(-2x)\, y_n + {}^nC_2(-2)\, y_{n-1} = y_n$$

or $\quad (1 - x^2)\, y_{n+1} - (2nx + 1)\, y_n - n(n-1)\, y_{n-1} = 0$

Replacing n by $n - 1$ here, gives

$$(1 - x^2)\, y_n - \{2(n-1)x + 1\}\, y_{n-1} - (n-1)(n-2) y_{n-2} = 0.$$ **Hence the desired result.**

EXAMPLE 1.18. If $y = e^{m \sin^{-1} x}$, show that $(1-x^2)xy_{n+2} - (2n+1)\, y_{n+1} - (m^2+n^2)y_n = 0$ and hence find $y_n\,(0)$. **(UPTU 2009–10)**

SOLUTION: Given $y = e^{m \sin^{-1} x}$ or $\log y = m \sin^{-1} x$. Differentiating w.r.t. x, we get.

$$\frac{1}{y}\, y_1 = \frac{m}{\sqrt{1-x^2}} \quad \text{or} \quad y^2_1\,(1-x^2) = m^2\, y^2. \qquad \text{... (1)}$$

On further differentiation, we get

$$(1-x^2)\, 2y_1y_2 - 2xy_1{}^2 = 2m^2\, y\, y_1 \quad \text{or} \quad (1-x^2)\, y_2 - xy_1 - m^2y = 0 \qquad \text{... (2)}$$

Differentiating the above relation n times using Lebneitz's theorem gives

$$(1-x^2)y_{n+2} + n(-2x)y_{n+1} + \frac{n(n-1)}{2}(-2)y_n - xy_{n+1} - n.1y_n - m^2y_n = 0$$

or $$(1-x^2)y_{n+2} - (2n+1)xy_{n+1} - (m^2+n^2)\, y_n = 0. \qquad \text{... (3)}$$

At $x = 0$ the above relation becomes

$$y_{n+2}\,(0) = (m^2+n^2)\, y_n\,(0)$$

$$\Rightarrow \quad y_n\,(0) = [m^2 + (n-2)^2]\, y_{n-2}\,(0) = [m^2 + (n-2)^2]\,[m^2 + (n-4)^2]\, y_{n-4}\,(0) = \ldots$$

$$= \begin{cases} [m^2+(n-2)^2][m^2+(n-4)^2][m^2+(n-6)^2]\ldots[m^2+1]m & \text{when } n \text{ is odd} \\ [m^2+(n-2)^2][m^2+(n-4)^2][m^2+(n-6)^2]\ldots[m^2+2^2]\, y_2(0) & \text{when } n \text{ is even} \end{cases}$$

But from (1) $y_1(0) = my(0) = m$ and from (2), $y_2(0) = m^2y(0) = m^2$.

$$\therefore \quad y_n\,(0) = \begin{cases} [m^2+(n-2)][m^2+(n-4)^2][m^2+(n-6)^2]\ldots[m^2+1]m & \text{when } n \text{ is odd} \\ [m^2+(n-2)^2][m^2+(n-4)^2][m^2+(n-6)^2]\ldots[m^2+2^2]\, y_2(0) & \text{when } n \text{ is even} \end{cases}$$ **Ans.**

EXAMPLE 1.19. If $y = (\sin^{-1}x)^2$, prove that $(1-x^2)\, y_{n+2} - (2n+1)xy_{n+1} - n^2y_n = 0$ and hence find $y_n\,(0)$. **(UPTU 2005-06; UPTU 2008-09)**

SOLUTION: We are given $y = (\sin^{-1}x)^2$ hence $y_1 = \dfrac{2\sin^{-1} x}{\sqrt{1-x^2}}$ ...(1)

or $$(1-x^2)\, y^2_1 = 4y \quad \text{which on differentiation again, gives}$$

$$(1-x^2)\, 2y_1y_2 - 2xy^2_1 = 4y_1 \quad \text{or} \quad (1-x^2)\, y_2 - xy_1 = 2 \qquad \text{... (2)}$$

Now differentiating (2) n times w.r.t. x using Lebneitz's theorem, gives

$$(1-x^2)\, y_{n+2} - n(-2x)\, y_{n+1} + \frac{n(n-1)}{2}.\,(-2)\, y_n - xy_{n+1} - n\,.\,y_n = 0$$

or $$(1-x^2)\, y_{n+2} - (2n+1)\, xy_{n+1} - n^2y_n = 0 \qquad \text{... (3)}$$

Next, at $x = 0$ relation (3) becomes $y_{n+2}\,(0) = n^2\, y_n\,(0)$ ... (4)

or $$y_n(0) = (n-2)^2\, y_{n-2}\,(0) = (n-2)^2\,(n-4)^2\, y_{n-4}\,(0)$$

$$= (n-2)^2\,(n-6)^2\, y_{n-6}\,(0) = \ldots.$$

$$= \begin{cases} (n-2)^2\,(n-4)^2\,(n-6)^2 \,.....\, 2^2\, y_2(0) & \text{when } n \text{ is even} \\ (n-2)^2\,(n-4)^2\,(n-6)^2 \,.....\, 1^2\, y_1(0) & \text{when } n \text{ is odd} \end{cases}$$

But from (1), $y_1(0) = 0$ and from (2), $y_2(0) = 2$, therefore

$$y_n(0) = \begin{cases} 0 & \text{when } n \text{ is odd} \\ (n-2)^2\,(n-4)^2 (n-6)^2 .....2^2.\,2 & \text{when } n \text{ is even.} \end{cases}$$ **Ans.**

EXAMPLE 1.20. If $y^{1/m} + y^{-1/m} = 2x$ show that

$(x^2 - 1)\, y_{n+2} + (2n + 1)\, x\, y_{n+1} + (n^2 - m^2)\, y_n = 0.$ **(UPTU 2001, 2002)**

SOLUTION: The given relation can be written as

$$y^{2/m} - 2xy^{1/m} + 1 = 0 \text{ whose roots are } y^{1/m} = \frac{2x \pm \sqrt{4x^2 - 4}}{2}$$

$\therefore \quad y = (x \pm \sqrt{x^2 - 1})^m \quad$ or $\quad \log y = m \log\, ((x \pm \sqrt{x^2 - 1})$

Differentiating *w.r.t.* x, gives

$$\frac{1}{y}\, y_1 = \frac{m}{x \pm \sqrt{x^2 - 1}} \left[1 \pm \frac{x}{\sqrt{x^2 - 1}} \right] = \pm \frac{m}{\sqrt{x^2 - 1}}$$

or $\quad (x^2 - 1)\, y_1^2 = m^2\, y^2$. This, on further differentiating *w.r.t.* x, gives

$(x^2 - 1)\, 2y_1\, y_2 + 2x\, y_1^2 = 2m^2\, yy_1 \quad$ or $\quad (x^2 - 1)y_2 + xy_1 = m^2 y.$

Differentiating the above relation n times, using Leibnitz's theorem, we get

$$(x^2 - 1)y_{n+2} + n.2x\, y_{n+1} + \frac{n(n-1)}{2}.2.\, y_n + x\, y_{n+1} + n.\, y_n = m^2\, y_n$$

or $\quad (x^2 - 1)\, y_{n+2} + (2n + 1)\, x\, y_{n+1} + (n^2 - m^2)\, y_n = 0 \quad$ **Hence the result.**

EXAMPLE 1.21. **(a)** If $y = \cos\,(m \sin^{-1} x)$ show that

$$(1 - x^2)\, y_{n+2} - (2n + 1)x\, y_{n+1} + (m^2 - n^2)\, y_n = 0$$

and hence find y_n (0). **(UPTU 2001, 2004-05)**

(b) If $y = \sin\,(a \sin^{-1} x)$, find $(y_n)_0$. **(UPTU 2008-09, 2010-11)**

SOLUTION: (a) $y = \cos\,(m \sin^{-1} x), \quad \therefore \quad y_1 = \sin\,(m \sin^{-1} x).\, \dfrac{-m}{\sqrt{1 - x^2}}$ ...(1)

or $\quad (1 - x^2)\, y_1^2 = m^2\, (1 - y^2)$. Again differentiation, gives

$(1 - x^2)\, 2y_1\, y_2 - 2xy_1^2 = -\, 2m^2 y\, y_1 \quad$ or $\quad (1 - x^2)\, y_2 - xy_1 + m^2 y = 0$ ...(2)

Differentiating (2) n times, using Leibnitz's theorem, gives

$$(1 - x^2)y_{n+2} + n\,(-\,2x)y_{n+1} + \frac{n(n-1)}{2}\,(-\,2)\, y_n - xy_{n+1} - n\,.\,1.\, y_n + m^2\, y_n = 0$$

or $\quad (1 - x^2)y_{n+2} - (2n + 1)\, xy_{n+1} + (m^2 - n^2)y_n = 0$ ...(3)

which is the required result.

Now to find y_n (0), we put $x = 0$ in (3) and get

$y_{n+2}\,(0) = (n^2 - m^2)\, y_n(0) \quad$ or $\quad y_n(0) = [(n - 2)^2 - m^2]\, y_{n-2}\,(0)$ ...(4)

$\therefore \qquad y_n(0) = [(m-2)^2 - m^2]\,[(n-4)^2 - m^2]\,y_{n-4}(0)$

$$= \begin{cases} [n-2)^2 - m^2][(n-4)^2 - m^2] \ldots (1^2 - m^2)\,y(0) & \text{if } n \text{ is odd} \\ [(n-2)^2 - m^2][(n-4)^2 - m^2] \ldots (2^2 - m^2)\,y_2(0) & \text{if } n \text{ is even} \end{cases}$$

From (1), we have $y_1(0) = 0$ and from (2), $y_2(0) = -m^2 y(0) = -m^2$.

Therefore, $y_n(0) = \begin{cases} 0 & \text{if } n \text{ is odd} \\ [(n-2)^2 - n^2][n-4)^2 - m^2] \ldots [2^2 - m^2](-m^2) & \text{if } n \text{ is even} \end{cases}$ **Ans.**

(b) $y = \sin(a \sin^{-1} x) \quad \therefore \quad y_1 = \cos(a \sin^{-1} x) \cdot \dfrac{a}{\sqrt{1-x^2}}$ ...(1)

or $(1 - x^2)\,y_1^2 = a^2(1 - y^2)$. Again differentiation, gives

$(1 - x^2)\,2y_1y_2 - 2xy_1^2 = -2a^2yy_1$

or $(1 - x^2)y_2 - xy_1 + a^2y = 0$ provided $y_1 \neq 0$ ...(2)

Differentiating (2) n times, using Leibnitz theorem, gives

$$(1 - x)^2y_{n+2} + n(-2x)y_{n+1} + \frac{n(n-1)}{2}(-2)\,y_n - xy_{n+1} - n \cdot 1 \cdot y_n + a^2y_n = 0$$

or $(1 - x^2)\,y_{n+2} - (2n + 1)\,xy_{n+1} + (a^2 - n^2)\,y_n = 0.$ ...(3)

Now to find $(y_n)_0$ we put $x = 0$ in (3) and get

$(y_{n+2})_0 = (n^2 - a^2)\,(y_n)_0$

or $(y_n)_0 = [(n-2)^2 - a^2]\,(y_{n-2})_0$ ...(4)

$$\therefore \qquad (y_n)_0 = \begin{cases} [(n-2)^2 - a^2][(n-4)^2 - a^2] \ldots (2^2 - a^2)\,y_0(0) & \text{if } n \text{ is even} \\ [(n-2)^2 - a^2][(n-4)^2 - a^2] \ldots (1^2 - a^2)\,y_1(0) & \text{if } n \text{ is odd} \end{cases}$$

from (1) we have $y(0) = 0$ and $y_1(0) = a$ and from (2) $y_2(0) = 0$

Therefore $(y_n)_0 = \begin{cases} 0 & \text{if } n \text{ is even} \\ [(n-2)^2 - a^2][(n-4)^2 - a^2] \ldots (1 - a^2)a & \text{if } n \text{ is odd} \end{cases}$ **Ans.**

EXAMPLE 1.22. If $x = \tan y$, prove that $(1 + x^2)\,y_{n+1} + 2nxy_n + n(n-1)\,y_{n-1} = 0.$

(UPTU 2006–07)

SOLUTION: $x = \tan y$ or $y = \tan^{-1} x$ hence $y_1 = \dfrac{1}{1+x^2}$ or $(1 + x^2)\,y_1 = 1.$

Differentiating both sides n times w.r.t. x, using Leibnitz theorem, we get

$$(1+x^2)\,y_{n+1} + n \cdot 2xy_n + \frac{n(n-1)}{2!} \cdot 2y_{n-1} = 0$$

or $\qquad (1+x^2)\,y_{n+1} + 2nxy_n + n(n-1)\,y_{n-1} = 0.$ **Hence Proved**

EXAMPLE 1.23. If $y=(x^2-1)^n$ prove that $(x^2-1)\,y_{n+2}+2xy_{n+1}-n\,(n+1)\,y_n=0.$

(UPTU 2000–01; 2002–03)

SOLUTION: $y=(x^2-1)^n$ hence $y_1=n\,(x^2-1)^{n-1}\cdot 2x$ or $(x^2-1)\,y_1=2nxy$

Differentiating $(n+1)$ times w.r.t. x on both sides and using Leibnitz theorem, we get

$$(x^2-1)\,y_{n+2}+(n+1)\,2x\,y_{n+1}+\frac{(n+1)\,n}{2!}\cdot 2y_n = 2nx\,y_{n+1}+(n+1)\,2n\,y_n$$

or $$(x^2-1)\,y_{n+2}+2xy_{n+1}+y_n\,(n^2+n-2n^2-2n)=0$$

or $$(x^2-1)\,y_{n+2}+2xy_{n+1}-n\,(n+1)\,y_n=0$$ **Hence Proved**

EXAMPLE 1.24. If $x=\tan(\log y)$ show that $(1+x^2)y_{n+1}+(2nx-1)y_n+n(n-1)y_{n-1}=0$

OR

If $y=e^{\tan^{-1}x}$, show that

$(1+x^2)y_{n+2}+\{2(n+1)x-1\}y_{n+1}+n(n+1)y_n=0$ **(UPTU 2003)**

SOLUTION: $\log y=\tan^{-1}x$, hence $\dfrac{1}{y}y_1=\dfrac{1}{1+x^2}$ or $(1+x^2)y_1=y.$

Differentiating it again, we get

$$(1+x^2)y_2+2xy_1=y_1 \quad \text{or} \quad (1+x^2)y_2+(2x-1)y_1=0$$

Now, differentiating n times w.r.t. x, using Leibnitz theorem, we get

$$(1+x^2)y_{n+2}+{}^nC_1\,2x\,y_{n+1}+{}^nC_2\cdot 2y_n+(2x-1)y_{n+1}+{}^nC_1\cdot 2y_n=0$$

or $$(1+x^2)y_{n+2}+\{2(n+1)x-1\}y_{n+1}+n(n+1)y_n=0.$$ **Hence Proved.**

EXAMPLE 1.25. Find the value of $D^n(x^{n-1}\log x)$ where $D\equiv\dfrac{d}{dx}$. **(UPTU 2009-10, 2011-12)**

SOLUTION: Using Leibnitz theorem we have

$$D^n(x^{n-1}\log x)=\log x\cdot D^n(x^{n-1})+{}^nC_1\cdot\frac{1}{x}D^{n-1}(x^{n-1})+{}^nC_2\cdot\left(\frac{-1}{x^2}\right)D^{n-2}(x^{n-1})+{}^nC_3\left(\frac{2}{x^3}\right)D^{n-3}(x^{n-1})+\ldots$$

$$+\frac{d^n}{dx^n}(\log x)\cdot x^{n-1}$$

$$=0+{}^nC_1\left(\frac{1}{x}\right)(n-1)!+{}^nC_2\left(\frac{-1}{x^2}\right)(n-1)!\,x+{}^nC_3\left(\frac{2}{x^3}\right)(n-1)!\,\frac{x^2}{2}+\ldots+\frac{(-1)^{n-1}(n-1)!}{x^n}\cdot x^{n-1}$$

$$=\frac{(n-1)!}{x}[{}^nC_1-{}^nC_2+{}^nC_3-\ldots+(-1)^{n-1}] = \frac{(n-1)!}{x}(1) \quad (\text{since } 1-{}^nC_1+{}^nC_2-{}^nC_3+\ldots=0).$$

Ans.

EXERCISE 1B

1. Find the nth derivative of

   (*i*) $x^2 \sin x$ **(UPTU 2008-09)** (*ii*) $e^x \log x$. **(UPTU 2001)**

2. If $y = x \log(x+1)$, show that $y_n = \dfrac{(-1)^{n-2}(n-2)!(x+n)}{(x+1)^n}$. **(UPTU 2006-07)**

3. If $y = a\cos(\log x) + b\sin(\log x)$, show that

   $x^2 y_{n+2} + (2n+1)x\, y_{n+1} + (n^2+1)\, y_n = 0$. **(UPTU 2003-04, 2011-12)**

4. If $f(x) = x^2 e^{-x/a}$, show that $f_n(0) = \dfrac{n(n-1)(-1)^n}{a^{n-2}}$.

5. If $y = \tan^{-1}\left(\dfrac{a+x}{a-x}\right)$, prove that $(a^2+x^2)\, y_{n+2} + 2(n+1)\, x\, y_{n+1} + n(n+1)\, y_n = 0$.

   (UPTU 2011-12)

6. Given that $y = \left\{\log\left(x + \sqrt{a^2+x^2}\right)\right\}^2$, show that

   (*i*) $(a^2+x^2)y_2 + xy_1 = 2$ (*ii*) $(a^2+x^2)\, y_{n+2} + (2n+1)\, xy_{n+1} + n^2 y_n = 0$.

7. Given that $y = \left(x - \sqrt{x^2-1}\right)^m$, show that

   $$(x^2-1)\, y_{n+2} + (2n+1)\, xy_{n+1} + (n^2 - m^2)\, y_n = 0.$$

8. If $f(x) = \cot x$, prove that

   ${}^nC_1 f_{n-1}(0) - {}^nC_3 f_{n-3}(0) + {}^nC_5 f_{n-5}(0) - \ldots = \cos n\pi/2$.

9. If $x = \sin t$, $y = \cos at$, show that

   $(1-x^2)\, y_1^2 = a^2(1-y^2)$ and hence deduce that

   $(1-x^2)\, y_{n+2} - (2n+1)\, xy_{n+1} - (n^2 - a^2)\, y_n = 0$.

10. If $y = (1+x^2)\sin(m\tan^{-1} x)$, find $y_n(0)$.

11. If $\cos^{-1}(y/b) = \log(x/n)^n$, show that $x^2 y_{n+2} + (2n+1)xy_{n+1} + 2n^2 y_n = 0$. **(UPTU 2005)**

12. Find the nth derivative of the function $y = x^2 \sin x$ at $x = 0$. **(UPTU 2008-09)**

❑❑❑

EXPANSION OF FUNCTIONS

MACLAURIN'S THEOREM

If a function $f(x)$ is differentiable any number of times and can be expanded in a convergent series of terms of positive integral powers of x, then

$$f(x) = f(0)+xf'(0)+\frac{x^2}{2!}f''(0)+\frac{x^3}{3!}f'''(0)+...+\frac{x^n}{n!}f^{(n)}(0)+...$$

PROOF: Let $f(x) = a_0 + a_1 x + a_2 x^2 + a_3 x^3 + ...+ a_n x^n + ...$ ... (1)

where $a_0, a_1, a_2, a_3, ...$ are to be determined.

Differentiating (1) successively, we get

$$f'(x) = a_1+2a_2x+3a_3x^2+....+na_nx^{n-1}+.....$$

$$f''(x) = 2a_2+3.2a_3x+4.3a_4x^2+....+n(n-1)a_nx^{n-2}+...$$

$$f'''(x) = 3\cdot2a_3+4\cdot3\cdot2a_4x+5\cdot4\cdot3a_5x^2+....+n(n-1)(n-2)a_nx^{n-3}+...$$

and, in general,

$$f^{(n)}(x) = n(n-1)(n-2)...3\cdot2\cdot1\cdot a_n + \text{term with positive powers of } x.$$

At $x=0$, we have

$$f(0) = a_0,\ f'(0)=a_1,\ f''(0)=2.1a_2,\ f'''(0)=3.2.1.\ a_3, ...$$

and, in general, $f^{(n)}(0) = n!\ a_n$.

Substituting these values of $a_0, a_1, a_2, ...$ in (1), we get

$$f(x) = f(0)+xf'(0)+\frac{x^2}{2!}f''(0)+\frac{x^3}{3!}f'''(0)+...+\frac{x^n}{n!}f^{(n)}(0)+...$$

or $$f(x) = \sum_{n=0}^{\infty}\frac{x^n}{n!}f^{(n)}(0)$$ ...(2)

which is the expansion of $f(x)$ in ascending powers of x and is known as Maclaurin's expansion of $f(x)$.

The conditions under which the expansion (2) is valid, are

(*i*) *$f(x)$ and its successive derivatives must be finite and continous in the range of x in which $f(x)$ is defined.*

(*ii*) *the series on the right hand side of (2) must be convergent.*

For the condition of convergence of the power series (2), if we write

$$f(x) = f(0)+xf'(0)+\frac{x^2}{2!}f''(0)+...+\frac{x^{n-1}}{(n-1)!}f^{(n-1)}(0)+R_n$$

then the remainder R_n should tend to 0 as n tends to ∞. The Lagrangian form of the remainder R_n is

$$R_n = \frac{x^n}{n!}f^{(n)}(\theta x) \quad \text{where} \quad 0<\theta<1.$$

Maclaurin's Expansion of Some Standard Functions

1. *Expansion of* $\sin x$

Let $f(x) = \sin x$, then $f^{(n)}(x) = \sin(x + n\pi/2)$ and $f^{(n)}(0) = \sin n\pi/2$

$$\therefore \quad \sin x = \sum_{1}^{\infty} \frac{x^n}{n!} \sin\frac{n\pi}{2}$$

or $$\mathbf{\sin x = x - \frac{x^3}{3!} + \frac{x^5}{5!} - \frac{x^7}{7!} + \ldots}$$

2. *Expansion of* $\cos x$

Since $\cos x = \dfrac{d}{dx}\sin x$, the expansion can also be obtained by differentiating the expansion of $\sin x$ term by term,

hence $$\mathbf{\cos x = 1 - \frac{x^2}{2!} + \frac{x^4}{4!} - \frac{x^6}{6!} + \ldots}$$

3. *Expansion of* e^x

If $f(x) = e^x$, $f^{(n)}(x) = e^x$ and $f^{(n)}(0) = 1$

therefore, $$\mathbf{e^x = 1 + x + \frac{x^2}{2!} + \frac{x^3}{3!} + \ldots\ldots + \frac{x^n}{n!} + \ldots\ldots}$$

4. *Expansion of* $\log(1 + x)$

If $f(x) = \log(1 + x)$, *then* $f^{(n)}(x) = \dfrac{(-1)^{n-1}(n-1)!}{(1+x)^n}$

hence $f^{(n)}(0) = (-1)^{n-1}(n-1)!$.

Thus, $$\log(1+x) = \log 1 + \sum_{1}^{\infty}(-1)^{n-1}(n-1)!\frac{x^n}{n!} = \sum_{1}^{\infty}\frac{(-1)^{n-1}x^n}{n}$$

or $$\mathbf{\log(1+x) = x - \frac{x^2}{2} + \frac{x^3}{3} - \frac{x^4}{4} + \ldots \ (|x| < 1)}$$

or $$\mathbf{\log(1-x) = -x - \frac{x^2}{2} - \frac{x^3}{3} - \frac{x^4}{4} - \ldots - \ (|x| < 1)}$$

5. *Expansion of* $\tan x$

If $f(x) = \tan x$, then $f'(x) = \sec^2 x$,

$f''(x) = 2\sec^2 x \tan x$, $f'''(x) = 2\sec^4 x + 4\sec^2 x \tan^2 x$,

hence $f(0) = 0$, $f'(0) = 1$, $f''(0) = 0$

and $f'''(0) = 2, f^{(iv)}(0) = 0$, $f^{(v)}(0) = 16$, and so on.

Therefore, $$\mathbf{\tan x = x + \frac{x^3}{3} + \frac{2}{15}x^5 + \ldots}$$

6. *Binomial Expansion of* $(1 + x)^n$ for $|x| < 1$.

$$\mathbf{(1+x)^n = 1 + nx + \frac{n(n-1)}{2!}x^2 + \frac{n(n-1)(n-2)}{3!}x^3 + \ldots \ (|x| < 1).}$$

TAYLOR'S THEOREM

Let $f(x)$ be a function of x and h be small. If the function $f(x+h)$ is capable of being expanded in a convergent series of terms of positive integral powers of h, then this expansion is given by

$$f(x+h) = f(x) + h f'(x) + \frac{h^2}{2!} f''(x) + \frac{h^3}{3!} f'''(x) + \ldots\ldots + \frac{h^n}{n!} f^{(n)}(x) + \ldots\ldots$$

PROOF: Assume that $f(x+h) = A_0 + A_1 h + A_2 h^2 + A_3 h^3 + \ldots + A_n h^n + \ldots$ ...(1)

where A's are functions of x

Differentiating successively w.r.t. h, we get

$$f'(x+h) = A_1 + 2A_2 h + 3A_3 h^2 + 4A_4 h^3 + \ldots + nA_n h^{n-1} + \ldots$$

$$f''(x+h) = 2\cdot A_2 + 3\cdot 2\cdot A_3 h + 4\cdot 3\cdot A_4 h^2 + \ldots + n(n-1) A_n h^{n-2} + \ldots$$

$$f'''(x+h) = 3\cdot 2\cdot A_3 + 4\cdot 3\cdot 2\cdot A_4 h + \ldots + n(n-1)(n-2) A_n h^{n-3} + \ldots$$

and, in general, $f^{(n)}(x+h) = n(n-1)(n-2) \ldots 3.2\, A_n$ + terms in ascending powers of h.

Putting $h = 0$, we get $f^{(n)}(x) = n!\, A_n$ so that $A_n = \dfrac{f^{(n)}(x)}{n!}$

Substituting these values of A_0, A_1, A_2, ... in (1), we get

$$f(x+h) = f(x) + \frac{h}{1!} f'(x) + \frac{h^2}{2!} f''(x) + \frac{h^3}{3!} f'''(x) + \ldots + \frac{h^n}{n!} f^{(n)}(x) + \ldots \quad \text{...(2)}$$

Its another form can be obtained by replacing x by a and h by $x - a$, so as to get

$$f(x) = f(a) + (x-a) f'(a) + \frac{(x-a)^2}{2!} f''(a) + \ldots + \frac{(x-a)^n}{n!} f^{(n)}(a) + \ldots \quad (3)$$

The conditions under which the above expansion is valid, are

(*i*) *the function $f(x)$ and its derivatives must be finite and continuous in the range of definition of $f(x)$.*

(*ii*) *the series on the right hand side of (2) must be convergent for which the remainder term*

$$R_n \to 0 \text{ as } n \to \infty \text{ where } R^n = \frac{h^n}{n!} f^{(n)}(x + \theta h) \text{ and } 0 < \theta < 1.$$

In the form (3) of Taylor's expansion, if we take $a = 0$ then we have

$$f(x) = f(0) + x f'(0) + \frac{x^2}{2!} f''(0) + \ldots$$

which is nothing but the **Maclaurin's expansion of $f(x)$**. Thus Maclurin's expansion is a particular case of Taylor's expansion. **With a slightly different approach we can show here that Taylor's series can be derived from Maclaurin's series.**

The Maclaurin's series for a function $g(x)$, is

$$g(x) = g(0) + x g'(0) + \frac{x^2}{2!} g''(0) + \ldots + \frac{x^n}{n!} g^{(n)}(0) + \ldots$$

If we replace here $g(x)$ by $f(x+h)$, then

$$g'(x) = f'(x+h),\quad g''(x) = f''(x+h),\quad \ldots\ g^{(n)}(x) = f^{(n)}(x+h), \ldots$$

Therefore, $g(0) = f(h),\ g'(0) = f'(h),\ g''(0) = f''(h), \ldots$

Thus, we get $$f(x+h) = f(h) + xf'(h) + \frac{x^2}{2!}f''(h) + \ldots + \frac{x^n}{n!}f^{(n)}(h) + \ldots$$

Now, in this relation we interchange x and h so as to get

$$f(x+h) = f(x) + hf'(x) + \frac{h^2}{2!}f''(x) + \ldots + \frac{h^n}{n!}f^{(n)}(x) + \ldots$$

which is Taylor's expansion of $f(x+h)$.

Thus, it can be concluded that Taylor's and Maclaurin's series are not essentially different.

EXAMPLE 1.26. ***(a)*** Expand $\tan^{-1}x$ in powers of $(x-1)$.

(b) Show that $\log(1+e^x) = \log 2 + \frac{x}{2} + \frac{x^2}{8} - \frac{x^4}{192} + \ldots$

SOLUTION: ***(a)*** The Taylor's expansion of $f(x)$ in powers of $(x-a)$, is

$$f(x) = f(a) + (x-a)f'(a) + \frac{(x-a)^2}{2!}f''(a) + \ldots + \frac{(x-a)^n}{n!}f^{(n)}(a) + \ldots$$

Here $f(x) = \tan^{-1}x$ and $a = 1$ therefore, $f(1) = \pi/4$,

and $f'(x) = \frac{1}{1+x^2}$ hence $f'(1) = \frac{1}{2}$.

Next $f''(x) = \frac{-2x}{(1+x^2)^2}$ hence $f''(1) = -\frac{1}{2}$,

and $$f'''(x) = \frac{(1+x^2)^2(-2) - (-2x)\cdot 2(1+x^2)\cdot 2x}{(1+x^2)^4} = \frac{6x^2-2}{(1+x^2)^3}, \qquad \therefore\ f'''(1) = \frac{1}{2}.$$

Therefore, we have

$$\tan^{-1}x = \frac{\pi}{4} + (x-1)\left(\frac{1}{2}\right) + \frac{(x-1)^2}{2!}\left(-\frac{1}{2}\right) + \frac{(x-1)^3}{3!}\left(\frac{1}{2}\right) + \ldots\ldots$$

or $$\tan^{-1}x = \frac{\pi}{4} + \frac{1}{2}(x-1) - \frac{1}{4}(x-1)^2 + \frac{1}{12}(x-1)^3 + \ldots\ldots$$ **Ans.**

(b) Let $f(x) = \log(1+e^x)$. By Maclaurin's theorem

$$f(x) = f(0) + xf'(0) + \frac{x^2}{2!}f''(0) + \frac{x^3}{3!}f'''(0) + \frac{x^4}{4!} + f^{\text{iv}}(0) + \ldots$$

Now $f(0) = \log 2,\quad f'(x) = \frac{e^x}{1+e^x} \quad \therefore\ f'(0) = \frac{1}{2}$

$$f''(x) = \frac{e^x}{(1+e^x)^2} \quad \therefore\ f''(0) = \frac{1}{4}$$

$$f'''(x) = \frac{e^x(e^x-1)}{(e^x+1)^3} \quad \therefore \quad f'''(0) = 0$$

and $$f^{iv}(x) = \frac{e^x(1+e^x)(2e^x-1) - 3e^{2x}(e^x-1)}{(e^x+1)^4} \quad \therefore \quad f^{iv}(0) = \frac{1}{8}, \ldots$$

Hence $$f(x) = \log(1+e^x) = \log 2 + \frac{x}{2} + \frac{x^2}{8} - \frac{x^4}{192} + \ldots$$ **Hence proved.**

EXAMPLE 1.27. **(a)** Expand the polynomial $2x^3 + 7x^2 + x - 1$ in powers of $(x-2)$.

(b) Obtain the series for $\log_e(1+x)$ and then find the series for $\log_e \frac{1+x}{1-x}$ and hence determine the value of $\log_e \frac{11}{9}$ upto five decimal places. **(UPTU 2011-12)**

SOLUTION: (a) Let $f(x) = 2x^3 + 7x^2 + x - 1$, then

$$f'(x) = 6x^2 + 14x + 1, \quad f''(x) = 12x + 14, \quad f'''(x) = 12.$$

Now, by Taylor's theorem for $f(x)$ about the point $x = 2$, we have

$$f(x) = f(2) + (x-2) f'(2) + \frac{(x-2)^2}{2!} f''(2) + \frac{(x-2)^3}{3!} f'''(2) + \ldots$$

Here $f(2) = 45$, $f'(2) = 53$, $f''(2) = 38$, $f'''(2) = 12$ hence

$$f(x) = 45 + 53(x-2) + 19(x-2)^2 + 2(x-2)^3.$$ **Ans.**

(b) We know, by Maclaurin's theorem, that

$$f(x) = f(0) + x f'(0) + \frac{x^2}{2!} f''(0) + \ldots$$

Let us take $f(x) = \log_e(1+x)$ then $f(0) = \log(1+0) = 0$,

$$f'(x) = \frac{1}{1+x}, \quad f''(x) = -\frac{1}{(1+x)^2}, \quad f'''(x) = \frac{2}{(1+x)^3}, \ldots$$

$$\therefore \quad f'(0) = 1, \quad f''(0) = -1, \quad f'''(0) = 2,$$

Therefore, we have $$\log_e(1+x) = x - \frac{x^2}{2} + \frac{x^3}{3} - \frac{x^4}{4} + \ldots, \quad \text{where } |x| < 1.$$

Also $$\log_e(1-x) = -x - \frac{x^2}{2} - \frac{x^3}{3} - \frac{x^4}{4} - \ldots \quad \text{where } |x| < 1.$$

$$\Rightarrow \quad \log_e\left(\frac{1+x}{1-x}\right) = \log_e(1+x) - \log_e(1-x)$$

$$= 2\left(x + \frac{x^3}{3} + \frac{x^5}{5} + \frac{x^7}{7} + \ldots\right)$$

Now, if we take $x = \frac{1}{10} = 0.1$ here, then we get

$$\log_e\left(\frac{1+\frac{1}{10}}{1-\frac{1}{10}}\right) = \log_e\left(\frac{11}{9}\right) = 2\left[0.1+\frac{1}{3}(0.1)^3+\frac{1}{5}(0.1)^5+...\right]$$

$$= 2[0.1 + 0.00033 + 0.000002 + ...]$$

$$= 0.200664.$$ **Ans.**

EXAMPLE 1.28. (*a*) Show that $\tan\left(\frac{\pi}{4}+x\right) = 1+2x+2x^2+\frac{8}{3}x^3+\frac{10}{3}x^4+...$

and hence calculate the value of tan 46° correct to four decimal places, given that $\pi = 3.14159$.

(*b*) Expand $\tan x$ in powers of $\left(x-\frac{\pi}{4}\right)$ upto first four terms.

SOLUTION: (*a*) Let $f(x) = \tan\left(\frac{\pi}{4}+x\right)$, then $f(0) = 1$.

Next, $f'(x) = \sec^2\left(\frac{\pi}{4}+x\right) = 1+\tan^2\left(\frac{\pi}{4}+x\right) = 1+f^2(x) \quad \therefore \quad f'(0) = 1+1 = 2$

Next $f''(x) = 2f(x)f'(x)$, hence $f''(0) = 2f(0)f'(0) = 4$.

and $f'''(x) = 2f(x)f''(x) + 2\{f'(x)\}^2$

hence $f'''(0) = 2f(0)f''(0) + 2\{f'(0)\}^2 = 16$

Now, $f^{iv}(x) = 2f(x)f'''(x) + 2f'(x)f''(x) + 4f'(x)f''(x)$,

hence $f^{iv}(0) = 2f(0)f'''(0) + 6f'(0)f''(0) = 80$

Therefore, $$\tan\left(\frac{\pi}{4}+x\right) = f(0)+xf'(0)+\frac{x^2}{2!}f''(0)+\frac{x^3}{3!}f'''(0)+\frac{x^4}{4!}f^{iv}(0)+...$$

$$= 1+2x+\frac{4}{2!}x^2+\frac{16}{3!}x^3+\frac{80}{4!}x^4+... = 1+2x+2x^2+\frac{8}{3}x^3+\frac{10}{3}x^4+...$$

Hence the result.

Taking $x = 1° = \frac{\pi}{180} = \frac{3.14159}{180} = 0.01745$

we have $x^2 = 0.000345$ and $x^3 = 0.000005, ...$

To achieve the desired accuracy we need consider only first four terms, hence

$$\tan 46° = 1 + 2(0.01745) + 2(0.000345) + \frac{8}{3}(0.000005)$$

$$= 1 + 0.03490 + 0.00069 = 1.0355$$

which is correct to four places of decimal.

(***b***) By Taylor's theorem the expansion of $f(x)$ about $x = a$, is

$$f(x) = f(a) + (x-a)f'(a) + \frac{(x-a)^2}{2!}f''(a) + \frac{(x-a)^3}{3!}f'''(a) + \ldots$$

Here $\quad f(x) = \tan x$ and 'a' $= \frac{\pi}{4}$ and $f\left(\frac{\pi}{4}\right) = \tan\frac{\pi}{4} = 1.$

$f'(x) = \sec^2 x$, $f''(x) = 2\sec^2 x \tan x$ and $f'''_{(x)} = 2\sec^4 x + 4\tan^2 x \sec^2 x$,

$$\therefore \quad f'\left(\frac{\pi}{4}\right) = \sec^2\frac{\pi}{4} = 2, \; f''\left(\frac{\pi}{4}\right) = 2\sec^2\frac{\pi}{4}\tan\frac{\pi}{4} = 4,$$

$$f'''\left(\frac{\pi}{4}\right) = 2\sec^4\frac{\pi}{4} + 4\tan^2\frac{\pi}{4}\sec^2\frac{\pi}{4} = 8 + 4\cdot 1\cdot 2 = 16$$

$$\therefore \quad \tan x = 1 + \left(x - \frac{\pi}{4}\right)2 + \frac{1}{2!}\left(x-\frac{\pi}{4}\right)^2\cdot 4 + \frac{1}{3!}\left(x-\frac{\pi}{4}\right)^3\cdot 16 \quad \text{upto first four terms}$$

$$= 1 + 2\left(x-\frac{\pi}{4}\right) + 2\left(x-\frac{\pi}{4}\right)^2 + \frac{8}{3}\left(x-\frac{\pi}{4}\right)^3. \quad \textbf{Ans.}$$

EXAMPLE 1.29. Show that $\sin^{-1}x = x + \frac{1^2}{3!}x^3 + \frac{1^2.3^2}{5!}x^5 + \frac{1^2.3^2.5^2}{7!}x^7 + \ldots$

and hence find the approximate value of π.

SOLUTION: Let $y = \sin^{-1}x$ hence $y_1 = \frac{1}{\sqrt{1-x^2}}$ or $(1-x^2)\,y_1^2 = 1.$

Differentiating both sides w.r.t. x, gives

$$(1-x^2)\,2y_1 y_2 - 2xy_1^2 = 0 \quad \text{or} \quad (1-x^2)\,y_2 - xy_1 = 0 \text{ as } y_1 \neq 0.$$

Differentiating the above relation n times using Leibnitz's theorem, we get

$$(1-x^2)\,y_{n+2} + {}^nC_1\,(-2x)\,y_{n+1} + {}^nC_2\,(-2)\,y_n - x\,y_{n+1} - {}^nC_1\,1.\,y_n = 0$$

or $\quad (1-x^2)\,y_{n+2} - (2n+1)x\,y_{n+1} - n^2\,y_n = 0$

At $x = 0$, we have $y_{n+2}(0) = n^2\,y_n(0)$ or $y_n(0) = (n-2)^2\,y_{n-2}(0)$.

and $\quad y_n(0) = (n-2)^2\,(n-4)^2\,y_{n-4}(0)$.

Continuing the process, we get

$$y_n(0) = (n-2)^2\,(n-4)^2\,(n-6)^2 \ldots 2^2\,y_2(0), \text{ when } n \text{ is even}$$
$$= (n-2)^2\,(n-4)^2\,(n-6)^2 \ldots 1^2\,y_1(0). \text{ when } n \text{ is odd.}$$

But $\quad y_1(0) = 1$ and $y_2(0) = 0$,

therefore, $\quad y_n(0) = 0$ when n is even

$$= 1^2\cdot 3^2\cdot 5^2\cdot \ldots (n-2)^2 \text{ when } n \text{ is odd.}$$

Now, by Maclaurin's theorem, we have

$$y(x) = y(0) + x\,y_1(0) + \frac{x^2}{2!}y_2(0) + \frac{x^3}{3!}y_3(0) + \frac{x^4}{4!}y_4(0) + \frac{x^5}{5!}y_5(0) + \ldots$$

or $$\sin^{-1} x = 0 + x + \frac{1^2}{3!}x^3 + \frac{1^2 \cdot 3^2}{5!}x^5 + \frac{1^2 \cdot 3^2 \cdot 5^2}{7!}x^7 + ...$$

$$= x + \frac{1}{2}\frac{x^3}{3} + \frac{1 \cdot 3}{2 \cdot 4}\frac{x^5}{5} + \frac{1 \cdot 3 \cdot 5}{2 \cdot 4 \cdot 6}\frac{x^7}{7} + ...$$

Here $-1 < x < 1$ and to obtain the value of π we put $x = 1/2$, to get

$$\frac{\pi}{6} = \frac{1}{2} + \frac{1}{6}\cdot\frac{1}{8} + \frac{3}{40}\cdot\frac{1}{32} + \frac{5}{112}\cdot\frac{1}{128} + ...$$

or $$\pi = 3 + \frac{1}{8} + \frac{9}{640} + \frac{30}{14336} = 3.141 \quad \text{approximately.}$$ **Ans.**

Aliter:

Let $y = \sin^{-1} x$ then $\frac{dy}{dx} = (1 - x^2)^{-1/2}$

Expanding by Binomial theorem, we get

$$\frac{dy}{dx} = 1 + \left(-\frac{1}{2}\right)\left(-x^2\right) + \frac{\left(-\frac{1}{2}\right)\left(-\frac{3}{2}\right)}{2!}\left(-x^2\right)^2 + \frac{\left(-\frac{1}{2}\right)\left(-\frac{3}{2}\right)\left(-\frac{5}{2}\right)}{3!}\left(-x^2\right)^3 + ...$$

$$= 1 + \frac{x^2}{2} + \frac{1 \cdot 3}{2 \cdot 4}x^4 + \frac{1 \cdot 3 \cdot 5}{2 \cdot 4 \cdot 6}x^6 + ...$$

Integrating both sides w.r.t. x, gives

$$y = \left(x + \frac{1}{2}\frac{x^3}{3} + \frac{1 \cdot 3}{2 \cdot 4}\frac{x^5}{5} + \frac{1 \cdot 3 \cdot 5}{2 \cdot 4 \cdot 6}\frac{x^7}{7} + ...\right) + C$$

Since $y = 0$ at $x = 0$ we have $C = 0$, hence

$$\sin^{-1} x = x + \frac{1}{2}\frac{x^3}{3} + \frac{1 \cdot 3}{2 \cdot 4}\frac{x^5}{5} + \frac{1 \cdot 3 \cdot 5}{2 \cdot 4 \cdot 6}\frac{x^7}{7} + ...$$ **Ans.**

EXAMPLE 1.30. Apply Taylor's theorem to show that

$$\tan^{-1}(x + h) = \tan^{-1} x + h \sin\alpha \cdot \frac{\sin\alpha}{1} - (h \sin\alpha)^2 \cdot \frac{\sin 2\alpha}{2} + (h \sin\alpha)^3 \cdot \frac{\sin 3\alpha}{3} - ...$$

where $\alpha = \cot^{-1} x$.

SOLUTION: Let $f(x) = \tan^{-1} x$,

hence $$f'(x) = \frac{1}{x^2 + 1} = \frac{1}{(x + i)(x - i)} = \frac{1}{2i}\left[\frac{1}{(x - i)} - \frac{1}{(x + i)}\right]$$

Differentiating w.r.t. x, $(n - 1)$ times, gives

$$f^{(n)}(x) = \frac{1}{2i}\left[\frac{(-1)^{n-1}(n-1)!}{(x - i)^n} - \frac{(-1)^{n-1}(n-1)!}{(x + i)^n}\right]$$

Since $\alpha = \cot^{-1} x$ or $x = \cot \alpha$, we have

$$f^{(n)}(x) = \frac{(-1)^{n-1}(n-1)!}{2i}\left[\frac{\sin^n \alpha}{(\cos\alpha - i\sin\alpha)^n} - \frac{\sin^n \alpha}{(\cos\alpha + i\sin\alpha)^n}\right]$$

$$= \frac{(-1)^{n-1}(n-1)!}{2i} \sin^n \alpha\ [(\cos n\alpha + i \sin n\alpha) - (\cos n\alpha - i \sin n\alpha)]$$

(using de Moivre's Theorem)

$$= (-1)^{n-1} (n-1)!\ \sin^n \alpha \sin n\alpha$$

We know, by Taylor's Theoem that

$$f(x+h) = f(x) + \frac{h}{1!}f'(x) + \frac{h^2}{2!}f''(x) + \frac{h^3}{3!}f'''(x) + \ldots + \frac{h^n}{n!}f^{(n)}(x) + \ldots$$

$\therefore$ $\tan^{-1}(x+h)$

$$= \tan^{-1}x + \frac{h}{1!}\sin\alpha\sin\alpha - \frac{h^2}{2!}\sin^2\alpha\sin 2\alpha + \frac{h^3}{3!}2!\cdot\sin^3\alpha\sin 3\alpha - \frac{h^4}{4!}3!\sin^4\alpha\sin 4\alpha + \ldots$$

$$= \tan^{-1}x + (h\sin\alpha)\frac{\sin\alpha}{1} - (h\sin\alpha)^2\frac{\sin 2\alpha}{2} + (h\sin\alpha)^3\frac{\sin 3\alpha}{3} - (h\sin\alpha)^4\frac{\sin 4\alpha}{4} + \ldots$$

Hence the result.

EXAMPLE 1.31. Estimate the value of $\sqrt{10}$ correct to four places of decimal using Taylor's theorem.

SOLUTION: Let $f(x) = \sqrt{x}$, then using Taylor's expansion

$$f(x+h) = f(x) + h\,f'(x) + \frac{h^2}{2!}f''(x) + \frac{h^3}{3!}f'''(x) + \ldots$$

we get $$(x+h)^{1/2} = x^{1/2} + h\frac{d}{dx}x^{1/2} + \frac{h^2}{2!}\frac{d^2}{dx^2}x^{1/2} + \frac{h^3}{3!}\frac{d^3}{dx^3}x^{1/2} + \ldots$$

Taking $x = 9$ and $h = 1$ in the above expansion, gives

$$\sqrt{10} = \left[3 + h\left(\frac{1}{2}x^{-1/2}\right) + \frac{h^2}{2!}\left(\frac{1}{2}\left(-\frac{1}{2}\right)x^{-3/2}\right) + \frac{h^3}{3!}\left(\frac{1}{2}\left(-\frac{1}{2}\right)\left(-\frac{3}{2}\right)x^{-5/2}\right) + \ldots\right]_{x=9,\ h=1}$$

$$= 3 + \frac{1}{2.3} - \frac{1}{8.27} + \frac{3}{3.2.8.243} - \ldots\ldots$$

$= 3.16227$ approximately. **Ans.**

EXAMPLE 1.32. Obtain the Maclaurin's expansion of $e^{x\cos x}$ upto first four terms.

SOLUTION: Let $y = e^{x\cos x}$, then $y(0) = 1$.

$\therefore$ $\log y = x\cos x$ which on differentiation w.r.t. x, gives $\frac{1}{y}\frac{dy}{dx} = 1.\cos x - x\sin x$

or $y_1 = y(\cos x - x\sin x)$ hence $y_1(0) = 1$.

further, $y_2 = y_1 (\cos x - x \sin x) + y (-\sin x - x\cos x - \sin x)$

on putting $x = 0$ we get $y_2(0) = 1$.

Differentiating again, gives

$$y_3 = y_2(\cos x - x\sin x) + y_1(-\sin x - x\cos x - \sin x) + y_1(-2\sin x - x\cos x) + y(-3\cos x + x\sin x)$$

$$= y_2(\cos x - x\sin x) - 2y_1(2\sin x + x\cos x) + y(x\sin x - 3\cos x)$$

At $x = 0$ we have $y_3(0) = 1\, y_2(0) - 0 - 3\, y(0) = -2$, etc.

Therefore, $y(x) = y(0) + x\, y_1(0) + \dfrac{x^2}{2!} y_2(0) + \dfrac{x^3}{3!} y_3(0) + \ldots = 1 + x + \dfrac{x^2}{2!} + \dfrac{x^3}{3!}(-2) + \ldots$

or $y(x) = 1 + x + \dfrac{x^2}{2} - \dfrac{x^3}{3}$ (retaining terms upto x^3). **Ans.**

Aliter:

The above result can also be obtained in a simpler way by expanding the standard functions in ascending powers of x. Let us put $x\cos x = t$, then

$$e^{x\cos x} = e^t = 1 + t + \frac{t^2}{2!} + \frac{t^3}{3!} + \ldots = 1 + (x\cos x) + \frac{1}{2!}(x\cos x)^2 + \frac{1}{3!}(x\cos x)^3 + \ldots..$$

But $x\cos x = x\left(1 - \dfrac{x^2}{2!} + \dfrac{x^4}{4!} - \ldots\right) = x - \dfrac{x^3}{2!} + \dfrac{x^5}{4!} - \dfrac{x^7}{6!} + \ldots$

$$\therefore e^{x\cos x} = 1 + \left(x - \frac{x^3}{2!} + \frac{x^5}{4!} - \frac{x^7}{6!} + \ldots\right) + \frac{1}{2!}\left(x - \frac{x^3}{2!} + \frac{x^5}{4!} - \frac{x^7}{6!} + \ldots\right)^2 + \frac{1}{3!}\left(x - \frac{x^3}{2!} + \frac{x^5}{4!} - \ldots\right)^3 + \ldots$$

$$= 1 + x + \frac{x^2}{2!} - x^3\left(-\frac{1}{2!} + \frac{1}{3!}\right) + \ldots....$$

or $e^{x\cos x} = 1 + x + \dfrac{x^2}{2} - \dfrac{x^3}{3}$ (on retaining terms upto x^3). **Ans.**

EXAMPLE 1.33. Prove that $\log(1 + x + x^2 + x^3 + x^4) = x + \dfrac{x^2}{2} + \dfrac{x^3}{3} + \dfrac{x^4}{4} - \dfrac{4}{5}x^5 + \dfrac{x^6}{6} + \ldots$

SOLUTION: The result will be derived by using the method of expansion of standard functions.

Since $1 + x + x^2 + x^3 + x^4 = \dfrac{1-x^5}{1-x}$ (as sum of a geometric series)

we have, $\log(1 + x + x^2 + x^3 + x^4) = \log(1 - x^5) - \log(1 - x)$

$$= \left[-x^5 - \frac{(x^5)^2}{2} - \frac{(x^5)^3}{3} - \frac{(x^5)^4}{4} - \ldots\right] - \left[-x - \frac{x^2}{2} - \frac{x^3}{3} - \frac{x^4}{4} - \frac{x^5}{5} - \frac{x^6}{6} - \ldots\right]$$

$$= x + \frac{x^2}{2} + \frac{x^3}{3} + \frac{x^4}{4} + x^5\left(\frac{1}{5} - 1\right) + \frac{x^6}{6} + \frac{x^7}{7} + \frac{x^8}{8} + \frac{x^9}{9} + x^{10}\left(\frac{1}{10} - \frac{1}{2}\right) + \ldots$$

or $\log(1 + x + x^2 + x^3 + x^4) = x + \dfrac{x^2}{2} + \dfrac{x^3}{3} + \dfrac{x^4}{4} - \dfrac{4x^5}{5} + \dfrac{x^6}{6} + \dfrac{x^7}{7} + \dfrac{x^8}{8} + \dfrac{x^9}{9} - \dfrac{2}{5}x^{10} + \ldots$

Hence Proved.

EXAMPLE 1.34. Show that $(1+x)^x = 1+x^2-\dfrac{x^3}{2}+\dfrac{5}{6}x^4-\dfrac{3}{4}x^5+\dfrac{33}{40}x^6+\ldots$

SOLUTION: We can write $(1+x)^x = e^{\log(1+x)^x} = e^{x\log(1+x)} = e^t$ where $t = x\log(1+x)$.

Now, $$t = x\log(1+x) = x\left(x-\frac{x^2}{2}+\frac{x^3}{3}-\frac{x^4}{4}+\ldots\right) = x^2-\frac{x^3}{2}+\frac{x^4}{3}-\frac{x^5}{4}+\ldots$$

$$\therefore \quad (1+x)^x = e^t = 1+t+\frac{t^2}{2!}+\frac{t^3}{3!}+\frac{t^4}{4!}+\ldots$$

$$= 1+\left(x^2-\frac{x^3}{2}+\frac{x^4}{3}-\frac{x^5}{4}+\frac{x^6}{5}-\ldots\right)+\frac{1}{2!}\left(x^2-\frac{x^3}{2}+\frac{x^4}{3}-\ldots\right)^2+\frac{1}{3!}\left(x^2-\frac{x^3}{2}+\frac{x^4}{3}-\ldots\right)^3+\ldots\ldots$$

$$= 1+x^2-\frac{x^3}{2}+x^4\left(\frac{1}{3}+\frac{1}{2}\right)+x^5\left(-\frac{1}{4}-\frac{1}{2}\right)+x^6\left(\frac{1}{5}+\frac{1}{8}+\frac{1}{3}+\frac{1}{6}\right)+\ldots$$

or $$(1+x)^x = 1+x^2-\frac{x^3}{2}+\frac{5}{6}x^4-\frac{3}{4}x^5+\frac{33}{40}x^6+\ldots$$ **Hence Proved.**

EXAMPLE 1.35. Expand $\cos^{-1}\left(\dfrac{x-x^{-1}}{x+x^{-1}}\right)$ in ascending powers of x. $(x>0)$

SOLUTION: Let us put $x = \cot\theta$, then

$$\cos^{-1}\left(\frac{x-x^{-1}}{x+x^{-1}}\right) = \cos^{-1}\left(\frac{\cot\theta-\tan\theta}{\cot\theta+\tan\theta}\right) = \cos^{-1}\left(\frac{\cos^2\theta-\sin^2\theta}{\cos^2\theta+\sin^2\theta}\right)$$

$$= \cos^{-1}(\cos 2\theta) = 2\theta = 2\cot^{-1}x = 2\left(\frac{\pi}{2}-\tan^{-1}x\right) = \pi-2\tan^{-1}x.$$

We know that $\tan^{-1}x = x-\dfrac{x^3}{3}+\dfrac{x^5}{5}-\dfrac{x^7}{7}+\ldots$

and general value of $\tan^{-1}x$ is written as

$$\tan^{-1}x = n\pi+x-\frac{x^3}{3}+\frac{x^5}{5}-\frac{x^7}{7}+\ldots$$

therefore, $$\cos^{-1}\left(\frac{x-x^{-1}}{x+x^{-1}}\right) = \pi-2\left(n\pi+x-\frac{x^3}{3}+\frac{x^5}{5}-\frac{x^7}{7}+\ldots\right)$$

$$= -(2n-1)\pi-2x+\frac{2}{3}x^3-\frac{2}{5}x^5+\frac{2}{7}x^7-\ldots\ldots$$ **Ans.**

APPROXIMATE ERROR

Let y be a function of x, given by $y = f(x)$. Now, if x suffers a small change δx, it is often required to find how much change takes place in y. Let this change in y be denoted by δy.

As such $y + \delta y = f(x + \delta x)$ or $\delta y = f(x + \delta x) - f(x)$

Using Taylor's expansion, we get

$$\delta y = \left\{ f(x) + \delta x\, f'(x) + \frac{\delta x^2}{2!} f''(x) + \ldots \right\} - f(x) = \delta x\, f'(x) + \frac{(\delta x)^2}{2!} f''(x) + \ldots$$

Here, δx is small and if we neglet its square and higher powers, then

$$\delta y = f'(x)\, \delta x = \frac{dy}{dx} \delta x \quad \text{approximately.}$$

Further, if δx is the error in x, then $\frac{\delta x}{x}$ is called *relative error* and $\frac{\delta x}{x} \times 100$ is called the *percentage error* in x.

EXAMPLE 1.36. Find the change in the total surface area of a right circular cone when

(*i*) the radius r is constant and the altitude h changes by a small amount δh.

(*ii*) the altitude is constant and the radius changes by a small amount δr.

SOLUTION: The total surfade area S of a right circular cone with radius of the base as r and altitude h, is given by

$$S = \pi r^2 + \pi r l = \pi r^2 + \pi r \sqrt{r^2 + h^2}$$

Now, (*i*) if r is constant and altitude h changes by δh, then

$$\frac{dS}{dh} = 0 + \frac{\pi r}{2} \left(r^2 + h^2 \right)^{-1/2} .2h = \frac{\pi r h}{\sqrt{r^2 + h^2}}$$

therefore, the consequential change δS in S, will be given by

$$\delta S = \frac{dS}{dh} \delta h = \frac{\pi r h}{\sqrt{r^2 + h^2}} \delta h \quad \text{approximately.} \qquad \textbf{Ans.}$$

(*ii*) if h is constant and the radius r changes by δr, then

$$\frac{dS}{dr} = 2\pi r + \pi \sqrt{r^2 + h^2} + \frac{\pi r . 2r}{2\sqrt{r^2 + h^2}} = 2\pi r + \frac{\pi \left(2r^2 + h^2 \right)}{\sqrt{r^2 + h^2}}$$

therefore, the resulting change δS in S, will be given by

$$\delta S = \left[2\pi r + \frac{\pi (2r^2 + h^2)}{\sqrt{r^2 + h^2}} \right] \delta r \quad \text{approximately.} \qquad \textbf{Ans.}$$

EXAMPLE 1.37. If Δ is the area of a triangle ABC having sides equal to a, b, c and S is the semi-perimeter, prove that the error $\delta \Delta$ in Δ resulting from a small error δc in the measurement of c, is given by

$$\delta \Delta = \frac{\Delta}{4} \left[\frac{1}{S} + \frac{1}{S - a} + \frac{1}{S - b} - \frac{1}{S - c} \right] \delta c .$$

SOLUTION: We know that $\Delta^2 = S(S-a)(S-b)(S-c)$ where $S = (a+b+c)/2$.

or $2\log\Delta = \log S + \log(S-a) + \log(S-b) + \log(S-c)$.

Differentiating both sides w.r.t. c, gives

$$\frac{2}{\Delta}\frac{d\Delta}{dc} = \frac{1}{S}\frac{dS}{dc} + \frac{1}{S-a}\frac{d}{dc}(S-a) + \frac{1}{S-b}\frac{d}{dc}(S-b) + \frac{1}{S-c}\frac{d}{dc}(S-c)$$

$$= \frac{1}{S}\cdot\frac{1}{2} + \frac{1}{2(S-a)} + \frac{1}{2(S-b)} + \frac{1}{(S-c)}\left(\frac{1}{2}-1\right)$$

or $$\frac{d\Delta}{dc} = \frac{\Delta}{4}\left[\frac{1}{S} + \frac{1}{S-a} + \frac{1}{S-b} - \frac{1}{S-c}\right]$$

Therefore the error $\delta\Delta$ in the area, is given by

$$\delta\Delta = \frac{d\Delta}{dc}\delta c = \frac{\Delta}{4}\left[\frac{1}{S} + \frac{1}{S-a} + \frac{1}{S-b} - \frac{1}{S-c}\right]\delta c.$$ **Hence Proved.**

EXAMPLE 1.38. A heavy string is suspended from two poles of equal height, taking the shape of a catenary with sag a and equation $y = a\cosh(x/a)$. If the absolute value of x is small, show that the shape of the string can be approximated by the parabola $y = a + x^2/(2a)$.

SOLUTION: Here $y = a\cosh(x/a)$, hence $y(0) = a$ and

$$y_1 = a\sinh(x/a)\,.\frac{1}{a} = \sinh(x/a), \text{ hence } y_1(0) = 0.$$

Further $y_2 = \frac{1}{a}\cos h(x/a)$, hence $y_2(0) = \frac{1}{a}$.

Again $y_3 = \frac{1}{a^2}\sin h(x/a)$, hence $y_3(0) = 0$

and $y_4 = \frac{1}{a^3}\cos h(x/a)$, hence $y_4(0) = \frac{1}{a^3}$, and so on.

Therefore, by Maclaurin's expansion of y, we have

$$y = y(0) + xy_1(0) + \frac{x^2}{2!}y_2(0) + \frac{x^3}{3!}y_3(0) + \frac{x^4}{4!}y_4(0) + \dots$$

$$= a + 0 + \frac{x^2}{2a} + 0 + \frac{x^4}{24a^3} + \dots$$

As $|x|$ is small, neglecting terms beyond x^2, we get

$$y = a + \frac{x^2}{2a}$$ which is the equation of a parabola. **Hence Proved.**

EXAMPLE 1.39. Apply Taylor's theorem to estimate the value of $f(11/10)$ where

$$f(x) = x^3 + 3x^2 + 15x - 10.$$

SOLUTION: By Taylor's expansion, we have

$$f(a+\delta x) = f(a) + \delta x\, f'(a) + \frac{(\delta x)^2}{2!}f''(a) + \frac{(\delta x)^3}{3!}f'''(a) + \dots \qquad ...(1)$$

Here $f(x) = x^3 + 3x^2 + 15x - 10$ and $a = 1$ and $\delta x = 0.1$

$\therefore$ $f(1) = 1^3 + 3.1^2 + 15.1 - 10 = 9$

Also $f'(x) = 3x^2 + 6x + 15$ $\therefore$ $f'(1) = 3 + 6 + 15 = 24$

and $f''(x) = 6x + 6,$ $\therefore$ $f''(1) = 12$

and $f'''(x) = 6$ $\therefore$ $f'''(1) = 6$

Using (1) here we have

$$\therefore \quad f\left(\frac{11}{10}\right) = f\left(1+\frac{1}{10}\right) = 9 + 0.1\ (24) + \frac{(0.1)^2}{2}12 + \frac{(0.1)^3}{3!}(6)$$

$$= 9 + 2.4 + 0.06 + 0.001 = 11.461. \quad \textbf{Ans.}$$

EXAMPLE 1.40. **(a)** Taylor's series becomes MaClaurin's series when

(i) $x = a,\ h = 0$ (ii) $x = 0,\ h = x$

(iii) $h = x,\ x = a$ (iv) none of these

(b) Which function cannot be expanded by MaClaurin's theorem.

(i) $\cos x$ (ii) $\sec x$

(iii) $\tan x$ (iv) $\cot x$

SOLUTION: **(a)** Answer is (ii)

(b) Answer is (iv).

EXAMPLE 1.41. Expand $\log\ (x+\sqrt{x^2+1})$ upto first four terms by Maclaurin's theorem. Putting $x = 0.75$ in the expansion, calculate the value of log 2 to four place of decimal and find the percentage error.

SOLUTION: $f(x) = \log\ (x+\sqrt{x^2+1})\ ,\ f(0) = 0.$

$$\therefore \quad f'(x) = \frac{1+\dfrac{1}{2}\dfrac{2x}{\sqrt{x^2+1}}}{x+\sqrt{x^2+1}} = \frac{1}{\sqrt{x^2+1}} \quad \text{hence } f'(0) = 1.$$

and $$f''(x) = \frac{-1}{2}(x^2+1)^{-3/2}\cdot 2x = -x\,(x^2+1)^{-3/2} \quad \text{hence } f''(0) = 0.$$

Also, $$f'''(x) = -1\cdot(x^2+1)^{-3/2} - x\left(\frac{-3}{2}\right)(x^2+1)^{-5/2}\cdot 2x$$

$$= -(x^2+1)^{-3/2} + 3x^2(x^2+1)^{-5/2}, \quad \text{hence } f'''(0) = -1.$$

By Maclaurin's theorem

$$f(x) = f(0) + x f'(0) + \frac{x^2}{2!}f''(0) + \frac{x^3}{3}f'''(0) \quad \text{(retaining upto four terms)} \quad ...(1)$$

Putting here $x = 0.75$ we get $f\left(\frac{3}{4}\right) = \log\left(\frac{3}{4}+\sqrt{\frac{9}{16}+1}\right) = \log 2.$

Thus from (1), we get

$$\therefore \quad \log(2) = \left[0 + x.1 + \frac{x^2}{2!}(0) + \frac{x^3}{3!}(-1)\right]_{x=0.75}$$

$$= 0.75 - \frac{1}{6}(0.75)^3 = \frac{3}{4} - \frac{1}{6}\cdot\frac{3}{4}\cdot\frac{3}{4}\cdot\frac{3}{4} = \frac{87}{128}$$

$= 0.6626$ approximately.

Actual value of $\log 2 = 0.6931$ hence error $= 0.0305$.

Therefore, the percentage error $= \dfrac{0.0305}{0.6931} \times 100 = 4.4\%$ **Ans.**

EXAMPLE 1.42. The angles of a triangle are calculated from the sides $a, b, c,$ if small errors $\delta a, \delta b, \delta c,$ are made in measuring the sides a, b, c respectively. Show that $\delta A = \dfrac{a}{2\Delta}[\delta a - \delta b \cos C - \delta c \cos B]$ approximately where Δ is the area of the triangle and hence verify that $\delta A + \delta B + \delta C = 0$.

SOLUTION: We know that $\cos A = \dfrac{b^2 + c^2 - a^2}{2bc}$. Since $\delta a, \delta b, \delta c$ are the errors made in measuring the sides a, b, c then as a result δA is the error made in measuring angle A.

$$\therefore \quad -\sin A\, \delta A = \frac{-2a\delta a}{2bc} + \delta b\left(\frac{1}{2c} - \frac{c}{2b^2} + \frac{a^2}{2cb^2}\right) + \delta c\left(\frac{1}{2b} - \frac{b}{2c^2} + \frac{a^2}{2bc^2}\right)$$

$$= \frac{-a}{bc}\delta a + \frac{\delta b}{2cb^2}(b^2 - c^2 + a^2) + \frac{\delta c}{2bc^2}(c^2 + a^2 - b^2)$$

$$= \frac{-a}{bc}\delta a + \frac{\delta b}{2cb^2}2ab\cos C + \frac{\delta c}{2bc^2}2ac\cos B$$

$$= \frac{-a}{bc}\delta a + \frac{a}{bc}\delta b\cos C + \frac{a}{bc}\delta c\cos B$$

or $$\frac{bc}{a}\sin A\,\delta A = \delta a - \delta h \cos C - \delta c \cos B$$

But $\Delta = \dfrac{1}{2}bc \sin A$ hence we have

$$\frac{2\Delta}{a}\delta A = \delta a - \delta b \cos C - \delta c \cos B$$

or $$\delta A = \frac{a}{2\Delta}[\delta a - \delta b \cos C - \delta c \cos B]$$

$$\therefore \quad \delta A + \delta B + \delta C = \frac{1}{2\Delta}[a\delta a - a\cos c\,\delta b - a\cos B\,\delta c + b\delta b$$
$$- b\cos A\,\delta c - b\cos C\,\delta a + c\delta c - c\cos A\,\delta b - c\cos B\,\delta a]$$

$$= \frac{1}{2\Delta}[(a - b\cos c - c\cos B)\,\delta a + (b - c\cos A - a\cos C)\,\delta b$$
$$+ (c - a\cos B - b\cos A)\,\delta c]$$

$= 0.$ **Hence the result.**

EXERCISE 1C

1. Use Maclaurin's theorem to show that $e^x \cos x = 1+x-\frac{2x^3}{3!}-\frac{2^2x^4}{4!}-\frac{2^2x^5}{5!}+\frac{2^3x^7}{7!}+\ldots$

2. Show that $\sqrt{1+x+2x^2} = 1+\frac{x}{2}+\frac{7}{8}x^2-\frac{7}{16}x^3+\ldots$

3. Expand $\sin x$ in powers of $(x-\pi/2)$.

4. Employing Maclaurin's theorem, show that

$$e^{a\sin^{-1}x} = 1+ax+\frac{(ax)^2}{2!}+\frac{a(1^2+a^2)x^3}{3!}+\frac{a^2(2^2+a^2)x^4}{4!}+\ldots$$

and hence deduce that

$$e^{\theta} = 1+\sin\theta+\frac{\sin^2\theta}{2!}+\frac{2}{3!}\sin^3\theta+\frac{5}{4!}\sin^4\theta+\ldots$$

5. Show that $\sin(x+h)$ differs from $\sin x + h\cos x$ by not more than $h^2/2$.

6. If $|x| < a$, then show that within $(x/a)^2$, we have

$$e^{x/a} = \sqrt{\frac{a+x}{a-x}} \text{ approximately.}$$

7. The pressure p and volume v of a gas are connected by the relation $pv^{1.4} = c$ where c is constant. Find the percentage increase in the pressure corresponding to a decrease of 1/2% in the volume.

8. A soap bubble of radius 2 cm shrinks to radius 1.9 cm. Estimate the decrease in (*i*) Volume (*ii*) Surface area.

9. If ABCD is a rectangular protector in which AB = 6 cm, BC = 2 cm and O is the mid point of AB. An angle BOP is indicated by a mark P on the edge CD. If, in setting off an angle θ degrees, the mark is made $\frac{1}{100}$ of a cm away along the edge. Show that the error in the angle is $\frac{9}{10\pi}\sin^2\theta$ degrees.

10. Prove that $f(mx) = f(x) + (m-1)xf'(x) + \frac{(m-1)^2}{2!}x^2 f''(x) + \ldots$

11. If $x = y-\frac{1}{2}y^2+\frac{1}{3}y^3-\frac{1}{4}y^4 \ldots\infty$, show that $y = x+\frac{x^2}{2!}+\frac{x^3}{3!}+\frac{x^4}{4!}+\ldots\infty$.

12. Show that $\log\sec x = \frac{x^2}{2}+\frac{1}{3}\frac{x^4}{4}+\frac{2}{15}\frac{x^6}{6}+\ldots$

13. Compute to four decimal places the value of $\cos 32^\circ$ using Taylor's theorem.

14. Using Taylor's series compute $\sin 31^\circ$ to four decimal places.

15. Prove that $\frac{e^x}{e^x+1} = \frac{1}{2}+\frac{x}{4}-\frac{x^3}{48}+\ldots$

16. Given $\log_{10}4 = 0.6021$ calculate apprrixmately $\log_{10}404$.

❑❑❑

CHAPTER

2 Asymptotes, Curvature and Curve Tracing

Asymptotes for Cartesian Coordinates. Curve Tracing Cartesian and Polar Coordinates

DEFINITION OF ASYMPTOTE

We shall consider here only those curves whose branches extend upto infinity, like parabola and hyperbola. At a point on the curve, tangent is drawn and then the point of contact is moved farther and farther away from the origin. If this tangent in the process tends to some definite straight line, it is called ***asymptote*** of the curve. In other words, **the *limiting position of the tangent line as the point of contact recedes to infinity is known as asymptote*.** In some typical cases these asymptotes are curves also. However, we shall consider here only those curves which have straight lines as asymptotes. Some asymptotes are parallel to x-axis or to y-axis and those asymptotes which are not parallel to any of the coordinate axes are called *oblique asymptotes*.

DETERMINING THE ASYMPTOTES

Let the equation of the curve be algebraic in x and y of degree n, its general form being,

$$(a_0 x^n + a_1 x^{n-1} y + a_2 x^{n-2} y^2 + \ldots + a_n y^n) + (b_1 x^{n-1} + b_2 x^{n-2} y + \ldots + b_n y^{n-1}) +$$
$$(c_2 x^{n-2} + c_3 x^{n-3} y + \ldots + c_n y^{n-2}) + \ldots + (l_{n-1} x + l_n y) + k_n = 0. \qquad \ldots(1)$$

Since the expressions in brackets are homogeneous, the above equation can be written as

$$x^n \phi_n (y/x) + x^{n-1} \phi_{n-1} (y/x) + x^{n-2} \phi_{n-2} (y/x) + \ldots + x \phi_1 (y/x) + k_n = 0$$

where $\phi_n (y/x)$ is a function of (y/x) of degree n, etc.

or $$\phi_n(y/x) + \frac{1}{x} \phi_{n-1}(y/x) + \frac{1}{x^2} \phi_{n-2}(y/x) + \ldots + \frac{1}{x^{n-1}} \phi_1(y/x) + \frac{k_n}{x^n} = 0 \qquad \ldots(2)$$

Now, let $y = mx + c$ or $y/x = m + \dfrac{c}{x}$ ...(3)

be an asymptote to (1), then $\lim\limits_{x \to \infty} (y/x) = m$.

Taking limit in (2) as $x \to \infty$, we get $\phi_n(m) = 0$. ...(4)

The real roots of the equation (4) in m give the slopes of the asymptotes.

Using (3), the equation (2) becomes

$$\phi_n\left(m+\frac{c}{x}\right)+\frac{1}{x}\phi_{n-1}\left(m+\frac{c}{x}\right)+\frac{1}{x^2}\phi_{n-2}\left(m+\frac{c}{x}\right)+\ldots\ldots = 0.$$

Since x is very large, $\frac{c}{x}$ will be very small and we can expand $\phi_n\left(m+\frac{c}{x}\right)$, $\phi_{n-1}\left(m+\frac{c}{x}\right)$, ...

by Taylor's theorem, to get

$$\left[\phi_n(m)+\frac{c}{x}\phi_n{}'(m)+\frac{c^2}{2!\,x^2}\phi_n{}''(m)+\ldots\right]+\frac{1}{x}\left[\phi_{n-1}(m)+\frac{c}{x}\phi_{n-1}{}'(m)+\frac{c^2}{2!\,x^2}\phi_{n-1}{}''(m)+\ldots\right]$$

$$+\frac{1}{x^2}\left[\phi_{n-2}(m)+\frac{c}{x}\phi_{n-2}{}'(m)+\frac{c^2}{2!\,x^2}\phi_{n-2}{}''(m)+\ldots\right] = 0. \qquad \ldots(5)$$

But, by (4), $\phi_n(m) = 0$, hence the above equation (5) on multiplying throughout by x, becomes

$$\left[c\,\phi_n{}'(m)+\phi_{n-1}(m)\right]+\frac{1}{x}\left[\frac{c^2}{2!}\phi_n{}''(m)+c\,\phi_{n-1}{}'(m)+\phi_{n-2}(m)\right]+$$

$$\frac{1}{x^2}\left[\frac{c^3}{3!}\phi_n{}'''(m)+\frac{c^2}{2!}\phi_{n-1}{}''(m)+c\,\phi_{n-2}{}'(m)+\phi_{n-3}(m)\right]+\ldots = 0$$

As $x \to \infty$, we get $c\,\phi_n{}'(m) + \phi_{n-1}(m) = 0$

or
$$c = \frac{-\phi_{n-1}(m)}{\phi_n'(m)}. \qquad \ldots(6)$$

Now, if the roots of the equation $\phi_n(m) = 0$ are real and distinct as $m = m_1, m_2, \ldots, m_n$ the corresponding values of c as $c_1, c_2, \ldots, c_n$ are easily obtained from (6). However, in case two roots of (4) are equal, say $m_1 = m_2$, then it is a case of two parallel asymptotes and two values of c cannot be obtained from (6) since $\phi_{n-1}(m_1) = 0$ and $\phi'_n(m_1) = 0$.

In such a case we take into account $\frac{c^2}{2!}\phi_n''(m) + c\phi_{n-1}'(m) + \phi_{n-2}(m) = 0$

which gives two values of c, say c_1 and c_2 for one value of m.

Similarly, in the case when three roots of equation (4) are equal we get three parallel asymptotes and the corresponding three values of c are obtained from the equation

$$\frac{c^3}{3!}\phi_n'''(m) + \frac{c^2}{2!}\phi_{n-1}''(m) + \frac{c}{1!}\phi_{n-2}'(m) + \phi_{n-3}(m) = 0$$

To sum up, the working procedure for obtaining the equations of the asymptotes, will be as follows: In the given algebraic equation we first replace y by $(mx + c)$, and equate to zero the coefficient of the highest degree term in x to get the values of m, the slopes of the asymptotes. Then equate to zero the coefficient of the next to the highest degree term in x to get 'c' of the asymptote.

As a further short cut, the polynomials $\phi_n(m)$, $\phi_{n-1}(m)$,... are obtained by replacing x by 1 and y by m in the nth degree term, then in the $(n-1)^{th}$ degree term, and so on, respectively in the equation of the curve.

Asymptotes Parallel to Coordinate Axes

The asymptotes parallel to x-axis can be obtained by the method just explained but the asymptotes parallel to y-axis cannot be determined by this method simply because the slope m of such asymptotes is infinite. For the asymptotes, parallel to y-axis we make slight change in the approach as follows.

The general algebraic equation of nth degree can be written as

$$y^n \phi_0(x) + y^{n-1} \phi_1(x) + y^{n-2} \phi_2(x) + ... = 0 \qquad ...(1)$$

where $\phi_0(x)$, $\phi_1(x)$, $\phi_2(x)$,... are polynomials in x of degree 0, 1, 2, ... respectively.

Dividing throughout by y^n, (1) becomes

$$\phi_0(x)+\frac{1}{y}\phi_1(x)+\frac{1}{y^2}\phi_2(x)+.... = 0. \qquad ...(2)$$

Now we take limit as y tends to infinity and x tends to some finite value k, then we get

$\phi_0(k) = 0$ where k is the root of the equation $\phi_0(x) = 0$.

If the roots of the equation $\phi_0(x) = 0$ are k_1, k_2,... then the asymptotes parallel to y-axis are

$x = k_1$, $x = k_2$,...

Thus, **the *asymptotes parallel to y-axis are obtained by equating to zero the coefficient of the highest degree term in y*** and if the cofficient of the highest degree term in y is constant, then obviously, there is no asymptote parallel to y-axis.

Following on the similar lines we can conclude **that the *asymptotes parallel to x-axis can be easily obtained by equating to zero the cofficient of the highest degree term in x*.** However, if the coefficient of the highest degree term in x is constant there is no asymptote parallel to x-axis.

EXAMPLE 2.1. **(a)** Find the asymptotes of the curve $x^2 y^2 - a^2(x^2 + y^2) = 0$.
(b) What is the asymptote of the curve $y^2(2a - x) = x^3$?

SOLUTION: (a) The equation of the curve is of fourth degree therefore, there cannot be more than four asymptotes of the curve.

First, let us find the asymptotes parallel to x-axis for which we rewrite the given equation as

$$x^2(y^2 - a^2) - a^2 y^2 = 0$$

Equating to zero the cofficient of the highest degree term in x, we get

$$y^2 - a^2 = 0 \quad \text{or} \quad y = \pm a$$

Thus there are two asymptotes of the curve parallel to x-axis as $y = \pm a$. Similarly, to find the asymptotes parallel to y-axis we rewrite the given equation as

$$y^2(x^2 - a^2) - a^2x^2 = 0.$$

Equating to zero the coefficient of the highest degree term in y, we get

$$x^2 - a^2 = 0 \quad \text{or} \quad x = \pm a.$$

Therefore, there are two asymptotes of the curve parallel to y-axis as $x = \pm a$. **Ans.**

(b) For asymptote parallel to y-axis we equate to zero the coefficient of highest degree term in y and get

$$2a - x = 0 \quad \text{or} \quad x = 2a. \quad \textbf{Ans.}$$

EXAMPLE 2.2. Find the asymptotes of the curve $y^3 - x^2y + 2y^2 + 4y + x = 0$.

SOLUTION : The given equation of the curve is of degree 3. Replacing y by m and x by 1 in the third degree terms in the equation of the curve, we get

$$\phi_3(m) = m^3 - m = m(m^2 - 1).$$

Therefore, $\phi_3(m) = 0$ gives the slopes of the asymptotes as $m = 0, 1, -1$ which are all distinct.

Next, replacing y by m and x by 1 in the second degree terms of the equation of the curve, we get $\phi_2(m) = 2m^2$.

The 'c' of the asymptotes are given by

$$c = -\frac{\phi_2(m)}{\phi_3'(m)} = -\frac{2m^2}{3m^2 - 1}$$

Hence, for $m = 0$ we have $c = 0$, and for $m = \pm 1$ we have $c = -1$ and thus there are three asymptotes of the curve whose equations are $y = 0$, $y = x - 1$ and $y = -x - 1$. **Ans.**

It is to be observed here that the asymptote $y = 0$ parallel to x-axis could be obtained directly also, by equating to zero the coefficient of the highest degree term in x. The highest degree term in x here is x^2 whose coefficient y when equated to 0 gives $y = 0$, the x-axis itself. Further, since the cofficient of the highest degree term in y is constant there is no asymptote parallel to y-axis.

EXAMPLE 2.3. Find the asymptotes of the curve $y^3 - 6xy^2 + 11x^2y - 6x^3 + x + y = 0$

SOLUTION: The given equation is of degree three hence not more than three asymptotes can be there for this curve. Here

$$\phi_3(m) = m^3 - 6m^2 + 11m - 6 = (m-1)(m^2 - 5m + 6) = (m-1)(m-2)(m-3),$$

$$\phi_2(m) = 0 \quad \text{and} \quad \phi_1(m) = 1 + m.$$

The slopes of the asymptotes are the roots of the equation $\phi_3(m) = 0$ which are $m = 1, 2, 3$.

The 'c' of the asymptotes are given by

$$c = -\frac{\phi_2(m)}{\phi_3'(m)} = 0 \quad \text{as} \quad \phi_2(m) = 0.$$

Therefore, there are three asymptotes of the curve, with equations $y = x$, $y = 2x$, $y = 3x$. **Ans.**

EXAMPLE 2.4. Determine the asymptotes, if any, of the curve

$$4x^3 - 3xy^2 - y^3 + 2x^2 - xy - y^2 - 1 = 0.$$

SOLUTION: The coefficient of the highest degree term in x, that is, x^3 is 4 hence no asymptote is there parallel to x-axis. Similarly, the coefficient of the highest degree term in y, is y^3 whose coefficient is -1, hence no asymptote parallel to y-axis.

Next, here $\phi_3(m) = 4 \cdot 1^3 - 3 \cdot 1 \cdot m^2 - m^3 = -m^3 - 3m^2 + 4$

and $\phi_2(m) = 2 \cdot 1^2 - 1 \cdot m - m^2 = -m^2 - m + 2$ and $\phi_1(m) = 0$.

The slopes of the asymptotes are given by the equation

$$\phi_3(m) = 0 \quad \text{or} \quad m^3 + 3m^2 - 4 = 0.$$

or $\quad (m-1)(m+2)^2 = 0$ hence $m = 1, -2, -2$.

The 'c' of the asymptote corresponding to $m = 1$, is given by

$$c = -\frac{\phi_2(m)}{\phi_3'(m)} = \left(\frac{m^2+m-2}{-3m^2-6m}\right)_{m=1} = 0.$$

Therefore one asymptote has equation $y = x$. Since two roots of the equation $\phi_3(m) = 0$, are equal, this is the case of a pair of parallel asymptotes. The values of 'c' for these parallel asymptotes are given by

$$\frac{c^2}{2!}\phi_3''(m) + \frac{c}{1!}\phi_2'(m) + \phi_1(m) = 0$$

or $$\frac{c^2}{2}(-6m-6) + c(-2m-1) + 0 = 0 \quad \text{or} \quad 3c^2(m+1) + c(2m+1) = 0$$

Putting $m = -2$, in this equation, we get

$-3c^2 - 3c = 0$ which gives $c = 0, -1$.

Therefore, the equations of the parallel asymptotes are

$y = -2x + 0$ and $y = -2x - 1$

Thus, the given curve has three asymptotes with equations

$y = x$, $y = -2x$ and $y = -2x - 1$. **Ans.**

EXAMPLE 2.5. Find the asymptotes of the curve given by the equation

$$y^4 - 2xy^3 + 2x^3y - x^4 - 3x^3 + 3x^2y + 3xy^2 - 3y^3 - 2x^2 + 2y^2 - 1 = 0$$

SOLUTION: The equation of the curve is of degree 4 hence it cannot have more than four asymptotes. From the equation of the curve, we have

$$\phi_4(m) = m^4 - 2\cdot 1\, m^3 + 2\cdot 1^3 m - 1^4 = m^4 - 2m^3 + 2m - 1$$
$$= (m-1)(m^3 - m^2 - m + 1) = (m-1)^2(m^2-1) = (m-1)^3(m+1)$$

and $$\phi_3(m) = -3\cdot 1^3 + 3\cdot 1^2\cdot m + 3\cdot 1\cdot m^2 - 3m^3 = -3m^3 + 3m^2 + 3m - 3$$

and $$\phi_2(m) = -2\cdot 1^2 + 2m^2 = 2(m^2-1), \quad \phi_1(m) = 0.$$

For the slopes of the asymptote, we have the equation

$$\phi_4(m) = 0 \quad \text{or} \quad (m-1)^3(m+1) = 0$$

whose roots are $m = 1, 1, 1$ and -1. Thus, of the four asymptotes, three are parallel.

The 'c' of the asymptote with slope -1, is given by

$$c = -\frac{\phi_3(m)}{\phi_4'(m)} = \frac{-3(-m^3+m^2+m-1)}{3(m-1)^2(m+1) + (m-1)^3\cdot 1}$$

$$= \frac{-3\,[-(-1)^3 + (-1)^2 - 1 - 1]}{0 + (-2)^3} = 0 \text{ at } m = -1.$$

Therefore, the asymptote corresponding to $m = -1$ is $y = -x$.

To find the values of 'c' for the three parallel asymptotes corresponding to $m = 1$, we have

$$\frac{c^3}{3!}\phi_4'''(m)+\frac{c^2}{2!}\phi_3''(m)+c\,\phi_2'(m)+\phi_1(m)=0$$

or $$\frac{c^3}{6}(24m-12)+\frac{c^2}{2}(-18m+6)+c\cdot 4m+0=0$$

or $$c\,[(4m-2)\,c^2+(-9m+3)\,c+4m]=0$$

Putting $m=1$ we get $c=0$ and $2c^2-6c+4=0$

or $(c-1)(c-2)=0$

Thus, the three values of c for $m = 1$, are $c = 0, 1, 2.$

Therefore, the four asymptotes are

$$y=-x,\ y=x,\ y=x+1 \text{ and } y=x+2$$ **Ans.**

EXAMPLE 2.6. An asymptote touches the curve in at least (i) one point (ii) two points (iii) three points (iv) one of these

SOLUTION: Since the asymptote touches the curve at infinity hence the correct answer is (iv).

EXAMPLE 2.7. Find the asymptote of the curve $(x + y)^2 (x + 2y + 2) = x + 9y - 2$.

SOLUTION: Equation of the curve can be written as $(x + y)^2 (x + 2y) + 2 (x + y)^2 - x - 9y + 2 = 0$

Here $\phi_3(m) = (1 + m)^2 (1 + 2m)$, $\phi_2(m) = 2 (1 + m)^2$ and $\phi_1(m) = -1 - 9m$

The slopes of the asymptotes are obtained from $\phi_3(m) = 0$

$\Rightarrow$ $(1 + m)^2 (1 + 2m) = 0$ hence $m = -1, -1, -1/2.$

For $m = -1/2$, we have $c\,\phi'_3(m) + \phi_2(m) = 0$

or $$c = \left[\frac{-\phi_2(m)}{\phi_3'(m)}\right]_{m=\frac{-1}{2}} = \left[\frac{-2(1+m)^2}{2(1+m)^2+2(1+m)(1+2m)}\right]_{m=\frac{-1}{2}}$$

$$= \left[\frac{-(1+m)}{(1+m)+(1+2m)}\right]_{m=\frac{-1}{2}} = -1$$

For $m = -1$, which is the repeated root, we have $\frac{c^2}{2}\phi_3''(m) + c\phi'_2(m) + \phi_1(m) = 0$

or $$\frac{c^2}{2}\phi_3''(-1) + c\phi'_2(-1) + \phi_1(-1) = 0 \quad \text{or} \quad \frac{c^2}{2}\,2(-3+2) + c\,4\,(1+m) - 1 + 9 = 0$$

or $$-c^2 + 0 + 8 = 0,\quad c = \pm\, 2\sqrt{2}$$

Hence the asymptote corresponding to $m = \frac{-1}{2}$ is $y = \frac{-1}{2}x - 1$ and the two asymptotes corresponding to $m = -1$, are $y = -x \pm 2\sqrt{2}$. **Ans.**

EXAMPLE 2.8. Find the asymptotes, if any, of the curve $\dfrac{a^2}{x^2}+\dfrac{b^2}{y^2}=1$

SOLUTION: Equation of the curve is $\dfrac{a^2}{x^2}+\dfrac{b^2}{y^2}=1$ or $x^2y^2 - a^2y^2 - b^2x^2 = 0$ ...(1)

For asymptotes parallel to X-axis, we equate to zero the coefficient of x^2, to get

$$y^2 - b^2 = 0 \quad \text{or} \quad y = \pm b.$$

For asymptotes parallel to Y-axis, we equate to zero the coefficient of y^2, to get

$$x^2 - a^2 = 0 \quad \text{or} \quad x = \pm a.$$

Thus, we have four asymptotes of the curve as $y = \pm b$ and $x = \pm a$. **Ans.**

EXAMPLE 2.9. Find the asymptotes of the curve ***(a)*** $x^3 + 3x^2y - 4y^3 - x + y + 3 = 0$.
(b) $y^3 - x^2y - 2xy^2 + 2x^3 - 7xy + 3y^2 + 2x^2 + 2x + 2y + 1 = 0$. **[UPTU 2001-12]**

SOLUTION: (a) Equation of the curve is $x^3 + 3x^2y - 4y^3 - x + y + 3 = 0$.

Here $\phi_3(m) = 1 + 3m - 4m^3,\quad \phi_2(m) = 0,\quad \phi_1(m) = -1 + m,\quad \phi_0(m) = 3.$

To find the slopes 'm' of asymptotes, we write $\phi_3(m) = 0$

or $4m^3 - 3m - 1 = 0$ or $(m-1)(4m^2 + 4m + 1) = 0$

which gives $m = 1, \dfrac{-1}{2}, \dfrac{-1}{2}.$

For 'c' of the asymptote corresponding to $m = 1$, we have

$$c\phi_3'(m) + \phi_2(m) = 0 \quad \text{or} \quad c(3 - 12m^2) - 0 = 0 \quad \text{hence } c = 0$$

For $m = \dfrac{-1}{2}$, we have two parallel asymptotes whose 'c' is given by

$$\frac{c^2}{2!}\phi_3''(m) + \frac{c}{1!}\phi_2'(m) + \phi_1(m) = 0$$

or $$\frac{c^2}{2}(-24m) + c(0) + (-1 + m) = 0$$

or $$-c^2(12m) - 1 + m = 0$$

For $m = \dfrac{-1}{2}$ we have $6c^2 - \dfrac{3}{2} = 0$ or $c = \pm\dfrac{1}{2}$

Thus we have three asymptotes as

$$y = x,\quad y = \frac{-1}{2}x + \frac{1}{2} \quad \text{and} \quad y = \frac{-1}{2}x - \frac{1}{2}. \qquad \textbf{Ans.}$$

(b) In the equation of the given curve

$\phi_3(m) = m^3 - 2m^2 - m + 2,\quad \phi_2(m) = -7m + 3m^2 + 2,\quad \phi_1(m) = 2 + 2m.$

For slopes of the asymptotes we have $\phi_3(m) = 0$ or $m^3 - 2m^2 - m + 2 = 0$

or $(m+1)(m^2 - 3m + 2) = 0$ $\therefore$ $m = -1, 1, 2.$

For 'c' of the asymptotes, we have $c = \dfrac{-\phi_2(m)}{\phi_3'(m)} = \dfrac{-(3m^2 - 7m + 2)}{3m^2 - 4m - 1}$

When $m = -1$, $c = \dfrac{-(3+7+2)}{3+4-1} = -2$

When $m = 1$, $c = \dfrac{-(3-7+2)}{3-4-1} = -1$

and when, $m = 2$, $c = -\dfrac{(12-14+2)}{12-8-1} = 0$

$\therefore$ Asymptotes are $y = -x - 2$, $y = x - 1$ and $y = 2x$. **Ans.**

EXERCISE 2A

Find the asymptotes of the following curves

1. $x^3 + 2x^2y - xy^2 - 2y^3 + 4y^2 + 2xy + y - 1 = 0.$
2. $x^3 - 4y^3 + 3x^2y + y - x + 3 = 0.$
3. $y^3 + 2xy^2 + x^2y - y + 1 = 0.$
4. $\dfrac{x^2}{a^2} - \dfrac{y^2}{b^2} = 1.$
5. $(x^3 + a^3)\, y = bx^3.$
6. $y\,(x - y)^3 = y\,(x - y) + 2.$
7. $x^3 - 5x^2y + 8xy^2 - 4y^3 + x^2 - 3xy + 2y^2 - 1 = 0.$
8. Find all the asymptotes of the curve $y^2\,(x - 2a) = x^3 - a^3.$
9. Find all the asymptotes of the curve $x^2y^2 - x^2y - xy^2 + x + y + 1 = 0.$
10. Find the asymptotes of the curve $y^3 - xy^2 - x^2y + x^3 + x^2 - y^2 = 0.$

❑❑❑

DERIVATIVE OF ARC LENGTH IN CARTESIAN FORM

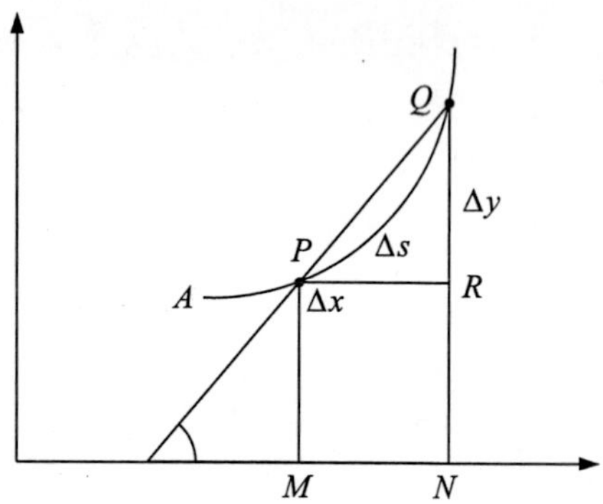

Let $P(x, y)$ and $Q(x+\Delta x,\ y+\Delta y)$ be the two neighbouring points on the curve $y = f(x)$. Let arc $AP = s$ and arc $AQ = s + \Delta s$, where A is a fixed point on the curve. Then arc $PQ = \Delta s$.

From the right angled triangle PQR.

$$PQ^2 = PR^2 + RQ^2$$
$$= (\Delta x)^2 + (\Delta y)^2 \qquad \text{...(1)}$$

Dividing by $(\Delta x)^2$,

$$\left(\frac{PQ}{\Delta x}\right)^2 = \left(\frac{PQ}{\Delta s}\right)^2\left(\frac{\Delta s}{\Delta x}\right)^2 = 1+\left(\frac{\Delta y}{\Delta x}\right)^2 \qquad \text{..(2)}$$

where $Q \to P$, $\dfrac{PQ}{\Delta x} = \left(\dfrac{\text{chord } PQ}{\text{arc } PQ}\right) \to 1$ and $\Delta x \to 0$

Also $\lim\limits_{\Delta x \to 0} \dfrac{\Delta s}{\Delta x} = \dfrac{ds}{dx} = \lim\limits_{\Delta x \to 0} \dfrac{\Delta y}{\Delta x} = \dfrac{dy}{dx}$

Thus as $Q \to P$, Eq. (2) reduces to

$$\left(\frac{ds}{dx}\right)^2 = 1+\left(\frac{dy}{dx}\right)^2$$

$$\frac{ds}{dx} = \sqrt{1+\left(\frac{dy}{dx}\right)^2}$$

Similarly, didividng Eq. (1) by $(\Delta y)^2$ and taking limits $Q \to P$, i.e., $\Delta y \to 0$

$$\frac{ds}{dy} = \sqrt{1+\left(\frac{dx}{dy}\right)^2}$$

Corollary: If equation of the curve is given in parametric form $x = x(t)$, $y = y(t)$

$$\frac{ds}{dt} = \frac{ds}{dx}\cdot\frac{dx}{dt}$$

$$= \sqrt{1+\left(\frac{dy}{dx}\right)^2}\cdot\frac{dx}{dt} = \sqrt{\left(\frac{dx}{dt}\right)^2+\left(\frac{dy}{dx}\right)^2\cdot\left(\frac{dx}{dt}\right)^2}$$

$$= \sqrt{\left(\frac{dx}{dt}\right)^2+\left(\frac{dy}{dt}\right)^2}$$

CURVATURE

Of any two curves one may be bending more sharply than the other. Also the bending may be different at different points of curve. For example, in case of parabola the bending is different at its different points, being maximum at its vertex. Actually, the curvature is a numerical measure of bending of the curve.

Let P be any point of a curve and s be the length of the arc from a fixed point A of the curve to the point P. Let the tangent at P be inclined at an angle ψ with the axis of x and arc AP = s. (See the figure). The direction of the curve at a point is determined by the slope of the tangent at that point. Let Q be another point on the curve close to P so that arc AQ = $s + \delta s$, and the tangent at Q make an angle $\psi + \delta\psi$ with the axis of x. The measure of the rapidity with which the curve is turning (or bending) at the point is the rate of change of ψ with respect to s. This rate of change is called *the* ***curvature*** *at the particular point.* Thus, the curvature at P

$$= \lim_{Q \to P} \frac{\delta\psi}{\delta s} = \lim_{\delta s \to 0} \frac{\delta\psi}{\delta s} = \frac{d\psi}{ds}.$$

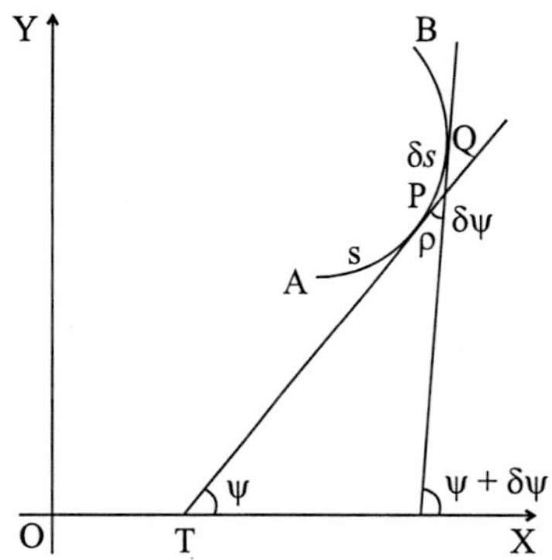

For example, consider a circle of radius a and centre on the y-axis at $(0, a)$, then we have $s = a\psi$ and therefore the curvature $= \frac{d\psi}{ds} = \frac{1}{a}$.

This shows that the curvature of a circle is same at every point of its circumference and is reciprocal of the radius of the circle. Idea is extended to any curve and this is the reason that the reciprocal of the curvature of a curve at a point is defined as its **radius of curvature** at that point and is denoted by ρ. Therefore, we have

$$\frac{1}{\rho} = \frac{d\psi}{ds} \quad \text{or} \quad \rho = \frac{ds}{d\psi}$$

Obviously, the radius of curvature of circle at any point is same as its radius. **Radius of Curvature** (in Cartesian Coordinates).

We know that $\tan\psi = \frac{dy}{dx}$. ...(1)

Differentiating on both sides with respect to s, gives

$$\sec^2\psi \frac{d\psi}{ds} = \frac{d}{ds}\left(\frac{dy}{dx}\right) = \frac{d^2y}{dx^2} \cdot \frac{dx}{ds}$$

or $$\frac{ds}{d\psi} = \frac{\sec^2\psi}{\frac{d^2y}{dx^2}} \cdot \frac{ds}{dx} = \frac{(1 + \tan^2\psi)}{\frac{d^2y}{dx^2}} \cdot \sqrt{1 + \left(\frac{dy}{dx}\right)^2} \quad \text{since} \quad \frac{ds}{dx} = \sqrt{1 + \left(\frac{dy}{dx}\right)^2}.$$

Now using the relation (1), we have

$$\rho = \frac{ds}{d\psi} = \frac{\left[1 + \left(\frac{dy}{dx}\right)^2\right]^{3/2}}{\frac{d^2y}{dx^2}} \quad \text{where} \quad \frac{d^2y}{dx^2} \neq 0.$$

In case the tangent to the curve at the point under consideration is parallel to y-axis then $\frac{dy}{dx}$ will be infinite, hence $\frac{dx}{dy} = 0$ then we consider $\frac{dx}{dy} = \cot \psi$.

Differentiating both sides w.r.t. s, gives

$$\frac{d}{ds}\left(\frac{dx}{dy}\right) = \frac{d}{ds} \cot \psi$$

or $$\frac{d}{dy}\left(\frac{dx}{dy}\right)\frac{dy}{ds} = -\operatorname{cosec}^2 \psi \frac{d\psi}{ds}$$ or $$\frac{d^2x}{dy^2}\frac{dy}{ds} = -(1+\cot^2\psi)\frac{d\psi}{ds}$$

or $$\rho = \left|\frac{ds}{d\psi}\right| = \left|\frac{-(1+\cot^2\psi)}{\frac{d^2x}{dy^2}}\frac{ds}{dy}\right| = \left|\frac{-\left[1+\left(\frac{dx}{dy}\right)^2\right]\sqrt{1+\left(\frac{dx}{dy}\right)^2}}{\frac{d^2x}{dy^2}}\right|$$

or $$\rho = \frac{\left[1+\left(\frac{dx}{dy}\right)^2\right]^{\frac{3}{2}}}{\frac{d^2x}{dy^2}} \quad \text{where} \quad \frac{d^2x}{dy^2} \neq 0.$$

RADIUS OF CURVATURE FOR PARAMETRIC CURVES

Let the equation of the curve be $x = f(t)$ and $y = g(t)$ where t is the parameter,

then $$\frac{dy}{dx} = \frac{dy}{dt}\bigg/\frac{dx}{dt} = \frac{y'}{x'}$$ where the dashe represents the derivative w.r.t. t.

Further, $$\frac{d^2y}{dx^2} = \frac{d}{dx}\left(\frac{y'}{x'}\right) = \frac{d}{dt}\left(\frac{y'}{x'}\right)\frac{dt}{dx} = \frac{x'y''-y'x''}{x'^2\cdot\frac{dx}{dt}}$$

$$\therefore \quad \rho = \frac{\left[1+\left(\frac{dy}{dx}\right)^2\right]^{\frac{3}{2}}}{\frac{d^2y}{dx^2}} = \frac{\left(1+\frac{y'^2}{x'^2}\right)^{\frac{3}{2}}}{\frac{x'y''-y'x''}{x'^3}} = \frac{\left(x'^2+y'^2\right)^{\frac{3}{2}}}{x'y''-y'x''}.$$

RADIUS OF CURVATURE AT THE ORIGIN BY NEWTON'S METHOD

This is a very fundamental method but has limited use because it is applicable *only when the curve passes through the origin and has x-axis or y-axis as the tangent there.*

Let us first consider the case where the curve passes through the origin and x-axis is the tangent to the curve at the origin. Thus, $y_1 = 0$ at the origin and hence $\rho = \frac{\left(1+y_1^2\right)^{3/2}}{y_2} = \frac{1}{y_2}$.

$\therefore$ Let us consider $\lim_{x\to 0} \dfrac{x^2}{2y}$ which is of the form $\dfrac{0}{0}$

Applying L′ Hospital's rule, we get

$$\lim_{x\to 0} \frac{x^2}{2y} = \lim_{x\to 0} \frac{2x}{2y_1} \quad \text{which is again of } \frac{0}{0} \text{ form}$$

Again applying L′ Hospital's rule $= \lim_{x\to 0} \dfrac{2}{2y_2} = \rho$ {where $y_2 \neq 0$ at (0, 0))

Therefore, we have $\boldsymbol{\rho = \lim_{x\to 0} \dfrac{x^2}{2y}}$, when x-axis is tangent to curve at the origin.

Similarly, in the case when curve passes through the origin and y-axis is the tangent to the curve at the origin, we have $\boldsymbol{\rho = \lim_{y\to 0} \dfrac{y^2}{2x}}$.

CENTRE OF CURVATURE AND CIRCLE OF CURVATURE

Let APB be a portion of the curve and P (x, y) be any point on it and PT the tangent to the curve at P. If we draw a circle passing through P on the same side of the tangent as the curve, with centre at a point C on the normal to the curve at P so that the radius CP = ρ the radius of curvature at P, then such a circle is known as *circle of curvature* at P (See figure on next page).

The point C is called the *centre of curvature*. Let C = (α, β) then from the figure, we have

$$\alpha = x - \rho \sin\psi \quad \text{and} \quad \beta = y + \rho \cos\psi.$$

Since $\tan\psi = \dfrac{dy}{dx} = y_1$, we have $\sin\psi = \dfrac{y_1}{\sqrt{1+y_1^2}}$ and $\cos\psi = \dfrac{1}{\sqrt{1+y_1^2}}$,

Already we have $\rho = \dfrac{\left(1+y_1^2\right)^{3/2}}{y_2}$

therefore $\boldsymbol{\alpha} = x - \dfrac{\left(1+y_1^2\right)^{3/2}}{y_2}\dfrac{y_1}{\sqrt{1+y_1^2}} = \boldsymbol{x - \dfrac{y_1\left(1+y_1^2\right)}{y_2}}$

and $\boldsymbol{\beta} = y + \dfrac{\left(1+y_1^2\right)^{3/2}}{y_2}\dfrac{1}{\sqrt{1+y_1^2}} = \boldsymbol{y + \dfrac{\left(1+y_1^2\right)}{y_2}}$

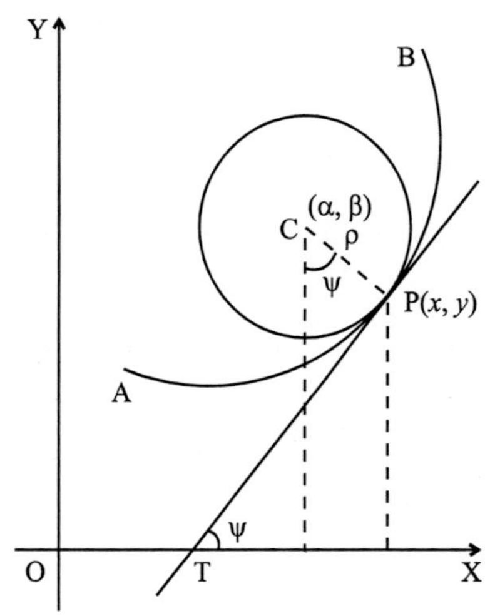

Therefore, the circle of curvature has the equation $\boldsymbol{(x-\alpha)^2 + (y-\beta)^2 = \rho^2}$.

For parametric curves

$$\boldsymbol{\alpha = x - \frac{y'(x'^2+y'^2)}{x'y''-y'x''}, \quad \beta = y + \frac{x'(x'^2+y'^2)}{x'y''-y'x''}}$$

Further, the locus of the centre of curvature of a circle is called its EVOLUTE and the curve is referred to as INVOLUTE of the evolute.

EXAMPLE 2.10. Find the radius of curvature at the origin for the following curves

(*i*) $x^2 = 4ay$ (*ii*) $y^2(a^2 - x^2) = a^3x$.

SOLUTION: (*i*) The parabola $x^2 = 4ay$ passes through the origin and tangent at the origin is $y = 0$, the axis of x. Therefore, by Newton's method, $\rho = \lim\limits_{x\to 0} \dfrac{x^2}{2y} = \lim\limits_{x\to 0} \dfrac{4ay}{2y} = 2a$ **Ans.**

(*ii*) The given curve $y^2(a^2 - x^2) = a^3x$ passes through the origin and by equating to 0 the lowest degree term we get the tangent at the origin as $x = 0$, the axis of y. Therefore, by Newton's method,

$$\rho = \lim_{y\to 0} \frac{y^2}{2x} = \lim_{\substack{y\to 0\\ x\to 0}} \frac{a^3 x}{2\left(a^2 - x^2\right)\cdot x} = \lim_{\substack{y\to 0\\ x\to 0}} \frac{a^3}{2a^2 - 2x^2} = \frac{a}{2}. \qquad \textbf{Ans.}$$

EXAMPLE 2.11. If ρ is the radius of curvature at any point P on the parabola $y^2 = 4ax$ and S is its focus then show that ρ^2 varies as SP^3. Also, show that the radius of curvature at the vertex is equal to the length of the semi-latus rectum.

SOLUTION: Differentiating $y^2 = 4ax$ we get $2y\, y_1 = 4a$ or $y_1 = \dfrac{2a}{y}$.

and $$y_2 = \frac{-2a}{y^2}\cdot\frac{dy}{dx} = \frac{-2a}{y^2}\cdot\frac{2a}{y} = \frac{-4a^2}{y^3}$$

$$\therefore \quad \rho = \frac{\left(1+y_1^2\right)^{3/2}}{y_2} = \frac{\left(1+4\dfrac{a^2}{y^2}\right)^{3/2}}{\dfrac{-4a^2}{y^3}} = \frac{\left(y^2+4a^2\right)^{3/2}}{-4a^2}$$

$$= \frac{\left(4ax+4a^2\right)^{3/2}}{4a^2} = \frac{(4a)^{3/2}}{4a^2}(x+a)^{3/2} = \frac{2}{\sqrt{a}}(x+a)^{3/2} \quad \text{(in absolute value).}$$

Since the focus S is at $(a, 0)$ and $P(x, y)$ is on the parabola, hence

$$SP = \sqrt{(x-a)^2 + (y-0)^2} = \sqrt{(x-a)^2 + 4ax} = x + a$$

$$\therefore \quad \rho = \frac{2}{\sqrt{a}} SP^{3/2} \quad \text{or} \quad \rho^2 = \frac{4}{a} SP^3$$

This establishes that ρ^2 varies as SP^3.

Next, to obtain ρ at the vertex $(0, 0)$, we find that $\dfrac{dy}{dx}$ is infinite at the vertex since tangent there, is parallel to y-axis. Therefore using the formula

$$\rho = \frac{\left[1+\left(\frac{dx}{dy}\right)^2\right]^{3/2}}{\frac{d^2x}{dy^2}}$$

and the equation of the curve, we have

$$\frac{dx}{dy} = \frac{y}{2a} = 0 \quad \text{at} \quad (0,0) \quad \text{and} \quad \frac{d^2x}{dy^2} = \frac{1}{2a}$$

therefore, $\rho\,(0,\,0) = \dfrac{(1+0)^{3/2}}{\left(\dfrac{1}{2a}\right)} = 2a =$ length of semi latus rectum. **Hence Proved.**

EXAMPLE 2.12. Find the radius of curvature at any point 'θ' of the cycloid

$$x = a\,(\theta + \sin\theta), \quad y = a\,(1 - \cos\theta).$$

SOLUTION: $x' = \dfrac{dx}{d\theta} = a\,(1+\cos\theta), \quad y' = \dfrac{dy}{d\theta} = a\sin\theta$

$$x'' = \frac{d^2x}{d\theta^2} = -a\sin\theta, \quad y'' = \frac{d^2y}{d\theta^2} = a\cos\theta$$

For parametric equation of the curve, we have

$$\rho = \frac{(x'^2+y'^2)^{3/2}}{x'\,y''-y'\,x''} = \frac{\{a^2(1+\cos\theta)^2 + a^2\sin^2\theta\}^{3/2}}{a^2(1+\cos\theta)\cos\theta + a^2\sin^2\theta}$$

$$= \frac{a\,(1+\cos^2\theta+2\cos\theta+\sin^2\theta)^{3/2}}{\cos\theta+\cos^2\theta+\sin^2\theta} = \frac{a\,(2+2\cos\theta)^{3/2}}{1+\cos\theta}$$

$$= 2\sqrt{2}\,a\sqrt{1+\cos\theta} = 2\sqrt{2}\,a\sqrt{2\cos^2\frac{\theta}{2}} = 4a\cos\frac{\theta}{2}$$

Therefore at the point 'θ' we have $\rho = 4a\cos\dfrac{\theta}{2}$. **Ans.**

EXAMPLE 2.13. **(a)** Show that the radius of curvature at any point (x, y) of the rectangular hyperbola $xy = c^2$, is given by $\rho = \dfrac{\left(x^2+y^2\right)^{3/2}}{2c^2}$.

(b) If ρ_1 and ρ_2 are the radii of curvature at the extremities of a focal chord of a parabola, show that $\rho_1^{-2/3} + \rho_2^{-2/3}$ is a constant.

SOLUTION: **(a)** The equation of the curve can be written as $y = \dfrac{c^2}{x}$

$$\therefore \quad y_1 = \frac{-c^2}{x^2} \quad \text{and} \quad y_2 = \frac{2c^2}{x^3}.$$

$$\therefore \quad \rho = \frac{\left(1+y_1^2\right)^{3/2}}{y_2} = \frac{\left(1+\dfrac{c^4}{x^4}\right)^{3/2}}{\dfrac{2c^2}{x^3}} = \frac{\left(x^4+c^4\right)^{3/2} x^3}{x^6 \cdot 2c^2}$$

$$= \frac{(x^4+c^4)^{3/2}}{2c^2 x^3} = \frac{(x^4+x^2 y^2)^{3/2}}{2c^2 x^3} = \frac{(x^2+y^2)^{3/2}}{2c^2} \quad (\text{using } c^2 = xy)$$ **Hence Proved.**

(b) Let the parabola be $y^2 = 4ax$, any point on it is $(at^2, 2at)$.

We know that if $(at_1^2, 2at_1)$ and $(at_2^2, 2at_2)$ are the extremities of a focal chord of the parabola then $t_1 t_2 = -1$.

We have $x = at^2,\ y = 2at$ hence $x' = 2at,\ y' = 2a$ and $x'' = 2a,\ y'' = 0$ therefore at the point $(at^2, 2at)$

$$\rho = \frac{(x'^2+y'^2)^{3/2}}{x'y''-y'x''} = \frac{(4a^2t^2+4a^2)^{3/2}}{0-4a^2} = 2a\,(1+t^2)^{3/2} \quad (\text{not taking} - \text{sign}).$$

or $$\rho^{-2/3} = (2a)^{-2/3}(1+t^2)^{-1}$$

Therefore $$\rho_1^{-2/3}+\rho_2^{-2/3} = (2a)^{-2/3}\left(\frac{1}{1+t_1^2}+\frac{1}{1+t_2^2}\right) = (2a)^{-2/3}\,\frac{(2+t_1^2+t_2^2)}{(1+t_1^2+t_2^2+t_1^2t_2^2)}$$

$$= (2a)^{-2/3} \quad \text{since } t_1 t_2 = -1 = \text{Constant.}$$ **Ans.**

EXAMPLE 2.14. If ρ and ρ' are the radii of curvature at the extremities of two conjugate diameters of the ellipse $\dfrac{x^2}{a^2}+\dfrac{y^2}{b^2} = 1$, prove that $(\rho^{2/3} + \rho'^{2/3})\,(ab)^{2/3} = a^2 + b^2$.

[GGSIPU I Sem End Term 2003]

SOLUTION: Let C P and C D be two conjugate semi diameters of the given ellipse. Parametric equation of the ellipse is $x = a\cos\theta,\ y = b\sin\theta$.

Let the coordinates of the point P be $(a\cos\theta, b\sin\theta)$ then those of the point D are obtained by replacing 'θ' by $\dfrac{\pi}{2}+\theta$, so as to get D = $\{a\cos(\theta+\pi/2),\ b\sin(\theta+\pi/2)\}$.

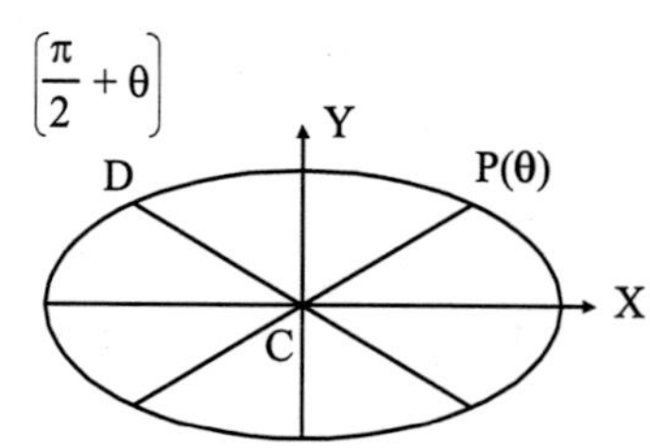

To find the radius of curvature at P, the point 'θ' of the ellipse, we have

$$x' = -a\sin\theta,\ y' = b\cos\theta \quad \text{and} \quad x'' = -a\cos\theta,\ y'' = -b\sin\theta$$

$$\therefore \quad \rho = \frac{(x'^2+y'^2)^{3/2}}{x'y''-y'x''} = \frac{(a^2\sin^2\theta+b^2\cos^2\theta)^{3/2}}{(-a\sin\theta)(-b\sin\theta)-(b\cos\theta)(-a\cos\theta)}$$

or $$\rho = \frac{(a^2\sin^2\theta+b^2\cos^2\theta)^{3/2}}{ab} \qquad \text{...(1)}$$

For the radius of curvature ρ' at D we have to replace θ by $(\theta + \pi/2)$ in (1) and get

$$\rho' = \frac{(a^2 \cos^2 \theta + b^2 \sin^2 \theta)^{3/2}}{ab}$$

$$\therefore \quad \rho^{2/3} + \rho'^{2/3} = \frac{(a^2 \sin^2 \theta + b^2 \cos^2 \theta)}{(ab)^{2/3}} + \frac{(a^2 \cos^2 \theta + b^2 \cos^2 \theta)}{(ab)^{2/3}} = \frac{a^2 + b^2}{(ab)^{2/3}}$$

or $\quad (ab)^{2/3} (\rho^{2/3} + \rho'^{2/3}) = a^2 + b^2.$ **Hence Proved.**

EXAMPLE 2.15. Find the radius of curvature at the origin for the curve

$$2x^3 - 3x^2 y + 4y^3 + y^2 - 3x = 0.$$

SOLUTION: The given curve is passing through the origin and on equating to zero the lowest degree term we get the equation of the tangent at the origin as $x = 0$, the axis of y.

Therefore, by Newton's method $\rho = \lim\limits_{y \to 0} \dfrac{y^2}{2x}$.

Dividing the given equation throughout by x, we get

$$2x^2 - 3xy + 4y \cdot \frac{y^2}{x} + \frac{y^2}{x} - 3 = 0. \quad \text{At the origin, we have}$$

$$0 + 0 + 0 + 2\rho - 3 = 0 \quad \text{or} \quad \rho = \frac{3}{2}. \qquad \textbf{Ans.}$$

EXAMPLE 2.16. Find the point on the curve $y = e^x$ at which the curvature is maximum and show that the tangent to the curve at that point forms with the coordinate axes, a triangle whose sides are in the ratio $1 : \sqrt{2} : \sqrt{3}$.

SOLUTION: We have $y = e^x$ hence $y_1 = e^x = y_2 = y$.

$$\therefore \quad \rho = \frac{\left(1 + y_1^2\right)^{3/2}}{y_2} \quad \text{or} \quad \rho^2 = \frac{\left(1 + y_1^2\right)^3}{y_2^2}.$$

Curvature is maximum when radius of curvature is minimum, *i.e.*, when $\dfrac{\left(1 + y_1^2\right)^3}{y_2^2}$ is minimum.

If we write $y^2 = t$ then $y_1^2 = y_2^2 = t$, hence

$$\rho^2 = \frac{(1 + t)^3}{t} = t^2 + 3t + 3 + \frac{1}{t}$$

Now, $\dfrac{d\rho^2}{dt} = 0$ gives $2t + 3 - \dfrac{1}{t^2} = 0$ or $2t^3 + 3t^2 - 1 = 0$

or $(t + 1)(2t^2 + t - 1) = 0$ or $(t + 1)^2 (2t - 1) = 0.$

$\Rightarrow \quad t = -1$ and $t = \dfrac{1}{2}$ or $y^2 = -1, \dfrac{1}{2}$

Obviously, $y^2 \neq -1$ as $y^2 \geq 0$ hence $y^2 = \dfrac{1}{2}$ or $y = \pm \dfrac{1}{\sqrt{2}}$.

But $y\ (= e^x)$ cannot be negative, hence $y = \dfrac{1}{\sqrt{2}}$.

For ρ^2 to be minimum $\frac{d\rho^2}{dt} = 0$ and $\frac{d^2\rho}{dt^2} > 0$

$$\frac{d^2\rho^2}{dt^2} = 2 + \frac{2}{t^3} = 2 + \frac{2}{y^6} > 0 \quad \text{hence } \rho^2 \text{ is minimum at } y = \frac{1}{\sqrt{2}}.$$ **Ans.**

For $y = \frac{1}{\sqrt{2}}$ we have $x = \log y = -\frac{1}{2}\log 2$

Thus, the required point is $\left(-\frac{1}{2}\log 2, \frac{1}{\sqrt{2}}\right)$ where the curvature is maximum.

The tangent at the point $\left(-\frac{1}{2}\log 2, \frac{1}{\sqrt{2}}\right)$ has the equation

$$y - \frac{1}{\sqrt{2}} = \frac{1}{\sqrt{2}}\left(x + \frac{1}{2}\log 2\right) \quad \text{or} \quad \frac{y}{1} - \frac{x}{\sqrt{2}} = \frac{1}{\sqrt{2}}\left(1 + \frac{1}{2}\log 2\right) = k, \text{ say.}$$

Then intercepts on the coordinate axes are $k\sqrt{2}$ and k therefore the tangent is of length

$$= \sqrt{\left(k\sqrt{2}\right)^2 + k^2} = k\sqrt{3}.$$

Hence the sides of the triangle are in the ratio $1 : \sqrt{2} : \sqrt{3}$. **Hence Proved.**

EXAMPLE 2.17. Show that the radius of curvature at any point of the curve $x^{2/3} + y^{2/3} = a^{2/3}$ is equal to three times the length of the perpendicular from the origin to the tangent at that point.

SOLUTION: The given curve in the parametric form, is $x = a\cos^3\theta$, $y = a\sin^3\theta$ hence $x' = -3a\cos^2\theta\sin\theta$, $y' = 3a\sin^2\theta\cos\theta$

and $x'' = 6a\cos\theta\sin^2\theta - 3a\cos^3\theta$, $y'' = 6a\sin\theta\cos^2\theta - 3a\sin^3\theta$.

$$\therefore \quad \rho = \frac{(x'^2 + y'^2)^{3/2}}{x'y'' - y'x''} = \frac{27a^3\cos^3\theta\sin^3\theta}{-9a^2\sin^2\theta\cos^2\theta} = 3a\sin\theta\cos\theta \text{ (on simplification).}$$

The equation of the tangent at $(a\cos^3\theta, a\sin^3\theta)$ to the curve, is

$$y - a\sin^3\theta = \frac{-\sin\theta}{\cos\theta}(x - a\cos^3\theta) \quad \text{or} \quad x\sin\theta + y\cos\theta = a\sin\theta\cos\theta$$

$\therefore$ The length of the perpendicular from the origin upon this perpendicular is $p = a\sin\theta\cos\theta$

Therefore we have $\rho = 3p$, **Hence the result.**

EXAMPLE 2.18. The radius of curvature of a circle at any point is same as its
(i) diameter (ii) radius (iii) circumference (iv) square of radius

SOLUTION: Ans is (ii).

EXAMPLE 2.19. Find the curvature of $x = 4\cos t, y = 3\sin t$. At what point on this ellipse does the curvature have the greatest and the least values? What are magnitudes?

SOLUTION: The curve is $x = 4\cos t,\quad y = 3\sin t,\quad$ hence $x' = -4\sin t,\ y' = 3\cos t,$

$$x'' = -4\cos t, \quad y'' = -3\sin t.$$

Therefore the radius of curvature ρ at any point 't' on the curve, is given by

$$\rho = \frac{(x'^2 + y'^2)^{3/2}}{x'y'' - y'x''} = \frac{(16\sin^2 t + 9\cos^2 t)^{3/2}}{-4\sin t(-3\sin t) - 3\cos t(-4\cos t)} = \frac{1}{12}(9\cos^2 t + 16\sin^2 t)^{3/2}$$

or $$\rho^{2/3} = \frac{1}{12^{2/3}}(9\cos^2 t + 16\sin^2 t)$$

Since the curvature is reciprocal of the radius of curvature, the curvature will be greatest and least when the radius of curvature is least and greatest respectively, in other words, when $(12\rho)^{2/3}$ is least and greatest respectively.

For maximum and minimum of $(12\rho)^{2/3}$ we have $\dfrac{d}{dt}(16\sin^2 t + 9\cos^2 t) = 0$

or $\quad 14\sin t\cos t = 0 \quad$ which gives $\quad t = 0$ and $t = \dfrac{\pi}{2}$.

Hence $(12\rho)^{2/3} = 9$ at $t = 0$, *i.e.* at $(4, 0)$ and is equal to 16 at $t = \pi/2$, *i.e.* at $(0, 3)$

Thus the curvature is least at $(0, 3)$ and greatest at $(4, 0)$, and least value of curvature is $3/16$ and greatest is $4/9$. **Ans.**

EXAMPLE 2.20. Find the radius of curvature of the curve $y^2 = \dfrac{4a^2(2a - x)}{x}$ at a point where the curve meets the axis of x.

SOLUTION: The given curve is $y^2 = \dfrac{4a^2(2a - x)}{x}$ which meets the axis of x at the point $(2a, 0)$. Since the tangent to the curve at $(2a, 0)$ is parallel to y-axis hence $y_1 = \infty$ at this point. Therefore we consider the other formula $\rho = \dfrac{(1 + x_1^2)^{3/2}}{x_2}$ for finding the radius if curvature at $(2a, 0)$

The equation of the curve can be written as $\quad x = \dfrac{8a^3}{y^2 + 4a^2}$

Hence $$x_1 = \frac{-16a^3 y}{(y^2 + 4a^2)^2} \quad \text{which is } 0 \text{ at } (2a, 0)$$

Also $$x_2 = \frac{-16a^3(y^2 + 4a^2)^2 + 16a^3 y \cdot 2(y^2 + 4a^2)\cdot 2y}{(y^2 + 4a^2)^4}$$

$$= \frac{-16a^3(y^2 + 4a^2) + 64a^3 y^2}{(y^2 + 4a^2)^3} = \frac{1}{a} \quad \text{at } (2a, 0)$$

Therefore $$\rho = \frac{(1 + x_1^2)^{3/2}}{x_2} = a \quad \text{at } (2a, 0).$$ **Ans.**

EXAMPLE 2.21. Find the centre of curvature of the parabola

$x = at^2, \ y = 2at$ at the point 't' and hence find its evolute.

SOLUTION: As $x = at^2, y = 2at$ we have $y_1 = \dfrac{dy}{dt} \Big/ \dfrac{dx}{dt} = \dfrac{2a}{2at} = \dfrac{1}{t}$

and $y_2 = \dfrac{d}{dx}\left(\dfrac{1}{t}\right) = -\dfrac{1}{t^2} \cdot \dfrac{dt}{dx} = -\dfrac{1}{t^2 \cdot 2at} = -\dfrac{1}{2at^3}$

The coordinates of the centre of curvature $(\bar{x}, \bar{y})$ are given by

$$\bar{x} = x - \frac{y_1(1+y_1^2)}{y_2} \quad \text{and} \quad \bar{y} = y + \frac{1+y_1^2}{y_2}$$

or $$\bar{x} = at^2 - \frac{1}{t}\left(1+\frac{1}{t^2}\right)(-2at^3) = at^2 + 2a(1+t^2) = a(2+3t^2) \quad \text{...(1)}$$

and $$\bar{y} = 2at + \left(1+\frac{1}{t^2}\right)(-2at^3) = 2at - 2at(1+t^2) = -2at^3 \quad \text{...(2)}$$

Thus, the centre of curvature $= (\bar{x}, \bar{y}) = \{a(2+3t^2), -2at^3\}$.

For evolute, we find the locus of the centre of curvature, for which, eliminating t in (1) and (2), gives

$$(\bar{x}-2a)^3 = (3a)^3\left(\frac{\bar{y}}{-2a}\right)^2 \quad \text{or} \quad 4(\bar{x}-2a)^3 = 27a\,\bar{y}^2$$

$\therefore$ Evolute is $4(x-2a)^3 = 27ay^2$. **Ans.**

EXAMPLE 2.22. Find the centre of curvature of the hypocycloid $x^{2/3} + y^{2/3} = a^{2/3}$ and also find the evolute of the curve.

SOLUTION: The parametric form of the equation of the given curve is $x = a\cos^3\theta, y = a\sin^3\theta$.

$$\therefore \quad y_1 = \frac{dy}{dx} = \frac{3a\sin^2\theta\cos\theta}{-3a\cos^2\theta\sin\theta} = -\tan\theta.$$

and $$y_2 = \frac{d^2y}{dx^2} = -\sec^2\theta\frac{d\theta}{dx} = \frac{-\sec^2\theta}{-3a\cos^2\theta\sin\theta} = \frac{1}{3a\cos^4\theta\sin\theta}.$$

Hence, the centre of curvature $(\bar{x}, \bar{y})$ at the point 'θ', is given by

$$\bar{x} = x - \frac{y_1(1+y_1^2)}{y_2} = a\cos^3\theta + \tan\theta(1+\tan^2\theta)\,3a\cos^4\theta\sin\theta$$
$$= a(\cos^3\theta + 3\sin^2\theta\cos\theta)$$

$$\bar{y} = y + \frac{1+y_1^2}{y_2} = a\sin^3\theta + (1+\tan^2\theta)\,3a\cos^4\theta\sin\theta = a(\sin^3\theta + 3\cos^2\theta\sin\theta)$$

Further $\quad \bar{x} + \bar{y} = a(\cos\theta + \sin\theta)^3$

and $\quad \bar{x} - \bar{y} = a(\cos\theta - \sin\theta)^3$

or $\quad (\bar{x}+\bar{y})^{2/3} = a^{2/3}(\cos\theta + \sin\theta)^2$ and $(\bar{x}-\bar{y})^{2/3} = a^{2/3}(\cos\theta - \sin\theta)^2$

$\therefore \quad (\bar{x}+\bar{y})^{2/3} + (\bar{x}-\bar{y})^{2/3} = 2a^{2/3}$

Therefore the evolute of the given curve, as locus of $(\bar{x}, \bar{y})$, is

$(x+y)^{2/3} + (x-y)^{2/3} = 2a^{2/3}$. **Ans.**

EXERCISE 2B

1. Find the radius of curvature at the point (x, y) of the curves

   (*i*) $y = c \cosh (x/c)$ (*ii*) $y = a \log \sec (x/a)$.

2. Obtain the radius of curvature at the point (x, y) of the curves

   (*i*) $x^{2/3} + y^{2/3} = a^{2/3}$. (*ii*) $\sqrt{x} + \sqrt{y} = \sqrt{a}$

3. Find the radius of curvature at any point '*t*' on the curve

$$x = a (\cos t + \log \tan t/2), \quad y = a \sin t$$

4. The tangents at two points P and Q on the cycloid

$$x = a (\theta - \sin \theta), \quad y = a (1 - \cos \theta)$$

   are at right angles. If ρ_1 and ρ_2 are the radii of curvature at these points then show that

$$\rho_1^2 + \rho_2^2 = 16a^2.$$

5. Apply Newton's method to find the radius of curvature at the origin for the curve

$$x^3 + y^3 - 2x^2 + 6y = 0.$$

6. Find the radius of curvature at the origin for the curve $x^3 + y^3 = 3axy.$

7. Show that the radius of curvature at the end of the major axis of the ellipse

$$\frac{x^2}{a^2} + \frac{y^2}{b^2} = 1$$

   is equal to the semi-latus rectum of the ellipse.

8. Apply Newton's method to obtain the radius of curvature of the curve $y^2 = (2a - x)/x$ at the point $(2a, 0)$.

9. Find the radius of curvature of the curve $y^2 = x^2 (a + x)/(a - x)$ at the origin.

10. For the curve $s = ce^{x/c}$ show that $c\rho = s (s^2 - c^2)^{1/2}$

11. For the curve $y = \dfrac{ax}{a + x}$, prove that

$$\left(\frac{2\rho}{a}\right)^{2/3} = \left(\frac{y}{x}\right)^2 + \left(\frac{x}{y}\right)^2$$

   where ρ is the radius of curvature of the curve at its point (x, y).

12. Show that the ratio of radii of curvature at the points on the curves $xy = c^2$, $x^3 = 3c^2y$ which have same abscissa, varies as the square root of the ratio of the ordinates.

13. Find the radius of curvature at the origin for the curve

$$x^3y - xy^3 + 2x^2y + xy - y^2 + 2x = 0$$

14. For the ellipse $\dfrac{x^2}{a^2} + \dfrac{y^2}{b^2} = 1$, prove that $\rho = \dfrac{a^2b^2}{p^3}$, p being the perpendicular distance from the centre on the tangent at (x, y).

15. If ρ_1 and ρ_2 are the radii of curvature at the ends of a focal chord of the parabola $y^2 = 4ax$ then show that $\rho_1^{-2/3} + \rho_2^{-2/3} = (2a)^{-2/3}$.

16. Show that the radius of curvature at any point '*t*' of the curve

$$x = a(\cos t + t \sin t), \quad y = a(\sin t - t \cos t) \quad \text{is given by } \rho = at.$$

17. Find the centre of curvature at the point (x, y) of the hyperbola $\frac{x^2}{a^2} - \frac{y^2}{b^2} = 1$ and also obtain its evolute.

19. Find the radius of curvature and the centre of curvature for the curve $y = \tan x$ at the point where $x = \frac{\pi}{4}$.

19. Show that the centre of curvature at any point on the ellipse

$$\frac{x^2}{a^2} + \frac{y^2}{b^2} = 1 \quad \text{is} \quad \left(\frac{a^2 - b^2}{a^4} x^3, \frac{a^2 - b^2}{-b^4} y^3 \right).$$

CURVE TRACING

For computing the areas, lengths, volume of solids of revolution and surface of solids of revolution it is very important and useful to know the shape of the curve represented by the given equation. Curve tracing tit bits help us in obtaining the rough shape of the curve without plotting a large number of points on it. In addition to it, the properties of some of the standard curves which frequently occur in engineering applications, will be discussed here.

Tracing of Cartesian Curves

We give below some of the main rules which definitely help in determining the shape of the curves from their equations.

1. Symmetry

(*a*) If powers of y in the equation of the curve, **are all even**, then the curve is **symmetrical about *x*-axis**, *i.e.*, whatever is the shape of the curve on one side of x-axis, the mirror image of the shape should be on the other side of it, because for one value of x we get two values of y which differ in sign only. For example, the curve $y^2 = 4ax$ is **symmetrical about *x*-axis.**

(*b*) If all the powers of x in the equation **are even**, the curve is **symmetrical about *y*-axis**, *i.e.*, whatever is the shape of the curve on one side of y-axis the mirror image of the shape should be on the other side of it, because for one value of y we get two values of x which differ in sign only. For example, the curve $x^2 = 4ay$ is **symmetrical about *y*-axis.**

(*c*) If on interchanging x and y, the equation of the curve remains unchanged, the curve is **symmetrical about the line $y = x$.** For example, the curves $xy = c^2$ and $x^3 + y^3 = 3\,axy$ both are symmetrical about the line $y = x$.

(*d*) If x is changed to $-x$ and y to $-y$ and the equation of the curve remains unchanged, then the curve is **symmetrical in the opposite quadrants**, *i.e.*, whatever is the shape of the curve in the first quadrant mirror image is in the third quadrant and similarly, for the second and fourth quadrants.

For example, the curves $xy = c^2$ and $\frac{x^2}{a^2} + \frac{y^2}{b^2} = 1$ are symmetrical in the opposite quadrants.

2. The Origin

(*a*) If there is no constant term in the equation of the curve, the curve passes through the origin.

(*b*) If the curve is passing through the origin, we can find the equation of the tangent to the curve at the origin, by equating to zero the lowest degree term in the equation. For example, the curve $y^2 = 4ax$ has y-axis as tangent at the origin and the curve $y = x^3$ has x-axis as tangent to the curve at the origin.

(*c*) If there are two or more tangents to the curve at the origin, it is called a **multiple point**. Further, the origin is called a **node**, a **cusp** or an **isolated point according** as the tangents there are real and distinct, real and coincident or imaginary. For example, in the curve $y^2 (a^2 + x^2) + x^2(a^2 - x^2) = 0$, the origin is an isolated point and for the curve $y^2 (a - x) = x^3$ the origin is a cusp.

3. Intersection with the Coordinate Axes

(a) Find the points where the curve intersects the x-axis and the y-axis separately by putting $y = 0$ and $x = 0$ respectively.

(b) Find the tangent to the curve at its point of intersection with the coordinate axes by first shifting the origin to this point and then equating to zero the lowest degree term.

4. Region (s) of Absence of Curve

If possible, express y in terms of x or x in terms of y explicitly, and examine how y varies as x varies and how x varies as y varies. On solving for y, say, in terms of x, suppose we find that between $x = a$ and $x = b$, y is imaginary, then the curve does not exist in the region between the lines $x = a$ and $x = b$. For example, in the curve $y^2 (1 - x) = x^3$, y is imaginary when $x < 0$ and also when $x > 1$ hence no curve lies on the right of the line $x = 1$ and on the left of y-axis.

5. Maxima, Minima and Points of Inflexion.

Find $\frac{dy}{dx}$ and equate it to zero to get the critical points at which y may be maximum or minimum.

Also find $\frac{d^2y}{dx^2}$ and equate it to zero to get the possible points of inflexion.

6. Asymptotes

Find the asymptotes parallel to x-axis (or y-axis) by equating to zero the cofficient of the highest degree term in x (or y). For example, in the curve $y^2 (1 - x) = x^3$ the asymptote parallel to y-axis is the line $x = 1$.

To obtain oblique asymptotes put $y = mx + c$ in the equation of the curve and then equate to zero the coefficients of highest and the next highest degree terms in x, this will give equations to find m and c of the equations of asymptotes.

7. Conversion to Polar Form

If needed, transform the given equation into polar coordinates r and θ by putting $x = r\cos\theta$, $y = r\sin\theta$. The polar form of some curves is more convinient to handle than their cartesian form.

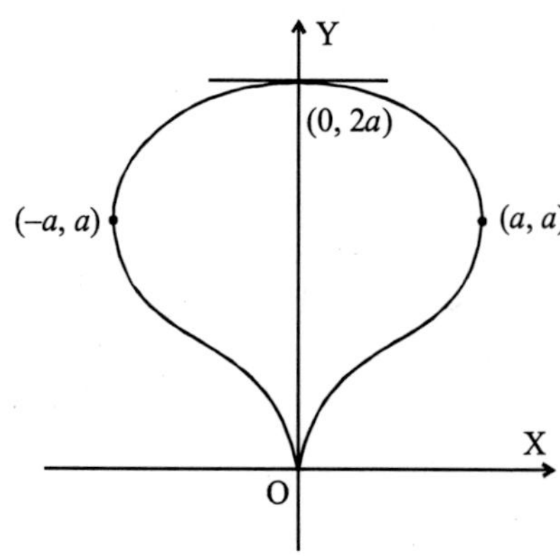

EXAMPLE 2.23. Trace the curve $a^2x^2 = y^3(2a - y)$.

SOLUTION: The equation of the curve is $a^2x^2 = y^3 (2a - y)$.

(i) In the given equation of the curve all the powers of x are even, hence the curve is symmetrical about y-axis.

(ii) Since there is no constant term in the equation, the curve passes through the origin. Now, equating to zero the lowest degree term, gives the tangents to the curve at the origin as $x^2 = 0$ which are real and coincident *i.e.*, y-axis. Here origin is a **cusp**.

(*iii*) The curve meets the y-axis at $(0, 0)$ and at $(0, 2a)$. Shifting the origin to the point $(0, 2a)$ the equation of the curve becomes $a^2x^2 = -(y + 2a)^3 y$. Equating to zero the lowest degree term gives $y = 0$. Therefore, tangent at $(0, 2a)$ is parallel to x-axis.

(*iv*) Since $x^2 < 0$ for $y > 2a$ and for $y < 0$, there is no portion of the curve below the x-axis and above the line $y = 2a$.

(*v*) Putting $y = a$ we get $x = \pm a$. Giving suitable values to y like $y = \frac{a}{2}$ and $\frac{3a}{2}$ we can find corresponding values of x and such points give the shape of the curve as shown in the above Figure.

EXAMPLE 2.24. Trace the curve $\boxed{4ay^2 = x(x-2a)^2}$.

SOLUTION: The equation of the curve is $4ay^2 = x(x - 2a)^2$.

(*i*) Since all powers of y in the given equation are even, the curve is symmetrical about the x-axis.

(*ii*) There is no constant term in the equation hence the curve passes through the origin. Equating to zero the lowest degree term we get $x = 0$ as the tangent to the curve at the origin, *i.e.*, y-axis.

(*iii*) The curve meets the y-axis at the origin only and the x-axis at A $(2a, 0)$. Shifting the origin to the point $(2a, 0)$ the equation of the curve becomes $4ay^2 = (x + 2a)x^2$ and now equating the lowest degree term to 0, we get $y = \pm x/\sqrt{2}$ as the tangents to the curve at the point A.

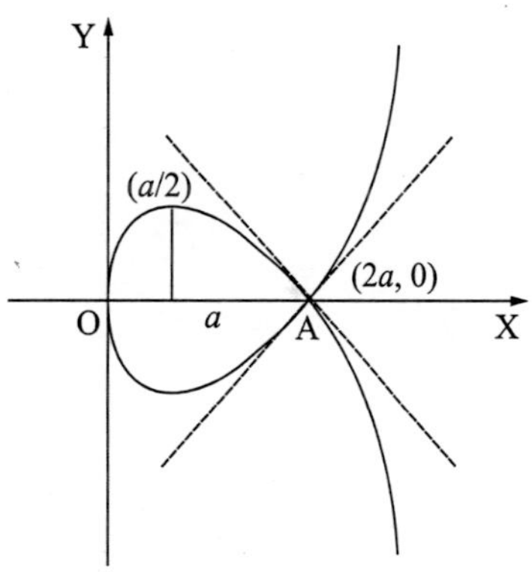

(*iv*) From the given equation of the curve, we note that y is imaginary when x is negative hence no portion of the curve lies on the left of y-axis.

(*v*) From the equation of the curve, we have

$$y = \pm \frac{(x-2a)}{2}\sqrt{\frac{x}{a}}$$

therefore,
$$\frac{dy}{dx} = \pm\left[\frac{x-2a}{2}\frac{1}{2\sqrt{ax}} + \frac{1}{2}\sqrt{\frac{x}{a}}\right]$$

$$= \pm\frac{1}{4\sqrt{ax}}(3x-2a).$$

Now, $\frac{dy}{dx} = 0$ at $x = \frac{2a}{3}$

for which y is maximum. The maximum value of y is given by

$$y = \frac{2a}{3}\sqrt{\frac{2}{3}} = \frac{a\sqrt{24}}{9}.$$

(*vi*) Beyond $x = 2a$, y increases monotonically as x increases.

(*vii*) Give certain suitable values to x like $x = a$, $3a$, $4a$ and find the corresponding values of y and plot such points. The rough shape of the curve is given in the above Figure.

EXAMPLE 2.25. Trace the curve $y^2(x+a)=x^2(3a-x)$. **(UPTU 2008-09)**

SOLUTION: The equation of the curve is $y^2(x+a)=x^2(3a-x)$.

(*i*) The curve is symmetrical about the axis of x.

(*ii*) The curve passes through the origin. Equating to 0 the lowest degree term, we get $ay^2=3ax^2$ or $y = \pm x\sqrt{3}$ as two tangents to the curve at the origin.

(*iii*) The curve meets the y-axis only at the origin while it meets x-axis at $(3a, 0)$ also. Shifting the origin to this point the equation of the curve becomes

$$y^2(x+4a) = -(x+3a)^2 x,$$

and now equating to 0 the lowest degree term, we get $x=0$, the y-axis, as the tangent there. Thus, at $(3a, 0)$ tangent is parallel to y-axis.

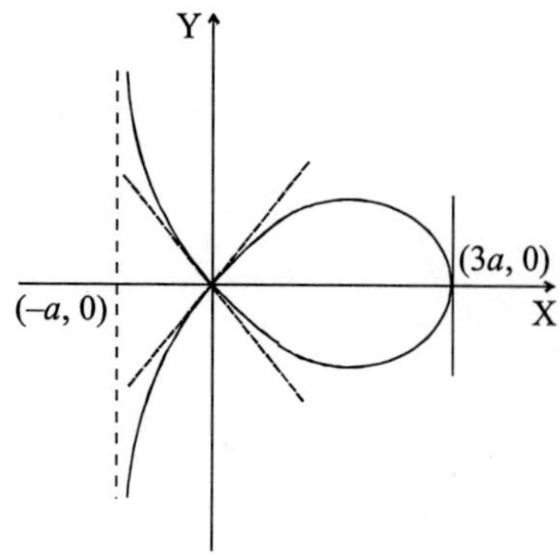

(*iv*) y is imaginary for $x<-a$ and also for $x>3a$, therefore there is no portion of the curve on the left of the line $x=-a$ and on the right of the line $x=3a$.

(*v*) There is no asymptote parallel to x-axis, and for asymptote parallel to y-axis, we equate to zero the coefficient of the highest degree term in y so as to get $x+a=0$ as the asymptote parallel to y-axis.

(*vi*) Differentiating the given equation, we get

$$2y\frac{dy}{dx}(x+a) + y^2\cdot 1 = 6ax-3x^2$$

or $$2y(x+a)^2\frac{dy}{dx}+y^2(x+a) = (6ax-3x^2)(x+a)$$

or $$2y(x+a)^2\frac{dy}{dx} = 3x(2a-x)(a+x)-x^2(3a-x) = 2x(3a^2-x^2).$$

For maxima $\frac{dy}{dx}=0$ and we get $x=a\sqrt{3}$ and $x=0$.

Thus, y is maximum at $x=a\sqrt{3}$.

(*vii*) Give suitable values to x like $x=-a/2,\ a,\ 2a$ and find the corresponding values of y, plot such points and get the rough shape of the curve as shown in the above Figure.

EXMAPLE 2.26. Trace the curve $ay^2 = x(a^2-x^2)$.

SOLUTION: The equation of the curve is $ay^2=x(a^2-x^2)$.

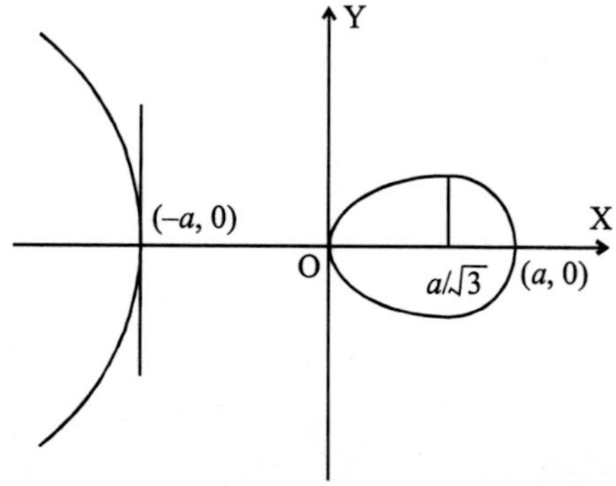

(*i*) The curve is symmetrical about x-axis.

(*ii*) It passes through the origin and equating to 0 the lowest degree term, we get $x=0$, the y-axis as the tangent to the curve at the origin.

(*iii*) The curve does not meet the y-axis except at the origin. It meets the x-axis at $(-a, 0)$ and at $(a, 0)$.

Shifting the origin to the point $(a, 0)$, the equation of the curve becomes $ay^2 = (x + a)(-2ax - x^2)$. Now, equating to 0 the lowest degree term, we get $x = 0$, therefore the tangent at $(a, 0)$ is parallel to y-axis.

Similarly, shifting the origin to the point $(-a, 0)$, the equation of the curve becomes $ay^2 = (x - a)(2ax - x^2)$ and equating to 0 the lowest degree term here, we get $x = 0$ hence tangent to the curve at $(-a, 0)$ is also parallel to y-axis.

(*iv*) Since y is imaginary for $x > a$ and also for $-a < x < 0$, hence there is no portion of the curve on the right of the line $x = a$ and between the lines $x = -a$ and $x = 0$.

(*v*) Differentiating the given equation, gives

$$2ay\frac{dy}{dx} = a^2 - 3x^2$$

And, $\frac{dy}{dx} = 0$ gives $x = \frac{a}{\sqrt{3}}$ which is the maxima.

(*vi*) Now, give suitable values to x like $x = \frac{a}{2}, -2a, -3a$ and find the corresponding values of y and plot these points and the shape of the curves is as shown in the above Figure.

EXAMPLE 2.27. Trace the curve $y^2(a^2 - x^2) = a^3x$.

SOLUTION: The equation of the curve is $y^2(a^2 - x^2) = a^3x$.

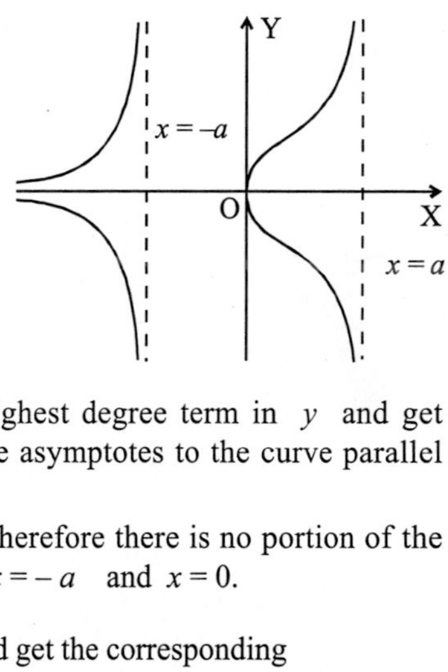

(*i*) The curve is symmetrical about the axis of x.

(*ii*) It passes through the origin and the tangent to the curve at the origin is $x = 0$, the y-axis.

(*iii*) The curve does not meet the coordinate axes except at the origin.

(*iv*) For the asymptotes parallel to x-axis we equate to zero the coefficient of the highest degree term in x and get $y^2 = 0$, hence x-axis itself is the asymptote. For asymptotes parallel to y-axis, we equate to zero the cofficient of the highest degree term in y and get $a^2 - x^2 = 0$ therefore both the lines $x = a$ and $x = -a$ are asymptotes to the curve parallel to y-axis.

(*v*) Value of y is imaginary for $x > a$ and for $-a < x < 0$, therefore there is no portion of the curve on the right of the line $x = a$ and between the lines $x = -a$ and $x = 0$.

(*vi*) Give the suitable values to x like $x = -3a, -2a, -\frac{3a}{2}, \frac{a}{2}$ and get the corresponding values of y and plot the points to get the rough shape of the curve as shown in the above Figure.

EXAMPLE 2.28. Trace the curve $x^2(x^2 + y^2) = a^2(x^2 - y^2)$.

SOLUTION:

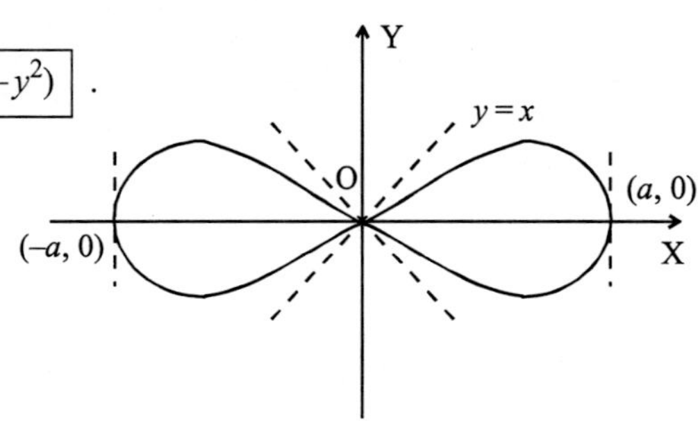

(*i*) The curve is symmetrical about both the axes.

(*ii*) It passes through the origin and equating to zero the lowest degree term we get $x^2 - y^2 = 0$, or $y = \pm x$ as tangents to the curve at the origin.

(*iii*) The curve does not meet the y-axis except at the origin but meets x-axis at the $(a, 0)$ and at $(-a, 0)$. Shifting the origin to these points we find that the tangents at $(\pm a, 0)$ are parallel to y-axis.

(*iv*) We can rewrite the equation of the curve as $y^2(a^2 + x^2) = x^2(a^2 - x^2)$. Now y^2 is negative for $x > a$ and for $x < -a$, therefore, there is no portion of the curve on the right of the line $x = a$ and on the left of the line $x = -a$.

(*v*) Give suitable values to x like $x = \frac{a}{3}, \frac{a}{2}, \frac{2a}{3}$ and get the corresponding values of y and plot the points to get the shape of the curve as given in the above Figure having two loops.

EXAMPLE 2.29. Trace the curve $xy^2 + (x + a)^2 (x + 2a) = 0$.

SOLUTION:

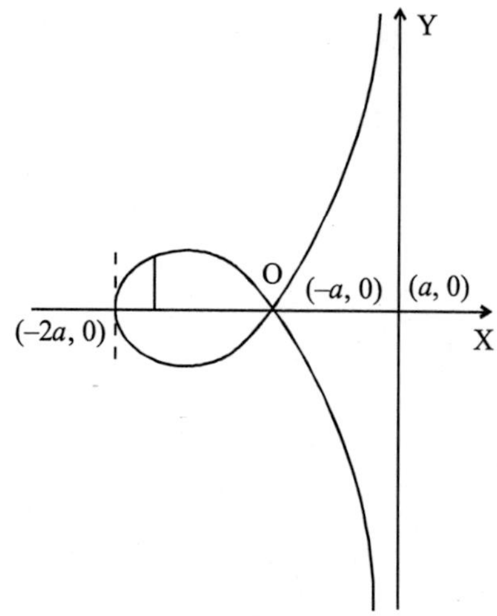

(*i*) The curve is symmetrical about x-axis.

(*ii*) The curve does not pass through the origin and meets the x-axis at $(-2a, 0)$ and at $(-a, 0)$.

Shifting the origin to $(-a, 0)$ the equation of the curve becomes $(x - a)y^2 + x^2(x + a) = 0$ and equating to zero the lowest degree term, we get $y = \pm x$ as tangents at the new origin. Similarly, shifting the origin to the point $(-2a, 0)$, the given equation of curve becomes $(x - 2a)y^2 + (x - a)^2 x = 0$ and then equating to zero the lowest degree term, we get $x = 0$ the new y-axis as tangent there.

(*iii*) For asymptotes parallel to y-axis we equate to zero the coefficient of the highest degree term in y and get y-axis as the asymptote.

(*iv*) From the given equation of the curve we observe that $y = \pm (x + a)\sqrt{-\frac{(x+2a)}{x}}$ and hence y is imaginary when $x < -2a$ and also when $x > 0$. Thus, there is no portion of the curve on the left of $x = -2a$ and on the right of y-axis.

(*v*) Differentiating the given equation and equating $\frac{dy}{dx}$ to zero, we get maxima at $x = \frac{-a\left(1+\sqrt{5}\right)}{2}$.

(*vi*) Give suitable values to x, like $x = -3a/2, -a/2$ and find the corresponding points on the curve and get the shape of the curve as shown in the above Figure.

EXAMPLE 2.30. Trace the curve $y(x^2 + a^2) = a^3$.

SOLUTION:

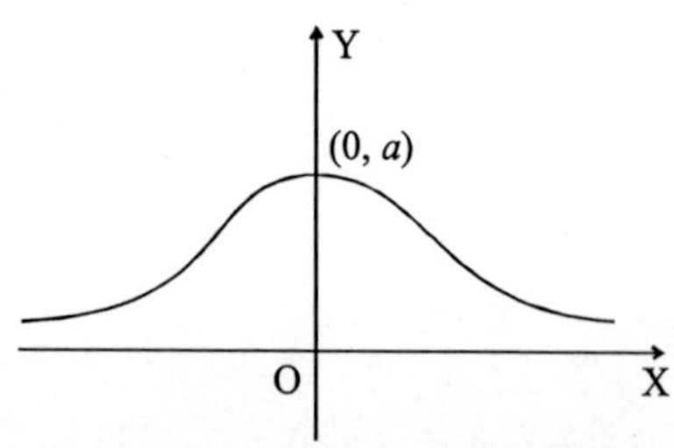

(*i*) The curve is symmetrical about y-axis.

(*ii*) The curve does not meet x-axis but meets y-axis at $(0, a)$. Shifting the origin to $(0, a)$ the equation becomes $(y + a)(x^2 + a^2) = a^3$ or $y(x^2 + a^2) + ax^2 = 0$ and now equating to zero the lowest degree term, we get $y = 0$, the x-axis as the tangent there.

(*iii*) Solving the given equation for x, we get, $x = \pm a\sqrt{\frac{a-y}{y}}$ which shows that x is imaginary when $y < 0$ and also when $y > a$. Thus, there is no curve below x-axis and above the line $y = a$.

(*iv*) Equating to 0 the coefficient of the highest degree term in x we get $y = 0$, the x-axis, as the asymptote of the curve.

(*v*) Give suitable values to x, like $x = \frac{a}{2}$, a, $2a$ and obtain the corresponding points on the curve to get the rough shape of the curve as shown in the above Figure.

EXAMPLE 2.31. Trace the **hypocycloid** $\boxed{x^{2/3} + y^{2/3} = a^{2/3}}$. **(UPTU 2005-06)**

SOLUTION:

(*i*) If we write the equation of the curve as $(x^{1/3})^2 + (y^{1/3})^2 = a^{2/3}$ it is clear that the curve is symmetrical about x-axis and y-axis both.

(*ii*) The curve does not pass through the origin but meets the x-axis at $(\pm a, 0)$ and the y-axis at $(0, \pm a)$.

(*iii*) Differentiating the given equation, we get

$$\frac{2}{3}x^{-1/3} + \frac{2}{3}y^{-1/3}\frac{dy}{dx} = 0 \quad \text{or} \quad \frac{dy}{dx} = -\left(\frac{y}{x}\right)^{1/3}.$$

At $(\pm a, 0)$ the slope of the tangent is 0 hence x-axis is the tangent at these points. At $(0, \pm a)$ the slope of the tangent is infinity, hence y-axis is tangent at these points.

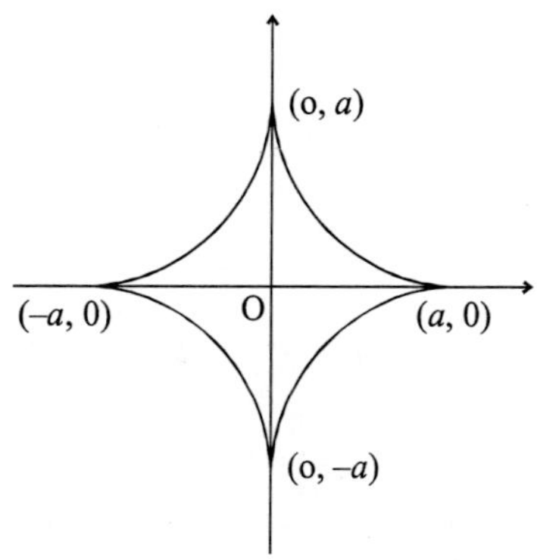

(*iv*) We can write the given equation of the curve in parametric form as $x = a\cos^3 t$, $y = a\sin^3 t$ which shows that $-a \le x \le a$ and $-a \le y \le a$. Therefore, no point of the curve lies outside the square within the lines $x = \pm a$, $y \pm a$.

(*v*) Give suitable values to x like $x = a/3$, $a/2$, $2a/3$ and find points on the curve and get the shape of the curve as shown in the adjoining Figure.

CURVES IN PARAMETRIC FORM

Following few points will help in tracing the curves whose equations are given in parametric form or can be expressed in parametric form: $x = f_1(t)$, $y = f_2(t)$.

1. (*a*) If $x = f_1(t)$ is an even function of t and $y = f_2(t)$ is an odd function of t, the curve is **symmetrical about x-axis.**

   (*b*) If $x = f_1(t)$ is an odd function of t and $y = f_2(t)$ is an even function of t, the curve is **symmetrical about y-axis**, because if (x, y) is a point on the curve then $(-x, y)$ also lies on the curve.

2. If, for some value of t both x and y become zero then the curve **passes through the origin**.

3. (*a*) Find the value of t for which $f_2(t) = 0$ and for this value of t find $x = f_1(t)$ and get the point where the **curve meets the x-axis.**

   (*b*) Find the value of t for which $f_1(t) = 0$ and for this value of t find $y = f_2(t)$ and get the point where the **curve meets the y-axis.**

4. If possible find the greatest and the least values of x and y which give the region or regions in which the curve lies.

5. Find $\frac{dy}{dx}\left(=\frac{dy}{dt}\Big/\frac{dx}{dt}\right)$ and find the points on the curve where the tangent to the curve is parallel to either coordinate axis.

6. Find the range of the parameter t for which x or y cannot be obtained and this will indicate the region in which the curve does not lie.

7. Find the possible asymptotes of the curve, parallel to the coordinate axes.

EXAMPLE 2.32. Trace the **cycloid** $\boxed{x = a(\theta + \sin\theta),\ y = a(1 - \cos\theta)}$. **(UPTU 2001)**

SOLUTION:

(*i*) Here θ is the parameter and x is an odd function of θ, and y an even function of θ hence the curve is symmetrical about y-axis.

(*ii*) For $\theta = 0$ we have $x = 0$ and $y = 0$ hence the curve passes through the origin.

(*iii*) The curve meets y-axis at the origin only while it meets the x-axis where $1 - \cos\theta = 0$ *i.e.*, at $\theta = 0, \pm 2\pi, \pm 4\pi, \pm 6\pi, \ldots$ Thus the curve meets the x-axis at $(0, 0)$, $(\pm 2a\pi, 0)$ $(\pm 4a\pi, 0)$ $(\pm 6a\pi, 0)$, etc.

(*iv*) Observe here that $0 \le y \le 2a$ therefore the curve lies only within the strip $y = 0$ and $y = 2a$.

(*v*) From the equation of the curve, we have $\frac{dx}{d\theta} = a(1 + \cos\theta)$, $\frac{dy}{d\theta} = a\sin\theta$

hence $\frac{dy}{dx} = \tan\theta/2$. Now $\frac{dy}{dx} = 0$ at $\theta = 0, \pm 2\pi, \pm 4\pi, \ldots$

hence x-axis is tangent to the curve at its points $(0, 0)$, $(\pm 2a\pi, 0)$ $(\pm 4a\pi, 0)$, ...

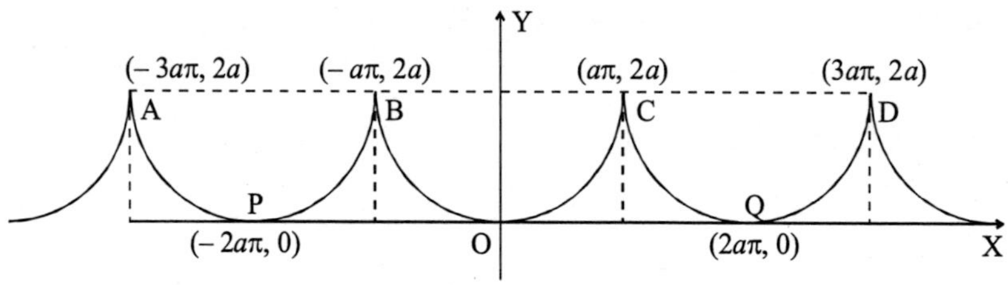

At $\theta = \pm\pi, \pm 3\pi, \pm 5\pi, \ldots$ $\frac{dy}{dx}$ is infinite, hence at the points $(\pm a\pi, 2a)$, $(\pm 3a\pi, 2a)$, $(\pm 5a\pi, 2a)$, ... tangents are parallel to y-axis.

Curve APB is one cycloid, BOC is another cycloid, CQD is yet another cycloid, and so on as shown in the above Figure.

Similarly, the shape of the **cycloid** $\boxed{x = a(\theta - \sin\theta),\ y = a(1 - \cos\theta)}$ is as shown in Figure given below and these are called **inverted cycloids**.

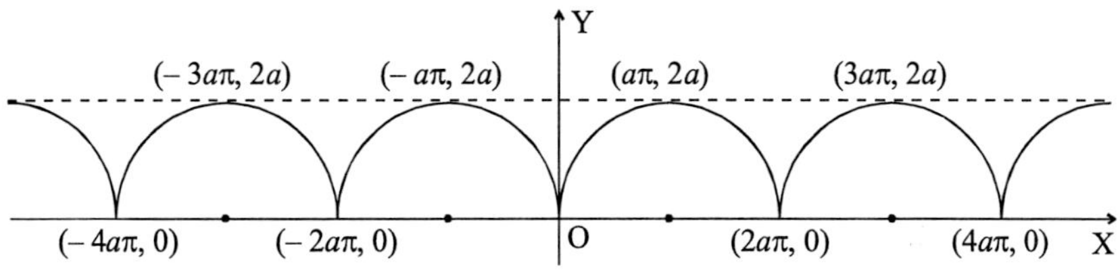

Also, the shape of the **cycloid** $x = a(\theta - \sin\theta),\ \ y = a(1 + \cos\theta)$ is shown in the Figure given below.

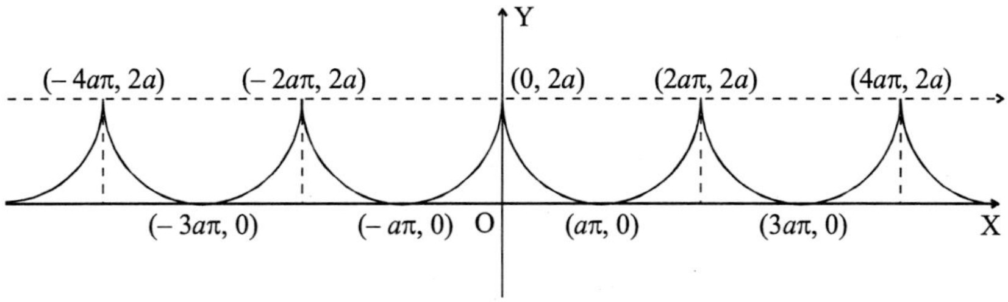

And the shape of the **cycloid** $x = a(\theta + \sin\theta),\ \ y = a(1 + \cos\theta)$ is as shown in the Figure given below.

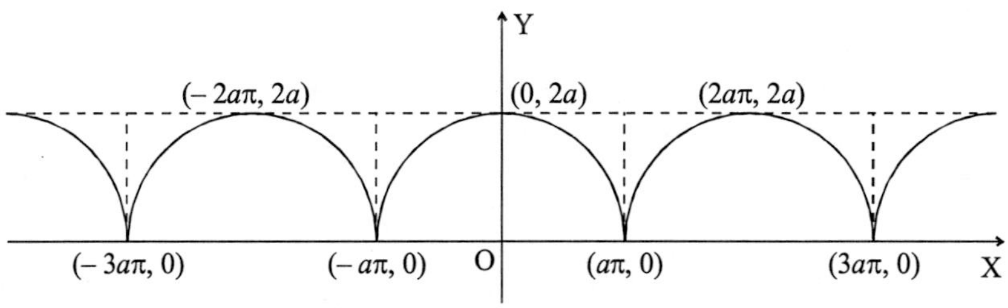

EXAMPLE 2.33. Trace the **hypocycloid** $\left(\dfrac{x}{a}\right)^{2/3} + \left(\dfrac{y}{b}\right)^{2/3} = 1$.

SOLUTION: The given equation of the curve can be written in parametric form as

$\boldsymbol{x = a\cos^3\theta,\ y = b\sin^3\theta}$ where θ is the parameter.

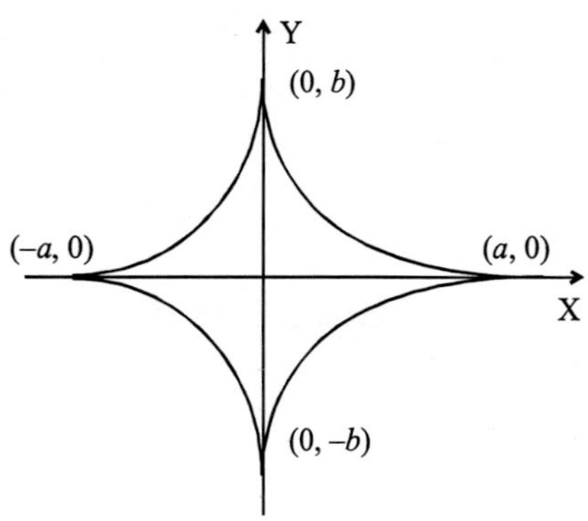

(*i*) Here x is an even function of θ and y is an odd function of θ hence the curve is symmetrical about x-axis.

Also, if θ is changed to $(\pi - \theta)$, y remains the same but x becomes $-x$, therefore the curve is symmetical about y-axis.

The symmetry of the curve about both the axes can be observed also by rewriting the given equation as

$$\left[(x/a)^{1/3}\right]^2 + \left[(y/b)^{1/3}\right]^2 = 1.$$

(*ii*) Since $-a \le x \le a$ and $-b \le y \le b$, the entire curve lies within the rectangle $x = \pm a$, $y = \pm b$.

(*iii*) The curve does not pass through the origin and meets the x-axis at $(\pm a, 0)$ and y-axis at $(0, \pm b)$.

(*iv*) $$\frac{dy}{dx} = \frac{dy}{d\theta}\bigg/\frac{dx}{d\theta} = \frac{3b \sin^2 \theta \cos \theta}{-3a \cos^2 \theta \sin \theta} = \frac{-b}{a} \tan \theta.$$

which is 0 at $(\pm a, 0)$ and is infinite at $(0, \pm b)$. Hence x-axis is the tangent to the curve at $(\pm a, 0)$ and y-axis is the tangent to the curve at $(0, \pm b)$.

(*v*) Next, give suitable values to the parameter θ as $\theta = \pi/6, \pi/4, \pi/3$ and get the corresponding points on the curve and get the rough shape of the curve as given above.

EXAMPLE 2.34. Trace the **tractrix** $x = a(\cos t + \log|\tan(t/2|), \; y = a \sin t$.

SOLUTION: We can rewrite the given equation as $x = a\left(\cos t + \frac{1}{2}\log \tan^2 \frac{t}{2}\right)$, $y = a \sin t$.

(*i*) The curve is symmetrical about x-axis as x is an even function of t and y is an odd function of the parameter t.

(*ii*) When we put $t = \pm \pi/2$, we get $x = 0$ and $y = \pm a$, hence the curve meets the y-axis at the points $(0, \pm a)$.

(*iii*) Since $-a \leq y \leq a$, therefore the curve lies wholly within the strip $y = \pm a$.

(*iv*) As t approaches to 0, x tends to infinity and y tends to 0, therefore the x-axis is an asymptote to the curve.

(*v*) Next, $$\frac{dx}{dt} = a\left(-\sin t + \frac{\sec^2 t/2}{2 \tan t/2}\right) = a\left(-\sin t + \frac{1}{\sin t}\right) = \frac{a \cos^2 t}{\sin t} \quad \text{and} \quad \frac{dy}{dt} = a \cos t$$

hence $$\frac{dy}{dx} = \frac{dy}{dt}\bigg/\frac{dx}{dt} = \frac{a \cos t}{a \cos^2 t} \sin t = \tan t.$$

Also, at $t = \pm \pi / 2$. that is, at the points $(0 \pm a)$, $\frac{dy}{dx}$ is infinite hence y-axis is the tangent there.

And, $$\frac{dy}{dx} = 0 \text{ at } t = 0,$$

that is, at $y = 0$, $x = \infty$ which further establishes that x-axis is an asymptote to the curve.

(*vi*) Give suitable values to t and get the corresponding points on the curve and the rough shape of the curve comes out as shown in the Figure.

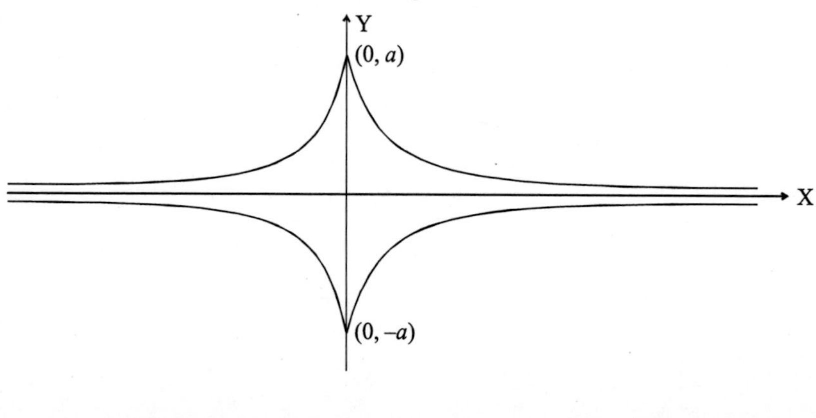

EXAMPLE 2.35. Trace the **catenary** $\boxed{y = c\cosh(x/c)}$.

SOLUTION: The equation of the curve can also be written as $y = \frac{c}{2}\left(e^{x/c} + e^{-x/c}\right)$.

(*i*) Changing x to $-x$ the equation remains unchanged, hence the curve is symmetrical about y-axis.

(*ii*) The curve does not meet the axis of x but meets the y-axis at $(0, c)$.

(*iii*) Differentiating the given equation, we get

$$\frac{dy}{dx} = \frac{c}{2}\left(\frac{e^{x/c}}{c} - \frac{e^{-x/c}}{c}\right) = \sinh x/c.$$

As $\frac{dy}{dx} = 0$ at $x = 0$, the tangent to the curve at $(0, c)$ is parallel to x-axis.

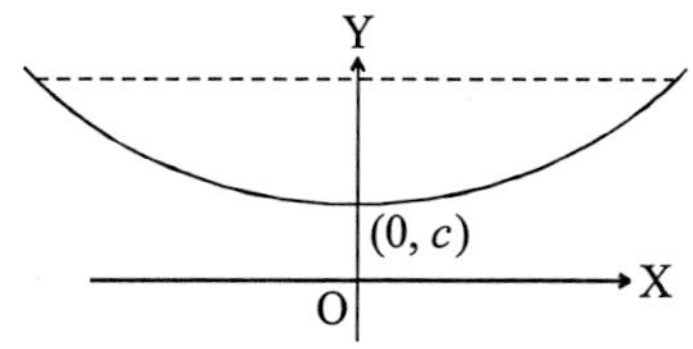

(*iv*) Since $\cosh(x/c) \geq 1$ there is no part of the curve below the line $y = c$.

(*v*) Give suitable values to x and find the corresponding values of y and plot the points on the curve and get the shape of the curve as shown in the Figure.

As physical illustration, consider a chain hanging from two fixed points at the same horizontal level. The shape of the chain is nothing but the above catenary.

EXAMPLE 2.36. Trace the curve $\boxed{y^2(2a - x) = x^3}$. **(UPTU 2004–05, 2006–07, 2010–11)**

SOLUTION: Equation is $y^2(2a - x) = x^3$. The curve is symmetrical about X-axis. It passes through the origin and the tangent to the curve at the origin is X-axis itself. The curve has asymptote parallel to Y-axis, as $x = 2a$. There is no curve on the left of Y-axis and on the right of the line $x = 2a$. The shape of the curve is as given in the adjoining Figure.

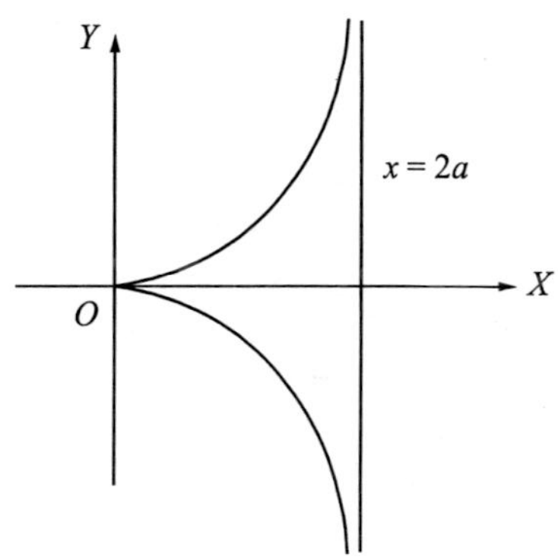

EXAMPLE 2.37. Trace the curve $\boxed{x^{1/2} + y^{1/2} = a^{1/2}}$.

SOLUTION: Curve is symmetrical about the line $y = x$. The curve meets x-axis at $(a, 0)$ and y-axis at $(0, a)$.

Other points on the curve are $(a/4, a/4)$. At $x = a./2$, we have

$$y = a\left(\frac{\sqrt{2} - 1}{\sqrt{2}}\right)^2 \quad \text{and at} \quad x = 3a/4, \quad y = a\left(\frac{2 - \sqrt{3}}{2}\right)^2$$

Shape of the curve is shown in the Figure.

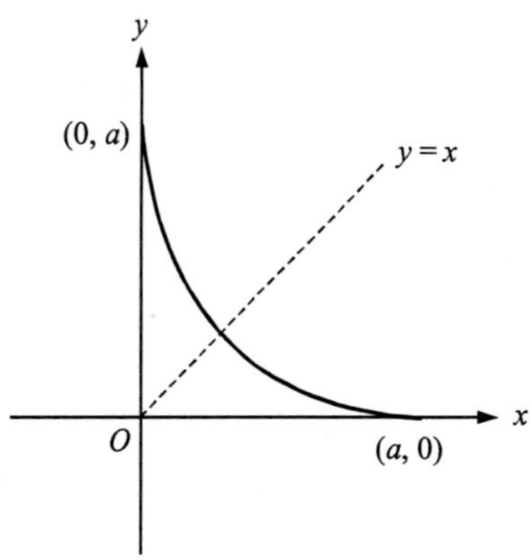

EXAMPLE 2.38. Trace the curve $x^2y^2 = x^2 + a^2$.

SOLUTION: Curve is symmetrical about both the axes. At $x = \pm a$ we have $y = \pm\sqrt{2}$ and at $x = \pm 2a$ we have $y = \pm\sqrt{5}/2$, etc. Further, there is asymptote parallel to X-axis as $y = \pm 1$ and asymptote parallel to Y-axis, is $x = 0$, that is, Y-axis itself. Shape of the curve is given in the adjoining Figure.

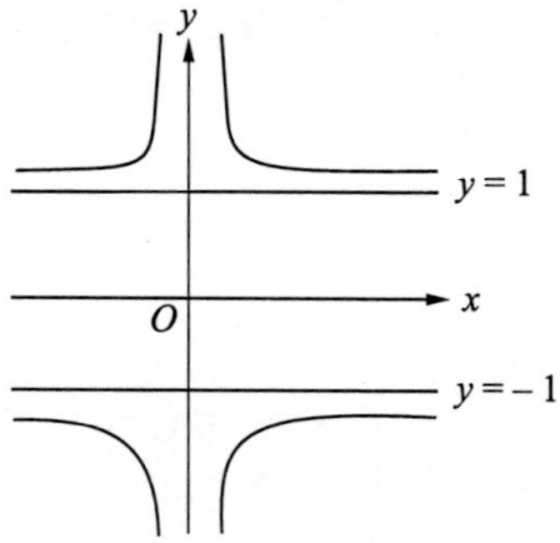

EXERCISE 2C

Trace the following curves:

1. $y = x^3$ (cubical parabola).
2. $y^2 = (x - a)^3$ (Semi cubical parabola).
3. $3ay^2 = x^2 (a - x)$.
4. $ay^2 = x (a^2 + x^2)$.
5. $a^2y^2 = x^2 (2a - x) (x - a)$.
6. $y^2x = a^2 (a - x)$.
7. $y(x^2 + 4a^2) = 8a^3$.
8. $x^2y^2 = a^2 (y^2 - x^2)$.
9. $a^2y^2 = x^3 (2a - x)$.
10. $y = \dfrac{x}{1+x^2}$.
11. $x = a \cos \theta, y = b \sin \theta$.
12. $x = a \sec t, y = b \tan t$.
13. $x = a \cos^3 \theta, y = a \sin^3 \theta$. **(UPTU 2005-06)**
14. $y^2x = a^2 (a - y)$.

❑❑❑

TRACING OF POLAR CURVES

Curves in polar coordinates r and θ are traced basically in the same manner as the cartesian curves. The hints given below would definitely help in sketching the polar curves.

1. Symmetry

(*a*) If θ is changed to $-\theta$ and the equation remains unchanged, the curve is **symmetrical about the initial line (x-axis in cartesian)** $\theta = 0$.

(*b*) If θ is changed to $\pi - \theta$ and the equation remains unchanged, the curve is **symmetrical about the line $\theta = \pi/2$ (*y*-axis in cartesian).**

(*c*) If θ is changed to $\pi + \theta$ and the equation remains unchanged, the curve is symmetrical about the pole **(same as symmetry in opposite quadrants in cartesian).** In other words, we can say that if r is changed to $-r$ and the equation remains unchanged, the curve is **symmetrical about the pole.**

(*d*) If θ is changed to $(\pi/2 - \theta)$ and the equation remains unaltered, the curve is **symmetrical about the line $\theta = \pi/4$ (same as symmetry about the line $y = x$ in cartesian).**

2. Pole

We find whether the pole lies on the curve or not. If $r = 0$ for $\theta = \alpha$, then the curve passes through the pole and the tangent to the curve at the pole is the line $\theta = \alpha$.

3. Direction of the tangent

Differentiate the equation of the curve with respect to θ and using $\tan\phi = r\dfrac{d\theta}{dr}$ we obtain ϕ which is the angle between the tangent to the curve at a point and the radius vector of this point. This angle ϕ gives the direction of the tangent to the curve at the point.

4. Regions

Solve the given equation of the curve for r in terms of θ and if we can find the range of θ, say $\alpha < \theta < \beta$ in which r becomes imaginary then we conclude that no portion of the curve lies in the triangular strip $\theta = \alpha$ and $\theta = \beta$.

Also, if possible, observe the greatest and the least values of r. If the least value of r is a then no part of the curve lies within the circle with radius a and centre at the pole. Similarly, if the greatest value of r is b then no part of the curve lies outside the circle with centre at the pole and radius b.

5. Asymptotes

If r becomes infinite for $\theta = \theta_1$, say, then there may exist asymptote of the curve. The method to obtain asymptotes for polar curves has already been discussed in the early part of this chapter.

6. Intersection with axes

Determine where the curve meets the initial line $\theta = 0$ and the lines $\theta = \pi/2$, $\theta = \pi$, $\theta = 3\pi/2$, etc. and obtain the tangents, at these points. Then give some other suitable values to θ and find the corresponding points on the curve and the tangents there.

The above mentioned points would enable us to determine rough shape of the curve.

EXAMPLE 2.39. Trace the curve $r = a \sin 3\theta$ **(three leaved rose).** **(UPTU 2001-02)**

SOLUTION:

(*i*) On changing θ to $(\pi - \theta)$ the equation of the curve remains unchanged hence the curve is symmetrical about the line $\theta = \pi/2$.

(*ii*) The greatest value of r is 'a' as the maximum value of $\sin 3\theta$ is 1 therefore the curve wholly lies within a circle of radius a and centre at the pole.

(*iii*) The curve meets the pole for $\theta = 0, \pi/3, 2\pi/3, \pi$. The variation of r with the variation of θ is shown in the following Table:

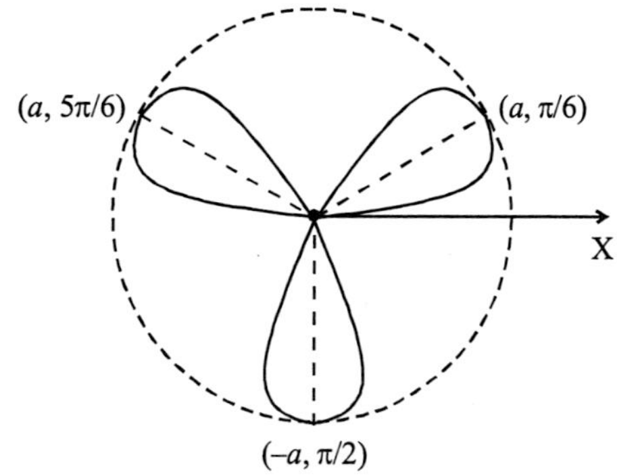

θ	0	$\pi/6$	$\pi/3$	$\pi/2$	$2\pi/3$	$5\pi/6$	π
r	0	a	0	$-a$	0	a	0

Thus, the tangents to the curve at the pole are having the equations. $\theta = 0$, $\theta = \pi/3$, $\theta = 2\pi/3$ and $\theta = \pi$.

(*iv*) We shall get three loops as shown in the adjoining Figure. It is more popularly known as **a rose with three leaves.**

EXAMPLE 2.40. Trace the curve $r = a \sin 2\theta$ **(four leaved rose).** **(UPTU 2001)**

SOLUTION:

(*i*) On changing θ to $\pi + \theta$ the equation of the curve remains unaltered therefore, it is symmetrical about the pole.

(*ii*) On changing θ to $(\pi/2 - \theta)$ the equation remains unchanged hence the curve is symmetrical about the line $\theta = \pi/4$.

(*iii*) The curve meets the pole at $\theta = 0, \pi/2, \pi, 3\pi/2$. Therefore, the tangents to the curve at the pole are the lines $\theta = 0$, $\theta = \pi/2$, $\theta = \pi$ and $\theta = 3\pi/2$.

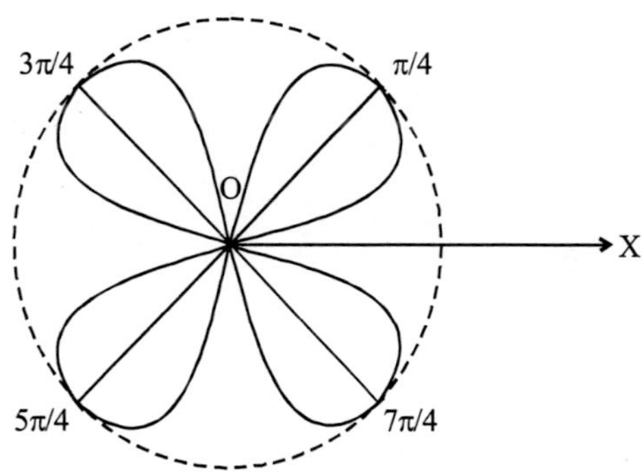

(*iv*) The maximum value of r is 'a' as the maximum value of $\sin 2\theta$ is 1, hence the curve wholly lies within a circle with centre at the pole and radius a. The variation of r with the variation of θ is shown in the following table.

θ	0	$\pi/4$	$\pi/2$	$3\pi/4$	π	$5\pi/4$	$3\pi/2$	$7\pi/4$	2π
r	0	a	0	$-a$	0	a	0	$-a$	0

Thus, we get four loops as shown in the above Figure. This is more popularly known as **four leaved rose.**

EXAMPLE 2.41. Trace the **cardioide** $r = a(1 - \cos\theta)$.

SOLUTION:

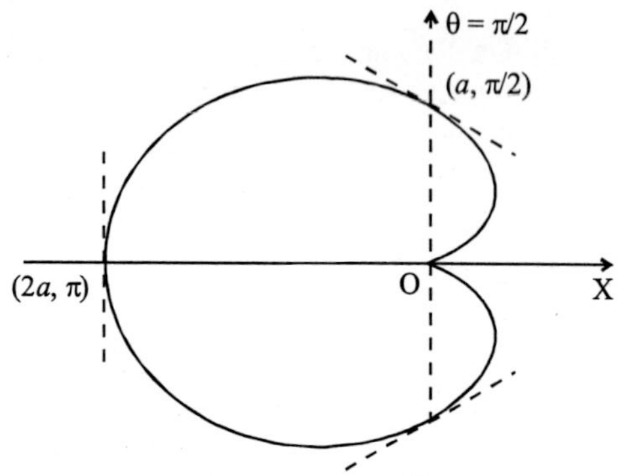

(*i*) When θ is changed to $-\theta$ the equation of the curve remains unaltered, hence there is symmetry about the initial line and no other point of symmetry is satisfied here.

(*ii*) When $\theta = 0$, $r = 0$ hence the curve passes through the pole and the tangent to the curve at the pole is the initial line.

(*iii*) The minimum value of $\cos\theta$ is -1 hence the maximum value of r is $2a$ and as such the whole curve lies within the circle with centre at the pole and radius $2a$.

(*iv*) From $r = a(1 - \cos\theta)$ we have $\dfrac{dr}{d\theta} = a \sin\theta$.

Hence $\tan\phi = r\dfrac{d\theta}{dr} = \dfrac{a(1-\cos\theta)}{a\sin\theta} = \tan\theta/2$, so $\phi = \theta/2$.

Now, at $\theta = \pi/2$, $\phi = \pi/4$, hence the tangent at the point $(a, \pi/2)$ makes an angle $\pi/4$ with line $\theta = \pi/2$. At $\theta = \pi$, $\phi = \pi/2$ hence at the point $(2a, \pi)$ the tangent is perpendicular to the line $\theta = \pi$.

(*v*) Give suitable values to θ, like $\theta = \pi/4, 3\pi/4$ and find the corresponding points on the curve. The rough shape of the curve is as given in the above Figure.

EXAMPLE 2.42. Trace the **Leminiscate of Bernoulli** $r^2 = a^2 \cos 2\theta$.

(UPTU 2000-01, 2008-09, 2011-12)

SOLUTION:

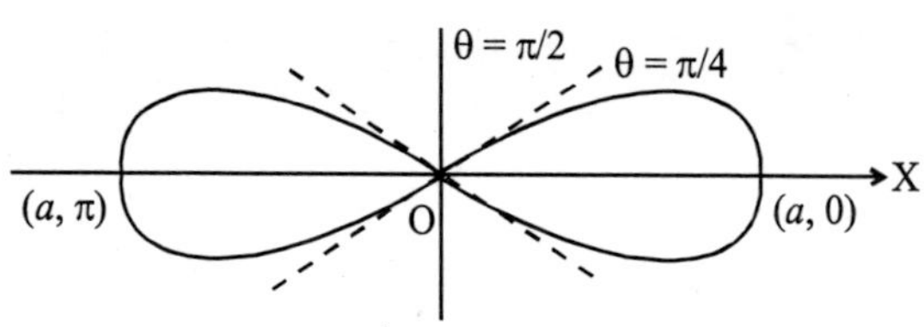

(*i*) On changing θ to $-\theta$, the equation remains unchanged hence the curve is symmetrical about the initial line.

(*ii*) When θ is changed to $\pi - \theta$ the equation remains the same hence there is symmetry about the line $\theta = \pi/2$.

(*iii*) When θ is changed to $\pi + \theta$ the equation remains unchanged hence the curve is also symmetrical about the pole.

(*iv*) $r = 0$ when $\theta = \pm\pi/4$, therefore curve passes through the pole and the tangents to the curve at the pole are the lines $\theta = \pm\pi/4$.

(*v*) The maximum value of r is 'a' when $\theta = 0$ and when $\theta = \pi$, hence the curve lies wholly within the circle with centre at the pole and radius a.

(*vi*) When θ increases from $\pi/4$, r^2 becomes negative hence r is imaginary, and it remains so till θ becomes $3\pi/4$. Thus, there is no portion of the curve in the region between the lines $\theta = \pi/4$ and $\theta = 3\pi/4$.

(*vii*) Differentiating the equation of the curve, we have

$$2r\frac{dr}{d\theta} = -2a^2 \sin 2\theta$$

hence $\tan\phi = r\dfrac{d\theta}{dr} = \dfrac{r^2}{-a^2 \sin 2\theta} = \dfrac{a^2 \cos 2\theta}{-a^2 \sin 2\theta} = -\cot 2\theta = \tan\left(\dfrac{\pi}{2}+2\theta\right)$

Thus, $\phi = \dfrac{\pi}{2}+2\theta.$

Therefore, $\phi = \pi/2$ when $\theta = 0$, and the tangent to the curve at the point $(a, 0)$ is perpendicular to the initial line. Similarly, when $\theta = \pi$, we have $\phi = 5\pi/2$ hence the tangent to the curve at the point (a, π) is perpendicular to the line $\theta = \pi$.

(*viii*) Give suitable values to θ like $\theta = \pm\, \pi/6$ and get the corresponding points on the curve. The final shape of the curve is as shown in the above Figure.

EXAMPLE 2.43. Trace the curve $\boxed{r = a + b\cos\theta}$ **(lamicon)**, a and b being positive. **(UPTU 2005)**

SOLUTION: When θ is changed to $-\theta$, the equation remains unaltered hence the curve is symmetrical about the initial line. Here we shall get three types of curve in the following three cases :

CASE I. $a > b$. Here r is never negative.

At $\theta = 0$, $r = a + b$ and at $\theta = \pi$, $r = a - b$.

Also, $\dfrac{dr}{d\theta} = -\,b\sin\theta$, hence $\tan\phi = r\dfrac{d\theta}{dr} = \dfrac{a + b\cos\theta}{-b\sin\theta}$.

When $\theta = 0$, we have $\phi = \pi/2$.

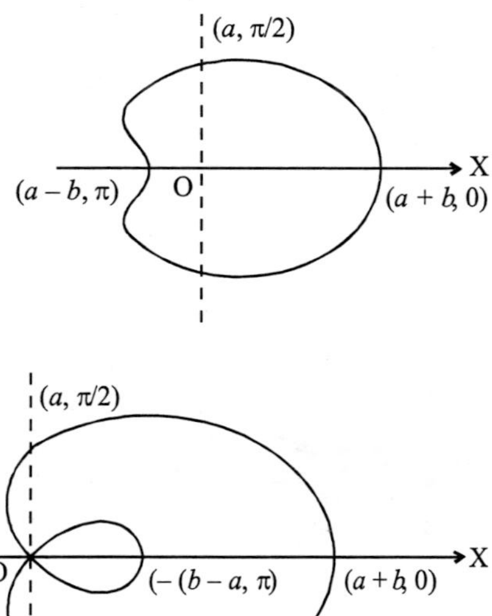

and when $\theta = \pi$, again $\phi = \pi/2$, therefore, tangents to the curve at $(a + b, 0)$ and $(a - b, \pi)$ are perpendicular to the initial line. Then give suitable values to θ like $\theta = \pi/3, \pi/2, 2\pi/3$ and get corresponding points on the curve. The shape of the curve is given in the above Figure.

CASE II. $a < b$.

Here r is negative for some values of θ and positive for other values of θ. Actually, we get one small loop inside a bigger loop as shown in the adjoining Figure.

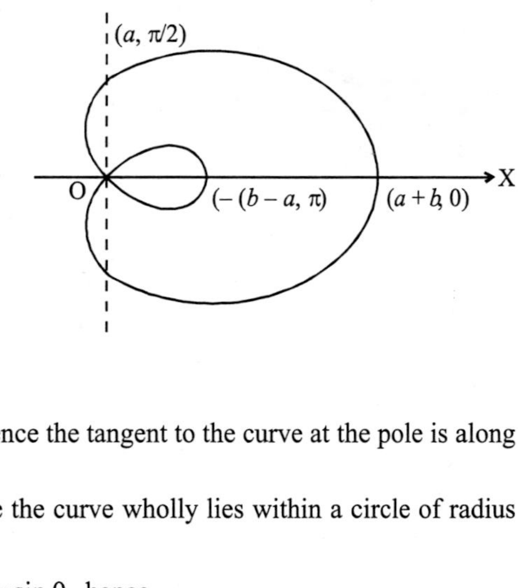

CASE III. $a = b$.

The equation of the curve here becomes $r = a\,(1 + \cos\theta)$ which is a cardioide.

(*i*) The curve passes through the pole when $\theta = \pi$ hence the tangent to the curve at the pole is along the negative direction of the initial line.

(*ii*) Maximum value of r is $2a$ at $\theta = 0$, therefore the curve wholly lies within a circle of radius $2a$ and centre at the pole.

(*iii*) Differentiating the given equation, we get $\dfrac{dr}{d\theta} = -\,a\sin\theta$, hence

$$\tan\phi = r\frac{d\theta}{dr} = \frac{a(1+\cos\theta)}{-a\sin\theta} = -\cot\theta/2 = \tan\left(\frac{\pi}{2}+\frac{\theta}{2}\right)$$

Therefore, $\phi = \pi/2 + \theta/2$.

Thus, at the point $(2a, 0)$. $\phi = \pi/2$ hence the tangent there is perependicular to the initial line and at the point $(a, \pi/2)$, $\phi = 3\pi/4$ and the tangent there would make an angle $3\pi/4$ with the line $\theta = \pi/2$.

(*iv*) Finally, give suitable values to θ like $\theta = \pi/6$, $\pi/3$, $2\pi/3$ and get the corresponding points on the curve. The shape of the curve is given in the adjoining Figure.

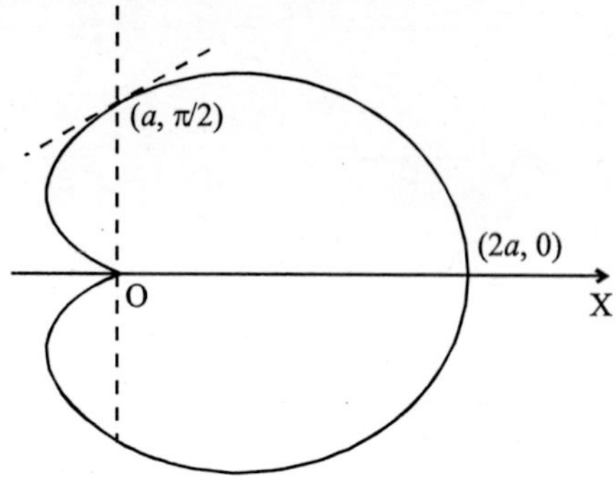

EXAMPLE 2.44. Trace the **equiangular spiral** $\boxed{r = ae^{\theta \cot \alpha}}$.

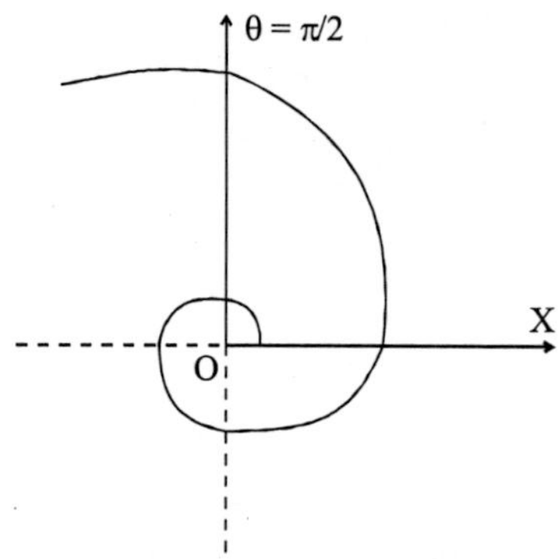

SOLUTION: There is no symmetry in this curve.

At $\theta = 0$, we have $r = a$.

The shape of this curve is mainly guided by an important characteristic property it possesses.

Here $\dfrac{dr}{d\theta} = a\,e^{\theta \cot \alpha} \cot \alpha = r \cot \alpha$.

Hence $\tan \phi = r\dfrac{d\theta}{dr} = \dfrac{r}{r \cot \alpha} = \tan \alpha$.

Thus, $\phi = \alpha$, meaning thereby that the tangent to the curve at any point of it, always makes a constant angle α with the radius vector of that point. Give suitable value to θ, like $\theta = \pi/2, \pi, 3\pi/2, 2\pi, 5\pi/2$ and get corresponding points on the curve. The bend of the curve at every point is guided by the above mentioned property. The shape of the curve is as shown in the adjoining Figure.

EXAMPLE 2.45. Trace the **spiral** $\boxed{r\theta = a}$.

SOLUTION: The equation of the curve is $r = a/\theta$.

(*i*) The curve is symmetrical about the line $\theta = \pi/2$ because when θ is changed to $-\theta$ and r to $-r$ equation of the curve remains unchanged, that is, for every point (r, θ) on the curve we shave $(-r, -\theta)$ also on the curve.

(*ii*) As $\theta \to 0$, $r \to \infty$ hence an asymptote may exist here. We know that the equation of asymptote for the polar curve $\dfrac{1}{r} = f(\theta)$, is

$$r \sin(\theta - \alpha) = \frac{1}{f'(\alpha)} \quad \text{and} \quad r \to \infty \text{ as } \theta \to \alpha.$$

In the present case, $\alpha = 0$ and $f(\theta) = \dfrac{\theta}{a}$

$\therefore \quad f'(\theta) = \dfrac{1}{a}$, therefore, the asymptote here is

$r \sin(\theta - 0) = \dfrac{1}{1/a}$ or $r \sin \theta = a$

or $y = a$ (in cartesian coordinates).

Further, since $\sin\theta < \theta$, r for the curve is always less than the r for the asymptote. Thus, the curve lies completely below the asymptote.

(*iii*) The way r changes with the change in θ, is shown in the Table below.

θ	0	$\pi/2$	π	$3\pi/2$	2π	∞
r	∞	$2a/\pi$	a/π	$2a/3\pi$	$a/2\pi$	0

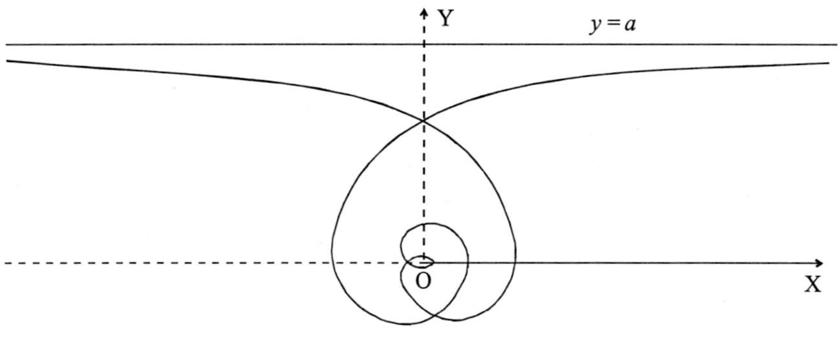

The shape of the curve is as shown in the above Figure. It is in two parts, one when θ and r both are positive and the other when θ and r both are negative. Both the parts are symmetrical about the line $\theta = \pi/2$.

EXAMPLE 2.46. Trace the **Foluim of Descartes** $\boxed{x^3 + y^3 = 3axy}$. **(UPTU 2002-03)**

SOLUTION:

(*i*) On interchanging x and y the equation remains unchanged, hence the curve is symmetrical about the line $y = x$.

(*ii*) The curve passes through the origin and equating to zero the lowest degree term, we find that both x-axis and y-axis are tangents to the curve at the origin.

(*iii*) The curve does not meet the coordinate axes except at the origin.

(*iv*) The curve meets the line of symmetry $y = x$ at the point $(3a/2, 3a/2)$. To find the tangent to the curve at the point $(3a/2, 3a/2)$. shift the origin to this point and equation becomes

$$(x + 3a/2)^3 + (y + 3a/2)^3 = 3a\,(x + 3a/2)\,(y + 3a/2)$$

or $$x^3 + y^3 - 3axy + \frac{9a}{2}\left(x^2 + y^2\right) + \frac{27a^2}{4}(x + y) - \frac{9a^2}{2}(x + y) = 0$$

or $$x^3 + y^3 - 3axy + \frac{9a}{2}\left(x^2 + y^2\right) + \frac{9a^2}{4}(x + y) = 0$$

Equating to 0 the lowest degree term we get the tangent to the curve at the new origin as $x + y = 0$.

Therefore, the tangent to the given curve at $(3a/2, 3a/2)$ is $x + y = 3a$.

(*v*) No asymptote parallel to x-axis or y-axis exists. However an oblique asymptote does exist. By the method discussed in the early part of this chapter we get the equation of this asymptote as $x + y + a = 0$.

(*vi*) Observe here that x and y cannot be simultaneously negative hence no portion of the curve lies in the third quadrant.

(*vii*) Now, for convenience, we convert the given equation to polar coordinates so as to get

$$r^3 \cos^3 \theta + r^3 \sin^3 \theta = 3a\, r^2 \cos \theta \sin \theta$$

or $$r = \frac{3a \sin\theta \cos\theta}{\sin^3\theta + \cos^3\theta}$$

Next, we give suitable values to θ and find the corresponding values of r and put them in tabular form as

θ	0	$\pi/4$	$\pi/3$	$\pi/2$	$2\pi/3$	π
r	0	$\frac{3a\sqrt{2}}{2}$	$\frac{6a\sqrt{3}}{1+3\sqrt{3}}$	0	$\frac{-6a\sqrt{3}}{3\sqrt{3}-1}$	0

We observe here that as θ increases from 0 to $\pi/4$, r increases, and when θ goes from $\pi/4$ to $\pi/2$, r decreases. When θ goes from $\pi/2$ to π, r is negative.

The shape of the curve is as shown in the Figure given below.

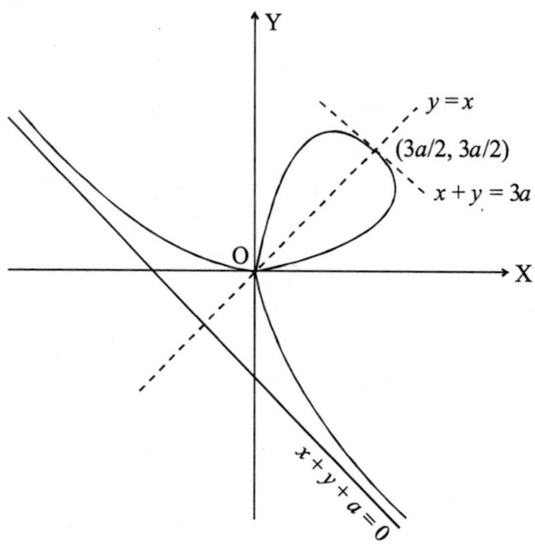

EXERCISE 2D

Trace the following curves:

1. (*i*) $r = 2a \cos \theta$ (*ii*) $r = 2a \cos (\theta - \alpha)$.

2. $r = a\theta$ **(spiral of Archamedes)**.

3. $r = a \cos 2\theta$. **(UPTU 2003-04)**

4. $r = a \cos 3\theta$.

5. $r^2 = a^2 \sin 2\theta$.

6. $r = a \sin 5\theta$.

7. $r^2 \cos 2\theta = a^2$.

8. $x^6 + y^6 = a^2 x^2 y^2$.

9. $x^5 + y^5 = 5ax^2y^2$.

❑❑❑

UNIT-II
DIFFERENTIAL CALCULUS-II

Partial Differentiation
Euler's Theorem
Change of Variables
Expansion of Functions of Several Variables
Jacobian
Approximation of Errors
Extrema of Functions of Several Variables
Lagrange's Method of Multipliers with Simple Applications

CHAPTER

3 Partial Differentiation

Partial Differentiation, Euler's Theorem, Total Derivative and Change of Independent variables

FUNCTIONS OF SEVERAL VARIABLES

In engineering problems one frequently comes across a variable quantity which depends for its values on two or more independent variables. For instance, the area of a rectangle depends upon its length and breadth and we say that this area is a function of two independent variables, the length and the breadth.

We use the notation $f(x, y)$ or $F(x, y)$, etc. to denote the value of the function f at (x, y) and write $z = f(x, y)$ or $z = F(x, y)$, etc., Also, we may write $z = z(x, y)$ where it should be clearly understood that in this notation z is used in two senses, namely, as a function and as a variable. The concept can be easily extended to the functions of three or more variables. Thus, $\omega = f(x, y, z)$ denotes the value of the function f at (x, y, z), a point in three dimensional space.

PARTIAL DERIVATIVES

Let us extend the notion of ordinary derivative of the function of one variable to the derivative of a function f (or z) of two independent variables x and y. Now, question arises, should we differentiate f with respect to x or y? The answer is simple: treat y as constant while differentiating f with respect to x and treat x as constant while differentiating f with respect to y. This way we define two different derivatives and call them partial derivatives to distinguish them from the ordinary derivative of a function of a single independent variable.

We denote the partial derivative of z (or f) with respect to x by

$$\frac{\partial z}{\partial x} \text{ or } \frac{\partial f}{\partial x} \text{ or } f_x \text{ or } z_x \text{ or } \left(\frac{\partial z}{\partial x}\right)_{(\text{treating } y \text{ as constant})}.$$

and the partial derivative of z (or f) w.r.t. y by $\frac{\partial z}{\partial y}$ or $\frac{\partial f}{\partial y}$ or z_y or f_y or $\left(\frac{\partial z}{\partial y}\right)_{(\text{treating } x \text{ as constant})}$.

Thus, $$\boxed{\frac{\partial f}{\partial x} = \lim_{\delta x \to 0} \frac{f(x+\delta x, y) - f(x, y)}{\delta x} = \left(\frac{\partial f}{\partial x}\right)_y}$$

$$= \text{value of } \frac{\partial f}{\partial x} \text{ when } y \text{ is kept constant.}$$

Similarly, $$\boxed{\frac{\partial f}{\partial y} = \lim_{\delta y \to 0} \frac{f(x, y+\delta y) - f(x, y)}{\delta y} = \left(\frac{\partial f}{\partial y}\right)_x}$$

$$= \text{value of } \frac{\partial f}{\partial y} \text{ when } x \text{ is kept constant.}$$

Geometrically, $\frac{\partial z}{\partial x}\left(\text{or } \frac{\partial f}{\partial x}\right)$ is the slope of the tangent drawn to the curve of intersection of the surface $z = f(x, y)$ with a plane parallel to the plane $y = 0$, *i.e.*, zox- plane.

Similarly, for $\frac{\partial z}{\partial y}\left(\text{or } \frac{\partial f}{\partial y}\right)$.

In general, $\frac{\partial f}{\partial x}$ and $\frac{\partial f}{\partial y}$ are functions of both x and y and therefore, we can obtain the higher order partial derivatives of $f(x, y)$ as

$$\boxed{\frac{\partial^2 f}{\partial x^2} = \frac{\partial}{\partial x}\left(\frac{\partial f}{\partial x}\right) = f_{xx}}, \quad \boxed{\frac{\partial^2 f}{\partial x\, \partial y} = \frac{\partial}{\partial x}\left(\frac{\partial f}{\partial y}\right) = f_{xy}},$$

$$\boxed{\frac{\partial^2 f}{\partial y\, \partial x} = \frac{\partial}{\partial y}\left(\frac{\partial f}{\partial x}\right) = f_{yx}}, \quad \boxed{\frac{\partial^2 f}{\partial y^2} = \frac{\partial}{\partial y}\left(\frac{\partial f}{\partial y}\right) = f_{yy}}, \quad \text{etc.}$$

In $\frac{\partial^2 f}{\partial x\, \partial y}$ we first differentiate f partially with respect to y and then partially with respect to x, It must be noted that $\frac{\partial^2 f}{\partial x\, \partial y} = \frac{\partial^2 f}{\partial y\, \partial x}$, **in general**, meaning thereby, that **the order of partial differentiation is generally immaterial.**

EXAMPLE 3.1. If $z = f(x + ay) + \phi(x - ay)$, then show that $z_{yy} = a^2 z_{xx}$. **(UTPU 2005)**

SOLUTION: From $z = f(x + a y) + \phi(x - a y)$, we get

$$z_x = \frac{\partial z}{\partial x} = f'(x + ay)\frac{\partial}{\partial x}(x + a y) + \phi'(x - a y)\frac{\partial}{\partial x}(x - ay)$$

$$= f'(x + ay) + \phi'(x - ay)$$

where f' and ϕ' mean ordinary derivatives of f and ϕ with respect to $x + ay$ and $x - ay$ respectively.

Similarly, $$z_y = \frac{\partial z}{\partial y} = f'(x + a y)\frac{\partial}{\partial y}(x + a y) + \phi'(x - ay)\frac{\partial}{\partial y}(x - ay)$$

$$= a f'(x + a y) - a\,\phi'(x - a y)$$

Now, differentiating z_x and z_y again partially with respect to x and y respectively, gives

$$z_{xx} = \frac{\partial}{\partial x}\left(\frac{\partial z}{\partial x}\right) = \frac{\partial}{\partial x}\, f'(x + a y) + \frac{\partial}{\partial x}\, \phi'(x - a y)$$

$$= f''\frac{\partial}{\partial x}(x + a y) + \phi''\frac{\partial}{\partial x}(x - a y) = f''(x + ay) + \phi''(x - ay).$$

and
$$z_{yy} = \frac{\partial}{\partial y}\left(\frac{\partial z}{\partial y}\right) = \frac{\partial}{\partial y}[a f'(x + a y)] - \frac{\partial}{\partial y}[a\, \phi'(x - a y)]$$

$$= a f''\frac{\partial}{\partial y}(x + a y) - a\,\phi''\frac{\partial}{\partial y}(x - a y)$$

$$= a^2 f''(x + a y) + (-a)^2\, \phi''(x - a y)$$

which clearly implies that $z_{yy} = a^2 z_{xx}$. **Hence Proved.**

EXAMPLE 3.2. Show that, at a point on the surface $x^x\, y^y\, z^z = c$ where $x = y = z$, we have $\dfrac{\partial^2 z}{\partial x\, \partial y} = \dfrac{-1}{x\, \log(ex)}$. **(UTPU 2001 Q.B.)**

SOLUTION: From the given relation $x^x\, y^y\, z^z = c$ ...(1)

it is clear that we can take x as a function of y and z, or y as a function of z and x, or z as a function of x and y. As we are to calculate $\dfrac{\partial^2 z}{\partial x \partial y}$ we shall take here z as a function of x and y.

Taking logarithm in (1), gives

$$x \log x + y \log y + z \log z = \log c \qquad \text{... (2)}$$

Differentiating (2) partially with respect to y, gives

$$0 + y \cdot \frac{1}{y} + 1.\ \log y + z \cdot \frac{1}{z}\frac{\partial z}{\partial y} + \log z \cdot \frac{\partial z}{\partial y} = 0 \quad \text{or} \quad \frac{\partial z}{\partial y} = \frac{-(1+\log y)}{1+\log z} \qquad \text{... (3)}$$

Similarly, differentiating (2) partially *w.r.t.* x, gives $\dfrac{\partial z}{\partial x} = \dfrac{-(1+\log x)}{1+\log z}$ ... (4)

Next, differentiating (3) partially *w.r.t* x, we get

$$\frac{\partial^2 z}{\partial x\, \partial y} = -(1 + \log y)\, \frac{\partial}{\partial x}\frac{1}{(1+\log z)}$$

$$= -(1+\log y)\frac{(-1)}{(1+\log z)^2}\frac{\partial}{\partial x}(1+\log z) = \frac{1+\log y}{(1+\log z)^2}\frac{1}{z}\frac{\partial z}{\partial x}$$

Using (4) here, gives

$$\frac{\partial^2 z}{\partial x\, \partial y} = \frac{1+\log y}{(1+\log z)^2}\, \frac{(-1)(1+\log x)}{z\,(1+\log z)} = \frac{-(1+\log x)(1+\log y)}{z\,(1+\log z)^3}$$

At the point $x = y = z$, we have

$$\frac{\partial^2 z}{\partial x\, \partial y} = \frac{-1}{x(1+\log x)} = \frac{-1}{x(\log e+\log x)} = \frac{-1}{x\log(ex)}.$$ **Hence Proved.**

EXAMPLE 3.3. If $u = \log \dfrac{x^2+y^2}{xy}$, verify that $\dfrac{\partial^2 u}{\partial y\, \partial x} = \dfrac{\partial^2 u}{\partial x\, \partial y}$.

SOLUTION: We can write $u = \log (x^2 + y^2) - \log x - \log y$

Hence $\dfrac{\partial u}{\partial x} = \dfrac{2x}{x^2+y^2} - \dfrac{1}{x}$ and $\dfrac{\partial u}{\partial y} = \dfrac{2y}{x^2+y^2} - \dfrac{1}{y}$

Further $$\frac{\partial^2 u}{\partial x \partial y} = \frac{\partial}{\partial x}\left(\frac{\partial u}{\partial y}\right) = \frac{\partial}{\partial x}\left[\frac{2y}{x^2+y^2} - \frac{1}{y}\right] = \frac{-4xy}{(x^2+y^2)^2} - 0$$

and $$\frac{\partial^2 u}{\partial y \partial x} = \frac{\partial}{\partial y}\left(\frac{\partial u}{\partial x}\right) = \frac{\partial}{\partial y}\left[\frac{2x}{x^2+y^2} - \frac{1}{x}\right] = \frac{-2x.2y}{(x^2+y^2)^2} - 0$$

Thus, $\dfrac{\partial^2 u}{\partial x \partial y} = \dfrac{\partial^2 u}{\partial y \partial x}$ **is verified.**

EXAMPLE 3.4. If $\dfrac{x^2}{a^2+u} + \dfrac{y^2}{b^2+u} + \dfrac{z^2}{c^2+u} = 1$, show that

$$u_x^2 + u_y^2 + u_z^2 = 2\,(xu_x + yu_y + zu_z).$$ **(UPTU 2002-03)**

SOLUTION: Presence of u_x, u_y and u_z indicates that we have to take u as a function of x, y and z.

Differentiating the given relation partially *w.r.t.* x, we get

$$\frac{2x}{a^2+u} - \frac{x^2 u_x}{(a^2+u)^2} - \frac{y^2 u_x}{(b^2+u)^2} - \frac{z^2 u_x}{(c^2+u)^2} = 0$$

or $u_x = \dfrac{2x}{K(a^2+u)}$ where $K = \dfrac{x^2}{(a^2+u)^2} + \dfrac{y^2}{(b^2+u)^2} + \dfrac{z^2}{(c^2+u)^2}$

Similarly, differentiating the given relation partially *w.r.t.* y and z separately, gives

$$u_y = \frac{2y}{K(b^2+u)} \quad \text{and} \quad u_z = \frac{2z}{K(c^2+u)}.$$

Therefore, $$u_x^2 + u_y^2 + u_z^2 = \frac{4}{K^2}\left[\frac{x^2}{(a^2+u)^2} + \frac{y^2}{(b^2+u)^2} + \frac{z^2}{(c^2+u)^2}\right]$$

$$= \frac{4}{K^2}.K = \frac{4}{K}.$$

And $$xu_x + yu_y + zu_z = \frac{2}{K}\left[\frac{x^2}{a^2+u} + \frac{y^2}{b^2+u} + \frac{z^2}{c^2+u}\right]$$

$$= \frac{2}{K} \cdot 1 \quad \text{(using the given relation.)}$$

This establishes that $u_x^2 + u_y^2 + u_z^2 = 2\,(xu_x + yu_y + zu_z)$. **Hence Proved.**

EXAMPLE 3.5. If $u = \log(x^3 + y^3 - x^2y - y^2x)$, show that

$$\left(\frac{\partial}{\partial x}+\frac{\partial}{\partial y}\right)^2 u = -\frac{4}{(x+y)^2}.$$

SOLUTION: Differentiating the given relation *w.r.t.* x and y separately, we get

$$\frac{\partial u}{\partial x} = \frac{3x^2-2xy-y^2}{x^3+y^3-x^2y-y^2x}, \quad \frac{\partial u}{\partial y} = \frac{3y^2-x^2-2xy}{x^3+y^3-x^2y-y^2x}$$

$$\therefore \quad \frac{\partial u}{\partial x}+\frac{\partial u}{\partial y} = \frac{2x^2+2y^2-4xy}{x^3+y^3-xy\,(x+y)} = \frac{2\,(x-y)^2}{(x+y)\,[x^2+y^2-xy-xy]} = \frac{2}{x+y}$$

Now, $$\left(\frac{\partial}{\partial x}+\frac{\partial}{\partial y}\right)^2 u = \left(\frac{\partial}{\partial x}+\frac{\partial}{\partial y}\right)\left(\frac{\partial u}{\partial x}+\frac{\partial u}{\partial y}\right) = \left(\frac{\partial}{\partial x}+\frac{\partial}{\partial y}\right)\frac{2}{x+y} = 2\left[\frac{\partial}{\partial x}\frac{1}{x+y}+\frac{\partial}{\partial y}\frac{1}{x+y}\right]$$

$$= 2\left[\frac{-1}{(x+y)^2}+\frac{-1}{(x+y)^2}\right] = \frac{-4}{(x+y)^2}.$$ **Hence Proved.**

EXAMPLE 3.6. If $\theta = t^n e^{-\frac{r^2}{4t}}$ find the value of n for which

$$\frac{1}{r^2}\frac{\partial}{\partial r}\left(r^2\frac{\partial \theta}{\partial r}\right) = \frac{\partial \theta}{\partial t}.$$ **(UPTU 2001 Q.B.)**

SOLUTION: From the given relation $\theta = t^n e^{-\frac{r^2}{4t}}$ it is clear that θ is a function of r and t.
Taking logarithm on the both sides, we get

$$\log \theta = n \log t - \frac{r^2}{4t}. \quad \ldots (1)$$

Differentiating (1) partially *w.r.t.* r, gives $\frac{1}{\theta}\frac{\partial \theta}{\partial r} = 0 - \frac{2r}{4t}$

Hence $$r^2\frac{\partial \theta}{\partial r} = -\frac{r^3\theta}{2t} \quad \ldots (2)$$

Differentiating (2) partially *w.r.t.* r, gives

$$\frac{\partial}{\partial r}\left(r^2\frac{\partial \theta}{\partial r}\right) = -\frac{3r^2\theta}{2t} - \frac{r^3}{2t}\frac{\partial \theta}{\partial r} = -\frac{3r^2\theta}{2t}+\frac{r^4\theta}{4t^2} \quad \text{[using (2)]}$$

or $$\frac{1}{r^2}\frac{\partial}{\partial r}\left(r^2\frac{\partial \theta}{\partial r}\right) = \left(-\frac{3}{2t}+\frac{r^2}{4t^2}\right)\theta.$$

Next, differentiating (1) partially *w.r.t.* t, we get $\frac{1}{\theta}\frac{\partial \theta}{\partial t} = \frac{n}{t}+\frac{r^2}{4t^2}$

Given that $\frac{1}{r^2}\frac{\partial}{\partial r}\left(r^2\frac{\partial \theta}{\partial r}\right) = \frac{\partial \theta}{\partial t}$, we get

$$\left(-\frac{3}{2t}+\frac{r^2}{4t^2}\right)\theta = \left(\frac{n}{t}+\frac{r^2}{4t^2}\right)\theta$$ which gives $n = -3/2$. **Ans.**

EXAMPLE 3.7. If $z = e^{ax+by} f(ax - by)$ show that $b\dfrac{\partial z}{\partial x} + a\dfrac{\partial z}{\partial y} = 2abz.$ **(UPTU 2001 Q.B.)**

SOLUTION: Given that $z = e^{ax+by} f(ax - by)$, we have

$$\frac{\partial z}{\partial x} = ae^{ax+by} f(ax-by) + ae^{ax+by} f'(ax-by)$$

and
$$\frac{\partial z}{\partial y} = be^{ax+by} f(ax-by) - be^{ax+by} f'(ax-by)$$

or
$$a\frac{\partial z}{\partial y} = abe^{ax+by} [f(ax-by) - f'(ax-by)]$$

Therefore
$$b\frac{\partial z}{\partial x} + a\frac{\partial z}{\partial y} = abe^{ax+by} [2f(ax-by)] = 2abz.$$
Hence the result.

EXAMPLE 3.8. If $V = (x^2 + y^2 + z^2)^{m/2}$ find the value of $m(\neq 0)$ which will make

$$\frac{\partial^2 V}{\partial x^2} + \frac{\partial^2 V}{\partial y^2} + \frac{\partial^2 V}{\partial z^2} = 0.$$

SOLUTION: Given $V = (x^2 + y^2 + z^2)^{m/2}$ hence $\dfrac{\partial V}{\partial x} = \dfrac{m}{2}(x^2+y^2+z^2)^{\frac{m}{2}-1} \cdot 2x$

$$\therefore \quad \frac{\partial^2 V}{\partial x^2} = mx\left(\frac{m}{2}-1\right)(x^2+y^2+z^2)^{\frac{m}{2}-2} \cdot 2x + m(x^2+y^2+z^2)^{\frac{m}{2}-1}$$

$$= m(x^2+y^2+z^2)^{\frac{m}{2}-2} [(m-2)\,x^2 + (x^2+y^2+z^2)]$$

Similarly for $\dfrac{\partial^2 V}{\partial y^2}$ and $\dfrac{\partial^2 V}{\partial z^2}$.

Therefore
$$\frac{\partial^2 V}{\partial x^2} + \frac{\partial^2 V}{\partial y^2} + \frac{\partial^2 V}{\partial z^2} = m(x^2+y^2+z^2)^{\frac{m}{2}-2} [(m-2)\,(x^2+y^2+z^2) + 3(x^2+y^2+z^2)]$$

$$= m(x^2+y^2+z^2)^{\frac{m}{2}-2} (m+1)\,(x^2+y^2+z^2)$$

$$= m(m+1)\,(x^2+y^2+z^2)^{\left(\frac{m}{2}-1\right)}$$

$$\Rightarrow \quad \frac{\partial^2 V}{\partial x^2} + \frac{\partial^2 V}{\partial y^2} + \frac{\partial^2 V}{\partial z^2} = 0 \text{ when } m = 0 \text{ or } m = -1.$$

Therefore, the required non-zero value of m is -1. **Ans.**

EXAMPLE 3.9. If $u = \log(x^3 + y^3 + z^3 - 3xyz)$, prove that

$$\left(\frac{\partial}{\partial x}+\frac{\partial}{\partial y}+\frac{\partial}{\partial z}\right)^2 u = \frac{-9}{(x+y+z)^2}.$$ **(UPTU 2003-04)**

SOLUTION: Given that $u = \log(x^3 + y^3 + z^3 - 3xyz)$

Hence $\dfrac{\partial u}{\partial x} = \dfrac{3x^2 - 3yz}{x^3 + y^3 + z^3 - xyz}$, $\dfrac{\partial u}{\partial y} = \dfrac{3y^2 - 3xz}{x^3 + y^3 + z^3 - 3xyz}$, $\dfrac{\partial u}{\partial z} = \dfrac{3z^2 - 3xy}{x^3 + y^3 + z^3 - 3xyz}$

$$\therefore \qquad \left(\frac{\partial}{\partial x}+\frac{\partial}{\partial y}+\frac{\partial}{\partial z}\right)u = \frac{3(x^2+y^2+z^2-xy-yz-zx)}{x^3+y^3+z^3-3xyz} = \frac{3}{x+y+z}$$

Next
$$\left(\frac{\partial}{\partial x}+\frac{\partial}{\partial y}+\frac{\partial}{\partial z}\right)^2 u = \left(\frac{\partial}{\partial x}+\frac{\partial}{\partial y}+\frac{\partial}{\partial z}\right)\frac{3}{(x+y+z)}$$

$$= \frac{-3}{(x+y+z)^2} + \frac{-3}{(x+y+z)^2} + \frac{-3}{(x+y+z)^2} = \frac{-9}{(x+y+z)^2}.$$

Hence proved.

EXAMPLE 3.10. If $x = \dfrac{\cos\theta}{u}$, $y = \dfrac{\sin\theta}{u}$, show that

$$\left(\frac{\partial x}{\partial u}\right)_\theta \left(\frac{\partial u}{\partial x}\right)_y + \left(\frac{\partial y}{\partial u}\right)_\theta \left(\frac{\partial u}{\partial y}\right)_x = 1.$$

SOLUTION: From $x = \dfrac{\cos\theta}{u}$, $y = \dfrac{\sin\theta}{u}$, we can get

$$\left(\frac{\partial x}{\partial u}\right)_\theta = -\frac{\cos\theta}{u^2} \quad \text{and} \quad \left(\frac{\partial y}{\partial u}\right)_\theta = -\frac{\sin\theta}{u^2}.$$

Next, to find $\left(\dfrac{\partial u}{\partial x}\right)_y$ and $\left(\dfrac{\partial u}{\partial y}\right)_x$ we have to express u as a function of x and y.

Eliminating θ in the given relations, we get

$$x^2u^2 + y^2u^2 = 1 \quad \text{or} \quad u^2 = \frac{1}{x^2+y^2}$$

Therefore, $$2u\left(\frac{\partial u}{\partial x}\right)_y = \frac{-2x}{(x^2+y^2)^2} = -2xu^4$$

or $$\left(\frac{\partial u}{\partial x}\right)_y = -xu^3 = -u^2\cos\theta$$

Similarly $$\left(\frac{\partial u}{\partial y}\right)_x = -yu^3 = -u^2\sin\theta.$$

$$\therefore \qquad \left(\frac{\partial x}{\partial u}\right)_\theta \left(\frac{\partial u}{\partial x}\right)_y + \left(\frac{\partial y}{\partial u}\right)_\theta \left(\frac{\partial u}{\partial y}\right)_x = \frac{-\cos\theta}{u^2}(-u^2\cos\theta) - \frac{\sin\theta}{u^2}(-u^2\sin\theta) = 1.$$

Hence Proved.

EXAMPLE 3.11. If $u = f(r)$ where $r^2 = x^2 + y^2 + z^2$, show that

$$u_{xx} + u_{yy} + u_{zz} = f''(r) + \frac{2}{r} f'(r).$$

SOLUTION: Since $u = f(r)$ we have $u_x = f'(r) \frac{\partial r}{\partial x}$

Differentiating the relation $r^2 = x^2 + y^2 + z^2$ partially w.r. t. x, we get

$$2r \frac{\partial r}{\partial x} = 2x \quad \text{or} \quad \frac{\partial r}{\partial x} = \frac{x}{r}.$$

Similarly, we can have $\frac{\partial r}{\partial y} = \frac{y}{r}$ and $\frac{\partial r}{\partial z} = \frac{z}{r}$.

$\therefore$ $u_x = f'(r) \frac{x}{r}$. Differentiating again *w.r.t.* x partially, we get

$$u_{xx} = \frac{\partial}{\partial x}\left[f'(r)\frac{x}{r}\right] = f'(r)\frac{\partial}{\partial x}\left(\frac{x}{r}\right) + \frac{x}{r} f''(r) \frac{\partial r}{\partial x}$$

$$= f'(r)\left(\frac{1}{r} - \frac{x}{r^2} \cdot \frac{\partial r}{\partial x}\right) + \frac{x^2}{r^2} f''(r) = \frac{f'(r)}{r} - f'(r) \cdot \frac{x}{r^2} \cdot \frac{x}{r} + \frac{x^2}{r^2} f''(r)$$

$$= \frac{f'(r)}{r} + x^2\left(\frac{1}{r^2} f''(r) - \frac{f'(r)}{r^3}\right)$$

Similarly, we can get

$$u_{yy} = \frac{f'(r)}{r} + y^2\left(\frac{1}{r^2} f''(r) - \frac{f'(r)}{r^3}\right) \quad \text{and} \quad u_{zz} = \frac{f'(r)}{r} + z^2\left(\frac{1}{r^2} f''(r) - \frac{f'(r)}{r^3}\right).$$

$$\therefore \quad u_{xx} + u_{yy} + u_{zz} = \frac{3f'(r)}{r} + (x^2 + y^2 + z^2)\left\{\frac{1}{r^2} f''(r) - \frac{f'(r)}{r^3}\right\}$$

$$= \frac{3f'(r)}{r} + r^2\left\{\frac{1}{r^2} f''(r) - \frac{1}{r^3} f'(r)\right\} = f''(r) + \frac{2}{r} f'(r).$$ **Hence Proved.**

EXAMPLE 3.12. If $f(x, y) = (1 - 2xy + y^2)^{-1/2}$, show that

$$\frac{\partial}{\partial x}\left[(1-x^2)\frac{\partial f}{\partial x}\right] + \frac{\partial}{\partial y}\left[y^2 \frac{\partial f}{\partial y}\right] = 0.$$

SOLUTION: Let $V = (1 - 2xy + y^2)$ then $f = V^{-1/2}$.

$$\therefore \quad \frac{\partial f}{\partial x} = -\frac{1}{2} V^{-3/2} \frac{\partial V}{\partial x} = -\frac{1}{2} V^{-3/2}(-2y) = yV^{-3/2}$$

and

$$\frac{\partial f}{\partial y} = -\frac{1}{2} V^{-3/2} \frac{\partial V}{\partial y} = -\frac{1}{2} V^{-3/2}(2y - 2x) = (x - y)V^{-3/2}$$

Also,

$$\frac{\partial^2 f}{\partial x^2} = \frac{-3}{2} V^{-5/2} y(-2y) = 3y^2 V^{-5/2}$$

and $$\frac{\partial^2 f}{\partial y^2} = (x-y)\left(\frac{-3}{2}\right)V^{-5/2}(-2x+2y) - V^{-3/2} = V^{-3/2}\left[3(x-y)^2\frac{1}{V} - 1\right].$$

$$\therefore \quad \frac{\partial}{\partial x}\left[(1-x^2)\frac{\partial f}{\partial x}\right]+\frac{\partial}{\partial y}\left[y^2\frac{\partial f}{\partial y}\right]$$

$$= (1-x^2)\frac{\partial^2 f}{\partial x^2}-2x\frac{\partial f}{\partial x}+y^2\frac{\partial^2 f}{\partial y^2}+2y\frac{\partial f}{\partial y}$$

$$= 3(1-x^2)y^2V^{-5/2}-2xyV^{-3/2}+y^2\left[3(x-y)^2\cdot\frac{1}{V}-1\right]V^{-3/2}+2y(x-y)V^{-3/2}$$

$$= yV^{-3/2}\left[\frac{3(1-x^2)y}{V}-2x+\frac{3y(x-y)^2}{V}-y+2(x-y)\right]$$

$$= yV^{-5/2}\,[3y(1-x^2)+3y(x-y)^2-3y(1-2xy+y^2)]$$

$$= 3y^2V^{-5/2}\,[1-x^2+(x-y)^2-1+2xy-y^2] = 0.$$ **Hence Proved.**

EXAMPLE 3.13. If $u=f(r)$ and $x = r\cos\theta$, $y = r\sin\theta$, prove that

$$\frac{\partial^2 u}{\partial x^2}+\frac{\partial^2 u}{\partial y^2} = f''(r)+\frac{1}{r}f'(r).$$ **(UPTU 2000–01, 2005–06)**

SOLUTION: Given $u=f(r)$ where $r^2=x^2+y^2$ $\therefore$ $\frac{\partial r}{\partial x}=\frac{x}{r}$ and $\frac{\partial r}{\partial y}=\frac{y}{r}$

Now $$\frac{\partial u}{\partial x} = \frac{du}{dr}\cdot\frac{\partial r}{\partial x} = f'(r)\frac{x}{r} \quad\text{and}\quad \frac{\partial u}{\partial y} = \frac{du}{dr}\cdot\frac{\partial r}{\partial y} = f'(r)\frac{y}{r}$$

Next $$\frac{\partial^2 u}{\partial x^2} = \frac{\partial}{\partial x}\left[x\cdot\frac{1}{r}f'(r)\right] = 1\cdot\frac{1}{r}f'(r)+x\frac{d}{dr}\left[\frac{1}{r}f'(r)\right]\cdot\frac{\partial r}{\partial x}$$

$$= \frac{1}{r}f'(r)+x\cdot\frac{x}{r}\left[\frac{1}{r}f''(r)-\frac{1}{r^2}f'(r)\right] = \frac{1}{r}f'(r)\left(1-\frac{x^2}{r^2}\right)+\frac{x^2}{r^2}f''(r)$$

Similarly $$\frac{\partial^2 u}{\partial y^2} = \frac{1}{r}f'(r)\left(1-\frac{y^2}{r^2}\right)+\frac{y^2}{r^2}f''(r)$$

$$\therefore \quad \frac{\partial^2 u}{\partial x^2}+\frac{\partial^2 u}{\partial y^2} = \frac{1}{r}f'(r)\left[2-\frac{x^2+y^2}{r^2}\right]+\frac{x^2+y^2}{r^2}f''(r) = f''(r)+\frac{1}{r}f'(r).$$ **Hence Proved.**

EXERCISE 3A

1. If $u = \dfrac{e^{x+y+z}}{e^x + e^y + e^z}$, show that $\dfrac{\partial u}{\partial x} + \dfrac{\partial u}{\partial y} + \dfrac{\partial u}{\partial z} = 2u$.

2. If $V = \log(\tan x + \tan y + \tan z)$ find the value of

$$\sin 2x \frac{\partial V}{\partial x} + \sin 2y \frac{\partial V}{\partial y} + \sin 2z \frac{\partial V}{\partial z}.$$

(UPTU 2011-12)

3. Given that $z = x^2 \tan^{-1}(y/x) - y^2 \tan^{-1}(x/y)$, find the value of $\dfrac{\partial^2 z}{\partial y \partial x}$

and verify that $\dfrac{\partial^2 z}{\partial x \partial y} = \dfrac{\partial^2 z}{\partial y \partial x}$. **(UPTU 2001 SP. Exam.)**

4. If $u^2 = x^2 + y^2 + z^2$ find the value of $\dfrac{\partial^2 u}{\partial x^2} + \dfrac{\partial^2 u}{\partial y^2} + \dfrac{\partial^2 u}{\partial z^2}$.

5. If $u(x, t) = A e^{-gx} \sin(nt - gx)$ where A, g and n are constants, and $\dfrac{\partial u}{\partial t} = a^2 \dfrac{\partial^2 u}{\partial x^2}$,

show that $g = \dfrac{1}{a}\sqrt{\dfrac{n}{2}}$.

6. If $u = e^{r \cos\theta} \cos(r \sin\theta)$ and $v = e^{r\cos\theta} \sin(r \sin\theta)$, show that

$$\frac{\partial u}{\partial r} = \frac{1}{r}\frac{\partial v}{\partial \theta} \quad \text{and} \quad \frac{\partial v}{\partial r} = -\frac{1}{r}\frac{\partial u}{\partial \theta}.$$

7. If $u = r^n$ and $r^2 = x^2 + y^2 + z^2$, prove that

$u_{xx} + u_{yy} + u_{zz} = n(n+1) r^{n-2}$.

8. Given that $u = x \log(x + r) - r$ where $r^2 = x^2 + y^2$, evaluate $u_{xx} + u_{yy}$ in the simplified form.

9. Let $u = lx + my$ and $v = mx - ly$ where l and m are constants, show that

$$\left(\frac{\partial u}{\partial x}\right)_y \left(\frac{\partial x}{\partial u}\right)_v = \frac{l^2}{l^2 + m^2} = \left(\frac{\partial y}{\partial v}\right)_u \left(\frac{\partial v}{\partial y}\right)_x.$$

10. If $u = \log(\tan x + \tan y)$, evaluate $\sin 2x \dfrac{\partial u}{\partial x} + \sin 2y \dfrac{\partial u}{\partial y}$.

11. Let $xu + yv = 0$ and $\dfrac{u}{x} + \dfrac{v}{y} = 1$, then show that

$$\left(\frac{\partial u}{\partial x}\right)_y - \left(\frac{\partial v}{\partial y}\right)_x = \frac{y^2 + x^2}{y^2 - x^2}.$$

12. If $z = (x + y)\left[1 + \phi\left(\dfrac{y}{x}\right)\right]$ prove that

$$x\left(\frac{\partial^2 z}{\partial x^2} - \frac{\partial^2 z}{\partial x \partial y}\right) = y\left(\frac{\partial^2 z}{\partial y^2} - \frac{\partial^2 z}{\partial y \partial x}\right).$$

13. If $u = e^{xyz} f\left(\frac{xz}{y}\right)$ show that

$$x\frac{\partial u}{\partial x} + y\frac{\partial u}{\partial y} = 2\,xyzu = y\frac{\partial u}{\partial y} + z\frac{\partial u}{\partial z}$$

and deduce that $x\frac{\partial^2 u}{\partial x\,\partial z} = y\frac{\partial^2 u}{\partial y\,\partial z}$.

14. Given $x = \cos\theta - r\sin\theta$, $y = \sin\theta + r\cos\theta$, show that

(*i*) $\frac{\partial \theta}{\partial x} = \frac{-\cos\theta}{r}$, $\frac{\partial r}{\partial x} = \frac{x}{r}$

(*ii*) $\frac{\partial^2 \theta}{\partial x^2} = \frac{\cos\theta}{r^3}(\cos\theta - 2r\sin\theta)$.

❑❑❑

HOMOGENEOUS FUNCTIONS

An expression of the form $a_0 x^n + a_1 x^{n-1} y + a_2 x^{n-2} y^2 + ... + a_n y^n$ in which all the terms are of same degree n is called **homogeneous** function in x and y of degree n. The definition can be extended to functions of three or more variables. The above expression can also be written as

$$x^n [a_0 + a_1 (y/x) + a_2 (y/x)^2 + ... + a_n (y/x)^n] = x^n f(y/x)$$

where $f(y/x)$ is a nth degree polynomial in (y/x). It is important to note here that we are now in a position to generalise $f(y/x)$ to include trigonometric, exponential and other functions.

Thus, $x^n f(y/x)$ defines a homogeneous function in x and y, of degree n, whatever may be the functional nature of f. For example, $x^4 \tan (y/x)$ is homogeneous in x and y of degree 4. Similarly, the expression $\dfrac{x^{1/3}+y^{1/3}}{x^{1/2}+y^{1/2}}$ is a homogeneous function of degree $-\dfrac{1}{6}$ because it can be written as

$$\frac{x^{1/3}\left\{1+(y/x)^{1/3}\right\}}{x^{1/2}\left\{1+(y/x)^{1/2}\right\}} = x^{\frac{1}{3}-\frac{1}{2}} g(y/x) = x^{-1/6} g(y/x).$$

In this connection a very useful theorem follows:

EULER'S THEOREM ON HOMOGENEOUS FUNCTIONS

(UPTU 2005-06, 2006-07)

If u is a homogeneous function in x and y, of degree n, then

$$x\frac{\partial u}{\partial x} + y\frac{\partial u}{\partial y} = nu.$$

PROOF: As explained above, we can write $u = x^n f(y/x)$

or $u = x^n f(t)$ where $t = y/x$.

Then $$\frac{\partial u}{\partial x} = nx^{n-1} f(t) + x^n f'(t)\frac{\partial t}{\partial x} = nx^{n-1} f(t) + x^n f'(t)(-y/x^2)$$

$$= nx^{n-1} f(t) - x^{n-2} y f'(t)$$

and $$\frac{\partial u}{\partial y} = x^n f'(t)\frac{\partial t}{\partial y} = x^n f'(t)\frac{1}{x} = x^{n-1} f'(t)$$

Therefore, $$x\frac{\partial u}{\partial x} + y\frac{\partial u}{\partial y} = nx^n f(t) - x^{n-1} y f'(t) + x^{n-1} y f'(t) = nu.$$

Some Useful Deductions of Euler's Theorem

I. If u is a homogeneous function in x, y and z of degree n, we can write $u = x^n f(y/x, z/x)$ and then it can be easily proved that

$$x\frac{\partial u}{\partial x} + y\frac{\partial u}{\partial y} + z\frac{\partial u}{\partial z} = nu.$$

II. If u is a homogeneous function in x and y of degree n, then

$$\boxed{x^2\frac{\partial^2 u}{\partial x^2}+2xy\frac{\partial^2 u}{\partial x\,\partial y}+y^2\frac{\partial^2 u}{\partial y^2}=n\,(n-1)\,u}\,.$$

Proof : We already have $x\dfrac{\partial u}{\partial x}+y\dfrac{\partial u}{\partial y}=nu.$ ... (1)

Differentiating (1) partially *w.r.t.* x and y separately, we get

$$x\frac{\partial^2 u}{\partial x^2}+\frac{\partial u}{\partial x}.1+y\frac{\partial^2 u}{\partial x\,\partial y}=n\frac{\partial u}{\partial x} \qquad \text{... (2)}$$

and $$x\frac{\partial^2 u}{\partial y\,\partial x}+y\frac{\partial^2 u}{\partial y^2}+1.\frac{\partial u}{\partial y}=n\frac{\partial u}{\partial y} \qquad \text{... (3)}$$

Multiplying (2) by x and (3) by y and adding these, gives

$$x^2\frac{\partial^2 u}{\partial x^2}+x\frac{\partial u}{\partial x}+xy\frac{\partial^2 u}{\partial x\,\partial y}+xy\frac{\partial^2 u}{\partial y\,\partial x}+y^2\frac{\partial^2 u}{\partial y^2}+y\frac{\partial u}{\partial y}=nx\frac{\partial u}{\partial x}+ny\frac{\partial u}{\partial y}$$

Using (1) here and the fact that $\dfrac{\partial^2 u}{\partial y\,\partial x}=\dfrac{\partial^2 u}{\partial x\,\partial y}$, we get

$$x^2\frac{\partial^2 u}{\partial x^2}+2xy\frac{\partial^2 u}{\partial x\,\partial y}+y^2\frac{\partial^2 u}{\partial y^2}=(n-1)\,x\frac{\partial u}{\partial x}+(n-1)\,y\frac{\partial u}{\partial y}=n(n-1)u$$

III. If z is homogeneous in x and y of degree n and z is a function of u as $z=f(u)$ then we have

$$\boxed{(i)\;\; x\frac{\partial u}{\partial x}+y\frac{\partial u}{\partial y}=\frac{n\,f(u)}{f'(u)}}$$

$$\boxed{(ii)\;\; x^2\frac{\partial^2 u}{\partial x^2}+2xy\frac{\partial^2 u}{\partial x\,\partial y}+y^2\frac{\partial^2 u}{\partial y^2}=g\,(u)\,[g'\,(u)-1]\;\text{ where }\; g\,(u)=\frac{n\,f(u)}{f'(u)}}\,.$$

Proof: ***(i)*** Since z is homogenous in x and y of degree n, we have $x\dfrac{\partial z}{\partial x}+y\dfrac{\partial z}{\partial y}=nz.$ ... (4)

Also, since $z=f(u)$, we have $\dfrac{\partial z}{\partial x}=f'(u)\dfrac{\partial u}{\partial x},\quad \dfrac{\partial z}{\partial y}=f'(u)\dfrac{\partial u}{\partial y}$

Therefore, (4) becomes $x\,f'(u)\dfrac{\partial u}{\partial x}+y\,f'(u)\dfrac{\partial u}{\partial y}=n\,f(u)$

or $$x\frac{\partial u}{\partial x}+y\frac{\partial u}{\partial y}=\frac{n\,f(u)}{f'(u)}=g\,(u) \quad\text{where}\quad g\,(u)=\frac{n\,f(u)}{f'(u)}. \qquad \text{... (5)}$$

(ii) Differentiating (5) partially *w.r.t.* x and y separately, gives

$$x\frac{\partial^2 u}{\partial x^2}+1.\frac{\partial u}{\partial x}+y\frac{\partial^2 u}{\partial x\,\partial y}=g'(u)\frac{\partial u}{\partial x} \qquad \text{... (6)}$$

and $$x\frac{\partial^2 u}{\partial y\,\partial x}+y\frac{\partial^2 u}{\partial y^2}+1.\frac{\partial u}{\partial y}=g'(u)\frac{\partial u}{\partial y} \qquad \text{... (7)}$$

Multiplying (6) by x and (7) by y and adding, we get

$$x^2 \frac{\partial^2 u}{\partial x^2} + 2xy \frac{\partial^2 u}{\partial x \partial y} + y^2 \frac{\partial^2 u}{\partial y^2} + x \frac{\partial u}{\partial x} + y \frac{\partial u}{\partial y} = g'(u) \left(x \frac{\partial u}{\partial x} + y \frac{\partial u}{\partial y} \right)$$

or $$x^2 \frac{\partial^2 u}{\partial x^2} + 2xy \frac{\partial^2 u}{\partial x \partial y} + y^2 \frac{\partial^2 u}{\partial y^2} = \left(x \frac{\partial u}{\partial x} + y \frac{\partial u}{\partial y} \right) (g'(u) - 1) = g(u)(g'(u) - 1).$$

where $$g(u) = \frac{n f(u)}{f'(u)}.$$

EXAMPLE 3.14. If $u = x^4 \log \left(\frac{\sqrt[3]{y} - \sqrt[3]{x}}{\sqrt[3]{y} + \sqrt[3]{x}} \right)$ evaluate $x \frac{\partial u}{\partial x} + y \frac{\partial u}{\partial y}$.

SOLUTION : The expression $\frac{\sqrt[3]{y} - \sqrt[3]{x}}{\sqrt[3]{y} + \sqrt[3]{x}}$ is homogeneous in x and y of degree 0 therefore $x^4 \log \left(\frac{\sqrt[3]{y} - \sqrt[3]{x}}{\sqrt[3]{y} + \sqrt[3]{x}} \right)$ is homogeneous in x and y of degree 4.

Applying Euler's theorem, we get $x \frac{\partial u}{\partial x} + y \frac{\partial u}{\partial y} = 4u$. **Ans.**

EXAMPLE 3.15. **(a)** If $z = x^4 y^2 \sin^{-1} \frac{x}{y} + \log x - \log y$, show that

$$x \frac{\partial z}{\partial x} + y \frac{\partial z}{\partial y} = 6x^4 y^2 \sin^{-1} \frac{x}{y}.$$

(b) If $u = x^2yz - 4y^2z^2 + 2xz^3$, find the value of $x \frac{\partial u}{\partial x} + y \frac{\partial u}{\partial y} + z \frac{\partial u}{\partial z}$.

[UPTU 2011-12]

SOLUTION: **(a)** Let us write $z = u + v$

where $u = x^4 y^2 \sin^{-1} \frac{x}{y}$ and $v = \log x - \log y = \log \left(\frac{x}{y} \right)$.

Observe here that $\frac{x}{y}$ is homogenous of degree 0, so is $\sin^{-1} \frac{x}{y}$ and, in turn, u is homogenous in x, y of degree $4 + 2\ (= 6)$. Also $\log \left(\frac{x}{y} \right)$ is homogenous in x and y of degree 0, therefore by Euler's theorem, we have

$$x \frac{\partial u}{\partial x} + y \frac{\partial u}{\partial y} = 6u \quad \text{and} \quad x \frac{\partial v}{\partial x} + y \frac{\partial v}{\partial y} = 0.$$

Adding these, gives $x \frac{\partial}{\partial x}(u + v) + y \frac{\partial}{\partial y}(u + v) = 6u + 0$

or $$x \frac{\partial z}{\partial x} + y \frac{\partial z}{\partial y} = 6x^4 y^2 \sin^{-1} \frac{x}{y}.$$ **Hence Proved.**

(b) Here $u = x^2yz - 4y^2z^2 + 2xz^3$ is homogeneous function in x, y, z of degree 4, hence, by Euler's theorem, we get

$$x\frac{\partial u}{\partial x} + y\frac{\partial u}{\partial y} + z\frac{\partial u}{\partial z} = 4u. \quad \textbf{Ans.}$$

EXAMPLE 3.16. **(a)** If $u = \cos^{-1}\left(\dfrac{x+y}{\sqrt{x}+\sqrt{y}}\right)$, show that $x\dfrac{\partial u}{\partial x} + y\dfrac{\partial u}{\partial y} + \dfrac{1}{2}\cot u = 0$.

(b) If $u = \sin^{-1}\left[\dfrac{x^{1/3}+y^{1/3}}{x^{1/2}-y^{1/2}}\right]^{1/2}$ show that $x\dfrac{\partial u}{\partial x} + y\dfrac{\partial u}{\partial y} = -\dfrac{1}{12}\tan u$.

[UPTU 2011-12]

SOLUTION: (a) Given that $u = \cos^{-1}\left(\dfrac{x+y}{\sqrt{x}+\sqrt{y}}\right)$ we have $\cos u = \dfrac{x+y}{\sqrt{x}+\sqrt{y}}$.

Obviously $\dfrac{x+y}{\sqrt{x}+\sqrt{y}}$ is homogeneous is x and y of degree 1/2 hence cos u is homogeneous of degree 1/2. $\therefore$ By Euler's theorem

$$x\frac{\partial}{\partial x}\cos u + y\frac{\partial}{\partial y}\cos u = \frac{1}{2}\cos u$$

or $$-x\sin u\frac{\partial u}{\partial x} - y\sin u\frac{\partial u}{\partial y} = \frac{1}{2}\cos u.$$

or $$x\frac{\partial u}{\partial x} + y\frac{\partial u}{\partial y} = -\frac{1}{2}\frac{\cos u}{\sin u} \quad \text{or} \quad x\frac{\partial u}{\partial x} + y\frac{\partial u}{\partial y} + \frac{1}{2}\cot u = 0.$$ **Hence the result.**

(b)
$$\sin u = \left(\frac{x^{1/3}+y^{1/3}}{x^{1/2}-y^{1/2}}\right)^{1/2} = \frac{(x^{1/3})^{1.2}\left[1+\left(\dfrac{y}{x}\right)^{1/3}\right]^{1/2}}{(x^{1/2})^{1/2}\left[1-\left(\dfrac{y}{x}\right)^{1/2}\right]^{1/2}}$$

Let $t = \sin u$, then $$t = x^{-1/12}\left[\frac{1+\left(\dfrac{y}{x}\right)^{1/3}}{1-\left(\dfrac{y}{x}\right)^{1/2}}\right]^{1/2} = x^{-1/12}\phi\left(\frac{y}{x}\right) \text{ (say)}$$

where $\phi\left(\dfrac{y}{x}\right)$ is of degree zero.

Thus, t is a homogeneous function in x and y of degree $-1/12$. Therefore, by Euler's theorem

$$x\frac{\partial t}{\partial x} + y\frac{\partial t}{\partial y} = -\frac{1}{12}t$$

Since $t = \sin u$, we have $$x\cos u\frac{\partial u}{\partial x} + y\cos u\frac{\partial u}{\partial y} = -\frac{1}{12}\sin u$$

or $$x\frac{\partial u}{\partial x} + y\frac{\partial u}{\partial y} = -\frac{1}{12}\tan u$$ **Hence Proved.**

EXAMPLE 3.17. If $u = \sin^{-1}\left(\dfrac{x^2+y^2}{x+y}\right)$ prove that $x\dfrac{\partial u}{\partial x}+y\dfrac{\partial u}{\partial y} = \tan u.$

SOLUTION: Given $u = \sin^{-1}\left(\dfrac{x^2+y^2}{x+y}\right)$ we have $\sin u = \dfrac{x^2+y^2}{x+y}.$

Clearly, $\sin u$ is a homogeneous function in x and y of degree one, hence by Euler's theorem

$$x\frac{\partial}{\partial x}\sin u+y\frac{\partial}{\partial y}\sin u = 1 \sin u$$

or $$x\cos u\frac{\partial u}{\partial x}+y\cos u\frac{\partial u}{\partial y} = \sin u \quad \text{or} \quad x\frac{\partial u}{\partial x}+y\frac{\partial u}{\partial y} = \tan u.$$ **Hence the result.**

EXAMPLE 3.18. If $u = \tan^{-1}\left(\dfrac{x^3+y^3}{x-y}\right)$ prove that

(i) $x\dfrac{\partial u}{\partial x}+y\dfrac{\partial u}{\partial y} = \sin 2u$

(ii) $x^2\dfrac{\partial^2 u}{\partial x^2}+2xy\dfrac{\partial^2 u}{\partial x\partial y}+y^2\dfrac{\partial^2 u}{\partial y^2} = \sin 4u-\sin 2u.$ **(UPTU 2001 Q.B.)**

SOLUTION: Give that $u = \tan^{-1}\left(\dfrac{x^3+y^3}{x-y}\right)$ we have $\tan u = \dfrac{x^3+y^3}{x-y} = f(u).$

Clearly, $\tan u$ is a homogeneous function in x and y of degree 2, hence by Euler's theorem

$$x\frac{\partial u}{\partial x}+y\frac{\partial u}{\partial y} = \frac{nf(u)}{f'(u)} = \frac{2\tan u}{\sec^2 u} = \sin 2u = g(u), \text{ say.}$$ **Hence Proved.**

Also by Euler's theorem (Deduction III), we have

$$x^2\frac{\partial^2 u}{\partial x^2}+2xy\frac{\partial^2}{\partial x\partial y}+y^2\frac{\partial^2 u}{\partial y^2} = g(u)\,[.g'(u)-1]$$

$$= \sin 2u\ [2\cos 2u-1] = 2\sin 2u\cos 2u - \sin 2u$$

$$= \sin 4u - \sin 2u.$$ **Hence the result.**

EXAMPLE 3.19. If $u = \sin^{-1}\left(\dfrac{x+y}{\sqrt{x}+\sqrt{y}}\right)$ prove that

$$x^2\frac{\partial^2 u}{\partial x^2}+2xy\frac{\partial^2 u}{\partial x\partial y}+y^2\frac{\partial^2 u}{\partial y^2} = -\frac{\sin u\cos 2u}{4\cos^3 u}.$$ **(UPTU 2001 Q.B.)**

SOLUTION: Since $u = \sin^{-1}\left(\dfrac{x+y}{\sqrt{x}+\sqrt{y}}\right)$ we have $\sin u = \dfrac{x+y}{\sqrt{x}+\sqrt{y}} = f(u)$, say.

Clearly $\sin u$ is homogeneous in x and y of degree 1/2 hence by Euler's theorem we have

$$x\frac{\partial u}{\partial x}+y\frac{\partial u}{\partial y} = \frac{nf(u)}{f'(u)} = \frac{1}{2}\frac{\sin u}{\cos u} = \frac{1}{2}\tan u = g\ (u), \text{ say}$$

Also by Euler's theorem (Deduction III), we know that

$$x^2\frac{\partial^2 u}{\partial x^2}+2xy\frac{\partial^2 u}{\partial x\partial y}+y^2\frac{\partial^2 u}{\partial y^2} = g(u)[g'(u)-1] = \frac{1}{2}\tan u\left[\frac{1}{2}\sec^2 u-1\right] = \frac{1}{4}\tan u(\sec^2 u-2)$$

$$= \frac{1}{4}\frac{\sin u}{\cos u}\left[\frac{1}{\cos^2 u}-2\right] = -\frac{1}{4}\frac{\sin u}{\cos u}\cdot\frac{2\cos^2 u-1}{\cos^2 u}$$

$$= \frac{-1}{4}\frac{\sin u\cos 2u}{\cos^3 u}.$$ **Hence the result.**

EXAMPLE 3.20. If $u = \sin^{-1}(x^2+y^2)^{1/5}$, prove that

$$x^2 u_{xx} + 2xy\, u_{xy} + y^2 u_{yy} = \frac{2}{25}\tan u\,(2\tan^2 u-3).$$

SOLUTION: Since $u = \sin^{-1}(x^2+y^2)^{1/5}$ we have $\sin u = (x^2+y^2)^{1/5} = f(u)$, say.

Since $\sin u$ is homogeneous in x and y of degree 2/5, we have, by Euler's theorem

$$xu_x + yu_y = \frac{nf(u)}{f'(u)} = \frac{2}{5}\frac{\sin u}{\cos u} = \frac{2}{5}\tan u = g(u).$$

Again by Euler's theorem (Deduction III), we have

$$x^2 u_{xx} + 2xy\, u_{xy} + y^2 u_{yy} = g(u)\,[g'(u)-1]$$

$$= \frac{2}{5}\tan u\left[\frac{2}{5}\sec^2 u-1\right] = \frac{2}{25}\tan u\,[2\tan^2 u-3].$$ **Hence the result.**

EXAMPLE 3.21. If $u = \dfrac{(x^2+y^2)^n}{2n(2n-1)} + x f\left(\dfrac{y}{x}\right) + \phi\left(\dfrac{x}{y}\right)$,

evaluate $x^2\dfrac{\partial^2 u}{\partial x^2}+2xy\dfrac{\partial^2 u}{\partial x\partial y}+y^2\dfrac{\partial^2 u}{\partial y^2}$.

SOLUTION: Let $u = u_1 + u_2 + u_3$ where $u_1 = \dfrac{(x^2+y^2)^n}{2n(2n-1)}$, $u_2 = x f\left(\dfrac{y}{x}\right)$, $u_3 = \phi\left(\dfrac{x}{y}\right)$.

Clearly u_1 is homogeneous in x and y of degree $2n$, u_2 is homogeneous in x and y of degree 1 and u_3 is homogenous in x and y of degree 0. Thus applying Euler's theorem separately to u_1, u_2, u_3, we get

$$x^2\frac{\partial^2 u_1}{\partial x^2}+2xy\frac{\partial^2 u_1}{\partial x\partial y}+y^2\frac{\partial^2 u_1}{\partial y^2} = 2n(2n-1)\,u_1$$

$$x^2\frac{\partial^2 u_2}{\partial x^2}+2xy\frac{\partial^2 u_2}{\partial x\partial y}+y^2\frac{\partial^2 u_2}{\partial y^2} = 1(1-1)u_2 = 0$$

$$x^2\frac{\partial^2 u_3}{\partial x^2}+2xy\frac{\partial^2 u_3}{\partial x\partial y}+y^2\frac{\partial^2 u_3}{\partial y^2} = 0(0-1)\,u_3 = 0$$

Adding these, gives

$$x^2\frac{\partial^2 u}{\partial x^2}+2xy\frac{\partial^2 u}{\partial x\partial y}+y^2\frac{\partial^2 u}{\partial y^2} = 2n(2n-1)\,u_1 + 0 + 0$$

$$= 2n(2n-1)\frac{(x^2+y^2)^n}{2n(2n-1)} = (x^2+y^2)^n.$$ **Ans.**

EXAMPLE 3.22. If $u = x^n f\left(\frac{y}{x}\right) + y^{-n} \phi\left(\frac{x}{y}\right)$, show that

$$x^2 u_{xx} + 2xy\, u_{xy} + y^2 u_{yy} + xu_x + yu_y = n^2 u.$$

SOLUTION: Let $u = u_1 + u_2$ where

$$u_1 = x^n f\left(\frac{y}{x}\right) \quad \text{and} \quad u_2 = y^{-n} \phi\left(\frac{x}{y}\right).$$

Here u_1 is homogeneous in x and y of degree n while u_2 is homogenous in x and y of degree $-n$. Making use of Euler's theorem to u_1 and u_2 separately, we get

$$x\frac{\partial u_1}{\partial x} + y\frac{\partial u_1}{\partial y} = nu_1, \qquad \text{...(1)}$$

$$x\frac{\partial u_2}{\partial x} + y\frac{\partial u_2}{\partial y} = -nu_2. \qquad \text{...(2)}$$

And
$$x^2\frac{\partial^2 u_1}{\partial x^2} + 2xy\frac{\partial^2 u_1}{\partial x \partial y} + y^2\frac{\partial^2 u_1}{\partial y^2} = n(n-1)u_1, \qquad \text{... (3)}$$

$$x^2\frac{\partial^2 u_2}{\partial x^2} + 2xy\frac{\partial^2 u_2}{\partial x \partial y} + y^2\frac{\partial^2 u_2}{\partial y^2} = -n(-n-1)u_2. \qquad \text{... (4)}$$

Adding (1), (2), (3) and (4), gives

$$x^2\frac{\partial^2}{\partial x^2}(u_1+u_2) + 2xy\frac{\partial^2}{\partial x \partial y}(u_1+u_2) + y^2\frac{\partial^2}{\partial y^2}(u_1+u_2) + x\frac{\partial}{\partial x}(u_1+u_2) + y\frac{\partial}{\partial y}(u_1+u_2)$$

$$= n(n-1)u_1 + (-n)(-n-1)u_2 + nu_1 - nu_2$$

or
$$x^2\frac{\partial^2 u}{\partial x^2} + 2xy\frac{\partial^2 u}{\partial x \partial y} + y^2\frac{\partial^2 u}{\partial y^2} + x\frac{\partial u}{\partial x} + y\frac{\partial u}{\partial y} = u_1(n^2 - n + n) + u_2(n^2 + n - n)$$

$$= n^2(u_1 + u_2) = n^2 u.$$ **Hence Proved.**

EXAMPLE 3.23. If $u = \operatorname{cosec}^{-1}\left(\sqrt{\frac{x^{1/2}+y^{1/2}}{x^{1/3}+y^{1/3}}}\right)$ evaluate $x^2 u_{xx} + 2xy\, u_{xy} + y^2 u_{yy}$.

SOLUTION: From the given equation we can write

$$\operatorname{cosec} u = \sqrt{\frac{x^{1/2}+y^{1/2}}{x^{1/3}+y^{1/3}}} = z, \text{ say}$$

or
$$z = \sqrt{\frac{x^{1/2}\left(1+(y/x)^{1/2}\right)}{x^{1/3}\left(1+(y/x)^{1/3}\right)}} = x^{1/12}\sqrt{\frac{1+(y/x)^{1/2}}{1+(y/x)^{1/3}}}$$

Thus z is a homogeneous function in x and y of degree $\frac{1}{12}$ and z is a function of u as $z = f(u) = \operatorname{cosec} u$ therefore by Euler's theorem (Deduction III), we have

$$x^2 \frac{\partial^2 u}{\partial x^2} + 2xy \frac{\partial^2 u}{\partial x \partial y} + y \frac{\partial^2 u}{\partial y^2} = g(u)\,[\,g'(u) - 1]. \qquad \text{... (1)}$$

where $g(u) = \dfrac{n f(u)}{f'(u)} = \left(\dfrac{1}{12}\right)\left(\dfrac{\operatorname{cosec} u}{-\operatorname{cosec} u \cot u}\right) = -\dfrac{1}{12} \tan u.$

As such $x^2 u_{xx} + 2xy\, u_{xy} + y^2 u_{yy} = \left(-\dfrac{1}{12} \tan u\right)\left(-\dfrac{1}{12} \sec^2 u - 1\right)$

$$= \frac{1}{12} \tan u \left(\frac{13}{12} + \frac{1}{12} \tan^2 u\right). \qquad \textbf{Ans.}$$

EXAMPLE 3.24. If $u = \sin^{-1}\left(\dfrac{x^3 + y^3 + z^3}{ax + by + cz}\right)$ prove that $x\dfrac{\partial u}{\partial x} + y\dfrac{\partial u}{\partial y} + z\dfrac{\partial u}{\partial z} = 2 \tan u.$

(UPTU 2002-03)

SOLUTION: Here u is not homogeneous function

but $f(u) = \sin u = \dfrac{x^3 + y^3 + z^3}{ax + by + cz}$ is homogeneous in x, y, z of degree 2.

$$\therefore\ x\frac{\partial u}{\partial x} + y\frac{\partial u}{\partial y} + z\frac{\partial u}{\partial z} = \frac{2f(u)}{f'(u)} = \frac{2 \sin u}{\cos u} = 2 \tan u. \qquad \textbf{Hence Proved}$$

EXAMPLE 3.25. If $z = x^2 \tan^{-1}\left(\dfrac{y}{x}\right) - y^2 \tan^{-1}\left(\dfrac{x}{y}\right)$, then evaluate $x^2 z_{xx} + 2xy\, z_{xy} + y^2 z_{yy}$.

SOLUTION: Here $\tan^{-1}(y/x)$ and $\tan^{-1}(x/y)$ are both of degree zero and hence behave like constants. Therefore z is homogeneous function in x and y of degree two. Applying Euler's theorem, we get

$$x^2 z_{xx} + 2xy\, z_{xy} + y^2 z_{yy} = 2(2 - 1)\, z = 2z. \qquad \textbf{Ans.}$$

EXAMPLE 3.26. If $u = \log\left(\dfrac{x^4 + y^4}{x + y}\right)$ show that $x\dfrac{\partial u}{\partial x} + y\dfrac{\partial u}{\partial y} = 3.$ **(UPTU 2000-01)**

SOLUTION: Given $e^u = \dfrac{x^4 + y^4}{x + y}$. Here e^u is homogeneous function of x and y of degree 3.

Hence, using Euler' theorem on e^u, we get

$$x\frac{\partial}{\partial x} e^u + y\frac{\partial}{\partial y} e^u = 3e^u \quad \text{or} \quad xe^u \frac{\partial u}{\partial x} + ye^u \frac{\partial u}{\partial y} = 3e^u \;\Rightarrow\; x\frac{\partial u}{\partial x} + y\frac{\partial u}{\partial y} = 3. \qquad \textbf{Hence Proved}$$

EXAMPLE 3.27. If $u(x, y) = \sin^{-1}\left(\frac{x}{y}\right) + \tan^{-1}\left(\frac{y}{x}\right)$, prove that $x\frac{\partial u}{\partial x} + y\frac{\partial u}{\partial y} = 0.$ **(UPTU 2006–07)**

SOLUTION: Let $u = A(x,y) + B(x,y)$ where $A = \sin^{-1}\left(\frac{x}{y}\right)$ and $B = \tan^{-1}\left(\frac{y}{x}\right)$

Hence $\sin A = \frac{x}{y}$ and $\tan B = \frac{y}{x}$.

Now, x/y is homogeneous in x and y of degree zero and y/x is also homogeneous in x and y of degree zero. Therefore, by Euler's theorem

$$x\frac{\partial}{\partial x}\sin A + y\frac{\partial}{\partial y}\sin A = 0 \quad \text{and} \quad x\frac{\partial}{\partial x}\tan B + y\frac{\partial}{\partial y}\tan B = 0$$

$$\Rightarrow \quad x\cos A\frac{\partial A}{\partial x} + y\cos A\frac{\partial A}{\partial y} = 0 \quad \text{and} \quad x\sec^2 B\frac{\partial B}{\partial x} + y\sec^2 B\frac{\partial B}{\partial y} = 0$$

or $$x\frac{\partial A}{\partial x} + y\frac{\partial A}{\partial y} = 0 \quad \text{and} \quad x\frac{\partial B}{\partial x} + y\frac{\partial B}{\partial y} = 0.$$

Adding these two, we get

$$x\frac{\partial (A+B)}{\partial x} + y\frac{\partial}{\partial y}(A+B) = 0 \quad \text{or} \quad x\frac{\partial u}{\partial x} + y\frac{\partial u}{\partial y} = 0.$$ **Hence Proved**

EXAMPLE 3.28. If $u = \sin^{-1}\left(\frac{x+2y+3z}{\sqrt{x^8+y^8+z^8}}\right)$ show that

$$x\frac{\partial u}{\partial x} + y\frac{\partial u}{\partial y} + z\frac{\partial u}{\partial z} + 3\tan u = 0.$$ **(UPTU 2003–04)**

SOLUTION: Here $\sin u = \frac{x+2y+3z}{\sqrt{x^8+y^8+z^8}}$ which is homogeneous in x, y and z of degree (-3).

Applying Euler's theorem on $\sin u$, we get

$$x\frac{\partial}{\partial x}(\sin u) + y\frac{\partial}{\partial y}(\sin u) + z\frac{\partial}{\partial z}(\sin u) = (-3)\sin u$$

or $$x\cos u\frac{\partial u}{\partial x} + y\cos u\frac{\partial u}{\partial y} + z\cos u\frac{\partial u}{\partial z} = -3\sin u$$

or $$x\frac{\partial u}{\partial x} + y\frac{\partial u}{\partial y} + z\frac{\partial u}{\partial z} + 3\tan u = 0$$ **Hence Proved**

EXAMPLE 3.29. If $u = \cos^{-1}\left(\frac{x+y}{\sqrt{x}+\sqrt{y}}\right)$ show that

$$x\frac{\partial u}{\partial x} + y\frac{\partial u}{\partial y} + \frac{1}{2}\cot u = 0.$$ **(UPTU 2003–04)**

SOLUTION: We have $\cos u = \frac{x+y}{\sqrt{x}+\sqrt{y}}$. Here $\cos u$ is homogeneous in x and y of degree $\frac{1}{2}$.

Hence, by Euler's theorem

$$x\frac{\partial u}{\partial x} + y\frac{\partial u}{\partial y} = \frac{n f(u)}{f'(u)} = \frac{1}{2}\frac{\cos u}{(-\sin u)} = -\frac{1}{2}\cot u.$$ **Hence Proved.**

EXERCISE 3B

1. If $u = \dfrac{x^3 + y^3}{y\sqrt{x}} + \dfrac{1}{x^7}\sin^{-1}\left(\dfrac{x^2+y^2}{x^2+2xy}\right)$ find the value of $x^2 u_{xx} + 2xy\, u_{xy} + y^2 u_{yy} + x\, u_x + y\, u_y$ at the point (1, 2).

2. If $u = \sin^{-1}\left(\dfrac{\sqrt{x}-\sqrt{y}}{\sqrt{x}+\sqrt{y}}\right)$ show that $\dfrac{\partial u}{\partial x} = \dfrac{-y}{x}\dfrac{\partial u}{\partial y}$.

3. If $z = x^3 e^{-x/y}$ prove that $x^2 z_{xx} + 2xy\, z_{xy} + y^2 z_{yy} = 6z$.

4. If $u = \sin^{-1}\left(\dfrac{x^{1/4}+y^{1/4}}{x^{1/6}+y^{1/6}}\right)$, evaluate $x^2\dfrac{\partial^2 u}{\partial x^2} + 2xy\dfrac{\partial^2 u}{\partial x \partial y} + y^2\dfrac{\partial^2 u}{\partial y^2}$.

5. If $y = x \sin u$, find the value of $x^2 u_{xx} + 2xy\, u_{xy} + y^2 u_{yy}$.

6. If $f(x, y) = \dfrac{1}{x^2} + \dfrac{1}{xy} + \dfrac{\log x - \log y}{x^2+y^2}$, show that $x\dfrac{\partial f}{\partial x} + y\dfrac{\partial f}{\partial y} + 2f = 0$.

7. Determine the value of $x^2 u_{xx} + 2xy\, u_{xy} + y^2 u_{yy}$ if $u = \sin^{-1}(x^3 + y^3)^{2/5}$.

8. If $u = \sec^{-1}\left(\dfrac{x^3 - y^3}{x - y}\right)$, then evaluate $x^2 u_{xx} + 2xy\, u_{xy} + y^2 u_{yy}$.

9. If $u = \dfrac{x^2 y^2 z^2}{x^2+y^2+z^2} + \cos\dfrac{2xy+yz+zx}{x^2+y^2+z^2}$ evaluate $x\, u_x + y\, u_y + z\, u_z$.

10. If $u = \log\left(\dfrac{x+y}{\sqrt{x^2+y^2}}\sin^{-1}\left(\dfrac{x+y}{\sqrt{x}+\sqrt{y}}\right)\right)$, show that

$$x^2\frac{\partial^2 u}{\partial x^2} + 2xy\frac{\partial^2 u}{\partial x \partial y} + y^2\frac{\partial^2 u}{\partial y^2} = \frac{1}{2}\tan\theta\left(\frac{1}{2}\sec^2\theta - 1\right)\log\left(\frac{x+y}{\sqrt{x^2+y^2}}\right)$$

where $\theta = \sin^{-1}\dfrac{x+y}{\sqrt{x}+\sqrt{y}}$.

11. If $u = \dfrac{1}{r}f(\theta)$ where $x = r\cos\theta$, $y = r\sin\theta$, then evaluate $x\dfrac{\partial u}{\partial x} + y\dfrac{\partial u}{\partial y}$.

12. If $u = \log\sqrt{x^2+y^2}$, evaluate $x^2\dfrac{\partial^2 u}{\partial x^2} + 2xy\dfrac{\partial^2 u}{\partial x \partial y} + y^2\dfrac{\partial^2 u}{\partial y^2}$.

13. If $x = e^u \tan v$, $y = e^u \sec v$ find the value of

$$\left(x\frac{\partial u}{\partial x} + y\frac{\partial u}{\partial y}\right)\left(x\frac{\partial v}{\partial x} + y\frac{\partial v}{\partial y}\right)$$

14. If $u = \log\left(\dfrac{x^2+y^2}{x+y}\right)$ prove that $x\dfrac{\partial u}{\partial x} + y\dfrac{\partial u}{\partial y} = 1$. **(UPTU 2008-09)**

15. Verify Euler's theorem for $z = \dfrac{x^{1/3}+y^{1/3}}{x^{1/2}+y^{1/2}}$. **(UPTU 2009-10)**

❑❑❑

TOTAL DERIVATIVE

If $u = f(x, y)$ is a function of two independent variables x and y, and x and y are separately functions of a single independent variable t then u can be expressed as a function of t alone and then we can find the ordinary derivative $\frac{du}{dt}$ which is called the **total derivative** of u w.r.t. t, to distinguish it from the partial derivatives $\frac{\partial u}{\partial x}$ and $\frac{\partial u}{\partial y}$.

Let us be interested in finding $\frac{du}{dt}$ without actually substituting the values of x and y in terms of t in $f(x, y)$. We derive below the relation between the total derivative $\frac{du}{dt}$ and the partial derivatives $\frac{\partial u}{\partial x}$ and $\frac{\partial u}{\partial y}$. When t is given an increment δt, suppose x and y get increments δx and δy respectively and this, in turn, causes u to get an increment of δu. In other words, when t becomes $t + \delta t$ let x become $x + \delta x$ and y become $y + \delta y$ and consequently, u becomes $u + \delta u$. Thus, δx and δy both tend to 0 as δt tends to 0 and then by definition,

$$\frac{du}{dt} = \lim_{\delta t \to 0} \frac{f(x+\delta x, y+\delta y) - f(x, y)}{\delta t}$$

$$= \lim_{\delta t \to 0} \frac{f(x+\delta x, y+\delta y) - f(x, y+\delta y) + f(x, y+\delta y) - f(x, y)}{\delta t}$$

$$= \lim_{\delta t \to 0} \frac{f(x+\delta x, y+\delta y) - f(x, y+\delta y)}{\delta x} \cdot \frac{\delta x}{\delta t} + \lim_{\delta t \to 0} \frac{f(x, y+\delta y) - f(x, y)}{\delta y} \cdot \frac{\delta y}{\delta t}$$

$$= \lim_{\delta x \to 0} \frac{f(x+\delta x, y+\delta y) - f(x, y+\delta y)}{\delta x} \cdot \left(\frac{dx}{dt}\right) + \lim_{\delta y \to 0} \frac{f(x, y+\delta y) - f(x, y)}{\delta y} \cdot \left(\frac{dy}{dt}\right)$$

$$= \lim_{\delta x \to 0} \frac{\delta f(x, y+\delta y)}{\delta x} \cdot \frac{dx}{dt} + \frac{\partial f}{\partial y} \cdot \frac{dy}{dt}$$

$$= \frac{\partial f}{\partial x}\frac{dx}{dt} + \frac{\partial f}{\partial y}\frac{dy}{dt} = \frac{\partial u}{\partial x}\frac{dx}{dt} + \frac{\partial u}{\partial y}\frac{dy}{dt} \quad \text{since } u = f(x, y).$$

Thus, we have

$$\boxed{\frac{du}{dt} = \frac{\partial u}{\partial x}\frac{dx}{dt} + \frac{\partial u}{\partial y}\frac{dy}{dt}} \qquad \text{...(1)}$$

and in terms of differentials this result can be better written as

$$\boxed{du = \frac{\partial u}{\partial x}\, dx + \frac{\partial u}{\partial y}\, dy} \qquad \text{...(2)}$$

where du is called the **total derivative of u.**

This is a very important result and will be used very frequently in partial differentiation.

On extending this result to functions of three variables, we have

$$\frac{du}{dt} = \frac{\partial u}{\partial x}\frac{dx}{dt} + \frac{\partial u}{\partial y}\frac{dy}{dt} + \frac{\partial u}{\partial z}\frac{dz}{dt}$$

or
$$du = \frac{\partial u}{\partial x}dx + \frac{\partial u}{\partial y}dy + \frac{\partial u}{\partial z}dz.$$

Some Deductions

If $f(x, y) = 0$ is an implicit relation between x and y, *i.e.*, y is an implicit function of x, then we have

$$df = \frac{\partial f}{\partial x}dx + \frac{\partial f}{\partial y}dy = 0 \quad \text{... (1)}$$

Therefore, $\frac{dy}{dx}$ interms of partial derivatives of f w.r.t. x and y, is given by

$$\frac{dy}{dx} = -\frac{\partial f}{\partial x}\Big/\frac{\partial f}{\partial y} \quad \text{... (2)}$$

In this case, we can also calculate $\frac{d^2y}{dx^2}$ in terms of partial derivatives. But, before that let us introduce the following conventional notations:

$$\frac{\partial f}{\partial x} = p,\ \frac{\partial f}{\partial y} = q,\ \frac{\partial^2 f}{\partial x^2} = \frac{\partial p}{\partial x} = r,\ \frac{\partial^2 f}{\partial y^2} = \frac{\partial q}{\partial y} = t,\ \frac{\partial^2 f}{\partial x \partial y} = \frac{\partial p}{\partial y} = \frac{\partial q}{\partial x} = s.$$

Thus, we can write $\frac{dy}{dx} = \frac{-p}{q}.$

Hence
$$\frac{d^2y}{dx^2} = -\frac{q\frac{dp}{dx} - p\frac{dq}{dx}}{q^2} \quad \text{... (3)}$$

Now, using (3) we can write

$$\frac{dp}{dx} = \frac{\partial p}{\partial x}.1 + \frac{\partial p}{\partial y}.\frac{dy}{dx} = r + s\left(\frac{-p}{q}\right) = \frac{qr - ps}{q}$$

and
$$\frac{dq}{dx} = \frac{\partial q}{\partial x} + \frac{\partial q}{\partial y}\frac{dy}{dx} = s + t\left(\frac{-p}{q}\right) = \frac{qs - pt}{q}$$

Putting the values of $\frac{dp}{dx}$ and $\frac{dq}{dx}$ in (3), we get

$$\frac{d^2y}{dx^2} = -\frac{1}{q^2}\left[q.\frac{qr - ps}{q} - p.\frac{qs - pt}{q}\right] = -\frac{1}{q^3}\left[q^2r - 2\,pqs + p^2t\right]$$

Thus, we have $\frac{d^2y}{dx^2} = -\frac{[q^2r - 2pqs - p^2t]}{q^3}$

CHANGE OF INDEPENDENT VARIABLES

Let $u = f(x, y)$ where $x = f_1(t_1, t_2)$ and $y = f_2(t_1, t_2)$ ... (1)

It is frequently necessiated to change the expressions involving $u, x, y, \frac{\partial u}{\partial x}, \frac{\partial u}{\partial y}$ to the expressions involving $u, t_1, t_2, \frac{\partial u}{\partial t_1}, \frac{\partial u}{\partial t_2}$, etc. In this constext, the earlier result

$\frac{du}{dt} = \frac{\partial u}{\partial x}\frac{dx}{dt} + \frac{\partial u}{\partial y}\frac{dy}{dt}$ can now be extended to the following

$$\boxed{\frac{\partial u}{\partial t_1} = \frac{\partial u}{\partial x}\frac{\partial x}{\partial t_1} + \frac{\partial u}{\partial y}\cdot\frac{\partial y}{\partial t_1}} \quad \text{and} \quad \boxed{\frac{\partial u}{\partial t_2} = \frac{\partial u}{\partial x}\frac{\partial x}{\partial t_2} + \frac{\partial u}{\partial y}\cdot\frac{\partial y}{\partial t_2}}.$$

Next, if instead of (1) we are given that

$$u = u(t_1, t_2), \quad t_1 = \phi_1(x, y) \quad \text{and} \quad t_2 = \phi_2(x, y).$$

Then the equations of transformation become

$$\frac{\partial u}{\partial x} = \frac{\partial u}{\partial t_1}\cdot\frac{\partial t_1}{\partial x} + \frac{\partial u}{\partial t_2}\cdot\frac{\partial t_2}{\partial x}$$

and
$$\frac{\partial u}{\partial y} = \frac{\partial u}{\partial t_1}\cdot\frac{\partial t_1}{\partial y} + \frac{\partial u}{\partial t_2}\cdot\frac{\partial t_2}{\partial y}.$$

Conversion from cartesian to polar coordinates and *vice versa* are frequently encountered examples. Thus, if $u = f(x, y)$ where $x = r\cos\theta$, $y = r\sin\theta$, then

$$\boxed{\frac{\partial u}{\partial r} = \frac{\partial u}{\partial x}\frac{\partial x}{\partial r} + \frac{\partial u}{\partial y}\frac{\partial y}{\partial r} \quad \text{and} \quad \frac{\partial u}{\partial \theta} = \frac{\partial u}{\partial x}\frac{\partial x}{\partial \theta} + \frac{\partial u}{\partial y}\cdot\frac{\partial y}{\partial \theta}}$$

and other way round, we have the formulae

$$\boxed{\frac{\partial u}{\partial x} = \frac{\partial u}{\partial r}\frac{\partial r}{\partial x} + \frac{\partial u}{\partial \theta}\frac{\partial \theta}{\partial x} \quad \text{and} \quad \frac{\partial u}{\partial y} = \frac{\partial u}{\partial r}\frac{\partial r}{\partial y} + \frac{\partial u}{\partial \theta}\frac{\partial \theta}{\partial y}}.$$

EXAMPLE 3.30. If $(\cos x)^y - (\sin y)^x = 0$, find $\frac{dy}{dx}$.

SOLUTION: The given relation is $(\cos x)^y = (\sin y)^x$

Taking logarithm on both sides, gives $y \log(\cos x) = x \log(\sin y)$

Now, let us take $f(x, y) = y \log(\cos x) - x \log(\sin y) = 0$

So, y can be taken as an imlicit function of x or *vice versa*.

Here $\dfrac{\partial f}{\partial x} = \dfrac{y(-\sin x)}{\cos x} - 1.\log(\sin y) = -(y \tan x + \log \sin y)$

and $\dfrac{\partial f}{\partial y} = \log \cos x \,.\, 1 - \dfrac{x \cos y}{\sin y} = \log \cos x - x \cot y.$

Therefore, $\dfrac{dy}{dx} = -\dfrac{\partial f}{\partial x} \Big/ \dfrac{\partial f}{\partial y} = \dfrac{(y \tan x + \log \sin y)}{\log \cos x - x \cot y}.$ **Ans.**

EXAMPLE 3.31. If $x^n + y^n = a^n$, find $\dfrac{d^2y}{dx^2}$.

SOLUTION: Let $f(x, y) = x^n + y^n - a^n = 0$

then $\dfrac{\partial f}{\partial x} = p = n\, x^{n-1}$, $\dfrac{\partial f}{\partial y} = q = n\, y^{n-1}$,

$\dfrac{\partial^2 f}{\partial x^2} = r = n\,(n-1)\, x^{n-2}$, $s = \dfrac{\partial^2 f}{\partial y \partial x} = \dfrac{\partial p}{\partial y} = 0$ and $\dfrac{\partial^2 f}{\partial y^2} = t = n\,(n-1)\, y^{n-2}$.

Since, $f(x, y) = 0$ we can take y as an implicit function of x

$$\therefore \quad \frac{d^2y}{dx^2} = -\frac{1}{q^3}\,[\,q^2 r - 2pqs + p^2 t\,]$$

$$= -\frac{1}{n^3 y^{3n-3}}\,[n^2 y^{2n-2} \cdot n\,(n-1)\, x^{n-2} - 2\, n^2 x^{n-1} y^{n-1} \cdot 0 + n^2 x^{2n-2} \cdot n\,(n-1)\, y^{n-2}]$$

$$= -\frac{n^3 (n-1)}{n^3 y^{3n-3}}\,[\,x^{n-2} y^{2n-2} + y^{n-2} x^{2n-2}]$$

$$= -\frac{(n-1)}{y^{3n-3}}\, x^{n-2} y^{n-2}\,[\,y^n + x^n\,] = -\frac{(n-1)\, x^{n-2} \,.\, a^n}{y^{2n-1}}. \qquad \textbf{Ans.}$$

EXAMPLE 3.32. If $f(x, y) = 0$ and $\phi(y, z) = 0$, show that

$$\frac{\partial f}{\partial y} \cdot \frac{\partial \phi}{\partial z}\frac{dz}{dx} = \frac{\partial f}{\partial x} \cdot \frac{\partial \phi}{\partial y}.$$ **(UPTU 2005)**

SOLUTION: Since $f(x, y) = 0$ we have

$$df = \frac{\partial f}{\partial x} dx + \frac{\partial f}{\partial y} dy = 0 \qquad \therefore \quad \frac{dy}{dx} = -\frac{\partial f}{\partial x} \Big/ \frac{\partial f}{\partial y}.$$

And since $\phi(y, z) = 0$, we have

$$d\phi = \frac{\partial \phi}{\partial y} dy + \frac{\partial \phi}{\partial z} dz = 0 \qquad \therefore \quad \frac{dz}{dy} = -\frac{\partial \phi}{\partial y} \Big/ \frac{\partial \phi}{\partial z}$$

$$\therefore \quad \frac{dy}{dx} \cdot \frac{dz}{dy} = (-1)^2 \frac{\dfrac{\partial f}{\partial x}\,\dfrac{\partial \phi}{\partial y}}{\dfrac{\partial f}{\partial y}\,\dfrac{\partial \phi}{\partial z}} \quad \text{or} \quad \frac{\partial f}{\partial y} \cdot \frac{\partial \phi}{\partial z}\frac{dz}{dx} = \frac{\partial f}{\partial x}\,\frac{\partial \phi}{\partial y}.$$ **Hence Proved.**

EXAMPLE 3.33. If $y^x + x^y = (x + y)^{(x + y)}$ find $\frac{dy}{dx}$.

SOLUTION: Let $f(x, y) = y^x + x^y - (x + y)^{(x + y)}$

$$\therefore \quad \frac{\partial f}{\partial x} = y^x \log y + y \cdot x^{y-1} - (x + y)^{x+y} [1 + \log (x + y)]$$

and $$\frac{\partial f}{\partial y} = x y^{x-1} + x^y \log x - (x + y)^{x+y} [1 + \log (x + y)]$$

Hence $$\frac{dy}{dx} = -\frac{\partial f/\partial x}{\partial f/\partial y} = -\left[\frac{y^x \log y + y x^{y-1} - (x+y)^{x+y}(1+\log(x+y))}{x y^{x-1} + x^y \log x - (x+y)^{x+y}(1+\log(x+y))}\right].$$ **Ans.**

EXAMPLE 3.34. If $z = xy f(y/x)$ and z is constant, show that

$$\frac{f'(y/x)}{f(y/x)} = \frac{x\left(y + x\frac{dy}{dx}\right)}{y\left(y - x\frac{dy}{dx}\right)}.$$

SOLUTION: Since z is constant we write $xy f(y/x) = z = \text{constant}$

$$\therefore \quad \frac{dy}{dx} = \frac{-\frac{\partial}{\partial x}[xyf(y/x)]}{\frac{\partial}{\partial y}[xyf(y/x)]} = \frac{-[yf(y/x) + xyf'(y/x)(-y/x^2)]}{[xf(y/x).xyf'(y/x)(1/x)]}$$

or $$\frac{x\frac{dy}{dx}}{y} = \frac{\frac{y}{x}f'(y/x) - f(y/x)}{\frac{y}{x}f'(y/x) + f(y/x)}$$

Applying componendo and dividendo here, we get

$$\frac{y + x\frac{dy}{dx}}{y - x\frac{dy}{dx}} = \frac{2\frac{y}{x}f'(y/x)}{2f(y/x)} = \frac{yf'(y/x)}{xf(y/x)}$$

or $$\frac{f'(y/x)}{f(y/x)} = \frac{x\left(y + x\frac{dy}{dx}\right)}{y\left(y - x\frac{dy}{dx}\right)}.$$ **Hence the result.**

EXAMPLE 3.35. If $u = f(x - y, y - z, z - x)$ show that $\frac{\partial u}{\partial x} + \frac{\partial u}{\partial y} + \frac{\partial u}{\partial z} = 0.$

(UPTU 2002-03)

SOLUTION: Let $x' = x - y$, $y' = y - z$, $z' = z - x$ then we have $u = f(x', y', z')$

$$\therefore \quad \frac{\partial u}{\partial x} = \frac{\partial u}{\partial x'}\frac{\partial x'}{\partial x} + \frac{\partial u}{\partial y'}\cdot\frac{\partial y'}{\partial x} + \frac{\partial u}{\partial z'}\cdot\frac{\partial z'}{\partial x}$$

But $\dfrac{\partial x'}{\partial x} = 1, \quad \dfrac{\partial y'}{\partial x} = 0 \quad$ and $\quad \dfrac{\partial z'}{\partial x} = -1$

Hence $\dfrac{\partial u}{\partial x} = \dfrac{\partial u}{\partial x'} - \dfrac{\partial u}{\partial z'}$

Similarly, $\dfrac{\partial u}{\partial y} = \dfrac{\partial u}{\partial x'}\dfrac{\partial x'}{\partial y} + \dfrac{\partial u}{\partial y'}.\dfrac{\partial y'}{\partial y} + \dfrac{\partial u}{\partial z'}.\dfrac{\partial z'}{\partial y} = -\dfrac{\partial u}{\partial x'} + \dfrac{\partial u}{\partial y'} + 0$

And $\dfrac{\partial u}{\partial z} = \dfrac{\partial u}{\partial x'}\dfrac{\partial x'}{\partial z} + \dfrac{\partial u}{\partial y'}.\dfrac{\partial y'}{\partial z} + \dfrac{\partial u}{\partial z'}.\dfrac{\partial z'}{\partial z} = 0 - \dfrac{\partial u}{\partial y'} + \dfrac{\partial u}{\partial z'}$

Therefore, $\dfrac{\partial u}{\partial x} + \dfrac{\partial u}{\partial y} + \dfrac{\partial u}{\partial z} = \dfrac{\partial u}{\partial x'} - \dfrac{\partial u}{\partial z'} - \dfrac{\partial u}{\partial x'} + \dfrac{\partial u}{\partial y'} - \dfrac{\partial u}{\partial y'} + \dfrac{\partial u}{\partial z'} = 0.$ **Hence Proved.**

EXAMPLE 3.36. If $z = f(u, v)$ where $u = x\cos\alpha - y\sin\alpha$ and $v = x\sin\alpha + y\cos\alpha$ show that $x\dfrac{\partial z}{\partial x} + y\dfrac{\partial z}{\partial y} = u\dfrac{\partial z}{\partial u} + v\dfrac{\partial z}{\partial v}$, α being constant.

SOLUTION: Here $z = f(u, v)$ where u and v are functions of x, y.

$$\therefore \quad \frac{\partial z}{\partial x} = \frac{\partial z}{\partial u}.\frac{\partial u}{\partial x} + \frac{\partial z}{\partial v}.\frac{\partial v}{\partial x} = \cos\alpha\frac{\partial z}{\partial u} + \sin\alpha\frac{\partial z}{\partial v}$$

and
$$\frac{\partial z}{\partial y} = \frac{\partial z}{\partial u}.\frac{\partial u}{\partial y} + \frac{\partial z}{\partial v}.\frac{\partial v}{\partial y} = -\sin\alpha\frac{\partial z}{\partial u} + \cos\alpha\frac{\partial z}{\partial v}$$

Therefore,
$$x\frac{\partial z}{\partial x} + y\frac{\partial z}{\partial y} = x\left(\cos\alpha\frac{\partial z}{\partial u} + \sin\alpha\frac{\partial z}{\partial v}\right) + y\left(-\sin\alpha\frac{\partial z}{\partial u} + \cos\alpha\frac{\partial z}{\partial v}\right)$$

$$= \frac{\partial z}{\partial u}(x\cos\alpha - y\sin\alpha) + \frac{\partial z}{\partial v}(x\sin\alpha + y\cos\alpha) = u\frac{\partial z}{\partial u} + v\frac{\partial z}{\partial v}.$$

Hence Proved.

EXAMPLE 3.37. If $z = f(x, y)$ and $x = r\cos\theta$, $y = r\sin\theta$, show that

$$\left(\frac{\partial z}{\partial x}\right)^2 + \left(\frac{\partial z}{\partial y}\right)^2 = \left(\frac{\partial z}{\partial r}\right)^2 + \frac{1}{r^2}\left(\frac{\partial z}{\partial \theta}\right)^2.$$

SOLUTION: Here $\dfrac{\partial z}{\partial r} = \dfrac{\partial z}{\partial x}.\dfrac{\partial x}{\partial r} + \dfrac{\partial z}{\partial y}.\dfrac{\partial y}{\partial r} = \dfrac{\partial z}{\partial x}\cos\theta + \dfrac{\partial z}{\partial y}\sin\theta$ ... (1)

and
$$\frac{\partial z}{\partial \theta} = \frac{\partial z}{\partial x}\frac{\partial x}{\partial \theta} + \frac{\partial z}{\partial y}\frac{\partial y}{\partial \theta} = \frac{\partial z}{\partial x}(-r\sin\theta) + \frac{\partial z}{\partial y}(r\cos\theta)$$

or $\dfrac{1}{r}\dfrac{\partial z}{\partial \theta} = -\sin\theta\dfrac{\partial z}{\partial x} + \cos\theta\dfrac{\partial z}{\partial y}$ ...(2)

Squaring and adding (1) and (2), we get

$$\left(\frac{\partial z}{\partial r}\right)^2 + \frac{1}{r^2}\left(\frac{\partial z}{\partial \theta}\right)^2 = \left(\frac{\partial z}{\partial x}\right)^2 \cos^2\theta + \left(\frac{\partial z}{\partial y}\right)^2 \sin^2\theta + 2\sin\theta\cos\theta\,\frac{\partial z}{\partial x}\frac{\partial z}{\partial y}$$

$$+\left(\frac{\partial z}{\partial x}\right)^2 \sin^2\theta + \left(\frac{\partial z}{\partial y}\right)^2 \cos^2\theta - 2\sin\theta\cos\theta\,\frac{\partial z}{\partial x}\frac{\partial z}{\partial y}$$

$$= \left(\frac{\partial z}{\partial x}\right)^2 + \left(\frac{\partial z}{\partial y}\right)^2.$$ **Hence Proved.**

EXAMPLE 3.38. Let $z = f(x, y)$ where $x = e^u \cos v$, $y = e^u \sin v$ then show that

(*i*) $y\frac{\partial z}{\partial u} + x\frac{\partial z}{\partial v} = e^{2u}\frac{\partial z}{\partial y}$

(*ii*) $\left(\frac{\partial f}{\partial x}\right)^2 + \left(\frac{\partial f}{\partial y}\right)^2 = e^{-2u}\left[\left(\frac{\partial f}{\partial u}\right)^2 + \left(\frac{\partial f}{\partial v}\right)^2\right].$ **(UPTU 2001 Q.B.)**

SOLUTION: (*i*) We have $z = f(x, y)$ and x and y are functions of u and v.

Hence $$\frac{\partial z}{\partial u} = \frac{\partial z}{\partial x}\frac{\partial x}{\partial u} + \frac{\partial z}{\partial y}\frac{\partial y}{\partial u} = e^u\cos v\frac{\partial z}{\partial x} + e^u \sin v\frac{\partial z}{\partial y} = x\frac{\partial z}{\partial x} + y\frac{\partial z}{\partial y} \quad \text{...(1)}$$

And $$\frac{\partial z}{\partial v} = \frac{\partial z}{\partial x}\frac{\partial x}{\partial v} + \frac{\partial z}{\partial y}\frac{\partial y}{\partial v} = -e^u\sin v\frac{\partial z}{\partial x} + e^u \cos v\frac{\partial z}{\partial y} = -y\frac{\partial z}{\partial x} + x\frac{\partial z}{\partial y} \quad \text{...(2)}$$

$$\therefore \quad y\frac{\partial z}{\partial u} + x\frac{\partial z}{\partial v} = xy\frac{\partial z}{\partial x} + y^2\frac{\partial z}{\partial y} - xy\frac{\partial z}{\partial x} + x^2\frac{\partial z}{\partial y}$$

$$= (x^2 + y^2)\frac{\partial z}{\partial y} = (e^{2u}\cos^2 v + e^{2u}\sin^2 v)\frac{\partial z}{\partial y} = e^{2u}\frac{\partial z}{\partial y}.$$ **Hence Proved.**

(*ii*) Squaring and adding (1) and (2), gives

$$\left(\frac{\partial z}{\partial u}\right)^2 + \left(\frac{\partial z}{\partial v}\right)^2 = x^2\left(\frac{\partial z}{\partial x}\right)^2 + 2xy\frac{\partial z}{\partial x}\frac{\partial z}{\partial y} + y^2\left(\frac{\partial z}{\partial y}\right)^2 + y^2\left(\frac{\partial z}{\partial x}\right)^2 - 2xy\frac{\partial z}{\partial x}\frac{\partial z}{\partial y} + x^2\left(\frac{\partial z}{\partial y}\right)^2$$

$$= (x^2 + y^2)\left(\frac{\partial z}{\partial x}\right)^2 + (x^2 + y^2)\left(\frac{\partial z}{\partial y}\right)^2 + 0$$

$$= (x^2+y^2)\left[\left(\frac{\partial z}{\partial x}\right)^2+\left(\frac{\partial z}{\partial y}\right)^2\right] = (e^{2u}\cos^2 v + e^{2u}\sin^2 v)\left[\left(\frac{\partial z}{\partial x}\right)^2+\left(\frac{\partial z}{\partial y}\right)^2\right]$$

$$= e^{2u}\left[\left(\frac{\partial z}{\partial x}\right)^2+\left(\frac{\partial z}{\partial y}\right)^2\right]$$

or $$e^{-2u}\left[\left(\frac{\partial z}{\partial u}\right)^2+\left(\frac{\partial z}{\partial v}\right)^2\right] = \left(\frac{\partial z}{\partial x}\right)^2+\left(\frac{\partial z}{\partial y}\right)^2.$$ **Hence the result.**

EXAMPLE 3.39. By changing the independent variables u and v to x and y, by means of the relations $x = u\cos\alpha - v\sin\alpha$, $y = u\sin\alpha + v\cos\alpha$,

show that $\frac{\partial^2 z}{\partial u^2}+\frac{\partial^2 z}{\partial v^2}$ transforms into $\frac{\partial^2 z}{\partial x^2}+\frac{\partial^2 z}{\partial y^2}$.

SOLUTION: Since $x = u\cos\alpha - v\sin\alpha$, $y = u\sin\alpha + v\cos\alpha$ we have

$$\frac{\partial z}{\partial u} = \frac{\partial z}{\partial x}\frac{\partial x}{\partial u}+\frac{\partial z}{\partial y}.\frac{\partial y}{\partial u} = \cos\alpha\frac{\partial z}{\partial x}+\sin\alpha\frac{\partial z}{\partial y}$$

and $$\frac{\partial z}{\partial v} = \frac{\partial z}{\partial x}\frac{\partial x}{\partial v}+\frac{\partial z}{\partial y}\frac{\partial y}{\partial v} = -\sin\alpha\frac{\partial z}{\partial x}+\cos\alpha\frac{\partial z}{\partial y}.$$

Next $$\frac{\partial^2 z}{\partial u^2} = \frac{\partial}{\partial x}\left(\cos\alpha\frac{\partial z}{\partial x}+\sin\alpha\frac{\partial z}{\partial y}\right).\frac{\partial x}{\partial u}+\frac{\partial}{\partial y}\left(\cos\alpha\frac{\partial z}{\partial x}+\sin\alpha\frac{\partial z}{\partial y}\right)\frac{\partial y}{\partial u}$$

$$= \left(\cos\alpha\frac{\partial^2 z}{\partial x^2}+\sin\alpha\frac{\partial^2 z}{\partial x\partial y}\right)\cos\alpha+\left(\cos\alpha\frac{\partial^2 z}{\partial x\partial y}+\sin\alpha\frac{\partial^2 z}{\partial y^2}\right)\sin\alpha$$

$$= \cos^2\alpha\frac{\partial^2 z}{\partial x^2}+2\sin\alpha\cdot\cos\alpha\frac{\partial^2 z}{\partial x\partial y}+\sin^2\alpha\frac{\partial^2 z}{\partial y^2}$$

and $$\frac{\partial^2 z}{\partial v^2} = \frac{\partial}{\partial x}\left(-\sin\alpha\frac{\partial z}{\partial x}+\cos\alpha\frac{\partial z}{\partial y}\right)\frac{\partial x}{\partial v}+\frac{\partial}{\partial y}\left(-\sin\alpha\frac{\partial z}{\partial x}+\cos\alpha\frac{\partial z}{\partial y}\right)\cdot\frac{\partial y}{\partial v}$$

$$= \left(-\sin\alpha\frac{\partial^2 z}{\partial x^2}+\cos\alpha\frac{\partial^2 z}{\partial x\partial y}\right)(-\sin\alpha)+\left(-\sin\alpha\frac{\partial^2 z}{\partial x\partial y}+\cos\alpha\frac{\partial^2 z}{\partial y^2}\right)\cos\alpha$$

$$= \sin^2\alpha\frac{\partial^2 z}{\partial x^2}-2\sin\alpha\cos\alpha\frac{\partial^2 z}{\partial x\partial y}+\cos^2\alpha\frac{\partial^2 z}{\partial y^2}$$

$\therefore$ $$\frac{\partial^2 z}{\partial u^2}+\frac{\partial^2 z}{\partial v^2} = (\cos^2\alpha+\sin^2\alpha)\frac{\partial^2 z}{\partial x^2}+0\frac{\partial^2 z}{\partial x\partial y}+(\cos^2\alpha+\sin^2\alpha)\frac{\partial^2 z}{\partial y^2}$$

$$= \frac{\partial^2 z}{\partial x^2}+\frac{\partial^2 z}{\partial y^2}.$$ **Hence the result.**

EXAMPLE 3.40. If z is a function of x and y, and $u = lx + my$ and $v = ly - mx$,

then prove that $\dfrac{\partial^2 z}{\partial x^2} + \dfrac{\partial^2 z}{\partial y^2} = (l^2 + m^2)\left(\dfrac{\partial^2 z}{\partial u^2} + \dfrac{\partial^2 z}{\partial v^2}\right)$. **(UPTU 2001 Q.B.)**

SOLUTION: Let $z = z(u, v)$ where $u = lx + my$, $v = ly - mx$.

Thus, z is a composite function of x and y.

$$\therefore \quad \frac{\partial z}{\partial x} = \frac{\partial z}{\partial u}\frac{\partial u}{\partial x} + \frac{\partial z}{\partial v}.\frac{\partial v}{\partial x} = l\frac{\partial z}{\partial u} - m\frac{\partial z}{\partial v} = \left(l\frac{\partial}{\partial u} - m\frac{\partial}{\partial v}\right)z$$

$$\Rightarrow \quad \frac{\partial}{\partial x} \equiv l\frac{\partial}{\partial u} - m\frac{\partial}{\partial v}.$$

$$\therefore \quad \frac{\partial^2 z}{\partial x^2} = \frac{\partial}{\partial x}\left(\frac{\partial z}{\partial x}\right) = \left(l\frac{\partial}{\partial u} - m\frac{\partial}{\partial v}\right)\left(l\frac{\partial z}{\partial u} - m\frac{\partial z}{\partial v}\right)$$

$$= l\frac{\partial}{\partial u}\left(l\frac{\partial z}{\partial u} - m\frac{\partial z}{\partial v}\right) - m\frac{\partial}{\partial v}\left(l\frac{\partial z}{\partial u} - m\frac{\partial z}{\partial v}\right)$$

$$= l^2\frac{\partial^2 z}{\partial u^2} - lm\frac{\partial^2 z}{\partial u \partial v} - lm\frac{\partial^2 z}{\partial v \partial u} + m^2\frac{\partial^2 z}{\partial v^2} \qquad \text{... (1)}$$

$$\text{Next,} \quad \frac{\partial z}{\partial y} = \frac{\partial z}{\partial u}\frac{\partial u}{\partial y} + \frac{\partial z}{\partial v}\frac{\partial v}{\partial y} = m\frac{\partial z}{\partial u} + l\frac{\partial z}{\partial v} = \left(m\frac{\partial}{\partial u} + l\frac{\partial}{\partial v}\right)z$$

$$\Rightarrow \quad \frac{\partial}{\partial y} \equiv m\frac{\partial}{\partial u} + l\frac{\partial}{\partial v}.$$

$$\therefore \quad \frac{\partial^2 z}{\partial y^2} = \frac{\partial}{\partial y}\left(\frac{\partial z}{\partial y}\right) = \left(m\frac{\partial}{\partial u} + l\frac{\partial}{\partial v}\right)\left(m\frac{\partial z}{\partial u} + l\frac{\partial z}{\partial v}\right)$$

$$= m^2\frac{\partial^2 z}{\partial u^2} + ml\frac{\partial^2 z}{\partial u \partial v} + lm\frac{\partial^2 z}{\partial v \partial u} + l^2\frac{\partial^2 z}{\partial v^2} \qquad \text{...(2)}$$

Adding (1) and (2), we get

$$\frac{\partial^2 z}{\partial x^2} + \frac{\partial^2 z}{\partial y^2} = (l^2 + m^2)\frac{\partial^2 z}{\partial u^2} + 0 + (m^2 + l^2)\frac{\partial^2 z}{\partial v^2}$$

$$= (l^2 + m^2)\left(\frac{\partial^2 z}{\partial u^2} + \frac{\partial^2 z}{\partial v^2}\right).$$ **Hence Proved.**

EXAMPLE 3.41. Transform the equation $\dfrac{\partial^2 u}{\partial x^2}+\dfrac{\partial^2 u}{\partial y^2}=0$ into polar co-ordinates r and θ.

(UPTU 2001 Q.B.)

SOLUTION: Since $x = r\cos\theta$, $y = r\sin\theta$ we have $r^2 = x^2 + y^2$ and $\theta = \tan^{-1}\dfrac{y}{x}$.

We have $\dfrac{\partial u}{\partial x} = \dfrac{\partial u}{\partial r}\cdot\dfrac{\partial r}{\partial x}+\dfrac{\partial u}{\partial \theta}\cdot\dfrac{\partial \theta}{\partial x}$ and $\dfrac{\partial u}{\partial y} = \dfrac{\partial u}{\partial r}\cdot\dfrac{\partial r}{\partial y}+\dfrac{\partial u}{\partial \theta}\cdot\dfrac{\partial \theta}{\partial y}$.

From $r^2 = x^2+y^2$, we have $2r\dfrac{\partial r}{\partial x} = 2x$ or $\dfrac{\partial r}{\partial x} = \dfrac{x}{r} = \cos\theta$.

and $\dfrac{\partial r}{\partial y} = \dfrac{y}{r} = \sin\theta$.

Similarly, from $\theta = \tan^{-1}\dfrac{y}{x}$ we have $\dfrac{\partial \theta}{\partial x} = \dfrac{1}{1+\dfrac{y^2}{x^2}}\cdot\left(\dfrac{-y}{x^2}\right) = \dfrac{-y}{x^2+y^2} = \dfrac{-\sin\theta}{r}$

and $\dfrac{\partial \theta}{\partial y} = \dfrac{1}{1+\dfrac{y^2}{x^2}}\cdot\left(\dfrac{1}{x}\right) = \dfrac{x}{x^2+y^2} = \dfrac{\cos\theta}{r}$.

Therefore, $\dfrac{\partial u}{\partial x} = \cos\theta\dfrac{\partial u}{\partial r} - \dfrac{\sin\theta}{r}\dfrac{\partial u}{\partial \theta}$

and $\dfrac{\partial u}{\partial y} = \sin\theta\dfrac{\partial u}{\partial r} + \dfrac{\cos\theta}{r}\dfrac{\partial u}{\partial \theta}$.

Thus, in terms of operators, we can write

$$\frac{\partial}{\partial x} \equiv \cos\theta\frac{\partial}{\partial r} - \frac{\sin\theta}{r}\frac{\partial}{\partial \theta} \quad \text{and} \quad \frac{\partial}{\partial y} \equiv \sin\theta\frac{\partial}{\partial r} + \frac{\cos\theta}{r}\frac{\partial}{\partial \theta}$$

Therefore, $\dfrac{\partial^2 u}{\partial x^2} = \dfrac{\partial}{\partial x}\left(\dfrac{\partial u}{\partial x}\right) = \left(\cos\theta\dfrac{\partial}{\partial r} - \dfrac{\sin\theta}{r}\dfrac{\partial}{\partial \theta}\right)\left(\cos\theta\dfrac{\partial u}{\partial r} - \dfrac{\sin\theta}{r}\dfrac{\partial u}{\partial \theta}\right)$

$$= \cos\theta\frac{\partial}{\partial r}\left(\cos\theta\frac{\partial u}{\partial r} - \frac{\sin\theta}{r}\frac{\partial u}{\partial \theta}\right) - \frac{\sin\theta}{r}\frac{\partial}{\partial \theta}\left(\cos\theta\frac{\partial u}{\partial r} - \frac{\sin\theta}{r}\frac{\partial u}{\partial \theta}\right)$$

$$= \cos\theta\left[\cos\theta\frac{\partial^2 u}{\partial r^2} + \frac{\sin\theta}{r^2}\frac{\partial u}{\partial \theta} - \frac{\sin\theta}{r}\frac{\partial^2 u}{\partial r\,\partial \theta}\right]$$

$$-\frac{\sin\theta}{r}\left[-\sin\theta\frac{\partial u}{\partial r} + \cos\theta\frac{\partial^2 u}{\partial \theta\,\partial r} - \frac{\cos\theta}{r}\frac{\partial u}{\partial \theta} - \frac{\sin\theta}{r}\frac{\partial^2 u}{\partial \theta^2}\right]$$

$$= \cos^2\theta\frac{\partial^2 u}{\partial r^2} - \frac{2\sin\theta\cos\theta}{r}\frac{\partial^2 u}{\partial \theta\,\partial r} + \frac{\sin^2\theta}{r^2}\frac{\partial^2 u}{\partial \theta^2} + \frac{2\sin\theta\cos\theta}{r^2}\frac{\partial u}{\partial \theta} + \frac{\sin^2\theta}{r}\frac{\partial u}{\partial r}$$

and $$\frac{\partial^2 u}{\partial y^2} = \frac{\partial}{\partial y}\left(\frac{\partial u}{\partial y}\right) = \left(\sin\theta\frac{\partial}{\partial r} + \frac{\cos\theta}{r}\frac{\partial}{\partial\theta}\right)\left(\sin\theta\frac{\partial u}{\partial r} + \frac{\cos\theta}{r}\frac{\partial u}{\partial\theta}\right)$$

$$= \sin\theta\frac{\partial}{\partial r}\left(\sin\theta\frac{\partial u}{\partial r} + \frac{\cos\theta}{r}\frac{\partial u}{\partial\theta}\right) + \frac{\cos\theta}{r}\frac{\partial}{\partial\theta}\left(\sin\theta\frac{\partial u}{\partial r} + \frac{\cos\theta}{r}\frac{\partial u}{\partial\theta}\right)$$

$$= \sin\theta\left[\sin\theta\frac{\partial^2 u}{\partial r^2} - \frac{\cos\theta}{r^2}\frac{\partial u}{\partial\theta} + \frac{\cos\theta}{r}\frac{\partial^2 u}{\partial r\,\partial\theta}\right]$$

$$+ \frac{\cos\theta}{r}\left[\sin\theta\frac{\partial^2 u}{\partial\theta\,\partial r} + \cos\theta\frac{\partial u}{\partial r} - \frac{\sin\theta}{r}\frac{\partial u}{\partial\theta} + \frac{\cos\theta}{r}\frac{\partial^2 u}{\partial\theta^2}\right]$$

$$= \sin^2\theta\frac{\partial^2 u}{\partial r^2} + \frac{2\sin\theta\cos\theta}{r}\frac{\partial^2 u}{\partial r\,\partial\theta} + \frac{\cos^2\theta}{r^2}\frac{\partial^2 u}{\partial\theta^2} - \frac{2\sin\theta\cos\theta}{r^2}\frac{\partial u}{\partial\theta} + \frac{\cos^2\theta}{r}\frac{\partial u}{\partial r}$$

Therefore, $$\frac{\partial^2 u}{\partial x^2} + \frac{\partial^2 u}{\partial y^2} = (\sin^2\theta + \cos^2\theta)\frac{\partial^2 u}{\partial r^2} + \frac{\sin^2\theta + \cos^2\theta}{r^2}\frac{\partial^2 u}{\partial\theta^2} + \frac{\sin^2\theta + \cos^2\theta}{r}\frac{\partial u}{\partial r}$$

$$= \frac{\partial^2 u}{\partial r^2} + \frac{1}{r^2}\frac{\partial^2 u}{\partial\theta^2} + \frac{1}{r}\frac{\partial u}{\partial r}$$

Thus, the Laplace equation in cartesian form $\frac{\partial^2 u}{\partial x^2} + \frac{\partial^2 u}{\partial y^2} = 0$

gets transformed into polar form as $\frac{\partial^2 u}{\partial r^2} + \frac{1}{r}\frac{\partial u}{\partial r} + \frac{1}{r^2}\frac{\partial^2 u}{\partial\theta^2} = 0$. **Ans.**

EXAMPLE 3.42. Let f be a composite function of u and v and u and v be functions of x and y, given by $u = x^2 - y^2$, $v = 2xy$, then show that

$$\frac{\partial^2 f}{\partial x^2} + \frac{\partial^2 f}{\partial y^2} = 4\,(x^2 + y^2)\left(\frac{\partial^2 f}{\partial u^2} + \frac{\partial^2 f}{\partial v^2}\right).$$

SOLUTION: Given that $f = f(u, v)$ and $u = x^2 - y^2$, $v = 2xy$, we have

$$\frac{\partial f}{\partial x} = \frac{\partial f}{\partial u}\frac{\partial u}{\partial x} + \frac{\partial f}{\partial v}\frac{\partial v}{\partial x} = \frac{\partial f}{\partial u}(2x) + \frac{\partial f}{\partial v}(2y)$$

$$\Rightarrow \quad \frac{\partial}{\partial x} \equiv 2\left(x\frac{\partial}{\partial u} + y\frac{\partial}{\partial v}\right) \qquad \text{... (1)}$$

Similarly, $$\frac{\partial f}{\partial y} = \frac{\partial f}{\partial u}\frac{\partial u}{\partial y} + \frac{\partial f}{\partial v}\frac{\partial v}{\partial y} = \frac{\partial f}{\partial u}(-2y) + \frac{\partial f}{\partial v}(2x)$$

$$\Rightarrow \quad \frac{\partial}{\partial y} \equiv 2\left(-y\frac{\partial}{\partial u} + x\frac{\partial}{\partial v}\right) \qquad \text{... (2)}$$

$$\therefore \quad \frac{\partial^2 f}{\partial x^2} = \frac{\partial}{\partial x}\left(\frac{\partial f}{\partial x}\right) = \frac{\partial}{\partial x}\left(2x\frac{\partial f}{\partial u} + 2y\frac{\partial f}{\partial v}\right) = 2\left[x\frac{\partial}{\partial x}\left(\frac{\partial f}{\partial u}\right) + 1.\frac{\partial f}{\partial u} + y\frac{\partial}{\partial x}\left(\frac{\partial f}{\partial v}\right)\right]$$

$$= 2\left[x.2\left(x\frac{\partial}{\partial u} + y\frac{\partial}{\partial v}\right)\left(\frac{\partial f}{\partial u}\right) + \frac{\partial f}{\partial u} + y.2\left(x\frac{\partial}{\partial u} + y\frac{\partial}{\partial v}\right)\left(\frac{\partial f}{\partial v}\right)\right]$$

$$= 4\left[x\left(x\frac{\partial^2 f}{\partial u^2} + y\frac{\partial^2 f}{\partial u\,\partial v}\right) + \frac{1}{2}\frac{\partial f}{\partial u} + y\left(x\frac{\partial^2 f}{\partial u\,\partial v} + y\frac{\partial^2 f}{\partial v^2}\right)\right]$$

$$= 4x^2\frac{\partial^2 f}{\partial u^2} + 8xy\frac{\partial^2 f}{\partial u\,\partial v} + 4y^2\frac{\partial^2 f}{\partial v^2} + 2\frac{\partial f}{\partial u}. \qquad \text{... (3)}$$

Similarly,

$$\frac{\partial^2 f}{\partial y^2} = \frac{\partial}{\partial y}\left(\frac{\partial f}{\partial y}\right) = \frac{\partial}{\partial y}\left(-2y\frac{\partial f}{\partial u} + 2x\frac{\partial f}{\partial v}\right) = -2y\frac{\partial}{\partial y}\left(\frac{\partial f}{\partial u}\right) - 2.1\frac{\partial f}{\partial u} + 2x\frac{\partial}{\partial y}\left(\frac{\partial f}{\partial v}\right)$$

$$= -2y\left(-2y\frac{\partial}{\partial u} + 2x\frac{\partial}{\partial v}\right)\left(\frac{\partial f}{\partial u}\right) - 2\frac{\partial f}{\partial u} + 2x\left(-2y\frac{\partial}{\partial u} + 2x\frac{\partial}{\partial v}\right)\left(\frac{\partial f}{\partial v}\right)$$

$$= 4y^2\frac{\partial^2 f}{\partial u^2} - 4xy\frac{\partial^2 f}{\partial u\,\partial v} - 2\frac{\partial f}{\partial u} - 4xy\frac{\partial^2 f}{\partial u\,\partial v} + 4x^2\frac{\partial^2 f}{\partial v^2}. \qquad \text{...(4)}$$

Adding (3) and (4), gives

$$\frac{\partial^2 f}{\partial x^2} + \frac{\partial^2 f}{\partial y^2} = 4(x^2 + y^2)\frac{\partial^2 f}{\partial u^2} + 0 + 4(y^2 + x^2)\frac{\partial^2 f}{\partial v^2} + 0 = 4(x^2 + y^2)\left(\frac{\partial^2 f}{\partial u^2} + \frac{\partial^2 f}{\partial v^2}\right)$$

Hence Proved.

EXAMPLE 3.43. By changing the independent variables x and y to u and v, by means of the relations $u = x - ay$, $v = x + ay$, show that the expression $a^2\frac{\partial^2 z}{\partial x^2} - \frac{\partial^2 z}{\partial y^2}$ transforms into $4a^2\frac{\partial^2 z}{\partial u\,\partial v}$.

SOLUTION:

$$\frac{\partial z}{\partial x} = \frac{\partial z}{\partial u}\cdot\frac{\partial u}{\partial x} + \frac{\partial z}{\partial v}\cdot\frac{\partial v}{\partial x} = \frac{\partial z}{\partial u} + \frac{\partial z}{\partial v}$$

and

$$\frac{\partial z}{\partial y} = \frac{\partial z}{\partial u}\cdot\frac{\partial u}{\partial y} + \frac{\partial z}{\partial v}\cdot\frac{\partial v}{\partial y} = -a\frac{\partial z}{\partial u} + a\frac{\partial z}{\partial v}$$

Next,

$$\frac{\partial^2 z}{\partial x^2} = \frac{\partial}{\partial u}\left(\frac{\partial z}{\partial u} + \frac{\partial z}{\partial v}\right)\frac{\partial u}{\partial x} + \frac{\partial}{\partial v}\left(\frac{\partial z}{\partial u} + \frac{\partial z}{\partial v}\right)\frac{\partial v}{\partial x} = \frac{\partial^2 z}{\partial u^2} + \frac{2\partial^2 z}{\partial u\,\partial v} + \frac{\partial^2 z}{\partial v^2}$$

and $$\frac{\partial^2 z}{\partial y^2} = \frac{\partial}{\partial u}\left(-a\frac{\partial z}{\partial u}+a\frac{\partial z}{\partial v}\right)\frac{\partial u}{\partial y}+\frac{\partial}{\partial v}\left(-a\frac{\partial z}{\partial u}+a\frac{\partial z}{\partial v}\right)\frac{\partial v}{\partial y}$$

$$= -a\frac{\partial}{\partial u}\left(\frac{\partial z}{\partial u}-\frac{\partial z}{\partial v}\right)(-a)-a\frac{\partial}{\partial v}\left(\frac{\partial z}{\partial u}--\frac{\partial z}{\partial v}\right)(a)$$

$$= a^2\frac{\partial^2 z}{\partial u^2}+a^2\frac{\partial^2 z}{\partial v^2}-2a^2\frac{\partial^2 z}{\partial u\,\partial v}$$

Therefore $a^2\dfrac{\partial^2 z}{\partial x^2}-\dfrac{\partial^2 z}{\partial y^2} = 4a^2\dfrac{\partial^2 z}{\partial u\,\partial v}$. **Hence Proved.**

EXAMPLE 3.44. If $u=f(r, s, t)$ where $r=\dfrac{x}{y},\ s=\dfrac{y}{z},\ t=\dfrac{z}{x}$ prove that $x\dfrac{\partial u}{\partial x}+y\dfrac{\partial u}{\partial y}+z\dfrac{\partial u}{\partial z} = 0.$

SOLUTION: Given $u = f(r, s, t)$ and $r=\dfrac{x}{y},\ s=\dfrac{y}{z},\ t=\dfrac{z}{x}.$

$$\therefore \quad \frac{\partial u}{\partial x} = \frac{\partial u}{\partial r}\frac{\partial r}{\partial x}+\frac{\partial u}{\partial s}\frac{\partial s}{\partial x}+\frac{\partial u}{\partial t}\frac{\partial t}{\partial x} = \frac{1}{y}\frac{\partial u}{\partial r}+0-\frac{z}{x^2}\frac{\partial u}{\partial t},$$

$$\frac{\partial u}{\partial y} = \frac{\partial u}{\partial r}\frac{\partial r}{\partial y}+\frac{\partial u}{\partial s}\frac{\partial s}{\partial y}+\frac{\partial u}{\partial t}\frac{\partial t}{\partial y} = \frac{-x}{y^2}\frac{\partial u}{\partial r}+\frac{1}{z}\frac{\partial u}{\partial s}+0,$$

$$\frac{\partial u}{\partial z} = \frac{\partial u}{\partial r}\frac{\partial r}{\partial z}+\frac{\partial u}{\partial s}\frac{\partial s}{\partial z}+\frac{\partial u}{\partial t}\frac{\partial t}{\partial z} = 0-\frac{y}{z^2}\frac{\partial u}{\partial s}+\frac{1}{x}\frac{\partial u}{\partial t}$$

Hence $x\dfrac{\partial u}{\partial x}+y\dfrac{\partial u}{\partial y}+z\dfrac{\partial u}{\partial z} = \dfrac{x}{y}\dfrac{\partial u}{\partial r}-\dfrac{z}{x}\dfrac{\partial u}{\partial t}-\dfrac{x}{y}\dfrac{\partial u}{\partial r}+\dfrac{y}{z}\dfrac{\partial u}{\partial s}-\dfrac{y}{z}\dfrac{\partial u}{\partial s}+\dfrac{z}{x}\dfrac{\partial u}{\partial t} = 0.$ **Hence Proved.**

EXAMPLE 3.45. If $u = x\log(xy)$ where $x^3+y^3+3xy = 1$, find $\dfrac{du}{dx}$. **(UPTU 2005–06)**

SOLUTION: Given $u = x(\log x+\log y)$ hence $du = \dfrac{\partial u}{\partial x}dx+\dfrac{\partial u}{\partial y}dy$

or $$\frac{du}{dx} = \frac{\partial u}{\partial x}+\frac{\partial u}{\partial y}\frac{dy}{dx} = x\cdot\frac{1}{x}+(\log x+\log y)+\frac{x}{y}\frac{dy}{dx} = 1+\log(xy)+\frac{x}{y}\frac{dy}{dx}$$

Now, since $x^3+y^3+3xy = 1$ we have

$$\frac{dy}{dx} = \frac{-\dfrac{\partial}{\partial x}(x^3+y^3+3xy-1)}{\dfrac{\partial}{\partial y}(x^3+y^3+3xy-1)}$$

or $$\frac{dy}{dx} = \frac{-(3x^2+3y)}{(3y^2+3x)} = \frac{-(x^2+y)}{y^2+x}$$

Therefore $$\frac{du}{dx} = 1+\log(xy)-\frac{x}{y}\left(\frac{x^2+y}{y^2+x}\right)$$

$$= 1+\log(xy)-\frac{x^3+xy}{y^3+xy}. \quad \textbf{Ans.}$$

EXAMPLE 3.46. If $f(x, y, z) = 0$ show that $\left(\frac{\partial y}{\partial z}\right)_x \left(\frac{\partial z}{\partial x}\right)_y \left(\frac{\partial x}{\partial y}\right)_z = -1.$ **(UPTU 2003–04)**

SOLUTION: Treating x as constant in the given relation, we have

$$\left(\frac{\partial y}{\partial z}\right)_x = -\frac{\partial f}{\partial z}\Big/\frac{\partial f}{\partial y}$$

Similarly treating y as constant in the given relation, we have

$$\left(\frac{\partial z}{\partial x}\right)_y = -\frac{\partial f}{\partial x}\Big/\frac{\partial f}{\partial z}$$

and treating z as constant in the given relation, we have

$$\left(\frac{\partial x}{\partial y}\right)_z = -\frac{\partial f}{\partial y}\Big/\frac{\partial f}{\partial x}$$

Multiplying the above relations, we get

$$\left(\frac{\partial y}{\partial z}\right)_x \left(\frac{\partial z}{\partial x}\right)_y \left(\frac{\partial x}{\partial y}\right)_z = -1. \quad \textbf{Hence Proved.}$$

EXAMPLE 3.47. If $u = u\left(\frac{y-x}{xy}, \frac{z-x}{xz}\right)$ prove that $x^2\frac{\partial u}{\partial x}+y^2\frac{\partial u}{\partial y}+z^2\frac{\partial u}{\partial z} = 0.$

(UPTU 2003–04)

SOLUTION: Let $u = u(s, t)$ where $s = \frac{y-x}{xy} = \frac{1}{x}-\frac{1}{y}$ and $t = \frac{z-x}{xz} = \frac{1}{x}-\frac{1}{z}.$

Then $$\frac{\partial u}{\partial x} = \frac{\partial u}{\partial s}\frac{\partial s}{\partial x}+\frac{\partial u}{\partial t}\frac{\partial t}{\partial x} = \frac{\partial u}{\partial s}\left(-\frac{1}{x^2}\right)+\frac{\partial u}{\partial t}\left(-\frac{1}{x^2}\right)$$

and $$\frac{\partial u}{\partial y} = \frac{\partial u}{\partial s}\frac{\partial s}{\partial y}+\frac{\partial u}{\partial t}\cdot\frac{\partial t}{\partial y} = \frac{1}{y^2}\frac{\partial u}{\partial s}+\frac{\partial u}{\partial t}(0) = \frac{1}{y^2}\frac{\partial u}{\partial s}.$$

and $$\frac{\partial u}{\partial z} = \frac{\partial u}{\partial s}\cdot\frac{\partial s}{\partial z}+\frac{\partial u}{\partial t}\frac{\partial t}{\partial z} = 0\frac{\partial u}{\partial s}+\frac{1}{z^2}\frac{\partial u}{\partial t} = \frac{1}{z^2}\frac{\partial u}{\partial t}$$

$$\Rightarrow \quad x^2\frac{\partial u}{\partial x}+y^2\frac{\partial u}{\partial y}+z^2\frac{\partial u}{\partial z} = -\frac{\partial u}{\partial s}-\frac{\partial u}{\partial t}+\frac{\partial u}{\partial s}+\frac{\partial u}{\partial t} = 0. \quad \textbf{Hence proved.}$$

EXAMPLE 3.48. If $x + y = 2e^{\theta} \cos \phi$ and $x - y = 2\, ie^{\theta} \sin \phi$, then

show that $\dfrac{\partial^2 f}{\partial \theta^2} + \dfrac{\partial^2 f}{\partial \phi^2} = 4xy \dfrac{\partial^2 f}{\partial x\, \partial y}$. **(UPTU 2002, 2003–04)**

SOLUTION: From $x + y = 2\, e^{\theta} \cos \phi,\ x - y = 2ie^{\theta} \sin \phi$, we have $x = e^{\theta} (\cos \phi + i \sin \phi) = e^{\theta + i\phi}$ and $y = e^{\theta} (\cos \phi - i \sin \phi) = e^{\theta - i\phi}$.

$$\therefore \quad \frac{\partial x}{\partial \theta} = e^{\theta + i\phi} = x, \quad \frac{\partial y}{\partial \theta} = e^{\theta - i\phi} = y$$

and
$$\frac{\partial x}{\partial \phi} = ie^{\theta + i\phi} = ix, \quad \frac{\partial y}{\partial \phi} = -ie^{\theta - i\phi} = -iy$$

Now
$$\frac{\partial f}{\partial \theta} = \frac{\partial f}{\partial x}\frac{\partial x}{\partial \theta} + \frac{\partial f}{\partial y}\frac{\partial y}{\partial \theta} = x\frac{\partial f}{\partial x} + y\frac{\partial f}{\partial y}$$

and
$$\frac{\partial f}{\partial \phi} = \frac{\partial f}{\partial x}\frac{\partial x}{\partial \phi} + \frac{\partial f}{\partial y}\cdot\frac{\partial y}{\partial \phi} = i\left(x\frac{\partial f}{\partial x} - y\frac{\partial f}{\partial y}\right)$$

These, in operator form, can be written as

$$\frac{\partial}{\partial \theta} \equiv x\frac{\partial}{\partial x} + y\frac{\partial}{\partial y} \quad \text{and} \quad \frac{\partial}{\partial \phi} \equiv i\left(x\frac{\partial}{\partial x} - y\frac{\partial}{\partial y}\right).$$

Next
$$\frac{\partial^2 f}{\partial \theta^2} + \frac{\partial^2 f}{\partial \phi^2} = \left(\frac{\partial}{\partial \theta} + \frac{i\partial}{\partial \phi}\right)\left(\frac{\partial}{\partial \theta} - i\frac{\partial}{\partial \phi}\right) f$$

$$= \left[x\frac{\partial}{\partial x} + y\frac{\partial}{\partial y} + i^2\left(x\frac{\partial}{\partial x} - y\frac{\partial}{\partial y}\right)\right]\left[x\frac{\partial}{\partial x} + y\frac{\partial}{\partial y} - i^2\left(x\frac{\partial}{\partial x} - y\frac{\partial}{\partial y}\right)\right] f$$

$$= \left[2y\frac{\partial}{\partial y}\right]\left[2x\frac{\partial}{\partial x}\right] f = 4xy\frac{\partial^2 f}{\partial x\, \partial y}.$$ **Hence proved.**

EXERCISE 3C

1. If $x^3 + 3x^2 y + 6xy^2 + y^3 = 1$, find $\dfrac{dy}{dx}$.

2. If $ax^2 + 2hxy + by^2 + 2gx + 2fy + c = 0$ obtain $\dfrac{d^2y}{dx^2}$.

3. If $x^4 + y^4 + 4a^2 xy = 0$, prove that $(y^3 + a^2 x)^3 \dfrac{d^2y}{dx^2} = 2a^2 xy\,(x^2 y^2 + 3a^4)$.

4. If $u = x \log (x y)$ where $x^3 + y^3 + 3xy = 1$, find $\dfrac{du}{dx}$ at $(1, 1)$.

(UPTU 2001 Sp. Exam.)

5. The relations $f(x, y, z) =$ constant and $xyz =$ constant together define y as a function of x, then show that

$$\frac{dy}{dx} = \frac{-y\left(x\dfrac{\partial f}{\partial x} - z\dfrac{\partial f}{\partial z}\right)}{x\left(y\dfrac{\partial f}{\partial y} - z\dfrac{\partial f}{\partial z}\right)}.$$

6. If $u = \log (\tan x + \tan y + \tan z)$, find the value of $\sin 2x \dfrac{\partial u}{\partial x} + \sin 2y \dfrac{\partial u}{\partial y} + \sin 2z \dfrac{\partial u}{\partial z}$.

(UPTU 2000-01)

7. If $u = f\left(\dfrac{x}{y}, \dfrac{y}{z}, \dfrac{z}{x}\right)$ prove that $x\dfrac{\partial u}{\partial x} + y\dfrac{\partial u}{\partial y} + z\dfrac{\partial u}{\partial z} = 0$.

8. If $u = x + ay$ and $v = x + by$ transforms the equation $2\dfrac{\partial^2 z}{\partial x^2} - 5\dfrac{\partial^2 z}{\partial x \partial y} + 3\dfrac{\partial^2 z}{\partial y^2} = 0$ into $\dfrac{\partial^2 z}{\partial u \partial v} = 0$, find the values of a and b.

9. If $z = f(x, y)$ where $x = \log u$, $y = \log v$ then show that

(*i*) $\dfrac{\partial^2 z}{\partial x^2} + \dfrac{\partial^2 z}{\partial y^2} = u^2 \dfrac{\partial^2 z}{\partial u^2} + v^2 \dfrac{\partial^2 z}{\partial v^2} + u\dfrac{\partial z}{\partial u} + v\dfrac{\partial z}{\partial v}$.

(*ii*) $\dfrac{\partial^2 z}{\partial x \partial y} = uv \dfrac{\partial^2 z}{\partial u \partial v}$.

10. If $z = f(u, v)$ where $u = x + y$, $v = x - y$ show that $\dfrac{\partial z}{\partial x} + \dfrac{\partial z}{\partial y} = 2\dfrac{\partial z}{\partial u}$.

11. If $z = f(x, y)$ where $x = u \cosh v$ and $y = u \sinh v$, then show that

$$\left(\frac{\partial z}{\partial x}\right)^2 - \left(\frac{\partial z}{\partial y}\right)^2 = \left(\frac{\partial z}{\partial u}\right)^2 - \frac{1}{u^2}\left(\frac{\partial z}{\partial v}\right)^2.$$

12. If $z = f(x, y)$ where $x = uv$ and $y = \dfrac{u}{v}$ then show that

(i) $\dfrac{\partial z}{\partial x} = \dfrac{1}{2v}\dfrac{\partial z}{\partial u} + \dfrac{1}{2u}\dfrac{\partial z}{\partial v}$

(ii) $\dfrac{\partial z}{\partial y} = \dfrac{v}{2}\dfrac{\partial z}{\partial u} - \dfrac{v^2}{2u}\dfrac{\partial z}{\partial v}$.

13. A function $f(x, y)$ is expressed in terms of new variables u and v given by $u = e^x \cos y$ and $v = e^x \sin y$ show that

(i) $\dfrac{\partial f}{\partial x} = u\dfrac{\partial f}{\partial u} + v\dfrac{\partial f}{\partial v}$

(ii) $\dfrac{\partial f}{\partial y} = -v\dfrac{\partial f}{\partial u} + u\dfrac{\partial f}{\partial v}$

(iii) $\dfrac{\partial^2 f}{\partial x^2} + \dfrac{\partial^2 f}{\partial y^2} = (u^2 + v^2)\left(\dfrac{\partial^2 f}{\partial u^2} + \dfrac{\partial^2 f}{\partial v^2}\right)$

14. If $u = f(ax^2 + 2hxy + by^2)$ and $v = \phi\,(ax^2 + 2hxy + by^2)$ prove that

$$\frac{\partial}{\partial y}\left(u\frac{\partial v}{\partial x}\right) = \frac{\partial}{\partial x}\left(u\frac{\partial v}{\partial y}\right).$$

15. If $V = f(2x - 3y, 3y - 4z, 4z - 2x)$ compute the value of $6V_x + 4V_y + 3V_z$.

(UPTU 2008-09)

❑❑❑

CHAPTER

4 Expansion of Functions of Several Variables and Jacobians

Expansion of Functions of Two and Three Variables, Jacobians and Functional Dependence

TAYLOR'S THEOREM FOR FUNCTIONS OF TWO VARIABLES

Let us first recall the **Taylor's expansion** of function of one variable as

$$f(x+h) = f(x) + hf'(x) + \frac{h^2}{2!} f''(x) + \ldots \qquad \ldots (1)$$

where h is small. Consider the extension of above result to function of two independent variables, *i.e.* to obtain the expansion of $f(x+h, y+k)$ in powers of h and k, both h and k being small. We shall prove that

$$f(x+h,\ y+k) = f(x, y) + \left(h\frac{\partial}{\partial x} + k\frac{\partial}{\partial y}\right)f + \frac{1}{2!}\left(h\frac{\partial}{\partial x} + k\frac{\partial}{\partial y}\right)^2 f + \ldots$$

In $f(x+h,\ y+k)$ first take $x+h$ as constant and expand $f(x+h,\ y+k)$ in powers of k using the relation (1), as

$$f(x+h, y+k) = f(x+h, y) + k\frac{\partial}{\partial y} f(x+h, k) + \frac{k^2}{2!}\frac{\partial^2}{\partial y^2} f(x+h, y) + \ldots . \qquad \ldots (2)$$

Now expanding each term on the R.H.S. of (2) in powers of h by Taylor's theorem of one variable considering y as constant, we have

$$f(x+h, y+k) = \left[f(x,y) + h\frac{\partial f}{\partial x} + \frac{h^2}{2!}\frac{\partial^2 f}{\partial x^2} + \ldots\right] + k\frac{\partial}{\partial y}\left[f(x,y) + h\frac{\partial f}{\partial x} + \frac{h^2}{21}\frac{\partial^2 f}{\partial x^2} + \ldots\right]$$

$$+ \frac{k^2}{2!}\frac{\partial^2}{\partial y^2}\left[f(x,y) + h\frac{\partial f}{\partial x} + \frac{h^2}{2!}\frac{\partial^2 f}{\partial x^2} + \ldots\right] + \ldots$$

or $$\boldsymbol{f(x+h, y+k) = f(x,y) + \left(h\frac{\partial f}{\partial x} + k\frac{\partial f}{\partial y}\right) + \frac{1}{2!}\left(h^2\frac{\partial^2 f}{\partial x^2} + 2hk\frac{\partial^2 f}{\partial x\,\partial y} + k^2\frac{\partial^2 f}{\partial x^2}\right) + \ldots}$$

This extension of Taylor's theorem is useful when we discuss the maxima and minima of function of two variables. If we put $x = 0$, $y = 0$ and write x for h and y for k we arrive at the extension of **Maclaurin's theorem for two independent variables** as

$$f(x, y) = f(0, 0) + x\left(\frac{\partial f}{\partial x}\right)_{(0,0)} + y\left(\frac{\partial f}{\partial y}\right)_{(0,0)}$$

$$+\frac{1}{2!}\left\{x^2\left(\frac{\partial^2 f}{\partial x^2}\right)_{(0,0)} + 2xy\left(\frac{\partial^2 f}{\partial x\,\partial y}\right)_{(0,0)} + y^2\left(\frac{\partial^2 f}{\partial y^2}\right)_{(0,0)}\right\} + \dots .$$

EXAMPLE 4.1. Expand $f(x, y) = e^{xy}$ about $(1, 1)$ upto second degree terms.

SOLUTION: $$f(x, y) = f(1, 1) + \left[(x-1)\frac{\partial f}{\partial x} + (y-1)\frac{\partial f}{\partial y}\right]_{(1,1)}$$

$$+\frac{1}{2!}\left[(x-1)^2\frac{\partial^2 f}{\partial x^2} + 2(x-1)(y-1)\frac{\partial^2 f}{\partial x\,\partial y} + (y-1)^2\frac{\partial^2 f}{\partial y^2}\right]_{(1,1)} + \dots$$

$$= e + (x-1)(ye^{xy})_{(1,1)} + (y-1)(xe^{xy})_{(1,1)} + \frac{1}{2}[(x-1)^2 e + 2(x-1)(y-1)e + (y-1)^2 e] + \dots$$

$$= e(x+y-1) + \frac{e}{2}(x+y-2)^2 \text{ (on simplification).}$$ **Ans.**

EXAMPLE 4.2. Find the expansion of $\cos x \cos y$ in powers of x, y upto fourth order terms.

SOLUTION: Expanding the function $f(x, y) = \cos x \cos y$ about the origin by Taylor's theorem and retaining terms upto fourth order, we have

$$f(x, y) = f(0, 0) + (xf_x + yf_y) + \frac{1}{2!}(x^2 f_{xx} + 2xyf_{xy} + y^2 f_{yy})$$

$$+\frac{1}{3!}[x^3 f_{xxx} + 3x^2 y\, f_{xxy} + 3xy^2\, f_{xyy} + y^3\, f_{yyy}]$$

$$+\frac{1}{4!}[x^4 f_{xxxx} + 4x^3 y f_{xxxy} + 6x^2 y^2 f_{xxyy} + 4xy^3 f_{xyyy} + y^4 f_{yyyy}]$$

where derivatives are taken at the origin.

Now, $f(0, 0) = \cos 0 \cos 0 = 1$, $f_x = -\sin x \cos y = 0$ at $(0, 0)$, $f_y = -\cos x \sin y = 0$ at $(0,0)$

$f_{xx} = -\cos x \cos y = -1$ at $(0, 0)$ $f_{xy} = \sin x \sin y = 0$ at $(0, 0)$,

$f_{yy} = -\cos x \cos y = -1$ at $(0, 0)$.

$f_{xxx} = \sin x \sin y = 0$ at $(0, 0)$, $f_{xxy} = +\cos x \sin y = 0$ at $(0, 0)$,

$f_{xyy} = \sin x \cos y = 0$ at $(0, 0)$

$f_{yyy} = +\cos x \sin y = 0$ at $(0, 0)$, $f_{xxxx} = \cos x \cos y =$ at $(0, 0)$,

$f_{xxxy} = -\sin x \sin y = 0$ at $(0, 0)$, $f_{xxyy} = \cos x \cos y = 1$ at $(0, 0)$,

$$f_{xyyy} = -\sin x \sin y = 0 \text{ at } (0, 0) \quad f_{yyyy} = \cos x \cos y = 1 \text{ at } (0, 0)$$

$$\therefore \cos x \cos y = 1 + (0.x + 0.y) + \frac{1}{2!}(-1x^2 + 0.2xy - 1y^2) + \frac{1}{3!}[0\,x^3 + 0.3x^2y + 0.3xy^2 + 0.y^3]$$
$$+ \frac{1}{4!}[1.x^4 + 0.4x^3y + 1.6x^2y^2 + 0.4xy^3 + 1.y^4]$$

$$= 1 - \frac{x^2}{2!} - \frac{y^2}{2!} + \frac{x^4}{4!} + \frac{6x^2y^2}{4!} + \frac{1}{4!}y^4$$

$$= 1 - \frac{1}{2}(x^2 + y^2) + \frac{1}{24}(x^4 + 6x^2y^2 + y^4). \qquad \textbf{Ans.}$$

ALITER

$$\cos x \cos y = [1 - \frac{x^2}{2!} + \frac{x^4}{4!} - \frac{x^6}{6!} + ...][1 - \frac{y^2}{2!} + \frac{y^4}{4!} - \frac{y^6}{6!} + ...]$$

$$= 1 - \frac{x^2}{2!} - \frac{y^2}{2!} + \frac{x^4}{4!} + \frac{y^4}{4!} + \frac{x^2y^2}{2!2!} \quad \text{keeping terms upto 4th orders}$$

$$= 1 - \frac{1}{2}(x^2 + y^2) + \frac{x^4}{24} + \frac{x^2y^2}{4} + \frac{y^4}{24}. \qquad \textbf{Ans.}$$

EXAMPLE 4.3. Expand $e^x \log(1 + y)$ in the neighbourhood of the origin retaining terms upto second degree in x and y.

SOLUTION: We know that $e^x = 1 + x + \frac{x^2}{2!} + \frac{x^3}{3!} + ...$

and $\log(1 + y) = y - \frac{y^2}{2} + \frac{y^3}{3} - \frac{y^4}{4} + ...$

Therefore $e^x \log(1 + y) = \left[1 + x + \frac{x^2}{2!} + \frac{x^3}{3!} + ...\right]\left[y - \frac{y^2}{2} + \frac{y^3}{3} ...\right]$

$$= y - \frac{y^2}{2} + xy \quad \text{retaining terms upto second degree.} \qquad \textbf{Ans.}$$

EXAMPLE 4.4. Expand $\sin(xy)$ is powers of $(x - 1)$ and $(y - \pi/2)$ upto and including the second degree terms.

SOLUTION: $f(x, y) = \sin(xy)$, $f_x = y\cos(xy)$, $f_y = x\cos(xy)$,
$f_{xx} = -y^2\sin(xy)$, $f_{xy} = 1\cos(xy) - xy\sin(xy)$, $f_{yy} = -x^2\sin(xy)$.

The expansion or $f(x, y)$ about $(1, \pi/2)$ by, Taylor's theorem, is

$$f(x, y) = f(1, \pi/2) + [(x - 1)f_x + (y - \pi/2)f_y] + \frac{1}{2!}\left[(x-1)^2 f_{xx} + 2(x-1)\left(y - \frac{\pi}{2}\right)f_{xy} + \left(y - \frac{\pi}{2}\right)^2 f_{yy}\right]$$

retaining terms upto second degree and taking derivatives at $(1, \pi/2)$.

Here $f(1, \pi/2) = 1, \quad f_x = \frac{\pi}{2} \cos \frac{\pi}{2}, \quad f_y = 1 \cos \pi/2$

and $\quad f_{xx} = -\left(\frac{\pi}{2}\right)^2 .1, \quad f_{xy} = 0 - \frac{\pi}{2} . 1 = -\frac{\pi}{2}, \quad f_{yy} = -1$ at $(1, \pi/2)$

$$\therefore \quad \sin(xy) = 1 + [0(x-1) + 0(y - \pi/2)] + \frac{1}{2!}\left[-\frac{\pi^2}{4}(x-1)^2 - 2\left(\frac{-\pi}{2}\right)(x-1)\left(y-\frac{\pi}{2}\right) - \left(y-\frac{\pi}{2}\right)^2\right]$$

$$= 1 - \frac{\pi}{8}(x-1)^2 + \pi(x-1)(y-\pi/2) - (y-\pi/2)^2. \qquad \textbf{Ans.}$$

EXAMPLE 4.5. Obtain the Taylor's expansion of the function $f(x, y) = \tan^{-1}\left(\frac{y}{x}\right)$ about $(1, 1)$, upto including second degree terms and compute $f(1.1, 0.9)$.

(UPTU 2002-03; UPTU 2005-06)

SOLUTION: Given $f(x, y) = \tan^{-1}\left(\frac{y}{x}\right)$. Taylor's expansion of $f(x, y)$ about $(1, 1)$ is

$$f(x, y) = f(1, 1) + (x-1)\left(\frac{\partial f}{\partial x}\right)_{(1,1)} + (y-1)\left(\frac{\partial f}{\partial y}\right)_{(1,1)}$$

$$+\frac{1}{2!}\left[(x-1)^2\left(\frac{\partial^2 f}{\partial x^2}\right)_{(1,1)} + 2(x-1)(y-1)\left(\frac{\partial^2 f}{\partial x \partial y}\right)_{(1,1)} + (y-1)^2\left(\frac{\partial^2 f}{\partial y^2}\right)_{(1,1)}\right] + \ldots.$$

Here $\quad \frac{\partial f}{\partial x} = \frac{-y}{x^2+y^2} = -\frac{1}{2}$ at $(1, 1)$, $\quad \frac{\partial f}{\partial y} = \frac{x}{x^2+y^2} = \frac{1}{2}$ at $(1, 1)$,

$\frac{\partial^2 f}{\partial x^2} = \frac{2xy}{(x^2+y^2)^2} = \frac{1}{2}$ at $(1, 1)$, $\quad \frac{\partial^2 f}{\partial y^2} = \frac{-2xy}{(x^2+y^2)^2} = -\frac{1}{2}$ at $(1, 1)$

and $\quad \frac{\partial^2 f}{\partial x \partial y} = \frac{y^2 - x^2}{(x^2+y^2)} = 0$ at $(1, 1)$.

$$\therefore \quad \tan^{-1}\left(\frac{y}{x}\right) = \frac{\pi}{4} + (x-1)\left(-\frac{1}{2}\right) + (y-1)\left(\frac{1}{2}\right) + \frac{1}{2!}\left[(x-1)^2\left(\frac{1}{2}\right) + 2(x-1)(y-1)(0) + (y-1)^2\left(-\frac{1}{2}\right)\right]$$

Next, $\quad x - 1 = 1.1 - 1 = 0.1, \quad y - 1 = 0.9 - 1 = -0.1$

$$\therefore \quad f(1.1, 0.9) = \frac{\pi}{4} - \frac{1}{2}(0.1) + \frac{1}{2}(-0.1) + \frac{1}{4}(0.1)^2 - \frac{1}{4}(-0.1)^2 = 0.7862. \qquad \textbf{Ans.}$$

EXAMPLE 4.6. Expand $e^{ax}\sin by$ in powers of x and y upto third degree terms.

(UPTU 2010-11)

SOLUTION: $f(x,y) = e^{ax}\sin by$. Its expansion in powers of x and y, is

$$f(x,y) = f(0,0) + xf_x(0,0) + yf_y(0,0) + \frac{1}{2!}[x^2 f_{xx}(0,0) + 2xyf_{xy}(0,0) + y^2 f_{yy}(0,0)]$$

$$+ \frac{1}{3!}[x^3 f_{xxx}(0,0) + 3x^2 yf_{xxy}(0,0) + 3xy^2 f_{xyy}(0,0) + y^3 f_{yyy}(0,0)] + ...$$

Now, $f_x = ae^{ax}\sin by,\ f_y = be^{ax}\cos by,\ f_{xx} = a^2e^{ax}\sin by,\ f_{xy} = abe^{ax}\cos by,\ f_{yy} = -b^2e^{ax}\sin by$,

$$f_{xxx} = a^3e^{ax}\sin by,\ f_{xxy} = a^2be^{ax}\cos by,\ f_{xyy} = -ab^2e^{ax}\sin by,\ f_{yyy} = -b^3e^{ax}\cos by.$$

$$f_x(0,0) = f_{xx}(0,0) = f_{yy}(0,0) = f_{xxx}(0,0) = f_{xyy}(0,0) = 0,\ f_y(0,0) = b,\ f_{xy}(0,0) = ab,$$

$$f_{xxy}(0,0) = a^2b \text{ and } f_{yyy}(0,0) = -b^3.$$

Hence $\quad f(x,y) = by + \frac{1}{2!}2xy\cdot ab + \frac{1}{3!}[3x^2ya^2b - y^3b^3].$ **Ans.**

EXAMPLE 4.7. **(a)** Expand $x^2y + 3y - 2$ in powers of $x-1$ and $y+2$ using Taylor's theorem.

(UPTU 2001 Q.B.)

(b) Find the Taylor series expansion of $f(x, y) = x^3 + xy^2$ about the point (2, 1). **[UPTU 2011-12]**

SOLUTION: **(a)** Expansion of $f(x, y)$ in powers of $x - a$ and $y - b$ is

$$f(x,y) = f(a,b) + (x-a)f_x(a,b) + (y-b)f_y(a,b)$$

$$+ \frac{1}{2!}[(x-a)^2 f_{xx}(a,b) + 2(x-a)(y-b)f_{xy}(a,b) + (y-b)^2 f_{yy}(a,b)] + ...$$

Here $\quad f(x,y) = x^2y + 3y - 2, \quad \therefore \quad f(1,-2) = -10.$

$$f_x = 2xy \ \therefore\ f_x(1,-2) = -4,\ f_y = x^2 + 3 \ \therefore\ f_y(1,-2) = 4$$

$$f_{xx} = 2y \ \therefore\ f_{xx}(1,-2) = -4,\ f_{xy} = 2x,\ f_{xy}(1,-2) = 2,\ f_{yy} = 0.$$

Therefore, the required expansion is

$$x^2y + 3y - 2 = -10 - 4(x-1) + 4(y+2) + \frac{1}{2!}[-4(x-1)^2 + 4(x-1)(y+2) + (y+2)^2(0)] + ...$$

$$= -10 - 4(x-1) + 4(y+2) - 2(x-1)^2 + 2(x-1)(y+2) + ... \quad \textbf{Ans.}$$

(b) The Taylor's expansion of $f(x, y)$ about the point (2, 1) is

$$f(x,y) = f(2,1) + \left[(x-2)\frac{\partial f}{\partial x} + (y-1)\frac{\partial f}{\partial y}\right] + \frac{1}{2!}\left[(x-2)^2\frac{\partial^2 f}{\partial x^2} + 2(x-2)(y-1)\frac{\partial^2 f}{\partial x \partial y} + (y-1)^2\frac{\partial^2 f}{\partial y^2}\right] + ...$$

where derivatives of f are to be taken at (2, 1).

$$f(x, y) = x^3 + xy^2 \quad \therefore \quad f(2, 1) = 8 + 2 = 10$$

$$\frac{\partial f}{\partial x} = 3x^2 + y^2 = 13 \text{ at } (2, 1), \quad \frac{\partial f}{\partial y} = 2xy = 4 \text{ at } (2, 1);$$

$$\frac{\partial^2 f}{\partial x^2} = 6x = 12 \text{ at } (2, 1), \quad \frac{\partial^2 f}{\partial x \partial y} = 2y = 2 \text{ at } (2, 1), \quad \frac{\partial^2 f}{\partial y^2} = 2x = 4 \text{ at } (2, 1)$$

and so on.

$$\therefore \quad x^3 + xy^2 = 10 + [13(x-2) + 4(y-1)] + \frac{1}{2!}[12(x-2)^2 + 4(x-2)(y-1) + 4(y-1)^2] + \ldots$$ **Ans.**

EXAMPLE 4.8. Find Taylor series of the function $f(x, y) = e^{-x^2-y^2} \cos xy$ about the point $x_0 = 0,\ y_0 = 0$ upto three terms. **(UPTU 2006–07)**

SOLUTION: $f(x, y) = e^{-x^2-y^2} \cos xy$ hence $f(0, 0) = 1$

Now, $f_x = -(2x \cos xy + y \sin xy)\, e^{-x^2-y^2}$ and $f_y = -(2y \cos xy + x \sin xy)\, e^{-x^2-y^2}$
hence $f_x(0, 0) = f_y(0, 0) = 0.$

Next, $f_{xx} = -(-2x)(2x \cos xy + y \sin xy)\, e^{-x^2-y^2} - e^{-x^2-y^2}(2 \cos xy - 2xy \sin xy + y^2 \cos xy)$
hence $f_{xx}(0, 0) = -2.$

and $f_{yy} = -(-2y)(2y \cos xy + x \sin xy)\, e^{-x^2-y^2} - e^{-x^2-y^2}(2 \cos xy - 2yx \sin xy + x^2 \cos xy)$
hence $f_{yy}(0, 0) = -2.$

Next, $f_{xy} = -(-2y)(2x \cos xy + y \sin xy)\, e^{-x^2-y^2} - e^{-x^2-y^2}(2x^2 \sin xy + xy \cos xy + \sin xy)$
hence $f_{xy}(0, 0) = 0.$

We know that $f(x, y) = f(0, 0) + [xf_x(0, 0) + yf_y(0, 0)] + \frac{1}{2!}[x^2 f_{xx}(0, 0) + 2xyf_{xy}(0, 0) + y^2 f_{yy}(0, 0)]$

Hence $f(x, y) = 1 + 0 + 0 + \frac{1}{2!}[-2x^2 + 0 - 2y^2] + \ldots$

$= 1 - x^2 - y^2$ (upto three terms). **Ans.**

EXAMPLE 4.9. Expand $e^{a \sin^{-1} x}$ in ascending powers of x as far upto x^4 by Maclaurins series. **(UPTU 2001–02)**

SOLUTION: Let $y = e^{a \sin^{-1} x}$ then $y_1 = e^{a \sin^{-1} x} \frac{a}{\sqrt{1-x^2}}$ or $(1 - x^2)\, y_1^2 = y^2 a^2.$

Differentiating again, we get $(1 - x^2)\, 2y_1 y_2 - 2xy_1^2 = 2a^2 yy_1$ or $(1 - x^2)\, y_2 - xy_1 - a^2 y = 0,$ if $y_1 \neq 0$

Differentiating n times the above relation w.r.t. x and using Leibnitz's theorem, we get

$$(1-x^2)\,y_{n+2}+n\,(-2x)\,y_{n+1}+\frac{n(n-1)}{2}(-2)\,y_n-xy_{n+1}-n\cdot 1\cdot y_n-a^2y_n=0$$

or $\qquad (1-x^2)\,y_{n+2}-x(2n+1)\,y_{n+1}-(n^2+a^2)\,y_n=0$

Putting $x = 0$, we get $y_{n+2}(0) = (n^2 + a^2)\,y_n(0)$

Clearly $y(0) = 1$, $\quad y_1(0) = a$, $\quad y_2(0) = a^2$

By Maclaurin's theorem $\quad y = e^{a\sin^{-1}x} = y(0) + xy_1(0) + \frac{x^2}{2!}y_2(0)+\frac{x^3}{3!}y_3(0)+\frac{x^4}{4!}y_4(0)+...$

Also, from the previous relations, $y_3(0) = (a^2 + 1)\,a$ and $y_4(0) = (a^2 + 2^2)\,a^2$.

Thus, $\qquad e^{a\sin^{-1}x} = 1+ax+\frac{a^2x^2}{2!}+\frac{(a^2+1)a}{3!}x^3+\frac{(a^2+4)a^2}{4!}x^4+...\,.$ **Ans.**

EXAMPLE 4.10. Expand x^y in powers of $(x - 1)$ and $(y - 1)$ upto third degree terms.

(UPTU 2003–04)

SOLUTION: We know, by Taylor's theorem, that $f(x, y)$

$$= f(1,1)+(x-1)f_x(1,1)+(y-1)f_y(1,1)+\frac{1}{2!}[(x-1)^2 f_{xx}(1,1)+2(x-1)(y-1)f_{xy}(1,1)+(y-1)^2 f_{yy}(1,1)]$$

$$+\frac{1}{3!}[(x-1)^3 f_{xxx}(1,1)+3(x-1)^2(y-1)f_{xxy}(1,1)+3(x-1)(y-1)^2 f_{xyy}(1,1)+(y-1)^3 f_{yyy}(1,1)]+...$$

Now $\qquad f(x, y) = x^y$ hence $f(1, 1) = 1$

Next, $\qquad f_x = yx^{y-1}$ and $f_y = x^y \log x$ hence $f_x(1, 1) = 1$ and $f_y(1, 1) = 0$

Further, $\qquad f_{xx} = y(y-1)\,x^{y-2}$ and $f_{yy} = x^y(\log x)^2$ and $f_{xy} = yx^{y-1}\log x + x^y x^{y-1}$

Hence $\qquad f_{xx}(1, 1) = 0$, $f_{yy}(1, 1) = 0$ and $f_{xy}(1, 1) = 1$

Next $\qquad f_{xxx} = y(y-1)(y-2)\,x^{y-3}$, $f_{xxy} = (y^2 - y)\,x^{y-2}\log x + (2y - 1)\,x^{y-2}$,

$\qquad f_{xyy} = \log x\,[\,yx^{y-1}\log x + x^{y-1}] + x^{y-1}\log x$ and $f_{yyy} = (\log x)^3\,x^y$

$\therefore \qquad f_{xxx}(1, 1) = 0$, $f_{xxy}(1, 1) = 1$, $f_{xyy} = 0$ and $f_{yyy} = 0$

$$\therefore \qquad x^y = 1+1(x-1)+0(y-1)+\frac{1}{2!}\left[0(x-1)^2+2(x-1)(y-1)\cdot 1+0(y-1)^2\right]$$

$$\frac{1}{3!}\left[(x-1)^3(0)+3(x-1)^2(y-1)\cdot 1+3(x-1)(y-1)^2(0)+(y-1)^3(0)\right]+...$$

$$= 1+(x-1)+(x-1)(y-1)+\frac{1}{2}(x-1)(y-1)^2$$ (upto third degree terms). **Ans.**

JACOBIANS

Let $u = u(x, y)$ and $v = v(x, y)$ be two continuous functions of the independent variables x and y such that $\frac{\partial u}{\partial x}, \frac{\partial u}{\partial y}, \frac{\partial v}{\partial x}, \frac{\partial v}{\partial y}$ are also continuous in x and y. Then the determinant $\begin{vmatrix} \frac{\partial u}{\partial x} & \frac{\partial u}{\partial y} \\ \frac{\partial v}{\partial x} & \frac{\partial v}{\partial y} \end{vmatrix}$ is called the Jacobian of u and v with respect to x and y, and is denoted by $\mathbf{J}\left(\frac{u, v}{x, y}\right)$ or $\frac{\partial (u, v)}{\partial (x, y)}$.

Similarly, if u, v, w are functions of x, y, z then Jacobian of u, v, w with respect to x, y, z is defined as

$$\begin{vmatrix} \frac{\partial u}{\partial x} & \frac{\partial u}{\partial y} & \frac{\partial u}{\partial z} \\ \frac{\partial v}{\partial x} & \frac{\partial v}{\partial y} & \frac{\partial v}{\partial z} \\ \frac{\partial w}{\partial x} & \frac{\partial w}{\partial y} & \frac{\partial w}{\partial z} \end{vmatrix} = \mathbf{J}\left(\frac{u, v, w}{x, y, z}\right) = \frac{\partial (u, v, w)}{\partial (x, y, z)}.$$

CHAIN RULE FOR JACOBIANS

If u, v are functions of x, y and x, y are themselves functions of r, s then

$$\frac{\partial (u, v)}{\partial (x, y)} \cdot \frac{\partial (x, y)}{\partial (r, s)} = \frac{\partial (u, v)}{\partial (r, s)}$$

Now,
$$\frac{\partial (u, v)}{\partial (x, y)} \cdot \frac{\partial (x, y)}{\partial (r, s)} = \begin{vmatrix} \frac{\partial u}{\partial x} & \frac{\partial u}{\partial y} \\ \frac{\partial v}{\partial x} & \frac{\partial v}{\partial y} \end{vmatrix} \begin{vmatrix} \frac{\partial x}{\partial r} & \frac{\partial x}{\partial s} \\ \frac{\partial y}{\partial r} & \frac{\partial y}{\partial s} \end{vmatrix}$$

$$= \begin{vmatrix} \frac{\partial u}{\partial x}\frac{\partial x}{\partial r} + \frac{\partial u}{\partial y}\frac{\partial y}{\partial r} & \frac{\partial u}{\partial x}\frac{\partial x}{\partial s} + \frac{\partial u}{\partial y}\frac{\partial y}{\partial s} \\ \frac{\partial v}{\partial x}\frac{\partial x}{\partial r} + \frac{\partial v}{\partial y}\frac{\partial y}{\partial r} & \frac{\partial v}{\partial x}\frac{\partial x}{\partial s} + \frac{\partial v}{\partial y}\frac{\partial y}{\partial s} \end{vmatrix} \quad \text{... (1)}$$

But since $u = u(x, y)$, $v = v(x, y)$ and $x = x(r, s)$, $y = y(r, s)$, we have

$$\frac{\partial u}{\partial r} = \frac{\partial u}{\partial x}\frac{\partial x}{\partial r} + \frac{\partial u}{\partial y}\frac{\partial y}{\partial r}, \quad \frac{\partial u}{\partial s} = \frac{\partial u}{\partial x}\frac{\partial x}{\partial s} + \frac{\partial u}{\partial y}\frac{\partial y}{\partial s},$$

$$\frac{\partial v}{\partial r} = \frac{\partial v}{\partial x}\frac{\partial x}{\partial r} + \frac{\partial v}{\partial y}\frac{\partial y}{\partial r}, \quad \frac{\partial v}{\partial s} = \frac{\partial v}{\partial x}\frac{\partial x}{\partial s} + \frac{\partial v}{\partial y}\frac{\partial y}{\partial s}.$$

Therefore (1) becomes
$$\frac{\partial (u, v)}{\partial (x, y)} \cdot \frac{\partial (x, y)}{\partial (r, s)} = \begin{vmatrix} \frac{\partial u}{\partial r} & \frac{\partial u}{\partial s} \\ \frac{\partial v}{\partial r} & \frac{\partial v}{\partial s} \end{vmatrix} = \frac{\partial (u, v)}{\partial (r, s)}.$$

In general
$$\frac{\partial (x_1, x_2, \ldots, x_n)}{\partial (u_1, u_2, \ldots, u_n)} \cdot \frac{\partial (u_1, u_2, \ldots, u_n)}{\partial (v_1, v_2, \ldots, v_n)} = \frac{\partial (x_1, x_2, \ldots, x_n)}{\partial (v_1, v_2, \ldots, v_n)}.$$

Corollary of Chain Rule

(UPTU 2005-06)

$$\frac{\partial (u, v)}{\partial (x, y)} \cdot \frac{\partial (x, y)}{\partial (u, v)} = 1.$$

In the previous result if we replace r, s by u, v this corollary immediately follows.

If $\mathbf{J} = \dfrac{\partial (u, v)}{\partial (x, y)}$ then the Jacobian $\dfrac{\partial (x, y)}{\partial (u, v)}$ is denoted by J′, that is,

$$\mathbf{J}' = \frac{\partial (x, y)}{\partial (u, v)} \text{ and we have } \mathbf{JJ}' = 1.$$ **(UPTU 2001 Q.B.)**

Also, in general, we have

$$\frac{\partial (x_1, x_2, \ldots, x_n)}{\partial (u_1, u_2, \ldots, u_n)} \cdot \frac{\partial (u_1, u_2, \ldots, u_n)}{\partial (x_1, x_2, \ldots, x_n)} = 1.$$

JACOBIAN OF IMPLICIT FUNCTIONS

If u and v are implicit functions of the variables x and y, connected by the relations

$$f_1 (u, v, x, y) = 0 \quad \text{and} \quad f_2 (u, v, x, y) = 0.$$

Differentiating the above relations partially *w.r.t.* x and y separately, we get

$$\frac{\partial f_1}{\partial x} + \frac{\partial f_1}{\partial u}\frac{\partial u}{\partial x} + \frac{\partial f_1}{\partial v}\frac{\partial v}{\partial x} = 0, \quad \frac{\partial f_1}{\partial y} + \frac{\partial f_1}{\partial u}\frac{\partial u}{\partial y} + \frac{\partial f_1}{\partial v}\frac{\partial v}{\partial y} = 0 \qquad \ldots (1)$$

and

$$\frac{\partial f_2}{\partial x} + \frac{\partial f_2}{\partial u}\frac{\partial u}{\partial x} + \frac{\partial f_2}{\partial v}\frac{\partial v}{\partial x} = 0, \quad \frac{\partial f_2}{\partial y} + \frac{\partial f_2}{\partial u}\frac{\partial u}{\partial y} + \frac{\partial f_2}{\partial v}\frac{\partial v}{\partial y} = 0 \qquad \ldots (2)$$

Now consider

$$\frac{\partial (f_1, f_2)}{\partial (u, v)} \cdot \frac{\partial (u, v)}{\partial (x, y)} = \begin{vmatrix} \dfrac{\partial f_1}{\partial u} & \dfrac{\partial f_1}{\partial v} \\ \dfrac{\partial f_2}{\partial u} & \dfrac{\partial f_2}{\partial v} \end{vmatrix} \begin{vmatrix} \dfrac{\partial u}{\partial x} & \dfrac{\partial u}{\partial y} \\ \dfrac{\partial v}{\partial x} & \dfrac{\partial v}{\partial y} \end{vmatrix}$$

$$= \begin{vmatrix} \dfrac{\partial f_1}{\partial u}\dfrac{\partial u}{\partial x} + \dfrac{\partial f_1}{\partial v}\dfrac{\partial v}{\partial x} & \dfrac{\partial f_1}{\partial u}\dfrac{\partial u}{\partial y} + \dfrac{\partial f_1}{\partial v}\dfrac{\partial v}{\partial y} \\ \dfrac{\partial f_2}{\partial u}\dfrac{\partial u}{\partial x} + \dfrac{\partial f_2}{\partial v}\dfrac{\partial v}{\partial x} & \dfrac{\partial f_2}{\partial u}\dfrac{\partial u}{\partial y} + \dfrac{\partial f_2}{\partial v}\dfrac{\partial v}{\partial y} \end{vmatrix}$$

$$= \begin{vmatrix} -\dfrac{\partial f_1}{\partial x} & -\dfrac{\partial f_1}{\partial y} \\ -\dfrac{\partial f_2}{\partial x} & -\dfrac{\partial f_2}{\partial y} \end{vmatrix} = (-1)^2 \frac{\partial (f_1, f_2)}{\partial (x, y)} \qquad \text{[using (1) and (2)]}$$

Thus, we have

$$\boxed{\frac{\partial (u, v)}{\partial (x, y)} = (-1)^2 \frac{\dfrac{\partial (f_1, f_2)}{\partial (x, y)}}{\dfrac{\partial (f_1, f_2)}{\partial (u, v)}}}.$$

For the case when u, v, w are implicit functions of $x, y, z,$ given by the relation

$$f_1(u, v, w, x, y, z) = 0$$
$$f_2(u, v, w, x, y, z) = 0$$
$$f_3(u, v, w, x, y, z) = 0$$

we shall have
$$\boxed{\frac{\partial(u,v,w)}{\partial(x,y,z)} = (-1)^3 \frac{\dfrac{\partial(f_1,f_2,f_3)}{\partial(x,y,z)}}{\dfrac{\partial(f_1,f_2,f_3)}{\partial(u,v,w)}}}.$$

EXAMPLE 4.11. (a) If $r^2 = x^2 + y^2$, $\theta = \tan^{-1}(y/x)$ evaluate $\dfrac{\partial(r,\theta)}{\partial(x,y)}$. **(UPTU 2001 Q.B.)**

(b) If $u = e^x \sin y$ and $v = e^x \cos y$ evaluate $\dfrac{\partial(u,v)}{\partial(x,y)}$. **[UPTU 2011-12]**

SOLUTION: (a) Since $r^2 = x^2 + y^2$ we have $\dfrac{\partial r}{\partial x} = \dfrac{x}{r} = \cos\theta$ and $\dfrac{\partial r}{\partial y} = \dfrac{y}{r} = \sin\theta$

and as $\theta = \tan^{-1}(y/x)$, we have
$$\frac{\partial\theta}{\partial x} = \frac{-y/x^2}{1+\dfrac{y^2}{x^2}} = \frac{-y}{x^2+y^2} = \frac{-r\sin\theta}{r^2} = \frac{-\sin\theta}{r}$$

and
$$\frac{\partial\theta}{\partial y} = \frac{1/x}{1+\dfrac{y^2}{x^2}} = \frac{x}{x^2+y^2} = \frac{r\cos\theta}{r^2} = \frac{\cos\theta}{r}.$$

Therefore,
$$\frac{\partial(r,\theta)}{\partial(x,y)} = \begin{vmatrix} \dfrac{\partial r}{\partial x} & \dfrac{\partial\theta}{\partial x} \\ \dfrac{\partial r}{\partial y} & \dfrac{\partial\theta}{\partial y} \end{vmatrix} = \begin{vmatrix} \cos\theta & \dfrac{-\sin\theta}{r} \\ \sin\theta & \dfrac{\cos\theta}{r} \end{vmatrix} = \frac{1}{r}[\cos^2\theta + \sin^2\theta) = \frac{1}{r}. \quad \textbf{Ans.}$$

(b) Here $u = e^x \sin y$, $v = e^x \cos y$. Hence

$$\frac{\partial u}{\partial x} = e^x \sin y, \quad \frac{\partial u}{\partial y} = e^x \cos y, \quad \frac{\partial v}{\partial x} = e^x \sin y, \quad \frac{\partial v}{\partial y} = -e^x \sin y$$

$$\therefore \quad \frac{\partial(u,v)}{\partial(x,y)} = \begin{vmatrix} \dfrac{\partial u}{\partial x} & \dfrac{\partial u}{\partial y} \\ \dfrac{\partial v}{\partial x} & \dfrac{\partial v}{\partial y} \end{vmatrix} = \begin{vmatrix} e^x \sin y & e^x \cos y \\ e^x \cos y & -e^x \sin y \end{vmatrix}$$

$$= (e^x)^2 \begin{vmatrix} \sin y & \cos y \\ \cos y & -\sin y \end{vmatrix} = e^{2x}(-\sin^2 y - \cos^2 y) = -e^{2x}. \quad \textbf{Ans.}$$

EXAMPLE 4.12. Find the value of the Jacobian $\frac{\partial(u,v)}{\partial(r,\theta)}$ where $u = x^2 - y^2$, $v = 2xy$ and $x = r\cos\theta$, $y = r\sin\theta$.

SOLUTION: We know that $\frac{\partial(u,v)}{\partial(r,\theta)} = \frac{\partial(u,v)}{\partial(x,y)} \cdot \frac{\partial(x,y)}{\partial(r,\theta)}$

Now,
$$\frac{\partial(u,v)}{\partial(x,y)} = \begin{vmatrix} \frac{\partial u}{\partial x} & \frac{\partial u}{\partial y} \\ \frac{\partial v}{\partial x} & \frac{\partial v}{\partial y} \end{vmatrix} = \begin{vmatrix} 2x & -2y \\ 2y & 2x \end{vmatrix} = 4(x^2 + y^2) = 4r^2$$

and
$$\frac{\partial(x,y)}{\partial(r,\theta)} = \begin{vmatrix} \frac{\partial x}{\partial r} & \frac{\partial x}{\partial \theta} \\ \frac{\partial y}{\partial r} & \frac{\partial y}{\partial \theta} \end{vmatrix} = \begin{vmatrix} \cos\theta & -r\sin\theta \\ \sin\theta & r\cos\theta \end{vmatrix} = r$$

Therefore,
$$\frac{\partial(u,v)}{\partial(r,\theta)} = 4r^2.r = 4r^3 \quad \textbf{Ans.}$$

PARTIAL DERIVATIVES FROM IMPLICIT FUNCTIONS USING JACOBIANS

Suppose u, v are implicit functions of the independent variables x, y connected by the functional relations

$$f_1(u, v, x, y) = 0 \qquad \text{... (1)}$$

and
$$f_2(u, v, x, y) = 0. \qquad \text{... (2)}$$

To obtain partial derivatives $\frac{\partial u}{\partial x}, \frac{\partial u}{\partial y}, \frac{\partial v}{\partial x}, \frac{\partial v}{\partial y}$ we partially differentiate (1) and (2) *w.r.t.* x and y separately and get

$$\left.\begin{aligned} \frac{\partial f_1}{\partial x}.1 + \frac{\partial f_1}{\partial u}.\frac{\partial u}{\partial x} + \frac{\partial f_1}{\partial v}.\frac{\partial v}{\partial x} = 0 \\ \frac{\partial f_2}{\partial x}.1 + \frac{\partial f_2}{\partial u}\frac{\partial u}{\partial x} + \frac{\partial f_2}{\partial v}\frac{\partial v}{\partial x} = 0 \end{aligned}\right\} \qquad \text{... (3)}$$

and
$$\left.\begin{aligned} \frac{\partial f_1}{\partial y}.1 + \frac{\partial f_1}{\partial u}.\frac{\partial u}{\partial y} + \frac{\partial f_1}{\partial v}.\frac{\partial v}{\partial y} = 0 \\ \frac{\partial f_2}{\partial y}.1 + \frac{\partial f_2}{\partial u}\frac{\partial u}{\partial y} + \frac{\partial f_2}{\partial v}\frac{\partial v}{\partial y} = 0 \end{aligned}\right\} \qquad \text{... (4)}$$

Solving (3) for $\frac{\partial u}{\partial x}$ and $\frac{\partial v}{\partial x}$, we get

$$\boxed{\frac{\partial u}{\partial x} = -\begin{vmatrix} \frac{\partial f_1}{\partial x} & \frac{\partial f_1}{\partial v} \\ \frac{\partial f_2}{\partial x} & \frac{\partial f_2}{\partial v} \end{vmatrix} \Bigg/ \begin{vmatrix} \frac{\partial f_1}{\partial u} & \frac{\partial f_1}{\partial v} \\ \frac{\partial f_2}{\partial u} & \frac{\partial f_2}{\partial v} \end{vmatrix} = -\frac{\frac{\partial(f_1, f_2)}{\partial(x, v)}}{\frac{\partial(f_1, f_2)}{\partial(u, v)}}}$$

and $$\frac{\partial v}{\partial x} = -\begin{vmatrix} \frac{\partial f_1}{\partial u} & \frac{\partial f_1}{\partial x} \\ \frac{\partial f_2}{\partial u} & \frac{\partial f_2}{\partial x} \end{vmatrix} \Bigg/ \begin{vmatrix} \frac{\partial f_1}{\partial u} & \frac{\partial f_1}{\partial v} \\ \frac{\partial f_2}{\partial u} & \frac{\partial f_2}{\partial v} \end{vmatrix} = -\frac{\frac{\partial (f_1, f_2)}{\partial (u, x)}}{\frac{\partial (f_1, f_2)}{\partial (u, v)}}$$

Similarly, solving (4) for $\frac{\partial u}{\partial y}$ and $\frac{\partial v}{\partial y}$, we get

$$\frac{\partial u}{\partial y} = -\frac{\frac{\partial (f_1, f_2)}{\partial (y, v)}}{\frac{\partial (f_1, f_2)}{\partial (u, v)}} \quad \text{and} \quad \frac{\partial v}{\partial y} = -\frac{\frac{\partial (f_1, f_2)}{\partial (u, y)}}{\frac{\partial (f_1, f_2)}{\partial (u, v)}}.$$

EXAMPLE 4.13. If $u^2 + xv^2 = x + y, \quad v^2 + yu^2 = x - y,$ find $\frac{\partial u}{\partial x}, \frac{\partial v}{\partial y}$.

SOLUTION: The functional relations are:

$f_1(u, v, x, y) = u^2 + xv^2 - x - y = 0$ and $f_2(u, v, x, y) = v^2 + yu^2 - x + y = 0.$

Now, $$\frac{\partial u}{\partial x} = -\frac{\frac{\partial (f_1, f_2)}{\partial (x, v)}}{\frac{\partial (f_1, f_2)}{\partial (u, v)}} = -\begin{vmatrix} \frac{\partial f_1}{\partial x} & \frac{\partial f_1}{\partial v} \\ \frac{\partial f_2}{\partial x} & \frac{\partial f_2}{\partial v} \end{vmatrix} \Bigg/ \begin{vmatrix} \frac{\partial f_1}{\partial u} & \frac{\partial f_1}{\partial v} \\ \frac{\partial f_2}{\partial u} & \frac{\partial f_2}{\partial v} \end{vmatrix}$$

$$= -\begin{vmatrix} v^2 - 1 & 2xv \\ -1 & 2v \end{vmatrix} \Bigg/ \begin{vmatrix} 2u & 2xv \\ 2uy & 2v \end{vmatrix} = -\frac{1}{2}\begin{vmatrix} v^2 - 1 & xv \\ -1 & v \end{vmatrix} \Bigg/ \begin{vmatrix} u & xv \\ uy & v \end{vmatrix}$$

$$= -\frac{1}{2}\frac{(v^3 - v + xv)}{uv - uvxy} = \frac{-1(v^2 - 1 + x)}{2u(1 - xy)}.$$

and $$\frac{\partial v}{\partial y} = \frac{-\frac{\partial (f_1, f_2)}{\partial (u, y)}}{\frac{\partial (f_1, f_2)}{\partial (u, v)}} = -\begin{vmatrix} 2u & -1 \\ 2yu & 1 \end{vmatrix} \Bigg/ \begin{vmatrix} 2u & 2xv \\ 2uy & 2v \end{vmatrix}$$

$$= \frac{-1}{2}\begin{vmatrix} u & -1 \\ yu & 1 \end{vmatrix} \Bigg/ \begin{vmatrix} u & xv \\ uy & v \end{vmatrix} = \frac{-1}{2}\frac{(u + yu)}{(uv - xyuv)} = \frac{-(1 + y)}{2v(1 - xy)}.$$ **Ans.**

EXAMPLE 4.14. If $u = x + y^2, \quad v = y + z^2, \quad w = z + x^2,$ prove that $\frac{\partial x}{\partial u} = -(1 + 8xyz)^{-1}$.

SOLUTION: We are given three functional relations in $u, v, w, x, y, z,$ as

$$f_1 \equiv u - x - y^2 = 0.$$
$$f_2 \equiv v - y - z^2 = 0.$$
$$f_3 \equiv w - z - x^2 = 0.$$

$$\frac{\partial x}{\partial u} = \frac{\dfrac{\partial(f_1, f_2, f_3)}{\partial(u, y, z)}}{\dfrac{\partial(f_1, f_2 f_3)}{\partial(x, y, z)}} = \frac{\begin{vmatrix} 1 & -2y & 0 \\ 0 & -1 & -2z \\ 0 & 0 & -1 \end{vmatrix}}{\begin{vmatrix} -1 & -2y & 0 \\ 0 & -1 & -2z \\ -2x & 0 & -1 \end{vmatrix}} = \frac{1}{-1+2y(-4xz)} = \frac{-1}{1+8xyz}.$$

JACOBIANS TO DETERMINE FUNCTIONAL DEPENDENCE

Jacobian is also used in determining whether or not two functions are functionally dependent. Two functions $f(x, y)$ and $\phi(x, y)$ are called functionally dependent if they are functions of each other.

Assume that $f(x, y)$ and $\phi(x, y)$ are functionally dependent then there exists a relation of the type $F(f, \phi) = 0$.

Differentiating it partially *w.r.t.* x and y, we get

$$\frac{\partial F}{\partial f}\frac{\partial f}{\partial x} + \frac{\partial F}{\partial \phi}\frac{\partial \phi}{\partial x} = 0 \quad \text{and} \quad \frac{\partial F}{\partial f}\frac{\partial f}{\partial y} + \frac{\partial F}{\partial \phi}\frac{\partial \phi}{\partial y} = 0.$$

These are homogeneous equations. Their non-trivial solution would exist only when

$$\begin{vmatrix} \dfrac{\partial f}{\partial x} & \dfrac{\partial \phi}{\partial x} \\ \dfrac{\partial f}{\partial y} & \dfrac{\partial \phi}{\partial y} \end{vmatrix} = 0, \quad i.e. \quad \frac{\partial(f, \phi)}{\partial(x, y)} = 0.$$

Hence $f(x, y)$ **and** $\phi(x, y)$ **are functionally dependent if their Jacobian vanishes identically.**

The idea can be easily extended to three functions and, in general, to n functions.

EXAMPLE 4.15. Show that the functions $u = x + y + z$, $v = x^3 + y^3 + z^3 - 3xyz$, $w = x^2 + y^2 + z^2 - xy - yz - zx$ are functionally dependent and find the relation between them.

SOLUTION: The functions u, v, w are functionally dependent if $\dfrac{\partial(u, v, w)}{\partial(x, y, z)} = 0$

$$\frac{\partial(u, v, w)}{\partial(x, y, z)} = \begin{vmatrix} \dfrac{\partial u}{\partial x} & \dfrac{\partial u}{\partial y} & \dfrac{\partial u}{\partial z} \\ \dfrac{\partial v}{\partial x} & \dfrac{\partial v}{\partial y} & \dfrac{\partial v}{\partial z} \\ \dfrac{\partial w}{\partial x} & \dfrac{\partial w}{\partial y} & \dfrac{\partial w}{\partial z} \end{vmatrix} = \begin{vmatrix} 1 & 1 & 1 \\ 3(x^2 - yz) & 3(y^2 - xz) & 3(z^2 - xy) \\ 2x - y - z & 2y - z - x & 2z - x - y \end{vmatrix}$$

(Applying $C_2 \to C_2 - C_1$ and $C_3 \to C_3 - C_1$)

$$= 3\begin{vmatrix} 1 & 0 & 0 \\ x^2 - yz & y^2 - x^2 + z(y - x) & z^2 - x^2 + y(z - x) \\ 2x - y - z & 3(y - x) & 3(z - x) \end{vmatrix}$$

$$= 6\begin{vmatrix} (y-x)(x+y+z) & (z-x)(x+y+z) \\ y-x & z-x \end{vmatrix}$$

$$= 6(y-x)(z-x)(x+y+z)\begin{vmatrix} 1 & 1 \\ 1 & 1 \end{vmatrix} = 0$$

Therefore u, v, w are functionally dependent. Further, since

$$u = x^3 + y^3 + z^3 - 3xyz = (x+y+z)(x^2+y^2+z^2-xy-yz-zx) = uw$$

Hence $u = vw$ is the desired relation. **Ans.**

EXAMPLE 4.16. Find the Jacobian of the following transformation

$$u = a\cosh x\cos y, \quad v = a\sinh x\sin y.$$

SOLUTION: The required Jacobian J is given by

$$J = \frac{\partial(u,v)}{\partial(x,y)} = \begin{vmatrix} \frac{\partial u}{\partial x} & \frac{\partial u}{\partial y} \\ \frac{\partial v}{\partial x} & \frac{\partial v}{\partial y} \end{vmatrix} = \begin{vmatrix} a\sinh x\cos y & -a\cosh x\sin y \\ a\cosh x\sin y & a\sinh x\cos y \end{vmatrix}$$

$$= a^2(\sinh^2 x\cos^2 y + \cosh^2 x\sin^2 y) = \frac{a^2}{2}[\sinh^2 x(1+\cos 2y) + \cosh^2 x(1-\cos 2y)]$$

$$= \frac{a^2}{2}[\sinh^2 x + \cosh^2 x - \cos 2y(\cosh^2 x - \sinh^2 x)] = \frac{a^2}{2}[\cosh 2x - \cos 2y]. \quad \textbf{Ans.}$$

EXAMPLE 4.17. If $y_1 = \frac{x_2 x_3}{x_1}$, $y_2 = \frac{x_3 x_1}{x_2}$, $y_3 = \frac{x_1 x_2}{x_3}$, find the value of $\frac{\partial(y_1,y_2,y_3)}{\partial(x_1,x_2,x_3)}$.

(UPTU 2001, Sp. Exam. 2004-05, 2010-11)
(UPTU 2011-12)

SOLUTION:
$$J = \frac{\partial(y_1,y_2,y_3)}{\partial(x_1,x_2,x_3)} = \begin{vmatrix} \frac{\partial y_1}{\partial x_1} & \frac{\partial y_1}{\partial x_2} & \frac{\partial y_1}{\partial x_3} \\ \frac{\partial y_2}{\partial x_1} & \frac{\partial y_2}{\partial x_2} & \frac{\partial y_2}{\partial x_3} \\ \frac{\partial y_3}{\partial x_1} & \frac{\partial y_3}{\partial x_2} & \frac{\partial y_3}{\partial x_3} \end{vmatrix} = \begin{vmatrix} -\frac{x_2x_3}{x_1^2} & \frac{x_3}{x_1} & \frac{x_2}{x_1} \\ \frac{x_3}{x_2} & -\frac{x_3x_1}{x_2^2} & \frac{x_1}{x_2} \\ \frac{x_2}{x_3} & \frac{x_1}{x_3} & -\frac{x_1x_2}{x_3^2} \end{vmatrix}$$

$$= -\frac{x_2x_3}{x_1^2}\left(\frac{x_1^2}{x_2x_3} - \frac{x_1^2}{x_2x_3}\right) + \frac{x_3}{x_1}\left(\frac{x_1x_2}{x_2x_3} + \frac{x_1x_2x_3}{x_2x_3^2}\right) + \frac{x_2}{x_1}\left(\frac{x_3x_1}{x_2x_3} + \frac{x_1x_2x_3}{x_3x_2^2}\right)$$

$$= 0 + \frac{x_3}{x_1}\left(\frac{x_1}{x_3} + \frac{x_1}{x_3}\right) + \frac{x_2}{x_1}\left(\frac{x_1}{x_2} + \frac{x_1}{x_2}\right) = 2 + 2 = 4. \quad \textbf{Ans.}$$

EXAMPLE 4.18. For the transformation $x = a(u + v)$, $y = b(u - v)$ and

$$u = r^2 \cos 2\theta, \quad v = r^2 \sin 2\theta \quad \text{find} \quad \frac{\partial(x, y)}{\partial(r, \theta)}.$$

SOLUTION: From the relations $x = a(u + v)$, $y = b(u - v)$, we have

$$\frac{\partial(x, y)}{\partial(u, v)} = \begin{vmatrix} \frac{\partial x}{\partial u} & \frac{\partial x}{\partial v} \\ \frac{\partial y}{\partial u} & \frac{\partial y}{\partial v} \end{vmatrix} = \begin{vmatrix} a & a \\ b & -b \end{vmatrix} = -2ab$$

And from the relations $u = r^2 \cos 2\theta$, $v = r^2 \sin 2\theta$, we have

$$\frac{\partial(u, v)}{\partial(r, \theta)} = \begin{vmatrix} \frac{\partial u}{\partial r} & \frac{\partial u}{\partial \theta} \\ \frac{\partial v}{\partial r} & \frac{\partial v}{\partial \theta} \end{vmatrix} = \begin{vmatrix} 2r\cos 2\theta & -2r^2 \sin 2\theta \\ 2r \sin 2\theta & 2r^2 \cos 2\theta \end{vmatrix} = 4r^3 \begin{vmatrix} \cos 2\theta & -\sin 2\theta \\ \sin 2\theta & \cos 2\theta \end{vmatrix} = 4r^3$$

By chain rule, $$\frac{\partial(x, y)}{\partial(r, \theta)} = \frac{\partial(x, y)}{\partial(u, v)} \cdot \frac{\partial(u, v)}{\partial(r, \theta)} = -2ab \cdot 4r^3 = -8abr^3.$$ **Ans.**

EXAMPLE 4.19. If $x + y + z = u$, $y + z = uv$, $z = uvw$, find $\frac{\partial(x, y, z)}{\partial(u, v, w)}$.

(UPTU 2003-04)

SOLUTION: The given relations can be written as

$$z = uvw, \quad y = uv - z = uv - uvw = uv(1 - w)$$

and $$x = u - (y + z) = u - uv = u(1 - v).$$

Therefore, $$\frac{\partial(x, y, z)}{\partial(u, v, w)} = \begin{vmatrix} \frac{\partial x}{\partial u} & \frac{\partial x}{\partial v} & \frac{\partial x}{\partial w} \\ \frac{\partial y}{\partial u} & \frac{\partial y}{\partial v} & \frac{\partial y}{\partial w} \\ \frac{\partial z}{\partial u} & \frac{\partial z}{\partial v} & \frac{\partial z}{\partial w} \end{vmatrix}$$

$$= \begin{vmatrix} 1-v & -u & 0 \\ v(1-w) & u(1-w) & -uv \\ vw & uw & uv \end{vmatrix}$$ (applying $R_2 \to R_2 + R_3$)

$$= \begin{vmatrix} 1-v & -u & 0 \\ v & u & 0 \\ vw & uw & uv \end{vmatrix}$$

$$= uv[u(1 - v) + uv] = u^2 v.$$ **Ans.**

EXAMPLE 4.20. If $u = xyz$, $v = x^2 + y^2 + z^2$, $w = x + y + z$ find $\dfrac{\partial(x, y, z)}{\partial(u, v, w)}$.

(UPTU 2002-03)

SOLUTION: Let us first calculate the value of:

$$J = \frac{\partial(u, v, w)}{\partial(x, y, z)} = \begin{vmatrix} \dfrac{\partial u}{\partial x} & \dfrac{\partial u}{\partial y} & \dfrac{\partial u}{\partial z} \\ \dfrac{\partial v}{\partial x} & \dfrac{\partial v}{\partial y} & \dfrac{\partial v}{\partial z} \\ \dfrac{\partial w}{\partial x} & \dfrac{\partial w}{\partial y} & \dfrac{\partial w}{\partial z} \end{vmatrix}$$

$$= \begin{vmatrix} yz & zx & xy \\ 2x & 2y & 2z \\ 1 & 1 & 1 \end{vmatrix} = \begin{vmatrix} yz & z(x-y) & y(x-z) \\ 2x & 2(y-x) & 2(z-x) \\ 1 & 0 & 0 \end{vmatrix}$$

$$= 2z\,(x - y)\,(z - x) - 2y\,(y - x)\,(x - z)$$

$$= -\,2\,(x - y)\,(z - x)\,(y - z)$$

Using the fact that $JJ' = 1$, we have

$$\frac{\partial(x, y, z)}{\partial(u, v, w)} = J' = \frac{1}{J} = \frac{-1}{2(x-y)\,(y-z)\,(z-x)}. \qquad \textbf{Ans.}$$

EXAMPLE 4.21. If $u^3 + v + w = x + y^2 + z^2$, $u + v^3 + w = x^2 + y + z^2$ and

$u + v + w^3 = x^2 + y^2 + z$, find $\dfrac{\partial(u, v, w)}{\partial(x, y, z)}$.

SOLUTION: The given relations can be written as implicit functions as

$$f_1\,(u, v, w, x, y, z) = u^3 + v + w - x - y^2 - z^2 = 0$$

$$f_2\,(u, v, w, x, y, z) = u + v^3 + w - x^2 - y - z^2 = 0$$

$$f_3\,(u, v, w, x, y, z) = u + v + w^3 - x^2 - y^2 - z = 0$$

Now, $$\frac{\partial(f_1, f_2, f_3)}{\partial(u, v, w)} = \begin{vmatrix} 3u^2 & 1 & 1 \\ 1 & 3v^2 & 1 \\ 1 & 1 & 3w^2 \end{vmatrix}$$

$$= 3u^2\,(9v^2w^2 - 1) - 1\,(3w^2 - 1) + 1\,(1 - 3v^2)$$

$$= 27\,u^2 v^2 w^2 - 3u^2 - 3v^2 - 3\,w^2 + 2$$

and $$\frac{\partial(f_1, f_2, f_3)}{\partial(x, y, z)} = \begin{vmatrix} -1 & -2y & -2z \\ -2x & -1 & -2z \\ -2x & -2y & -1 \end{vmatrix}$$

$$= -1\,(1 - 4yz) + 2y\,(2x - 4xz) - 2z\,(4xy - 2x)$$
$$= -1 + 4\,(xy + yz + zx) - 16\,xyz$$

Therefore,

$$\frac{\partial(u,v,w)}{\partial(x,y,z)} = (-1)^3\, \frac{\partial(f_1,f_2,f_3)}{\partial(x,y,z)} \Big/ \frac{\partial(f_1,f_2,f_3)}{\partial(u,v,w)}$$

$$= \frac{1-4(xy+yz+zx)+16xyz}{2-3(u^2+v^2+w^2)+27u^2v^2w^2}. \qquad \textbf{Ans.}$$

EXAMPLE 4.22. Show that the functions

$f_1(x, y) = x^2 + y^2 + xy = 0$ and

$f_2(x, y) = x^4 + 2x^3y + 3x^2y^2 + 2xy^3 + y^4 = 0$

are functionally dependent.

SOLUTION: We have $\dfrac{\partial(f_1,f_2)}{\partial(x,y)} = \begin{vmatrix} 2x+y & 2y+x \\ 4x^3+6x^2y+6xy^2+2y^3 & 2x^3+6x^2y+6xy^2+4y^3 \end{vmatrix}$

$$= (2x+y)\,(2x^3 + 6x^2y + 6xy^2 + 4y^3) - (2y+x)\,(4x^3 + 6x^2y + 6xy^2 + 2y^3)$$
$$= 0 \quad \text{(on simplification).}$$

which implies that f_1 and f_2 are functionally dependent. **Hence Proved.**

EXAMPLE 4.23. If $u = \dfrac{y^2}{2x}$ and $v = \dfrac{x^2+y^2}{2x}$, find $\dfrac{\partial(u,v)}{\partial(x,y)}$.

SOLUTION: Given $u = \dfrac{y^2}{2x}$ and $v = \dfrac{x^2+y^2}{2x}$ we have

$$\frac{\partial(u,v)}{\partial(x,y)} = \begin{vmatrix} \dfrac{\partial u}{\partial x} & \dfrac{\partial u}{\partial y} \\ \dfrac{\partial v}{\partial x} & \dfrac{\partial v}{\partial y} \end{vmatrix} = \begin{vmatrix} -\dfrac{y^2}{2x^2} & \dfrac{y}{x} \\ \dfrac{1}{2} - \dfrac{y^2}{2x^2} & \dfrac{y}{x} \end{vmatrix}.$$ Now applying $R_2 \to R_2 - R_1$, we get

$$\frac{\partial(u,v)}{\partial(x,y)} = \begin{vmatrix} \dfrac{-y^2}{2x^2} & \dfrac{y}{x} \\ \dfrac{1}{2} & 0 \end{vmatrix} = \frac{-y}{2x}. \qquad \textbf{Ans.}$$

EXAMPLE 4.24. If $u^3 + v^3 = x + y$, $u^2 + v^2 = x^3 + y^3$ show that $\dfrac{\partial(u,v)}{\partial(x,y)} = \dfrac{y^2-x^2}{2uv\,(u-v)}$.

(UPTU 2006–07)

SOLUTION: The given relations are $f_1(u, v, x, y) = 0$ and $f_2(u, v, x, y) = 0$

where $f_1 \equiv u^3 + v^3 - x - y$ and $f_2 \equiv u^2 + v^2 - x^3 - y^3$

Since,
$$\frac{\partial(u,v)}{\partial(x,y)} = \frac{(-1)^2 \dfrac{\partial(f_1,f_2)}{\partial(x,y)}}{\dfrac{\partial(f_1,f_2)}{\partial(u,v)}} = (-1)^2 \frac{\begin{vmatrix} \dfrac{\partial f_1}{\partial x} & \dfrac{\partial f_1}{\partial y} \\ \dfrac{\partial f_2}{\partial x} & \dfrac{\partial f_2}{\partial y} \end{vmatrix}}{\begin{vmatrix} \dfrac{\partial f_1}{\partial u} & \dfrac{\partial f_1}{\partial v} \\ \dfrac{\partial f_2}{\partial u} & \dfrac{\partial f_2}{\partial v} \end{vmatrix}}$$

$$= \frac{\begin{vmatrix} -1 & -1 \\ -3x^2 & -3y^2 \end{vmatrix}}{\begin{vmatrix} 3u^2 & 3v^2 \\ 2u & 2v \end{vmatrix}} = \frac{3(y^2-x^2)}{6(u^2v-uv^2)} = \frac{y^2-x^2}{2uv(u-v)}.$$ **Hence Proved.**

EXAMPLE 4.25. If u, v, w are the roots of the equation

$$(x-a)^3+(x-b)^3+(x-c)^3=0, \text{ find } \frac{\partial(u,v,w)}{\partial(a,b,c)}.$$ **(UPTU 2001–02, 2010-11)**

SOLUTION: $(x-a)^3+(x-b)^3+(x-c)^3=0$

or $3x^3 - 3(a+b+c)x^2 + 3(a^2+b^2+c^2)x - (a^3+b^3+c^3) = 0$ whose roots are u, v, w.

Therefore we have

$$u+v+w = a+b+c, \quad uv+vw+wu = a^2+b^2+c^2$$

and $$uvw = \frac{1}{3}(a^3+b^3+c^3).$$

Let us write $f_1 \equiv (u+v+w-a-b-c)$, $f_2 = (uv+vw+wu-a^2-b^2-c^2)$

and $$f_3 \equiv uvw - \frac{1}{3}(a^3+b^3+c^3)$$

Next,
$$\frac{\partial(f_1,f_2,f_3)}{\partial(a,b,c)} = \begin{vmatrix} -1 & -1 & -1 \\ -2a & -2b & -2c \\ -a^2 & -b^2 & -c^2 \end{vmatrix} = \begin{vmatrix} -1 & 0 & 0 \\ -2a & 2(a-b) & 2(a-c) \\ -a^2 & a^2-b^2 & a^2-c^2 \end{vmatrix}$$

$$= -2\begin{vmatrix} a-b & a-c \\ a^2-b^2 & a^2-c^2 \end{vmatrix} = -2(a-b)(a-c)\begin{vmatrix} 1 & 1 \\ a+b & a+c \end{vmatrix}$$

$$= -2(a-b)(b-c)(c-a).$$

and
$$\frac{\partial(f_1,f_2,f_3)}{\partial(u,v,w)} = \begin{vmatrix} 1 & 1 & 1 \\ v+w & w+u & u+v \\ vw & wu & uv \end{vmatrix} = \begin{vmatrix} 1 & 0 & 0 \\ v+w & u-v & u-w \\ vw & w(u-v) & v(u-w) \end{vmatrix}$$

$$= \begin{vmatrix} u-v & u-w \\ w(u-v) & v(u-w) \end{vmatrix} = (u-v)(u-w)\begin{vmatrix} 1 & 1 \\ \omega & v \end{vmatrix}$$

$$= -(u-v)(v-w)(w-u)$$

$$\Rightarrow \quad \frac{\partial(u,v,w)}{\partial(a,b,c)} = (-1)^3 \frac{\partial(f_1,f_2,f_3)}{\partial(a,b,c)} \Big/ \frac{\partial(f_1,f_2,f_3)}{\partial(u,v,w)} = \frac{2(a-b)(b-c)(c-a)}{-(u-v)(v-w)(w-u)}. \quad \textbf{Ans.}$$

EXAMPLE 4.26. (*a*) Show that the functions $u = x^2 + y^2 + z^2$, $v = x + y + z$, $w = yz + zx + xy$ are not independent of one another.

(*b*) Verify the chain rule for Jacobians if $x = u$, $y = u \tan v$, $z = w$.

(UPTU 2008-09)

SOLUTION: (*a*)

$$\frac{\partial(u,v,w)}{\partial(x,y,z)} = \begin{vmatrix} \frac{\partial u}{\partial x} & \frac{\partial u}{\partial y} & \frac{\partial u}{\partial z} \\ \frac{\partial v}{\partial x} & \frac{\partial v}{\partial y} & \frac{\partial v}{\partial z} \\ \frac{\partial w}{\partial x} & \frac{\partial w}{\partial y} & \frac{\partial w}{\partial z} \end{vmatrix} = \begin{vmatrix} 2x & 2y & 2z \\ 1 & 1 & 1 \\ y+z & z+x & x+y \end{vmatrix} = 2\begin{vmatrix} x & y & z \\ 1 & 1 & 1 \\ y+z & z+x & x+y \end{vmatrix}.$$

Applying $R_3 \to R_3 + R_1$

$$= 2\begin{vmatrix} x & y & z \\ 1 & 1 & 1 \\ x+y+z & x+y+z & x+y+z \end{vmatrix} = 2(x+y+z)\begin{vmatrix} x & y & z \\ 1 & 1 & 1 \\ 1 & 1 & 1 \end{vmatrix} = 0$$

Hence u, v, w are not independent of each other. **Hence Proved.**

(*b*)

$$\frac{\partial(x,y,z)}{\partial(u,v,w)} = \begin{vmatrix} \frac{\partial x}{\partial u} & \frac{\partial x}{\partial v} & \frac{\partial x}{\partial w} \\ \frac{\partial y}{\partial u} & \frac{\partial y}{\partial v} & \frac{\partial y}{\partial w} \\ \frac{\partial z}{\partial u} & \frac{\partial z}{\partial v} & \frac{\partial z}{\partial w} \end{vmatrix} = \begin{vmatrix} 1 & 0 & 0 \\ 0 & u\sec^2 v & 0 \\ 0 & 0 & 1 \end{vmatrix} = u\sec^2 v$$

Also, we have $u = x$, $v = \tan^{-1}\left(\frac{y}{x}\right)$ and $w = z$, therefore

$$\frac{\partial(u,v,w)}{\partial(x,y,z)} = \begin{vmatrix} \frac{\partial u}{\partial x} & \frac{\partial u}{\partial y} & \frac{\partial u}{\partial z} \\ \frac{\partial v}{\partial x} & \frac{\partial v}{\partial y} & \frac{\partial v}{\partial z} \\ \frac{\partial w}{\partial x} & \frac{\partial w}{\partial y} & \frac{\partial w}{\partial z} \end{vmatrix} = \begin{vmatrix} 1 & 0 & 0 \\ \frac{-y}{x^2+y^2} & \frac{x}{x^2+y^2} & 0 \\ 0 & 0 & 1 \end{vmatrix}$$

$$= \frac{x}{x^2+y^2} = \frac{u}{u^2+u^2\tan^2 v} = \frac{u}{u^2\sec^2 v} = \frac{1}{u\sec^2 v}$$

Hence $\dfrac{\partial(x,y,z)}{\partial(u,v,w)} \cdot \dfrac{\partial(u,v,w)}{\partial(x,y,z)} = 1.$ **Hence Verified.**

EXERCISE 4A

1. Find the Jacobian of the following transformations
$u = \cos x, \ v = \sin x \cos y, \ w = \sin x \sin y \cos z.$

2. If (x, y, z) and (r, θ, ϕ) are respectively the cartesian and spherical polar coordinates of a point then show that $\dfrac{\partial (x, y, z)}{\partial (r, \theta, \phi)} = r^2 \sin \theta.$ **(UPTU 2000-01)**

3. For the transformation $x = e^u \cos v, \ y = e^u \sin v$ show that $\dfrac{\partial(u, v)}{\partial(x, y)} = e^{-2u}.$ **(UPTU 2011-12)**

4. Verify $JJ' = 1$ when $x = v^2 + w^2, \ y = w^2 + u^2, \ z = u^2 + v^2.$ where $J = \dfrac{\partial(x, y, z)}{\partial(u, v, w)}.$

5. If $x^2 + y^2 + u^2 - v^2 = 0$ and $uv + xy = 0$ show that $\dfrac{\partial(u, v)}{\partial(x, y)} = \dfrac{x^2 - y^2}{u^2 + v^2}.$

6. If $u^3 + v^3 + w^3 = x + y + z, \ u^2 + v^2 + w^2 = x^3 + y^3 + z^3$ and $u + v + w = x^2 + y^2 + z^2$, find $\dfrac{\partial(u, v, w)}{\partial(x, y, z)}.$

7. If $u = xyz, \ v = x^2 + y^2 + z^2, \ w = x + y + z,$ find $\dfrac{\partial x}{\partial u}.$

8. If $x + y + z + u + v = a$ and $x^2 + y^2 + z^2 + u^2 + v^2 = b$ where a and b are constants, find $\left(\dfrac{\partial v}{\partial y}\right)_{(x,u)}$ and $\left(\dfrac{\partial y}{\partial v}\right)_{(x,z)}.$

9. If $u = x + y + z, \ v = x^2 + y^2 + z^2, \ w = xy + yz + zx$ investigate whether u, v, w are functionally dependent on $x, y, z,$ or not. If yes, find the relation between them. **(UPTU 2009)**

10. If $u = \sin^{-1} x + \sin^{-1} y, \ v = x\sqrt{1 - y^2} + y\sqrt{1 - x^2}$ find out if they are functionally dependent and if so, find the relation between them. **(UPTU 2001 Q.B.)**

11. Investigate if $u = \dfrac{x - y}{x + z}, \ v = \dfrac{x + z}{y + z}$ are functionally dependent and if so, find the functional relation between them. **(UPTU 2011-12)**

12. If u and v are functions of x and y then show that $\dfrac{\partial(u, v)}{\partial(x, y)} \cdot \dfrac{\partial(x, y)}{\partial(u, v)} = 1.$

13. Given that $x = u + v + w, \ y = u^2 + v^2 + w^2, \ z = u^3 + v^3 + w^3$ find $\dfrac{\partial u}{\partial x}.$

14. Show that the functions $u = x + y + z, \ v = x^3 + y^3 + z^3 - 3xyz,$
$w = x^2 + y^2 + z^2 - xy - yz - zx$ are functionally dependent. Also find the relation between them.

15. If $x = e^v \sec u, \ y = e^v \tan u,$ evaluate $\dfrac{\partial(x, y)}{\partial(u, v)}.$ **(UPTU 2009-10)**

16. If $u = \dfrac{x + y}{z}, \ v = \dfrac{y + z}{x}, \ w = \dfrac{y(x + y + z)}{xz}$ then show that u, v, w are not independent and find the relation between them. **(UPTU 2009-10)**

❑❑❑

CHAPTER

5 Applications of Partial Differentiation

Approximation of Errors. Extrema (Maxima and Minima) of Function of two and Three variables. Lagrange's method of Undetermined Multipliers.

ERRORS AND APPROXIMATIONS

Let u be a function of two variables x and y and let δx and δy be small changes made in x and y respectively and let the resulting change in u be δu, then

$$\begin{aligned}\delta u &= u\,(x+\delta x, y+\delta y) - u\,(x, y)\\ &= [u\,(x+\delta x, y+\delta y) - u\,(x, y+\delta y)] + [u\,(x, y+\delta y) - u\,(x, y)]\\ &= \frac{u(x+\delta x, y+\delta y)-u(x, y+\delta y)}{\delta x}\,\delta x + \frac{u(x, y+\delta y)-u(x,y)}{\delta y}\,\delta y.\end{aligned}$$

Since, $$\lim_{\delta y\to 0}\frac{u\,(x, y+\delta y)-u(x,y)}{\delta y} = \frac{\partial u}{\partial y}$$

and $$\lim_{\delta x\to 0}\frac{u(x+\delta x, y+\delta y)-u(x, y+\delta y)}{\delta x} = \frac{\partial u}{\partial x}$$

therefore, we can write

$$\boxed{\delta u = \frac{\partial u}{\partial x}\,\delta x + \frac{\partial u}{\partial y}\,\delta y} \quad \text{approximately.} \qquad ...(1)$$

If δx and δy are considered as small errors in the measurement of x and y then δu, given by the above equation (1), represents the error in the calculated value of u.

EXAMPLE 5.1. Compute the approximate value of $(1.04)^{3.01}$.

SOLUTION: Let $f(x, y) = x^y$. Here $x = 1$, $y = 3$ and $\delta x = 0.04$, $\delta y = 0.01$.

Therefore $$\delta f = \frac{\partial f}{\partial x}\,\delta x + \frac{\partial f}{\partial y}\,\delta y \text{ approximately.}$$

$$= yx^{y-1}.\delta x + x^y \log x\,.\,\delta y \;=\; 3(1^{3-1})\,(0.04)$$

$\therefore$ $(1.04)^{3.1} = 1.12$ approximately. **Ans.**

EXAMPLE 5.2. Compute the value of $\left[(3.8)^2 + 2\,(2.1)^3\right]^{1/5}$ using the theory of approximation.

[UPTU 2011-12]

SOLUTION: Consider the function $u = (x^2 + 2y^3)^{\frac{1}{5}}$

Let us take $x = 4$ and $y = 2$ so that

$$x + \delta x = 3.8 \text{ and } y + \delta y = 2.1 \quad \Rightarrow \quad \delta x = -0.2 \text{ and } \delta y = 0{\cdot}1\,.$$

The corresponding error δu in u, is given by

$$\delta u = \frac{\partial u}{\partial x}\delta x + \frac{\partial u}{\partial y}\delta y = \frac{1}{5}(x^2 + 2y^3)^{-\frac{4}{5}}.2x\,\delta x + \frac{1}{5}(x^2 + 2y^3)^{-\frac{4}{5}}.\,6y^2\,\delta y$$

$$= \frac{1}{5}\left(x^2 + 2y^3\right)^{\frac{-4}{5}}(2x\,\delta x + 6y^2\,\delta y)$$

$$= \frac{1}{5}\left(4^2 + 2(2)^3\right)^{-\frac{4}{5}}[\,2(4)\,(-0.2) + 6(2)^2\,(0.1)] \quad \text{at } x = 4, y = 2$$

$$= \frac{1}{5}(32)^{-\frac{4}{5}}[-1.6 + 2.4] = \frac{1}{5}\,2^{-4}\,(0.8) = \frac{0.8}{80} = 0.01$$

Also, at $x = 4$, $y = 2$ we have

$$u = (x^2 + 2y^3)^{\frac{1}{5}} = \left(4^2 + 2(2^3)\right)^{\frac{1}{5}} = 32^{\frac{1}{5}} = 2$$

$\therefore$ Required approximate value $= u + \delta u = 2 + 0.01 = 2.01$. **Ans.**

EXAMPLE 5.3. If the sides and angles of a triangle ABC vary in such a way that its circum-radius remains constant, prove that $\frac{da}{\cos A} + \frac{db}{\cos B} + \frac{dc}{\cos C} = 0$.

SOLUTION: By sine rule in triangle ABC, we have

$$\frac{a}{\sin A} = \frac{b}{\sin B} = \frac{c}{\sin C} = 2R \quad \text{(which is constant)}$$

$$\Rightarrow \qquad a = 2R\sin A, \quad b = 2R\sin B, \quad c = 2R\sin C$$

hence $\quad da = 2R\cos A\,dA, \quad db = 2R\cos B\,dB, \quad dc = 2R\cos C\,dc$

or $\quad \frac{da}{\cos A} + \frac{db}{\cos B} + \frac{dc}{\cos C} = 2R(dA + dB + dC)$

Since $A + B + C = \pi$ we have $dA + dB + dC = 0$

Therefore $\frac{da}{\cos A} + \frac{db}{\cos B} + \frac{dc}{\cos C} = 0$. **Hence Proved.**

EXAMPLE 5.4. The acceleration of a piston is calculated from the formula $f = w^2 r\left(\cos\theta + \frac{r}{l}\cos 2\theta\right)$ where r and l are constants. If ω and θ suffer changes $\delta\omega$ and $\delta\theta$, then show that the resulting change δf in f, is given by

$$\frac{\delta f}{f} = 2\frac{\delta\omega}{\omega} - \left(\frac{\sin\theta + \frac{2r}{l}\sin 2\theta}{\cos\theta + \frac{r}{l}\cos 2\theta}\right)\delta\theta$$

If $\theta = 30^\circ$ and $\frac{r}{l} = \frac{1}{4}$ and θ and ω are each 1% less, find the corresponding percentage change in f.

SOLUTION: In the given relation, taking logarithm, we get

$$\log f = 2\log\omega + \log r + \log\left(\cos\theta + \frac{r}{l}\cos 2\theta\right)$$

$$\therefore \quad \frac{\delta f}{f} = 2\frac{\delta\omega}{\omega} + 0 + \frac{-\sin\theta - \frac{2r}{l}\sin 2\theta}{\cos\theta + \frac{r}{l}\cos 2\theta}\delta\theta \quad \textbf{hence the first result.}$$

$$\text{or} \quad \frac{\delta f}{f}\times 100 = 2\frac{\delta\omega}{\omega}\times 100 - \frac{\sin\theta + \frac{2r}{l}\sin 2\theta}{\cos\theta + \frac{r}{l}\cos 2\theta}\frac{\delta\theta\times 100}{\theta}\cdot\theta.$$

Taking $\theta = (\pi/6)$, $\frac{r}{l} = \frac{1}{4}$, $\frac{\delta\theta}{\theta}\times 100 = -1$ and $\frac{\delta\omega}{\omega}\times 100 = -1$, we get

$$\frac{\delta f}{f}\times 100 = 2(-1) - \frac{\sin 30^\circ + 2\left(\frac{1}{4}\right)\sin 60^\circ}{\cos 30^\circ + \left(\frac{1}{4}\right)\cos 60^\circ}.(-1)\left(\frac{\pi}{6}\right)$$

$$= -2 + \frac{\frac{1}{2} + \frac{2}{4}\frac{\sqrt{3}}{2}}{\frac{\sqrt{3}}{2} + \frac{1}{8}}.\frac{\pi}{6} = -2 + \frac{4 + 2\sqrt{3}}{4\sqrt{3} + 1}.\frac{\pi}{6} = 1{\cdot}5 \text{ approximately.}$$

$\therefore$ Error in f is 1.5 % approximately. **Ans.**

EXAMPLE 5.5. The angles of a triangle are calculated from the sides a, b, c. If small changes $\delta a, \delta b, \delta c$ are made in measuring the sides a, b, c, show that the change in the angle A, is given by **(UPTU 2002, 2011-12)**

$$\delta A = \frac{a}{2\Delta}[\delta a - \delta b \cos C - \delta c \cos B]$$ where Δ is the area of the triangle.

Also, verify that $\delta A + \delta B + \delta C = 0$.

SOLUTION: We know that $\cos A = \frac{b^2 + c^2 - a^2}{2bc}$

$$\therefore \quad -\sin A\,\delta A = \frac{2bc\,(2b\delta b + 2c\delta c - 2a\delta a) - (b^2 + c^2 - a^2)(2b\delta c + 2c\delta b)}{(2bc)^2}$$

or $\quad -2\sin A\,\delta A = \dfrac{1}{(bc)^2}\,[2b^2c\,\delta b + 2bc^2\,\delta c - 2abc\,\delta a - (b^2+c^2-a^2)(b\delta c + c\delta b)$

$$= \frac{1}{(bc)^2}[\delta b\,(2b^2c - b^2c - c^3 + ca^2) + \delta c\,(2bc^2 - b^3 - bc^2 + ba^2) - 2abc\,\delta a]$$

$$= \frac{1}{(bc)^2}\,[c\delta b\,(b^2 + a^2 - c^2) + b\delta c\,(c^2 + a^2 - b^2) - 2abc\,\delta a]$$

$$= \frac{\delta b}{b}\left(\frac{b^2 + a^2 - c^2}{bc}\right) + \frac{\delta c}{c}\left(\frac{c^2 + a^2 - b^2}{bc}\right) - \frac{2a\,\delta a}{bc}$$

Since, $\quad \Delta = \dfrac{1}{2}\,bc\sin A$

$$\therefore \quad \delta A = -\frac{1.bc}{4\Delta}\left[\frac{\delta b}{b}\,\frac{a^2+b^2-c^2}{bc} + \frac{\delta c}{c}\,\frac{c^2+a^2-b^2}{bc} - \frac{2a\,\delta a}{bc}\right]$$

$$= -\frac{1}{4\Delta}\left[\frac{\delta b}{b}\cos C\,.\,2ab + \frac{\delta c}{c}\cos B\,.\,2ac - 2a\,\delta a\right]$$

$$\left(\text{since in } \Delta ABC,\quad \cos C = \frac{a^2+b^2-c^2}{2ab} \text{ and } \cos B = \frac{c^2+a^2-b^2}{2ac}\right)$$

$$= -\frac{2a}{4\Delta}\,[\cos C\,\delta b + \cos B\,\delta c - \delta a]$$

or $\quad \delta A = \dfrac{a}{2\Delta}\,[\delta a - \delta b\cos C - \delta c\cos B]$ which is the required result.

Similarly, we can get

$$\delta B = \frac{b}{2\Delta}\,[\,\delta b - \delta a\cos C - \delta c\cos A] \quad \text{and} \quad \delta C = \frac{c}{2\Delta}\,[\delta c - \delta a\cos B - \delta b\cos A]$$

Adding these, we get

$$\delta A + \delta B + \delta C = \frac{1}{2\Delta}\,[a\delta a - a\delta b\cos C - a\delta c\cos B + b\delta b - b\delta a\cos C - b\delta c\cos A + c\delta c - c\delta a\cos B - c\delta b\cos A]$$

$$= \frac{1}{2\Delta}\,[\delta a\,(a - b\cos C - c\cos B) + \delta b\,(\,b - a\cos C - c\cos A) + \delta c\,(c - a\cos B - b\cos A)]$$

$$= \frac{1}{2\Delta}\,.\,(0) = 0$$

(using the projection rule for Δ ABC). **Hence Proved.**

EXAMPLE 5.6. **(a)** The height h and the semi vertical angle a of a cone are measured and from these, A, the total area of the cone including the base, is calculated. If h and α are in error by small quantities δh and $\delta\alpha$ respectively, find the corresponding error in A. Also show that, if $\alpha = \dfrac{\pi}{6}$, an error of + 1% in h will be compensated by an error of $-0.33°$ in α. **(UPTU 2009)**

(b) If the base radius and height of a cone measured as 4 cm and 8 cm with a possible error of 0.04 and 0.08 cm respectively, calculate the percentage error in calculating volume of the cone. **[UPTU 2011-12]**

SOLUTION: **(a)** Let r be the radius of the base and l the slant height of the cone, then $\tan\alpha = \dfrac{r}{h}$

or $r = h\tan\alpha$ and $\sec\alpha = \dfrac{l}{h}$ or $l = h\sec\alpha$

$\therefore$ Total surface area $= A = \pi rl + \pi r^2 = \pi h^2\tan\alpha\sec\alpha + \pi h^2\tan^2\alpha$

$$= \pi h^2(\tan\alpha\sec\alpha + \tan^2\alpha)$$

Hence $\delta A = \dfrac{\partial A}{\partial h}\delta h + \dfrac{\partial A}{\partial \alpha}\delta\alpha$

$$= 2\pi h(\tan\alpha\sec\alpha + \tan^2\alpha)\,\delta h + \pi h^2(\tan^2\alpha\sec\alpha + \sec^3\alpha + 2\tan\alpha\sec^2\alpha)\,\delta\alpha$$

or $\delta A = 2\pi h\tan\alpha(\sec\alpha + \tan\alpha)\,\delta h + \pi h^2\sec\alpha(\sec\alpha + \tan\alpha)^2\,\delta\alpha$

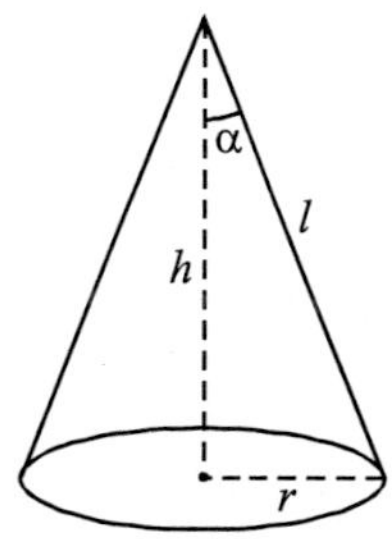

Putting $\alpha = \dfrac{\pi}{6}$ and $\dfrac{\delta h}{h}\times 100 = 1$, we get

$$\delta A = \frac{2\pi h^2}{100}\frac{1}{\sqrt{3}}\left(\frac{2}{\sqrt{3}} + \frac{1}{\sqrt{3}}\right) + \pi h^2\frac{2}{\sqrt{3}}\left(\frac{1}{3} + \frac{4}{3} + \frac{4}{\sqrt{3}}\frac{1}{\sqrt{3}}\right)\delta\alpha$$

$$= 2\pi h^2(0.01) + \pi h^2\, 2\sqrt{3}\,\delta\alpha$$

However, as given $\delta A = 0$, therefore

$$2\pi h^2(0.01) + 2\sqrt{3}\,\pi h^2\,\delta\alpha = 0$$

or $0.01 + \sqrt{3}\,\delta\alpha = 0 \Rightarrow \delta\alpha = -\dfrac{0.01}{\sqrt{3}}$ radians

or $\delta\alpha = -\dfrac{0.01}{\sqrt{3}}\times\dfrac{180}{\pi}$ (in degrees) $= -0.33°$ **Ans.**

(b) Let x be the base radius and y the height of the cone.

Volume V of the cone is given by $V = \dfrac{1}{3}\pi x^2 y$

Let error in $V = \delta V$ as a result of $\delta x = 0.04$ cm, $\delta y = 0.08$ cm.

Hence $$\frac{\delta V}{V}\times 100 = \frac{\pi}{3V}[2xy\,\delta x + x^2\delta y]\times 100$$

$$= \frac{\pi}{3}\cdot\frac{3}{\pi x^2 y}[2xy\,\delta x + x^2\delta y]\times 100$$

$$= \left[\frac{2}{x}\delta x + \frac{1}{y}\delta y\right]\times 100 = 2\left(\frac{\delta x}{x}\times 100\right) + \left(\frac{\delta y}{y}\times 100\right)$$

$$= 2\left(\frac{0.04}{4}\times 100\right) + \frac{0.08}{8}\times 100$$

$= 2\% + 1\% = 3\%$ **Ans.**

EXAMPLE 5.7. In a plane triangle ABC, if the sides a and b are kept constant, show that the variations of its angles are given by the relation

$$\frac{dA}{\sqrt{a^2 - b^2 \sin^2 A}} = \frac{dB}{\sqrt{b^2 - a^2 \sin^2 B}} = -\frac{dC}{c}.$$

SOLUTION: By the sine formula we have $\frac{a}{\sin A} = \frac{b}{\sin B}$ or $a \sin B = b \sin A$. ...(1)

Taking differentials on both sides, we get $a \cos B\, dB = b \cos A\, dA$

$$\Rightarrow \quad \frac{dA}{a \cos B} = \frac{dB}{b \cos A} = \frac{dA + dB}{a \cos B + b \cos A} \quad \text{(by componendo and dividendo)} \qquad ...(2)$$

Next, $a \cos B = a\sqrt{1 - \sin^2 B} = \sqrt{a^2 - a^2 \sin^2 B} = \sqrt{a^2 - b^2 \sin^2 A}$ using (1),

and $b \cos A = b\sqrt{1 - \sin^2 A} = \sqrt{b^2 - b^2 \sin^2 A} = \sqrt{b^2 - a^2 \sin^2 B}$. using (1).

Also, by projection rule in Δ ABC we have $a \cos B + b \cos A = c$,

and since $A + B + C = \pi$ we have $dA + dB + dC = 0$ or $dA + dB = -dC$.

As such (2) becomes $\frac{dA}{\sqrt{a^2 - b^2 \sin A}} = \frac{dB}{\sqrt{b^2 - a^2 \sin^2 B}} = \frac{-dC}{c}$. **Hence Proved.**

EXAMPLE 5.8. Find the possible percentage error in computing the parallel resistance r of the three resistances r_1, r_2, and r_3 from formula

$\frac{1}{r} = \frac{1}{r_1} + \frac{1}{r_2} + \frac{1}{r_3}$ if r_1, r_2, r_3 are each in error by $+ 1.2$ %.

SOLUTION : Using the differentials in $\frac{1}{r} = \frac{1}{r_1} + \frac{1}{r_2} + \frac{1}{r_3}$, we have

$$-\frac{1}{r^2}\delta r = -\frac{1}{r_1^2}\delta r_1 - \frac{1}{r_2^2}\delta r_2 - \frac{1}{r_3^2}\delta r_3$$

and since $\frac{\delta r_1}{r_1} \times 100 = \frac{\delta r_2}{r_2} \times 100 = \frac{\delta r_3}{r_3} \times 100 = 1.2$, we get

$$\frac{1}{r^2}\delta r = \frac{1.2}{r_1 \times 100} + \frac{1.2}{r_2 \times 100} + \frac{1.2}{r_3 \times 100}$$

$$= \frac{1.2}{100}\left(\frac{1}{r_1} + \frac{1}{r_2} + \frac{1}{r_3}\right) = \frac{1.2}{r \times 100}$$

$$\Rightarrow \quad \frac{\delta r}{r} \times 100 = \frac{1.2\, r}{r} = 1.2$$

Therefore, the resultant resistance is also having an error of + 1.2 %. **Ans.**

EXAMPLE 5.9. A balloon in the form of a right cylinder of radius 1.5 m and height 4 m is surmounted by hemispherical ends. If the radius is increased by 0.01 m and the length by 0.05 m how much change (%) will be in the volume of balloon?

SOLUTION : Volume of the baloon $= V = \pi r^2 h + \frac{2}{3}\pi r^3 + \frac{2}{3}\pi r^3 = \pi r^2 h + \frac{4}{3}\pi r^3.$

Hence $\delta V = \frac{\partial V}{\partial r}\partial r + \frac{\partial V}{\partial h}\partial h = \pi\, 2r \cdot \delta r \cdot h + \pi r^2 \delta h + \frac{4}{3} \cdot \pi\, 3r^2 \delta r$

$$\therefore \quad \frac{\delta V}{V} = \frac{\pi r[2h\,\delta r + r\delta h + 4r\delta r]}{\pi r^2 h + \frac{4}{3}\pi r^3} = \frac{2\delta r \cdot h + r \cdot \delta h + 4r \cdot \delta r}{rh + \frac{4}{3}r^2}$$

$$= \frac{2(0.01)(4) + 1.5(0.05) + 4(1.5)\,0.01}{(1.5)4 + \frac{4}{3}(1.5)^2}$$

$$= \frac{0.08 + 0.075 + 0.06}{6 + 3} = \frac{0.215}{9}$$

and hence percentage error in volume

$$= \frac{\delta V}{V} \cdot 100 = 2.389\%. \quad \textbf{Ans.}$$

EXAMPLE 5.10. (*a*) In estimating the number of bricks in a pile which is measured to be (5 m × 10 m × 5 m) the count of bricks is 100 bricks per m^3. Find the error in the cost when the tape is stretched 2% beyond its standard length. The cost of bricks is Rs. 2000 per thousand bricks. **(UPTU 2000–01)**

(*b*) Find the percentage of error in calculating the area of an ellipse $\frac{x^2}{a^2} + \frac{y^2}{b^2} = 1$, when error of +1% is made in measuring the major and minor axes. **(UPTU 2010-11)**

SOLUTION: (*a*) Volume $V = xyz$ or $\log V = \log x + \log y + \log z$

$$\Rightarrow \quad \frac{\delta V}{V} = \frac{\delta x}{x} + \frac{\delta y}{y} + \frac{\delta z}{z} \quad \Rightarrow \quad \frac{\delta V}{V} \times 100 = 2 + 2 + 2 = 6.$$

$$\text{or} \quad \delta V = \frac{6V}{100} = \frac{6(5 \times 10 \times 5)}{100} = 15 \text{ m}^3$$

Now, the number of bricks in δV, the error in V, = 15 × 100 = 1500

Therefore the error in the cost $= \frac{1500 \times 2000}{1000}$ = Rs. 3000. **Ans.**

(*b*) If x and y are semi-major and semi-minor axes of the ellipse, then its area A is given by

$$A = \pi xy$$

$$\Rightarrow \quad \log A = \log \pi + \log x + \log y.$$

Taking differentials we get $\frac{dA}{A} = 0 + \frac{dx}{x} + \frac{dy}{y}$

$$\text{or} \quad \frac{dA}{A} \times 100 = \frac{dx}{x} \times 100 + \frac{dy}{y} \times 100 = 1 + 1 = 2$$

$\therefore$ Error in area = 2%. **Ans.**

EXAMPLE 5.11. The power P required to propel a steamer of length l at a speed u, is given by $P = \lambda u^3 l^3$ where λ is constant. If u is increased by 3% and l is decreased by 1%, find the corresponding increase in P. **(UPTU 2009-10)**

SOLUTION: $P = \lambda u^3 l^3$. Here change in P is given by $\delta P = \frac{\partial P}{\partial u}\delta u + \frac{\partial P}{\partial l}\delta l$ approx.

or $$\delta P = 3\lambda u^2 l^3 \delta u + 3\lambda u^3 l^2 \delta l$$

$\therefore$ $$\frac{\delta P}{P} = \frac{3\lambda u^2 l^3}{\lambda u^3 l^3}\delta u + \frac{3\lambda u^3 l^2}{\lambda u^3 l^3}\delta l = \frac{3}{4}\delta u + \frac{3}{l}\delta l$$

$\therefore$ $$\frac{\delta P}{P}\times 100 = 3\frac{\delta u}{u}\times 100 + 3\frac{\delta l}{l}\times 100 = 3(3) + 3(-1) = 6\%.$$ **Ans.**

EXERCISE 5A

1. Find the percentage error in the calculated area of rectangle when an error of + 1 % is made in measuring its length and breadth.

2. The period of a simple pendulum is given by $T = 2\pi\sqrt{\frac{l}{g}}$. Find the maximum percentage error in T due to possible errors up to 1 % in l and 2.5 % in g.

(UPTU 2003-04)

3. In computing the volume of a right circular cone, errors of 2 % and 1 % are made in the height and the radius of the base respectively. Calculate the resulting percentage error in the volume.

4. A pile of bricks measures $2\,m \times 20\,m \times 6\,m$ and the count is 430 bricks to $1\,m^3$. Find the error in the cost when the tape is stretched by 1% beyond its standard length, the cost of bricks being Rs. 400 per thousand.

5. Find the approximate value of $[0.98^2 + 2.01^2 + 1.94^2]^{\frac{1}{2}}$.

6. The area of a triangle ABC is computed from the formula $\Delta = \frac{1}{2} bc \sin A$. Errors of 1%, 2% and 3% are made in measuring b, c and A. If A = 45°, find the percentage error in the calculated value of Δ.

7. The focal length of a mirror is found from the formula $\frac{2}{f} = \frac{1}{v} - \frac{1}{u}$. Find the percentage error in f if u and v are both in error by 2%.

8. The resistance R of a circuit is found by the formula $I = \frac{E}{R}$. If there is an error of 0.1 ampear in reading I and 0.5 volts in E, find the corresponding possible percentage error in R when I = 15 amp and E = 100 volts.

9. The resonant frequency f in a circuit is given by $f = \frac{1}{2\pi\sqrt{LC}}$. If the measurements of inductance L and capacitance C are in error by + 2% and – 1% respectively, find the percentage error in the calculated value of f.

10. The inner dimensions of a rectangular box are measured to be 1, 1.5 and 3 metres subject to an error of ± 0.2 metres in each. Within what limits must the diagonal of the box lie?

11. At a distance of 120 ft. from the foot of a tower, the elevation of its top is 60°, if the possible error in measuring the distance and the elevation are 1 inch and 1 minute, find the approximate error in the calculated height.

12. The area of a triangle is calculated from the lengths a, b, c of its sides by using the formula $\Delta = \sqrt{s(s-a)(s-b)(s-c)}$. If a is diminished and b is increased by the same small amount k, show that the resulting change in the area is given by

$$\frac{\delta\Delta}{\Delta} = 2(a-b)\,k[c^2 - (a-b)^2]^{-1/2}$$

13. Find the approximate value of $\sqrt{24}\,(18)^{1/4}\,(30)^{1/3}$ using the concept of total derivative.

❑❑❑

MAXIMA AND MINIMA OF A FUNCTION OF TWO OR MORE VARIABLES

We shall simply extend the definition of maxima and minima of functions of one variable to functions of two variables. The function $f(x, y)$ has a maximum value for a certain pair of values of x and y, if this value, say f_{max}, is greater than the values of $f(x, y)$ for all values of x and y in the small neighbourhood of the particular pair of values, *i.e.* $f(x, y) > f(x+h, y+k)$ where h and k are small. Similary, a minimum value of $f(x, y)$ is defined.

Thus, the quantity $f(x, y) - f(x+h, y+k)$ must retain a constant sign for small variations in h and k. When this constant sign is positive there exists a maximum value of $f(x, y)$ and when it is negative there exists a **minimum value** of $f(x, y)$. Clearly, if the sign does not remain constant, there will be **neither a maximum nor a minimum** and then the point is called **saddle point**.

To obtain the criterion for maxima and minima, recall the Taylor's expansion of $f(x+h, y+k)$ as

$$f(x+h, y+k) = f(x, y) + \left(h\frac{\partial f}{\partial x} + k\frac{\partial f}{\partial y}\right) + \frac{1}{2!}\left(h^2\frac{\partial^2 f}{\partial x^2} + 2hk\frac{\partial^2 f}{\partial x \partial y} + k^2\frac{\partial^2 f}{\partial y^2}\right) + \ldots$$

or $$f(x+h, y+k) - f(x, y) = (hf_x + kf_y) + \frac{1}{2!}(h^2 f_{xx} + 2hk f_{xy} + k^2 f_{yy}) + \ldots \quad \ldots(1)$$

When h and k are small the first term on the R.H.S. of (1) governs the sign of the R.H.S. and hence, for constant sign on the R.H.S., we must have

$$f_x = f_y = 0 \quad \text{or} \quad p = q = 0 \quad \ldots(2)$$

as preliminary condition for the existence of maxima and minima. A point satisfying the conditions in (2), is called a **STATIONARY POINT OR CRITICAL POINT**. Then in that case eqn. (1) becomes

$$f(x+h, y+k) - f(x, y) = \frac{1}{2!}(h^2 f_{xx} + 2hk f_{xy} + k^2 f_{yy}) + \ldots \quad \ldots(3)$$

Again since h and k are small, the sign of R.H.S. of (3) will be governed by the sign of its first term. Now, consider

$$h^2 f_{xx} + 2hk f_{xy} + k^2 f_{yy} = \frac{1}{f_{xx}}\left[h^2 f_{xx}^2 + 2hk\, f_{xy}\, f_{xx} + k^2 f_{xx}\, f_{yy}\right]$$
$$= \frac{1}{f_{xx}}\left[(h\, f_{xx} + k\, f_{xy})^2 + k^2(f_{xx}\, f_{yy} - f_{xy}^2)\right]$$

If $f_{xx}f_{yy} - f_{xy}^2 > 0$ or $rt - s^2 \geq 0$, the above expression in the square brackets is always positive and there will be **maxima** if $f_{xx} (= r)$ is negative and **minima** if f_{xx} $(= r)$ is positive. Whereas, if $f_{xx} f_{yy} - f_{xy}^2 < 0$ the expression in the square brackets can change in sign and therefore there will be neither maxima nor minima. Thus, provided that f_{xx}, f_{yy} and f_{xy} are not all zero, the criteria for the existence of maxima and minima runs as follows:

(i) $f_x = f_y = 0$, or $p = q = 0$ the solution of which gives pairs of values of x and y called **CRITICAL POINTS OR STATIONARY POINTS**.

(ii) Compute $f_{xx} (= r)$, $f_{yy} (= t)$ and $f_{xy} (= s)$ for each pair (x, y). If $f_{xx}\, f_{yy} - f_{xy}^2 > 0$, or $rt - s^2 > 0$, there will be a **maxima if f_{xx} (or f_{yy}) < 0** and **minima if f_{xx} (or f_{yy}) > 0.**

(iii) If $f_{xx}\, f_{yy} - f_{xy}^2 < 0$ there will be neither a maxima nor minima and points are called **SADDLE POINTS.**

EXAMPLE 5.12. Find the maximum and minimum values of the function $x^3 + y^3 - 3axy$.

(UPTU 2004-05; UPTU 2009-10; UTPU 2011-12)

SOLUTION: Let $f(x, y) = x^3 + y^3 - 3\,axy$ hence

$$\frac{\partial f}{\partial x} = 3x^2 - 3ay \quad \text{and} \quad \frac{\partial f}{\partial y} = 3y^2 - 3\,ax$$

For stationary points $\frac{\partial f}{\partial x} = 0$ and $\frac{\partial f}{\partial y} = 0$, which gives $x^2 = ay$ and $y^2 = ax$

thus the stationary points are (0, 0) and (a, a).

Next, $r = \frac{\partial^2 f}{\partial x^2} = 6x$, $s = \frac{\partial^2 f}{\partial x \partial y} = -3a$ and $t = \frac{\partial^2 f}{\partial y^2} = 6y$

At (0, 0), $rt - s^2 = 0 - 9a^2 < 0$.

Therefore, at (0, 0) $f(x, y)$ is neither maximum nor minimum.

At (a, a) we have $rt - s^2 = 36a^2 - 9a^2 = 27a^2$ which is positive. Now, the sign of r (or t) decides the existence of maxima or minima at the point (a, a) Since $r = 6a = t$ at (a, a) which is positive, hence (a, a) is minima and the minimum value of the function is $a^3 + a^3 - 3a^3 = -a^3$. **Ans.**

EXAMPLE 5.13. Find the dimensions of a rectangular box (without top) with a given volume so that the material used is minimum. **(UPTU 2001 Q.B., 2005-06)**

SOLUTION: Let the length, breadth and height of the box be x, y and z respectively and V be the given volume, then $V = xyz$ = constant

The surface area $S = xy + 2yz + 2zx$ ... (1)

Substituting $z = \frac{V}{xy}$ in (1), we get $S = xy + 2(x + y)\frac{V}{xy} = xy + \frac{2V}{y} + \frac{2V}{x}$ ... (2)

Therefore, $\frac{\partial S}{\partial x} = y - \frac{2V}{x^2}$ and $\frac{\partial S}{\partial y} = x - \frac{2V}{y^2}$.

At stationary points $\frac{\partial S}{\partial x} = 0$, $\frac{\partial S}{\partial y} = 0$, *i.e.*, $x^2 y = 2V$ and $xy^2 = 2V$

$\Rightarrow$ $xy\,(x - y) = 0$ which gives $x = 0$, $y = 0$, $x = y$

$\therefore$ Stationary points are (0, 0) and $\left((2V)^{1/3}, (2V)^{1/3}\right)$.

Next, $\frac{\partial^2 S}{\partial x^2} = \frac{4V}{x^3}$, $\frac{\partial^2 S}{\partial x \partial y} = 1$, $\frac{\partial^2 S}{\partial y^2} = \frac{4V}{y^3}$

At (0, 0) $\frac{\partial^2 s}{\partial x^2}$ and $\frac{\partial^2 s}{\partial y^2}$ are not defined, therefore, neither maxima nor minima at (0, 0).

At $\left((2V)^{1/3}, (2V)^{1/3}\right)$ $\frac{\partial^2 S}{\partial x^2} = 2$, $\frac{\partial^2 S}{\partial y^2} = 2$ hence

$$\frac{\partial^2 S}{\partial x^2}\frac{\partial^2 S}{\partial y^2} - \left(\frac{\partial^2 S}{\partial x \partial y}\right)^2 = 4 - 1 > 0 \quad \text{and} \quad \frac{\partial^2 S}{\partial x^2} > 0$$

hence S has minimum value at $x = (2V)^{1/3}$, $y = (2V)^{1/3}$.

At this point $z = \frac{V}{xy} = \frac{V}{(2V)^{1/3}.(2V)^{1/3}} = \frac{V^{1/3}}{2^{2/3}} = \frac{x}{2}$.

Therefore, the box should have a square base and the height should be half the length of the base, for material being used to be least in making the box. **Ans.**

EXAMPLE 5.14. Examine the function $f(x, y) = \sin x + \sin y + \sin(x + y)$ for maximum and minimum values. **(UPTU 2001 Q.B.)**

SOLUTION: $f(x, y) = \sin x + \sin y + \sin(x + y)$

Hence $\dfrac{\partial f}{\partial x} = \cos x + \cos(x + y) = 2\cos\left(x+\dfrac{y}{2}\right)\cos\dfrac{y}{2}$

and $\dfrac{\partial f}{\partial y} = \cos y + \cos(x + y) = 2\cos\left(y+\dfrac{x}{2}\right)\cos\dfrac{x}{2}$

For f to be maximum or minimum $\dfrac{\partial f}{\partial x} = 0 = \dfrac{\partial f}{\partial y}$, that is,

$$\cos\left(x+\frac{y}{2}\right)\cos\frac{y}{2} = 0 \quad \text{and} \quad \cos\left(y+\frac{x}{2}\right)\cos\frac{x}{2} = 0$$

which gives $x = \pi$, $y = \pi$; $x = \pi/3$, $y = \pi/3$.

Thus, the critical points are $\left(\dfrac{\pi}{3},\dfrac{\pi}{3}\right)$ and (π,π) if x and $y \in [0, \pi]$

Next, $\dfrac{\partial^2 f}{\partial x^2} = r = -\sin x - \sin(x+y)$, $\dfrac{\partial^2 f}{\partial y^2} = t = -\sin y - \sin(x+y)$

and $\dfrac{\partial^2 f}{\partial x \partial y} = s = -\sin(x+y)$.

At (π, π), $r = s = t = 0$. Hence saddle point.

At $\left(\dfrac{\pi}{3},\dfrac{\pi}{3}\right)$, $r = -2\sin\pi/3 = -\sqrt{3}$, $t = -\sqrt{3}$, and $s = -\sqrt{3}/2$

$\therefore \quad rt - s^2 = (\sqrt{3})^2 - \dfrac{3}{4} = \dfrac{9}{4} > 0$ and $r < 0$.

Hence $f(x, y)$ is maximum at $(\pi/3, \pi/3)$ and (π, π) is saddle point. **Ans.**

EXAMPLE 5.15. Find the maximum and minimum values of $f(x, y) = x^3 + 3xy^2 - 15x^2 - 15y^2 + 72x$.

SOLUTION: $f(x, y) = x^3 + 3xy^2 - 15x^2 - 15y^2 + 72x$.

Hence $\dfrac{\partial f}{\partial x} = 3x^2 + 3y^2 - 30x + 72$ and $\dfrac{\partial f}{\partial y} = 6xy - 30y$.

Now for f to have maximum or minimum $\dfrac{\partial f}{\partial x} = 0 = \dfrac{\partial f}{\partial y}$.

or $x^2 + y^2 - 10x + 24 = 0$ and $(x - 5)y = 0$

$\Rightarrow$ when $y = 0$, $(x - 4)(x - 6) = 0$. Hence two critical points are $(4, 0)$ and $(6, 0)$ and when $x = 5$

we have $y^2 - 1 = 0$, i.e., $y = \pm 1$, hence two more critical points are $(5, 1)$ and $(5, -1)$.

Next, $r = \dfrac{\partial^2 f}{\partial x^2} = 6x - 30$, $\dfrac{\partial^2 f}{\partial x \partial y} = s = 6y$ and $t = \dfrac{\partial^2 f}{\partial y^2} = 6x - 30$

Now at $(4, 0)$ $r = t = -6$ and $s = 0$ and $rt - s^2 > 0$, hence maxima at $(4, 0)$.

At $(6, 0)$, $r = t = 6$, $s = 0$ and $rt - s^2 > 0$, hence minima at $(6, 0)$.

At (5, 1), $r = t = 0$ and $s = 6$ and $rt - r^2 < 0$, so it is a saddle point.
and at (5, –1), $r = t = 0$ and $s = -6$ and $rt - r^2 < 0$, so it is also a saddle point.
Therefore, max. $f(x, y) = 64 - 16\,(15) + 228 = 112$
and min. $f(x, y) = 36\,(6) - 15\,(36) + 72\,(6) = 108$
and at $(5, \pm 1)$ we have saddle points. **Ans.**

EXAMPLE 5.16. Examine the function $f(x, y) = x^3 - 3x^2 - 4y^2 + 1$ for maximum and minimum.

SOLUTION: $f(x, y) = x^3 - 3x^2 - 4y^2 + 1$. Here $\dfrac{\partial f}{\partial x} = 3x^2 - 6x, \ \dfrac{\partial f}{\delta y} = -8y$

For $f(x, y)$ to have maximum or minimum, $\dfrac{\partial f}{\partial x} = 0 = \dfrac{\partial f}{\delta y}$.

Thus, the critical points are (0, 0) and (2, 0).

Next, $r = \dfrac{\partial^2 f}{\partial x^2} = 6x - 6, \quad s = \dfrac{\partial^2 f}{\partial x \partial y} = 0, \quad t = \dfrac{\partial^2 f}{\partial y^2} = -8.$

At (0, 0), $r < 0, t < 0$ and $rt - s^2 = 48 > 0$ hence maxima.
At (2, 0), $r = 6, s = 0, t = -8$ and $rt - s^2 < 0$ hence at (2, 0) we have a saddle point. **Ans.**

EXAMPLE 5.17. Discuss the maxima and minima of $x^3y^2\,(1 - x - y)$.

SOLUTION: Given function is $f = x^3y^2\,(1 - x - y)$.

Therefore $\dfrac{\partial f}{\partial x} = -3x^2y^2\,(x + y - 1) - x^3y^2 = -x^2y^2\,[4x + 3y - 3]$

and $\dfrac{\partial f}{\partial y} = -2x^3y\,(x + y - 1) - x^3y^2 = -x^3y\,[2x + 3y - 2]$

For f to be maximum or minimum $\dfrac{\partial f}{\partial x} = 0 = \dfrac{\partial f}{\partial y}$

$\Rightarrow \quad x = 0, \quad y = 0 \quad$ and $\quad x = \dfrac{1}{2}, \quad y = \dfrac{1}{3}$.

Thus the critical points are (0, 0) and $\left(\dfrac{1}{2}, \dfrac{1}{3}\right)$.

Next, $r = \dfrac{\partial^2 f}{\partial x^2} = -2xy^2(4x+3y-3) - 4x^2y^2, \quad s = \dfrac{\partial^2 f}{\partial x \partial y} = -2x^2y(4x+3y-3) - 3x^2y^2,$

$t = \dfrac{\partial^2 f}{\partial y^2} = -x^3(2x+3y-2) - 3x^3y.$

At (0, 0) $r = s = t = 0$ hence neither minima nor maxima here,

and at $\left(\dfrac{1}{2}, \dfrac{1}{3}\right)$, $r = -\dfrac{1}{9}, \ s = \dfrac{-1}{12}, \ t = \dfrac{-1}{8}$ hence $rt - s^2 = \dfrac{1}{72} - \dfrac{1}{144}$.

Since, $rt - s^2 > 0$ and r and t are negative we have a maxima at $\left(\dfrac{1}{2}, \dfrac{1}{3}\right)$. **Ans.**

LAGRANGE'S METHOD OF UNDETERMINED MULTIPLIERS

At occasions it would be necessary to find the maxima or minima of a function, subject to one or two conditions (or constraints) being satisfied. Suppose we are to maximise or minimise the function $u = f(x, y)$ subject to the condition $g(x, y) = 0$. If we can solve the latter equation for y in terms of x and substitute it in the function $f(x, y)$, then the problem reduces to that of finding the maxima or minima of a function of a single variable x. However, usually it is not possible and we take recourse to **Lagrange's method of undetermined multipliers** as explained below:

Since, $u = f(x, y)$ we have $\dfrac{du}{dx} = \dfrac{\partial f}{\partial x}.1 + \dfrac{\partial f}{\partial y}\dfrac{dy}{dx}$ ... (1)

and from the relation $g(x, y) = 0$ we have $\dfrac{\partial g}{\partial x}.1 + \dfrac{\partial g}{\partial y}\dfrac{dy}{dx} = 0$ ... (2)

From (2) we have $\dfrac{dy}{dx} = \dfrac{-\partial g/\partial x}{\partial g/\partial y} = -\dfrac{g_x}{g_y}$ on the curve $g(x, y) = 0$.

But on $g(x, y) = 0$ the function $u = f(x, y)$ is only a function of x as mentioned above, hence the stationary points are given by $\dfrac{du}{dx} = 0$, that is,

$$\frac{\partial f}{\partial x} + \frac{\partial f}{\partial y}\frac{dy}{dx} = 0 \quad \text{or} \quad \frac{\partial f}{\partial x} + \frac{\partial f}{\partial y}\left(\frac{-\partial g/\partial x}{\partial g/\partial y}\right) = 0$$

or $\quad f_x g_y - f_y g_x = 0$ ...(3)

which can be solved subject to the satisfaction of the condition $g(x, y) = 0$. ...(4)

Algebraically, this is equivalent to finding the stationary points of a function F given by

$$F(x, y) = f(x, y) + \lambda g(x, y) \quad ..(5)$$

where λ is a parameter to be determined.

The stationary points of $F(x, y)$ are given by the equations

$$f_x + \lambda g_x = 0 \quad \text{and} \quad f_y + \lambda g_y = 0$$

which will have a solution if the condition (3) is satisfied. The maxima or minima can then be obtained by examining the condition in the neighbourhood of the stationary points.

Further, suppose it is required to find the stationary values of a function of three variables, say

$$u = f(x, y, z) \quad ...(1)$$

subject to conditions $\phi(x, y, z) = 0$ and $\psi(x, y, z) = 0$. ... (2)

In this case we construct the function F as

$$F = f + \lambda_1 \phi + \lambda_2 \psi \quad ...(3)$$

where λ_1, λ_2 are called non-zero Lagrange's undetermined multipliers.

Next, form the equations $\dfrac{\partial F}{\partial x} = 0, \quad \dfrac{\partial F}{\partial y} = 0, \quad \dfrac{\partial F}{\partial z} = 0.$ ... (4)

Eliminating $x, y, z, \lambda_1, \lambda_2$ between (1), (2) and (4) we shall get an equation in u, the roots of which give the stationary values of $u = f(x, y, z)$.

The method would be clear through the following examples.

EXAMPLE 5.18. Show that the rectangular solid of maximum volume that can be inscribed in a given sphere, is a cube. **(UPTU 2004)**

SOLUTION: Let R be the radius of the given sphere and $2x, 2y, 2z$ be the length, breadth and height of the rectangular solid, then

$$V = 8\,xyz \quad \text{and} \quad x^2 + y^2 + z^2 = R^2.$$

Consider a function $F = 8\,xyz + \lambda(x^2 + y^2 + z^2 - R^2)$ and use Lagrange's method.

For stationary values $dF = 0$ hence $\dfrac{\partial F}{\partial x} = \dfrac{\partial F}{\partial y} = \dfrac{\partial F}{\partial z} = 0$

$\Rightarrow$ $8\,yz + 2\,\lambda x = 0,\quad 8\,zx + 2\,\lambda y = 0,\quad 8\,xy + 2\,\lambda z = 0$

or $2\,\lambda x^2 = -8\,xyz,\quad 2\,\lambda y^2 = -8xyz,\quad 2\,\lambda z^2 = -8\,xyz$

$\Rightarrow$ $x^2 = y^2 = z^2$. Hence the rectangular solid is a cube. **Hence proved.**

EXAMPLE 5.19. Determine the maxima of the function u given by

$$u = (x + 1)\,(y + 1)\,(z + 1)$$

subject to the condition $a^x\,b^y\,c^z = k.$

SOLUTION: From the given function and the condition, we can write

$$\log u = \log (x + 1) + \log (y + 1) + \log (z + 1) \qquad ...(1)$$

and $$x \log a + y \log b + z \log c = \log k. \qquad ...(2)$$

Now, when u is maximum $\log u$ will also be maximum. For finding critical points, we differetiate (1) and (2) and get

$$\frac{dx}{x+1} + \frac{dy}{y+1} + \frac{dz}{z+1} = 0 \qquad ... (3)$$

and $$\log a\, dx + \log b\, dy + \log c\, dz = 0 \qquad ... (4)$$

Multiplying (4) by λ and then adding to (3) and equating to zero the co-efficients of $dx, dy, dz,$ we get

$$\frac{1}{x+1} + \lambda \log a = 0, \quad \frac{1}{y+1} + \lambda \log b = 0, \quad \frac{1}{z+1} + \lambda \log c = 0$$

or
$$1 + \lambda\, x \log a + \lambda \log a = 0$$
$$1 + \lambda\, y \log b + \lambda \log b = 0 \qquad ...(5)$$
$$1 + \lambda\, z \log c + \lambda \log c = 0$$

Adding these, we get

$$3 + \lambda\,(x \log a + y \log b + z \log c) + \lambda\,(\log a + \log b + \log c) = 0$$

or $$3 + \lambda \log k + \lambda \log (abc) = 0 \quad \Rightarrow \quad \lambda = \frac{-3}{\log (k\,abc)}.$$

Substituting this value of λ in (5), we get

$$x + 1 = \frac{-1}{\lambda \log a} = \frac{\log (k\,abc)}{3 \log a}, \quad y + 1 = \frac{\log (k\,abc)}{3 \log b}, \quad z + 1 = \frac{\log (k\,abc)}{3 \log c}$$

Thus, the maxima is given by $x = \dfrac{\log\left(k\,\dfrac{bc}{a^2}\right)}{3 \log a}, \quad y = \dfrac{\log\left(k\,\dfrac{ca}{b^2}\right)}{3 \log b}, \quad z = \dfrac{\log\left(k\,\dfrac{ab}{c^2}\right)}{3 \log c}.$ **Ans.**

EXAMPLE 5.20. Find the stationary value of $a^3x^2 + b^3y^2 + c^3z^2$ subject to the fulfilment of the condition $\frac{1}{x}+\frac{1}{y}+\frac{1}{z} = 1.$

SOLUTION: Let $u = f(x, y, z) = a^3x^2 + b^3y^2 + c^3z^2$

and $\phi(x, y, z) = \frac{1}{x}+\frac{1}{y}+\frac{1}{z} - 1$

Now, consider the function $F = f + \lambda\,\phi$ where λ is an unknown non-zero quantity

or $$F = a^3x^2 + b^3y^2 + c^3z^2 + \lambda\left(\frac{1}{x}+\frac{1}{y}+\frac{1}{z}-1\right)$$

For stationary points we have $\frac{\partial F}{\partial x} = \frac{\partial F}{\partial y} = \frac{\partial F}{\partial z} = 0.$

or $$2a^3x - \frac{\lambda}{x^2} = 0, \quad 2b^3y - \frac{\lambda}{y^2} = 0, \quad 2c^3z - \frac{\lambda}{z^2} = 0.$$

or $$2a^3x^3 = \lambda, \quad 2b^3y^3 = \lambda, \quad 2c^3z^3 = \lambda.$$

or $ax = by = cz = K$, say

$\Rightarrow$ $$x = \frac{K}{a}, \quad y = \frac{K}{b}, \quad z = \frac{K}{c}$$

Substituting these in $\frac{1}{x}+\frac{1}{y}+\frac{1}{z} = 1$, we get

$$\frac{a}{K}+\frac{b}{K}+\frac{c}{K} = 1 \quad \text{or} \quad K = a + b + c.$$

Hence the stationary point is given by $x = \frac{a+b+c}{a}$, $y = \frac{a+b+c}{b}$, $z = \frac{a+b+c}{c}$

$\therefore$ Stationary value of $$f = \frac{a^3(a+b+c)^2}{a^2} + \frac{b^3(a+b+c)^2}{b^2} + \frac{c^3(a+b+c)^2}{c^2}$$
$$= (a + b + c)^3.$$ **Ans.**

EXAMPLE 5.21. Find the volume of the greatest rectangular parallelopiped that can be inscribed inside the ellipsoid $\frac{x^2}{a^2}+\frac{y^2}{b^2}+\frac{z^2}{c^2} = 1.$ **(UPTU 2000-01; 2002-03)**

SOLUTION: Let the edges of the parallelopiped be $2x$, $2y$ and $2z$ parallel to the co-ordinate axes. The volume V of the parallelopiped is given by $V = 8\,xyz$ which is to be maximised subject to the condition

$$\frac{x^2}{a^2}+\frac{y^2}{b^2}+\frac{z^2}{c^2} = 1.$$

Consider the function $F = 8\,xyz + \lambda\left(\frac{x^2}{a^2}+\frac{y^2}{b^2}+\frac{z^2}{c^2}-1\right)$ where λ is unknown multiplier.

For stationary values $\frac{\partial F}{\partial x} = \frac{\partial F}{\partial y} = \frac{\partial F}{\partial z} = 0$

or $$8yz + \frac{2\lambda x}{a^2} = 0, \quad 8zx + \frac{2\lambda y}{b^2} = 0, \quad 8xy + \frac{2\lambda z}{c^2} = 0$$

Eliminating λ between the above equations, taken in pairs, we get $\frac{x^2}{a^2} = \frac{y^2}{b^2} = \frac{z^2}{c^2}$.

Substituting these in the condition $\frac{x^2}{a^2} + \frac{y^2}{b^2} + \frac{z^2}{c^2} = 1$, we get

$$\frac{x^2}{a^2} = \frac{y^2}{b^2} = \frac{z^2}{c^2} = \frac{1}{3}$$

$\Rightarrow \quad x = \frac{a}{\sqrt{3}},\ y = \frac{b}{\sqrt{3}},\ z = \frac{c}{\sqrt{3}}$. since x, y, z cannot be $-ve$.

When $x = 0$ the parallelopiped becomes a rectangular sheet and, hence, the volume V is 0. Note that, as x increases volume also increases.

Hence V is maximum when $x = \frac{a}{\sqrt{3}},\ y = \frac{b}{\sqrt{3}},\ z = \frac{c}{\sqrt{3}}$.

$\therefore \quad$ Maximum volume $= 8xyz = \frac{8abc}{3\sqrt{3}}$. **Ans.**

EXAMPLE 5.22. Show that the stationary values of $u = \frac{x^2}{a^4} + \frac{y^2}{b^4} + \frac{z^2}{c^4}$ where $lx + my + nz = 0$ and $\frac{x^2}{a^2} + \frac{y^2}{b^2} + \frac{z^2}{c^2} = 1$, are roots of equation $\frac{l^2a^4}{1-a^2u} + \frac{m^2b^4}{1-b^2u} + \frac{n^2c^4}{1-c^2u} = 0$.

SOLUTION: Let us write $u = f(x, y, z) = \frac{x^2}{a^4} + \frac{y^2}{b^4} + \frac{z^2}{c^4}$, ... (1)

$\phi(x, y, z) \equiv lx + my + nz = 0$ ... (2)

and $\psi(x, y, z) \equiv \frac{x^2}{a^2} + \frac{y^2}{b^2} + \frac{z^2}{c^2} - 1 = 0$ ... (3)

and construct the function $F = f + \lambda_1 \phi + \lambda_2 \psi$, ... (4)

where λ_1, λ_2 are non-zero Lagrangian multipliers. For stationary values, we have

$$\frac{\partial F}{\partial x} = 0, \quad \frac{\partial F}{\partial y} = 0, \quad \frac{\partial F}{\partial z} = 0, \quad \text{that is,}$$

$$\frac{2x}{a^4} + \lambda_1 l + \lambda_2 \frac{2x}{a^2} = 0 \quad \text{...(5)}$$

$$\frac{2y}{b^4} + \lambda_1 m + \lambda_2 \frac{2y}{b^2} = 0 \quad \text{...(6)}$$

$$\frac{2z}{c^4} + \lambda_1 n + \lambda_2 \frac{2z}{c^2} = 0 \quad \text{...(7)}$$

Multiplying (5), (6), (7) by x, y, z respectively and adding, gives

$2u + 0 \cdot \lambda_1 + 2\lambda_2 = 0$ hence $\lambda_2 = -u$.

Substituting this value of λ_2 in (5), (6) and (7), we get

$$x\left(\frac{2}{a^4} - \frac{2u}{a^2}\right) = -\lambda_1 l \text{ or } x = \frac{-a^4 \lambda_1 l}{2(1-a^2u)} \text{ and } y = \frac{-b^4 \lambda_1 m}{2(1-b^2u)}, \quad z = \frac{-c^4 \lambda_1 n}{2(1-c^2u)}.$$

These values must satisfy (2), hence

$$-\frac{\lambda_1 a^4 l^2}{2(1-a^2 u)}-\frac{\lambda_1 b^4 m^2}{2(1-b^2 u)}-\frac{\lambda_1 c^4 n^2}{2(1-c^2 u)}=0$$

Since, $\lambda_1 \neq 0$ we have $\dfrac{a^4 l^2}{1-a^2 u}+\dfrac{b^4 m^2}{1-b^2 u}+\dfrac{c^4 n^2}{1-c^2 u}=0.$

which gives the stationary values of u. **Ans.**

EXAMPLE 5.23. Find the largest and the smallest distances from the origin to the ellipsoid

$$\frac{x^2}{4}+\frac{y^2}{5}+\frac{z^2}{25}=1 \quad \text{and the plane} \quad z=x+y.$$

SOLUTION: Let $u = x^2 + y^2 + z^2$ be the square of the distance of any point on the ellipsoid from the origin. We are to maximise and minimise u, subject to the constraints

$$\phi_1 \equiv \frac{x^2}{4}+\frac{y^2}{5}+\frac{z^2}{25}-1=0 \qquad \text{...(1)}$$

and $\phi_2 \equiv x + y - z = 0.$ ...(2)

Now, consider the function $F = u + \lambda_1 \phi_1 + \lambda_2 \phi_2$

$$= x^2 + y^2 + z^2 + \lambda_1\left(\frac{x^2}{4}+\frac{y^2}{5}+\frac{z^2}{25}-1\right) + \lambda_2 (x + y - z).$$

For critical points we have $\dfrac{\partial F}{\partial x}=0,\ \dfrac{\partial F}{\partial y}=0,\ \dfrac{\partial F}{\partial z}=0.$

Hence $2x+\dfrac{\lambda_1 x}{2}+\lambda_2=0,\quad 2y+\dfrac{2\lambda_1 y}{5}+\lambda_2=0$ and $2z+\dfrac{2\lambda_1 z}{25}-\lambda_2=0.$

Above equations when solved for $x, y, z,$ give

$$x=\frac{-2\lambda_2}{\lambda_1+4},\quad y=\frac{-5\lambda_2}{2\lambda_1+10},\quad z=\frac{25\lambda_2}{2\lambda_1+50} \qquad \text{... (3)}$$

Substituting these in the constraint $x + y - z = 0$ and on dividing by λ_2 ($\neq 0$), we get

$$\frac{2}{\lambda_1+4}+\frac{5}{2\lambda_1+10}+\frac{25}{2\lambda_1+50}=0$$

or $2(2\lambda_1+10)(2\lambda_1+50)+5(\lambda_1+4)(2\lambda_1+50)+25(\lambda_1+4)(2\lambda_1+10)=0$

or $17\lambda_1^2+245\lambda_1+750=0$ or $(17\lambda_1+75)(\lambda_1+10)=0$

which gives $\lambda_1 = -10,\ -\dfrac{75}{17}.$

When $\lambda_1 = -10$ we have, from (3) $x=\dfrac{\lambda_2}{3},\quad y=\dfrac{\lambda_2}{2},\quad z=\dfrac{5}{6}\lambda_2$

These values should satisfy (1), hence

$$\lambda_2^2\left(\frac{1}{36}+\frac{1}{20}+\frac{1}{36}\right)=1 \quad \text{or} \quad \lambda_2=\pm 6\sqrt{\frac{5}{19}}.$$

This gives two critical points $\left(2\sqrt{\frac{5}{19}}, 3\sqrt{\frac{5}{19}}, 5\sqrt{\frac{5}{19}}\right), \left(-2\sqrt{\frac{5}{19}}, -3\sqrt{\frac{5}{19}}, -5\sqrt{\frac{5}{19}}\right)$.

The stationary value of $x^2 + y^2 + z^2$ corresponding to these critical points, is

$$= \frac{(20+45+125)}{19} = 10.$$

Next, when $\lambda_1 = -\frac{75}{17}$ we have from (3). $x = \frac{34}{7}\lambda_2, \quad y = -\frac{17}{4}\lambda_2, \quad z = \frac{17}{28}\lambda_2$

Substituting these in (1), gives $\lambda_2 = \pm\frac{140}{17\sqrt{646}}$ which gives the critical points as

$$\left(\frac{40}{\sqrt{646}}, \frac{-35}{\sqrt{646}}, \frac{5}{\sqrt{646}}\right) \text{ and } \left(\frac{-40}{\sqrt{646}}, \frac{35}{\sqrt{646}}, \frac{-5}{\sqrt{646}}\right)$$

The stationary value of $x^2 + y^2 + z^2$ corresponding to these critical points, is

$$= \frac{(1600+1225+25)}{646} = \frac{75}{17}.$$

Thus, the required maximum value is 10 and the minimum value is $\frac{75}{17}$. **Ans.**

EXAMPLE 5.24. Find a point upon the plane $ax + by + cz = p$ at which the function $f = x^2 + y^2 + z^2$ has a minimum value and find the minimum f.

(UPTU 2004-05; UPTU 2009-10)

SOLUTION: We are to minimise $f = x^2 + y^2 + z^2$ such that $ax + by + cz = p$

We can write $f = x^2 + y^2 + \frac{1}{c^2}(p - ax - by)^2$

$$= \left(1+\frac{a^2}{c^2}\right)x^2 + \left(1+\frac{b^2}{c^2}\right)y^2 - \frac{2ap}{c^2}x - \frac{2bp}{c^2}y + \frac{2abxy}{c^2} + \frac{p^2}{c^2}$$

Hence $\frac{\partial f}{\partial x} = 2\left(1+\frac{a^2}{c^2}\right)x - \frac{2ap}{c^2} + \frac{2aby}{c^2}$ and $\frac{\partial f}{\partial y} = 2\left(1+\frac{b^2}{c^2}\right)y - \frac{2bp}{c^2} + \frac{2abx}{c^2}$.

For maxima and minima of f, we have $\frac{\partial f}{\partial x} = \frac{\partial f}{\partial y} = 0.$

Thus $(a^2 + c^2)\,x + aby = ap$ and $(b^2 + c^2)\,y + abx = bp$

which on solving for x and y, give

$$x = \frac{ap}{a^2+b^2+c^2} \quad \text{and} \quad y = \frac{bp}{a^2+b^2+c^2}$$

which defines a critical point $\left(\frac{ap}{a^2+b^2+c^2}, \frac{bp}{a^2+b^2+c^2}\right)$.

Next, $r = \dfrac{\partial^2 f}{\partial x^2} = 2\left(1+\dfrac{a^2}{c^2}\right)$, $s = \dfrac{\partial^2 f}{\partial x \partial y} = \dfrac{2ab}{c^2}$ and $t = \dfrac{\partial^2 f}{\partial y^2} = 2\left(1+\dfrac{b^2}{c^2}\right)$.

$$\therefore \quad rt - s^2 = \frac{4}{c^4}(a^2+c^2)(b^2+c^2) - \frac{4a^2b^2}{c^4} = \frac{4}{c^4}[c^4 + a^2c^2 + b^2c^2] = \frac{4}{c^2}[a^2+b^2+c^2) > 0$$

Here $r > 0$ hence at the critical point, function f has minima.

When $x = \dfrac{ap}{a^2+b^2+c^2}$, $y = \dfrac{bp}{a^2+b^2+c^2}$, we have

$$z = \frac{1}{c}[p - ax - by] = \frac{1}{c}\left[p - \frac{a^2 p}{a^2+b^2+c^2} - \frac{b^2 p}{a^2+b^2+c^2}\right]$$

$$= \frac{p}{c(a^2+b^2+c^2)}[a^2+b^2+c^2-a^2-b^2] = \frac{pc}{a^2+b^2+c^2}$$

$$\therefore \quad \text{Min. } f = x^2+y^2+z^2 = \frac{p^2}{(a^2+b^2+c^2)^2}[a^2+b^2+c^2] = \frac{p^2}{a^2+b^2+c^2}. \qquad \textbf{Ans.}$$

EXAMPLE 5.25. The temperature T at any point (x, y, z) of space is given by $T = 400\,xyz^2$ find the highest temperature at the surface of the sphere $x^2 + y^2 + z^2 = 1$.

(UPTU 2008-09)

SOLUTION: $T = 400\,xyz^2$ where $x^2 + y^2 + z^2 = 1$. Then $T = xyz^2 = xy\,(1 - x^2 - y^2)$

Hence $\dfrac{\partial T}{\partial x} = y\,(1 - x^2 - y^2) + xy\,(-2x) = -\,y\,(3x^2 + y^2 - 1)$

and $\dfrac{\partial T}{\partial y} = x(1 - x^2 - y^2) + xy\,(-2y) = -\,x\,(3y^2 + x^2 - 1)$.

For T to be maximum or minimum $\dfrac{\partial T}{\partial x} = \dfrac{\partial T}{\partial y} = 0$.

Thus, $y\,(3x^2 + y^2 - 1) = 0$ and $x\,(3y^2 + x^2 - 1) = 0$.

The above relations, when solved for x and y, give

$x = 0$, $y = 0$ and $x^2 = \dfrac{1}{4}$, $y^2 = \dfrac{1}{4}$. Thus the critical points are:

$(0, 0)$, $\left(\dfrac{1}{2}, \dfrac{1}{2}\right)$, $\left(\dfrac{1}{2}, \dfrac{-1}{2}\right)$, $\left(\dfrac{-1}{2}, \dfrac{1}{2}\right)$, $\left(-\dfrac{1}{2}, \dfrac{-1}{2}\right)$.

Next, $r = \dfrac{\partial^2 T}{\partial x^2} = -6xy$, $t = \dfrac{\partial^2 T}{\partial y^2} = -6xy$ and $s = \dfrac{\partial^2 T}{\partial x \partial y} = -3x^2 - 3y^2 + 1$.

At $(0, 0)$ $r = t = 0,$ $s = 1.$ so $rt < s^2$. Hence $(0, 0)$ is a saddle point.

At $\left(\frac{1}{2},\frac{1}{2}\right)$ and $\left(\frac{-1}{2},\frac{-1}{2}\right)$, $r = -\frac{3}{4},$ $t = \frac{-3}{4},$ $s = -\frac{-1}{2}$ $\therefore$ $rt - s^2 = \frac{5}{16} > 0$ and $r < 0$.

and at $\left(\frac{1}{2},\frac{-1}{2}\right)$ and $\left(-\frac{1}{2},\frac{1}{2}\right)$, $r = \frac{3}{4} = t$ and $s = \frac{-1}{2}$ $\therefore$ $rt - s^2 = \frac{5}{16} > 0$ and $r > 0$.

Therefore $(0, 0)$ is a saddle point, $\left(\frac{1}{2},\frac{1}{2}\right)$ and $\left(\frac{-1}{2},\frac{-1}{2}\right)$ are maxima and $\left(\frac{1}{2},\frac{-1}{2}\right)$ and $\left(\frac{-1}{2},\frac{1}{2}\right)$ are minima.

Maximum $T = 400.\ \frac{1}{2}.\frac{1}{2}.\left(1-\frac{1}{4}-\frac{1}{4}\right) = 100.$ **Ans.**

EXAMPLE 5.26. Divide $2u$ into three parts such that the continued product of the first, square of the second and cube of the third may be maximum.

SOLUTION: Let x, y, z be three parts, then $x + y + z = 2u$ and the product $P = xy^2z^3$.

Consider $F = xy^2z^3 + \lambda\,(x + y + z - 2u)$ then

$$\frac{\partial F}{\partial x} = y^2z^3 + \lambda = 0, \quad \frac{\partial F}{\partial y} = 2xyz^3 + \lambda = 0, \quad \frac{\partial F}{\partial z} = 3xy^2z^2 + \lambda = 0.$$

Solving these, we get $y = 2x,$ $2z = 3y$. Putting these in $x + y + z = 2u,$ we get $6x = 2u$ hence $x = u/3,$ $y = 2u/3,$ $z = u.$ **Ans.**

EXAMPLE 5.27. Find the points on the surface $z^2 = xy + 1$ nearest to the origin.

SOLUTION: Let (x, y, z) be a point on the surface, at minimum distance from $(0, 0)$.

$$\text{Distance } d = \sqrt{x^2 + y^2 + z^2} \qquad \text{so} \qquad d^2 = x^2 + y^2 + z^2$$

Since $z^2 = xy + 1$ we have $d^2 = x^2 + y^2 + xy + 1 = D,$ say.

When d is least, d^2 will be least. For maxima, minima of D

$$\frac{\partial D}{\partial x} = 2x + y = 0 \quad \text{and} \quad \frac{\partial D}{\partial y} = 2y + x = 0$$

which give $x = y = 0$ and hence $z^2 = 1$ so $z = \pm 1$.

Next $$r = \frac{\partial^2 D}{\partial x^2} = 2, \quad s = \frac{\partial^2 D}{\partial x \partial y} = 1 \quad \text{and} \quad t = \frac{\partial^2 D}{\partial y^2} = 2$$

$rt - s^2 = 4 - 1 = 3 > 0$ and $r > 0$ hence D^2 is minimum at $(0, 0, \pm 1)$. **Ans.**

EXAMPLE 5.28. **(a)** Find the minimum and maximum distances of the point (3, 4, 12) from the sphere $x^2 + y^2 + z^2 = 1$. **(UPTU 2001 Sp. Exam.)**

(b) Using the Lagrange's method, find the maximum and minimum distances from the origin to the curve $3x^2 + 4xy + 6y^2 = 140$. **(UPTU 20011-12)**

SOLUTION: Let A be the point (3, 4, 12) and $P(x, y, z)$ be any point on the sphere.

$\therefore \quad AP^2 = (x-3)^2 + (y-4)^2 + (z-12)^2 = f(x, y, z).$

and let $\phi(x, y, z) \equiv x^2 + y^2 + z^2 - 1 = 0.$

Now consider
$$F = f(x, y, z) + \lambda\phi(x, y, z)$$
$$= (x-3)^2 + (y-4)^2 + (z-12)^2 + \lambda(x^2 + y^2 + z^2 - 1)$$

$$\therefore \quad \frac{\partial F}{\partial x} = 2(x-3) + 2\lambda x, \quad \frac{\partial F}{\partial y} = 2(y-4) + 2\lambda y, \quad \frac{\partial F}{\partial z} = 2(z-12) + 2\lambda z.$$

For maxima and minima $\frac{\partial F}{\partial x} = \frac{\partial F}{\partial y} = \frac{\partial F}{\partial z} = 0$

which can be written as

$$-\lambda = \frac{x-3}{x} = \frac{y-4}{y} = \frac{z-12}{z} = \frac{\sqrt{(x-3)^2 + (y-4)^2 + (z-12)^2}}{\sqrt{x^2 + y^2 + z^2}} = \pm\sqrt{f}$$

$$\Rightarrow \quad x = \frac{3}{1+\lambda}, \; y = \frac{4}{1+\lambda}, \; z = \frac{12}{1+\lambda} \quad \text{or} \quad x = \frac{3}{1\pm\sqrt{f}}, \; y = \frac{4}{1\pm\sqrt{f}}, \; z = \frac{12}{1\pm\sqrt{f}}.$$

$$\therefore \quad x^2 + y^2 + z^2 = 1 \text{ gives } \frac{3^2 + 4^2 + 12^2}{(1\pm\sqrt{f})^2} = 1 \text{ or } 1\pm\sqrt{f} = 13 \Rightarrow f = 12^2, 14^2$$

$$\therefore \quad \sqrt{f} = 13 \pm 1 = 12, 14$$

$\therefore$ Max. distance = 14 and the Min. distance = 12. **Ans.**

(b) Let (x, y) be any point on the curve. Distance of the point (x, y) from origin is given by $d = \sqrt{(x-0)^2 + (y-0)^2}$.

If the distance is maximum or minimum it's square will also be.

Let $f(x, y) = (x-0)^2 + (y-0)^2 = x^2 + y^2$ ...(1)

and the condition is $f(x, y, z) = 3x^2 + 4xy + 6y^2 - 140 = 0.$ ...(2)

Consider
$$F(x, y, z) = f + \lambda\phi = (x^2 + y^2) + \lambda(3x^2 + 4xy + 6y^2)$$

For staionary values of F $\quad dF = 0$ hence $\frac{\partial F}{\partial x} = 0, \quad \frac{\partial F}{\partial y} = 0.$

$$\Rightarrow \quad 2x + \lambda(6x + 4y) = 0 \quad ...(3)$$

and
$$2y + \lambda(4x + 12y) = 0. \quad ...(4)$$

Multiplying (3) by $x/2$ and (4) by $y/2$ and adding, we get

$$\Rightarrow \quad x^2 + y^2 + \lambda(3x^2 + 4xy + 6y^2) = 0.$$

Using (1) and (2) here, we get

$$f(x, y) + \lambda 140 = 0 \quad \therefore \quad \lambda = -\frac{f}{140}.$$

Hence (3) and (4) become

$$x - \frac{f}{140}(3x + 2y) = 0 \quad \text{and} \quad y - \frac{f}{140}(2x + 6y) = 0$$

or $$(140 - 3f)x = 2fy \quad \text{and} \quad f_x = (70 - 3f)y.$$

Eliminating $\frac{x}{y}$ in the above equations

we get $$\frac{2f}{140 - 3f} = \frac{70 - 3f}{f}$$

or $$7f^2 - 630f + 9800 = 0 \quad \text{or} \quad f^2 - 90f + 1400 = 0$$

or $$(f - 70)(f - 20) = 0 \quad \therefore \quad f = 20, 70$$

Hence minimum distance = 20. and maximum distance = 70. **Ans.**

EXAMPLE 5.29. Find the minimum distance from the point (1, 2, 0) to the cone $z^2 = x^2 + y^2$.

(UPTU 2006–07)

SOLUTION: Cone is $z^2 = x^2 + y^2$. Let $P(x, y, z)$ be any point on the cone nearest to $A(1, 2, 0)$.

Distance AP = d, then

$$d^2 = (x - 1)^2 + (y - 2)^2 + (z - 0)^2 = x^2 + y^2 - 2x - 4y + 1 + 4 + z^2$$

or $$d^2 = 2x^2 + 2y^2 - 2x - 4y + 5. \quad (\text{since } z^2 = x^2 + y^2)$$

For d^2 to be minimum $\frac{\partial (d^2)}{\partial x} = \frac{\partial}{\partial y}(d^2) = 0$

or $$4x - 2 = 0, \quad 4y - 4 = 0 \quad \text{hence} \quad x = \frac{1}{2}, \quad y = 1 \quad \therefore \quad z = \sqrt{\frac{1}{4} + 1} = \pm\frac{\sqrt{5}}{2}.$$

Next, $$r = \frac{\partial^2}{\partial x^2}(d^2) = 4, \quad s = \frac{\partial^2 (d^2)}{\partial x \partial y} = 0, \quad t = \frac{\partial^2 (d^2)}{\partial y^2} = 4.$$

Hence $rt - s^2 = 16 > 0$ and $r > 0$

Hence at the point $\left(\frac{1}{2}, 1, \frac{\sqrt{5}}{2}\right)$, d^2 is minimum.

$\therefore$ Minimum $$d = \sqrt{\left(\frac{1}{2} - 1\right)^2 + (1 - 2)^2 + \frac{5}{4}} = \frac{\sqrt{5}}{2}.$$ **Ans.**

EXAMPLE 5.30. Find the shortest distance from (1, 2, –1) to the sphere $x^2 + y^2 + z^2 = 24$.

(UPTU 2009-10)

SOLUTION: The point $P(1, 2, -1)$ lies inside the sphere $x^2 + y^2 + z^2 = 24$ since $1^2 + 2^2 + (-1)^2 < 24$. Shortest distance of the point P from the sphere

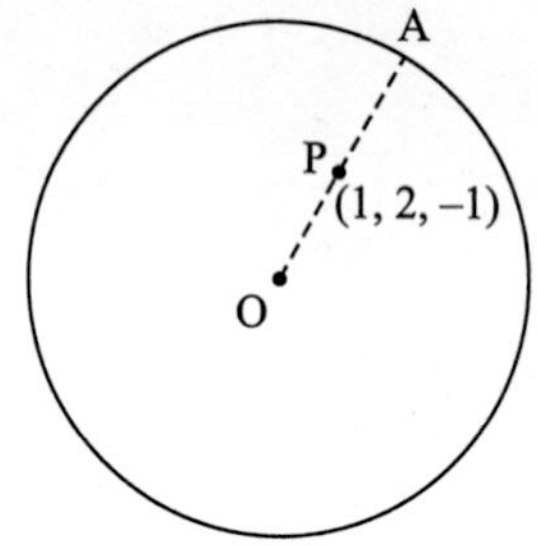

$$= PA = OA - OP$$

$$= \sqrt{24} - \sqrt{1^2 + 2^2 + (-1)^2}$$

$$= 2\sqrt{6} - \sqrt{6} = \sqrt{6} \quad \textbf{Ans.}$$

EXAMPLE 5.31. Find the dimensions of the rectangular box of maximum capacity whose surface area is given, when

(*a*) box is open at the top

(*b*) box is closed. **(UPTU 2008-09)**

SOLUTION: $V = xyz$ is to be maximized and surface area S is given.

(*a*) $S = xy + 2yz + 2xz \quad \therefore \quad z = \dfrac{S - xy}{2x + 2y}$ ...(1)

then $$V = \frac{xy\,(S - xy)}{2x + 2y} \quad \text{or} \quad 2V = \frac{xy\,S - x^2y^2}{x + y}.$$

$$\Rightarrow \quad 2\frac{\partial V}{\partial x} = \frac{(x+y)\,(yS - 2xy^2) - 1\,(xy\,S - x^2y^2)}{(x+y)^2} = \frac{Sy^2 - x^2y^2 - 2xy^3}{(x+y)^2}$$

and $$2\frac{\partial V}{\partial y} = \frac{Sx^2 - x^2y^2 - 2yx^3}{(x+y)^2}.$$

For V to be Max $\dfrac{\partial V}{\partial x} = 0$ and $\dfrac{\partial V}{\partial y} = 0$

$$\therefore \quad y^2(S - x^2 - 2xy) = 0 \quad \text{and} \quad x^2(S - y^2 - 2xy) = 0$$

$$\Rightarrow \quad S = x^2 + 2xy \quad \text{and} \quad S = y^2 + 2xy \Rightarrow y = x$$

then $S = 3x^2$ and $S = 3y^2$ or $x = y = \sqrt{\dfrac{S}{3}}$.

Putting these values of x and y in (1) gives

$$z = \frac{S - \dfrac{S}{3}}{4\sqrt{\dfrac{S}{3}}} = \frac{1}{2}\sqrt{\frac{S}{3}} \quad \text{Thus, we have } x = y = 2z$$

as the required dimensions of the box. **Ans.**

(*b*) $S = 2(xy + yz + zx)$ $\therefore$ $z = \dfrac{\frac{S}{2} - xy}{x+y}$ and $V = xyz = \dfrac{xy\left(\frac{S}{2} - xy\right)}{x+y}$

Then $$\frac{\partial V}{\partial x} = \frac{(x+y)\left(\frac{S}{2}y - 2xy^2\right) - xy\left(\frac{S}{2} - xy\right)}{(x+y)^2} = \frac{S\frac{y^2}{2} - x^2y^2 - 2x^3}{(x+y)^2}$$

and $$\frac{\partial V}{\partial y} = \frac{(x+y)\left(\frac{S}{2}x - 2x^2y\right) - xy\left(\frac{S}{2} - xy\right)}{(x+y)^2} = \frac{S\frac{x^2}{2} - x^2y^2 - 2x^3y}{(x+y)^2}.$$

For V to be max. $\dfrac{\partial V}{\partial x} = 0$ and $\dfrac{\partial V}{\partial y} = 0$

$$\Rightarrow \qquad y^2\left(\frac{S}{2} - x^2 - 2xy\right) = 0 \quad \text{and} \quad x^2\left(\frac{S}{2} - y^2 - 2xy\right) = 0.$$

As $x, y \neq 0$ $\therefore$ $\dfrac{S}{2} = x^2 + 2xy$ and $\dfrac{S}{2} = y^2 + 2xy.$

Subtracting one from the other, gives $x = y$. Hence $\dfrac{S}{2} = 3x^2$ or $x = y = \sqrt{\dfrac{S}{6}}.$

Substituting these in $z = \dfrac{\frac{S}{2} - xy}{x+y}$, we get

$$z = \frac{\frac{S}{2} - \frac{S}{6}}{2\sqrt{\frac{S}{6}}} = \frac{\frac{4S}{6}}{2\sqrt{\frac{S}{6}}} = 2\sqrt{\frac{S}{6}} = 2x = 2y$$

$\therefore$ required dimensions are $x = y = \dfrac{z}{2}$. **Ans.**

EXAMPLE 5.32. The temperature T at any point (x, y, z) in space is $T(x, y, z) = K\,xyz^2$ where k is constant Find the highest temperature on the surface of the sphere $x^2 + y^2 + z^2 = a^2$. **(UPTU 2008-09)**

SOLUTION: Here $T = K\,xyz^2$, which is to be maximized.

The constraint is $\phi \equiv x^2 + y^2 + z^2 - a^2 = 0$

Consider the function $F = T + \lambda\phi$ ...(1)

$$\therefore \quad \frac{\partial F}{\partial x} = \frac{\partial T}{\partial x} + \lambda\frac{\partial \phi}{\partial x} = Kyz^2 + \lambda(2x) \qquad ...(2)$$

$$\frac{\partial F}{\partial y} = \frac{\partial T}{\partial y} + \lambda\frac{\partial \phi}{\partial y} = Kxz^2 + \lambda(2y), \qquad ...(3)$$

and $$\frac{\partial F}{\partial z} = \frac{\partial T}{\partial z} + \lambda\frac{\partial \phi}{\partial z} = 2\,Kxyz + \lambda(2z). \qquad ...(4)$$

For T to be maximum $\dfrac{\partial F}{\partial x} = \dfrac{\partial F}{\partial y} = \dfrac{\partial F}{\partial z} = 0$. Thus, we have

$$\Rightarrow \qquad Kyz^2 + \lambda(2x) = 0 \qquad \text{...(5)}$$

$$Kxz^2 + \lambda(2y) = 0 \qquad \text{...(6)}$$

$$2\,Kxyz + \lambda(2z) = 0 \qquad \text{...(7)}$$

Multiplying (5), (6), (7) by x, y, z respectively, we get

$T + 2\lambda x^2 = 0, \quad T + 2\lambda y^2 = 0, \quad 2T + 2\lambda z^2 = 0$

$$\Rightarrow \qquad x^2 = \frac{-T}{2\lambda}, \quad y^2 = \frac{-T}{2\lambda}, \quad z^2 = \frac{-T}{\lambda}.$$

Adding these we get $x^2 + y^2 + z^2 = \dfrac{-2T}{\lambda} = a^2 \quad \therefore \quad \dfrac{-T}{\lambda} = \dfrac{a^2}{2}$

Thus $x^2 = \dfrac{a^2}{4}, \quad y^2 = \dfrac{a^2}{4}, \quad z^2 = \dfrac{a^2}{2}$

Hence $x = \dfrac{a}{2}, \quad y = \dfrac{a}{2}, \quad z = \dfrac{a}{\sqrt{2}}$ is the point of maxima

$\therefore$ Highest temperature $= Kxyz^2 = K\dfrac{a}{2}\cdot\dfrac{a}{2}\cdot\dfrac{a^2}{2} = \dfrac{Ka^3}{8}$. **Ans.**

EXERCISE 5B

1. Find the maximum and minimum value of the function $f(x, y) = x^4 + y^4 - 2x^2 + 4xy - 2y^2$.
2. Show that $u = x^2 y^2 (x^2 + y^2 - 1)$ has a maximum value of $-\frac{1}{27}$ for four sets of values of x and y and a minimum value at $x = 0,\ y = 0$.
3. Find the maximum and minimum values of $\sin x \sin y \sin (x + y)$.
4. A tent on a square base of side $2a$ consists of four vertical sides of height b surmounted by a regular pyramid of height h. If the volume enclosed is V show that the area of the canvas in the tent is

   $\frac{2V}{a} - \frac{8}{3}ah + 4a\sqrt{h^2 + a^2}$. Also show that if a and h can both vary, the least area of the canvas corresponding to a given volume V is given by

   $$a = \frac{\sqrt{5}}{2}h, \quad b = \frac{h}{2}.$$ **(UPTU 2001 Q.B.)**
5. Find the extreme values of $x^2 + y^2 + z^2$ subject to the constraints

   $$x^2 + \frac{y^2}{2} + \frac{z^2}{3} = 2 \quad \text{and} \quad 3x + 2y + z = 0.$$
6. Find the minimum value of $x^2 + y^2$ subject to the condition $ax + by = c$.
7. Find the points on the ellipse $\frac{x^2}{4} + \frac{y^2}{1} = 1$ whose distances from the straight line $2x + 3y = 6$ are the greatest and the least.
8. Find the minimum value of $f(x, y, z) = x^2 + y^2 + z^2$ subject to the condition $x + y + z = 1$.
9. Determine the values of x, y, z that maximise the function $3x + 5y + z - x^2 - y^2 - z^2$ subject to the constraint $x + y + z = 6$.
10. Divide 120 into three parts in such a way that the sum of their products taken two at a time, is maximum.
11. Find a point within a given triangle so that the sum of the squares of its distances from the sides of the triangle is minimum.
12. If r is the distance of a point on the conic $ax^2 + by^2 + cz^2 = 1$, $lx + my + nz = 0$ from the origin, then show that the stationary values of r are given by

    $$\frac{l^2}{1 - ar^2} + \frac{m^2}{1 - br^2} + \frac{n^2}{1 - cr^2} = 0.$$
13. Find the extreme value of $x^l \cdot y^m \cdot z^n$ subject to the conditions $ax + by + cz = l + m + n$.
14. Find the semi-vertical angle of a cone of maximum volume and of a given slant height **(UPTU 2002)**
15. The sum of three positive number is constant. Prove that their product is maximum when they are equal. **(UPTU 2001 Q.B.)**
16. In a place triangle ABC, find the maximum value of $\cos A \cos B \cos C$. **(UPTU 2009-10)**

❑❑❑

UNIT-III
MULTIPLE INTEGRALS

Double and Triple Integration
Change of Order of Integration
Change of Variables
Application of Double and Triple Integration to Area and Volume
Beta and Gamma Functions
Dirichlet's Integral and Applications

CHAPTER 6

Double Integration and Change of Order of Integration

Multiple Integration: Double Integrals, Change of Order of Integration, Cartesian and Polar Coordinates.

DOUBLE INTEGRAL—DEFINITION AND EVALUATION

Let $f(x, y)$ be a continuous function of x and y, defined within and on a region R bounded by a closed curve C in the xy-plane. Let the region R be subdivided into n subregions of areas $\delta A_1, \delta A_2, ..., \delta A_n$. Let (x_r, y_r) be any point within the r'th subregion δA_r. Now consider the sum

$$\sum_{r=1}^{n} f(x_r, y_r)\, \delta A_r.$$

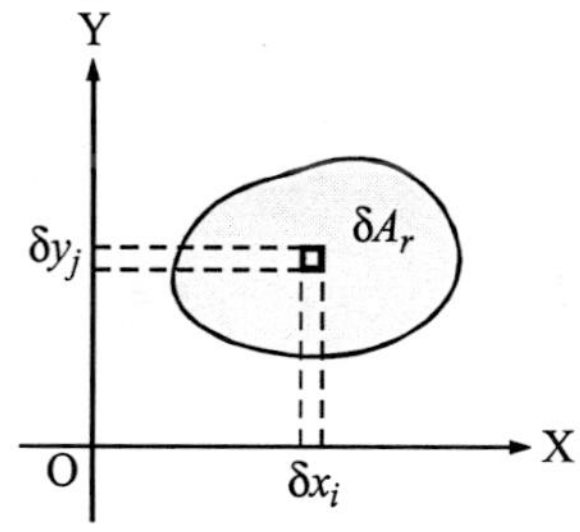

The limit of this sum as $n \to \infty$ and $\delta A_r \to 0$ $(r = 1, 2, ..., n)$, is defined as the double integral of $f(x, y)$ over the region R and is written as

$$\iint_R f(x, y)\, dA \qquad ...(1)$$

To evaluate the double integral, we subdivide R by lines parallel to the co-ordinate axes creating a rectangular grid. Since the area of the typical rectangle is $\delta A_r = \delta x_i\, \delta y_j$, we have

$$\iint_R f(x, y)\, dx\, dy = \lim_{n \to \infty} \sum_{i, j=1}^{n} f(x_i, y_j)\, \delta x_i\, \delta y_j\,.$$

The above definition can obviously be generalized to three dimensions giving rise to triple integrals.

The double integral over a region R is evaluated by two successive single integrals as described below:

Suppose the region R is described by the inequalities

$$\left.\begin{aligned} c \le y \le d \\ \text{and} \quad g_1(y) \le x \le g_2(y) \end{aligned}\right\} \qquad ...(2)$$

as shown in the adjoining figure,

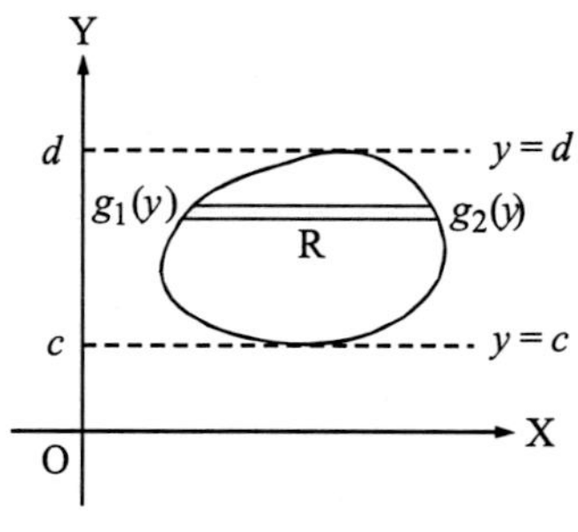

then $$\iint_R f(x, y)\, dx\, dy = \int_{y=c}^{y=d} \left[\int_{x=g_1(y)}^{x=g_2(y)} f(x, y)\, dx \right] dy \qquad ... (3)$$

The inner integral $\int_{x=g_1(y)}^{x=g_2(y)} f(x, y)\, dx$ is first evaluated with respect to x treating y as constant.

Since here the integrand and the limits are functions of y, the value of this integral would be a function of y, say $\phi(y)$,

hence $$\iint_R f(x,y)\, dx\, dy = \int_c^d \phi(y)\, dy \qquad ...(4)$$

which can be easily evaluated.

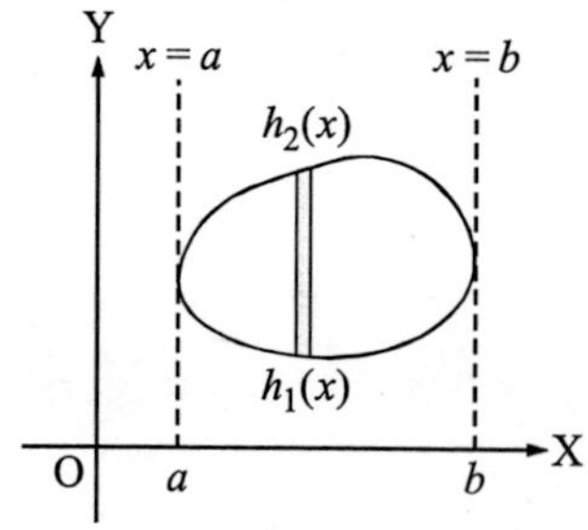

Alternatively, if the region R is described by the inequalities

and $$\left.\begin{aligned} a \le x \le b \\ h_1(x) \le y \le h_2(x) \end{aligned}\right\} \qquad ....(5)$$

as shown in the adjoining figure, then we write

$$\iint_R f(x,y)\, dx\, dy = \iint_R f(x,y)\, dy\, dx.$$

$$= \int_{x=a}^{x=b} \left[\int_{y=h_1(x)}^{y=h_2(x)} f(x, y)\, dy \right] dx. \qquad ...(6)$$

The inner integral $\int_{y=h_1(x)}^{y=h_2(x)} f(x, y)\, dy$ is a function of x, equal to F (x) say, then

$$\iint_R f(x, y)\, dx\, dy = \iint_R f(x, y)\, dy\, dx = \int_a^b F(x)\, dx \qquad ...(7)$$

which can be easily evaluated.

If the region R is a rectangle bounded by the lines $x = a, x = b;\ y = c, y = d$, then

$$\iint_R f(x, y)\, dx\, dy = \int_a^b \int_c^d f(x, y)\, dy\, dx = \int_c^d \int_a^b f(x, y)\, dx\, dy \qquad ...(8)$$

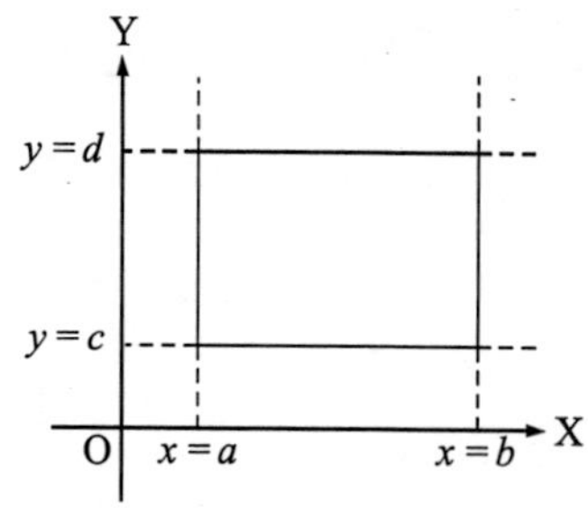

from which it is quite clear that the **order of integration is immaterial when limits of x and y are both constants.**

EXAMPLE 6.1. Evaluate $\int_0^1 \int_0^{\sqrt{1+x^2}} \frac{1}{1+x^2+y^2}\, dy\, dx$. **(UPTU 2001 Q.B)**

SOLUTION: Let $I = \int_0^1 \int_0^{\sqrt{1+x^2}} \frac{1}{1+x^2+y^2}\, dy\, dx = \int_0^1 \left[\int_0^{\sqrt{1+x^2}} \frac{1}{1+x^2+y^2}\, dy \right] dx.$

We first evalaute $\int_0^{\sqrt{1+x^2}} \frac{1}{1+x^2+y^2}\, dy$ where x has been treated as constant, hence

$$\int_0^{\sqrt{1+x^2}} \frac{1}{1+x^2+y^2}\, dy = \frac{1}{\sqrt{1+x^2}} \left[\tan^{-1} \frac{y}{\sqrt{1+x^2}} \right]_0^{\sqrt{1+x^2}}$$

$$= \frac{1}{\sqrt{1+x^2}} (\tan^{-1} 1 - \tan^{-1} 0) = \frac{\pi}{4\sqrt{1+x^2}}$$

$$\therefore \quad I = \int_0^1 \frac{\pi\, dx}{4\sqrt{1+x^2}} = \frac{\pi}{4} \left[\log\left(x + \sqrt{1+x^2} \right) \right]_0^1$$

$$= \frac{\pi}{4} \left[\log\left(1+\sqrt{2}\right) - \log 1 \right] = \frac{\pi}{4} \log\left(1+\sqrt{2}\right).$$ **Ans.**

EXAMPLE 6.2. **(a)** Evaluate $\int_0^\infty \int_0^\infty e^{-x^2(1+y^2)}\, x\, dx\, dy$.

(b) Find the value of the integral $\iint_R xy\, dx\, dy$ where R is the region bounded by the x-axis, the line $y = 2x$ and the parabola $x^2 = 4\,ay$. **(UPTU 2011-12)**

SOLUTION: **(a)** Both the limits of integration for x and y are constants hence we can integrate with respect to any variable first. Let us integrate, for convinience, first w.r.t. x.

Let $I = \int_0^\infty \left[\int_0^\infty e^{-x^2(1+y^2)}\, x\, dx \right] dy = \int_0^\infty dy \int_0^\infty e^{-x^2(1+y^2)}\, x\, dx$

Putting $x^2(1+y^2) = t$ in the inner integral treating y as constant, we get

$$\int_0^\infty e^{-x^2(1+y^2)}\, x\, dx = \int_0^\infty e^{-t} \frac{dt}{2(1+y^2)} = \frac{1}{2(1+y^2)} \int_0^\infty e^{-t}\, dt$$

$$= \frac{1}{2(1+y^2)} [-e^{-t}]_0^\infty = \frac{1}{2(1+y^2)}$$

$$\therefore \quad I = \int_0^\infty \frac{1}{2(1+y^2)}\, dy = \frac{1}{2} [\tan^{-1} y]_0^\infty = \frac{1}{2}\left(\frac{\pi}{2} - 0 \right) = \frac{\pi}{4}.$$ **Ans.**

(b) The points of intersection of the parabola $x^2 = 4ay$ and the line $y = 2x$ are (0, 0) and $(8a, 16a)$. Take horizontal strip PQ then the required integral will be

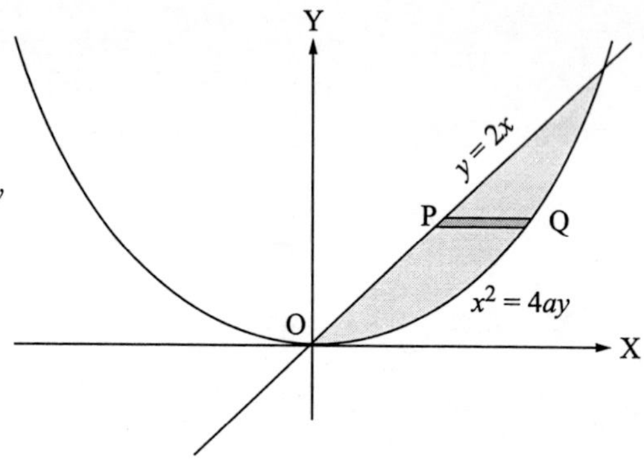

$$\int_{y=0}^{16a}\left(\int_{x=y/2}^{\sqrt{4ay}} x\,dx\right)y\,dy = \int_{y=0}^{16a}\left[\frac{x^2}{2}\right]_{y/2}^{\sqrt{4ay}} y\,dy$$

$$= \frac{1}{2}\int_{y=0}^{16a}\left[4ay - \left(\frac{y}{2}\right)^2\right]\cdot y\,dy$$

$$= \frac{1}{2}\int_{y=0}^{16a}\left(4ay^2 - \frac{y^3}{4}\right)dy$$

$$= \frac{1}{2}\left[\frac{4ay^3}{3} - \frac{y^4}{16}\right]_0^{16a}$$

$$= \frac{1}{2}\left[\frac{4a\cdot(16a)^3}{3} - \frac{(16a)^4}{16}\right]$$

$$= 16^3\cdot a^4\left[\frac{2}{3} - \frac{1}{2}\right]$$

$$= \frac{16^3 a^4}{6} \quad \textbf{Ans.}$$

EXAMPLE 6.3. Evaluate $\iint xy\,(x+y)\,dx\,dy$ over the region bounded by the line $y = x$ and the curve $y = x^2$.

SOLUTION: The region of integration is as shown in the adjoining figure.

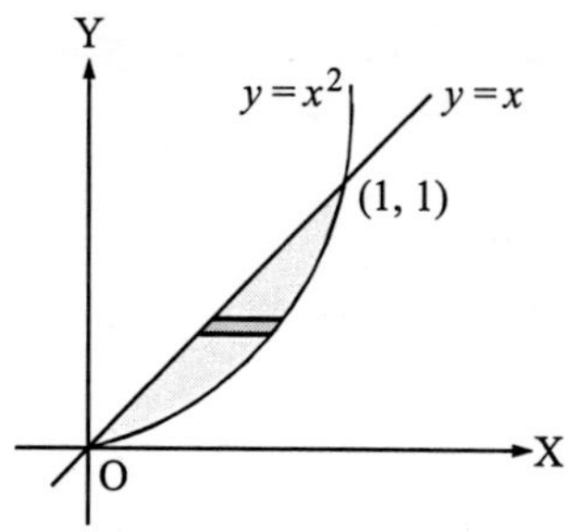

Let $\quad I = \iint xy\,(x+y)\,dx\,dy$

and let the order of integration be first with respect to x and then with respect to y. As shown in the adjoing figure we take an elementary strip parallel to x-axis and the limits for x are clearly from the line $x = y$ to the curve $x = \sqrt{y}$ and then the limits for y will be from $y = 0$ to $y = 1$.

$$\therefore \quad I = \int_0^1\int_y^{\sqrt{y}} xy\,(x+y)\,dx\,dy = \int_0^1\left[\frac{x^3}{3}y + \frac{x^2}{2}y^2\right]_{x=y}^{x=\sqrt{y}} dy = \int_0^1\left[\frac{y^2\sqrt{y}}{3} + \frac{y^3}{2} - \left(\frac{y^4}{3} + \frac{y^4}{2}\right)\right]dy$$

$$= \int_0^1\left(\frac{y^{\frac{5}{2}}}{3} + \frac{y^3}{2} - \frac{5}{6}y^4\right)dy = \left[\frac{y^{\frac{7}{2}}}{3}\cdot\frac{2}{7} + \frac{y^4}{8} - \frac{5}{6}\cdot\frac{y^5}{5}\right]_0^1 = \frac{2}{21} + \frac{1}{8} - \frac{1}{6} = \frac{3}{56}. \quad \textbf{Ans.}$$

CHANGE OF ORDER OF INTEGRATION

To evaluate a double integral we integrate first w.r.t. one variable (y or x depending upon the limits and the elementary strip) considering the other variable as constant and then integrate with respect to the remaining variable. However, if the limits are constant, the order of integration is immaterial, as stated earlier, and in that case we have

$$\int_a^b\int_c^d f(x, y)\, dy\, dx = \int_c^d\int_a^b f(x, y)\, dx\, dy$$

i.e. $$\int_a^b dx \int_c^d f(x, y)\, dy = \int_c^d dy \int_a^b f(x, y)\, dx.$$

But **if the limits are variable and the integrand $f(x, y)$ in the double integral is either difficult or even impossible to integrate in the given order then we reverse the order of integration and corresponding change is made in the limits of integration.** The new limits are obtained by geometrical considerations therefore a clear sketch of the curve is to be drawn. Sometimes in changing the order of integration we have to split up the region of integration and the new integral is expressed as the sum of a number of double integrals.

EXAMPLE 6.4. Change the order of integration and evaluate $\displaystyle\int_0^a\int_0^{a-\sqrt{a^2-y^2}} \frac{xy\log(x+a)}{(x-a)^2}\, dx\, dy$.

SOLUTION: In the given integral, as it stands, it is to be integrated first w.r.t. x and then w.r.t. y which is complicated, therefore it requires the change of order of integration, for which the region over which the integration is to be done, is to be found first.

The region is bounded by $x = 0,\quad x = a - \sqrt{a^2 - y^2}$

or $(x - a)^2 = a^2 - y^2$

or $x^2 + y^2 - 2ax = 0$ (circle) and $y = 0,\ y = a,$

as shown in the adjoing figure.

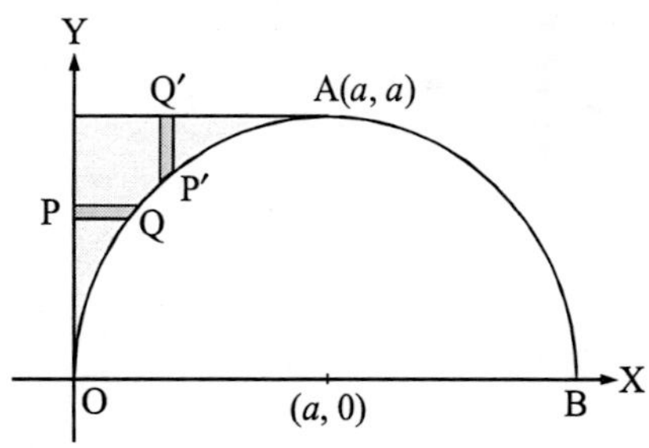

Actually from the equation of the circle we get $x = a \pm \sqrt{a^2 - y^2}$, *i.e.* for one value of y we have two values of x. Here $x = a - \sqrt{a^2 - y^2}$ represents the arc OA and $x = a + \sqrt{a^2 - y^2}$ the arc AB of the circle. Thus, as it is, we have elementary strip PQ parallel to X-axis as shown in the figure.

Thus when the order of integration is changed we take the strip P′Q′ parallel to Y-axis. Here y varies from the circle $y = \sqrt{2ax - x^2}$ to the line $y = a$ then the strip P′Q′ moves parallel to itself from $x = 0$ to $x = a$.

Therefore $$I = \int_0^a \int_{\sqrt{2ax-x^2}}^{a} \frac{xy\log(x+a)}{(x-a)^2}\, dy\, dx = \int_0^a \frac{x\log(x+a)}{(x-a)^2}\left[\int_{\sqrt{2ax-x^2}}^{a} y\, dy\right] dx$$

$$= \int_0^a \frac{x \log (x+a)}{2(x-a)^2}[a^2 - (2ax - x^2)]\, dx = \int_0^a \frac{x}{2} \log (x+a)\, dx \quad \text{(integrating it by parts)}$$

$$= \left[\log (x+a) \cdot \frac{x^2}{4}\right]_0^a - \int_0^a \frac{x^2\, dx}{4(x+a)} = \frac{a^2}{4} \log 2a - \frac{1}{4}\int_0^a \left(x - a + \frac{a^2}{x+a}\right) dx$$

$$= \frac{a^2}{4} \log 2a - \frac{1}{4}\left[\frac{x^2}{2} - ax + a^2 \log (x+a)\right]_0^a$$

$$= \frac{a^2}{4} \log 2a - \frac{1}{4}\left[\frac{a^2}{2} - a^2 + a^2 \log 2a - a^2 \log a\right] = \frac{a^2}{8} + \frac{a^2}{4} \log a = \frac{a^2}{8}[1 + 2 \log a]. \quad \textbf{Ans.}$$

EXAMPLE 6.5. Change the order of integration and hence evaluate $\int_0^1 \int_x^{2-x} \frac{x}{y}\, dy\, dx$.

SOLUTION: In the given integral, as it is, the elementary strip has to be taken parallel to y-axis. Region of integration is the shaded portion as shown in the adjoining figure bounded by the lines $y = x$, $y = 2 - x$ and $x = 0$, $x = 1$. On changing the order of integration the elementary strip has to be taken parallel to x-axis for which the region of integration, shown as shaded portion, has to be divided into two regions R_1 and R_2. Therefore

$$I = \int_0^1 \int_x^{2-x} \frac{x}{y}\, dy\, dx = \iint_{R_1} \frac{x}{y}\, dx\, dy + \iint_{R_2} \frac{x}{y}\, dx\, dy$$

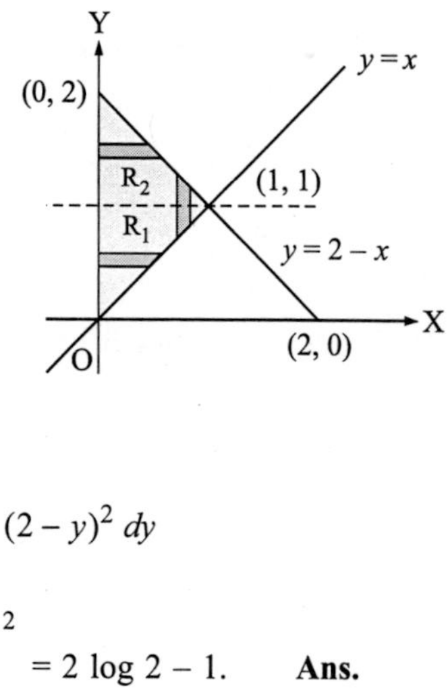

Region R_1 is bounded by $0 \le x \le y$, $0 \le y \le 1$ and R_2 is bounded by $0 \le x \le 2 - y$, $1 \le y \le 2$.

$$\therefore \quad I = \int_0^1 \int_0^y \frac{x}{y}\, dx\, dy + \int_1^2 \int_0^{2-y} \frac{x}{y}\, dx\, dy$$

$$= \int_0^1 \frac{1}{y} dy \int_0^y x\, dx + \int_1^2 \frac{1}{y} dy \int_0^{2-y} x\, dx$$

$$= \int_0^1 \frac{1}{y}\left[\frac{x^2}{2}\right]_0^y dy + \int_1^2 \frac{1}{y}\left[\frac{x^2}{2}\right]_0^{2-y} dy = \int_0^1 \frac{1}{y} \cdot \frac{y^2}{2} dy + \int_1^2 \frac{1}{2y}(2-y)^2\, dy$$

$$= \left[\frac{y^2}{4}\right]_0^1 + \int_1^2 \left(\frac{2}{y} - 2 + \frac{y}{2}\right) dy = \frac{1}{4} + \left[2 \log y - 2y + \frac{y^2}{4}\right]_1^2 = 2 \log 2 - 1. \quad \textbf{Ans.}$$

EXAMPLE 6.6. Evaluate $\int_0^a \int_0^{\sqrt{a^2-x^2}} \frac{dx\, dy}{(1+e^y)\sqrt{a^2 - x^2 - y^2}}$.

SOLUTION: The given integral is to be integrated first w.r.t. y which is not feasible, therefore we change the order of integration. As is clear from the given limits, the region of integration is the positive quadrant of the circle $x^2 + y^2 = a^2$. With the change of order of integration we have to take the elementary strip now parallel to x-axis and limits would change accordingly. (See figure.)

$$I = \int_0^a \int_0^{\sqrt{a^2-y^2}} \frac{dx\,dy}{(1+e^y)\sqrt{a^2-x^2-y^2}} = \int_0^a \frac{1}{1+e^y}\left[\int_0^{\sqrt{a^2-y^2}} \frac{dx}{\sqrt{a^2-y^2-x^2}}\right] dy$$

$$= \int_0^a \frac{1}{1+e^y}\left[\sin^{-1}\frac{x}{\sqrt{a^2-y^2}}\right]_{x=0}^{x=\sqrt{a^2-y^2}} dy$$

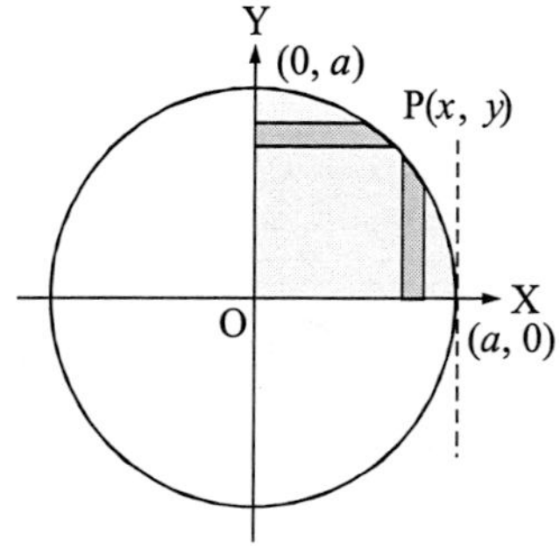

$$= \int_0^a \frac{1}{1+e^y}(\sin^{-1}1 - \sin^{-1}0)\,dy = \frac{\pi}{2}\int_0^a \frac{dy}{1+e^y}$$

Putting $e^y = t \quad \therefore\ e^y\,dy = dt$

$$I = \frac{\pi}{2}\int_1^{e^a} \frac{dt}{t(1+t)} = \frac{\pi}{2}\int_1^{e^a}\left(\frac{1}{t} - \frac{1}{t+1}\right) dt$$

$$= \frac{\pi}{2}\left[\log\frac{t}{t+1}\right]_1^{e^a} = \frac{\pi}{2}\left[\log\frac{e^a}{e^a+1} - \log\frac{1}{2}\right]$$

$$= \frac{\pi}{2}\log\frac{2e^a}{1+e^a}.$$ **Ans.**

EXAMPLE 6.7. Change the order of integration in the double integral $\int_0^{2a}\int_{\sqrt{2ax-x^2}}^{\sqrt{2ax}} f(x, y)\,dx\,dy.$

SOLUTION: The order of integration in the given integral is first w.r.t. y and then w.r.t x. To evaluate the given integral, as it stands, the elementary strip has to be taken parallel to Y-axis and its lower end is on $y = \sqrt{2ax - x^2}$, *i.e.* the circle $x^2 + y^2 - 2ax = 0$ and the upper end on $y = \sqrt{2ax}$, *i.e.* the parabola $y^2 = 2ax$. Then the strip is moved parallel to itself from $x = 0$ to $x = 2a$. Therefore the shaded portion between the parabola and the circle is the region of integration.

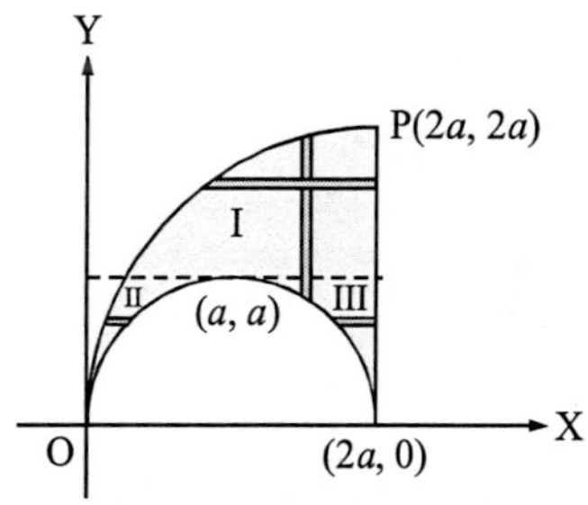

On changing the order of integration we first integrate w.r.t. x and then w.r.t. y. The elementary strip is taken parallel to X-axis. To cover the whole of the shaded area the region has to be divided into following three parts, as shown in the adjoining figure:

REGION I. The strip extends from the parabola $y^2 = 2ax$, *i.e.* $x = \frac{y^2}{2a}$, to the straight line $x = 2a$. The strip is then moved parallel to itself from $y = a$ to $y = 2a$ to cover the region I. Therefore, the part of double integral in this region is I_1 given by

$$I_1 = \int_a^{2a}\int_{\frac{y^2}{2a}}^{2a} f(x, y)\,dx\,dy.$$

REGION II. The strip extends from the parabola $y^2 = 2ax$, *i.e.* $x = \frac{y^2}{2a}$ to the circle $x^2 + y^2 - 2ax = 0$, *i.e.* $x = a - \sqrt{a^2 - y^2}$ as is clear from the figure. Then the strip is taken from $y = 0$ to $y = a$ to cover this region. Therefore the part of the integral in this region is I_2 given by

$$I_2 = \int_0^a \int_{\frac{y^2}{2a}}^{a-\sqrt{a^2-y^2}} f(x, y)\, dx\, dy.$$

REGION III. The strip extends from the circle $x^2 + y^2 - 2ax = 0$, *i.e.* $x = a + \sqrt{a^2 - y^2}$ to the line $x = 2a$. The strip is then taken from $y = 0$ to $y = a$ which covers this region. Denoting this part of integral by I_3 we have

$$I_3 = \int_0^a \int_{a+\sqrt{a^2-y^2}}^{2a} f(x, y)\, dx\, dy.$$

As such, the given integral is equal to $I_1 + I_2 + I_3$.

Thus, $$\int_0^{2a} \int_{\sqrt{2ax-x^2}}^{\sqrt{2ax}} f(x, y)\, dy\, dx = \int_a^{2a} \int_{\frac{y^2}{2a}}^{2a} f(x, y)\, dx\, dy + \int_0^a \int_{\frac{y^2}{2a}}^{a-\sqrt{a^2-y^2}} f(x, y)\, dx\, dy + \int_0^a \int_{a+\sqrt{a^2-y^2}}^{2a} f(x, y)\, dx\, dy.$$

Ans.

EXAMPLE 6.8. Express as a single integral $\int_0^{\frac{a}{\sqrt{2}}} \int_0^x x\, dy\, dx + \int_{\frac{a}{\sqrt{2}}}^a \int_0^{\sqrt{a^2-x^2}} x\, dy\, dx$ and evaluate it.

SOLUTION: Let $I_1 = \int_0^{\frac{a}{\sqrt{2}}} \int_0^x x\, dy\, dx$ and $I_2 = \int_{\frac{a}{\sqrt{2}}}^a \int_0^{\sqrt{a^2-x^2}} x\, dy\, dx$.

Let R_1 and R_2 be the regions over which I_1 and I_2 are being integrated respectively and are depicted by the shaded portion in Fig. I.

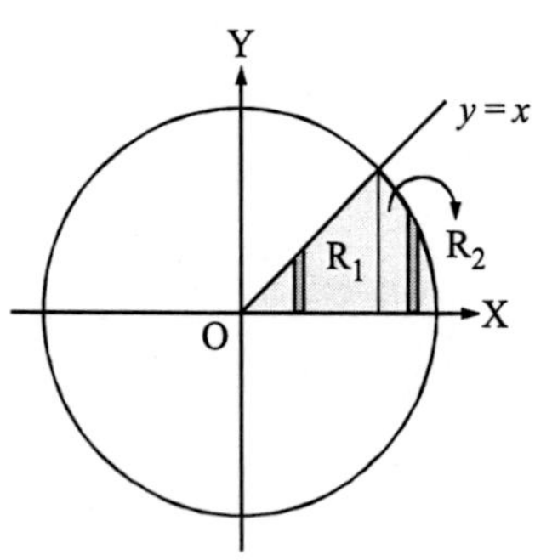

Fig. I

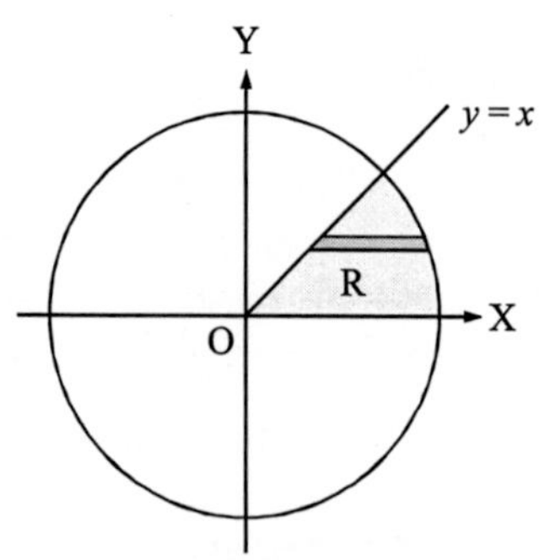

Fig. II

As is clear from Fig. II $R = R_1 + R_2$

$$\therefore \quad I = I_1 + I_2 = \iint_R x\,dx\,dy.$$

For evaluating I we change the order of integration hence the elementary strip has to be taken parallel to x-axis from $y = x$ to $y = \sqrt{a^2 - x^2}$, *i.e.* the circle $x^2 + y^2 = a^2$.

$$\therefore \quad I = \int_0^{\frac{a}{\sqrt{2}}} \int_y^{\sqrt{a^2-y^2}} x\,dx\,dy = \int_0^{\frac{a}{\sqrt{2}}} \left[\int_y^{\sqrt{a^2-y^2}} x\,dx \right] dy$$

$$= \int_0^{\frac{a}{\sqrt{2}}} \left[\frac{x^2}{2} \right]_y^{\sqrt{a^2-y^2}} dy \quad = \frac{1}{2} \int_0^{\frac{a}{\sqrt{2}}} (a^2 - y^2 - y^2)\,dy$$

$$= \frac{1}{2} \left[a^2 y - \frac{2y^3}{3} \right]_0^{\frac{a}{\sqrt{2}}} = \frac{1}{2} \left[\frac{a^3}{\sqrt{2}} - \frac{2}{3} \frac{a^3}{2\sqrt{2}} \right] = \frac{a^3}{3\sqrt{2}}. \qquad \textbf{Ans.}$$

EXAMPLE 6.9. Prove by the method of double integration that the area lying between the parabolas $y^2 = 4ax$ and $x^2 = 4ay$ is $\frac{16}{3} a^2$.

SOLUTION: The required area is

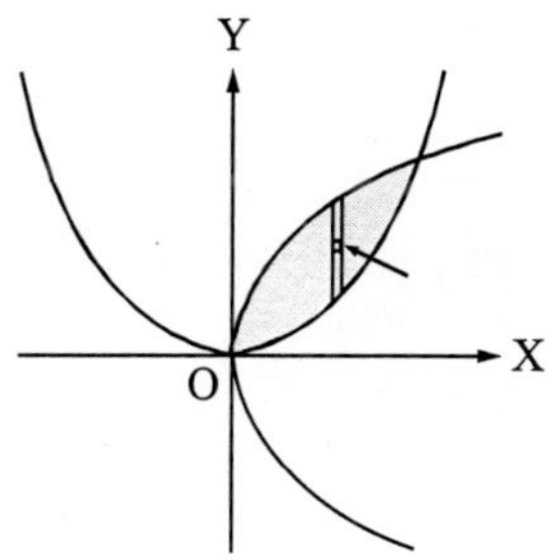

$$\int_0^{4a} \int_{x^2/4a}^{\sqrt{4ax}} dx\,dy = \int_0^{4a} [y]_{x^2/4a}^{\sqrt{4ax}} dx = \int_0^{4a} \left(\sqrt{4ax} - \frac{x^2}{4a} \right) dx$$

$$= \left[\sqrt{4a} \cdot \frac{2}{3} x^{3/2} - \frac{x^3}{3.4a} \right]_0^{4a}$$

$$= \frac{4}{3} \sqrt{a}\, 4a 2\sqrt{a} - \frac{4^3 a^3}{3.4a} = \frac{16}{3} a^2. \qquad \textbf{Hence Proved.}$$

EXAMPLE 6.10. Change the order of integration in the integral

$$\int_0^{a/\sqrt{2}} \int_y^{\sqrt{a^2-y^2}} x\,dx\,dy \quad \text{and then evaluate it.}$$

SOLUTION: The area of integration in the given integral $I = \int_0^{a/\sqrt{2}} \int_y^{\sqrt{a^2-y^2}} x\,dx\,dy$ is shaded portion in the figure given on next page.

Changing the order of integration the area has to be divided into two parts

by the line $x = a/\sqrt{2}$, we have

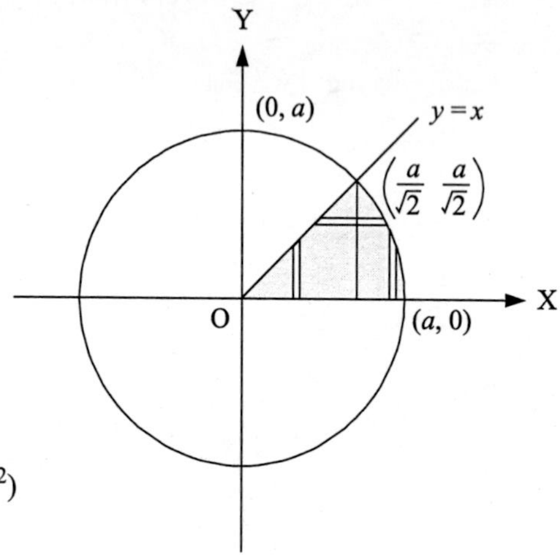

$$I = \int_0^{a/\sqrt{2}} \int_0^x x\,dy\,dx + \int_{a/\sqrt{2}}^{a} \int_0^{\sqrt{a^2-x^2}} x\,dy\,dx$$

$$= \int_0^{a/\sqrt{2}} x[y]_0^x\,dx + \int_{a/\sqrt{2}}^{a} x[y]_0^{\sqrt{a^2-x^2}}\,dx$$

$$= \int_0^{a/\sqrt{2}} x^2 dx + \int_{a/\sqrt{2}}^{a} x\sqrt{a^2-x^2}\,dx$$

$$= \left[\frac{x^3}{3}\right]_0^{a/\sqrt{2}} + \int_{a/\sqrt{2}}^{0} -t^2 dt$$

(On putting $a^2 - x^2 = t^2$)

$$= \frac{a^3}{3.2\sqrt{2}} + \left[\frac{t^3}{3}\right]_0^{a/\sqrt{2}}$$

$$= \frac{a^3}{6\sqrt{2}} + \frac{a^3}{6\sqrt{2}} = \frac{a^3}{3\sqrt{2}} = \frac{a^3\sqrt{2}}{6}. \quad \textbf{Ans.}$$

EXAMPLE 6.11. Sketch the region of integration and evaluate

$$\int_0^{\pi/2} dy \int_0^y \cos 2y\sqrt{1-a^2\sin^2 x}\,dx$$

SOLUTION: The given integral can be written as $\int_0^{\pi/2} dy \int_0^y \cos 2y\sqrt{1-a^2\sin^2 x}\;dx$

Now let us change the order of integration, then the given integral equals

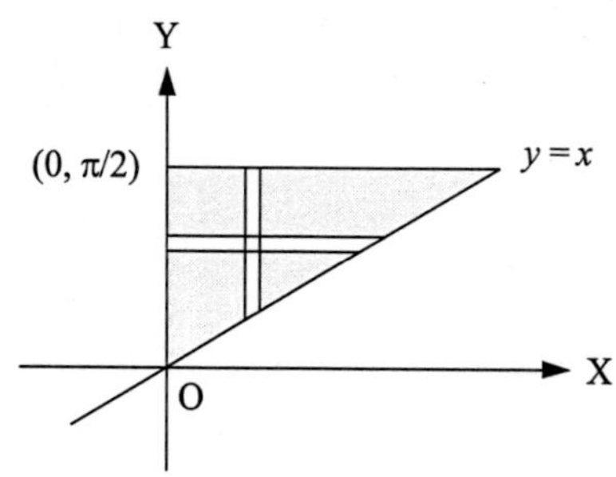

$$\int_0^{\pi/2} \int_{y=x}^{\pi/2} \cos 2y\sqrt{1-a^2\sin^2 x}\,dy\,dx$$

$$= \int_0^{\pi/2} \sqrt{1-a^2\sin^2 x}\left[\frac{1}{2}\sin 2y\right]_x^{\pi/2} dx$$

$$= \int_0^{\pi/2} \left(0 - \frac{1}{2}\sin 2x\right)\sqrt{1-a^2\sin^2 x}\,dx$$

(On putting $1 - a^2\sin^2 x = t^2$)

$$= \frac{1}{a^2}\int_1^{\sqrt{1-a^2}} t.t\,dt = \left[\frac{t^3}{3a^2}\right]_1^{\sqrt{1-a^2}} = \frac{1}{3a^2}[(1-a^2)^{3/2} - 1] \quad \textbf{Ans.}$$

EXAMPLE 6.12. (*a*) Evaluate the following integral by changing the order of integration

$$\int_0^3 \int_1^{\sqrt{4-y}} (x+y)\, dxdy.$$

(*b*) Change the order of integration in

$$I = \int_{x=0}^{2} \int_{y=x^2/4}^{3-x} xy\, dydx$$

and hence evaluate it. **(UPTU 2010-11)**

SOLUTION: (*a*) As given we are to integrate first w.r.t. x then w.r.t. y. The area of integration is the shaded portion. Changing the order of integration, the given integral equals

$$\int_1^2 \int_0^{4-x^2} (x+y)\, dydx = \int_1^2 \left[xy + \frac{y^2}{2} \right]_0^{4-x^2} dx$$

$$= \int_1^2 \left[x(4-x^2) + \frac{1}{2}(4-x^2)^2 \right] dx$$

$$= \int_1^2 \left(4x - x^3 + 8 - 4x^2 + \frac{x^4}{2} \right) dx$$

$$= \left[2x^2 - \frac{x^2}{4} + 8x - \frac{4}{3}x^3 + \frac{x^5}{10} \right]_1^2$$

$$= 8 - 1 + 16 - \frac{32}{3} + \frac{16}{5} - 2 + \frac{1}{4} - 8 + \frac{4}{3} - \frac{1}{10} = \frac{421}{60}. \quad \textbf{Ans.}$$

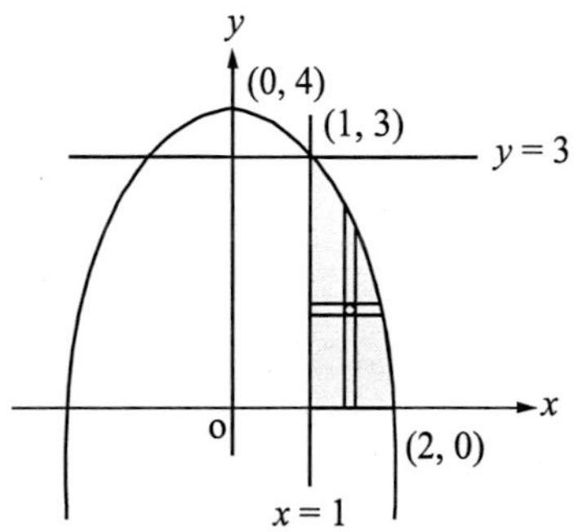

(*b*) As given we are to integrate first w.r.t. y then w.r.t. x.

The area of integration is the shaded portion.

If we change the order of integration.

$$I = \int_{y=0}^{1} \int_{x=0}^{2\sqrt{y}} xy\, dx\, dy + \int_{y=1}^{3} \int_{x=0}^{3-y} xy\, dx\, dy$$

$$= \int_{y=0}^{1} y \left[\frac{x^2}{2} \right]_{x=0}^{2\sqrt{y}} dy + \int_{y=1}^{3} y \left[\frac{x^2}{2} \right]_{x=0}^{3-y} dy$$

$$= \frac{1}{2} \int_{y=0}^{1} y \cdot (4y - 0)\, dy + \frac{1}{2} \int_{y=1}^{3} y[(3-y)^2 - 0]\, dy$$

$$= \int_{y=0}^{1} 2y^2 dy + \frac{1}{2} \int_{y=1}^{3} (y^3 - 6y^2 + 9y)\, dy$$

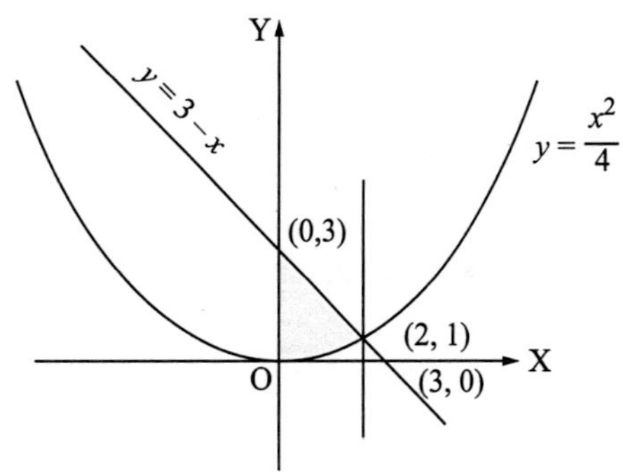

$$= 2\left[\frac{y^3}{3}\right]_{y=0}^{1} + \frac{1}{2}\left[\frac{y^4}{4} - \frac{6y^3}{3} + \frac{9y^2}{2}\right]_{y=1}^{3}$$

$$= \frac{2}{3} + \frac{1}{2}\left[\frac{81}{4} - 54 + \frac{81}{2} - \frac{1}{4} + 2 - \frac{9}{2}\right] = \frac{8}{3}. \quad \textbf{Ans.}$$

EXAMPLE 6.13. Evaluate the following integral by changing the order of integration

$$\int_0^\infty \int_0^x x e^{-x^2/y}\, dy dx.$$

(UPTU 2005-06)

SOLUTION: Area of integration is the shaded area. Changing the order of integration, the given

$$\text{integral} = \int_0^\infty \int_y^\infty x e^{-x^2/y}\, dx dy = \int_0^\infty \left[e^{-x^2/y}\left(\frac{-y}{2}\right)\right]_{x=y}^{\infty} dy$$

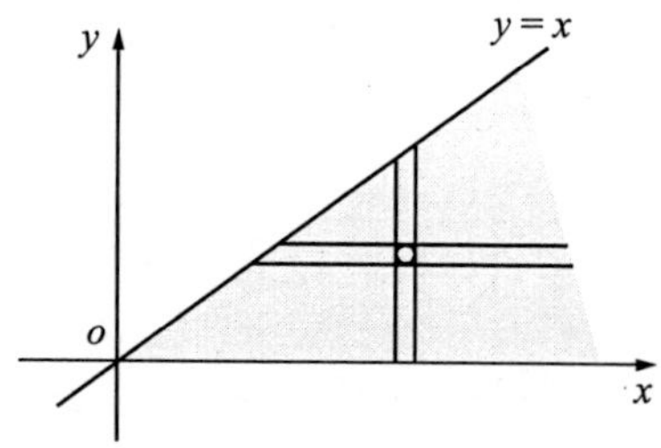

$$= \int_0^\infty 0 + \frac{1}{2}e^{-y} \cdot y\, dy$$

$$= \left[-\frac{y}{2}e^{-y}\right]_0^\infty - \int_0^\infty 1 \cdot \frac{e^{-y}}{-1 \cdot 2}\, dy$$

$$= 0 - \frac{1}{2}\left[e^{-y}\right]_0^\infty = \frac{1}{2}. \quad \textbf{Ans.}$$

EXAMPLE 6.14. By changing the order of integration in the integral $\int_0^\infty \int_0^\infty e^{-xy} \sin px\, dx dy$ show that

$$\int_0^\infty \frac{\sin px}{x}\, dx = \frac{\pi}{2}.$$

(UPTU. 2003-04, 2008-09)

SOLUTION: Let $\mathrm{I} = \int_0^\infty \int_0^\infty e^{-xy} \sin px\, dx dy = \int_0^\infty \sin px \int_{y=0}^\infty e^{-xy} dy \cdot dx$

$$= \int_0^\infty \sin px \left[\frac{e^{-xy}}{-x}\right]_{y=0}^\infty dx = \int_0^\infty \sin px\left(0 + \frac{1}{x}\right) dx = \int_0^\infty \frac{\sin px}{x}\, dx$$

Next, we change the order of integration, then

$$\mathrm{I} = \int_0^\infty \int_{x=0}^\infty e^{-xy} \sin px\, dx \cdot dy = \int_0^\infty \left[\frac{e^{-xy}}{p^2+y^2}(-p\cos px - y \sin px)\right]_{x=0}^\infty dy$$

$$= \int_0^\infty \left(0 + \frac{p}{p^2+y^2}\right) dy = \left[\tan^{-1}\frac{y}{p}\right]_0^\infty = \frac{\pi}{2}. \quad \text{Thus } \int_0^\infty \frac{\sin px}{x}\, dx = \frac{\pi}{2}. \quad \textbf{Ans.}$$

EXAMPLE 6.15. Evaluate $\iint (x+y)^2 dx\,dy$ over the area bounded by the ellipse $\dfrac{x^2}{a^2}+\dfrac{y^2}{b^2}=1.$

(UPTU 2003-04)

SOLUTION: The region R is the area bounded by $x=\pm a,\quad y=\pm\dfrac{b}{a}\sqrt{a^2-x^2}$.

$$\therefore\quad \iint_R (x+y)^2 dx\,dy = \int_{-a}^{a}\int_{-\frac{b}{a}\sqrt{a^2-x^2}}^{\frac{b}{a}\sqrt{a^2-x^2}} (x^2+y^2)dy\,dx + \int_{-a}^{a}\int_{-\frac{b}{a}\sqrt{a^2-x^2}}^{\frac{b}{a}\sqrt{a^2-x^2}} 2\,xy\,dy\,dx$$

$$= 2\int_{-a}^{a}\int_{0}^{\frac{b}{a}\sqrt{a^2-x^2}} (x^2+y^2)dy\,dx + 0 \qquad \text{(by property of definite integrals)}$$

$$= 2\int_{-a}^{a}\left[x^2y+\frac{y^3}{3}\right]_0^{\frac{b}{a}\sqrt{a^2-x^2}} dx$$

$$= 2\int_{-a}^{a}\left[x^2\frac{b}{a}\sqrt{a^2-x^2}+\frac{1}{3}\frac{b^3}{a^3}\left(a^2-x^2\right)^{3/2}\right]dx$$

$$= 4\frac{b}{a}\int_{0}^{a}\left[x^2\sqrt{a^2-x^2}+\frac{b^2}{3a^2}(a^2-x^2)^{3/2}\right]dx \qquad \text{(now putting } x=a\sin\theta\text{)}$$

$$= 4\frac{b}{a}\int_{0}^{\pi/2}\left[a^2\cdot\sin^2\theta\cdot a^2\cos^2\theta+\frac{b^2}{3a^2}a^3\cos^3\theta\cdot a\cos\theta\right]d\theta$$

$$= 4a^3b\int_{0}^{\pi/2}\left(\sin^2\theta\cos^2\theta+\frac{b^2}{3a^2}\cos^4\theta\right)d\theta$$

$$= 4a^3b\frac{\Gamma(3/2)\,\Gamma(3/2)}{2\Gamma(3)}+\frac{4ab^3}{3}\frac{\Gamma(5/2)\,\Gamma(1/2)}{2\Gamma(3)}$$

$$= a^3b\frac{\pi}{4}+ab^3\frac{\pi}{4} = \frac{\pi}{4}ab\left(a^2+b^2\right) \qquad \textbf{Ans.}$$

EXAMPLE 6.16. (*a*) If R is the region bounded by the curves $xy=16,\ x=y,\ y=0$ and $x=8$, evaluate the double integral $\iint_R x^2\,dx\,dy$ by expressing it as an appropriate integral. **(UPTU 2001-02)**

(*b*) Find the area of the region in the first quadrant bounded by the curves $y=x,\quad y=\dfrac{1}{x},\quad y=\dfrac{x}{4}$. **(UPTU 2008-09)**

SOLUTION: **(*a*)** Area R of integration is the shaded portion in the figure.

$$\therefore \quad \iint_R x^2\,dx\,dy = \int_0^4\int_0^x x^2 dy\,dx + \int_4^8\int_0^{16/x} x^2 dy\,dx$$

$$= \int_0^4 x^2 [y]_0^x\,dx + \int_4^8 x^2 [y]_0^{16/x}\,dx$$

$$= \int_0^4 x^3 dx + \int_4^8 16x\,dx$$

$$= \left[\frac{x^4}{4}\right]_0^4 + 16\left[\frac{x^2}{2}\right]_4^8$$

$$= 64 + 8(64 - 16) = 448. \qquad \textbf{Ans.}$$

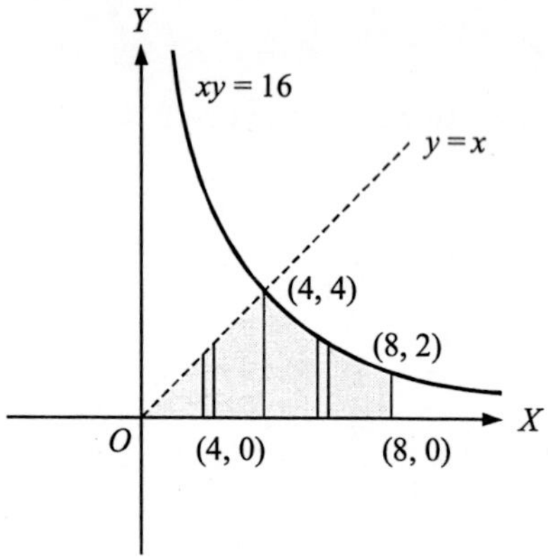

(*b*) Required area, as depicted in the adjoining figure, is equal to

$$\int_0^1\int_{x/4}^{x} dydx + \int_1^2\int_{x/4}^{1/x} dydx$$

$$= \int_0^1 [y]_{x/4}^x\,dx + \int_1^2 [y]_{x/4}^{1/x}\,dx$$

$$= \int_0^1\left(x - \frac{x}{4}\right)dx + \int_1^2\left(\frac{1}{x} - \frac{x}{4}\right)dx$$

$$= \left[\frac{3x^2}{8}\right]_0^1 + \left[\log x - \frac{x^2}{8}\right]_1^2$$

$$= \frac{3}{8} + \log 2 - \frac{1}{8}(4-1) = \log 2. \qquad \textbf{Ans.}$$

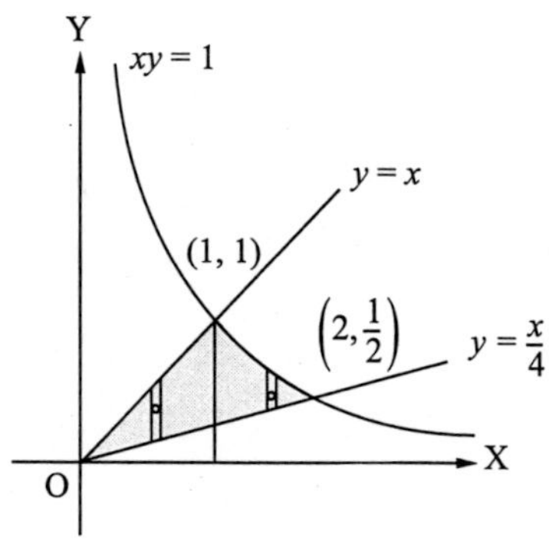

EXAMPLE 6.17. **(*a*)** Evaluate the following integral by changing the order of integration $\int_0^1\int_{e^x}^{e} \frac{dy\,dx}{\log y}$.

(UPTU 2001 Q.B.)

(*b*) Evaluate $\iint_R \left(1 - \frac{x^2}{a^2} + \frac{y^2}{b^2}\right)dx\,dy$ over the first quadrant of the ellipse $\frac{x^2}{a^2} + \frac{y^2}{b^2} = 1$.

(UPTU 2009-10)

SOLUTION: (*a*) Area of integration is shown as shaded portion in the figure.

Changing the order of integration

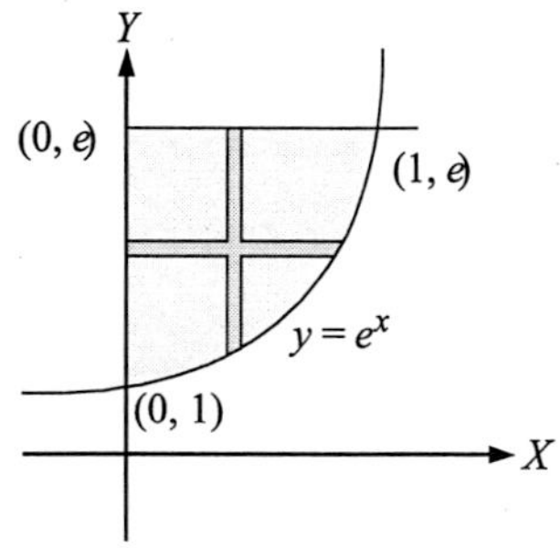

$$\therefore \quad \int_0^1 \int_{e^x}^{e} \frac{dy\,dx}{\log y} = \int_1^e \int_0^{\log y} \frac{dx\,dy}{\log y}$$

$$= \int_1^e \frac{1}{\log y}[x]_{x=0}^{\log y}\,dy$$

$$= \int_1^e 1\,dy = e - 1. \qquad \textbf{Ans.}$$

(*b*) Let $I = \iint_R \left(1 - \frac{x^2}{a^2} + \frac{y^2}{b^2}\right) dx\,dy$ where region $R: \frac{x^2}{a^2} + \frac{y^2}{b^2} = 1,\ x \geq 0,\ y \geq 0.$

Substituting, $x = aX,\ y = bY,$ we get

$$I = \iint_R (1 - X^2 + Y^2)\,dX\,dY \text{ where } R \text{ is region of } X^2 + Y^2 = 1,\ X \geq 0,\ Y \geq 0.$$

$$= ab \int_0^1 \int_0^{\sqrt{1-X^2}} (1 - X^2 + Y^2)\,dY dX = ab \int_0^1 \left[Y - YX^2 + \frac{Y^3}{3}\right]_0^{\sqrt{1-X^2}} dX$$

$$= ab \int_0^1 \left[(1-X^2)\sqrt{1-X^2} + \frac{1}{3}(1-X^2)^{3/2}\right] dX = \frac{4ab}{3}\int_0^1 (1-X^2)^{3/2}\,dX$$

$$= \frac{4}{3}ab \int_0^{\pi/2} (1-\sin^2\theta)^{3/2}\cos\theta\,d\theta \qquad (\text{on putting } X = \sin\theta)$$

$$= \frac{4}{3}ab \int_0^{\pi/2} \cos^4\theta\,d\theta = \frac{4ab}{3}\,\frac{\boxed{1/2}\,\boxed{5/2}}{2\boxed{3}} = \frac{ab}{3}\cdot\frac{3}{2}\cdot\frac{1}{2}\cdot\pi = \frac{\pi ab}{4}. \qquad \textbf{Ans.}$$

EXERCISE 6A

1. Evaluate $\int_0^1 \int_0^{x^2} e^{\frac{y}{x}}\, dy\, dx$.

2. Evaluate $\int_0^{a\sqrt{3}} \int_0^{\sqrt{x^2+a^2}} \frac{x\, dy\, dx}{y^2 + x^2 + a^2}$.

3. Evaluate the double integral $\iint \frac{dx\, dy}{x^4 + y^2}$ over the region bounded by $x = 1$ and $y = x^2$.

4. Evaluate $\iint \sin(ax + by)\, dx\, dy$ over the triangular area bounded by $x = 0, \quad y = 0, \quad ax + by = 1$.

5. Evaluate $\int_0^a \int_{\frac{y^2}{a}}^{y} \frac{y\, dx\, dy}{(a-x)\sqrt{ax - y^2}}$ by changing the order of integration.

(UPTU 2001 Sp. Exam.)

6. Change the order of integration in the double integral $\int_0^a \int_{\frac{x^2}{a}}^{2a-x} xy\, dy\, dx$ and the evaluate it.

(UPTU 2004-05)

7. Change the order of integration in the integral $\int_{-a}^{a} \int_0^{\sqrt{a^2-y^2}} f(x, y)\, dx\, dy$.

8. Evaluate (i) $\int_0^4 \int_0^{x^2} e^{\frac{y}{x}}\, dy\, dx$. (ii) $\int_0^1 \int_0^{\sqrt{1-x^2}} \frac{e^y}{(e^y + 1)\sqrt{1 - x^2 - y^2}}\, dx\, dy$ **(UPTU 2011-12)**

9. Change the order of integration in the integral $\int_0^1 \int_x^{\sqrt{2-x^2}} \frac{x\, dy\, dx}{\sqrt{x^2 + y^2}}$.

10. Evaluate $\int_0^\infty \int_x^\infty \frac{e^{-y}}{y}\, dy\, dx$. **(UPTU 2001 Q.B.)**

11. Evaluate $\int_0^1 dx \int_1^\infty e^{-y}\, y^x \log y\, dy$.

12. Evaluate, by changing the order of integration, the double integral $\int_0^1 \int_{e^x}^{e} \frac{dx\, dy}{y^2 \log y}$.

13. Sketch the region of integration and evaluate $\int_0^{\frac{\pi}{2}} \int_0^y \cos 2y \sqrt{1 - a^2 \sin^2 x}\, dx\, dy$.

14. Evaluate the double integral $\int_0^a \int_0^x \frac{f'(y)\, dy\, dx}{\sqrt{(a-x)(x-y)}}$.

❑❑❑

DOUBLE INTEGRALS IN POLAR CO-ORDINATES

In the polar co-ordinates we divide the region of integration into a mesh formed by circular arcs and straight lines, given by

r = constants (which are circles) and

θ = constants (which are straight lines)

as shown in the adjoining figure. Here the element in the mesh is $\delta r.\ r\ \delta\theta$.

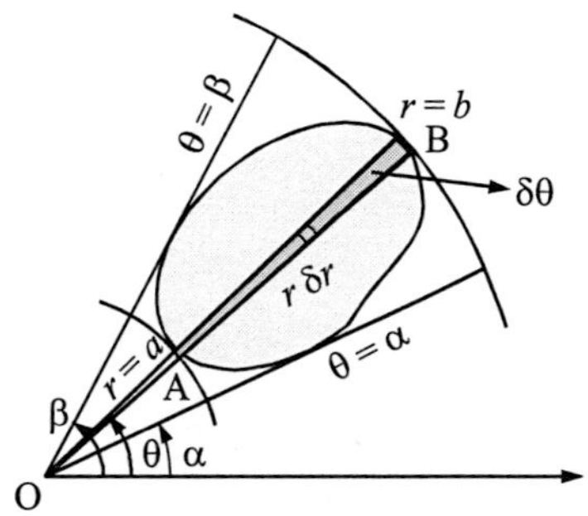

Now, if at the point $P(r, \theta)$ the function is $f(r, \theta)$ then , over the wedge AB, the sum

$$\underset{\delta r \to 0}{\text{Lt}}\ \delta\theta \sum_{A}^{B} f(r, \theta).\, r\, \delta r \quad \text{can be written as} \quad \delta\theta \int_{r_1(\theta)}^{r_2(\theta)} f(r, \theta)\, r\, dr.$$

where $r_1(\theta)$ and $r_2(\theta)$ arc the radii vectors of two curves and θ is kept constant while integrating w.r.t. r.

Finally, summing for all wedges between $\theta = \alpha$ and $\theta = \beta$, we get

$$\underset{\delta\theta \to 0}{\text{Lt}} \sum_{\alpha}^{\beta} \delta\theta \sum_{r_1(\theta)}^{r_2(\theta)} f(r, \theta)\, r\, \delta r = \int_{\theta=\alpha}^{\theta=\beta} d\theta \int_{r=r_1(\theta)}^{r=r_2(\theta)} f(r, \theta)\, r\, dr.$$

Further, the double integral in the cartesian form

$$\iint_R f(x, y)\, dx\, dy \qquad \text{becomes} \qquad \iint_R f(r\cos\theta, r\sin\theta)\, r\, dr\, d\theta$$

when changed to polar co-ordinates. In the latter integral the region R should be expressed in the polar co-ordinates. Here also, we can go in for changing the order of integration, if needed.

EXAMPLE 6.18. Evaluate $\displaystyle\int_0^{2a} \int_0^{\sqrt{2ax - x^2}} (x^2 + y^2)\, dx\, dy.$

SOLUTION: The region of integration is the upper half of the circle $x^2 + y^2 = 2ax$. If we transform to polar co-ordinates the region becomes $r = 2a\cos\theta$ from $\theta = 0$ to $\theta = \dfrac{\pi}{2}$, therefore the given integral becomes

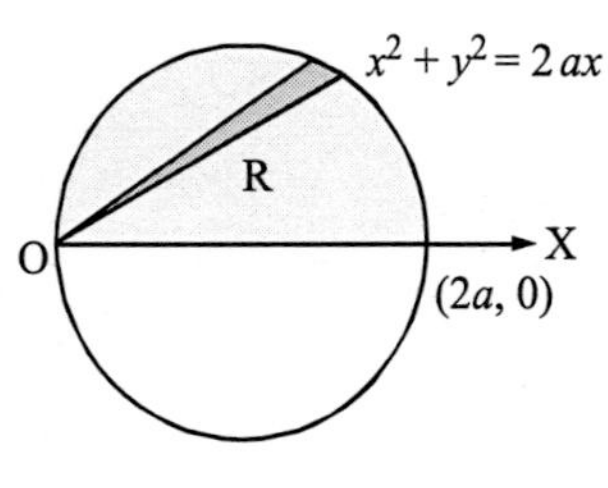

$$\int_0^{\pi/2} \int_0^{2a\cos\theta} r^2 .\, r\, dr\, d\theta = \int_0^{\pi/2} \left[\frac{r^4}{4}\right]_0^{2a\cos\theta} d\theta$$

$$= \frac{1}{4}(2a)^4 \int_0^{\pi/2} \cos^4\theta\, d\theta = \frac{3}{4}\cdot\frac{1}{2}\cdot\frac{\pi}{2}\cdot 4a^4 = \frac{3}{4}\pi a^4. \qquad \textbf{Ans.}$$

EXAMPLE 6.19. Evaluate the double integral $\iint r^3\, dr\, d\theta$ over the region bounded by the circles $r = 2\cos\theta$ and $r = 4\cos\theta$.

SOLUTION: The shaded area between the circles $r = 2\cos\theta$ and $r = 4\cos\theta$ is the region of integration. When we integrate w.r.t. r then r varies from P ($r = 2\cos\theta$) to Q ($r = 4\cos\theta$) and θ varies from $\theta = -\pi/2$ to $\theta = \pi/2$ so that the whole shaded area is covered. Then

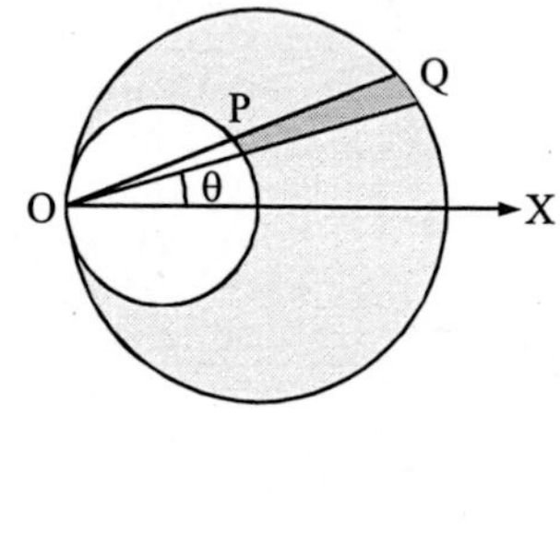

$$I = \int_{-\pi/2}^{\pi/2}\int_{2\cos\theta}^{4\cos\theta} r^3\, dr\, d\theta = \int_{-\pi/2}^{\pi/2}\left[\frac{r^4}{4}\right]_{2\cos\theta}^{4\cos\theta} d\theta$$

$$= \frac{1}{4}\int_{-\pi/2}^{\pi/2}[4^4\cos^4\theta - 2^4\cos^4\theta]\, d\theta$$

$$= \frac{20\times 12}{4}\int_{-\pi/2}^{\pi/2}\cos^4\theta\, d\theta = 60\times 2\int_0^{\pi/2}\cos^4\theta\, d\theta$$

$$= 120\times\frac{3}{4}\cdot\frac{1}{2}\cdot\frac{\pi}{2} = \frac{45\pi}{2}. \quad \textbf{Ans.}$$

EXAMPLE 6.20. (*a*) Evaluate $\iint r^2 dr\, d\theta$ over the area included between the circles $r = 2\sin\theta$ and $r = 4\sin\theta$. **(UPTU 2001 Q.B.)**

(*b*) Find the area of one loop of the lemniscate $r^2 = a^2\cos 2\theta$. **(UPTU 2010-11)**

SOLUTION: (*a*) Given integral is equal to

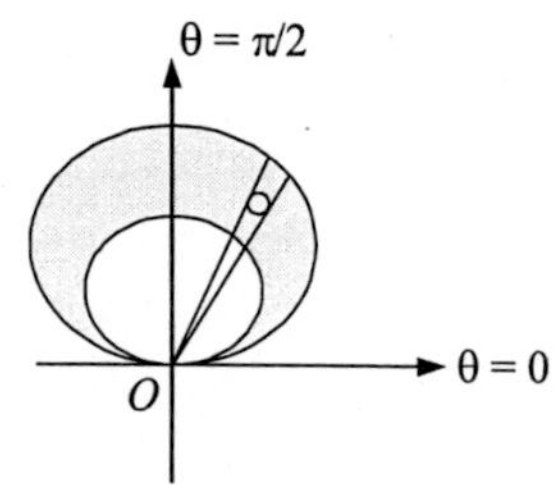

$$\int_0^{\pi}\int_{2\sin\theta}^{4\sin\theta} r^2 dr\, d\theta = \int_0^{\pi}\left[\frac{r^3}{3}\right]_{2\sin\theta}^{4\sin\theta} d\theta$$

$$= \frac{1}{3}\int_0^{\pi}(4^3 - 2^3)\sin^3\theta\, d\theta = \frac{56.(2)}{3}\int_0^{\pi/2}\sin^3\theta\, d\theta$$

$$= \frac{112}{3}\,\frac{\overline{|2}\,\overline{|1/2}}{2\overline{|5/2}} = \frac{56}{3\cdot\frac{3}{2}\cdot\frac{1}{2}} = \frac{224}{9}. \quad \textbf{Ans.}$$

(*b*) Area of one loop $= \int_{\theta=\pi/4}^{\theta=+\pi/4}\int_{r=0}^{a\sqrt{\cos 2\theta}} r\, dr\, d\theta$

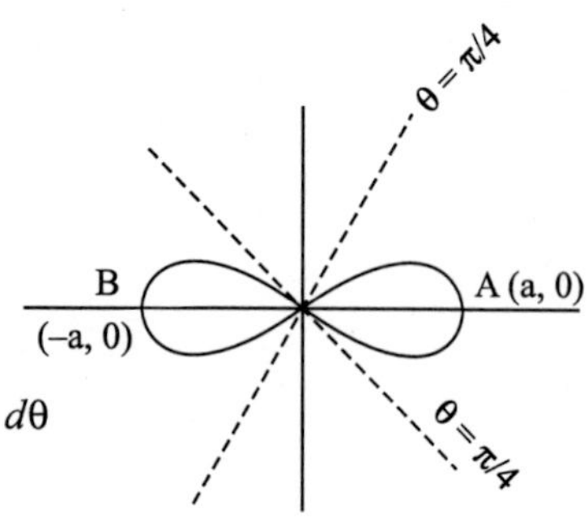

$$= 2\int_{\theta=0}^{\pi/4}\int_{r=0}^{a\sqrt{\cos 2\theta}} r\, dr\, d\theta = \int_{\theta=0}^{\pi/4}\left[r^2\right]_{r=0}^{a\sqrt{\cos 2\theta}} d\theta$$

$$= \int_{\theta=0}^{\pi/4} a^2\cos 2\theta\, d\theta = a^2\left[\frac{\sin 2\theta}{2}\right]_0^{\pi/4} = \frac{a^2}{2}. \quad \textbf{Ans.}$$

EXAMPLE 6.21. Evaluate $\iint \frac{r\,dr\,d\theta}{\sqrt{r^2+a^2}}$ over one loop of the leminiscate $r^2 = a^2 \cos 2\theta$.

SOLUTION: The loop of the curve $r^2 = a^2 \cos 2\theta$ is lying between $\theta = \pm\frac{\pi}{4}$. We integrate first w.r.t. r under the limits $r = 0$ to $r = a\sqrt{\cos 2\theta}$ and then to cover the entire loop, θ varies from $-\frac{\pi}{4}$ to $\frac{\pi}{4}$.

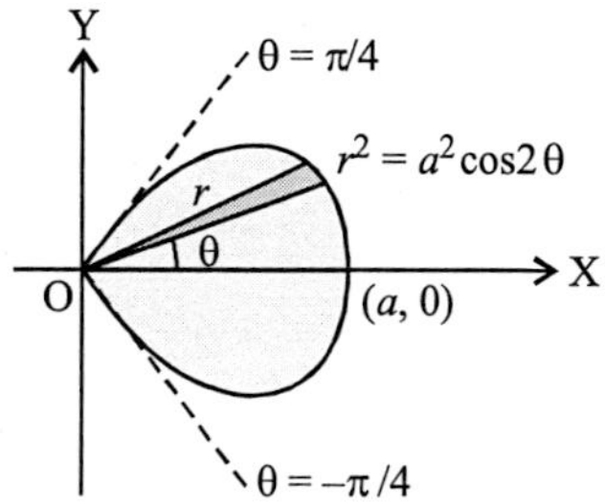

$$\text{Therefore, } I = \int_{-(\pi/4)}^{\pi/4}\int_0^{a\sqrt{\cos 2\theta}} \frac{r\,dr\,d\theta}{\sqrt{r^2+a^2}}$$

$$= \int_{-(\pi/4)}^{\pi/4} d\theta \int_0^{a\sqrt{\cos 2\theta}} \frac{r\,dr}{\sqrt{r^2+a^2}}$$

$$= \int_{-(\pi/4)}^{\pi/4} d\theta \left[\sqrt{r^2+a^2}\right]_0^{a\sqrt{\cos 2\theta}} = \int_{-(\pi/4)}^{\pi/4}\left[a\sqrt{1+\cos 2\theta} - a\right] d\theta$$

$$= 2a\int_0^{\pi/4}\left(\sqrt{1+\cos 2\theta} - 1\right) d\theta = 2a\int_0^{\pi/4}\left(\sqrt{2}\cos\theta - 1\right) d\theta$$

$$= 2a\,[\sqrt{2}\sin\theta - \theta]_0^{\pi/4} = 2a\left[\sqrt{2}\sin\frac{\pi}{4} - \frac{\pi}{4}\right] = 2a\left(1 - \frac{\pi}{4}\right). \quad \textbf{Ans.}$$

EXAMPLE 6.22. Evaluate $\iint y^2\,dx\,dy$ over the area outside the circle $x^2 + y^2 - ax = 0$ and inside the circle $x^2 + y^2 - 2ax = 0$.

SOLUTION: The region of integration is bounded by the circle $x^2 + y^2 - ax = 0$ or $r = a\cos\theta$ and the circle $x^2 + y^2 - 2ax = 0$ or $r = 2a\cos\theta$. The element $dx\,dy = r\,dr\,d\theta$. The elementary strip, as shown in the adjoining figure, extends from $r = a\cos\theta$ to $r = 2a\cos\theta$. To cover up the region of integration, θ varies from $\theta = -\frac{\pi}{2}$ to $\theta = \frac{\pi}{2}$. The integrand $= y^2 = (r\sin\theta)^2$

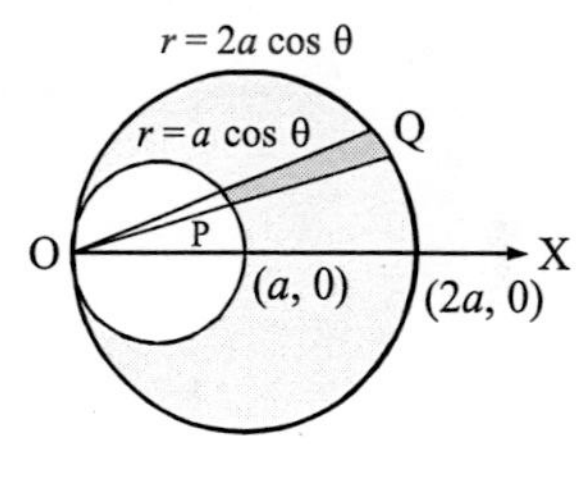

$$\therefore \quad \text{Given integral} = \int_{-(\pi/2)}^{\pi/2}\int_{a\cos\theta}^{2a\cos\theta} r^2\sin^2\theta\, r\,dr\,d\theta = \int_{-(\pi/2)}^{\pi/2}\sin^2\theta\,d\theta \int_{a\cos\theta}^{2a\cos\theta} r^3\,dr$$

$$= \int_{-(\pi/2)}^{\pi/2} \sin^2\theta\left[\frac{r^4}{4}\right]_{a\cos\theta}^{2a\cos\theta} d\theta$$

$$= \int_{-(\pi/2)}^{\pi/2}\frac{1}{4}\sin^2\theta\,[16a^4\cos^4\theta - a^4\cos^4\theta]\,d\theta = \frac{15}{4}a^4\int_{-(\pi/2)}^{\pi/2}\sin^2\theta\cos^4\theta\,d\theta$$

$$= \frac{15}{4}a^4\, 2\int_0^{\pi/2}\sin^2\theta\cos^4\theta\,d\theta = \frac{15a^4}{2}\frac{1}{6}.\frac{3}{4}.\frac{1}{2}.\frac{\pi}{2} = \frac{15\pi a^4}{64}. \quad \textbf{Ans.}$$

EXAMPLE 6.23. Change into polar co-ordinates and hence evaluate

$$\text{(i)} \int_0^a \int_0^{\sqrt{a^2-x^2}} \sqrt{x^2+y^2}\, dy\, dx. \qquad \text{(ii)} \int_0^a \int_0^{\sqrt{a^2-y^2}} (x^2+y^2)\, dx\, dy$$ **[UPTU 2011-12]**

SOLUTION: **(i)** The given integral is $I = \int_0^a \int_0^{\sqrt{a^2-x^2}} \sqrt{x^2+y^2}\, dy\, dx$

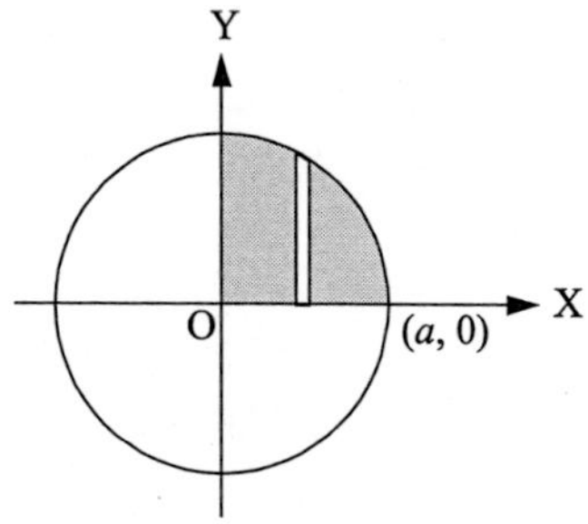

The area over which the integration is to be carried out is the shaded portion in the adjoining figure. Converting to polar co-ordinates by $x = r\cos\theta$, $y = r\sin\theta$ so that $r^2 = x^2 + y^2$ and $dxdy = r\, d\theta\, dr$, we get

$$I = \int_{\theta=0}^{\pi/2} \int_{r=0}^{a} r \cdot r d\theta\, dr$$

$$= \int_0^{\pi/2} \left[\frac{r^3}{3}\right]_0^a d\theta = \frac{a^3}{3} \cdot \frac{\pi}{2} = \pi a^3/6. \quad \textbf{Ans.}$$

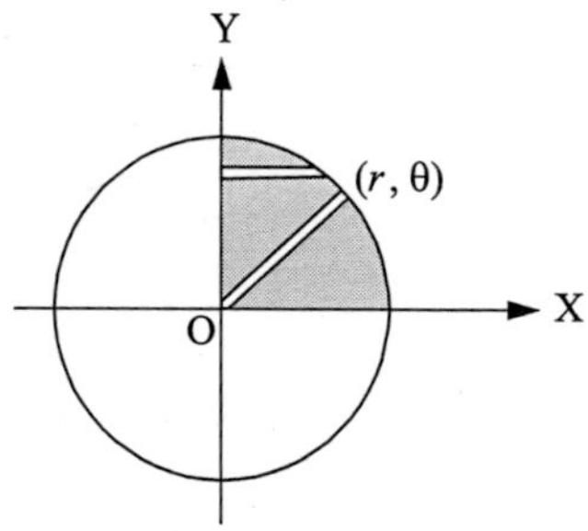

(ii) Let $I = \int_0^a \int_0^{\sqrt{a^2-y^2}} (x^2+y^2)\, dx\, dy$

Converting into polar coordinates $x = r\cos\theta$, $y = r\sin\theta$ so that $r^2 = x^2 + y^2$ and $dx\, dy = r\, dr\, d\theta$, we get

Area of integration is shaded portion in the figure in first quadrant.

Then $$I = \int_{\theta=0}^{\pi/2} \int_{r=0}^{a} r^2 \cdot r\, dr\, d\theta = \int_0^{\pi/2} \left[\frac{r^4}{4}\right]_0^a d\theta = \int_0^{\pi/2} \frac{a^4}{4} d\theta = \frac{\pi a^4}{8} \quad \textbf{Ans.}$$

EXAMPLE 6.24. Find by double integration the area lying inside the circle $r = a\sin\theta$ and outside the cardiode $r = a(1 - \cos\theta)$.

SOLUTION: The required area is the shaded portion in the adjacent figure and is equal to

$$\int_0^{\pi/2} \int_{a(1-\cos\theta)}^{a\sin\theta} r\, dr\, d\theta = \int_0^{\pi/2} \left[\frac{1}{2} r^2\right]_{a(1-\cos\theta)}^{a\sin\theta} d\theta$$

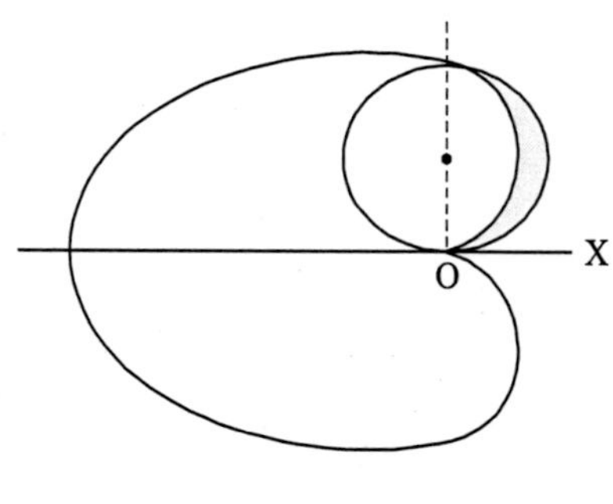

$$= \frac{1}{2} a^2 \int_0^{\pi/2} [\sin^2\theta - (1-\cos\theta)^2] d\theta$$

$$= \frac{a^2}{2} \int_0^{\pi/2} (-1 + 2\cos\theta - \cos 2\theta) d\theta$$

$$= \frac{a^2}{2}\left[-\theta + 2\sin\theta - \frac{1}{2}\sin 2\theta\right]_0^{\pi/2} = \frac{a^2}{2}\left[-\frac{\pi}{2} + 2\right] = a^2(1 - \pi/4). \quad \textbf{Ans.}$$

EXAMPLE 6.25. Evaluate the double integral

$$\int_0^1 x\, e^{-x^2} dx \int_{\sqrt{x-x^2}}^{\sqrt{1-x^2}} \frac{y\, e^{-y^2}}{x^2+y^2}\, dy.$$

SOLUTION: The region of integration is the shaded portion bounded by the upper halves of circles $x^2 + y^2 = 1$ and $x^2 + y^2 = x$ and by $x = 0$ and $x = 1$ as shown in the figure. On changing to polar co-ordinates the circle $x^2 + y^2 = x$ becomes $r = \cos\theta$ and $x^2 + y^2 = 1$ becomes $r = 1$, hence the given integral becomes

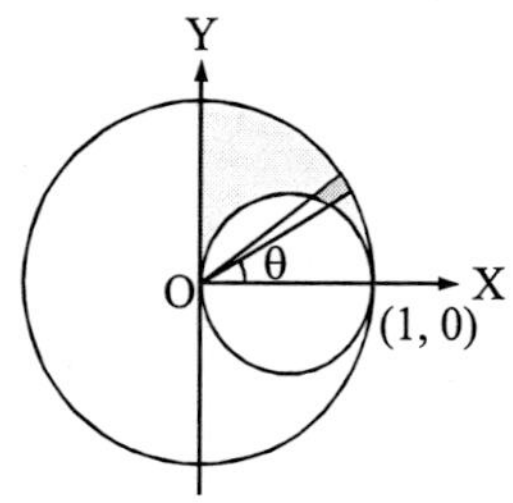

$$\text{I} = \int_0^{\pi/2} \int_{r=\cos\theta}^{1} \frac{r\cos\theta . r\sin\theta\, e^{-r^2}}{r^2} . r\, d\theta\, dr$$

$$= \int_0^{\pi/2} \sin\theta\cos\theta\, d\theta \int_{r=\cos\theta}^{1} r\, e^{-r^2}\, dr$$

$$= \int_0^{\pi/2} \sin\theta\cos\theta \left[-\frac{1}{2}e^{-r^2}\right]_{r=\cos\theta}^{1} d\theta$$

$$= \frac{1}{2}\int_0^{\pi/2} \left(e^{-\cos^2\theta} - e^{-1}\right)\sin\theta\cos\theta\, d\theta$$

$$= \frac{1}{2}\int_0^{\pi/2} e^{-\cos^2\theta}\sin\theta\cos\theta\, d\theta - \frac{e^{-1}}{2}\int_0^{\pi/2}\sin\theta\cos\theta\, d\theta$$

$$= \frac{1}{4}\left[e^{-\cos^2\theta}\right]_0^{\pi/2} - \frac{1}{4e}[\sin^2\theta]_0^{\pi/2}$$

$$= \frac{1}{4}(1-e^{-1}) - \frac{1}{4e} = \frac{1}{4}(1-2e^{-1}). \qquad \textbf{Ans.}$$

EXAMPLE 6.26. Evaluate $\iint \frac{dx\, dy}{\left(1+x^2+y^2\right)^{\frac{3}{2}}}$ over the area of the square bounded by $x = 0$, $y = 0$, $x = 1$, $y = 1$.

SOLUTION: Line $x = 1$ in polar co-ordinates is $r \cos\theta = 1$ or $r = \sec\theta$ and the line $y = 1$ is $r\sin\theta = 1$ or $r = \text{cosec}\,\theta$.

The region of integration is the square which is divided into two regions OACO and OCBO as shown in the adjoining figure.

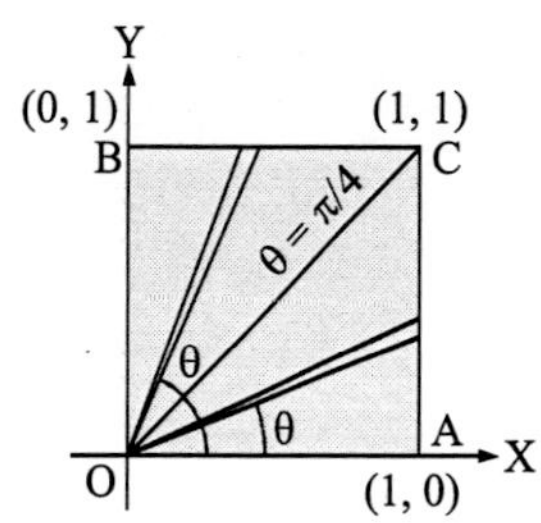

$$\text{I} = \iint \frac{dx\, dy}{\left(1+x^2+y^2\right)^{\frac{3}{2}}} = \iint_{\text{OACO}} \frac{r\, dr\, d\theta}{(1+r^2)^{\frac{3}{2}}} + \iint_{\text{OCBO}} \frac{r\, dr\, d\theta}{(1+r^2)^{\frac{3}{2}}}$$

$$= \int_{\theta=0}^{\frac{\pi}{4}} \int_0^{\sec\theta} \frac{r\, dr\, d\theta}{(1+r^2)^{\frac{3}{2}}} + \int_{\frac{\pi}{4}}^{\frac{\pi}{2}} \int_0^{\text{cosec}\,\theta} \frac{r\, dr\, d\theta}{(1+r^2)^{\frac{3}{2}}}$$

Putting $1 + r^2 = t^2$, so $r\,dr = t\,dt$, we have $\displaystyle\int \frac{r\,dr}{(1+r^2)^{\frac{3}{2}}} = \int \frac{t\,dt}{t^3} = \frac{-1}{\sqrt{1+r^2}}$

Therefore, $$I = \int_0^{\pi/4} \left[\frac{-1}{\sqrt{1+r^2}} \right]_0^{\sec\theta} d\theta + \int_{\pi/4}^{\pi/2} \left[\frac{-1}{\sqrt{1+r^2}} \right]_0^{\operatorname{cosec}\theta} d\theta$$

$$= \int_0^{\pi/4} \left[1 - \frac{1}{\sqrt{1+\sec^2\theta}} \right] d\theta + \int_{\pi/4}^{\pi/2} \left[1 - \frac{1}{\sqrt{1+\operatorname{cosec}^2\theta}} \right] d\theta$$

$$= \int_0^{\pi/4} d\theta - \int_0^{\pi/4} \frac{\cos\theta\,d\theta}{\sqrt{1+\cos^2\theta}} + \int_{\pi/4}^{\pi/2} d\theta - \int_{\pi/4}^{\pi/2} \frac{\sin\theta\,d\theta}{\sqrt{1+\sin^2\theta}}$$

$$= \int_0^{\pi/2} d\theta - \int_0^{\pi/4} \frac{\cos\theta\,d\theta}{\sqrt{2-\sin^2\theta}} - \int_{\pi/4}^{\pi/2} \frac{\sin\theta\,d\theta}{\sqrt{2-\cos^2\theta}} = \frac{\pi}{2} - I_1 - I_2$$

where $I_1 = \displaystyle\int_0^{\pi/4} \frac{\cos\theta\,d\theta}{\sqrt{2-\sin^2\theta}}$ and $I_2 = \displaystyle\int_{\pi/4}^{\pi/2} \frac{\sin\theta\,d\theta}{\sqrt{2-\cos^2\theta}}$.

In I_1 put $\sin\theta = u$ and in I_2 put $\cos\theta = v$

$$\therefore \quad I_1 = \int_0^{\frac{1}{\sqrt{2}}} \frac{du}{\sqrt{2-u^2}} = \left[\sin^{-1}\frac{u}{\sqrt{2}} \right]_0^{\frac{1}{\sqrt{2}}} = \sin^{-1}\frac{1}{2} = \frac{\pi}{6}$$

and $$I_2 = \int_{\frac{1}{\sqrt{2}}}^{0} \frac{-dv}{\sqrt{2-v^2}} = \left[\sin^{-1}\frac{v}{\sqrt{2}} \right]_0^{\frac{1}{\sqrt{2}}} = \sin^{-1}\frac{1}{2} - 0 = \frac{\pi}{6}$$

$$\therefore \quad I = \frac{\pi}{2} - \frac{\pi}{6} - \frac{\pi}{6} = \frac{\pi}{6}.$$ **Ans.**

EXAMPLE 6.27. Evaluate $\displaystyle\iint xy \left(\frac{x^2}{a^2} + \frac{y^2}{b^2} \right)^{m/2} dx\,dy$ over the positive quadrant of the ellipse

$$\frac{x^2}{a^2} + \frac{y^2}{b^2} = 1.$$

SOLUTION: The region of integration in the given double integral is the ellipse $\frac{x^2}{a^2} + \frac{y^2}{b^2} = 1$. Transforming these x, y to elliptical polar co-ordinates $x = ar\cos\theta$, $y = br\sin\theta$ so that $dx\,dy = abr\,d\theta\,dr$ and the ellipse $\frac{x^2}{a^2} + \frac{y^2}{b^2} = 1$ get transformed to the circle $r^2 = 1$, *i.e.*, $r = 1$, the circle with centre (0, 0) and radius 1 in the positive quadrant.

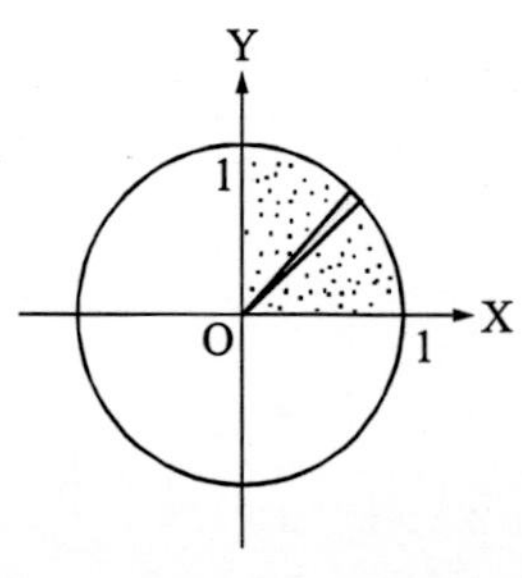

$$\therefore \quad I = \int_0^{\pi/2}\int_0^1 ar\cos\theta . br\sin\theta\,(r^2\cos^2\theta + r^2\sin^2\theta)^{m/2}\, ab\, r\, d\theta\, dr$$

$$= \frac{a^2 b^2}{2}\int_0^{\pi/2}\int_0^1 \sin 2\theta\ r^{m+3}\ d\theta\, dr$$

$$= \frac{a^2 b^2}{2}\left[-\frac{\cos 2\theta}{2}\right]_0^{\pi/2}\left[\frac{r^{m+4}}{m+4}\right]_0^1 = \frac{a^2 b^2}{2(m+4)}. \qquad \textbf{Ans.}$$

EXAMPLE 6.28. Evaluate $\iint \frac{(x^2+y^2)^2}{x^2y^2}\,dx\,dy$ over the area common to

$$x^2 + y^2 - 2ax = 0,\quad x^2 + y^2 - 2by = 0,\ a > b > 0.$$

SOLUTION: The region of integration is the common portion of the circle $x^2 + y^2 - 2ax = 0$ (centre $(a, 0)$, radius a) and the circle $x^2 + y^2 - 2by = 0$ (centre $(0, b)$, radius b) which on changing to polar co-ordinates, become $r = 2a\cos\theta$, $r = 2b\sin\theta$.

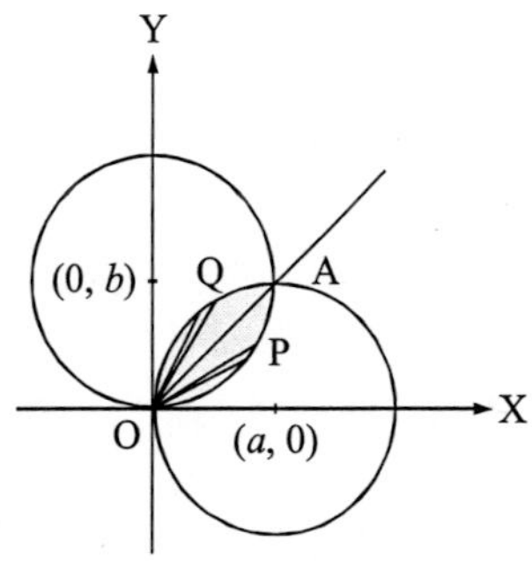

Their points of inter-section are O the pole and A $\left(\sqrt{a^2+b^2}, \alpha\right)$ where $\tan\alpha = \frac{a}{b}$. The region of integration, shown in the adjoining figure as shaded area, can be divided into two subregions OPAO and OAQO as shown in the figure. Then

$$I = \iint_{OPAQO} \frac{(x^2+y^2)^2\,dx\,dy}{x^2y^2} = \iint_{OPAO} \frac{r^4 . r\,dr\,d\theta}{r^4\sin^2\theta\cos^2\theta} + \iint_{OAQO} \frac{r^4 . r\,dr\,d\theta}{r^4\sin^2\theta\cos^2\theta}$$

$$= \int_0^{\alpha}\int_0^{2b\sin\theta} \frac{r\,dr\,d\theta}{\sin^2\theta\cos^2\theta} + \int_{\alpha}^{\pi/2}\int_0^{2a\cos\theta} \frac{r\,dr\,d\theta}{\sin^2\theta\cos^2\theta}$$

$$= \int_0^{\alpha} \frac{1}{\sin^2\theta\cos^2\theta}\left[\frac{r^2}{2}\right]_0^{2b\sin\theta} d\theta + \int_{\alpha}^{\pi/2} \frac{1}{\sin^2\theta\cos^2\theta}\left[\frac{r^2}{2}\right]_0^{2a\cos\theta} d\theta$$

$$= 2b^2\int_0^{\alpha}\sec^2\theta\,d\theta + 2a^2\int_{\alpha}^{\pi/2}\operatorname{cosec}^2\theta\,d\theta$$

$$= 2b^2[\tan\theta]_0^{\alpha} + 2a^2[-\cot\theta]_{\alpha}^{\pi/2}$$

$$= 2b^2\tan\alpha + 2a^2\cot\alpha = 2b^2\cdot\frac{a}{b} + 2a^2\cdot\frac{b}{a} = 4ab. \qquad \textbf{Ans.}$$

EXAMPLE 6.29. Evaluate $\int_0^2 \int_0^{\sqrt{2x-x^2}} \frac{x\,dydx}{\sqrt{x^2+y^2}}$ by changing to polar co-ordinates.

SOLUTION: Here y varies from $y = 0$ to $y = \sqrt{2x - x^2}$, i.e. $x^2 + y^2 - 2x = 0$ which is a circle with centre (1, 0) and radius 1. Here $y \geq 0$.

In polar co-ordinates $r^2 = 2r\cos\theta$ or $r = 2\cos\theta$.

Hence r varies from 0 to $2\cos\theta$ and θ varies from 0 to $\pi/2$

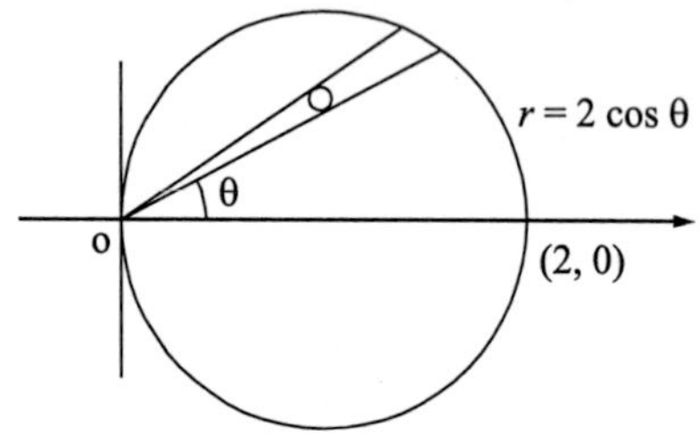

$$\text{Given Integral} = \int_0^{\pi/2} \int_0^{2\cos\theta} \frac{r\cos\theta \cdot r\,dr\,d\theta}{r}$$

$$= \int_0^{\pi/2} \cos\theta \left[\frac{r^2}{2}\right]_0^{2\cos\theta} d\theta = \frac{1}{2}\int_0^{\pi/2} 4\cos^3\theta\, d\theta = \frac{4}{3}. \quad \textbf{Ans.}$$

EXAMPLE 6.30. Change $\int_0^\infty \int_0^\infty e^{-(x^2+y^2)}\, dydx$ to polar co-ordinates and evaluate it.

SOLUTION: The region of integration being first quadrant of XY-plane hence $dxdy = r\,drd\theta$, r will vary from 0 to ∞ and θ from 0 to $\pi/2$.

$$\text{Given integral} = \int_0^\infty \int_0^\infty e^{-(x^2+y^2)}\, dxdy = \int_0^{\pi/2} \int_0^\infty e^{-r^2} \cdot r\, drd\theta = \int_0^{\pi/2} \left[-\frac{1}{2}e^{-r^2}\right]_0^\infty d\theta$$

$$= \int_0^{\pi/2} \left[0 - \left(-\frac{1}{2}\right)\right] d\theta = \frac{\pi}{4}. \quad \textbf{Ans.}$$

EXAMPLE 6.31. Evaluate $\iint r\sin\theta\, drd\theta$ over the area of the cardiode $r = a(1 + \cos\theta)$ above the initial line.

SOLUTION: For the cardiode above the initial line θ varies from 0 to π.

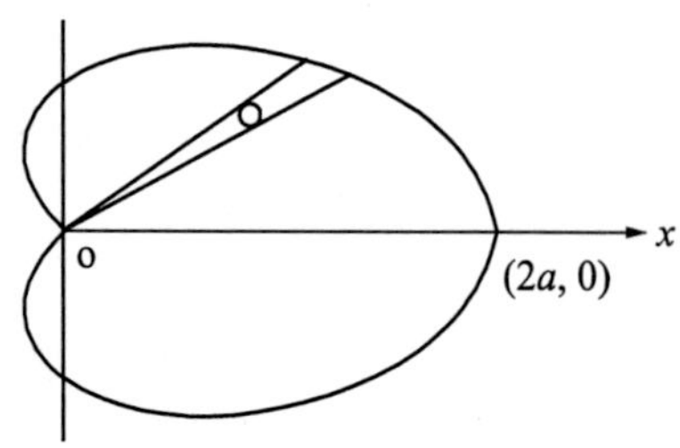

$$\text{Given integral} = \int_0^{\pi} \int_0^{a(1+\cos\theta)} r\sin\theta\, drd\theta$$

$$= \int_0^{\pi} \left[\frac{r^2}{2}\right]_0^{a(1+\cos\theta)} \sin\theta\, d\theta$$

$$= \int_0^{\pi} \frac{a^2}{2}(1+\cos\theta)^2 \sin\theta\, d\theta \quad \text{Putting } \cos\theta = t$$

$$= -\frac{a^2}{2}\int_1^{-1}(1+t)^2\, dt = \frac{a^2}{2}\int_{-1}^{1}(1+t^2+2t)\, dt = a^2\int_{-1}^{1}(1+t^2)\, dt + 0 = \frac{4a^2}{3}. \quad \textbf{Ans.}$$

EXERCISE 6B

1. Evaluate $\int_0^a \int_0^{\sqrt{a^2-x^2}} \sin\left(\frac{\pi}{a^2}\sqrt{a^2-x^2-y^2}\right) dx\, dy$.

2. Evaluate $\int_{-\infty}^{\infty}\int_{-\infty}^{\infty} \frac{dx\, dy}{(1+x^2+y^2)^{\frac{3}{2}}}$.

3. Find the value of $\iint \frac{x^2 y^2\, dx\, dy}{x^2+y^2}$ over the annular region between $x^2+y^2=a^2,\quad x^2+y^2=b^2\ (b>a>0)$.

4. Evaluate $\int_0^{4a}\int_{\frac{y^2}{4a}}^{y} \frac{x^2-y^2}{x^2+y^2}\, dy\, dx$.

5. Determine the value of $\int_0^a dx \int_{\sqrt{a^2-x^2}}^{\infty} \frac{x^{\frac{1}{3}}\, dx\, dy}{(x^2+y^2)^{\frac{3}{2}}}$.

6. Evaluate $\int_0^2 \int_{1-\sqrt{2x-x^2}}^{1+\sqrt{2x-x^2}} \frac{dy\, dx}{(x^2+y^2)^2}$.

7. Evaluate $\iint_R xy\, dx\, dy$ where R is the positive quadrant of the circle $x^2+y^2=a^2$.

(UPTU 2008-09)

8. Find $\int_0^{2a}\int_0^{\sqrt{2ax-x^2}} (x^2+y^2)\, dx\, dy$.

9. Evaluate $\int_0^{\infty}\int_0^{\infty} e^{-(x^2+y^2)}\, dx\, dy$ by transforming it into polar co-ordinates.

(UPTU 2001 Sp. Exam.) (UPTU 2002-03)

10. Evaluate $\int_0^1 \int_x^{\sqrt{2-x^2}} \frac{x\, dy\, dx}{\sqrt{x^2+y^2}}$.

11. Change the order of integration and evaluate $\int_0^a \int_{\sqrt{ax}}^{a} \frac{y^2 dx\, dy}{\sqrt{y^4-a^2x^2}}$.

12. Evaluate the integral $\int_0^{\infty}\int_0^{x} x \exp(-x^2/y) dy\, dx$.

❑❑❑

CHAPTER

7 Triple Integrals and Applications to Area and Volume

Applications of Double and Triple Integrals for Finding Area and Volume. Change of variables

TRIPLE INTEGRATION

In cartesian coordinates the triple integration of a function $f(x, y, z)$ defined over a closed parallelopiped, is expressed in the same manner as in the case of a double integral. Thus, we first obtain the sum

$$S = \sum_{i=1}^{n} f(x_i, y_j, z_k)\, \delta x\, \delta y\, \delta z$$

then the triple integral of $f(x, y, z)$ over a region V is the limiting value to which the sum S will tend to, as $n \to \infty$. Thus, $\iiint\limits_V f(x, y, z)\, dV = \lim\limits_{n\to\infty} \sum\limits_{i, j, k = 1}^{n} f(x_i, y_j, z_k)\, \delta x\, \delta y\, \delta z$

The triple integral

$$I = \int_{x=a}^{b} \int_{y=\phi_1(x)}^{\phi_2(x)} \int_{z=f_1(x,y)}^{f_2(x,y)} f(x, y, z)\, dx\, dy\, dz$$ is generally evaluated as follows.

$$I = \int_{x=a}^{b} \left[\int_{y=\phi_1(x)}^{\phi_2(x)} \left\{ \int_{z=f_1(x,y)}^{f_2(x,y)} f(x, y, z)\, dz \right\} dy \right] dx.$$

EXAMPLE 7.1. Evaluate $\int_0^a \int_0^x \int_0^{x+y} e^{x+y+z}\, dz\, dy\, dx$.

SOLUTION: We can write the given integral as

$$I = \int_0^a \int_0^x \int_0^{x+y} e^{x+y} . e^z\, dz\, dy\, dx = \int_0^a \int_0^x e^{x+y} \left[e^z\right]_0^{x+y} dy\, dx$$

$$= \int_0^a \int_0^x e^{x+y}(e^{x+y}-1)\,dy\,dx = \int_0^a \int_0^x (e^{2x}e^{2y} - e^x e^y)\,dy\,dx$$

$$= \int_0^a \left[e^{2x}\frac{e^{2y}}{2} - e^x.e^y \right]_{y=0}^{x} dx = \int_0^a \left[\frac{1}{2}e^{2x}(e^{2x}-1) - e^x(e^x-1) \right] dx$$

$$= \int_0^a \left[\frac{1}{2}e^{4x} - \frac{3}{2}e^{2x} + e^x \right] dx = \left[\frac{1}{8}e^{4x} - \frac{3}{4}e^{2x} + e^x \right]_0^a$$

$$= \frac{1}{8}(e^{4a}-1) - \frac{3}{4}(e^{2a}-1) + e^a - 1 = \frac{1}{8}e^{4a} - \frac{3}{4}e^{2a} + e^a - \frac{1}{8} + \frac{3}{4} - 1$$

Thus, $\quad I = \frac{1}{8}e^{4a} - \frac{3}{4}e^{2a} + e^a - \frac{3}{8}.$ **Ans.**

EXAMPLE 7.2. Evaluate $\displaystyle\int_0^{2a} \int_{-\sqrt{2ax-x^2}}^{\sqrt{2ax-x^2}} \int_0^{\sqrt{4a^2-x^2-y^2}} z\,dx\,dy\,dz.$

SOLUTION: $$I = \int_0^{2a} \int_{-\sqrt{2ax-x^2}}^{\sqrt{2ax-x^2}} \int_0^{\sqrt{4a^2-x^2-y^2}} z\,dz\,dy\,dx = \int_0^{2a} \int_{-\sqrt{2ax-x^2}}^{\sqrt{2ax-x^2}} \left[\frac{z^2}{2}\right]_0^{\sqrt{4a^2-x^2-y^2}} dy\,dx$$

$$= \int_0^{2a} \int_{-\sqrt{2ax-x^2}}^{\sqrt{2ax-x^2}} \frac{1}{2}[4a^2 - x^2 - y^2]\,dy\,dx = 2 \times \frac{1}{2}\int_0^{2a} \int_0^{\sqrt{2ax-x^2}} [4a^2 - x^2 - y^2]\,dy\,dx$$

$$= \int_0^{2a} \left[4a^2y - x^2y - \frac{y^3}{3} \right]_0^{\sqrt{2ax-x^2}} dx = \int_0^{2a} \left[(4a^2-x^2)\sqrt{2ax-x^2} - \frac{1}{3}(2ax-x^2)^{\frac{3}{2}} \right] dx$$

$$= \int_0^{2a} \sqrt{x(2a-x)} \left[4a^2 - x^2 - \frac{1}{3}(2ax-x^2) \right] dx$$

Putting $\quad x = 2a\sin^2\theta, \quad$ so $\quad dx = 4a\sin\theta\cos\theta\,d\theta, \quad$ we get

$$I = \int_0^{\pi/2} \sqrt{2a\sin^2\theta . 2a\cos^2\theta} \left[4a^2 - 4a^2\sin^4\theta - \frac{1}{3}(4a^2\sin^2\theta - 4a^2\sin^4\theta) \right] 4a\sin\theta\cos\theta\,d\theta$$

$$= 32a^4 \int_0^{\pi/2} \left[1 - \sin^4\theta - \frac{1}{3}\sin^2\theta(1-\sin^2\theta) \right] \sin^2\theta\cos^2\theta\,d\theta$$

$$= \frac{32}{3}a^4 \int_0^{\pi/2} \left[3(1+\sin^2\theta) - \sin^2\theta \right] \sin^2\theta\cos^4\theta\,d\theta$$

$$=\frac{32}{3}a^4\left[\int_0^{\pi/2}3\sin^2\theta\cos^4\theta\,d\theta+2\int_0^{\pi/2}\sin^4\theta\cos^4\theta\,d\theta\right]=\frac{32}{3}a^4\left[\frac{3.\Gamma\left(\frac{3}{2}\right).\Gamma\left(\frac{5}{2}\right)}{2\,\Gamma(4)}+\frac{2.\Gamma\left(\frac{5}{2}\right).\Gamma\left(\frac{5}{2}\right)}{2\,\Gamma(5)}\right]$$

$$=\frac{32}{3}a^4\left[\frac{3.\frac{3}{2}.\frac{1}{2}.\frac{1}{2}\pi}{12}+\frac{\frac{3}{2}.\frac{3}{2}.\frac{1}{2}.\frac{1}{2}\pi}{24}\right]=\frac{5\pi a^4}{4}.$$ **Ans.**

EXAMPLE 7.3. Evaluate $\int_0^1\int_0^{\sqrt{1-x^2}}\int_0^{\sqrt{1-x^2-y^2}}\frac{1}{\sqrt{1-x^2-y^2-z^2}}\,dz\,dy\,dx.$ **(UPTU 2003)**

SOLUTION: Given integral $=\int_0^1\int_0^{\sqrt{1-x^2}}\int_0^{\sqrt{1-x^2-y^2}}\frac{1}{\sqrt{(1-x^2-y^2-z^2)}}dzdydx$

$$=\int_0^1\int_0^{\sqrt{1-x^2}}\left[\sin^{-1}\frac{z}{\sqrt{1-x^2-y^2}}\right]_0^{\sqrt{1-x^2-y^2}}dydx$$

$$=\int_0^1\int_0^{\sqrt{1-x^2}}(\sin^{-1}1-\sin^{-1}0)dydx=\frac{\pi}{2}\int_0^1\sqrt{1-x^2}\,dx$$

$$=\frac{\pi}{2}\left[\frac{x}{2}\sqrt{1-x^2}+\frac{1}{2}\sin^{-1}x\right]_0^1=\frac{\pi}{2}\left[0+\frac{1}{2}\sin^{-1}1\right]=\frac{\pi}{2}.\left(\frac{\pi}{4}\right)=\frac{\pi^2}{8}.$$ **Ans.**

EXAMPLE 7.4. Evaluate $\iiint_R\frac{dx\,dy\,dz}{(x+y+z+1)^3}$ where R is the region bounded by the planes $x=0,\quad y=0,\ z=0$ and $x+y+z=1.$

SOLUTION: Given integal $=\int_0^1\int_0^{1-x}\int_0^{1-x-y}(x+y+z+1)^{-3}\,dz\,dy\,dx$

$$=\int_0^1\int_0^{1-x}\left[-\frac{1}{2}(x+y+z+1)^{-2}\right]_0^{1-x-y}dy\,dx=-\frac{1}{2}\int_0^1\int_0^{1-x}\left[\frac{1}{4}-\frac{1}{(x+y+1)^2}\right]dy\,dx$$

$$=-\frac{1}{2}\int_0^1\left[\frac{y}{4}+\frac{1}{x+y+1}\right]_0^{1-x}dx=-\frac{1}{2}\int_0^1\left(\frac{1-x}{4}+\frac{1}{2}-\frac{1}{1+x}\right)dx$$

$$=-\frac{1}{2}\left[\frac{x}{4}-\frac{x^2}{8}+\frac{x}{2}-\log(x+1)\right]_0^1=-\frac{1}{2}\left[\frac{1}{4}-\frac{1}{8}+\frac{1}{2}-\log 2\right]=\frac{1}{2}\left[\log 2-\frac{5}{8}\right].$$ **Ans.**

EXAMPLE. 7.5. Obtain the value of the triple integral $\iiint_V z^2\,dx\,dy\,dz$ over the volume V common to the sphere $x^2 + y^2 + z^2 = a^2$ and the cylinder $x^2 + y^2 = ax$.

SOLUTION: From the equation of the sphere we have $z = \pm\sqrt{a^2 - x^2 - y^2}$.

$$I = \iiint_V z^2\,dx\,dy\,dz = \iint\int_{-\sqrt{a^2-x^2-y^2}}^{\sqrt{a^2-x^2-y^2}} z^2\,dz\,dy\,dx = 2\iint\int_0^{\sqrt{a^2-x^2-y^2}} z^2\,dz\,dy\,dx$$

$$= 2\iint\left[\frac{z^3}{3}\right]_0^{\sqrt{a^2-x^2-y^2}} dy\,dx \quad = \frac{2}{3}\iint(a^2 - x^2 - y^2)^{\frac{3}{2}}\,dy\,dx.$$

Now x and y vary as per cylinder $x^2 + y^2 = ax$ or $r = a\cos\theta$. Converting to in polar co-ordinates, we get

$$I = \frac{2}{3}\int_{-(\pi/2)}^{\pi/2}\int_0^{a\cos\theta}(a^2 - r^2)^{\frac{3}{2}}\,r\,dr\,d\theta \quad \text{(now putting } a^2 - r^2 = t^2 \text{ so } -r\,dr = t\,dt)$$

$$= \frac{2}{3}\int_{-(\pi/2)}^{\pi/2}\int_a^{a\sin\theta}(-t^3.\,t\,dt)\,d\theta = -\frac{2}{3}\int_{-\pi/2}^{\pi/2}\left[\frac{t^5}{5}\right]_a^{a\sin\theta} d\theta = -\frac{2}{15}\int_{-(\pi/2)}^{\pi/2}[a^5\sin^5\theta - a^5]\,d\theta$$

$$= 0 + \frac{2}{15}a^5\,[\theta]_{-(\pi/2)}^{\pi/2} = \frac{2\pi a^5}{15}. \quad \text{(on using property of definite integrals)} \quad \textbf{Ans.}$$

EXAMPLE 7.6. Evaluate $\int_0^{\pi/2}\int_0^{a\sin\theta}\int_0^{(a^2-r^2)/a} r\,dz\,dr\,d\theta$.

SOLUTION: $$I = \int_0^{\pi/2}\int_0^{a\sin\theta}\int_0^{(a^2-r^2)/a} r\,dz\,dr\,d\theta$$

$$= \int_0^{\pi/2}\int_0^{a\sin\theta} r\,[z]_0^{(a^2-r^2)/a}\,dr\,d\theta = \int_0^{\pi/2}\int_0^{a\sin\theta} r\,\frac{(a^2 - r^2)}{a}\,dr\,d\theta$$

$$= \frac{1}{a}\int_0^{\pi/2}\left[a^2\frac{r^2}{2} - \frac{r^4}{4}\right]_0^{a\sin\theta} d\theta = \frac{a^3}{2}\int_0^{\pi/2}\sin^2\theta\,d\theta - \frac{a^3}{4}\int_0^{\pi/2}\sin^4\theta\,d\theta$$

$$= \frac{a^3}{2}\cdot\frac{1}{2}\int_0^{\pi/2} d\theta - \frac{a^3}{4}\cdot\frac{3}{4}\cdot\frac{1}{2}\int_0^{\pi/2} d\theta$$

$$= \frac{a^3\,\pi}{8} - \frac{3a^3\,\pi}{64} = \frac{5\pi\,a^3}{64}. \quad \textbf{Ans.}$$

CYLINDRICAL POLAR CO-ORDINATES

The use of cylindrical co-ordinates r, θ, z is particularly suited for problems in which there is an axis of symmetry of the solid. In place of the volume element $dx\,dy\,dz$ we use $r\,d\theta\,dr\,dz$ as element with cross-sectional area $r\,dr\,d\theta$ and height dz.

EXAMPLE 7.7. If a denotes the radius of the base and h the altitude of a right circular cone, express its volume as a triple integral and evaluate it using cylindrical co-ordinates.

SOLUTION: With cylindrical co-ordinates r, θ, z we have, from the adjoing figure

$$\frac{z}{r} = \frac{h}{a} \quad \text{or} \quad z = \frac{rh}{a}$$

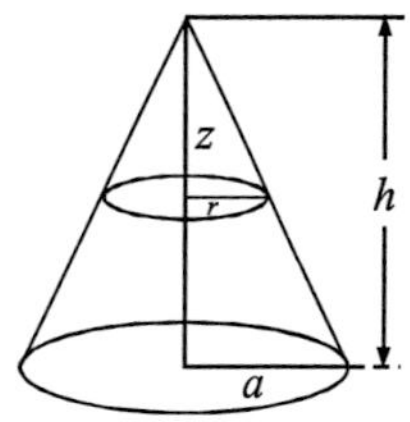

Hence the volume of the given come is

$$V = \int_{\theta=0}^{2\pi} \int_{r=0}^{a} \int_{z=\frac{rh}{a}}^{h} r\,dr\,d\theta\,dz = \int_0^{2\pi}\int_0^a [z]_{\frac{rh}{a}}^{h}\, r\,dr\,d\theta$$

$$= \int_0^{2\pi}\int_0^a \left(h - \frac{rh}{a}\right) r\,dr\,d\theta = h\int_0^{2\pi}\left[\frac{r^2}{2} - \frac{r^3}{3a}\right]_0^a d\theta = h\int_0^{2\pi}\left(\frac{a^2}{2} - \frac{a^3}{3a}\right)d\theta$$

$$= \pi\, a^2 h\left(-\frac{2}{3}+1\right) = \pi\, a^2 h/3. \quad \textbf{Ans.}$$

SPHERICAL POLAR CO-ORDIANTES

If there is symmetry about a point in any problem, it would be quite convenient to use spherical co-ordinates given by

$$\boldsymbol{x = r \sin\theta \cos\phi}$$

$$\boldsymbol{y = r \sin\theta \sin\phi}$$

$$\boldsymbol{z = r \cos\theta}$$

The volume element in spherical polar co-ordinates is

$$dV = dr \cdot r\,d\theta \cdot r\sin\theta\,d\phi = r^2 \sin\theta\,dr\,d\theta\,d\phi.$$

EXAMPLE 7.8. Evaluate $\iiint xyz\,dx\,dy\,dz$ over the positive octant of the sphere $x^2+y^2+z^2=a^2$.

SOLUTION: $$I = \iiint xyz\,dx\,dy\,dz = \int_{\theta=0}^{\pi/2}\int_{\phi=0}^{\pi/2}\int_{r=0}^{a} r\sin\theta\cos\phi\, r\sin\theta\sin\phi\, r\cos\theta \,.\, r^2\sin\theta\;\, dr\,d\theta\,d\phi$$

$$= \int_0^{\pi/2}\int_0^{\pi/2}\int_0^a r^5 \cos\phi\sin\phi\sin^3\theta\cos\theta\;\, dr\,d\phi\,d\theta$$

$$= \left[\frac{r^6}{6}\right]_0^a \int_0^{\pi/2}\cos\phi\sin\phi\,d\phi \int_0^{\pi/2}\sin^3\theta\cos\theta\,d\theta$$

$$= \frac{a^6}{6}\cdot\left[\frac{\sin^2\phi}{2}\right]_0^{\pi/2}\left[\frac{\sin^4\theta}{4}\right]_0^{\pi/2} = \frac{a^6}{48}. \quad \textbf{Ans.}$$

EXAMPLE 7.9. Evaluate the following integral, by changing to spherical polar co-ordinates,

$$\int_0^1 \int_0^{\sqrt{1-x^2}} \int_0^{\sqrt{1-x^2-y^2}} \frac{dxdydz}{\sqrt{1-x^2-y^2-z^2}}.$$ **(UPTU 2002-03)**

SOLUTION: The spherical polar co-ordinates are given by

$$x = r\sin\theta\cos\phi,\quad y = r\sin\theta\sin\phi,\quad z = r\cos\theta.$$

Here x, y, z are all positive, hence in the first octant only.

Also, we have $dxdydz = r^2\sin\theta\, drd\theta d\phi$.

$$\therefore \quad \text{Given integral} = \int_{\theta=0}^{\pi/2}\int_{\phi=0}^{\pi/2}\int_0^1 \frac{r^2\sin\theta\, drd\theta d\phi}{\sqrt{1-r^2}} = \int_{\phi=0}^{\pi/2} d\phi \int_0^{\pi/2}\sin\theta\, d\theta \int_0^1 \frac{r^2 dr}{\sqrt{1-r^2}}$$

$$= \frac{\pi}{2}\cdot 1\cdot \int_0^1 \frac{1-(1-r^2)\,dr}{\sqrt{1-r^2}} = \frac{\pi}{2}\int_0^1 \frac{dr}{\sqrt{1-r^2}} - \frac{\pi}{2}\int_0^1 \sqrt{1-r^2}\, dr$$

$$= \frac{\pi}{2}\left[\sin^{-1} r\right]_0^1 - \frac{\pi}{2}\left[\frac{r}{2}\sqrt{1-r^2} + \frac{1}{2}\sin^{-1} r\right]_0^1 = \frac{\pi}{2}\cdot\frac{\pi}{2} - \frac{\pi}{4}\cdot\frac{\pi}{2} = \frac{\pi^2}{8}.$$ **Ans.**

APPLICATIONS OF DOUBLE INTEGRALS

I. Area By Double Integration

(a) CARTESIAN CO-ORDINATES. Let us find the area enclosed between the curves $y = f_1(x)$ and $y = f_2(x)$ and the ordinates $x = a$, $x = b$.

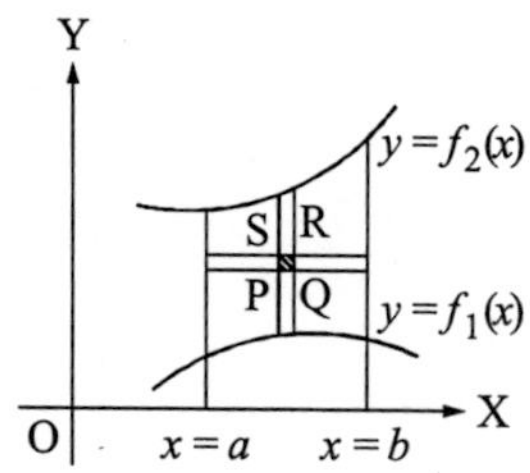

See adjoing figure. Let the area be divided into rectangular elements of the type PQRS of area $\delta x\, \delta y$ where $P \equiv (x, y)$ and $R \equiv (x + \delta x, y + \delta y)$. Now, when we move this element parallel to y-axis from $y = f_1(x)$ to $y = f_2(x)$, we get a vertical strip of area

$$\delta x \sum_{y=f_1(x)}^{y=f_2(x)} \delta y = \delta x \int_{y=f_1(x)}^{f_2(x)} dy$$

Adding up all such strips between the extreme ordinates at $x = a$ and $x = b$, we get the required area equal to

$$\lim_{\delta x \to 0} \sum_a^b \delta x \int_{f_1(x)}^{f_2(x)} dy = \int_a^b dx \int_{f_1(x)}^{f_2(x)} dy = \int_a^b \int_{f_1(x)}^{f_2(x)} dy\, dx$$

Similarly, the area enclosed between the curves $x = f_1(y)$ and $x = f_2(y)$ and the extreme abscissas $y = c$ and $y = d$ can be written as $\displaystyle\int_c^d \int_{f_1(y)}^{f_2(y)} dx\, dy$.

(b) POLAR CO-ORDINATES. As shown in the adjoining figure, the area of the element PQRS $= r\,\delta\theta\,\delta r$ (shaded). If the total area is divided into such curvelinear rectangles then the sum $\Sigma\Sigma\, r\,\delta\theta\,\delta r$ taken over all these rectangles gives the limiting value of area A, given by

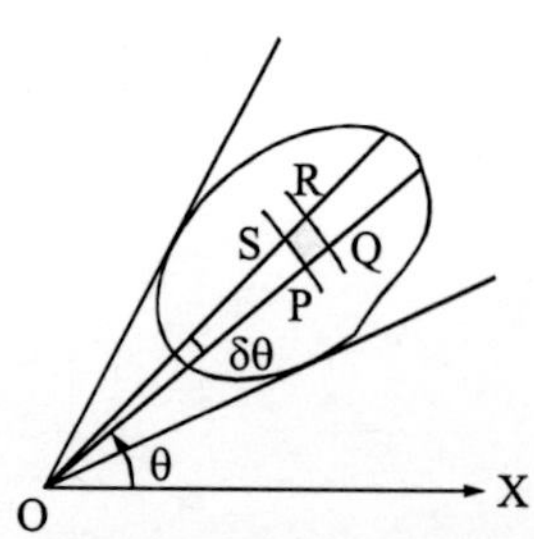

$$A = \Sigma\Sigma r\,\delta\theta\,\delta r = \iint r\, d\theta\, dr$$

where limits are to be chosen so as to cover the entire area.

EXAMPLE 7.10. Find the area bounded by the curve $(2a - x)\, y^2 = x^3$ and its asymptote using double integration concept.

SOLUTION: The shape of the curve is as shown in the adjoining figure. Curve $y^2\,(2a - x) = x^3$ has an asymptote parallel to y-axis at $x = 2a$.

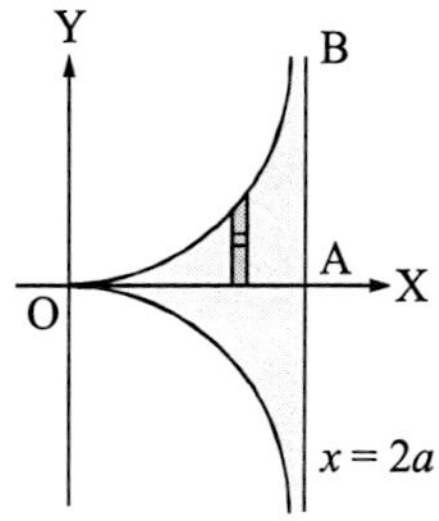

Required area, because of symmetry about the X-axis, is equal to

$$2 \iint\limits_{OABO} dx\, dy = 2 \int_0^{2a} \int_0^{\frac{x^{3/2}}{\sqrt{2a-x}}} dy\, dx = 2 \int_0^{2a} [y]_0^{\frac{x^{3/2}}{\sqrt{2a-x}}} dx = 2 \int_0^{2a} \frac{x^{3/2}\, dx}{\sqrt{2a-x}}$$

$$= 2 \int_0^{\pi/2} \frac{(2a \sin^2 \theta)^{3/2}}{\sqrt{2a - 2a \sin^2 \theta}}\, 4a \sin\theta \cos\theta\, d\theta \quad (\text{on putting } x = 2a \sin^2 \theta)$$

$$= 16a^2 \int_0^{\pi/2} \frac{\sin^3 \theta}{\cos\theta} \sin\theta \cos\theta\, d\theta = 16a^2 \int_0^{\pi/2} \sin^4 \theta\, d\theta$$

$$= 16a^2\, \frac{3}{4} \cdot \frac{1}{2} \cdot \frac{\pi}{2} \quad (\text{applying reduction formula}) = 3\pi\, a^2. \quad \textbf{Ans.}$$

EXAMPLE 7.11. **(a)** Find the area of the loop of the curve $y^2 = (x - a)\,(b - x)$, $0 < a < b$.

(b) Evaluate the area enclosed between the parabola $y = x^2$ and the straight line $y = x$. **[UPTU 2011-12]**

SOLUTION: **(a)** The given curve is symmetrical about x-axis and extends from $x = a$ to $x = b$. The shaded portion, as shown in the figure, is the required area

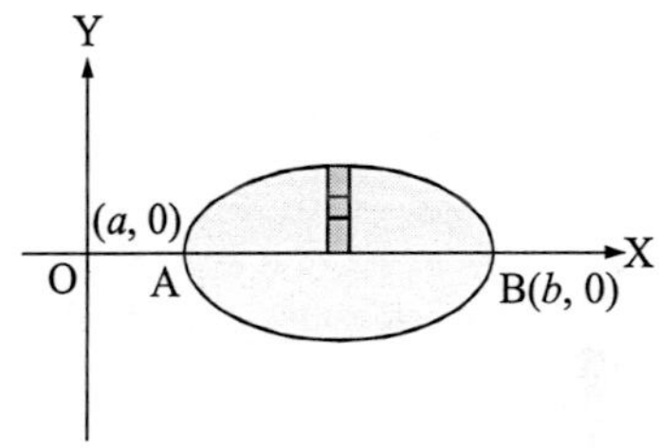

$$= 2 \int_a^b \int_0^{\sqrt{(x-a)(b-x)}} dx\, dy = 2 \int_a^b \sqrt{(x-a)(b-x)}\, dx$$

$$= 2 \int_a^b \sqrt{(b-a) \sin^2 t\, (b-a) \cos^2 t} \cdot 2\,(b-a) \sin t \cos t\, dt$$

(on putting $x = a \cos^2 t + b \sin^2 t$, so that $dx = 2\,(b - a) \sin t \cos t\, dt$)

$$\therefore \quad \text{The required area} = 4\,(b-a)^2 \int_0^{\pi/2} \sin^2 t \cos^2 t\, dt = 4\,(b-a)^2\, \frac{\Gamma\frac{3}{2} \cdot \Gamma\frac{3}{2}}{2\,\Gamma 3}$$

$$= 4\,(b-a)^2\, \frac{\left(\frac{1}{2} \cdot \sqrt{\pi}\right)^2}{2.2} = \frac{\pi}{4}(b-a)^2. \quad \textbf{Ans.}$$

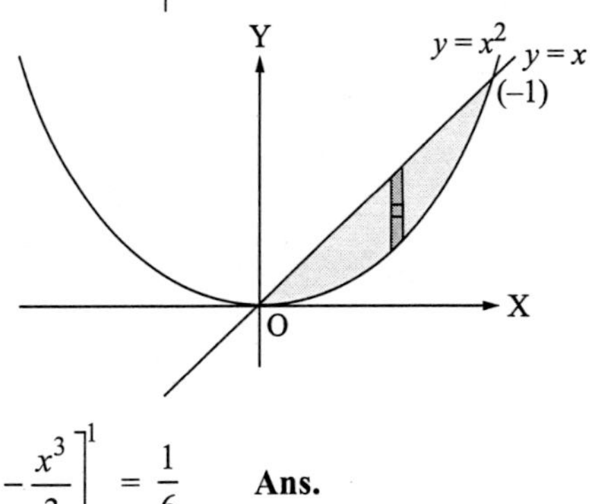

(b) The points of intersection of $y = x^2$ and $y = x$ are $(0, 0)$ and $(1, 1)$.

$$\text{Required area} = \int_{x=0}^{1} \int_{y=x^2}^{x} dy\, dx$$

$$= \int_0^1 [y]_{x^2}^{x}\, dx = \int_0^1 (x - x^2)\, dx = \left[\frac{x^2}{2} - \frac{x^3}{3}\right]_0^1 = \frac{1}{6} \quad \textbf{Ans.}$$

EXAMPLE 7.12. Find the area of the loop of the curve $x(x^2 + y^2) = a(x^2 - y^2)$ using double integration.

SOLUTION: The equation of the curve can be written as $y^2 = x^2 \dfrac{(a-x)}{a+x}$

The curve is symmetrical about x-axis. The shape of the curve is as shown in the adjoining figure. The required area A of the loop is equal to

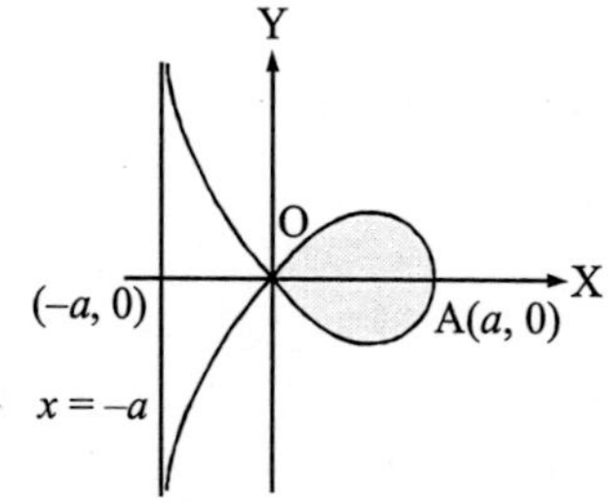

$$A = 2\int_0^a \int_0^{x\sqrt{\frac{a-x}{a+x}}} dy\, dx = 2\int_0^a x\sqrt{\frac{a-x}{a+x}}\, dx$$

$$= 2\int_0^a \frac{x(a-x)}{\sqrt{a^2 - x^2}}\, dx \qquad \text{(putting } x = a \sin\theta)$$

$$A = 2\int_0^{\pi/2} \frac{a\sin\theta\,(a - a\sin\theta)\, a\cos\theta\, d\theta}{\sqrt{a^2 - a^2\sin^2\theta}} = 2a^2 \int_0^{\pi/2} (\sin\theta - \sin^2\theta)\, d\theta$$

$$= 2a^2 \left[-\cos\theta\right]_0^{\pi/2} - a^2 \int_0^{\pi/2} (1 - \cos 2\theta)\, d\theta$$

$$= 2a^2(0+1) - a^2\left[\theta - \frac{1}{2}\sin 2\theta\right]_0^{\pi/2} = a^2\left(2 - \frac{\pi}{2}\right). \quad \textbf{Ans.}$$

EXAMPLE 7.13. Show that the area between one arch of the cycloid

$x = a(\theta + \sin\theta)$, $y = a(1 - \cos\theta)$ and its base is $3\pi a^2$.

SOLUTION: Required area A is the shaded portion as shown in the figure. By symmetry,

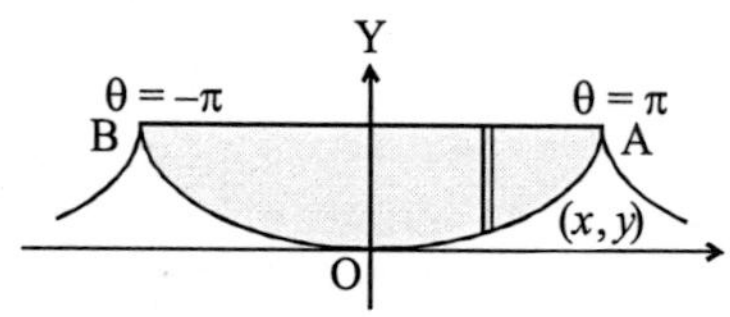

$$\text{the required area} = (\text{area AOBA}) = 2\int_0^{a\pi} \int_{a(1-\cos\theta)}^{2a} dy\, dx$$

$$= 2\int_0^{a\pi} \left[y\right]_{a(1-\cos\theta)}^{2a} dx$$

$$= 2\int_0^{a\pi} [2a - a(1 - \cos\theta)]\, dx \qquad (\text{here } dx = a(1 + \cos\theta)\, d\theta)$$

$$= 2a\int_0^{\pi} (1 + \cos\theta)\, a(1 + \cos\theta)\, d\theta$$

$$= 2a^2 \int_0^{\pi} \left(2\cos^2\frac{\theta}{2}\right)^2 d\theta \qquad \left(\text{Putting now } \frac{\theta}{2} = t\right)$$

$$= 2a^2 \int_0^{\pi/2} 4\cos^4 t \,.\, 2dt = 16 \cdot a^2 \frac{3}{4}\cdot\frac{1}{2}\cdot\frac{\pi}{2} = 3\pi a^2. \qquad \textbf{Hence Proved.}$$

EXAMPLE 7.14. Find the area enclosed by the leminiscate $r^2 = a^2 \cos 2\theta$ using double integration.

SOLUTION: The area as shown in the adjoining figure is symmetrical in all the four quadrants.

Therefore, the required area

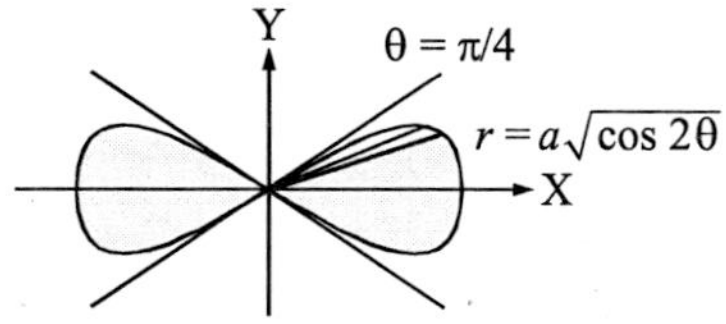

$$= 4\int_0^{\pi/4}\int_0^{a\sqrt{\cos 2\theta}} r\,dr\,d\theta = 4\int_0^{\pi/4}\left[\frac{r^2}{2}\right]_0^{a\sqrt{\cos 2\theta}} d\theta$$

$$= 2\int_0^{\pi/4} a^2 \cos 2\theta\,d\theta = 2a^2\left[\frac{1}{2}\sin 2\theta\right]_0^{\pi/4} = a^2. \quad \textbf{Ans.}$$

EXAMPLE 7.15. Find the area lying inside the circle $x^2 + y^2 - 2ax = 0$ and out side the circle $x^2 + y^2 = a^2$ using double integrtion.

SOLUTION: The centres of the two given circles are $(a, 0)$ and $(0, 0)$ and radii of both of them is a. The required area is the shaded portion as shown in the adjacent figure.

The two circles are $r = a$ and $r = 2a \cos\theta$ in polar co-ordinates and intersect in points B and B′ given by

$$2a\cos\theta = a \quad \text{or} \quad \cos\theta = \frac{1}{2} \quad \therefore\ \theta = \pm\frac{\pi}{3}$$

By symmetry the required area is equal to

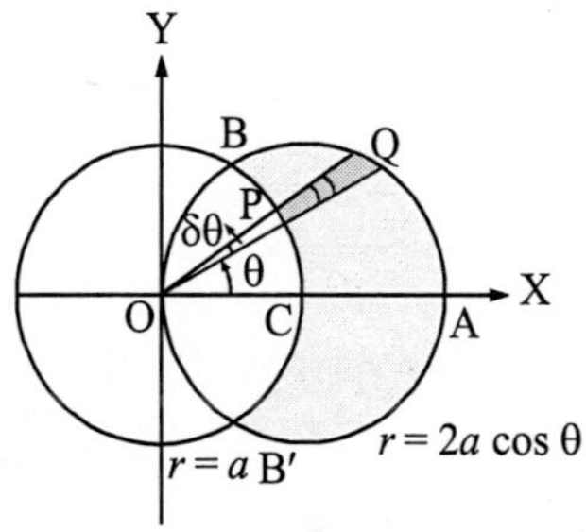

$$= 2\iint_{CABC} r\,dr\,d\theta = 2\int_0^{\pi/3}\int_a^{2a\cos\theta} r\,dr\,d\theta = 2\int_0^{\pi/3}\left[\frac{r^2}{2}\right]_a^{2a\cos\theta} d\theta$$

$$= \int_0^{\pi/3}(4a^2\cos^2\theta - a^2)\,d\theta = a^2\int_0^{\pi/3}[2(1+\cos 2\theta) - 1]\,d\theta$$

$$= a^2\left[\theta + \frac{2\sin 2\theta}{2}\right]_0^{\pi/3} = a^2\left[\frac{\pi}{3} + \sin\frac{2\pi}{3}\right] = a^2\left(\frac{\pi}{3} + \frac{\sqrt{3}}{2}\right). \quad \textbf{Ans.}$$

EXAMPLE 7.16. Find the whole area enclosed by the lamicon $r = a + b\cos\theta,\ a > b$.

SOLUTION: The curve is symmetrical about the x-axis as shown in the figure, hence the required area

$$= 2\int_0^{\pi}\int_0^{a+b\cos\theta} r\,dr\,d\theta = 2\int_0^{\pi}\left[\frac{r^2}{2}\right]_0^{a+b\cos\theta} d\theta$$

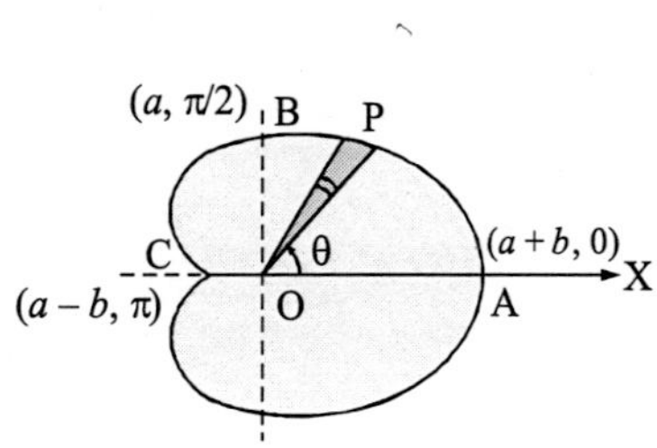

$$= \int_0^{\pi}(a + b\cos\theta)^2\,d\theta = \int_0^{\pi}(a^2 + 2ab\cos\theta + b^2\cos^2\theta)\,d\theta$$

$$= [a^2\theta + 2ab\sin\theta]_0^{\pi} + \frac{b^2}{2}\int_0^{\pi}(1 + \cos 2\theta)\,d\theta$$

$$= a^2\pi + 0 + \frac{b^2}{2}\left[\theta + \frac{1}{2}\sin 2\theta\right]_0^{\pi} = \pi\left(a^2 + \frac{b^2}{2}\right). \quad \textbf{Ans.}$$

EXAMPLE 7.17. Find area enclosed by the leaves of the following curves

(*i*) $r = a \sin 3\theta$. (*ii*) $r = a \cos 4\theta$.

SOLUTION: (*i*) We know that $r = a \sin n\,\theta$ has n leaves if n is odd and $2n$ leaves when n is even. Therefore $r = a \sin 3\theta$ has three symmetrical leaves and one leaf extends from $\theta = 0$ to $\theta = \frac{\pi}{3}$, as shown in the adjoing figure.

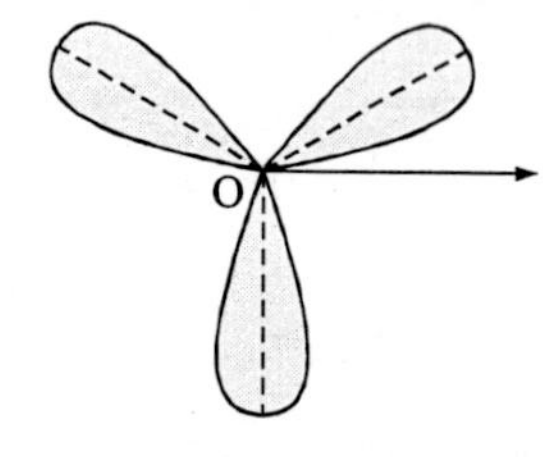

∴ Required area

$$= 3\int_0^{\pi/3}\int_0^{a\sin 3\theta} r\,d\theta\,dr = 3\int_0^{\pi/3}\left[\frac{r^2}{2}\right]_0^{a\sin 3\theta} d\theta = \frac{3}{2}\int_0^{\pi/3} a^2 \sin^2 3\theta\,d\theta$$

$$= \frac{3a^2}{4}\int_0^{\pi/3}(1-\cos 6\,\theta)\,d\theta = \frac{3a^2}{4}\left[\theta - \frac{1}{6}\sin 6\theta\right]_0^{\pi/3}$$

$$= \frac{3a^2}{4}\left[\frac{\pi}{3} - \frac{1}{6}\sin 2\pi\right] = \frac{\pi a^2}{4}. \quad \textbf{Ans.}$$

(*ii*) We know that $r = a \cos n\,\theta$ has n leaves if n is odd and $2n$ leaves when n is even.

Thus, $r = a \cos 4\,\theta$ will have eight symmetrical leaves. One leaf extends from $\theta = -\frac{\pi}{8}$ to $\theta = \frac{\pi}{8}$.

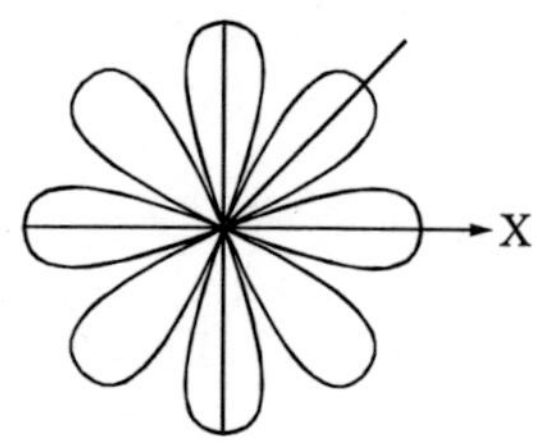

$$\text{Therefore, required area} = 8\int_{-(\pi/8)}^{\pi/8}\int_0^{a\cos 4\theta} r\,dr\,d\theta$$

$$= 8\int_{-(\pi/8)}^{\pi/8}\left[\frac{r^2}{2}\right]_0^{a\cos 4\theta} d\theta = \frac{8}{2}\int_{-(\pi/8)}^{\pi/8} a^2\cos^2 4\theta\,d\theta$$

$$= 4a^2\int_0^{\pi/8}(1+\cos 8\theta)\,d\theta = 4a^2\left[\theta + \frac{1}{8}\sin 8\theta\right]_0^{\pi/8}$$

$$= \frac{\pi a^2}{2}. \quad \textbf{Ans.}$$

EXAMPLE 7.18. Find the area bounded by the parabolas $y^2 = a - x$ and $y^2 = a - 4x$.

SOLUTION: The required area shown as shaded area, is

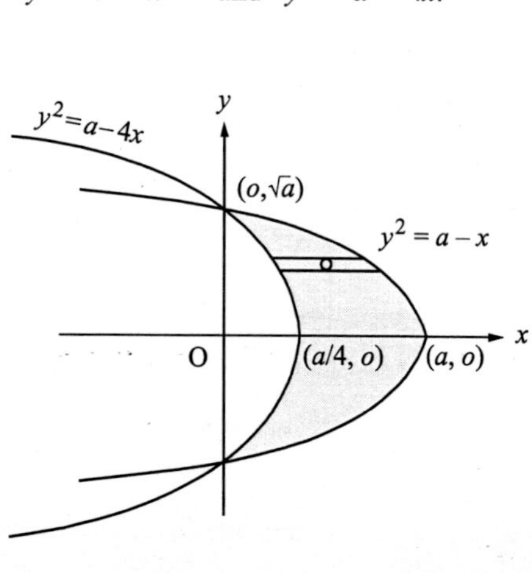

$$\int_{-\sqrt{a}}^{\sqrt{a}}\int_{(a-y^2)/4}^{a-y^2} dx\,dy = \int_{-\sqrt{a}}^{\sqrt{a}}\left[(a-y^2) - \frac{1}{4}(a-y^2)\right]dy$$

$$= 2\cdot\frac{3}{4}\int_0^{\sqrt{a}}(a-y^2)\,dy$$

$$= \frac{3}{2}\left[ay - \frac{y^3}{3}\right]_0^{\sqrt{a}}$$

$$= \frac{3}{2}\left[a\sqrt{a} - \frac{a\sqrt{a}}{3}\right] = a\sqrt{a}. \quad \textbf{Ans.}$$

EXAMPLE 7.19. Find the area of the region bounded by the curves

$$xy = 2,\ 4y = x^2,\quad y = 4.$$ **(UPTU 2002-03, 2009-10)**

SOLUTION: The shaded portion in the figure is the area of integration.

$$\text{Required area} = \int_1^4 \int_{2/y}^{2\sqrt{y}} dx\, dy = \int_1^4 \left(2\sqrt{y} - \frac{2}{y}\right) dy$$

$$= \left[2 \cdot \frac{2}{3} y^{3/2} - 2\log y\right]_1^4$$

$$= \frac{4}{3}(8-1) - 2\log 4 = \frac{28}{3} - 4\log 2.$$ **Ans.**

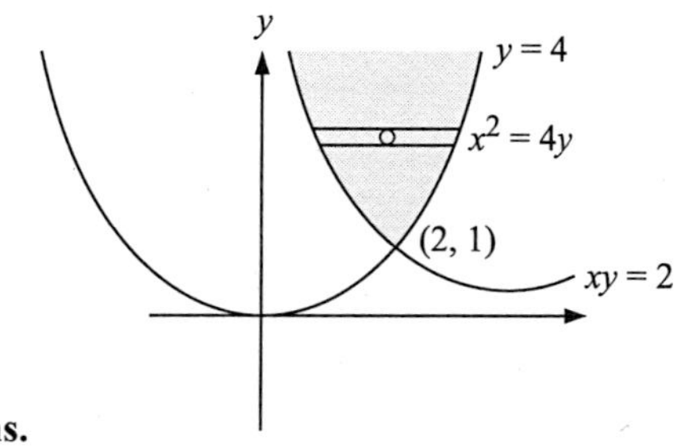

II. Volumes of solids by Double Integration

Consider a surface $z = f(x, y)$. Let S be a portion of this surface and S′ be its projection on the XY plane. We wish to show that the volume of the solid bounded by the surface $z = f(x, y)$ and the surrounding surface can be expressed as a double integral. Divide S′ into elementary rectangles of area $\delta x\, \delta y$ by drawing lines parallel to OX and OY. On each of these rectangles as base, erect a prism with its length parallel to z-axis. Volume of this typical elementary prism between S′ and the given surface $z = f(x, y)$, is $\delta V = z\, \delta x\, \delta y = f(x, y)\, \delta x\, \delta y$

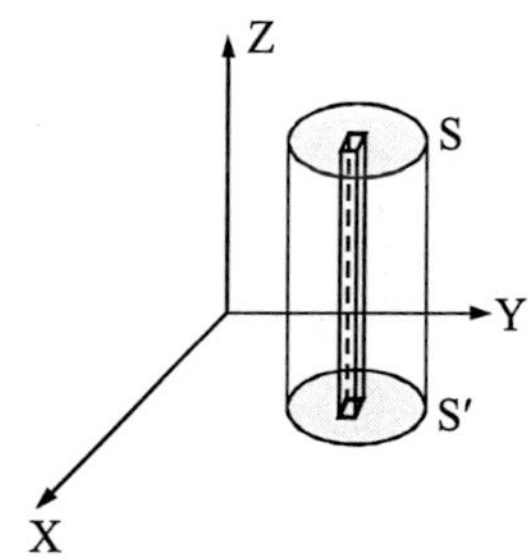

Therefore, the total volume is given by

$$V = \iint_{S'} f(x, y)\, dx\, dy$$

where S′ is the region of integration.

EXAMPLE 7.20. Obtain the volume bounded by the surface $z = c\left(1 - \frac{x}{a}\right)\left(1 - \frac{y}{b}\right)$ and the quadrant of the elliptic cylinder $\frac{x^2}{a^2} + \frac{y^2}{b^2} = 1.$

SOLUTION: In solid geometry $\frac{x^2}{a^2} + \frac{y^2}{b^2} = 1$ represents a cylinder whose axis is along z-axis and guiding curve is ellipse. Required volume V is given by

$$V = \iint z\, dx\, dy = \iint_C c\left(1 - \frac{x}{a}\right)\left(1 - \frac{y}{b}\right) dx\, dy \quad \text{where} \quad C: \frac{x^2}{a^2} + \frac{y^2}{b^2} = 1.$$

Let us use elliptic polar co-ordinates $x = a\, r\cos\theta,\ y = br\sin\theta$ so that $dx\, dy = ab\, r\, d\theta\, dr$

and $\frac{x^2}{a^2} + \frac{y^2}{b^2} = r^2$, hence

$$V = abc \int_0^{\pi/2} \int_0^1 (1 - r\cos\theta)(1 - r\sin\theta)\, r\, d\theta\, dr$$

$$= abc \int_0^{\pi/2} \int_0^1 [r - r^2 (\cos\theta + \sin\theta) + r^3 \cos\theta \sin\theta]\, dr\, d\theta$$

$$= abc \int_0^{\pi/2} \left[\frac{r^2}{2} - \frac{r^3}{3} (\cos\theta + \sin\theta) + \frac{r^4}{4} \sin\theta \cos\theta \right]_0^1 d\theta$$

$$= abc \int_0^{\pi/2} \left[\frac{1}{2} - \frac{1}{3} (\cos\theta + \sin\theta) + \frac{1}{4} \sin\theta \cos\theta \right] d\theta$$

$$= abc \left[\frac{\theta}{2} - \frac{1}{3} \sin\theta + \frac{1}{3} \cos\theta + \frac{1}{8} \sin^2\theta \right]_0^{\pi/2}$$

$$= abc \left[\frac{\pi}{4} - \frac{1}{3} + 0 + \frac{1}{8} - \frac{1}{3} \right] = abc \left(\frac{\pi}{4} - \frac{13}{24} \right).$$ **Ans.**

EXAMPLE 7.21. A triangular prism is formed by the planes $ay = bx, y = 0, x = a$. Find the volume of the prism between the planes $z = 0$ and the surface $z = c + xy$.

(UPTU 2002-03, 2009-10)

SOLUTION: Required volume $= \int_0^a \int_0^{bx/a} \int_0^{c+xy} dz\, dy\, dx$

$$= \int_0^a \int_0^{bx/a} (c + xy)\, dy\, dx = \int_0^a \left[cy + x\frac{y^2}{2} \right]_0^{bx/a} dx$$

$$= \int_0^a \left[c \cdot \frac{bx}{a} + \frac{x}{2} \cdot \frac{b^2 x^2}{a^2} \right] dx = \frac{bc}{a} \left[\frac{x^2}{2} \right]_0^a + \frac{b^2}{2a^2} \left[\frac{x^4}{4} \right]_0^a$$

$$= \frac{abc}{2} + \frac{b^2 a^2}{8} = \frac{ab}{8}(4c + ab).$$ **Ans.**

EXAMPLE 7.22. Find the volume common to the cylinders

$x^2 + y^2 = a^2$ and $x^2 + z^2 = a^2$. **(UPTU 2001 Q.B.; UPTU 2008-09)**

SOLUTION: Here z varies from $-\sqrt{a^2 - x^2}$ to $\sqrt{a^2 - x^2}$ and y also varies from $-\sqrt{a^2 - x^2}$ to $\sqrt{a^2 - x^2}$ and x varies from $-a$ to a.

$$\text{Required volume} = \int_{-a}^{a} \int_{-\sqrt{a^2 - x^2}}^{\sqrt{a^2 - x^2}} \int_{-\sqrt{a^2 - x^2}}^{\sqrt{a^2 - x^2}} dz\, dy\, dx$$

$$= \int_{-a}^{a} \int_{-\sqrt{a^2 - x^2}}^{\sqrt{a^2 - x^2}} 2\sqrt{a^2 - x^2}\, dy\, dx = \int_{-a}^{a} 2\sqrt{a^2 - x^2} \cdot 2\sqrt{a^2 - x^2}\, dx$$

$$= 8\int_0^a (a^2 - x^2)\, dx = 8\left[a^2 x - \frac{x^3}{3} \right]_0^a = \frac{16}{3} a^3.$$ **Ans.**

EXAMPLE 7.23. Find the volume bounded above by the sphere $x^2 + y^2 + z^2 = a^2$ and below by the cone $x^2 + y^2 = z^2$.

(UPTU 2001 Sp. Exam.)

SOLUTION: Here z varies from $\sqrt{x^2+y^2}$ to $\sqrt{a^2-x^2-y^2}$.

$$\text{Required volume} = \iint_A \int_{\sqrt{x^2+y^2}}^{\sqrt{a^2-x^2-y^2}} dz\, dx\, dy$$

The surfaces $x^2 + y^2 + z^2 = a^2$ and $x^2 + y^2 = z^2$ intersect on the cylinder $x^2+y^2+x^2+y^2=a^2$ or $x^2+y^2=a^2/2$.

Let us convert to polar co-ordinates for convenience. The area A is enclosed by $r = 0,\ r = a/\sqrt{2};\ \theta = 0,\ \theta = 2\pi$.

$$\therefore \quad \text{Volume} = \int_0^{2\pi}\int_0^{a/\sqrt{2}}\int_r^{\sqrt{a^2-r^2}} dz\ (r\,dr\,d\theta)$$

$$= \int_0^{2\pi}\int_0^{a/\sqrt{2}} \left(\sqrt{a^2-r^2}-r\right) r\,dr\,d\theta = \int_0^{2\pi}\left[-\frac{1}{3}(a^2-r^2)^{3/2}-\frac{r^3}{3}\right]_0^{a/\sqrt{2}} d\theta$$

$$= \frac{1}{3}\int_0^{2\pi}\left[-\left(\frac{a^2}{2}\right)^{3/2}+a^3-\frac{a^3}{2\sqrt{2}}\right]d\theta = \frac{2\pi}{3}a^3\left(1-\frac{1}{\sqrt{2}}\right) = \frac{\pi a^3}{3}\left(2-\sqrt{2}\right). \qquad \textbf{Ans.}$$

III. Volumes of Solids as Triple Integral

To express the volume of a solid as triple integral we note that the volume of an elementary cuboid with its faces parallel to the co-ordinate planes is $dx\, dy\, dz$ and therefore, the volume of the solid is given by

$$V = \iiint dx\, dy\, dz$$

where the limits of integration w.r.t. z (if we integrate first w.r.t. z) are z_1 and z_2 obtained from the equation of the top and the bottom of the given surface, and then the double integration is w.r.t. x and y carried out over the area of projection of the given solid on the XY-plane.

IV. Volume of Solid of Revolution

Consider the area A bounded by $y = f(x)$, the ordinates $x = a$ and $x = b$ and the x-axis. Let this area revolve about x-axis then a solid is generated whose volume is to be determined. Consider an elementary area. PQRS where $P \equiv (x, y)$ and $R\ (x + \delta x, y + \delta y)$. The area of the rectangle is $\delta x\ \delta y$. When revolved about x-axis the elementary area generates an elementary solid of volume $\pi\,(y + \delta y)^2\,\delta x - \pi y^2\,\delta x$

$$= \pi\,[(y+\delta y)^2 - y^2]\,\delta x = \pi\,[2y\,\delta y\,\delta x + \delta y^2\,.\,\delta x]$$

$$= 2\pi y\,\delta x\,\delta y \text{ neglecting the higher order infinitesimals.}$$

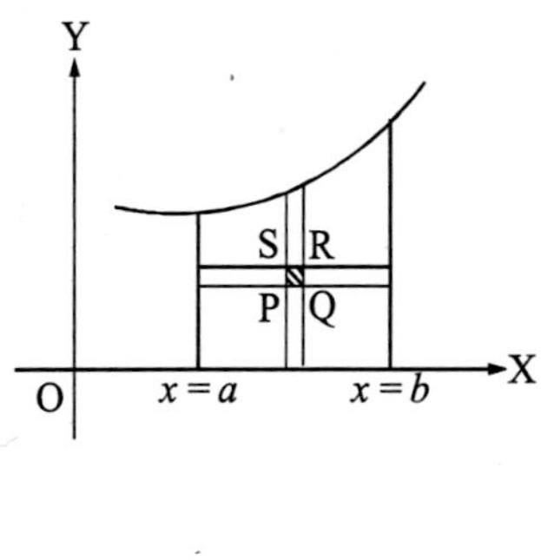

Therefore, the total volume of the solid of revolution of area A about X-axis, is equal to

$$\mathbf{V = 2\pi\int_a^b\int_0^{f(x)} y\,dy\,dx}$$

Similarly, if the area is revolved about Y-axis then the volume of the solid so generated, is

$$2\pi \iint_A x\, dx\, dy \quad \text{with appropriate limits.}$$

In **polar co-ordinates** the corresponding formula for **volume of solid of revolution of area A** about X-axis, becomes

$$\iint_A 2\pi r \sin\theta \,.\, r\, d\theta\, dr = 2\pi \iint_A r^2 \sin\theta\, d\theta\, dr$$

Similarly, the volume of solid of revolution of the area A about Y-axis $= 2\pi \iint_A r^2 \cos\theta\, d\theta\, dr$.

CHANGE OF VARIABLES USING JOCOBIANS

Sometimes it becomes difficult to evaluate double and triple integrals by usual methods, in such situations change of variables facilitates in transforming the given complicated integral into a simpler integral involving new variables.

For instance, let the variables x and y be replaced by new variables u and v by the transformation $x = f_1(u, v)$, $y = f_2(u, v)$ then

$$\iint_R f(x, y)\, dx\, dy = \iint_{R'} f\{f_1(u, v), f_2(u, v)\}\, |J|\, du\, dv \quad \text{where } J = \frac{\partial(x, y)}{\partial(u, v)} = \begin{vmatrix} \frac{\partial x}{\partial u} & \frac{\partial x}{\partial v} \\ \frac{\partial y}{\partial u} & \frac{\partial y}{\partial v} \end{vmatrix}.$$

Here the region of integration R in xy-plane get transformed to the region R′ in the uv-plane and J is known as Jacobian of transformation.

The idea can be extended to triple integral as given below

$$\iiint_R f(x, y, z)\, dx\, dy\, dz = \iiint f[f_1(u, v, w), f_2(u, v, w), f_3(u, v, w)]\, |J|\, du\, dv\, dw$$

where $x = f_1(u, v, w)$, $y = f_2(u, v, w)$, $z = f_3(u, v, w)$,

and
$$\boxed{J = \frac{\partial(x, y, z)}{\partial(u, v, w)} = \begin{vmatrix} \frac{\partial x}{\partial u} & \frac{\partial x}{\partial v} & \frac{\partial x}{\partial w} \\ \frac{\partial y}{\partial u} & \frac{\partial y}{\partial v} & \frac{\partial y}{\partial w} \\ \frac{\partial z}{\partial u} & \frac{\partial z}{\partial v} & \frac{\partial z}{\partial w} \end{vmatrix}}.$$

EXAMPLE 7.24. **(a)** Find the volume of the tetrahedron bounded by the co-ordinate planes and the plane $\frac{x}{a} + \frac{y}{b} + \frac{z}{c} = 1$. **(UPTU 2002-03, 2005-06)**

(b) Find the volume of the solid which is bounded by the surfaces $2z = x^2 + y^2$ and $z = x$. **[UPTU 2011-12]**

SOLUTION: **(a)** The region of integration R is enclosed by the co-ordinate planes $x = 0, y = 0, z = 0$ and the plane $\frac{x}{a} + \frac{y}{b} + \frac{z}{c} = 1$. Putting $x = a\text{X}$, $y = b\text{Y}$, $z = c\text{Z}$, the equation of the plane becomes

$$\text{X} + \text{Y} + \text{Z} = 1.$$

Then Z varies from Z = 0 to Z = (1 – X – Y)

Y varies from Y = 0 to Y = 1 – X

and X varies from X = 0 to X = 1.

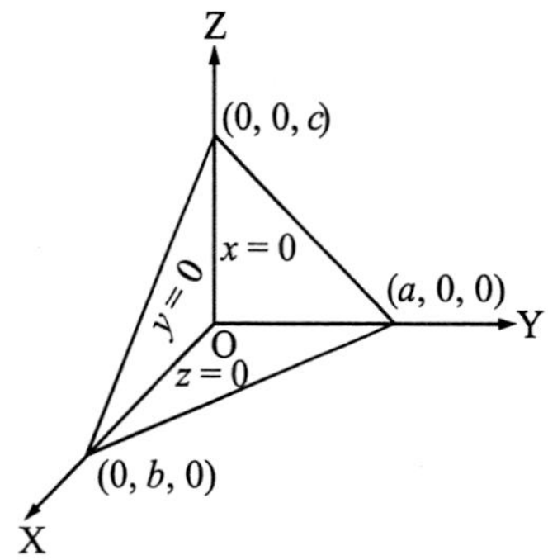

Volume V of the tetrahedron $= \iiint dx\, dy\, dx$

$$= abc \int_0^1 \int_0^{1-X} \int_0^{1-X-Y} dX\, dY\, dZ$$

or
$$V = abc \int_0^1 \int_0^{1-X} (1 - X - Y)\, dX\, dY$$

$$= abc \int_0^1 \left[\frac{(1-X-Y)^2}{-2} \right]_0^{1-X} dX = \frac{abc}{2} \int_0^1 (1-X)^2\, dX$$

$$= \frac{abc}{2} \left[\frac{(1-X)^3}{-3} \right]_0^1 = \frac{abc}{-6}(0-1) = \frac{abc}{6}. \quad \textbf{Ans.}$$

(b) Given $x^2 + y^2 = 2z, \quad z = x$

$\therefore$ $x^2 + y^2 = 2x$ is the curve of intersection,

or $(x-1)^2 + y^2 = 1.$

$$\text{Required volume } V = \int_{z=0}^{2} \int_{x=-\sqrt{2z}}^{+\sqrt{2z}} \int_{y=-\sqrt{2z-x^2}}^{y=+\sqrt{2z-x^2}} dy\, dx\, dz$$

$$= 4 \int_{z=0}^{2} \int_{x=0}^{\sqrt{2z}} \int_{y=0}^{\sqrt{2z-x^2}} dy\, dx\, dz$$

$$= 4 \int_{z=0}^{2} \int_{x=0}^{\sqrt{2z}} \sqrt{2z - x^2}\, dx\, dz$$

$$= 4 \int_{z=0}^{2} \left[\frac{x}{2}\sqrt{2z-x^2} + \frac{2z}{2} \sin^{-1} \frac{x}{\sqrt{2z}} \right]_{x=0}^{\sqrt{2z}}$$

$$= 4 \int_{z=0}^{2} \left[0 + z\left(\frac{\pi}{2}\right) \right] dz = 2\pi \left(\frac{z^2}{2} \right)_0^2$$

$= 4\pi$ **Ans.**

EXAMPLE 7.25. Evaluate $\iint_R \cos\frac{x-y}{x+y}\,dx\,dy$ where 'R' is bounded by $x=0,\ y=0,\ x+y=1$.

SOLUTION: Let $u=x-y$ and $v=x+y$ which gives $x=\frac{u+v}{2}$ and $y=\frac{v-u}{2}$.

Thus, the region $x=0$ means $u+v=0$ or $u=-v$

and the region $y=0$ means $v-u=0$ or $u=v$

and the region $x+y=1$ means $v=1$.

Region R is in the shaded area as shown in the figure.

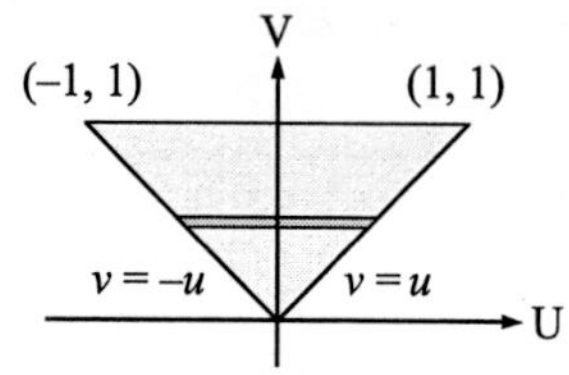

Next, $$dx\,dy=\frac{\partial(x,y)}{\partial(u,v)}\,du\,dv=\begin{vmatrix}\frac{1}{2} & \frac{1}{2}\\ -\frac{1}{2} & \frac{1}{2}\end{vmatrix}du\,dv=\frac{1}{2}\,du\,dv$$

$$\therefore\quad \iint_R \cos\frac{x-y}{x+y}\,dx\,dy=\frac{1}{2}\int_0^1\int_{-v}^{v}\cos\frac{u}{v}\,du\,dv=\frac{1}{2}\int_0^1\left[v\sin\frac{u}{v}\right]_{u=-v}^{v}dv$$

$$=\frac{1}{2}\int_0^1 v\,(\sin 1-\sin(-1)\,dv=\sin 1\left[\frac{v^2}{2}\right]_0^1=\frac{1}{2}\sin 1.\qquad \textbf{Ans.}$$

EXAMPLE 7.26. Find the volume bounded by the ellipsoid

$$(2x+y+z)^2+(3x-y+z)^2+4\,(x+2y)^2=16.$$

SOLUTION: Let us change the variables here as

$$X=2x+y+z,\quad Y=3x-y+z,\quad Z=2\,(x+2y).$$

Then the given ellipsoid becomes the sphere $X^2+Y^2+Z^2=16$

Now, required volume $$V=\iiint_{\text{ellipsoid}} dx\,dy\,dz=\iiint_{\text{sphere}}\frac{\partial(x,y,z)}{\partial(X,Y,Z)}\,dX\,dY\,dZ$$

Here $$\frac{\partial(X,Y,Z)}{\partial(x,y,z)}=\begin{vmatrix}\frac{\partial X}{\partial x} & \frac{\partial X}{\partial y} & \frac{\partial X}{\partial z}\\ \frac{\partial Y}{\partial x} & \frac{\partial Y}{\partial y} & \frac{\partial Y}{\partial z}\\ \frac{\partial Z}{\partial x} & \frac{\partial Z}{\partial y} & \frac{\partial Z}{\partial z}\end{vmatrix}=\begin{vmatrix}2 & 1 & 1\\ 3 & -1 & 1\\ 2 & 4 & 0\end{vmatrix}=8$$

$$\therefore\quad \frac{\partial(x,y,z)}{\partial(X,Y,Z)}=\frac{1}{8}$$

Therefore $$V=\iiint_R \frac{1}{8}\,dX\,dY\,dZ$$

where R, the region of integration, is the sphere $X^2 + Y^2 + Z^2 = 4^2$.

Now using spherical polar co-ordinates

$$X = r \sin\theta \cos\phi, \quad Y = r \sin\theta \sin\phi, \quad Z = r \cos\theta$$

Then $$dX\,dY\,dZ = \frac{\partial(X, Y, Z)}{\partial(r, \theta, \phi)} \cdot dr\,d\theta\,d\phi = r^2 \sin\theta\,dr\,d\theta\,d\phi$$

$$\therefore \quad V = \frac{1}{8}\int_{\theta=0}^{\pi}\int_{\phi=0}^{2\pi}\int_{r=0}^{4} r^2 \sin\theta\,dr\,d\theta\,d\phi = \frac{1}{8}\int_0^{\pi}\int_0^{2\pi} \sin\theta \left[\frac{r^3}{3}\right]_0^4 d\theta\,d\phi$$

$$= \frac{64}{8(3)}\int_0^{\pi}[\phi]_0^{2\pi} \sin\theta\,d\theta = \frac{16\pi}{3}\int_0^{\pi} \sin\theta\,d\theta = \frac{16\pi}{3}(-\cos\theta)_0^{\pi} = \frac{32\pi}{3}. \qquad \textbf{Ans.}$$

EXAMPLE 7.27. Evalvate $\iint_R (x+y)^2\,dxdy$ where R is the parallelogram in the xy-plane with vertices (1, 0), (3, 1), (2, 2), (0, 1) using the transformation

$$u = x + y, \; v = x - 2y.$$ **(UPTU 2003-04)**

SOLUTION: R is parallelogram $ABCD$ as shown in the adjacent figure. Equation of line AB is $x - 2y = 1$, Equation of line DC is $x - 2y = -2$, Equation of line DA is $x + y = 1$ and the equation of line CB is $x + y = 4$

The lines can also be represented as

$u = 1, \quad u = 4$ and $v = -2, \quad v = 1$

We also know by change of variables

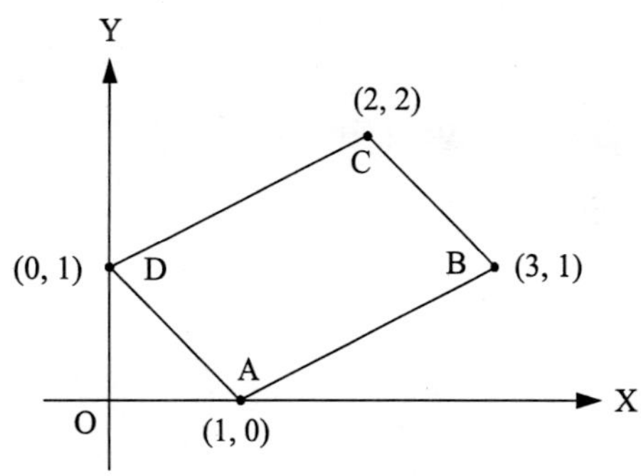

$$du\,dv = \frac{\partial(u,v)}{\partial(x,y)}\,dxdy$$

From $u = x + y, \; v = x - 2y,$ we have

$$\frac{\partial(u,v)}{\partial(x,y)} = \begin{vmatrix} \frac{\partial u}{\partial x} & \frac{\partial u}{\partial y} \\ \frac{\partial v}{\partial x} & \frac{\partial v}{\partial y} \end{vmatrix}$$

$$= \begin{vmatrix} 1 & 1 \\ 1 & -2 \end{vmatrix} = -3$$

$$\therefore \quad dxdy = \frac{\partial(x,y)}{\partial(u,v)}\,dudv = \frac{-1}{3}\,dudv$$

Therefore $$\iint_R (x+y)^2\,dxdy = \iint_R u^2\left(\frac{-1}{3}\right)dudv = \frac{-1}{3}\int_{-2}^{1}\int_1^4 u^2\,dudv = \frac{-1}{3}\int_{-2}^{1}\left[\frac{u^3}{3}\right]_1^4 dv$$

$$= -7\,(1+2) = -21. \qquad \textbf{Ans.}$$

EXAMPLE 7.28. (*a*) Find the volume bounded by the elliptic paraboloids $z = x^2 + 9y^2$ and $z = 18 - x^2 - 9y^2$. **(UPTU 2006-07, 2011-12)**

(*b*) Find the volume enclosed between the two surfaces $z = 8 - x^2 - y^2$ and $z = x^2 + 3y^2$. **(UPTU 2010-11)**

SOLUTION: (*a*) Required volume = $\iiint_V dx\,dy\,dz$ where V is the volume bounded by the paraboloids $z = x^2 + 9y^2$ and $z = 18 - x^2 - 9y^2$. Inter-section of these paraboloids is $x^2 + 9y^2 = 18 - x^2 - 9y^2$

or $x^2 + 9y^2 = 9$ or the ellipse $\frac{x^2}{9} + y^2 = 1$.

$$\therefore \quad \text{Volume} = \int_{-3}^{3}\int_{-\frac{1}{3}\sqrt{9-x^2}}^{\frac{1}{3}\sqrt{9-x^2}}\int_{x^2+9y^2}^{18-x^2-9y^2} dz\,dy\,dx = \int_{-3}^{3}\int_{-\frac{1}{3}\sqrt{9-x^2}}^{\frac{1}{3}\sqrt{9-x^2}} (18 - x^2 - 9y^2 - x^2 - 9y^2)\,dy\,dx$$

$$= 2\int_{-3}^{3}\left[18y - 2x^2y - 18\frac{y^3}{3}\right]_0^{\frac{1}{3}\sqrt{9-x^2}} dx$$

$$= 2\int_{-3}^{3}\left[\frac{1}{3}(18 - 2x^2)\sqrt{9 - x^2} - \frac{18}{81}(9 - x^2)^{3/2}\right]dx$$

$$= 4\left(\frac{2}{3} - \frac{2}{9}\right)\int_0^3 (9 - x^2)^{3/2}\,dx = \frac{16}{9}\int_0^3 (9 - x^2)^{3/2}\,dx \qquad \text{(On putting } x = 3\sin\theta\text{)}$$

$$= \frac{16}{9}\cdot 27\cdot 3\int_0^{\pi/2}\cos^4\theta\,d\theta = 27\pi \qquad \textbf{Ans.}$$

(*b*) The two surfaces intersect on the elliptic cylinder i.e. $z = 8 - x^2 - y^2 = x^2 + 3y^2$

or $\qquad 2x^2 + 4y^2 = 8 \quad$ or $\quad \frac{x^2}{4} + \frac{y^2}{2} = 1$

The projection of this volume onto *xy*-plane is the plane region enclosed by ellipse having the same equation $\frac{x^2}{2^2} + \frac{y^2}{(\sqrt{2})^2} = 1 \quad \therefore \quad y = \pm\sqrt{\frac{4 - x^2}{2}}$, as shown below.

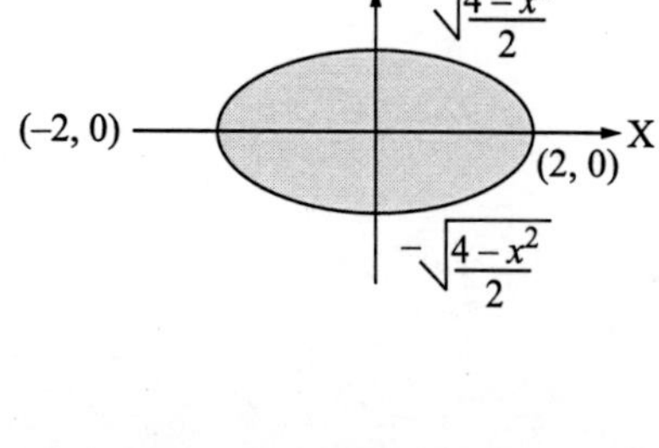

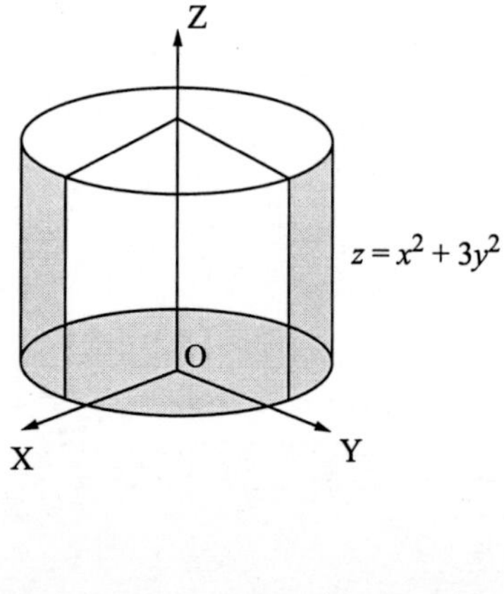

$$\therefore \qquad \text{Volume} = \int_{-2}^{2}\int_{-\sqrt{\frac{4-x^2}{2}}}^{\sqrt{\frac{4-x^2}{2}}}\int_{x^2+3y^2}^{8-x^2-y^2} dz\,dy\,dx = \int_{-2}^{2}\int_{-\sqrt{\frac{4-x^2}{2}}}^{\sqrt{\frac{4-x^2}{2}}} (z)_{x^2+3y^2}^{8-x^2-y^2}\,dy\,dx$$

$$= \int_{-2}^{2}\int_{-\sqrt{\frac{4-x^2}{2}}}^{\sqrt{\frac{4-x^2}{2}}} (8-x^2-y^2-x^2-3y^2)\,dy\,dx = 2\int_{-2}^{2}\int_{0}^{\sqrt{\frac{4-x^2}{2}}} (8-2x^2-4y^2)\,dy\,dx$$

$$= 2\int_{-2}^{2}\left[(8-2x^2)y-\frac{4y^3}{3}\right]_0^{\sqrt{\frac{4-x^2}{2}}} dx$$

$$= 2\int_{-2}^{2}\left[(8-2x^2)\sqrt{\frac{4-x^2}{2}}-\frac{4}{3}\left(\frac{4-x^2}{2}\right)^{3/2}\right] dx$$

$$= 4\int_{0}^{2}\left[(8-2x^2)\sqrt{\frac{4-x^2}{2}}-\frac{4}{3}\left(\frac{4-x^2}{2}\right)^{3/2}\right] dx$$

(put $x = 2\sin\theta$, $dx = 2\cos\theta\, d\theta$)

$$= 4\int_{0}^{\pi/2}\left[8\cdot(1-\sin^2\theta)\sqrt{2}\,(1-\sin^2\theta)^{1/2}-\frac{4}{3}\cdot\{2\,(1-\sin^2\theta)\}^{3/2}\right]\times 2\cos\theta\, d\theta$$

$$= 4\int_{0}^{\pi/2}\left[8\cdot\sqrt{2}\,(\cos^2\theta)^{3/2}-\frac{4}{3}(2)^{3/2}\,(\cos^2\theta)^{3/2}\right]\cdot 2\cos\theta\, d\theta$$

$$= 4\cdot 8\cdot\sqrt{2}\int_{0}^{\pi/2}\left[\left(1-\frac{1}{3}\right)\cos^3\theta\right]\cdot 2\cos\theta\, d\theta$$

$$= 64\sqrt{2}\cdot\frac{2}{3}\int_{0}^{\pi/2}\cos^4\theta\, d\theta = 64\cdot\sqrt{2}\cdot\frac{2}{3}\cdot\frac{3}{4}\cdot\frac{1}{2} = 8\sqrt{2}\,\pi \qquad \textbf{Ans.}$$

EXAMPLE 7.29. Transform $\int_{0}^{\pi/2}\int_{0}^{\pi/2}\sqrt{\frac{\sin\phi}{\sin\theta}}\,d\phi\, d\theta$ by the substitution $x = \sin\phi\cos\theta$, $y = \sin\phi\sin\theta$ and show that its value is π. **(UPTU 2001)**

SOLUTION: Since $x = \sin\phi\cos\theta$, $y = \sin\phi\sin\theta$ we have $x^2 + y^2 = \sin^2\phi$ and $\frac{y}{x} = \tan\theta$.

We know that $d\phi\, d\theta = \frac{\partial(\theta, \phi)}{\partial(x, y)} dx\, dy$

Here $$\frac{\partial(x, y)}{\partial(\theta, \phi)} = \begin{vmatrix} \frac{\partial x}{\partial\theta} & \frac{\partial x}{\partial\phi} \\ \frac{\partial y}{\partial\theta} & \frac{\partial y}{\partial\phi} \end{vmatrix} = \begin{vmatrix} -\sin\phi\sin\theta & \cos\phi\cos\theta \\ \sin\phi\cos\theta & \cos\phi\sin\theta \end{vmatrix} = -\sin\phi\cos\phi.$$

$$\therefore \quad \frac{\partial(\theta, \phi)}{\partial(x, y)} = \frac{-1}{\sin\phi\cos\phi}.$$

$$\therefore \quad d\theta d\phi = \frac{-1}{\sin\phi\cos\phi} dx\, dy$$

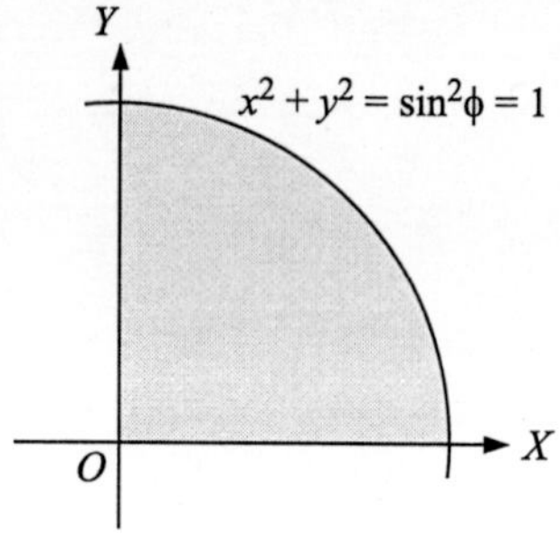

$$\text{Hence} \int_0^{\pi/2}\int_0^{\pi/2} \sqrt{\frac{\sin\phi}{\sin\theta}}\, d\phi\, d\theta = \int_0^1 \int_0^{\sqrt{1-y^2}} \sqrt{\frac{\sin\phi}{\sin\theta}} \frac{1}{\sin\phi\cos\phi} dx\, dy$$

$$= \int_0^1 \int_0^{\sqrt{1-y^2}} \frac{dx\, dy}{\cos\phi\sqrt{\sin\theta\sin\phi}} = \int_0^1 \int_0^{\sqrt{1-y^2}} \frac{dx\, dy}{\sqrt{y}\sqrt{1-x^2-y^2}}$$

$$= \int_0^1 \frac{dy}{\sqrt{y}} \int_0^{\sqrt{1-y^2}} \frac{dx}{\sqrt{1-y^2-x^2}} = \int_0^1 \frac{dy}{\sqrt{y}} \left[\sin^{-1}\frac{x}{\sqrt{1-y^2}}\right]_0^{\sqrt{1-y^2}}$$

$$= \int_0^1 \frac{\pi}{2}\frac{dy}{\sqrt{y}} = \frac{\pi}{2}\cdot\left[2\sqrt{y}\right]_0^1 = \pi.$$ **Hence Proved.**

EXAMPLE 7.30. Using the transformation $x + y = u$, $y = uv$, show that

$$\int_0^1 \int_0^{1-x} e^{y/(x+y)} dy\, dx = \frac{1}{2}(e-1).$$ **(UPTU 2001 Q.B.)**

SOLUTION: Since $y = uv$ and $x = u(1-v)$ we have

$$\frac{\partial(x, y)}{\partial(u, v)} = \begin{vmatrix} \frac{\partial x}{\partial u} & \frac{\partial x}{\partial v} \\ \frac{\partial y}{\partial u} & \frac{\partial y}{\partial v} \end{vmatrix} = \begin{vmatrix} 1-v & -u \\ v & u \end{vmatrix} = u$$

$$\therefore \quad dx\, dy = \frac{\partial(x, y)}{\partial(u, v)} du\, dv = u\, du\, dv$$

Here $x = 0$ gives $u(1-v) = 0 \Rightarrow u = 0, v = 1$

and $y = 0$ gives $uv = 0 \Rightarrow u = 0, v = 0$ and $x + y = 1$ gives $u = 1$.

$\therefore$ u varies from 0 to 1 and v also varies from 0 to 1.

$$\therefore \quad \int_0^1 \int_0^{1-x} e^{y/(x+y)} dy\, dx = \int_0^1 \int_0^1 e^{\frac{uv}{u}} \frac{\partial(x, y)}{\partial(u, v)} du\, dv$$

$$= \int_0^1 \int_0^1 u e^v\, du\, dv = \left[\frac{u^2}{2}\right]_0^1 \left[e^v\right]_0^1 = \frac{1}{2}(e-1).$$ **Hence Proved.**

EXAMPLE 7.31. If R is the region in the first quadrant bounded by $x = 0, y = 0$ and $x + y = 1$, change the variables x, y to u, v where $x + y = u, y = uv$ and evaluate

$$\iint_R xy\,(1 - x - y)^{1/2}\, dx\, dy.$$ **(UPTU 2002)**

SOLUTION: As in example 7.30 $\quad x = u(1 - v), \quad y = uv.$

$$\therefore \qquad \frac{\partial(x, y)}{\partial(u, v)} = u \quad \text{and} \quad dx\,dy = \frac{\partial(x, y)}{\partial(u, v)} du\,dv = u\,du\,dv$$

Here again u varies form 0 to 1 and v varies from 0 to 1.

$$\therefore \qquad \iint_D xy(1 - x - y)^{1/2} dx\,dy = \iint u(1 - v)uv(1 - u)^{1/2} \cdot u\,du\,dv$$

$$= \iint u^3 (1 - u)^{1/2} v(1 - v)\, du\, dv$$

$$= \int_0^1 u^3 (1 - u)^{1/2} du \int_0^1 v(1 - v) dv$$

$$= \beta(4, 3/2)\, \beta(2, 2) = \frac{\Gamma 4\, \Gamma 3/2}{\Gamma \frac{11}{2}} \cdot \frac{\Gamma 2\, \Gamma 2}{\Gamma 4} = \frac{16}{945}.$$ **Ans.**

ANOTHER APPLICATION – DETERMINATION OF MASS

(a) In case of **plane lamina,** if the density at any point $P(x, y)$ is $\rho(x, y)$ then the elementary mass at P is $\rho\, dx\, dy$, hence the total mass of the lamina = $\iint \rho(x, y)\, dx\, dy$ with appropriate limits so as to cover the complete lamina. In polar coordinates we take $\rho = \rho(r, \theta)$, at any point $P(r, \theta)$ then the total mass of the lamina = $\iint \rho(r, \theta)\, r\, dr\, d\theta$.

(b) In case of arc of a curve, the mass $= \int \rho\, ds$ where $\rho(x, y)$ is the density of the curve at $P(x, y)$.

(c) In case of solid, if the density at any point $P(x, y, z)$ is $\rho(x, y, z)$ then the total mass of the solid is $\iiint \rho(x, y, z)\, dx\, dy\, dz$ with appropriate limits so as to cover the complete solid.

EXAMPLE 7.32. Find the area and the mass contained in the first quadrant enclosed by the curve $\left(\frac{x}{a}\right)^{\alpha} + \left(\frac{y}{b}\right)^{\beta} = 1$ where $\alpha > 0,\ \beta > 0$ given that the density at any point $P(x, y)$ is $k\sqrt{xy}$. **(UPTU 2008-09)**

SOLUTION: Let us take the parametric equation of the curve as $x = a\cos^{2/\alpha}\theta$ and $y = b\sin^{2/\beta}\theta$. In the first quadrant, θ will vary from 0 to $\frac{\pi}{2}$ on the curve.

$$\therefore \quad \text{Area} = \int y\, dx = -\int_0^{\pi/2} b\sin^{2/\beta}\theta \cdot a \cdot \frac{2}{\alpha}\cos^{\frac{2}{\alpha}-1}\theta \cdot \sin\theta\, d\theta$$

$$= -\frac{2ab}{\alpha}\int_0^{\pi/2} \sin^{\frac{2}{\beta}+1}\theta\, \cos^{\frac{2}{\alpha}-1}\theta\, d\theta = \frac{-2ab}{\alpha} \cdot \frac{\Gamma\left(\left(\frac{2}{\beta}+2\right)\Big/2\right)\Gamma\left(\left(\frac{2}{\alpha}-1+1\right)\Big/2\right)}{2\,\Gamma\left(\left(\frac{2}{\beta}+\frac{2}{a}+2\right)\Big/2\right)}$$

$$= \frac{ab}{\alpha}\, \frac{\Gamma\left(\frac{1}{\beta}+1\right)\Gamma\left(\frac{1}{\alpha}\right)}{\Gamma\left(\frac{1}{\alpha}+\frac{1}{\beta}+1\right)}. \qquad \textbf{Ans.}$$

We know that Mass = Area × Density

$$\therefore \quad \text{required mass} = M = \int k\sqrt{xy} \cdot y\, dx = k\int x^{1/2} y^{3/2}\, dx.$$

$$= k\int_0^{\pi/2} a^{1/2}\cos^{1/\alpha}\theta \cdot b^{3/2}\sin^{3/\beta}\theta \cdot \frac{2a}{\alpha}\cos^{2/\alpha-1}\theta\, \sin\theta\, d\theta$$

$$= \frac{-k}{\alpha}(ab)^{3/2}\int_0^{\pi/2} \cos^{\frac{3}{\alpha}-1}\theta\, \sin^{\frac{3}{\beta}+1}\theta\, d\theta = \frac{-k}{\alpha}(ab)^{3/2}\, \frac{\Gamma\left(\frac{3}{2\alpha}\right)\Gamma\left(\frac{3}{2\beta}+1\right)}{2\,\Gamma\left(\frac{3}{2\alpha}+\frac{3}{2\beta}+1\right)}. \qquad \textbf{Ans.}$$

EXAMPLE 7.33. (a) Find the mass of the region bounded by the ellipsoid $\frac{x^2}{a^2}+\frac{y^2}{b^2}+\frac{z^2}{c^2}=1$ if the density varies as the square of the distance from the centre. **(UPTU 2009-10)**

(b) Find the mass of a plate which is bounded by the coordinate planes and the plane $\frac{x}{a}+\frac{y}{b}+\frac{z}{c}=1$, the density is given by $\rho = k\,xyz$. **[UPTU 2011-12]**

SOLUTION: (a) Since mass = density × volume and $\rho \propto (x^2+y^2+z^2)$ so $\rho = k(x^2+y^2+z^2)$.

$$M = k\iiint_R (x^2+y^2+z^2)\,dx\,dy\,dz\text{. Putting here } x=aX, y=bY, z=cZ.$$

$$= k\iiint (a^2X^2+b^2Y^2+c^2Z^2)\,abc\,dX\,dY\,dZ\,.$$

Converting to spherical polar co-ordinates
$X = r\sin\theta\cos\phi$, $Y = r\sin\theta\sin\phi$, $Z = r\cos\theta$, and taking into account

$$dXdYdZ = \frac{\partial(X,Y,Z)}{\partial(r,\theta,\phi)}\,drd\theta d\phi = r^2\sin\theta\,drd\theta d\phi, \text{ we get}$$

$$M = Kabc\int_0^1\int_{\theta=0}^{\pi}\int_{\phi=0}^{2\pi}(a^2r^2\sin^2\theta\cos^2\phi + b^2r^2\sin^2\theta\sin^2\phi + r^2c^2\cos^2\theta)\,r^2\sin\theta\,drd\theta d\phi$$

$$= Kabc\left[\frac{r^5}{5}\right]_0^1\int_0^{\pi}\left[a^2\sin^3\theta\int_0^{2\pi}\cos^2\phi\,d\phi + b^2\sin^3\theta\int_0^{2\pi}\sin^2\phi\,d\phi + c^2\cos^2\theta\sin\theta\int_0^{2\pi}d\phi\right]d\theta$$

$$= \frac{Kabc}{5}\int_0^{\pi}(\pi a^2\sin^3\theta + \pi b^2\sin^3\theta + 2\pi c^2\cos^2\theta\sin\theta)\,d\theta$$

$$= \frac{K\pi abc}{5}\left[(a^2+b^2)\frac{8}{3} + 2c^2\left[\frac{\cos^3\theta}{-3}\right]_0^{\pi}\right] = \frac{K\pi abc}{5\cdot 3}[8(a^2+b^2)+4c^2]$$

$$= \frac{4K\pi abc}{15}(a^2+b^2+2c^2). \qquad \textbf{Ans.}$$

(b) Let $u = \frac{x}{a}$, $v = \frac{y}{b}$ and $w = \frac{z}{c}$ then $u \ge 0$, $v \ge 0$, $w \ge 0$ and $u+v+w=1$. Also $dx = a\,dx$, $dy = b\,dv$, $dz = c\,dw$.

Required mass = volume × density

$$= \iiint k\,xyz\,dx\,dy\,dz = k\iiint (au)(bv)(cw)(abc)\,du\,dv\,dw$$

$$= k\,a^2b^2c^2\int_{u=0}^{1}\int_{v=0}^{1-u}\int_{w=0}^{1-u-v}uvw\,dw\,dv\,du = k\,a^2b^2c^2\int_0^1\int_0^{1-u}\frac{uv}{2}(1-u-v)^2\,dv\,dw$$

$$= ka^2b^2c^2\int_0^1\frac{u}{2}\int_0^{1-u}[v(1-u)^2 - 2v^2(1-u)+v^3]\,dv\,du = k\,a^2b^2c^2\int_0^1\frac{u}{24}(1-u)^4\,du$$

$$= \frac{1}{24}\int_0^1(1-u)u^4\,dx = \frac{k\,(abc)^2}{720}. \qquad \textbf{Ans.}$$

EXAMPLE 7.34. Find the volume to the cylinders $x^2 + y^2 = a^2$ and $x^2 + z^2 = a^2$.

(UPTU 2008-09)

SOLUTION: Required volume $= \int_{-a}^{a} \int_{-\sqrt{a^2-x^2}}^{\sqrt{a^2-x^2}} \int_{-\sqrt{a^2-x^2}}^{\sqrt{a^2-x^2}} dz\,dy\,dx$

$$= \int_{-a}^{a} \int_{-\sqrt{a^2-x^2}}^{\sqrt{a^2-x^2}} [z]_{-\sqrt{a^2-x^2}}^{\sqrt{a^2-x^2}} dydx = 2\int_{-a}^{a} \int_{-\sqrt{a^2-x^2}}^{\sqrt{a^2-x^2}} \sqrt{a^2-x^2}\, dydx$$

$$= 2\int_{-a}^{a} \sqrt{a^2-x^2}\,[y]_{-\sqrt{a^2-x^2}}^{\sqrt{a^2-x^2}} dy\,dx = 4\int_{-a}^{a} \sqrt{a^2-x^2}\sqrt{a^2-x^2}\,dx = 4\int_{-a}^{a} (a^2-x^2)dx$$

$$= 8\left[a^2x - \frac{x^3}{3}\right]_0^a = 8\left(a^3 - \frac{a^3}{3}\right) = \frac{16a^3}{3}.$$ **Ans.**

EXAMPLE 7.35. (a) Evaluate $\iiint (x-2y+z)\, dx\, dy\, dz$

where $R: 0 \le x \le 1,\ \ 0 \le y \le x^2,\ \ 0 \le z \le x+y.$ **(UPTU 2008-09)**

(b) Find the surface area of the plane $x + 2y + 2z = 12$ cut off by $x = 0,\ y = 0$ and $x^2 + y^2 = 16$. **[UPTU 2011-12]**

SOLUTION: (a) Given integral $= \int_0^1 \int_0^{x^2} \int_0^{x+y} (x-2y+z)\, dz\, dy\, dx$

$$= \int_0^1 \int_0^{x^2} \left[xz - 2yz + \frac{z^2}{2}\right]_0^{x+y} dydx = \int_0^1 \int_0^{x^2} \left[(x-2y)(x+y) + \frac{1}{2}(x+y)^2\right]_0 dy\,dx$$

$$= \int_0^1 \int_0^{x^2} \left(x^2 - xy - 2y^2 + \frac{1}{2}x^2 + xy + \frac{y^2}{2}\right) dy\,dx = \int_0^1 \int_0^{x^2} \left[\frac{3}{2}x^2 - \frac{3}{2}y^2\right] dy\,dx$$

$$= \int_0^1 \left[\frac{3}{2}x^2y - \frac{y^3}{2}\right]_0^{x^2} dx = \int_0^1 \left(\frac{3}{2}x^4 - \frac{x^6}{2}\right) dx = \left[\frac{3}{2}\cdot\frac{x^5}{5} - \frac{x^7}{14}\right]_0^1 = \frac{3}{10} - \frac{1}{14} = \frac{8}{35}.$$ **Ans.**

(b) Plane is $x + 2y + 2z = 12$ and circular cylinder is $x^2 + y^2 = 16$ whose axis is z-axis. Unit vector normal to the plane is

$$\hat{n} = \frac{i + 2j + 2k}{\sqrt{1+4+4}} = \frac{1}{3}(i + 2j + 2k)$$

Surface area of the plane cut off by $x = 0,\ y = 0$ and $x^2 + y^2 = 4^2$

= Projection of the circular area of the cylinder on the plane

$$= \pi(4^2)\hat{n}\cdot k = 16\pi\left[\frac{1}{3}(i + 2j + 2k)\cdot k\right]$$

$$= \frac{16\pi}{3}\times 2 = \frac{32\pi}{3}$$ **Ans.**

EXERCISE 7A

1. Evaluate $\displaystyle\int_0^1 \int_0^{\sqrt{1-x^2}} \int_0^{\sqrt{1-x^2-y^2}} xyz \, dz \, dy \, dx$.

2. Evaluate $\displaystyle\int_{-1}^1 \int_0^z \int_{x-z}^{x+z} (x+y+z) \, dy \, dx \, dz$.

3. Evaluate $\displaystyle\int_1^e \int_1^{\log y} \int_1^{e^x} \log z \, dz \, dx \, dy$.

4. Find volume of the ellipsoid $\dfrac{x^2}{a^2} + \dfrac{y^2}{b^2} + \dfrac{z^2}{c^2} = 1$.

5. A circular hole of radius b is made centrally through a sphere of radius a. Find the volume of the remaining portion of the sphere.

6. Evaluate $\displaystyle\int_0^a \int_y^{a+\sqrt{a^2-y^2}} \frac{dx \, dy}{(4a^2 + x^2 + y^2)^2}$.

7. Evaluate $\displaystyle\int_0^{\pi} \int_0^{a(1+\cos\theta)} \int_0^h 2r \left\{1 \frac{r}{a(1+\cos\theta)}\right\} dz \, dr \, d\theta$.

8. Evaluate $\displaystyle\iiint_R x^{l-1} y^{m-1} z^{n-1} \, dx \, dy \, dz$ over the region R bounded by

$$x > 0, \quad y > 0, \quad x + y + z < 1.$$

9. Evaluate $\iiint z^2 \, dx \, dy \, dz$ over the region common to the sphere $x^2 + y^2 + z^2 = a^2$ and the cylinder $x^2 + y^2 = ax$.

10. Determine the value of $\displaystyle\int_0^a \int_0^{\alpha} \int_0^{\beta} (x^2 + y^2 + z^2) \, dx \, dy \, dz$ where a is constant and

$$\alpha = \sqrt{a^2 - z^2}, \quad \beta = \sqrt{a^2 - y^2 - z^2}.$$

11. Evaluate $\displaystyle\iiint_R x^2 \, dx \, dy \, dz$ where R encloses the volume of the ellipsoid

$$\frac{x^2}{a^2} + \frac{y^2}{b^2} + \frac{z^2}{c^2} = 1.$$

12. Evaluate $\displaystyle\iint_R (x^2 + y^2) \, dx \, dy$ where R is bounded by the lines

$$x + y = 0, \quad x + y = 2, \quad x - y = 2.$$

13. Evaluate $\iiint x^2 y^2 z^2 \, dx \, dy \, dz$ over the volume bounded by

$$xy = 4, \quad xy = 9, \quad yz = 1, \quad yz = 4, \quad zx = 25 \quad \text{and} \quad zx = 49.$$

❑❑❑

CHAPTER

8 Beta and Gamma Functions, Dirichlet's Integral

Beta Function, Gamma Function, Dirichlet's Integral and Applications

BETA AND GAMMA FUNCTIONS

GAMMA FUNCTION of n, denoted by Γn, is defined as

$$\boxed{\Gamma n = \int_0^\infty e^{-x}\, x^{n-1}\, dx, \quad n > 0} \quad ...(1)$$

As such $\Gamma 1 = \int_0^\infty e^{-x}.1\, dx = -\left[e^{-x}\right]_0^\infty = 1.$

And, by definition $\Gamma(n+1) = \int_0^\infty e^{-x}\, x^n\, dx$

Integrating it by parts, we get

$$\Gamma(n+1) = -\left[x^n e^{-x}\right]_0^\infty + \int_0^\infty n\, x^{n-1} e^{-x}\, dx = -\lim_{x\to\infty} \frac{x^n}{e^x} + 0 + n\Gamma n$$

Applying L' Hospital's rule n times to $\lim_{x\to\infty} \frac{x^n}{e^x}$ its value comes out to be zero, therefore

$$\boxed{\Gamma(n+1) = n\,\Gamma n}.$$ **(UPTU 2004)**

If n is a positive integer, then

$$\Gamma(n+1) = n\Gamma n = n\,(n-1)\Gamma(n-1) = n\,(n-1)\,(n-2)\Gamma(n-2) = ...$$

$$= n(n-1)\,(n-2)...3.2.1\,\Gamma 1 = n\,(n-1)\,(n-2)...3.2.1 \quad \text{as} \quad \Gamma 1 = 1$$

$$= n! \quad (n \in \mathrm{N})$$

There is yet another representation of Γn as follows.

Putting $x = t^2$ in (1), gives

$$\Gamma n = \int_0^\infty e^{-t^2} . t^{2n-2} . 2t\, dt = 2\int_0^\infty e^{-t^2} . t^{2n-1} dt$$

or $$\boxed{\Gamma n = 2\int_0^\infty e^{-x^2} x^{2n-1}\, dx} \qquad ...(2)$$

Next, the **BETA FUNCTION,** denoted by $\beta\,(m, n)$, is defined as **(UPTU 2003)**

$$\boxed{\beta\,(m, n) = \int_0^1 x^{m-1}(1-x)^{n-1}\, dx} \qquad ...(3)$$

Applying the property of definite integrals. $\int_0^a f(x)\, dx = \int_0^a f(a-x)\, dx$ in (3), we get

$$\beta\,(m, n) = \int_0^1 (1-x)^{m-1} x^{n-1}\, dx = \beta\,(n, m)$$

Thus, we have $\boxed{\beta\,(m, n) = \beta\,(n, m)}$ ...(4)

There is yet another representation of $\beta\,(m, n)$. Putting $x = \dfrac{t}{1+t}$ or $t = \dfrac{x}{1-x}$ in (3), gives

$$\beta\,(m, n) = \int_0^\infty \left(\frac{t}{1+t}\right)^{m-1}\left(1-\frac{t}{1+t}\right)^{n-1}\frac{dt}{(1+t)^2} = \int_0^\infty \frac{t^{m-1}\, dt}{(1+t)^{m+n}}.$$

Thus, we can also write

$$\boxed{\beta\,(m, n) = \int_0^\infty \frac{t^{n-1}}{(1+t)^{m+n}}\, dt} \quad \text{on using (4).}$$

Further, in the relation $\beta\,(m, n) = \int_0^1 x^{m-1}(1-x)^{n-1}\, dx$,

if we substitute $x = \sin^2\theta$, so $dx = 2\sin\theta\cos\theta\, d\theta$, we get

$$\beta\,(m, n) = \int_0^{\pi/2} \sin^{2m-2}\theta\,(1-\sin^2\theta)^{n-1}\, 2\sin\theta\cos\theta\, d\theta$$

or $$\boxed{\beta\,(m, n) = 2\int_0^{\pi/2} \sin^{2m-1}\theta\, \cos^{2n-1}\theta\, d\theta}\,. \qquad ...(5)$$

RELATION BETWEEN BETA AND GAMMA FUNCTION

$$\boldsymbol{\beta(m, n) = \frac{\Gamma m \, \Gamma n}{\Gamma(m+n)}}$$ **(UPTU 2002, 2003)**

PROOF: We know that

$$\Gamma m = 2\int_0^\infty e^{-x^2} x^{2m-1}\, dx \quad \text{and} \quad \Gamma n = 2\int_0^\infty e^{-y^2} y^{2n-1}\, dy$$

$$\therefore \quad \Gamma m \, \Gamma n = 4\int_0^\infty e^{-x^2} x^{2m-1}\, dx \int_0^\infty e^{-y^2} y^{2n-1} dy = 4\int_0^\infty \int_0^\infty e^{-(x^2+y^2)} x^{2m-1} y^{2n-1} dx dy$$

Now, converting into polar co-ordinates, we get

$$\Gamma m \, \Gamma n = 4\int_0^{\pi/2} \int_0^\infty e^{-r^2} r^{2m+2n-2} \cos^{2m-1}\theta \sin^{2n-1}\theta \, r \, dr d\theta$$

$$= 4\int_0^\infty e^{-r^2} r^{2m+2n-1} dr \int_0^{\pi/2} \cos^{2m-1}\theta \sin^{2n-1}\theta \, d\theta$$

$$= \Gamma(m+n) \; \beta(m, n) \quad \text{on using (2) and (5).}$$

Hence $$\beta(m, n) = \frac{\Gamma m \, \Gamma n}{\Gamma(m+n)}. \qquad \text{...(6)}$$

In relation (5) writing $2m - 1 = p$ and $2n - 1 = q$ and using (6), we get

$$2\int_0^{\pi/2} \sin^p\theta \cos^q\theta \, d\theta = \frac{\Gamma\frac{p+1}{2}\,\Gamma\frac{q+1}{2}}{\Gamma\frac{p+q+2}{2}}$$

or $$\boldsymbol{\int_0^{\pi/2} \sin^p\theta \cos^q\theta \, d\theta = \frac{\Gamma\frac{p+1}{2}\,\Gamma\frac{q+1}{2}}{2\,\Gamma\frac{p+q+2}{2}}}. \qquad \text{... (7)}$$

This relation is a big facility and is used very frequently in the evaluation of the definite integrals. In (6) if we take $p = q = 0$, we get

$$\int_0^{\pi/2} \sin^0\theta \cos^0\theta \, d\theta = \frac{\Gamma 1/2 \, \Gamma 1/2}{2\,\Gamma 1} \quad \text{or} \quad \frac{\pi}{2} = \frac{\left(\Gamma 1/2\right)^2}{2} \quad \text{since } \Gamma 1 = 1$$

Therefore, we have yet another very important and useful result

$$\boldsymbol{\Gamma 1/2 = \sqrt{\pi}}$$ **(UPTU 2003)**

DUPLICATION FORMULA: $\boxed{\Gamma n\,\Gamma\left(n+\frac{1}{2}\right)=\frac{\sqrt{\pi}\,\Gamma 2n}{2^{2n-1}}}$ or $\boxed{\Gamma(n+1)\,\Gamma\left(n+\frac{1}{2}\right)=\frac{\sqrt{\pi}\,\Gamma(2n+1)}{2^{2n}}}$ **(UPTU 2001)**

In relation (6) putting $m = n$, we get

$$\beta(n, n) = \frac{\Gamma n\,\Gamma n}{\Gamma 2n} = \frac{\Gamma n\,\Gamma n}{(2n-1)(2n-2)(2n-3)\ldots 3.2.1}$$

$$\therefore \quad \beta(n, n) = \frac{\Gamma n\,\Gamma n}{2^{2n-1}\left(n-\frac{1}{2}\right)(n-1)\left(n-\frac{3}{2}\right)(n-2)\ldots 2\cdot\frac{3}{2}\cdot 1\cdot\frac{1}{2}}$$

$$= \frac{\Gamma n\,\Gamma n}{2^{2n-1}\left(n-\frac{1}{2}\right)\left(n-\frac{3}{2}\right)\ldots\frac{3}{2}.\frac{1}{2}(n-1)(n-2)\ldots 2.1}$$

or
$$\frac{\Gamma n\,\Gamma n}{\Gamma 2n} = \frac{\Gamma n\,\Gamma n\,\Gamma(1/2)}{2^{2n-1}\left(n-\frac{1}{2}\right)\left(n-\frac{3}{2}\right)\ldots\frac{3}{2}\cdot\frac{1}{2}\cdot\Gamma(1/2)\,\Gamma n} = \frac{\Gamma n\,\sqrt{\pi}}{2^{2n-1}\,\Gamma\left(n+\frac{1}{2}\right)} \quad (\text{as } \Gamma(1/2) = \sqrt{\pi})$$

$$\Rightarrow \quad \boxed{\Gamma n\,\Gamma\left(n+\frac{1}{2}\right) = \frac{\sqrt{\pi}\,\Gamma 2n}{2^{2n-1}}}$$

which is called the **duplication formula.**

replacing n by $n+\frac{1}{2}$, the above relation can also be written as

$$\boxed{\Gamma(n+1)\,\Gamma\left(n+\frac{1}{2}\right) = \frac{\sqrt{\pi}\,\Gamma(2n+1)}{2^{2n}}}.$$

EXAMPLE 8.1. Evaluate (*i*) $\int_0^\infty x^{1/4}e^{-\sqrt{x}}\,dx$ (*ii*) $\int_0^\infty x^2 e^{-a^2x^2}\,dx$.

SOLUTION: (*i*) Let $I = \int_0^\infty x^{1/4}e^{-\sqrt{x}}\,dx$. Substituting $x = t^2$, so $dx = 2t\,dt$, gives

$$I = \int_0^\infty t^{\frac{1}{2}}e^{-t}\,2t\,dt = 2\int_0^\infty e^{-t}\,t^{\frac{5}{2}-1}\,dt = 2\,\Gamma\frac{5}{2} = 2\cdot\frac{3}{2}\cdot\frac{1}{2}\,\Gamma\frac{1}{2} = \frac{3\sqrt{\pi}}{2}. \quad \textbf{Ans.}$$

(*ii*) Let $I = \int_0^\infty x^2 e^{-a^2x^2}\,dx$. Putting $x = \frac{\sqrt{t}}{a}$, so $dx = \frac{1}{2a\sqrt{t}}\,dt$, we have

$$I = \int_0^\infty \frac{t}{a^2}e^{-t}\frac{1}{2a\sqrt{t}}\,dt = \frac{1}{2a^3}\int_0^\infty e^{-t}\,t^{\frac{3}{2}-1}\,dt = \frac{1}{2a^3}\,\Gamma\frac{3}{2} = \frac{1}{2a^3}\cdot\frac{1}{2}\,\Gamma\frac{1}{2} = \frac{\sqrt{\pi}}{4a^3}. \quad \textbf{Ans.}$$

EXAMPLE 8.2. Evaluate (*i*) $\int_{-\infty}^{\infty} e^{-x^2}\, dx$ (*ii*) $\int_0^{\infty} \frac{x^a}{a^x}\, dx$. (*iii*) $\Gamma\left(-\frac{5}{2}\right)$.

(UPTU 2011-12)

SOLUTION: (*i*) Let $I = \int_{-\infty}^{\infty} e^{-x^2}\, dx = 2\int_0^{\infty} e^{-x^2}\, dx$ (by property of definite integrals)

Putting $x = \sqrt{t}$, so $dx = \frac{1}{2\sqrt{t}}\, dt$, we get

$$I = 2\int_0^{\infty} e^{-t}\frac{1}{2\sqrt{t}}\, dt = \int_0^{\infty} e^{-t}\, t^{\frac{1}{2}-1}\, dt = \Gamma\left(\frac{1}{2}\right) = \sqrt{\pi}. \quad \text{Thus} \quad \int_{-\infty}^{\infty} e^{-x^2}\, dx = \sqrt{\pi}. \qquad \textbf{Ans.}$$

(*ii*) Let $I = \int_0^{\infty} \frac{x^a}{a^x}\, dx$.

Substituting, $a^x = e^t$ or $x \log a = t$ so that $dx = \frac{dt}{\log a}$, we get

$$I = \int_0^{\infty} e^{-t}\left(\frac{t}{\log a}\right)^a \frac{dt}{\log a} = \frac{1}{(\log a)^{a+1}}\int_0^{\infty} e^{-t}\, t^{a+1-1}\, dt = \frac{\Gamma(a+1)}{(\log a)^{a+1}}. \qquad \textbf{Ans.}$$

(*iii*) Since $\Gamma(n+1) = n\Gamma(n)$ we have $\Gamma(n) = \frac{\Gamma(n+1)}{n}$ ...(1)

Taking $n = -\frac{5}{2}$ in (1), we get $\Gamma\left(-\frac{5}{2}\right) = \frac{\Gamma\left(-\frac{3}{2}\right)}{-\frac{5}{2}}$ and taking $x = \frac{-3}{2}$ in (1) gives $\Gamma\left(\frac{-3}{2}\right) = \frac{\Gamma\left(\frac{-3}{2}\right)}{\frac{-3}{2}}$

$$\therefore \quad \Gamma\left(-\frac{5}{2}\right) = \frac{1}{\left(-\frac{5}{2}\right)\left(-\frac{3}{2}\right)}\Gamma\left(-\frac{1}{2}\right)$$

Finally taking $n = -\frac{1}{2}$ in (1) we get $\Gamma\left(-\frac{1}{2}\right) = \frac{\Gamma\left(\frac{1}{2}\right)}{-\frac{1}{2}}$

$$\therefore \quad \Gamma\left(-\frac{5}{2}\right) = \frac{1}{\left(-\frac{5}{2}\right)\left(-\frac{3}{2}\right)}\frac{\Gamma\left(\frac{1}{2}\right)}{\left(-\frac{1}{2}\right)} = -\frac{8\sqrt{\pi}}{15} \qquad \textbf{Ans.}$$

EXAMPLE 8.3. Evaluate $\int_0^\infty x^{n-1} \cos ax\, dx$ and $\int_0^\infty x^{n-1} \sin ax\, dx$.

SOLUTION: Let us consider the integral

$$\mathrm{I} = \int_0^\infty x^{n-1} e^{-aix}\, dx = \int_0^\infty x^{n-1} (\cos ax - i \sin ax)\, dx \qquad \left(i = \sqrt{-1}\right)$$

Putting $aix = t$, $dx = \dfrac{dt}{ai}$, we get

$$\mathrm{I} = \int_0^\infty e^{-t} \left(\frac{t}{ai}\right)^{n-1} \frac{dt}{ai} = \frac{1}{a^n i^n} \int_0^\infty e^{-t}\, t^{n-1}\, dt = \frac{\overline{|n}}{a^n} (-i)^n$$

$$= \frac{\overline{|n}}{a^n} \left(\cos\frac{\pi}{2} - i \sin\frac{\pi}{2}\right)^n = \frac{\overline{|n}}{a^n} \left(\cos\frac{n\pi}{2} - i \sin\frac{n\pi}{2}\right)$$

Thus, $\displaystyle\int_0^\infty x^{n-1} (\cos ax - i \sin ax)\, dx = \frac{\overline{|n}}{a^n} \left(\cos\frac{n\pi}{2} - i \sin\frac{n\pi}{2}\right)$

Equating real and imaginary parts on both sides, we get

$$\int_0^\infty x^{n-1} \cos ax\, dx = \frac{\overline{|n}}{a^n} \cos\frac{n\pi}{2} \quad \text{and} \quad \int_0^\infty x^{n-1} \sin ax\, dx = \frac{\overline{|n}}{a^n} \sin\frac{n\pi}{2}. \qquad \textbf{Ans.}$$

EXAMPLE 8.4. Evaluate $\displaystyle\int_0^1 x^{m-1} \left(\log\frac{1}{x}\right)^{n-1} dx$, $m > 0$, $n > 0$.

SOLUTION: Let $\displaystyle\mathrm{I} = \int_0^1 x^{m-1} \left(\log\frac{1}{x}\right)^{n-1} dx$.

Putting $\log\dfrac{1}{x} = t$ or $x = e^{-t}$, so $dx = -e^{-t} dt$, we have

$$\mathrm{I} = \int_\infty^0 e^{-(m-1)t}\, t^{n-1} \left(-e^{-t}\right) dt = \int_0^\infty e^{-mt}\, t^{n-1}\, dt.$$

Again, putting $mt = y$, so $dt = \dfrac{1}{m} dy$, we get

$$\mathrm{I} = \int_0^\infty e^{-y} \left(\frac{y}{m}\right)^{n-1} \frac{1}{m} dy = \frac{1}{m^n} \int_0^\infty e^{-y}\, y^{n-1}\, dy = \frac{\overline{|n}}{m^n}. \qquad \textbf{Ans.}$$

EXAMPLE 8.5. Show that $n\,\beta\,(m+1,\,n) = m\,\beta\,(m,\,n+1)$.

SOLUTION: Using the property $\beta\,(m,n) = \dfrac{\Gamma m\,\Gamma n}{\Gamma (m+n)}$, we have

$$n\,\beta\,(m+1,n) = \frac{n\,\Gamma(m+1)\,\Gamma n}{\Gamma(m+n+1)} = \frac{n\cdot m\,\Gamma m\,\Gamma n}{\Gamma(m+n+1)} = \frac{m\,\Gamma m\,\Gamma(n+1)}{\Gamma(m+n+1)}$$

$= m\,\beta\,(m,\,n+1)$ by the above mentioned property. **Hence Proved.**

EXAMPLE 8.6. Evaluate $\displaystyle\int_0^{\pi} \sin^2\theta\,(1+\cos\theta)^4\,d\theta$.

SOLUTION: Let $\displaystyle I = \int_0^{\pi} \sin^2\theta\,(1+\cos\theta)^4\,d\theta$

$$= \int_0^{\pi}\left(2\sin\frac{\theta}{2}\cos\frac{\theta}{2}\right)^2\left(2\cos^2\frac{\theta}{2}\right)^4 d\theta$$

$$= 64\int_0^{\pi}\sin^2\frac{\theta}{2}\cos^{10}\frac{\theta}{2}\,d\theta.$$

Putting $\dfrac{\theta}{2} = t$ so $d\theta = 2dt$, we get

$$I = 64\int_0^{\pi/2}\sin^2 t\,\cos^{10} t.\,2dt$$

$$= 64.2\,\frac{\Gamma\frac{(2+1)}{2}\,\Gamma\frac{(10+1)}{2}}{2\,\Gamma\frac{(2+10+2)}{2}} = 64\,\frac{\Gamma\frac{3}{2}\,\Gamma\frac{11}{2}}{\Gamma 7}$$

$$= \frac{64\cdot\frac{9}{2}\cdot\frac{7}{2}\cdot\frac{5}{2}\cdot\frac{3}{2}\left(\Gamma\frac{3}{2}\right)^2}{6!} = \frac{4\cdot 9\cdot 7\cdot 5\cdot 3\left(\frac{1}{2}\Gamma\frac{1}{2}\right)^2}{6\cdot 5\cdot 4\cdot 3\cdot 2}$$

$$= \frac{21\,\pi}{16}$$

Thus, $\displaystyle\int_0^{\pi}\sin^2\theta\,(1+\cos\theta)^4\,d\theta = \frac{21\,\pi}{16}$. **Ans.**

EXAMPLE 8.7. Express the integral $\int_0^1 x^m (1-x^n)^p \, dx$ in terms of Gamma function and evaluate $\int_0^1 x^2 (1-x^2)^4 \, dx$.

SOLUTION: Let $I = \int_0^1 x^m (1-x^n)^p \, dx$, then putting $x^n = t$ so $nx^{n-1} dx = dt$, we get

$$I = \int_0^1 t^{\frac{m}{n}} (1-t)^p \frac{t^{1/n}\, dt}{nt} = \frac{1}{n}\int_0^1 t^{\frac{m+1}{n}-1} (1-t)^{p+1-1} \, dt$$

$$= \frac{1}{n}\beta\left(\frac{m+1}{n},\, p+1\right) = \frac{1}{n} \frac{\Gamma\left(\frac{m+1}{n}\right)\Gamma(p+1)}{\Gamma\left(\frac{m+1}{n}+p+1\right)}. \quad \textbf{Ans}$$

Now, taking $m = 2$, $n = 2$ and $p = 4$ in the above relation, we get

$$\int_0^1 x^2 \left(1-x^2\right)^4 dx = \frac{1}{2} \frac{\Gamma\left(\frac{2+1}{2}\right)\Gamma(4+1)}{\Gamma\left(\frac{2+1}{2}+4+1\right)} = \frac{1}{2}\frac{\Gamma\left(\frac{3}{2}\right)\Gamma(5)}{\Gamma\left(\frac{13}{2}\right)} = \frac{\Gamma\left(\frac{3}{2}\right)\cdot 4!}{2\cdot\frac{11}{2}\cdot\frac{9}{2}\cdot\frac{7}{2}\cdot\frac{5}{2}\cdot\frac{3}{2}\Gamma\left(\frac{3}{2}\right)} = \frac{16\cdot 4\cdot 3\cdot 2\cdot 1}{11\cdot 9\cdot 7\cdot 5\cdot 3}$$

or $\quad I = \dfrac{128}{3465}.$ **Ans.**

EXAMPLE 8.8. Show that $\int_a^b (x-a)^m (b-x)^n \, dx = (b-a)^{m+n+1} \beta(m+1, n+1)$.

SOLUTION: Let $I = \int_a^b (x-a)^m (b-x)^n \, dx$ Putting $x - a = y$, so $dx = dy$, we get

$I = \int_0^{b-a} y^m (b-a-y)^n \, dy$. Now putting $y = (b-a)\,t$, so $dy = (b-a)\,dt$, we get

$$I = \int_0^1 (b-a)^m t^m (b-a)^n (1-t)^n (b-a)\, dt$$

$$= (b-a)^{m+n+1}\int_0^1 t^m (1-t)^n \, dt = (b-a)^{m+n+1}\int_0^1 t^{m+1-1}(1-t)^{n+1-1}\, dt$$

$$= (b-a)^{m+n+1} \beta(m+1, n+1). \quad \textbf{Hence Proved.}$$

EXAMPLE 8.9. Show that $\displaystyle\int_0^{\pi/2} \sqrt{\tan\theta}\, d\theta = \int_0^{\pi/2} \sqrt{\cot\theta}\, d\theta = \frac{1}{2}\Gamma\left(\frac{1}{4}\right)\Gamma\left(\frac{3}{4}\right) = \frac{\pi}{\sqrt{2}}.$

SOLUTION: Using the property of definite integrals, we have

$$\int_0^{\pi/2} \sqrt{\tan\theta}\, d\theta = \int_0^{\pi/2} \sqrt{\tan\left(\frac{\pi}{2}-\theta\right)}\, d\theta = \int_0^{\pi/2} \sqrt{\cot\theta}\, d\theta$$

Next, $$\int_0^{\pi/2} \sqrt{\tan\theta}\, d\theta = \int_0^{\pi/2} \sin^{\frac{1}{2}}\theta \cos^{-\frac{1}{2}}\theta\, d\theta = \frac{\Gamma\left(\frac{(1/2)+1}{2}\right)\Gamma\left(\frac{(-1/2)+1}{2}\right)}{2\,\Gamma\left(\frac{(1/2)-(1/2)+2}{2}\right)}$$

$$= \frac{\Gamma\left(\frac{3}{4}\right)\cdot\Gamma\left(\frac{1}{4}\right)}{2\,\Gamma(1)} = \frac{\Gamma\left(\frac{3}{4}\right)\cdot\Gamma\left(\frac{1}{4}\right)}{2}, \quad \text{as } \Gamma(1) = 1.$$

Now, let us recall the duplication formula $2^{2n-1}\,\Gamma(n)\,\Gamma\left(n+\frac{1}{2}\right) = \sqrt{\pi}\,\Gamma(2n)$.

Taking $n = \frac{1}{4}$ here, we get

$$2^{\frac{-1}{2}}\,\Gamma\left(\frac{1}{4}\right)\Gamma\left(\frac{3}{4}\right) = \sqrt{\pi}\,\Gamma\left(\frac{1}{2}\right) \quad \text{or} \quad \Gamma\left(\frac{1}{4}\right)\Gamma\left(\frac{3}{4}\right) = \pi\sqrt{2} \quad \text{as} \quad \Gamma\left(\frac{1}{2}\right) = \sqrt{\pi}.$$

$$\therefore \quad \int_0^{\pi/2} \sqrt{\tan\theta}\, d\theta = \frac{1}{2}\pi\sqrt{2} = \frac{\pi}{\sqrt{2}} = \int_0^{\pi/2} \sqrt{\cot\theta}\, d\theta.$$ **Hence Proved.**

EXAMPLE 8.10. Show that $\displaystyle\int_0^1 \frac{x^{m-1} + x^{n-1}}{(1+x)^{m+n}}\, dx = \beta\,(m, n).$ **(UPTU 2001 Sp. Exam.) (UPTU 2011-12)**

SOLUTION: Let $\displaystyle \mathrm{I} = \int_0^1 \frac{x^{m-1}+x^{n-1}}{(1+x)^{m+n}}\, dx = \int_0^1 \frac{x^{m-1}}{(1+x)^{m+n}}\, dx + \int_0^1 \frac{x^{n-1}}{(1+x)^{m+n}}\, dx.$

Putting $x = \tan^2\theta$, $dx = 2\tan\theta\sec^2\theta\, d\theta$, in the above integrals, we get

$$\int_0^1 \frac{x^{m-1}}{(1+x)^{m+n}}\, dx = \int_0^{\pi/4} \frac{\tan^{2m-2}\theta\cdot 2\tan\theta\sec^2\theta\, d\theta}{\left(1+\tan^2\theta\right)^{m+n}} = 2\int_0^{\pi/4} \frac{\tan^{2m-1}\theta\sec^2\theta\, d\theta}{\sec^{2m+2n}\theta}$$

$$= 2\int_0^{\pi/4} \sin^{2m-1}\theta\, \cos^{2n-1}\theta\, d\theta$$

Similarly, $$\int_0^1 \frac{x^{n-1}\, dx}{(1+x)^{m+n}} = 2\int_0^{\pi/4} \sin^{2n-1}\theta\, \cos^{2m-1}\theta\, d\theta$$

In this integral, substituting $\theta = \frac{\pi}{2} - \phi$, so $d\theta = -d\phi$, we get

$$\int_0^1 \frac{x^{n-1}\,dx}{(1+x)^{m+n}} = -2\int_{\pi/2}^{\pi/4} \sin^{2n-1}\left(\frac{\pi}{2}-\phi\right)\cos^{2m-1}\left(\frac{\pi}{2}-\phi\right)d\phi = 2\int_{\pi/4}^{\pi/2}\cos^{2n-1}\phi\,\sin^{2m-1}\phi\,d\phi$$

$$= 2\int_{\pi/4}^{\pi/2}\cos^{2n-1}\theta\,\sin^{2m-1}\theta\,d\theta \quad \text{(by property of definite integrals replacing } \phi \text{ by } \theta\text{).}$$

$$\therefore \quad I = 2\int_0^{\pi/4}\sin^{2m-1}\theta\,\cos^{2n-1}\theta\,d\theta + 2\int_{\pi/4}^{\pi/2}\cos^{2n-1}\theta\,\sin^{2m-1}\,d\theta$$

$$= 2\int_0^{\pi/2}\sin^{2m-1}\theta\,\cos^{2n-1}\theta\,d\theta = 2\frac{\Gamma\left(\frac{2m-1+1}{2}\right)\Gamma\left(\frac{2n-1+1}{2}\right)}{2\,\Gamma\left(\frac{2m-1+2n-1+2}{2}\right)} = \frac{\Gamma(m)\Gamma(n)}{\Gamma(m+n)} = \beta(m,n). \quad \textbf{Hence Proved.}$$

EXAMPLE 8.11. Show that for a positive integer n, $\Gamma(n)$ can be expressed in the form

(i) $\Gamma(n) = 2a^n \int_0^\infty t^{2n-1} e^{-at^2}\,dt.$ (ii) $\Gamma(n) = a^n \int_0^\infty e^{-at} t^{n-1}\,dt$

SOLUTION: (i) We know that $\Gamma(n) = \int_0^\infty e^{-x} x^{n-1}\,dx.$

Here, put $x = at^2$ to get

$$\Gamma(n) = 2a^n \int_0^\infty e^{-at^2} t^{2n-1}\,dt. \quad \textbf{Hence Proved.}$$

(ii) Putting $x = at$ in (1), we get

$$\Gamma(n) = \int_0^\infty e^{-at}(at)^{n-1}\,a\,dt = a^n\int_0^\infty e^{-at}\,t^{n-1}\,dt. \quad \textbf{Hence Proved.}$$

EXAMPLE 8.12. Evaluate $\int_{-1}^{1}(1+x)^{p-1}(1-x)^{q-1}\,dx.$

SOLUTION: Let $I = \int_{-1}^{1}(1+x)^{p-1}(1-x)^{q-1}\,dx.$ Putting $1 + x = y$, gives

$I = \int_0^2 y^{p-1}(2-y)^{q-1}\,dy.$ Again putting $y = 2t$, we get

$$I = \int_0^1 2^{p-1}t^{p-1}\,2^{q-1}(1-t)^{q-1}\,2dt = 2^{p+q-1}\int_0^1 t^{p-1}(1-t)^{q-1}\,dt$$

$$= 2^{p+q-1}\beta(p,q) = 2^{p+q-1}\frac{\Gamma(p)\Gamma(q)}{\Gamma(p+q)} \quad \textbf{Ans.}$$

EXAMPLE 8.13. (*i*) Prove that $\displaystyle\int_0^{\frac{\pi}{2}} \sin^3 x \cos^{5/2} x\, dx = \frac{8}{77}.$

(*ii*) Show that $\Gamma(n)\,\Gamma(1-n) = \dfrac{\pi}{\sin n\pi}$ $(0 < n < 1)$. **(UPTU 2009-10)**

SOLUTION: (*i*) Using the result $\displaystyle\int_0^{\pi/2} \sin^p x \cos^q x dx = \frac{\Gamma\left(\frac{p+1}{2}\right)\Gamma\left(\frac{q+1}{2}\right)}{2\Gamma\left(\frac{p+q+2}{2}\right)}$ for $p = 3$, $q = 5/2$ we have

$$\int_0^{\pi/2} \sin^3 x \cos^{\frac{5}{2}} x\, dx = \frac{\Gamma\left(\frac{3+1}{2}\right)\Gamma\left(\frac{\left(\frac{5}{2}+1\right)}{2}\right)}{2\Gamma\left(\frac{3+\frac{5}{2}+2}{2}\right)} = \frac{\Gamma(2)\,\Gamma\left(\frac{7}{4}\right)}{2\Gamma\left(\frac{15}{4}\right)} = \frac{1\cdot\Gamma\left(\frac{7}{4}\right)}{2\cdot\frac{11}{4}\cdot\frac{7}{4}\cdot\frac{7}{4}\Gamma\left(\frac{7}{4}\right)} = \frac{8}{77}.$$ **Hence Proved.**

(*ii*) We know that $\displaystyle\beta(m, n) = \int_0^{\infty} \frac{x^{n-1} dx}{(1+x)^{m+n}} = \frac{\Gamma(m)\,\Gamma(n)}{\Gamma(m+n)}$

Now, if we write $m = 1 - n$, we get

$$\int_0^{\infty} \frac{x^{n-1} d x}{1+x} = \frac{\Gamma(n)\,\Gamma(1-n)}{1}.$$

We know that $\displaystyle\int_0^{\infty} \frac{t^{a-1} dt}{1+t} = \frac{\pi}{\sin a\pi}$ where $0 < a < 1$.

$$\Rightarrow \qquad \int_0^{\infty} \frac{x^{n-1} dx}{x+1} = \frac{\pi}{\sin n\pi}, \quad 0 < n < 1$$

$$\Rightarrow \qquad \Gamma(n)\,\Gamma(1-n) = \frac{\pi}{\sin n\pi}, \quad 0<x<1.$$ **Hence Proved.**

EXAMPLE 8.14. **(a)** Show that $\Gamma\left(\frac{3}{2}-x\right)\Gamma\left(\frac{3}{2}+x\right) = \left(\frac{1}{4}-x^2\right)\pi \sec \pi x$ provided $-1 < 2x < 1$.

(b) Using beta and gamma functions evaluate $\displaystyle\int_0^{\infty} \frac{dx}{1+x^4}$ **[UPTU 2011-12]**

SOLUTION: **(a)** L.H.S. $= \Gamma\left(\frac{3}{2}-x\right)\Gamma\left(\frac{3}{2}+x\right) = \left(\frac{1}{2}-x\right)\Gamma\left(\frac{1}{2}-x\right)\left(\frac{1}{2}+x\right)\Gamma\left(\frac{1}{2}+x\right) = \left(\frac{1}{4}-x^2\right)\Gamma\left(\frac{1}{2}+x\right)\Gamma\left(\frac{1}{2}-x\right)$

Now, we shall apply the well known result $\Gamma(n)\,\Gamma(n-1) = \dfrac{\pi}{\sin n\pi}$. Replacing n by $\left(\frac{1}{2}+x\right)$ here, we get

$$\left|\frac{1}{2}+x\right.\left|1-\left(\frac{1}{2}+x\right)\right. = \frac{\pi}{\sin\pi\left(\frac{1}{2}+x\right)} = \frac{\pi}{\sin\left(\frac{\pi}{2}+\pi x\right)}$$

or $$\left|\frac{1}{2}+x\right.\left|\frac{1}{2}-x\right. = \frac{\pi}{\cos\pi x} = \pi\sec\pi x.$$

Therefore, $\text{L.H.S.} = \left(\frac{1}{4}-x^2\right)\pi\sec\pi x = \text{R.H.S.}$ **Hence Proved.**

(b) Putting $x^4 = y$ $\therefore$ $x = y^{1/4}$ and $dx = \frac{1}{4}y^{-3/4}\,dy$, we have

$$I = \int_0^\infty \frac{dx}{1+x^4} = \frac{1}{4}\int_0^\infty \frac{y^{-3/4}dy}{1+y}$$

$$= \frac{1}{4}\int_0^\infty \frac{y^{1/4-1}dy}{1+y}$$

Now using the well known result $\int_0^\infty \frac{x^{n-1}\,dx}{1+x} = \frac{\pi}{\sin n\pi}$, we get

$$I = \frac{1}{4}\cdot\frac{\pi}{\sin(\pi/4)} = \frac{\pi\sqrt{2}}{4}$$ **Ans.**

EXAMPLE 8.15. Prove that $\beta\left(m,\frac{1}{2}\right) = 2^{2m-1}\,\beta(m, m)$.

SOLUTION: By definition, $\beta\left(m,\frac{1}{2}\right) = \dfrac{\left|m\right.\left|\frac{1}{2}\right.}{\left|m+\frac{1}{2}\right.}$.

Using the duplication formula $\left|m\right.\left|m+\frac{1}{2}\right. = \dfrac{\sqrt{\pi}\left|2m\right.}{2^{2m-1}}$ here, we get

$$\beta\left(m,\frac{1}{2}\right) = \frac{(\left|m\right.)^2\left|\frac{1}{2}\right.\times 2^{2m-1}}{\sqrt{\pi}\left|2m\right.}\left|m\right.$$

$$= \frac{\left|m\right.\left|m\right.\times 2^{2m-1}}{\left|2m\right.} \quad \text{since } \left|\frac{1}{2}\right. = \sqrt{\pi}.$$

$= 2^{2m-1}\,\beta(m, m) = \text{RHS}.$ **Hence Proved.**

EXAMPLE 8.16. Prove that $\int_0^1 \frac{dx}{\sqrt{1-x^4}} = \frac{\sqrt{\pi}\,\overline{|1/4}}{4\,\overline{|3/4}}$.

SOLUTION: Putting $x^4 = t$, the given integral becomes

$$I = \int_0^1 \frac{dx}{\sqrt{1-x^4}} = \int_0^1 \frac{\frac{1}{4}t^{-3/4}\,dt}{\sqrt{1-t}} = \frac{1}{4}\int_0^1 t^{\frac{1}{4}-1}(1-t)^{\frac{1}{2}-1}\,dt$$

$$= \frac{1}{4}\beta\left(\frac{1}{4},\frac{1}{2}\right) = \frac{1}{4}\frac{\overline{\left|\frac{1}{4}\right.}\,\overline{\left|\frac{1}{2}\right.}}{\overline{\left|\frac{1}{4}+\frac{1}{2}\right.}} = \frac{\sqrt{\pi}}{4}\frac{\overline{\left|\frac{1}{4}\right.}}{\overline{\left|\frac{3}{4}\right.}}.$$ **Hence Proved.**

EXAMPLE 8.17. Show that (i) $\int_0^\infty e^{-ax}x^{n-1}\cos bx\,dx = \frac{\overline{|n}\cos n\theta}{(a^2+b^2)^{n/2}}$

(ii) $\int_0^\infty e^{-ax}x^{n-1}\sin bx\,dx = \frac{\overline{|n}\sin n\theta}{(a^2+b^2)^{n/2}}$

where $\theta = \tan^{-1}(b/a)$. **(UPTU 2003-04)**

SOLUTION: Since $\overline{|n} = \int_0^\infty e^{-t}t^{n-1}\,dt$. Putting $t = ax$ we have $\int_0^\infty e^{-ax}x^{n-1}dx = \frac{\overline{|n}}{a^n}$.

Replacing a by $(a + ib)$, we get

$$\int_0^\infty e^{-(a+ib)x}x^{n-1}dx = \frac{\overline{|n}}{(a+ib)^n}.$$

Writing $a = r\cos\theta$, $b = r\sin\theta$ so that $\theta = \tan^{-1}(b/a)$, we get

$$\int_0^\infty e^{-ax}x^{n-1}(\cos bx - i\sin bx)\,dx = \frac{\overline{|n}}{r^n}(\cos\theta + i\sin\theta)^{-n}$$

$$= \frac{\overline{|n}}{(a^2+b^2)^{n/2}}(\cos n\theta - i\sin n\theta)$$

Equating real and imaginary parts on both sides, gives

$$\int_0^\infty e^{-ax}x^{n-1}\cos bx\,dx = \frac{\overline{|n}\cos x\theta}{(a^2+b^2)^{n/2}} \quad \text{and} \quad \int_0^\infty e^{-ax}x^{n-1}\sin bx\,dx\,dx = \frac{\overline{|n}\sin\theta}{(a^2+b^2)^{n/2}}.$$

Hence Proved.

EXAMPLE 8.18. Prove that $\beta(m, n) = \beta(m+1, n) + \beta(m, n+1)$ **(UPTU 2004-05, 2008-09)**

SOLUTION: We know that $\beta(m, n) = \dfrac{\Gamma(m)\,\Gamma(n)}{\Gamma(m+n)}$

$$\therefore \quad \text{RHS} = \beta(m+1, n) + \beta(m, n+1) = \frac{\Gamma(m+1)\,\Gamma(n)}{\Gamma(m+n+1)} + \frac{\Gamma(m)\,\Gamma(n+1)}{\Gamma(m+n+1)}$$

$$= \frac{m\Gamma(m)\,\Gamma(n) + \Gamma(m)\, n\Gamma(n)}{\Gamma(m+n+1)} = \frac{(m+n)\,\Gamma(m)\,\Gamma(n)}{(m+n)\,\Gamma(m+n)} = \beta(m, n)$$

Hence Proved.

DIRICHLET'S INTEGRAL

If V denotes the closed region in the first octanct bounded by the surface $(x/a)^p + (y/b)^q + (z/c)^r = 1$ and the co-ordinate planes, and if all the constants are positive, then

$$\iiint_V x^{\alpha-1} y^{\beta-1} z^{\gamma-1}\, dxdydz = \frac{a^\alpha b^\beta c^\gamma}{pqr} \frac{\Gamma\left(\frac{\alpha}{p}\right)\Gamma\left(\frac{\beta}{q}\right)\Gamma\left(\frac{\gamma}{r}\right)}{\Gamma\left(1+\frac{\alpha}{p}+\frac{\beta}{q}+\frac{\gamma}{r}\right)}.$$

This is known as Dirichlet's Integral.

As a particular case, in the above integral, if we take $\alpha = \beta = \gamma = 1$, then the volume V enclosed in the first octanct by the surface $(x/a)^p + (y/b)^q + (z/c)^r = 1$ and the co-ordinate planes, is given by

$$V = \iiint dxdydz = \frac{abc}{pqr} \frac{\Gamma\left(\frac{1}{p}\right)\Gamma\left(\frac{1}{q}\right)\Gamma\left(\frac{1}{r}\right)}{\Gamma\left(1+\frac{1}{p}+\frac{1}{q}+\frac{1}{r}\right)}.$$

EXAMPLE 8.19. Evaluate the integral $I = \iiint x^{\alpha-1} y^{\beta-1} z^{\gamma-1}\, dxdydz$ where V is the region in the first octanct bounded by the sphere $x^2 + y^2 + z^2 = 1$ and the co-ordinate planes where α, β, γ are positive. **(UPTU 2003)**

SOLUTION: Let us put $x^2 = u,\ y^2 = v,\ z^2 = w$ in the given integral, then

$$I = \iiint_R u^{(\alpha-1)/2} v^{(\beta-1)/2} w^{(\gamma-1)/2} \frac{du}{2\sqrt{u}} \cdot \frac{dv}{2\sqrt{v}} \cdot \frac{dw}{2\sqrt{w}}$$

$$= \frac{1}{8}\iiint_R u^{(\alpha/2-1)} . v^{(\beta/2-1)} . w^{(\gamma/2-1)}\, dudvdw$$

where R is the region in the uvw-space bounded by the plane $u + v + w = 1$ and the uv, vw, wu-planes. This is a Dirichlet's integral and has the value $\dfrac{\overline{|\frac{\alpha}{2}}\overline{|\frac{\beta}{2}}\overline{|\frac{\gamma}{2}}}{8\overline{\left|\left(\frac{\alpha}{2}+\frac{\beta}{2}+\frac{\gamma}{2}+1\right)\right.}}$.

Now let us derive it.

$$I = \frac{1}{8}\int_{u=0}^{1}\int_{v=0}^{1-u}\int_{w=0}^{1-u-v} u^{(\alpha/2-1)}.v^{(\beta/2-1)}.w^{(\gamma/2-1)}\,du dv dw$$

$$= \frac{1}{4\gamma}\int_{u=0}^{1}\int_{v=0}^{1-u} u^{(\alpha/2-1)}\,v^{(\beta/2-1)}\,(1-u-v)^{\gamma/2}\,du dv$$

$$= \frac{1}{4\gamma}\int_{u=0}^{1} u^{(\alpha/2-1)}\left[\int_{v=0}^{1-u} v^{(\beta/2-1)}\,(1-u-v)^{\gamma/2}\,dv\right]du$$

Substituting $v = (1-u)t$, we get

$$\int_{v=0}^{1-u} v^{(\beta/2-1)}(1-u-v)^{\gamma/2}\,dv = (1-u)^{(\beta+\gamma)/2}\int_{t=0}^{1} t^{(\beta/2-1)}(1-t)^{\gamma/2}\,dt$$

$$= (1-u)^{(\beta+\gamma)/2}\,\frac{\overline{|\beta/2}\,\overline{|\gamma/2+1}}{\overline{|(\beta+\gamma)/2+1}}$$

$$\therefore \qquad I = \frac{1}{4\gamma}\,\frac{\overline{|\beta/2}\,\overline{|\gamma/2+1}}{\overline{|(\beta+\gamma)/2+1}}\int_{u=0}^{1} u^{(\alpha/2)-1}(1-u)^{(\beta+\gamma)/2}\,du$$

$$= \frac{1}{4\gamma}\,\frac{\overline{|\beta/2}\,\overline{|\gamma/2+1}}{\overline{|(\beta+\gamma)/2+1}}\cdot\frac{\overline{|\alpha/2}\,\overline{|(\beta+\gamma)/2+1}}{\overline{|(\alpha+\beta+\gamma)/2+1}}$$

$$= \frac{\overline{|\alpha/2}\,\overline{|\beta/2}\,\overline{|\gamma/2}}{8\overline{|(\alpha+\beta+\gamma)/2+1}} \quad \left(\text{since } \overline{|\gamma/2+1} = (\gamma/2)\overline{|\gamma/2}\right). \qquad \textbf{Ans.}$$

EXAMPLE 8.20. Prove that $\iint_D x^{l-1}y^{m-1}\,dx\,dy = \dfrac{\overline{|l}\,\overline{|m}}{\overline{|l+m+1}}h^{l+m}$ where D is the domain $x \ge y \ge 0$ and $x + y \le h$. **(UPTU 2004-05)**

SOLUTION: Let $I = \iint_D x^{l-1}y^{m-1}\,dx\,dy$. Putting $x = Xh$ and $y = Yh$, then

$$I = \iint_{D'} (hX)^{l-1}\cdot(hY)^{m-1}\,h^2 dX\,dY \text{ where } D' \text{ is the domain } X \ge 0,\ Y \ge 0,\ X + Y \le 1$$

$$= h^{l+m}\iint_{D'} (X^{l-1}\,Y^{m-1}\,dX\,dY$$

$$= h^{l+m} \int_{X=0}^{1} \int_{Y=0}^{1-X} X^{l-1} Y^{m-1} \, dY dX$$

$$= h^{l+m} \int_0^1 X^{l-1} \left[\frac{Y^m}{m} \right]_0^{1-X} dX$$

$$= \frac{h^{l+m}}{m} \int_0^1 (1-X)^m \cdot X^{l-1} \, dX$$

$$= \frac{h^{l+m}}{m} \beta(l, m+1)$$

$$= \frac{h^{l+m}}{m} \frac{\Gamma l \, \Gamma (m+1)}{\Gamma (l+m+1)} = \frac{\Gamma l \, \Gamma m}{\Gamma (l+m+1)} h^{l+m}.$$ **Hence Proved.**

EXAMPLE 8.21. State Dirichlets's theorem for three variables. Hence evaluate $\iiint x^{l-1} y^{m-1} z^{n-1} \, dx\,dy\,dz$ where x, y, z are all positive but limited by the condition $\left(\frac{x}{a}\right)^p + \left(\frac{y}{b}\right)^q + \left(\frac{z}{c}\right)^r \le 1.$ **(UPTU 2005-06, 2010-11)**

SOLUTION: Dirichlets theorem: If l, m, n are all positive, then

$$\iiint_D x^{l-1} y^{m-1} z^{n-1} \, dx dy dz = \frac{\Gamma l \, \Gamma m \, \Gamma n}{\Gamma (l+m+n+1)}$$

where D is the domain $x \ge 0, y \ge 0, z \ge 0$ and $x + y + z \le 1$.

Let $$I = \iiint_D x^{l-1} y^{m-1} z^{n-1} \, dx dy dz \quad \text{where} \quad D : \left(\frac{x}{a}\right)^p + \left(\frac{y}{b}\right)^q + \left(\frac{z}{c}\right)^r \le 1.$$

Putting $\left(\frac{x}{a}\right)^p = X$, $\left(\frac{y}{b}\right)^q = Y$, $\left(\frac{z}{c}\right)^r = Z$ or $x = aX^{1/p}$, $y = bY^{1/q}$, $z = cZ^{1/r}$, we get

$$I = \iiint (aX^{1/p})^{l-1} (bY^{1/q})^{m-1} (cZ^{1/r})^{n-1} \, abc \frac{1}{p} \cdot \frac{1}{q} \cdot \frac{1}{r} X^{\frac{1-p}{p}} Y^{\frac{1-q}{q}} Z^{\frac{1-r}{r}} \, dX dY dZ$$

$$= \frac{a^l b^m c^n}{pqr} \iiint_D X^{\frac{l}{p}-1} Y^{\frac{m}{q}-1} Z^{\frac{n}{r}-1} \, dX dY dZ \quad \text{where} \quad D : X + Y + Z \le 1$$

$$= \frac{a^l b^m c^n}{pqr} \frac{\Gamma (l/p) \, \Gamma (m/q) \, \Gamma (n/r)}{\Gamma \left(\frac{l}{p} + \frac{m}{q} + \frac{n}{r} + 1\right)}.$$ **Ans.**

EXAMPLE 8.22. Show that, if l, m, n are all positive,

$$\iiint x^{l-1} y^{m-1} z^{n-1} dx\, dy\, dz = \frac{a^l b^m c^n}{8} \frac{\Gamma(l/2)\,\Gamma(m/2)\,\Gamma(n/2)}{\Gamma((l/2 + m/2 + m/2)}$$

where the triple integral is taken over the part of the ellipsoid $\dfrac{x^2}{a^2} + \dfrac{y^2}{b^2} + \dfrac{z^2}{c^2} = 1$ lying in the positive octant. **(UPTU 2001 Q.B.)**

SOLUTION: Let us write $X = \left(\dfrac{x}{a}\right)^2, \quad Y = \left(\dfrac{y}{b}\right)^2, \quad Z = \left(\dfrac{z}{c}\right)^2$

so $\quad x = a\sqrt{X}, \quad y = b\sqrt{Y}, \quad z = c\sqrt{Z}$ and $dx\, dy\, dz = \dfrac{\partial(x, y, z)}{\partial(X, Y, Z)} dX\, dY\, dZ = \dfrac{abc\, dX\, dY\, dZ}{8\sqrt{XYZ}}$

$$\therefore \quad \iiint x^{l-1} y^{n-1} z^{n-1}\, dx\, dy\, dz = \iiint a^{l-1} b^{m-1} c^{n-1} \left(\sqrt{X}\right)^{l-1} \left(\sqrt{Y}\right)^{m-1} \left(\sqrt{Z}\right)^{m-1} \frac{abc\, dX\, dY\, dZ}{8\sqrt{XYZ}}.$$

$$= \frac{a^l b^m c^n}{8} \iiint X^{(l/2-1)} Y^{(m/2-1)} Z^{(n/2-1)}\, dX\, dY\, dZ$$

$$= \frac{1}{8} a^l b^m c^n \frac{\Gamma(l/2)\,\Gamma(m/2)\,\Gamma(n/2)}{\Gamma(l/2 + m/2 + n/2)}$$ as Dirichlet's integral. **Hence Proved.**

EXERCISE 8A

1. Evaluate $\int_0^\infty x\, e^{-ax} \cos bx \, dx$.

2. Evaluate $\int_0^\infty \frac{x^4\, dx}{4^x}$.

3. Show that $\int_0^1 (x \log x)^3 \, dx = \frac{-3}{128}$.

4. Evaluate $\int_0^1 x^m (\log x)^n \, dx$.

5. Show that $\beta(m, n+1) + \beta(m+1, n) = \beta(m, n)$.

6. Evaluate $\int_0^\infty \frac{x^4 (1+x^5)\, dx}{(1+x)^{15}}$.

7. Show that $\int_0^1 \frac{x^{m-1}(1-x)^{n-1}\, dx}{(a+x)^{m+n}} = \frac{\beta(m,n)}{a^n (a+1)^m}$.

8. Evaluate $\int_0^1 \frac{x^{m-1}(1-x)^{n-1}}{(1+x)^{m+n}}\, dx$.

9. Show that $\int_1^\infty \frac{dx}{x^{m+1}(x-1)^n} = \beta(m+n, 1-n), -m < n < 1$.

10. If $\Gamma(n)\,\Gamma(1-n) = \frac{\pi}{\sin n\pi}$, show that $\int_0^\infty \frac{x^{n-1}\, dx}{(1+x)} = \frac{\pi}{\sin n\pi}$. **(UPTU 2009-10)**

11. Show that $\beta(m+1, n) = \frac{m}{m+n}\beta(m,n)$.

12. Prove that $\beta(n, n) = 2\int_0^{1/2} t^{n-1}(1-t)^{n-1}\, dt$.

13. If $I_n = \frac{\sqrt{\pi}\,\Gamma(n+1)}{2\,\Gamma(n/2+1)}$, show that $I_{n+2} = (n+1)\, I_n$.

14. Evaluate $\int_0^1 \left(1 - x^{1/n}\right)^{m-1} dx$.

15. The plane $\frac{x}{a}+\frac{y}{b}+\frac{z}{c} = 1$ meets the axes in A, B and C. Apply Dirichlet's integral to find the volume of the tetrahedron $OABC$, O being the origin. **(UPTU 2004)**

❑❑❑

UNIT-IV
VECTOR CALCULUS

Vector Differentiation:

Scalar and Vector Point Function

Gradient of Scalar Point Function

Divergence of Vector Point Function

Curl of Vector Point Function

Their Physical Interpretation.

Vector Integration:

Line, Surface and Volume Integrals

Statement of Green's, Theorem

Statement of Stoke's Theorem

Statement of Gauss Divergence Theorems and problems.

CHAPTER

9 Vector Differentiation

Scalar and vector Point Functions, Differentiation of Vector Function – Gradient, Divergence and Curl and their Physical Interpretations and Properties.

Elementary ideas of vector analysis in the form of vector algebra are presumed to be known. The differentiation and integration of vector fields has immense applications in a wide range of scientific and engineering disciplines. To mention a few we have rigid dynamics, fluid dynamics, heat transfer, continuum mechanics, electro magnetism, theory of relativity.

DIFFERENTIATION OF VECTOR FUNCTIONS

Let $\vec{V}$ be a vector continuous and single valued function of a scalar variable t. The ideas of limit, continuity and differentiability can easily be extended to the vector function $\vec{V}(t)$. Thus $\vec{V}(t)$ is continuous at $t = t_0$, if

$$\lim_{t \to t_0} \vec{V}(t) = \vec{V}(t_0).$$

Similarly, the derivative of $\vec{V}(t)$ with respect t is quite similar to that used for scalar functions. Thus corresponding to a small change δt in t let there be a change $\delta \vec{V}$ in $\vec{V}$, that is,

$$\delta \vec{V} = \vec{V}(t + \delta t) - \vec{V}(t)$$

then
$$\frac{d}{dt}\vec{V}(t) = \lim_{\delta t \to 0} \frac{\vec{V}(t+\delta t) - \vec{V}(t)}{\delta t} = \lim_{\delta t \to 0} \frac{\delta \vec{V}}{\delta t}.$$

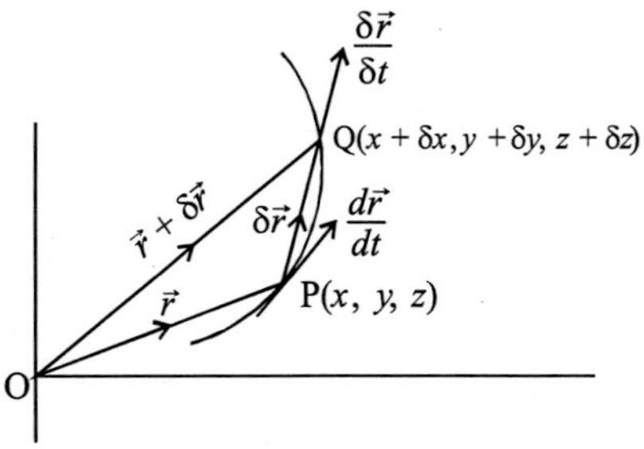

Geometrically, if a vector function $\vec{V}(t)$ represents a set of position vectors of points on some curve C, then $\vec{V}(t)$ defines a curve C with equation $\vec{r} = \vec{V}(t)$.

Now the increment $\delta \vec{r}$ in $\vec{r}$ corresponding to the increment δt in t is represented by the vector $\vec{PQ}$ in the adjoining figure. As $\delta t \to 0$ the point Q approaches to the point P. The chord $\vec{PQ}$ tends to coincide with the tangent at P. Then the limiting value of $\frac{\delta \vec{r}}{\delta t}$ as $\delta t \to 0$ will be a vector whose direction will be limiting direction of $d\vec{r}$, *i.e.*, tangent at P. This limiting value is called the **derivative of $\vec{r}$** w.r.t. t.

In particular, if the scalar variable t is taken to be the arc length s of the curve C measured from some fixed point on C, then we have

$$\lim_{\delta s \to 0} \left| \frac{\delta \vec{r}}{\delta s} \right| = \lim_{Q \to P} \frac{\text{chord PQ}}{\text{arc PQ}} = 1$$

Here $\frac{\delta \vec{r}}{\delta s}$ **is a unit vector along the tangent to the curve C at P.** The concepts of vector differentiation can easily be extended to the sum and product of vectors as follows, provided the original order of vectors is preserved in dealing with the vector product. Thus, for vectors $\vec{A}$, $\vec{B}$, $\vec{C}$ we have

General Rules of Vector Differentiation

	$\frac{d}{dt}(\vec{A} \pm \vec{B}) = \frac{d\vec{A}}{dt} \pm \frac{d\vec{B}}{dt}$
	$\frac{d}{dt}(u\vec{A}) = \frac{du}{dt}\vec{A} + u\frac{d\vec{A}}{dt}$ where u is a scalar function of t.
Also,	$\frac{d}{dt}(\vec{A} \cdot \vec{B}) = \frac{d\vec{A}}{dt} \cdot \vec{B} + \vec{A} \cdot \frac{d\vec{B}}{dt}$
and	$\frac{d}{dt}(\vec{A} \times \vec{B}) = \frac{d\vec{A}}{dt} \times \vec{B} + \vec{A} \times \frac{d\vec{B}}{dt}$
Next,	$\frac{d}{dt}(\vec{A} \cdot \vec{B} \times \vec{C}) = \frac{d\vec{A}}{dt} \cdot \vec{B} \times \vec{C} + \vec{A} \cdot \left(\frac{d\vec{B}}{dt} \times \vec{C}\right) + \vec{A} \cdot \left(\vec{B} \times \frac{d\vec{C}}{dt}\right)$
and	$\frac{d}{dt}\{\vec{A} \times (\vec{B} \times \vec{C})\} = \frac{d\vec{A}}{dt} \times (\vec{B} \times \vec{C}) + \vec{A} \times \left(\frac{d\vec{B}}{dt} \times \vec{C}\right) + \vec{A} \times \left(\vec{B} \times \frac{d\vec{C}}{dt}\right)$

If the vector $\vec{A}(t)$ is expressed in terms of its components as

$$\vec{A} = A_1(t)\, i + A_2(t)\, j + A_3(t)\, k$$

then
$$\frac{d\vec{A}}{dt} = \frac{dA_1}{dt} i + \frac{dA_2}{dt} j + \frac{dA_3}{dt} k.$$

Higher order derivatives of a vector function can be similarly defined.

Thus, if $\vec{r} = xi + yj + zk$ is the position vector of a particle at time t then its velocity $\vec{V}$ and acceleration $\vec{a}$ are given by

$$\vec{V} = \frac{d\vec{r}}{dt} = \frac{dx}{dt}i + \frac{dy}{dt}j + \frac{dz}{dt}k \qquad \text{and} \qquad \vec{a} = \frac{d^2x}{dt^2}i + \frac{d^2y}{dt^2}j + \frac{d^2z}{dt^2}k$$

where $\frac{d^2x}{dt^2}, \frac{d^2y}{dt^2}$ and $\frac{d^2z}{dt^2}$ are the components of acceleration $\vec{a}$ of the particle along the three co-ordinate axes.

If $\vec{V}$ is a vector function of three scalar variables x, y, z then the partial derivatives $\frac{\partial\vec{V}}{\partial x}, \frac{\partial\vec{V}}{\partial y}, \frac{\partial\vec{V}}{\partial z}$ are defined in the same way as in the case of scalar functions.

For example, $\vec{V}_x = \frac{\partial\vec{V}}{\partial x} = \lim_{\delta x\to 0} \frac{\vec{V}(x+\delta x, y, z) - \vec{V}(x,y,z)}{\delta x}$

provided the limit exists. Higher order partial derivatives also are defined analogously.

Two Important Deductions

> (1) **The necessary and sufficient condition for $\vec{V}$ (t) to have a constant magnitude, is**
>
> $$\vec{V} \cdot \frac{d\vec{V}}{dt} = 0.$$

To show that the condition is necessary, let $\vec{V}$ be a vector of constant magnitude then we have

$$\vec{V} \cdot \vec{V} = |\vec{V}|^2 = \text{constant}$$

$$\therefore \qquad \frac{d}{dt}(\vec{V} \cdot \vec{V}) = 0 \qquad \text{or} \qquad \frac{d\vec{V}}{dt} \cdot \vec{V} + \vec{V} \cdot \frac{d\vec{V}}{dt} = 0$$

$$\text{or} \qquad 2\vec{V} \cdot \frac{d\vec{V}}{dt} = 0 \qquad \text{or} \qquad \vec{V} \cdot \frac{d\vec{V}}{dt} = 0$$

which establishes the necessary part.

For sufficiency part, suppose $\vec{V} \cdot \frac{d\vec{V}}{dt} = 0 \Rightarrow \frac{d\vec{V}}{dt} \cdot \vec{V} + \vec{V} \cdot \frac{d\vec{V}}{dt} = 0$ or $\frac{d}{dt}(\vec{V} \cdot \vec{V}) = 0$

$\Rightarrow \quad \vec{V} \cdot \vec{V}$ = constant, hence $|\vec{V}|$ = constant.

Thus, the condition is necessary as well as sufficient.

> (2) **The necessary and sufficient condition for any vector $\vec{V}$ to have a constant direction, is**
>
> $$\vec{V} \times \frac{d\vec{V}}{dt} = 0.$$

First let us show the sufficiency part.

$\vec{V} = |\vec{V}|\,\hat{V} = V\,\hat{V}$ and denote $|\vec{V}|$ by V, then $\frac{d\vec{V}}{dt} = \frac{dV}{dt}\hat{V} + V\frac{d\hat{V}}{dt}$

Therefore $\vec{V} \times \dfrac{d\vec{V}}{dt} = V\hat{V} \times \left(\dfrac{dV}{dt}\hat{V} + V\dfrac{d\hat{V}}{dt}\right)$

$$= V\frac{dV}{dt}\hat{V} \times \hat{V} + V^2\,\hat{V} \times \frac{d\hat{V}}{dt}$$

$$= V^2\,\hat{V} \times \frac{d\hat{V}}{dt} \quad (\text{since } \hat{V} \times \hat{V} = 0)$$

If $\vec{V} \times \dfrac{d\vec{V}}{dt} = 0$ we have $\hat{V} \times \dfrac{d\hat{V}}{dt} = 0 \Rightarrow \dfrac{d\hat{V}}{dt}$ is parallel to $\hat{V}$.

Further since $\hat{V} \cdot \hat{V} = 1$ we have $\dfrac{d\hat{V}}{dt} \cdot \hat{V} + \hat{V} \cdot \dfrac{d\hat{V}}{dt} = 0 \Rightarrow \dfrac{d\hat{V}}{dt} \cdot \hat{V} = 0$

$\Rightarrow$ $\dfrac{d\hat{V}}{dt}$ is perpendicular to $\hat{V}$.

Thus, $\dfrac{d\hat{V}}{dt}$ is parallel as well as perpendicular to $\hat{V}$ which appears to be a contradiction in itself. Hence $\hat{V}$ is a constant vector which implies that $\vec{V}$ has constant direction and hence the condition is sufficient.

Next, to establish the necessary part, suppose $\vec{V}$ has a constant direction therefore $\hat{V}$ also has a constant direction and hence a constant vector.

$\Rightarrow$ $\dfrac{d\hat{V}}{dt} = 0$ and in turn $\hat{V} \times \dfrac{d\hat{V}}{dt} = 0.$

which establishes the necessary part.

EXAMPLE 9.1. A particle moves along a curve whose parametric equation is $x = 3t^2$, $y = t^2 - 2t$, $z = t^3$ where t is time. Find its velocity and acceleration at a time $t = 2$.

SOLUTION: If $\vec{r}$ is the position vector of the particle at any time t, then

$$\vec{r} = xi + yj + zk = 3t^2 i + (t^2 - 2t)\,j + t^3 k.$$

The velocity $\vec{V}$ of the particle at a time t, is given by

$$\vec{V} = \frac{d\vec{r}}{dt} = 6ti + (2t - 2)\,j + 3t^2 k$$

and the acceleration $\vec{a}$ of the particle at time t, is given by

$$\vec{a} = \frac{d^2\vec{r}}{dt^2} = 6i + 2j + 6tk$$

At time $t = 2$, we have

$$\left[\frac{d\vec{r}}{dt}\right]_{t=2} = 12i + 2j + 12k \quad \text{and} \quad \left[\frac{d^2\vec{r}}{dt^2}\right]_{t=2} = 6i + 2j + 12k. \quad \textbf{Ans.}$$

EXAMPLE 9.2. If $\frac{d\vec{A}}{dt} = \vec{C} \times \vec{A}$ and $\frac{d\vec{B}}{dt} = \vec{C} \times \vec{B}$, then show that

$$\frac{d}{dt}(\vec{A} \times \vec{B}) = \vec{C} \times (\vec{A} \times \vec{B}).$$

SOLUTION: $\frac{d}{dt}(\vec{A} \times \vec{B}) = \frac{d\vec{A}}{dt} \times \vec{B} + \vec{A} \times \frac{d\vec{B}}{dt}$

$$= (\vec{C} \times \vec{A}) \times \vec{B} + \vec{A} \times (\vec{C} \times \vec{B})$$

$$= (\vec{B} \cdot \vec{C})\,\vec{A} - (\vec{B} \cdot \vec{A})\,\vec{C} + (\vec{A} \cdot \vec{B})\,\vec{C} - (\vec{A} \cdot \vec{C})\,\vec{B}$$

$$= (\vec{B} \cdot \vec{C})\,\vec{A} - (\vec{A} \cdot \vec{C})\,\vec{B}$$

$= \vec{C} \times (\vec{A} \times \vec{B}).$ **Hence Proved**

EXAMPLE 9.3. Find the angle between the tangents to the curve

$\vec{r} = t^2 i - 2t j + t^3 k$ at the points $t = 1$ and $t = 2$.

SOLUTION: The tangent at any point 't' is given by

$$\frac{d\vec{r}}{dt} = \frac{d}{dt}(t^2 i - 2t j + t^3 k) = 2ti - 2j + 3t^2 k$$

Therefore the tangents $\vec{T}_1$ and $\vec{T}_2$ at $t = 1$ and $t = 2$ respectively, are given by

$$\vec{T}_1 = 2i - 2j + 3k \quad \text{and} \quad \vec{T}_2 = 4i - 2j + 12k.$$

If θ is the angle between $\vec{T}_1$ and $\vec{T}_2$ then

$$\cos\theta = \frac{\vec{T}_1 \cdot \vec{T}_2}{|\vec{T}_1|\,|\vec{T}_2|} = \frac{2(4) + (-2)(-2) + 3(12)}{\sqrt{2^2 + (-2)^2 + 3^2}\,\sqrt{4^2 + (-2)^2 + 12^2}} = \frac{48}{\sqrt{17}\,\sqrt{164}}$$

$\therefore$ $\theta = \cos^{-1}\left(\frac{24}{\sqrt{17}\,\sqrt{41}}\right)$ is the required angle. **Ans.**

EXAMPLE 9.4. Show that

$$\frac{d}{dt}\left(\vec{V} \cdot \frac{d\vec{V}}{dt} \times \frac{d^2\vec{V}}{dt^2}\right) = \vec{V} \cdot \left(\frac{d\vec{V}}{dt} \times \frac{d^3\vec{V}}{dt^3}\right)$$

SOLUTION: $\frac{d}{dt}\left(\vec{V} \cdot \frac{d\vec{V}}{dt} \times \frac{d^2\vec{V}}{dt^2}\right) = \frac{d\vec{V}}{dt} \cdot \left(\frac{d\vec{V}}{dt} \times \frac{d^2\vec{V}}{dt^2}\right) + \vec{V} \cdot \left(\frac{d^2\vec{V}}{dt^2} \times \frac{d^2\vec{V}}{dt^2} + \frac{d\vec{V}}{dt} \times \frac{d^3\vec{V}}{dt^3}\right)$

$$= 0 + \vec{V} \cdot \left(0 + \frac{d\vec{V}}{dt} \times \frac{d^3\vec{V}}{dt^3}\right) = \vec{V} \cdot \left(\frac{d\vec{V}}{dt} \times \frac{d^3\vec{V}}{dt^3}\right),$$

since a scalar triple product is zero if any two of its three vectors are identical or parallel. **Hence Proved.**

EXAMPLE 9.5. If $\vec{r}$ is the position vector of a particle of mass m at time t and $\vec{F}$ is the external force applied on the particle, then show that the moment M of the force $\vec{F}$ about its origin O, is given by

$\vec{M} = \dfrac{d\vec{H}}{dt}$ where $\vec{H} = \vec{r} \times m\vec{V}$ and $\vec{V}$ is the velocity of the particle at time t.

SOLUTION: Moment M of the force $\vec{F}$ about the origin O, is given by

$$\vec{M} = \vec{r} \times \vec{F} = \vec{r} \times \left(m\frac{d\vec{V}}{dt} \right)$$

Since, $\vec{H} = r \times m\vec{V}$, we have

$$\frac{d\vec{H}}{dt} = \frac{d}{dt}(\vec{r} \times m\vec{V}) = \frac{d\vec{r}}{dt} \times m\vec{V} + \vec{r} \times \frac{d}{dt}(m\vec{V})$$

$$= \vec{V} \times (m\vec{V}) + \vec{r} \times \frac{d}{dt}(m\vec{V}) = 0 + \vec{r} \times \left(m\frac{d\vec{V}}{dt} \right)$$

$\therefore$ $\vec{M} = \vec{r} \times m\dfrac{d\vec{V}}{dt} = \dfrac{d\vec{H}}{dt}.$ **Hence Proved.**

SCALAR AND VECTOR POINT FUNCTIONS AND LEVEL SURFACES

(UPTU 2001)

A variable quantity which depends for its value on its position only, that is, upon the co-ordinates of the points say (x, y, z) in space, is called a **point function**. There are two types of point functions.

***(a)* SCALAR POINT FUNCTION:** If a scalar quantity ϕ is a function of a point, that is, depends on its position (x, y, z) in space then $\phi\,(x, y, z)$ is called a scalar point function. The scalar point function does not depend upon the choice of co-ordinate system. For example, density of a body, the temperature of a body at any instant, potential of a body are all examples of scalar point function. Pressure in a fluid at a point usually depends on its depth hence it is also a scalar point function.

In dealing with the differentiation of a scalar point function $\phi\,(x, y, z)$, following formulae will be frequently used $d\phi = \dfrac{\partial\phi}{\partial x}dx + \dfrac{\partial\phi}{\partial y}dy + \dfrac{\partial\phi}{\partial z}dz$ and $\dfrac{\partial\phi}{\partial s} = \dfrac{\partial\phi}{\partial x}\dfrac{\partial x}{\partial s} + \dfrac{\partial\phi}{\partial y}\dfrac{\partial y}{\partial s} + \dfrac{\partial\phi}{\partial z}\dfrac{\partial z}{\partial s}.$

***(b)* VECTOR POINT FUNCTION:** If a vector quantity $\vec{V}$ depends for its value on its position (x, y, z) in space then $\vec{V}\,(x, y, z)$ is called a vector point function. Velocity of a moving fluid, force, gravitational force or electrical intensity are all examples of vector point function.

While differentiating a vector point function $\vec{V}\,(x, y, z)$ following formulae will be frequently used.

$$d\vec{F} = \frac{\partial\vec{F}}{\partial x}dx + \frac{\partial\vec{F}}{\partial y}dy + \frac{\partial\vec{F}}{\partial z}dz \quad \text{and} \quad \frac{\partial\vec{F}}{\partial s} = \frac{\partial\vec{F}}{\partial x}\frac{\partial x}{\partial s} + \frac{\partial\vec{F}}{\partial y}\frac{\partial y}{\partial s} + \frac{\partial\vec{F}}{\partial z}\frac{\partial z}{\partial s}.$$

LEVEL SURFACES: Let $\phi(x, y, z)$ be a scalar point function continuous in a region. The surface drawn in space containing all those points where $\phi(x, y, z)$ has same value, is called a **level surface**. Equipotential surfaces and isothermal surfaces are some examples of level surfaces.

GRADIENT OF A SCALAR POINT FUNCTION

The vector differential operator $i\frac{\partial}{\partial x}+j\frac{\partial}{\partial y}+k\frac{\partial}{\partial z}$ is denoted by the symbol ∇ and is called Del (or Nabla). It operates on any scalar point function $\phi(x, y, z)$ say, and we get a vector quantity $\nabla\phi = i\frac{\partial\phi}{\partial x}+j\frac{\partial\phi}{\partial y}+k\frac{\partial\phi}{\partial z}$ which is called the gradient of the scalar point function $\phi(x, y, z)$. This is also written as **gradient ϕ** or simply **grad ϕ.**

Actually, the operator ∇ is a differential operator but vector in nature.

Thus
$$\boxed{\text{grad } \phi = \nabla\phi = i\frac{\partial\phi}{\partial x}+j\frac{\partial\phi}{\partial y}+k\frac{\partial\phi}{\partial z}.}$$

PHYSICAL AND GEOMETRICAL SIGNIFICANCE OF GRADIENT

Let P (x, y, z) be the point with position vector $\vec{r}$, then $\vec{r} = xi + yj + zk$

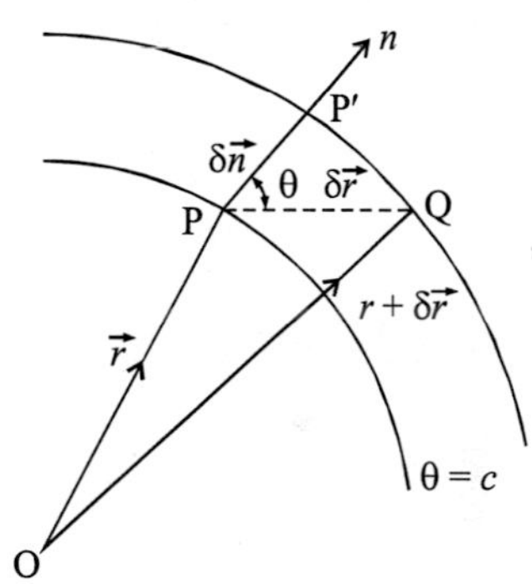

$$\therefore \qquad d\vec{r} = idx + jdy + kdz$$

Next,
$$\nabla\phi \cdot d\vec{r} = \left(i\frac{\partial\phi}{\partial x}+j\frac{\partial\phi}{\partial y}+k\frac{\partial\phi}{\partial z}\right)\cdot(idx + jdy + kdz)$$

$$= \frac{\partial\phi}{\partial x}dx+\frac{\partial\phi}{\partial y}dy+\frac{\partial\phi}{\partial z}dz = d\phi \qquad \text{...(1)}$$

Since $\nabla\phi$ is vector, we want to know its direction as well as magnitude. Let Q $(x + \delta x,\ y + \delta y, z + \delta z)$ be a neighbouring point with position vector $\vec{r} + \delta\vec{r}$. Consider the level surface through P at every point of which the value of the scalar point function is ϕ while the level surface through Q has value of the function as $\phi + \delta\phi$ throughout.

Thus, $\phi + \delta\phi = \phi(x + \delta x,\ y + \delta y, z + \delta z)$

$$= \phi(x, y, z) + \delta x\frac{\partial\phi}{\partial x} + \delta y\frac{\partial\phi}{\partial y} + \delta z\frac{\partial\phi}{\partial z} + \ldots \qquad \text{(by Taylor's series)}$$

$$\Rightarrow \qquad \delta\phi = \frac{\partial\phi}{\partial x}\delta x+\frac{\partial\phi}{\partial y}\delta y+\frac{\partial\phi}{\partial z}\delta z \qquad \text{neglecting higher order terms.}$$

$$= \left(i\frac{\partial\phi}{\partial x}+j\frac{\partial\phi}{\partial y}+k\frac{\partial\phi}{\partial z}\right)\cdot(i\,\delta x + j\,\delta y + k\,\delta z)$$

$$= \nabla\phi\cdot\delta\vec{r} \qquad \text{where} \qquad \delta\vec{r} = \vec{PQ}. \qquad \text{...(2)}$$

As the point Q approaches the point P, $\delta\phi \to 0$ and PQ becomes tangent to the level surface at P which clearly indicates that **the direction of $\nabla\phi$ at P is along the normal to the level surface of P.**

Further, let the normal at P meet the level surface of Q at P′ then $\overrightarrow{PP'} = \delta\vec{n}$ = projection of PQ along the normal. It is clear that $\dfrac{\partial\phi}{\partial r}$ is the rate of change of ϕ along $\overrightarrow{PQ}$ and $\dfrac{\partial\phi}{\partial n}$ is the rate of change of ϕ along PP′, *i.e.*, along the normal at P.

$$\therefore \quad \frac{\delta\phi}{\delta r} = \frac{\delta\phi}{\delta n}\frac{\delta n}{\delta r} = \frac{\delta\phi}{\delta n}\cos\theta \text{ (see the above figure)}$$

$$\text{As } Q \to P, \quad \lim\frac{\delta\phi}{\delta r} = \frac{\partial\phi}{\partial r} \quad \text{and} \quad \lim\frac{\delta\phi}{\delta n} = \frac{\partial\phi}{\partial n}$$

$$\Rightarrow \quad \frac{\partial\phi}{\partial r} = \frac{\partial\phi}{\partial n}\cos\theta \quad \text{hence} \quad \frac{\partial\phi}{\partial r} \le \frac{\delta\phi}{\delta n} \qquad ...(3)$$

Therefore **rate of change of ϕ is maximum in the direction of the normal.**

Since, δn is the projection of δr along the normal, we have $dn = \hat{n}\cdot d\vec{r}$ ...(4)

In relation (1) we had $d\phi = \nabla\phi\cdot d\vec{r}$ Let us write $d\phi = \dfrac{\partial\phi}{\partial n}dn$ ...(5)

Using (4) and (5) here we get $\nabla\phi\cdot d\vec{r} = \dfrac{\partial\phi}{\partial n}\hat{n}\cdot d\vec{r} \Rightarrow \nabla\phi = \dfrac{\partial\phi}{\partial n}\hat{n}$

which shows that grad ϕ at the point P represents maximum rate of change of ϕ which is along the outward drawn normal to the level surface at P. In other words, vector $\nabla\phi$ is in the direction in which ϕ increases most rapidly and $-\nabla\phi$ is in direction in which ϕ decreases most rapidly.

DIRECTIONAL DERIVATIVE

$\dfrac{\partial\phi}{\partial r}$ represents the rate of change of ϕ in any other direction and is called ***directional derivative.*** **The directional derivative along certain direction $\vec{a}$ is $\nabla\phi\cdot\hat{a}$. Among all the directional derivatives, $|\nabla\phi| = \dfrac{\partial\phi}{\partial n}$ has the maximum value.**

It is easy to derive the following corrollaries:

(1) $\nabla(u \pm v) = \nabla(u) \pm \nabla(v)$

(2) $\nabla(au) = a\nabla(u)$

(3) $\nabla(uv) = \nabla(u)\,v + u\nabla(v)$

(4) $\nabla[f(u)] = f'(u)\,\nabla u$

(5) $\nabla(\vec{a}\cdot\vec{r}) = \vec{a}$ where $\vec{a}$ is a constant vector. **(UPTU 2001 Q.B.)**

Let us prove the corollaries (4) and (5) as (1), (2) and (3) are obvious.

(4) Let $v = f(u)$ where u is a scalar function of x, y, z.

then $$\nabla(v) = i\frac{\partial v}{\partial x} + j\frac{\partial v}{\partial y} + k\frac{\partial v}{\partial z} = i\frac{\partial f(u)}{\partial x} + j\frac{\partial f(u)}{\partial y} + k\frac{\partial f(u)}{\partial z}$$

$$= i f'(u)\frac{\partial u}{\partial x} + j f'(u)\frac{\partial u}{\partial y} + k f'(u)\frac{\partial u}{\partial z} = f'(u)\left\{i\frac{\partial u}{\partial x} + j\frac{\partial u}{\partial y} + k\frac{\partial u}{\partial z}\right\}$$

or $$\boldsymbol{\nabla v = f'(u)\,\nabla u}.$$

(5) Let $\vec{a} = a_1 i + a_2 j + a_3 k$ and $\vec{r} = xi + yj + zk$

hence $\vec{a}\cdot\vec{r} = a_1 x + a_2 y + a_3 z$

then $$\nabla(\vec{a}\cdot\vec{r}) = \left(i\frac{\partial}{\partial x} + j\frac{\partial}{\partial y} + k\frac{\partial}{\partial z}\right)(a_1 x + a_2 y + a_3 z) = ia_1 + ja_2 + ka_3 = \vec{a}.$$

Thus, $\boldsymbol{\nabla(\vec{a}\cdot\vec{r}) = \vec{a}}$.

EXAMPLE 9.6. **(a)** Find a unit vector normal to the surface $x^3 + y^3 + 3xyz = 3$ at the point $(1, 2, -1)$ **(UPTU 2001 Q.B.)**

(b) Find the magnitude of the gradient of the functions $f = xyz^3$ at $(1, 0, 2)$. **[UPTU 2011-12]**

SOLUTION: (a) Equation of the given surface is $\phi \equiv x^3 + y^3 + 3xyz - 3 = 0$

Since, $\nabla\phi$ is a vector normal to the surface, we have

$$\nabla\phi = i\frac{\partial\phi}{\partial x} + j\frac{\partial\phi}{\partial y} + k\frac{\partial\phi}{\partial z} = i(3x^2 + 3yz) + j(3y^2 + 3xz) + k(3xy)$$

$$[\nabla\phi]_{(1,\,2,\,-1)} = i(3-6) + j(12-3) + k(6) = 3(-i + 3j + 2k)$$

Therefore unit vector along the normal to the given surface, is

$$\frac{(-i+3j+2k)}{\sqrt{1+9+4}} = \frac{1}{\sqrt{14}}(-i+3j+2k). \quad \textbf{Ans.}$$

(b) Gradient of $$f = \nabla f = i\frac{\partial f}{\partial x} + j\frac{\partial f}{\partial y} + k\frac{\partial f}{\partial z}$$

$$= i\frac{\partial}{\partial x}(xyz^3) + j\frac{\partial}{\partial y}(xyz^3) + k\frac{\partial}{\partial z}(xyz^3) = i(yz^3) + j(xz^3) + k(3xyz^2)$$

Now $(\nabla f)_{(1,\,0,\,2)} = i(0) + j(8) + k(0)$ and $|\nabla f| = 8$. **Ans.**

EXAMPLE 9.7. If $u = x + y + z$, $v = x^2 + y^2 + z^2$, $w = xy + yz + zx$, show that ∇u, ∇v, ∇w are coplanar vectors. **(UPTU 2002)**

SOLUTION: Here $$\nabla u = i\frac{\partial u}{\partial x} + j\frac{\partial u}{\partial y} + k\frac{\partial u}{\partial z} = i(1) + j(1) + k(1) = i + j + k,$$

$$\nabla v = i\frac{\partial v}{\partial x} + j\frac{\partial v}{\partial y} + k\frac{\partial v}{\partial z} = i(2x) + j(2y) + k(2z) = 2(xi + yj + zk),$$

$$\nabla w = i\frac{\partial w}{\partial x} + j\frac{\partial w}{\partial y} + k\frac{\partial w}{\partial z} = i(y+z) + j(z+x) + k(x+y).$$

Now, $\nabla u, \nabla v, \nabla w$ will be coplanar when $\nabla u \cdot (\nabla v \times \nabla w) = 0$

$$\nabla u \cdot (\nabla v \times \nabla w) = \begin{vmatrix} 1 & 1 & 1 \\ 2x & 2y & 2z \\ y+z & z+x & x+y \end{vmatrix} \quad \text{applying } R_3 \to R_3 + \frac{1}{2} R_2$$

$$= \begin{vmatrix} 1 & 1 & 1 \\ 2x & 2y & 2z \\ x+y+z & x+y+z & x+y+z \end{vmatrix} = (x+y+z) \begin{vmatrix} 1 & 1 & 1 \\ 2x & 2y & 2z \\ 1 & 1 & 1 \end{vmatrix} = 0$$

Therefore, $\nabla u, \nabla v, \nabla w$ are coplanar. **Hence Proved.**

EXAMPLE 9.8. If $r^2 = x^2 + y^2 + z^2$ then show that $\nabla f(r) = \frac{f'(r)}{r} \vec{r}$ and hence show that

(*i*) $\nabla (r) = \hat{r}$ (*ii*) $\nabla \left(\frac{1}{r}\right) = -\frac{1}{r^3} \vec{r}$

(*iii*) $\nabla (r^n) = n\, r^{n-2}\, \vec{r}$ (*iv*) $\nabla \int r^n dr = r^{n-1} \vec{r}$. **(UPTU 2003-04)**

SOLUTION: By definition

$$\nabla f(r) = i\frac{\partial}{\partial x} f(r) + j\frac{\partial}{\partial y} f(r) + k\frac{\partial}{\partial z} f(r) = i\, f'(r) \frac{\partial r}{\partial x} + j\, f'(r) \frac{\partial r}{\partial y} + k\, f'(r) \frac{\partial r}{\partial z}$$

$$= f'(r) \left(i\frac{\partial r}{\partial x} + j\frac{\partial r}{\partial y} + k\frac{\partial r}{\partial z} \right)$$

From $r^2 = x^2 + y^2 + z^2$ we have $2r \frac{\partial r}{\partial x} = 2x$, hence

$$\frac{\partial r}{\partial x} = \frac{x}{r}, \quad \frac{\partial r}{\partial y} = \frac{y}{r} \quad \text{and} \quad \frac{\partial r}{\partial z} = \frac{z}{r}.$$

$$\therefore \quad \nabla f(r) = f'(r) \left[i\frac{x}{r} + j\frac{y}{r} + k\frac{z}{r} \right] \quad \text{or} \quad \nabla f(r) = \frac{1}{r} f'(r)\, \vec{r}.$$ **Hence Proved.**

(*i*) $\nabla (r) = i\frac{\partial r}{\partial x} + j\frac{\partial r}{\partial y} + k\frac{\partial r}{\partial z} = i\frac{x}{r} + j\frac{y}{r} + k\frac{z}{r} = \frac{\vec{r}}{r} = \hat{r}.$

(*ii*) $\nabla \left(\frac{1}{r}\right) = i\frac{\partial}{\partial x}\left(\frac{1}{r}\right) + j\frac{\partial}{\partial y}\left(\frac{1}{r}\right) + k\frac{\partial}{\partial z}\left(\frac{1}{r}\right) = i\left(-\frac{1}{r^2}\right)\frac{\partial r}{\partial x} + j\left(-\frac{1}{r^2}\right)\frac{\partial r}{\partial y} + k\left(-\frac{1}{r^2}\right)\frac{\partial r}{\partial z}$

$$= -\frac{1}{r^2}\left(i\frac{\partial r}{\partial x} + j\frac{\partial r}{\partial y} + k\frac{\partial r}{\partial z} \right) = -\frac{1}{r^2}\left(i\frac{x}{r} + j\frac{y}{r} + k\frac{z}{r} \right) = -\frac{1}{r^3}\, \vec{r}.$$

(*iii*) Let us take $f(r) = r^n$ and using the result $\nabla f(r) = \frac{1}{r} f'(r)\, \vec{r}$

we get $\nabla (r^n) = \frac{1}{r}\, n\, r^{n-1}\, \vec{r} = n\, r^{n-2}\, \vec{r}$

(*iv*) We take $f(r) = \int r^n dr$ so that $f'(r) = r^n$

$$\therefore \quad \nabla \int r^n dr = \frac{1}{r} \cdot r^n\, \vec{r} = r^{n-1}\, \vec{r}.$$ **Hence Proved.**

EXAMPLE 9.9. Determine the constant 'a' such that at any point of inter-section of the two spheres $(x-a)^2+y^2+z^2=3$ and $x^2+(y-1)^2+z^2=1$ their tangent planes are perpendicular to each other.

SOLUTION: Let $\phi_1 = (x-a)^2 + y^2 + z^2$ and $\phi_2 = x^2 + (y-1)^2 + z^2$.

Then $$\nabla\phi_1 = i\frac{\partial\phi_1}{\partial x} + j\frac{\partial\phi_1}{\partial y} + k\frac{\partial\phi_1}{\partial z} = 2(x-a)\,i + 2yj + 2zk$$

and $$\nabla\phi_2 = i\frac{\partial\phi_2}{\partial x} + j\frac{\partial\phi_2}{\partial y} + k\frac{\partial\phi_2}{\partial z} = 2xi + 2(y-1)\,j + 2zk.$$

These vectors $\nabla\phi_1$ and $\nabla\phi_2$ are along the normals to the two spheres at a point (x, y, z) of their inter-section. The tangent planes to the two spheres will be perpendicular to each other when the normals are perpendicular to each other for which

$$[2(x-a)\,i + 2yj + 2zk]\cdot[2xi + 2(y-1) + 2zk] = 0 \quad \text{or} \quad x(x-2a) + y(y-1) + z^2 = 0$$

or $$x^2 + y^2 + z^2 - ax - y = 0 \qquad \text{...(1)}$$

At the point P (x, y, z) of inter-section of given spheres, we have

$$x^2 + y^2 + z^2 - 2ax + a^2 - 3 = 0 \quad \text{and} \quad x^2 + y^2 + z^2 - 2y = 0$$

Adding these two and dividing by 2, we get

or $$x^2 + y^2 + z^2 - ax - y + \frac{1}{2}a^2 - \frac{3}{2} = 0 \qquad \text{...(2)}$$

Comparing (1) and (2), we get $\frac{a^2}{2} - \frac{3}{2} = 0$ or $a = \pm\sqrt{3}$. **Ans.**

EXAMPLE 9.10. If $\vec{a}$ and $\vec{b}$ are constant vectors and $\vec{r} = xi + yj + zk$, show that

$$r^5\,[\vec{b}\cdot\nabla\{\vec{a}\cdot\nabla(1/r)\}] = 3(\vec{a}\cdot\vec{r})(\vec{b}\cdot\vec{r}) - (\vec{a}\cdot\vec{b})\,r^2.$$

SOLUTION: We know that $\nabla f(r) = \frac{f'(r)}{r}\vec{r}$ ...(1)

$$\Rightarrow \quad \nabla\left(\frac{1}{r}\right) = \frac{1}{r}\left(-\frac{1}{r^2}\right)\vec{r} = -\frac{1}{r^3}\vec{r} \quad \text{and} \quad \vec{a}\cdot\nabla\left(\frac{1}{r}\right) = -\frac{1}{r^3}(\vec{a}\cdot\vec{r})$$

$$\therefore \quad \nabla[\vec{a}\cdot\nabla(1/r)] = -\nabla\left[\frac{1}{r^3}(\vec{a}\cdot\vec{r})\right] = -\nabla\left(\frac{1}{r^3}\right)(\vec{a}\cdot\vec{r}) - \frac{1}{r^3}\nabla(\vec{a}\cdot\vec{r})$$

Using (1) we have $\nabla\left(\frac{1}{r^3}\right) = \frac{1}{r}\frac{d}{dr}\left(\frac{1}{r^3}\right)\vec{r} = -\frac{3}{r^5}\vec{r}$.

Also, we know that $\nabla(\vec{a}\cdot\vec{r}) = \vec{a}$ if $\vec{a}$ is a constant vector.

$$\therefore \quad \nabla[\vec{a}\cdot\nabla(1/r)] = \frac{3}{r^5}\vec{r}\,(\vec{a}\cdot\vec{r}) - \frac{1}{r^3}\vec{a}$$

Therefore $$r^5\,[\vec{b}\cdot\nabla\{\vec{a}\cdot\nabla(1/r)\}] = r^5\left[\frac{3}{r^5}(\vec{b}\cdot\vec{r})(\vec{a}\cdot\vec{r}) - \frac{1}{r^3}(\vec{b}\cdot\vec{a})\right]$$

$$= 3(\vec{a}\cdot\vec{r})(\vec{b}\cdot\vec{r}) - (\vec{a}\cdot\vec{b})\,r^2.$$ **Hence the result.**

EXAMPLE 9.11. If $\vec{A} = 2x^2i - 3yzj + xz^2k$ and $\phi = 2z - x^3y$, find $\vec{A}\cdot\nabla\phi$ and $\vec{A} \times \nabla\phi$ at the point (1, –1, 1).

SOLUTION: Given $\phi = 2z - x^3y$ as a scalar function and $\vec{A} = 2x^2i - 3yzj + xz^2k$ as vector function,

$$\nabla\phi = i\frac{\partial}{\partial x}(2z - x^3y) + j\frac{\partial}{\partial y}(2z - x^3y) + k\frac{\partial}{\partial z}(2z - x^3y)$$

$$= -3x^2yi - x^3i + 2k$$

At (1, – 1, 1), $\nabla\phi = 3i - j + 2k$ and $\vec{A} = 2i + 3j + k$

$\therefore$ $\vec{A}\,.\,\nabla\phi = 3(2) - 1(3) + 2(1) = 5.$

And $$\vec{A} \times \nabla\phi = \begin{vmatrix} i & j & k \\ 2 & 3 & 1 \\ 3 & -1 & 2 \end{vmatrix} = i(6+1) + j(3-4) + k(-2-9) = 7i - j - 11k$$

Thus $\vec{A}\,.\,\nabla\phi = 5$ and $\vec{A} \times \nabla\phi = 7i - j - 11k.$ **Ans.**

EXAMPLE 9.12. (*a*) Find the directional derivative of the scalar point function $\phi = 3e^{2x - y + z}$ at the point A (1, 1, – 1) in the direction towards the point B (– 3, 5, 6).

(*b*) Find the directional derivative of $\phi(x, y, z) = xy^2 + yz^3$ at the point (2, –1, 1) in the direction of the normal to the surface $x \log z - y^2 + 4 = 0$ at (2, –1, 1). **(UPTU 2010-11)**

SOLUTION: (*a*) $$\nabla\phi = i\frac{\partial\phi}{\partial x} + j\frac{\partial\phi}{\partial y} + k\frac{\partial\phi}{\partial z}$$

$$= 3\,[i\,2e^{2x-y+z} - j\,e^{2x-y+z} + ke^{2x-y+z}] = 3e^{2x-y+z}\,(2i - j + k).$$

$\therefore$ $[\nabla\phi]_{(1,\,1,\,-1)} = 3\,(2i - j + k)\,e^{2-1-1} = 6i - 3j + 3k.$

The directional derivative is the component of $\nabla\phi$ at the given point in the given direction

$$\overrightarrow{AB} = -3i + 5j + 6k - (i + j - k) = -4i + 4j + 7k$$

Therefore, the required directional derivative is the component of $6i - 3j + 3k$ in the direction of the vector $-4i + 4j + 7k$.

$$= (6i - 3j + 3k)\cdot\frac{(-4i + 4j + 7k)}{\sqrt{(-4)^2 + 4^2 + 7^2}}$$

$$= \frac{-24 - 12 + 21}{\sqrt{16 + 16 + 49}} = \frac{-5}{3}.$$ **Ans.**

(*b*)
$$\nabla\phi = i\frac{\partial}{\partial x}[xy^2 + yz^3] + j\frac{\partial}{\partial y}[xy^2 + yz^3] + k\frac{\partial}{\partial z}[xy^2 + yz^3]$$
$$= y^2 i + (2xy + z^3)\, j + 3yz^2 k$$
$$[\nabla\phi]_{(2,-1,1)} = i - 3j - 3k.$$

Vector normal to the surface ($x \log z - y^2 + 4 = 0$), is

$$\nabla(x \log z - y^2 + 4) = i\frac{\partial}{\partial x}[x\log z - y^2 + 4] + j\frac{\partial}{\partial y}[x\log z - y^2 + 4] + k\frac{\partial}{\partial z}[x\log z - y^2 + 4]$$
$$= \log z\; i - 2yj + \frac{x}{z}k$$
$$= 2j + 2k. \qquad \text{at point } (2, -1, 1)$$

Unit vector along the normal to the surface ($x \log z - y^2 + 4 = 0$) is $\hat{a}$, given by

$$\hat{a} = \frac{2j + 2k}{\sqrt{4+4}} = \frac{j+k}{\sqrt{2}},$$

$\therefore$ required directional derivative is

$$(\nabla\phi)_{(2,-1,1)} \cdot \hat{a} = (i - 3j - 3k)\cdot\frac{(j+k)}{\sqrt{2}}$$
$$= \frac{-6}{\sqrt{2}}. \qquad \textbf{Ans.}$$

EXAMPLE 9.13. Find the directional derivative of the scalar point function $\phi = x^2y + y^2z + z^2x$ at the point $(2, 2, 2)$ in the direction of the normal to the surface $4x^2y + 2z^2 = 2$ at the point $(2, -1, 3)$.

SOLUTION: We have
$$\nabla\phi = i\frac{\partial\phi}{\partial x} + j\frac{\partial\phi}{\partial y} + k\frac{\partial\phi}{\partial z}$$
$$= (2xy + z^2)\, i + (2yz + x^2)\, j + (2zx + y^2)\, k$$

Hence $[\nabla\phi]_{(2,2,2)} = 12i + 12j + 12k = 12\,(i + j + k)$

The given surface can be written as $u = 2$ where $u = 4x^2y + 2z^2$.

$\therefore$ $\nabla u = 8xyi + 4x^2j + 4zk$

and $[\nabla u]_{(2,-1,3)} = -16i + 16j + 12k$

which is the vector along the normal to the given surface at $(2, -1, 3)$. Therefore the required directional derivative is the component of $12\,(i + j + k)$ along $(-16i + 16j + 12k)$, equal to

$$= (12i + 12j + 12k)\cdot\frac{(-16i + 16j + 12k)}{\sqrt{(-16)^2 + 16^2 + 12^2}} = \frac{12(-16) + 12(16) + 12(12)}{4\sqrt{16+16+9}} = \frac{36}{\sqrt{41}}. \qquad \textbf{Ans.}$$

EXAMPLE 9.14. If the directional derivative of $\phi = ax^2y + by^2z + cz^2x$ at the point $(1, 1, 1)$ has maximum magnitude 15 in the direction parallel to the line

$$\frac{x-1}{2} = \frac{y-3}{-2} = \frac{z}{1},$$ find the values of a, b, c. **(UPTU 2001-02)**

SOLUTION: $\nabla\phi = i\frac{\partial\phi}{\partial x} + j\frac{\partial\phi}{\partial y} + k\frac{\partial\phi}{\partial z} = (2axy + cz^2)\, i + (2byz + ax^2)\, j + (2czx + by^2)\, k$

$\therefore \quad [\nabla\phi]_{(1,\,1,\,1)} = (2a + c)\, i + (2b + a)\, j + (2c + b)\, k$

This is along the normal to the surface $\phi(x, y, z) = \phi(1, 1, 1) = a + b + c$ and hence $[\nabla\phi]_{(1,\,1,\,1)}$ is the maximum directional derivative.

Since, this is given to be parallel to the line $\frac{x-1}{2} = \frac{y-3}{-2} = \frac{z}{1}$, we have

$$\frac{2a+c}{2} = \frac{2b+a}{-2} = \frac{2c+b}{1}$$

From the first two, we have $\quad 3a + 2b + c = 0$

and from the last two, we have $\quad a + 4b + 4c = 0$

Solving the above equations for $a, b, c,$ we get

$$\frac{a}{8-4} = \frac{b}{1-12} = \frac{c}{12-2} \quad \text{or} \quad \frac{a}{4} = \frac{b}{-11} = \frac{c}{10} = \lambda, \text{ say}$$

Hence $a = 4\lambda, \quad b = -11\lambda, \quad c = 10\lambda$

As we are given that $|\nabla\phi|_{(1,\,1,\,1)} = 15 \quad \Rightarrow \quad (2a + c)^2 + (2b + a)^2 + (2c + b)^2 = 15^2$

or $\quad (8\lambda + 10\lambda)^2 + (-22\lambda + 4\lambda)^2 + (20\lambda - 11\lambda)^2 = 15^2$

$$\therefore \quad \lambda^2 = \frac{15^2}{18^2 + 18^2 + 9^2} \quad \text{or} \quad \lambda = \pm\frac{5}{9}.$$

Therefore $\quad a = \pm\frac{20}{9}, \quad b = \pm\frac{55}{9}, \quad c = \pm\frac{50}{9}.$ **Ans.**

EXAMPLE 9.15. Find the directional derivative of the function $f(x, y, z) = 2xy + z^2$ at the point $(1, -1, 3)$ in the direction of the vector $i + 2j + 2k$.

SOLUTION: $\nabla f = i\frac{\partial}{\partial x}(2xy + z^2) + j\frac{\partial}{\partial y}(2xy + z^2) + k\frac{\partial}{\partial z}(2xy + z^2) = 2yi + 2xj + 2zk$

∇f at $(1, -1, 3)$ is equal to $-2i + 2j + 6k$.

Now, directional derivative of f at $(1, -1, 3)$ along the vector $(i + 2j + 2k)$

$$= (-2i + 2j + 6k).\frac{(i + 2j + 2k)}{\sqrt{1+4+4}} = \frac{1}{3}(-2+4+12) = \frac{14}{3}.$$ **Ans.**

EXAMPLE 9.16. Compute the directional derivative of $x^2 + y^2 + 4xz$ at $(1, -2, 2)$ in the direction of the vector $2i - 2j - k$.

SOLUTION: Let the scalar function $f = x^2 + y^2 + 4xz$.

Now $\quad \nabla f = i\frac{\partial f}{\partial x} + j\frac{\partial f}{\partial y} + k\frac{\partial f}{\partial z} = (2x + 4z)\, i + 2yj + 4xk$

$\therefore \quad$ grad f at $(1, -2, 2) = 10i - 4j + 4k$

The directional derivative of f at $(1, -2, 2)$ along the vector $2i - 2j - k$

$$= (10i - 4j + 4k)\cdot\frac{(2i - 2j - k)}{\sqrt{4+4+1}} = \frac{1}{3}(20 + 8 - 4) = 8.$$ **Ans.**

EXAMPLE 9.17. Find the directional derivative of the function $\phi = x^2 - y^2 + 2z^2$ at the point $P(1, 2, 3)$ in the direction of the line PQ when Q has co-ordinates (5, 0, 4).

(UPTU 2000-01)

SOLUTION: $\nabla \phi = i\dfrac{\partial \phi}{\partial x} + j\dfrac{\partial \phi}{\partial y} + k\dfrac{\partial \phi}{\partial z} = 2xi - 2yj + 4zk$

Hence $[\nabla \phi]_{(1,\ 2,\ 3)} = 2i - 4j + 12k$

Further, $\overrightarrow{PQ} = (5-1)i + (0-2)j + (4-3)k = 4i - 2j + k.$

$\therefore$ Directional derivative of ϕ along $\overrightarrow{PQ}$

$$= [\nabla \phi]\, \hat{PQ} = (2i - 4j + 12k).\ \frac{(4i - 2j + k)}{\sqrt{16+4+1}}$$

$$= \frac{1}{\sqrt{21}}[8+8+12] = \frac{28}{\sqrt{21}}. \qquad \textbf{Ans.}$$

EXAMPLE 9.18. Find the directional derivative of $\phi = xy^2 + yz^3$ at $(1, -1, 1)$ along the direction normal to the surface $x^2 + y^2 + z^2 = 9$ at $(1, 2, 2)$.

SOLUTION: $\nabla \phi = i\dfrac{\partial \phi}{\partial x} + j\dfrac{\partial \phi}{\partial y} + k\dfrac{\partial \phi}{\partial z} = i(y^2) + j(2xy + z^3) + k(3yz^2)$

$\therefore$ $[\nabla \phi]_{(1,\ -1,\ 1)} = i - j - 3k$

Next, vector along the normal to the surface $x^2 + y^2 + z^2 = 9$, is

$$\nabla (x^2 + y^2 + z^2) = i2x + 2yj + 2zk = 2i + 4j + 4k \text{ at } (1, 2, 2)$$

Therefore, the required directional derivative

$$= (i - j - 3k)\cdot \frac{(2i + 4j + 4k)}{\sqrt{4+16+16}}$$

$$= \frac{1}{3}(i - j - 3k).(i + 2j + 2k) = \frac{1}{3}(1 - 2 - 6) = -7/3. \qquad \textbf{Ans.}$$

EXAMPLE 9.19. Find the directional derivative of the scalar function $f(x, y, z) = x^2 + xy + z^2$ at the point $A\,(1, -1, -1)$ in the direction of the line AB where B has the co-ordinates (3, 2, 1)

SOLUTION: $f = x^2 + xy + z^2,\quad \therefore\quad \nabla f = i\dfrac{\partial f}{\partial x} + j\dfrac{\partial f}{\partial y} + k\dfrac{\partial f}{\partial z} = (2x + y)i + xj + 2zk$

$\therefore$ $(\nabla f)_A = [\nabla f]_{(1,\ -1,\ -1)} = (2-1)i + 1j - 2k = i + j - 2k.$

Also, $\overrightarrow{AB} = (3-1)i + (2+1)j + (1+1)k = 2i + 3j + 2k$

$\therefore$ The directional derivative of f at A along $\overrightarrow{AB}$

$$= [\nabla f]_{(1,\ -1,\ 1)} \cdot \hat{AB} = (i + j - 2k)\cdot \frac{(2i + 3j + 2k)}{\sqrt{2^2 + 3^2 + 2^2}}$$

$$= \frac{1}{\sqrt{17}}(2 + 3 - 4) = \frac{1}{\sqrt{17}}. \qquad \textbf{Ans.}$$

EXAMPLE 9.20. Find the directional derivative of $\phi = (x^2 + y^2 + z^2)^{-1/2}$ at the point (3, 1, 2) in the direction of the vector $yzi + zxj + xyk$. **(UPTU 2001 Sp. Exam.)**

SOLUTION: Let us write $r^2 = x^2 + y^2 + z^2$ then $\phi = \dfrac{1}{r}$

and therefore $$\nabla\phi = i\left(-\frac{1}{r^2}\right)\frac{\partial r}{\partial x} + j\left(-\frac{1}{r^2}\right)\frac{\partial r}{\partial y} + k\left(-\frac{1}{r^2}\right)\frac{\partial r}{\partial z} = \frac{-1}{r^2}\left(i\frac{x}{r} + j\frac{y}{r} + k\frac{z}{r}\right)$$

$$= \frac{-1}{r^3}(xi + yj + zk) = \frac{-\vec{r}}{r^3} = -\frac{(3i + j + 2k)}{14\sqrt{14}} \quad \text{at} \quad (3, 1, 2).$$

Unit vector $\hat{n}$ in the direction of $yzi + zxj + xyk$, is equal to

$$\hat{n} = \frac{yzi + zxj + xyk}{\sqrt{y^2z^2 + z^2x^2 + x^2y^2}} = \frac{2i + 6j + 3k}{\sqrt{2^2 + 6^2 + 3^2}} \quad \text{at} \quad (3, 1, 2).$$

$$= \frac{1}{7}(2i + 6j + 3k) \quad \text{at} \quad (3, 1, 2).$$

∴ Desired directional derivative

$$= \frac{-(3i + j + 2k)}{14\sqrt{14}} \cdot \frac{(2i + 6j + 3k)}{7}$$

$$= \frac{-1}{98\sqrt{14}}(6 + 6 + 6) = \frac{-9}{49\sqrt{14}}. \quad \textbf{Ans.}$$

EXAMPLE 9.21. Find the values of a, b, c so that the maximum value of the directional derivative of the scalar $\phi = axy^2 + byz + cz^2x^3$ at (1, 2, –1) has a magnitude 64 in the direction parallel to Z-axis. **(UPTU 2001 Q.B.)**

SOLUTION: $$\text{grad } \phi = \nabla\phi = \left(i\frac{\partial}{\partial x} + j\frac{\partial}{\partial y} + k\frac{\partial}{\partial z}\right)(axy^2 + byz + cz^2x^3)$$

$$= (ay^2 + 3cx^2z^2)i + (2axy + bz)j + (by + 2cx^3z)k$$

$$= (4a + 3c)i + (4a - b)j + (2b - 2c)k \quad \text{at} \quad (1, 2, -1).$$

We know that the directional derivative is maximum along the normal to the surface, that is, along $\nabla\phi$. But we are given that directional derivative is maximum in the direction parallel to Z-axis, hence co-efficients of i and j should be zero and co-efficient of k be 64.

$\Rightarrow$ $4a + 3c = 0$ and $4a - b = 0$ and $2b - 2c = 64$

Hence $a = 6$, $b = 24$ and $c = -8$. **Ans.**

EXAMPLE 9.22. Find the directional derivative of $\frac{1}{r}$ in the direction of $\vec{r}$ where $\vec{r} = xi + yj + zk$. **(UPTU 2002-03)**

SOLUTION: Here $\phi = \frac{1}{r}$ hence $\nabla\phi = \nabla\left(\frac{1}{r}\right) = -\frac{i}{r^2}\frac{\partial r}{\partial x} - \frac{i}{r^2}\frac{\partial r}{\partial y} - \frac{k}{r^2}\frac{\partial r}{\partial z}$

$$= -\frac{1}{r^2}\left[i\frac{x}{r} + j\frac{y}{r} + k\frac{z}{r}\right] = -\frac{1}{r^3}\vec{r}.$$

Unit vector along $\vec{r} = \hat{r} = \frac{\vec{r}}{r}$

$\therefore$ Directional derivative $= \left(-\frac{1}{r^3}\vec{r}\right)\cdot\left(\frac{1}{r}\vec{r}\right) = -\frac{1}{r^4}(\vec{r}\cdot\vec{r})$

$$= -\frac{r^2}{r^4} = -\frac{1}{r^2}.$$ **Ans.**

DIVERGENCE OF A VECTOR POINT FUNCTION

When the vector differential operator ∇ operates scalarly on a vector point function, the result is a scalar quantity called the divergence of the vector point function. Thus, if $\vec{F}(x, y, z)$ is a vector point function then $\nabla \cdot \vec{F}$ denotes the divergence of $\vec{F}$ and

$$\nabla \cdot \vec{F} = \left(i\frac{\partial}{\partial x} + j\frac{\partial}{\partial y} + k\frac{\partial}{\partial z}\right) \cdot (F_1 i + F_2 j + F_3 k) = \frac{\partial F_1}{\partial x} + \frac{\partial F_2}{\partial y} + \frac{\partial F_3}{\partial z} = \Sigma\frac{\partial F_1}{\partial x}$$

where F_1, F_2, F_3 are the components of $\vec{F}$.

The divergence of $\vec{F}$ can also be written as

$$\nabla \cdot \vec{F} = i\cdot\frac{\partial \vec{F}}{\partial x} + j\cdot\frac{\partial \vec{F}}{\partial y} + k\cdot\frac{\partial \vec{F}}{\partial z} = \Sigma i\cdot\frac{\partial \vec{F}}{\partial x}$$

As an illustration

$$\nabla \cdot \vec{r} = \left(i\frac{\partial}{\partial x} + j\frac{\partial}{\partial y} + k\frac{\partial}{\partial z}\right) \cdot (xi + yj + zk)$$

$$= \frac{\partial}{\partial x}(x) + \frac{\partial}{\partial y}(y) + \frac{\partial}{\partial z}(z) = 3.$$

In particular, **if $\nabla \cdot \vec{F} = 0$ the vector field $\vec{F}$ is called SOLENOIDAL.**

PHYSICAL INTERPRETATION OF DIVERGENCE OF A VECTOR FIELD **(UPTU 2009-10)**

To have a physical concept of the divergence of a vector point function, consider the flow of a fluid with velocity $\vec{V} = V_1 i + V_2 j + V_3 k$ at a point $A(x, y, z)$. Consider a small rectangular parallelopiped with edges $\delta x, \delta y, \delta z$ parallel to the co-ordinate axes OX, OY, OZ respectively in the fluid with one of its corners at $A(x, y, z)$.

Let us consider the fluid flow parallel to Y-axis entering the face ABCD and coming out of the face EFGH. (See adjoining figure.) Amount of the fluid entering the face ABCD in unit time = $V_2(x, y, z)\,\delta x\,\delta z$. Amount of the fluid coming out of the face EFGH in unit time $= V_2(x, y+\delta y, z)\,\delta x\,\delta z$.

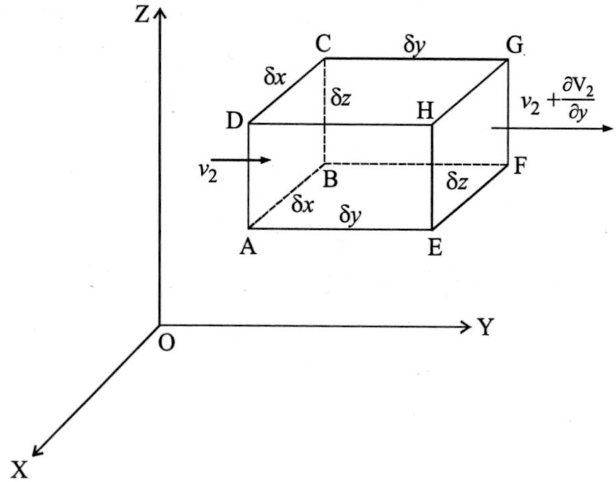

Thus, the rate of the fluid flow across the elementary volume along Y-axis

$$= V_2(x, y+\delta y, z)\,\delta x\,\delta z - V_2(x, y, z)\,\delta x\,\delta z$$

$$= \delta x\,\delta z\,[V_2(x, y+\delta y, z) - V_2(x, y, z)]$$

$$= \delta x\,\delta z\left[\left\{V_2(x, y, z) + \delta y\frac{\partial V_2}{\partial y} + \frac{(\delta y)^2}{2!}\cdot\frac{\partial^2 V_2}{\partial y^2} + \ldots\right\} - V_2(x, y, z)\right]$$

$$= \delta x\,\delta y\,\delta z\frac{\partial V_2}{\partial y} \qquad \text{neglecting the higher order terms.}$$

Similarly, the rate of fluid flow along X-axis is $\frac{\partial V_1}{\partial x}\delta x\delta y\delta z$ and along z-axis is $\frac{\partial V_3}{\partial z}\delta x\,\delta y\,\delta z$.

Thus, the total amount of the fluid crossing the elementary volume in unit time is

$$\left(\frac{\partial V_1}{\partial x} + \frac{\partial V_2}{\partial y} + \frac{\partial V_3}{\partial z}\right)\delta x\,\delta y\,\delta z$$

Therefore, the rate of fluid flow through unit volume

$$= \frac{\partial V_1}{\partial x} + \frac{\partial V_2}{\partial y} + \frac{\partial V_3}{\partial z} \quad (= \nabla \cdot \vec{V})$$

Thus the divergence of the fluid velocity $\vec{V}$ represents the rate of fluid flow through unit volume. It is also known as **FLUID FLUX.** Similarly, if $\vec{V}$ represents the electric flow then divergence $\vec{V}$ is the amount of electric flux which crosses a unit volume in unit time.

EXAMPLE 9.23. (a) Find the value of λ for which the vector $\vec{u} = (x + 3y)i + (y - 2z)j + (x + \lambda z)k$ is solenoidal vector.

(b) Show that the vector $\vec{V} = 3y^4z^2i + 4x^3z^2j - 3x^2y^2k$ is solenoidal.

[UPTU 2011-12]

SOLUTION: (a) For $\vec{u}$ to be solenoidal $\nabla \cdot \vec{u} = 0$

or $$\frac{\partial}{\partial x}(x+3y)+\frac{\partial}{\partial y}(y-2z)+\frac{\partial}{\partial z}(x+\lambda z) = 0$$

or $$1 + 1 + \lambda = 0$$

Hence $\lambda = -2$ for $\vec{u}$ to be solenoidal. **Ans.**

(b) $$\nabla\cdot\vec{V} = \frac{\partial}{\partial x}(3y^4x^2)+\frac{\partial}{\partial y}(4x^3z^2)+\frac{\partial}{\partial z}(-3x^2y^2)$$

$$= 0 + 0 + 0 = 0$$ **Hence proved**

EXAMPLE 9.24. If $u = x^2 + y^2 + z^2$ and $\vec{V} = xi + yj + zk$ show that $div\,(u\vec{V}) = 5u.$

SOLUTION: Given that $u = x^2 + y^2 + z^2$ and $\vec{V} = xi + yj + zk.$

$\therefore$ $\nabla u = 2xi + 2yj + 2zk$ and $\nabla.\vec{V} = 1 + 1 + 1 = 3.$

We know that
$$\nabla .(u\,\vec{V}) = (\nabla u).\vec{V} + u(\nabla .\vec{V})$$
$$= (2xi + 2yj + 2zk).(xi + yj + zk) + 3u$$
$$= 2x^2 + 2y^2 + 2z^2 + 3u = 2u + 3u = 5u.$$ **Hence the result.**

EXAMPLE 9.25. Show that

(*i*) $\nabla\cdot(r^n\,\vec{r}) = (n+3)\,r^n$ (*ii*) $\nabla\cdot(\phi\,\vec{F}) = \nabla\phi\cdot F + \phi\,(\nabla\cdot F)$

where ϕ is a scalar point function and $\vec{F}$ is a vector point function.

(UPTU 2000-01)

SOLUTION: (*i*)
$$\nabla\cdot(r^n\,\vec{r}) = \left(i\frac{\partial}{\partial x}+j\frac{\partial}{\partial y}+k\frac{\partial}{\partial z}\right)\cdot(r^n xi + r^n yj + r^n zk)$$
$$= \frac{\partial}{\partial x}(r^n x) + \frac{\partial}{\partial y}(r^n y) + \frac{\partial}{\partial z}(r^n z)$$
$$= r^n\frac{\partial x}{\partial x} + x\,nr^{n-1}\frac{\partial r}{\partial x} + r^n\frac{\partial y}{\partial y} + y\,nr^{n-1}\frac{\partial r}{\partial y} + r^n\frac{\partial z}{\partial z} + z\,nr^{n-1}\frac{\partial r}{\partial z}$$
$$= 3r^n + nr^{n-1}\left(x\frac{x}{r} + y\frac{y}{r} + z\frac{z}{r}\right) = (n+3)\,r^n.$$ **Hence Proved.**

(*ii*)
$$\nabla\cdot(\phi\,\vec{F}) = \left(i\frac{\partial}{\partial x}+j\frac{\partial}{\partial y}+k\frac{\partial}{\partial z}\right)\cdot(\phi F_1 i + \phi F_2 j + \phi F_3 k)$$

where F_1, F_2, F_3 are components of $\vec{F}$.

$$\therefore\quad \nabla\cdot(\phi\,\vec{F}) = \frac{\partial}{\partial x}(\phi F_1) + \frac{\partial}{\partial y}(\phi F_2) + \frac{\partial}{\partial z}(\phi F_3)$$

$$= \frac{\partial \phi}{\partial x} F_1 + \phi \frac{\partial F_1}{\partial x} + \frac{\partial \phi}{\partial y} F_2 + \phi \frac{\partial F_2}{\partial y} + \frac{\partial \phi}{\partial z} F_3 + \phi \frac{\partial F_3}{\partial z}$$

$$= \left(\frac{\partial \phi}{\partial x} F_1 + \frac{\partial \phi}{\partial y} F_2 + \frac{\partial \phi}{\partial z} F_3\right) + \phi\left(\frac{\partial F_1}{\partial x} + \frac{\partial F_2}{\partial y} + \frac{\partial F_3}{\partial z}\right)$$

$$= \left(i\frac{\partial \phi}{\partial x} + j\frac{\partial \phi}{\partial y} + k\frac{\partial \phi}{\partial z}\right) \cdot (F_1 i + F_2 j + F_3 k) + \phi\, \nabla \cdot \vec{F}$$

$= \nabla\phi \cdot \vec{F} + \phi\, \nabla \cdot \vec{F}$. **Hence Proved.**

EXAMPLE 9.26. If $\vec{a}$ is constant vector and $\vec{r} = xi + yj + zk$, show that

(i) $\text{div}\,(\vec{a} \times \vec{r}) = 0$ *(ii)* $\text{div}\,[(\vec{a} \cdot \vec{r})\,\vec{a}] = a^2$

(iii) $\text{div}\,[\vec{a} \times (\vec{r} \times \vec{a})] = 2a^2$. **(UPTU 2001-02)**

SOLUTION: *(i)* $\text{div}\,(\vec{a} \times \vec{r}) = \Sigma i \frac{\partial}{\partial x} \cdot (\vec{a} \times \vec{r}) = \Sigma i \cdot \frac{\partial}{\partial x}(\vec{a} \times \vec{r})$

$$= \Sigma i \cdot \left(\frac{\partial \vec{a}}{\partial x} \times \vec{r} + \vec{a} \times \frac{\partial \vec{r}}{\partial x}\right) = 0 + \Sigma i \cdot \left(\vec{a} \times \frac{\partial \vec{r}}{\partial x}\right) \quad \text{since } \vec{a} \text{ is constant.}$$

$= \Sigma i \cdot (\vec{a} \times i) = 0.$ **Hence Proved.**

(ii) $\text{div}\,\{(\vec{a} \cdot \vec{r})\,\vec{a}\} = \Sigma\, i \frac{\partial}{\partial x} \cdot \{(\vec{a} \cdot \vec{r})\,\vec{a}\} = \Sigma\, i \cdot \frac{\partial}{\partial x}\{(\vec{a} \cdot \vec{r})\,\vec{a}\}$

$$= \Sigma i \cdot \left\{\frac{\partial}{\partial x}(\vec{a} \cdot \vec{r})\vec{a} + (\vec{a} \cdot \vec{r})\frac{\partial \vec{a}}{\partial x}\right\} = \Sigma\, i \cdot \left\{\frac{\partial}{\partial x}(\vec{a} \cdot \vec{r})\right\}\vec{a} + 0$$

$$= \Sigma i \cdot \left(\frac{\partial \vec{a}}{\partial x} \cdot \vec{r} + \vec{a} \cdot \frac{\partial \vec{r}}{\partial x}\right) = 0 + \Sigma\, i \cdot (\vec{a} \cdot i)\,\vec{a} = \Sigma\,(\vec{a} \cdot i)\,(i \cdot \vec{a}) = \Sigma\, {a_1}^2$$

$= a_1^2 + a_2^2 + a_3^2 = a^2$ (where $\vec{a} = a_1 i + a_2 j + a_3 k$) **Hence Proved.**

(iii) $\text{div}\,\{\vec{a} \times (\vec{r} \times \vec{a})\} = \Sigma i \cdot \frac{\partial}{\partial x} \cdot \left\{(\vec{a} \cdot \vec{a})\,\vec{r} - (\vec{a} \cdot \vec{r})\,\vec{a}\right\}$

$$= \Sigma\, i \cdot \frac{\partial}{\partial x} \cdot \left\{(a^2)\,\vec{r}\right\} - \Sigma\, i \cdot \frac{\partial}{\partial x}\left\{(\vec{a} \cdot \vec{r})\,\vec{a}\right\}$$

$$= \Sigma\, i \cdot \left\{\left(\frac{\partial}{\partial x} a^2\right)\vec{r} + a^2 \frac{\partial \vec{r}}{\partial x}\right\} - \Sigma\, i \cdot \left\{\left(\frac{\partial}{\partial x}(\vec{a} \cdot \vec{r})\right)\vec{a} + (\vec{a} \cdot \vec{r})\frac{\partial \vec{a}}{\partial x}\right\}$$

$$= \Sigma i \cdot (0 + a^2 i) - \Sigma i \cdot \left(\frac{\partial \vec{a}}{\partial x} \cdot \vec{r} + \vec{a} \cdot \frac{\partial \vec{r}}{\partial x}\right)\vec{a} + 0$$

$= \Sigma\, a^2 - \Sigma\, i \cdot \{0 + \vec{a} \cdot i\}\,\vec{a} = 3a^2 - \Sigma\,(\vec{a} \cdot i)\,(i \cdot \vec{a}) = 3a^2 - \Sigma\, a_1^2$

$= 3a^2 - a^2 = 2a^2.$ **Hence Proved.**

THE CURL OF A VECTOR FIELD

When a vector point function is operated vectorially by the operator ∇ the result is a vector quantity called **curl** of the vector. Thus, if $\vec{F}(x, y, z)$ is the vector point function then

$$\text{curl } \vec{F} = \nabla \times \vec{F} = \sum i \times \frac{\partial \vec{F}}{\partial x} = \begin{vmatrix} i & j & k \\ \frac{\partial}{\partial x} & \frac{\partial}{\partial y} & \frac{\partial}{\partial z} \\ F_1 & F_2 & F_3 \end{vmatrix}$$

$$= i\left(\frac{\partial F_3}{\partial y} - \frac{\partial F_2}{\partial z}\right) + j\left(\frac{\partial F_1}{\partial z} - \frac{\partial F_3}{\partial x}\right) + k\left(\frac{\partial F_2}{\partial x} - \frac{\partial F_1}{\partial y}\right)$$

where F_1, F_2, F_3 are components of $\vec{F}$. **[UPTU 2011-12]**

> **A vector $\vec{F}$ is called IRROTATIONAL (or CONSERVATIVE) if its curl is zero, that is, if $\nabla \times \vec{F} = 0$.**

PHYSICAL INTERPRETATION OF CURL

(UPTU 2001 Sp. Exam.) (UPTU 2002-03; 2009-10)

Consider a rigid body rotating uniformly about a given axis, *i.e.* rotating with a constant angular velocity. If $\vec{V}$ is the linear velocity of a point P (x, y, z) of the body which is rotating with an angular velocity ω radians per second about a given axis (see the adjoining figure) then

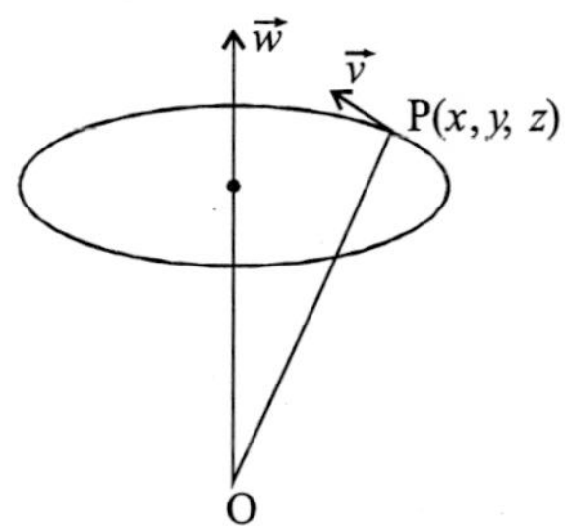

$$\vec{V} = \vec{\omega} \times \vec{r} \quad \text{where } \vec{r} = xi + yj + zk.$$

Let $\vec{\omega} = \omega_1 i + \omega_2 j + \omega_3 k$ where $\omega_1, \omega_2, \omega_3$ are constants.

Then $$\vec{V} = \vec{\omega} \times \vec{r} = \begin{vmatrix} i & j & k \\ \omega_1 & \omega_2 & \omega_3 \\ x & y & z \end{vmatrix} = (\omega_2 z - \omega_3 y)\, i + (\omega_3 x - \omega_1 z) j + (\omega_1 y - \omega_2 x)\, k$$

Therefore $$\text{curl } \vec{V} = \nabla \times \vec{V} = \begin{vmatrix} i & j & k \\ \frac{\partial}{\partial x} & \frac{\partial}{\partial y} & \frac{\partial}{\partial z} \\ \omega_2 z - \omega_3 y & \omega_3 x - \omega_1 z & \omega_1 y - \omega_2 x \end{vmatrix}$$

$$= i(\omega_1 + \omega_1) + j(\omega_2 + \omega_2) + k(\omega_3 + \omega_3) = 2\vec{\omega}$$

or $$\boxed{\vec{\omega} = \frac{1}{2} \text{curl } \vec{V} = \frac{1}{2}(\nabla \times \vec{V}).}$$

which shows that the angular velocity of a uniformly rotating body is one-half of the curl of the linear velocity of any point of the body.

An important point is note worthy here. **If $\vec{F}$ is irrotational (or conservative) there exists a scalar point function ϕ such that $\vec{F} = \nabla\phi$ and ϕ is called the SCALAR POTENTIAL of $\vec{F}$.** The formula for finding the scalar potential ϕ for an irrotational field $\vec{F}$, is given by

$$d\phi = F_1 dx + F_2 dy + F_3 dz \quad \text{which is directly integrable.}$$

EXAMPLE 9.27. If $\vec{r} = xi + yj + zk$ and $\vec{a}$ is a constant vector, show that

(*i*) $\vec{r}$ is irrotional, (*ii*) $\text{curl}(\vec{a} \times \vec{r}) = 2\vec{a}$. **(UPTU 2001-02)**

SOLUTION: (*i*) Curl $\vec{r} = \nabla \times \vec{r} = \begin{vmatrix} i & j & k \\ \frac{\partial}{\partial x} & \frac{\partial}{\partial y} & \frac{\partial}{\partial z} \\ x & y & z \end{vmatrix} = i\left(\frac{\partial z}{\partial y} - \frac{\partial y}{\partial z}\right) + j\left(\frac{\partial x}{\partial z} - \frac{\partial z}{\partial x}\right) + k\left(\frac{\partial y}{\partial x} - \frac{\partial x}{\partial y}\right) = 0.$

Hence $\vec{r}$ is irrotational.

(*ii*) Let $\vec{a} = a_1 i + a_2 j + a_3 k$ where a_1, a_2, a_3 are constants,

hence $\vec{a} \times \vec{r} = \begin{vmatrix} i & j & k \\ a_1 & a_2 & a_3 \\ x & y & z \end{vmatrix} = (a_2 z - a_3 y)\, i + (a_3 x - a_1 z) j + (a_1 y - a_2 x)\, k$

and $\nabla \times (\vec{a} \times \vec{r}) = \begin{vmatrix} i & j & k \\ \frac{\partial}{\partial x} & \frac{\partial}{\partial y} & \frac{\partial}{\partial z} \\ a_2 z - a_3 y & a_3 x - a_1 z & a_1 y - a_2 x \end{vmatrix}$

$= i\,(a_1 + a_1) + j\,(a_2 + a_2) + k\,(a_3 + a_3) = 2\vec{a}.$ **Hence Proved.**

EXAMPLE 9.28. **(a)** If $\vec{F} = (x + y + 1)\, i + j - (x + y)\, k,$ show that $\vec{F} \cdot \text{curl } \vec{F} = 0.$

(b) If $\vec{F} = (\vec{a} \cdot \vec{r})\vec{r}$ where $\vec{a}$ is a constant vector, find curl $\vec{F}$ and prove that it perpendicular to $\vec{a}$. **[UPTU 2011-12]**

SOLUTION: (a) Given that $\vec{F} = (x + y + 1)\, i + j - (x + y)\, k$

then curl $\vec{F} = \nabla \times \vec{F} = \begin{vmatrix} i & j & k \\ \frac{\partial}{\partial x} & \frac{\partial}{\partial y} & \frac{\partial}{\partial z} \\ x+y+1 & 1 & -(x+y) \end{vmatrix}$

$= i\,(-1 - 0) + j\,(0 + 1) + k\,(0 - 1) = -i + j - k$

$\therefore$ $\vec{F} \cdot \text{curl } \vec{F} = \{(x + y + 1)\, i + j - (x + y)\, k\} \cdot (-i + j - k)$

$= -(x + y + 1) + 1 + (x + y) = 0.$ **Hence Proved.**

(b) Let $\vec{a} = a_1 i + a_2 j + a_3 k$ be the constant vector and $\vec{r} = xi + yi + zk$

$\therefore \quad \vec{F} = (\vec{a} \cdot \vec{r})\vec{r} = (a_1 x + a^2 y + a^3 z)\,(xi + yj + zk)$

$$\text{and curl } \vec{F} = \nabla \times \vec{F} = \begin{vmatrix} i & j & k \\ \dfrac{\partial}{\partial x} & \dfrac{\partial}{\partial y} & \dfrac{\partial}{\partial z} \\ x(a_1 x + a_2 y + a_3 z) & y(a_1 x + a_2 y + a_3 z) & z(a_1 x + a_2 y + a_3 z) \end{vmatrix}$$

$$= i(a_2 z - a_3 y) + j(a_3 x - a_1 z) + k(a_1 y - a_2 x)$$

Then $\quad \text{curl } \vec{F} \cdot \vec{a} = [i(a_2 z - a_3 y) + j(a_3 x - a_1 z) + k(a_1 y - a_2 x)] \cdot (a_1 i + a_2 j + a_3 k)$

$$= a_1(a_2 z - a_3 y) + a_2(a_3 x - a_1 z) + a_3(a_1 y - a_2 x) = 0$$

$\therefore$ Curl $\vec{F}$ and $\vec{a}$ are mutually perpendicular. **Hence proved.**

EXAMPLE 9.29. If $\vec{r} = xi + yj + zk$ prove that curl $(r^n \vec{r}) = 0$ **(UPTU 2000-01)**

SOLUTION: $r^n \vec{r} = r^n xi + r^n yj + r^n zk$ where $r = |\vec{r}| = \sqrt{x^2 + y^2 + z^2}$.

$$\therefore \quad \text{curl } (r^n \vec{r}) = \vec{\nabla} \times (r^n \vec{r}) = \begin{vmatrix} i & j & k \\ \dfrac{\partial}{\partial x} & \dfrac{\partial}{\partial y} & \dfrac{\partial}{\partial z} \\ r^n x & r^n y & r^n z \end{vmatrix}$$

$$= i\left[\frac{\partial}{\partial y}(r^n z) - \frac{\partial}{\partial z}(r^n y)\right] + j\left[\frac{\partial}{\partial z}(r^n x) - \frac{\partial}{\partial x}(r^n z)\right] + k\left[\frac{\partial}{\partial x}(r^n y) - \frac{\partial}{\partial y}(r^n x)\right]$$

$$= \sum i\left[\frac{\partial}{\partial y}(r^n z) - \frac{\partial}{\partial z}(r^n y)\right] = \sum i\left[z.nr^{n-1}\frac{\partial r}{\partial y} - ynr^{n-1}\frac{\partial r}{\partial z}\right]$$

$$= \sum i\left[z.nr^{n-1}\frac{y}{r} - ynr^{n-1}\frac{z}{r}\right] = nr^{n-2}\sum i(yz - yz) = 0.$$ **Hence Proved.**

EXAMPLE 9.30. If $\vec{V} = (x^2 - y^2 + 2xz)\,i + (xz - xy - yz)\,j + (z^2 + x^2)\,k$ is a vector field, find curl $\vec{V}$. Show that the vectors given by curl $\vec{V}$ at $P_0(1, 2, -3)$ and $P_1(2, 3, -2)$ are orthogonal.

SOLUTION: $\vec{V} = (x^2 - y^2 + 2xz)\,i + (xz - xy - yz)\,j + (z^2 + x^2)\,k$

$$\text{Curl } \vec{V} = \nabla \times \vec{V} = \begin{vmatrix} i & j & k \\ \dfrac{\partial}{\partial x} & \dfrac{\partial}{\partial y} & \dfrac{\partial}{\partial z} \\ x^2 - y^2 + 2xz & xz - xy - yz & z^2 + x^2 \end{vmatrix}$$

$$= i\left[\frac{\partial}{\partial y}(z^2 + x^2) - \frac{\partial}{\partial z}(xz - xy - yz)\right] + j\left[\frac{\partial}{\partial z}(x^2 - y^2 + 2xz) - \frac{\partial}{\partial x}(z^2 + x^2)\right]$$

$$+ k\left[\frac{\partial}{\partial x}(xz - xy - yz) - \frac{\partial}{\partial y}(x^2 - y^2 + 2xz)\right]$$

$$= i\,[0 - (x - y)] + j\,[2x - 2x] + k\,[z - y) - (-2y + 0)] = (y - x)\,i + 0j + (z + y)\,k$$

Curl $\vec{V}$ at the point $P_0\,(1, 2, -3) = i - k$, and Curl $\vec{V}$ at the point $P_1\,(2, 3, -2) = i + k$. Since $(i - k).\,(i + k) = 0$, these are orthogonal. **Hence the result.**

VECTOR IDENTITIES

Given a scalar point function ϕ and vector point functions $\vec{F}$ and $\vec{G}$, following properties in the form of vector identities are important and very useful:

(1) $\text{div}\,(\phi\vec{F}) = \vec{F}\cdot\text{grad}\phi + \phi\,\text{div}\,\vec{F}$ or $\nabla\cdot(\phi\vec{F}) = (\nabla\phi)\bullet\vec{F} + \phi\,(\nabla\bullet\vec{F})$

(UPTU 2003-04; 2005-06)

Proof:

$$\nabla\cdot(\phi\vec{F}) = \sum i\frac{\partial}{\partial x}\bullet(\phi\vec{F}) = \sum i\bullet\frac{\partial}{\partial x}(\phi\vec{F}) = \sum i\bullet\left(\frac{\partial\phi}{\partial x}\vec{F}+\phi\frac{\partial\vec{F}}{\partial x}\right) = \sum i\bullet\left(\frac{\partial\phi}{\partial x}\vec{F}\right)+\phi\sum i\bullet\frac{\partial\vec{F}}{\partial x}$$

$$= \vec{F}\bullet\Sigma\, i\frac{\partial\phi}{\partial x}+\phi\Sigma\left(i\frac{\partial}{\partial x}\bullet\vec{F}\right) = \vec{F}\bullet\nabla\phi + \phi\,(\nabla\bullet\vec{F})$$

(2) $\text{Curl}\,(\phi\vec{F}) = (\text{grad}\,\phi)\times\vec{F} + \phi\,\text{curl}\,\vec{F}$ or $\nabla\times(\phi\vec{F}) = \nabla\phi\times\vec{F} + \phi\,(\nabla\times\vec{F})$

(UPTU 2003)

Proof:

$$\nabla\times(\phi\vec{F}) = \Sigma\, i\frac{\partial}{\partial x}\times(\phi\vec{F}) = \Sigma\, i\times\frac{\partial}{\partial x}(\phi\vec{F}) = \Sigma\, i\times\left(\frac{\partial\phi}{\partial x}\vec{F}+\phi\frac{\partial\vec{F}}{\partial x}\right)$$

$$= \Sigma\, i\times\left(\frac{\partial\phi}{\partial x}\vec{F}\right)+\Sigma\, i\times\left(\phi\frac{\partial\vec{F}}{\partial x}\right) = \Sigma\left(i\frac{\partial\phi}{\partial x}\right)\times\vec{F}+\phi\Sigma\, i\times\frac{\partial\vec{F}}{\partial x} = \nabla\phi\times\vec{F} + \phi\,\nabla\times\vec{F}$$

(3) $\text{div}\,(\vec{F}\times\vec{G}) = \vec{G}\cdot\text{Curl}\,\vec{F} - \vec{F}\cdot\text{curl}\,\vec{G}$ or $\nabla\cdot(\vec{F}\times\vec{G}) = \vec{G}\cdot(\nabla\times\vec{F}) - \vec{F}\cdot(\nabla\times\vec{G})$

(UPTU 2003, 2011-12)

Proof:

$$\nabla\cdot(\vec{F}\times\vec{G}) = \Sigma\, i\frac{\partial}{\partial x}\cdot(\vec{F}\times\vec{G}) = \Sigma\, i\cdot\frac{\partial}{\partial x}(\vec{F}\times\vec{G})$$

$$= \Sigma\, i\cdot\left(\frac{\partial\vec{F}}{\partial x}\times\vec{G}+\vec{F}\times\frac{\partial\vec{G}}{\partial x}\right) = \Sigma\, i\cdot\left(\frac{\partial\vec{F}}{\partial x}\times\vec{G}\right)-\Sigma\, i\cdot\left(\frac{\partial\vec{G}}{\partial x}\times\vec{F}\right)$$

$$= \Sigma\,\vec{G}\cdot\left(i\times\frac{\partial\vec{F}}{\partial x}\right)-\Sigma\,\vec{F}\cdot\left(i\times\frac{\partial\vec{G}}{\partial x}\right) = \vec{G}\cdot\Sigma\left(i\frac{\partial}{\partial x}\times\vec{F}\right)-\vec{F}\cdot\Sigma\left(i\frac{\partial}{\partial x}\times\vec{G}\right)$$

$$= \vec{G}\cdot(\nabla\times\vec{F}) - \vec{F}\cdot(\nabla\times\vec{G})$$

(4) Curl $(\vec{F} \times \vec{G}) = (\text{div } \vec{G})\, \vec{F} - (\text{div } \vec{F})\, \vec{G} + (\vec{G} \cdot \nabla)\, \vec{F} - (\vec{F} \cdot \nabla)\, \vec{G}$

or $\nabla \times (\vec{F} \times \vec{G}) = (\nabla \cdot \vec{G})\, \vec{F} - (\nabla \cdot \vec{F})\, \vec{G} + (\vec{G} \cdot \nabla)\, \vec{F} - (\vec{F} \cdot \nabla)\, \vec{G}$

Proof:

$$\text{Curl}\,(\vec{F} \times \vec{G}) = \nabla \times (\vec{F} \times \vec{G}) = \sum i \frac{\partial}{\partial x} \times (\vec{F} \times \vec{G}) = \sum i \times \frac{\partial}{\partial x}(\vec{F} \times \vec{G})$$

$$= \sum i \times \left(\frac{\partial \vec{F}}{\partial x} \times \vec{G} + \vec{F} \times \frac{\partial \vec{G}}{\partial x} \right) = \sum i \times \left(\frac{\partial \vec{F}}{\partial x} \times \vec{G} \right) - \sum i \times \left(\frac{\partial \vec{G}}{\partial x} \times \vec{F} \right)$$

Consider $$\sum i \times \left(\frac{\partial \vec{F}}{\partial x} \times \vec{G} \right) = \sum \left[(i \cdot \vec{G}) \frac{\partial \vec{F}}{\partial x} - \left(i \cdot \frac{\partial \vec{F}}{\partial x} \right) \vec{G} \right] = \sum (\vec{G} \cdot i) \frac{\partial \vec{F}}{\partial x} - \sum \left(i \cdot \frac{\partial \vec{F}}{\partial x} \right) \vec{G}$$

$$= \sum \left(\vec{G} \cdot i \frac{\partial}{\partial x} \right) \vec{F} - \sum \left(i \frac{\partial}{\partial x} \cdot \vec{F} \right) \vec{G} = (\vec{G} \cdot \nabla)\, \vec{F} - (\nabla \cdot \vec{F}) \vec{G}$$

Interchanging $\vec{F}$ and $\vec{G}$ in the above relation, we get

$$\sum i \times \left(\frac{\partial \vec{G}}{\partial x} \times \vec{F} \right) = (\vec{F} \cdot \nabla)\, \vec{G} - (\nabla \cdot \vec{G})\, \vec{F}$$

$\therefore \quad \nabla \times (\vec{F} \times \vec{G}) = \{(\vec{G} \cdot \nabla)\, \vec{F} - (\nabla \cdot \vec{F})\, \vec{G}\} - \{(\vec{F} \cdot \nabla)\, \vec{G} - (\nabla \cdot \vec{G})\, \vec{F}\}$

or $\quad \nabla \times (\vec{F} \times \vec{G}) = (\nabla \cdot \vec{G})\, \vec{F} - (\nabla \cdot \vec{F})\, \vec{G} + (\vec{G} \cdot \nabla)\, \vec{F} - (\vec{F} \cdot \nabla)\, \vec{G}.$

(5) Grad $(\vec{F} \cdot \vec{G}) = \vec{F} \times \text{Curl } \vec{G} + \vec{G} \times \text{Curl } \vec{F} + (\vec{F} \cdot \nabla)\, \vec{G} + (\vec{G} \cdot \nabla)\, \vec{F}$

or $\nabla (\vec{F} \cdot \vec{G}) = \vec{F} \times (\nabla \times \vec{G}) + \vec{G} \times (\nabla \times \vec{F}) + (\vec{F} \cdot \nabla)\, \vec{G} + (\vec{G} \cdot \nabla)\, \vec{F}$

Proof:

$$\nabla (\vec{F} \cdot \vec{G}) = \sum i \frac{\partial}{\partial x} (\vec{F} \cdot \vec{G}) = \sum i \left(\frac{\partial \vec{F}}{\partial x} \cdot \vec{G} + \vec{F} \cdot \frac{\partial \vec{G}}{\partial x} \right) = \sum i \left(\frac{\partial \vec{F}}{\partial x} \cdot \vec{G} \right) + \sum i \left(\frac{\partial \vec{G}}{\partial x} \cdot \vec{F} \right) \quad \text{...(1)}$$

Now consider

$$\vec{G} \times \left(i \times \frac{\partial \vec{F}}{\partial x} \right) = \left(\vec{G} \cdot \frac{\partial \vec{F}}{\partial x} \right) i - (\vec{G} \cdot i) \frac{\partial \vec{F}}{\partial x} \quad \text{hence} \quad \left(\vec{G} \cdot \frac{\partial \vec{F}}{\partial x} \right) i = \vec{G} \times \left(i \times \frac{\partial \vec{F}}{\partial x} \right) + (\vec{G} \cdot i) \frac{\partial \vec{F}}{\partial x}$$

$$\therefore \sum \left(\vec{G} \cdot \frac{\partial \vec{F}}{\partial x} \right) i = \sum \vec{G} \times \left(i \times \frac{\partial \vec{F}}{\partial x} \right) + \sum (\vec{G} \cdot i) \frac{\partial \vec{F}}{\partial x} = \sum \vec{G} \times \left(i \frac{\partial}{\partial x} \times \vec{F} \right) + \sum \left(\vec{G} \cdot i \frac{\partial}{\partial x} \right) \vec{F}$$

Thus $$\sum \left(\vec{G} \cdot \frac{\partial \vec{F}}{\partial x} \right) i = \vec{G} \times (\nabla \times \vec{F}) + (\vec{G} \cdot \nabla)\, \vec{F} \quad \text{...(2)}$$

Interchanging $\vec{F}$ and $\vec{G}$ in the last relation, we get

$$\Sigma\left(\vec{F}\cdot\frac{\partial \vec{G}}{\partial x}\right) i = \vec{F} \times (\nabla \times \vec{G}) + (\vec{F} \cdot \nabla)\, \vec{G} \qquad ...(3)$$

Using (2) and (3) in (1), we get

$$\nabla(\vec{F} \cdot \vec{G}) = \vec{F} \times (\nabla \times \vec{G}) + \vec{G} \times (\nabla \times \vec{F}) + (\vec{F} \cdot \nabla)\, \vec{G} + (\vec{G} \cdot \nabla)\, \vec{F}.$$

Following are the properties on vector differentiation dealing with the double operation of ∇.

(6) **Div (grad ϕ)** $= \dfrac{\partial^2\phi}{\partial x^2}+\dfrac{\partial^2\phi}{\partial y^2}+\dfrac{\partial^2\phi}{\partial z^2}$ **or** $\nabla \cdot (\nabla\phi) = \nabla^2\phi = \dfrac{\partial^2\phi}{\partial x^2}+\dfrac{\partial^2\phi}{\partial y^2}+\dfrac{\partial^2\phi}{\partial z^2}$

Proof:

$$\nabla^2\phi = (\nabla \cdot \nabla)\, \phi = \left(i\frac{\partial}{\partial x}+j\frac{\partial}{\partial y}+k\frac{\partial}{\partial z}\right) \cdot \left(i\frac{\partial}{\partial x}+j\frac{\partial}{\partial y}+k\frac{\partial}{\partial z}\right) \phi$$

$$= \left(\frac{\partial^2}{\partial x^2}+\frac{\partial^2}{\partial y^2}+\frac{\partial^2}{\partial z^2}\right) \phi = \frac{\partial^2\phi}{\partial x^2}+\frac{\partial^2\phi}{\partial y^2}+\frac{\partial^2\phi}{\partial z^2}.$$

(7) **Curl (grad ϕ) = 0 or $\nabla \times (\nabla\phi) = 0$** **(UPTU 2010-11)**

Proof:

$$\nabla \times (\nabla\phi) = \left(i\frac{\partial}{\partial x}+j\frac{\partial}{\partial y}+k\frac{\partial}{\partial z}\right)\times\left(i\frac{\partial\phi}{\partial x}+j\frac{\partial\phi}{\partial y}+k\frac{\partial\phi}{\partial z}\right)$$

$$= \begin{vmatrix} i & j & k \\ \partial/\partial x & \partial/\partial y & \partial/\partial z \\ \partial\phi/\partial x & \partial\phi/\partial y & \partial\phi/\partial z \end{vmatrix}$$

$$= i\left(\frac{\partial^2\phi}{\partial y\,\partial z}-\frac{\partial^2\phi}{\partial z\,\partial y}\right)+j\left(\frac{\partial^2\phi}{\partial z\,\partial x}-\frac{\partial^2\phi}{\partial x\,\partial z}\right)+k\left(\frac{\partial^2\phi}{\partial x\,\partial y}-\frac{\partial^2\phi}{\partial y\,\partial x}\right) = 0.$$

(8) **Div (Curl $\vec{F}$) = 0 or $\nabla \cdot (\nabla \times \vec{F}) = 0$** **(UPTU 2010-11)**

Proof:

Let $\vec{F} = F_1 i + F_2 j + F_3 k$ then

$$\nabla \times \vec{F} = \begin{vmatrix} i & j & k \\ \dfrac{\partial}{\partial x} & \dfrac{\partial}{\partial y} & \dfrac{\partial}{\partial z} \\ F_1 & F_2 & F_3 \end{vmatrix} = i\left(\frac{\partial F_3}{\partial y}-\frac{\partial F_2}{\partial z}\right)+j\left(\frac{\partial F_1}{\partial z}-\frac{\partial F_3}{\partial x}\right)+k\left(\frac{\partial F_2}{\partial x}-\frac{\partial F_1}{\partial y}\right)$$

$$\therefore \quad \nabla \cdot (\nabla \times \vec{F}) = \left(i\frac{\partial}{\partial x}+j\frac{\partial}{\partial y}+k\frac{\partial}{\partial z}\right)\cdot\left\{i\left(\frac{\partial F_3}{\partial y}-\frac{\partial F_2}{\partial z}\right)+j\left(\frac{\partial F_1}{\partial z}-\frac{\partial F_3}{\partial x}\right)+k\left(\frac{\partial F_2}{\partial x}-\frac{\partial F_1}{\partial y}\right)\right\}$$

$$= \frac{\partial}{\partial x}\left(\frac{\partial F_3}{\partial y}-\frac{\partial F_2}{\partial z}\right)+\frac{\partial}{\partial y}\left(\frac{\partial F_1}{\partial z}-\frac{\partial F_3}{\partial x}\right)+\frac{\partial}{\partial z}\left(\frac{\partial F_2}{\partial x}-\frac{\partial F_1}{\partial y}\right)$$

$$= \frac{\partial^2 F_3}{\partial x\partial y}-\frac{\partial^2 F_2}{\partial x\partial z}+\frac{\partial^2 F_1}{\partial y\partial z}-\frac{\partial^2 F_3}{\partial y\partial x}+\frac{\partial^2 F_2}{\partial z\partial x}-\frac{\partial^2 F_1}{\partial z\partial y} = 0.$$

(9) Curl (Curl $\vec{F}$) = grad (div $\vec{F}$) – $\nabla^2\vec{F}$

or $\nabla \times (\nabla \times \vec{F}) = \nabla(\nabla \cdot \vec{F}) - (\nabla \cdot \nabla)\vec{F}$ **(UPTU 2001 Sp. Exam.)**

(UPTU 2003)

Proof:

Let $\vec{F} = F_1 i + F_i j + F_3 k$ then

$$\nabla \times (\nabla \times \vec{F}) = \left(i\frac{\partial}{\partial x}+j\frac{\partial}{\partial y}+k\frac{\partial}{\partial z}\right)\times\left\{i\left(\frac{\partial F_3}{\partial y}-\frac{\partial F_2}{\partial z}\right)+j\left(\frac{\partial F_1}{\partial z}-\frac{\partial F_3}{\partial x}\right)+k\left(\frac{\partial F_2}{\partial x}-\frac{\partial F_1}{\partial y}\right)\right\}$$

$$= k\left(\frac{\partial^2 F_1}{\partial x\partial z}-\frac{\partial^2 F_3}{\partial x^2}\right)-j\left(\frac{\partial^2 F_2}{\partial x^2}-\frac{\partial^2 F_1}{\partial x\partial y}\right)-k\left(\frac{\partial^2 F_3}{\partial y^2}-\frac{\partial^2 F_2}{\partial y\partial z}\right)+i\left(\frac{\partial^2 F_2}{\partial y\partial x}-\frac{\partial^2 F_1}{\partial y^2}\right)$$
$$+ j\left(\frac{\partial^2 F_3}{\partial z\partial y}-\frac{\partial^2 F_2}{\partial z^2}\right)-i\left(\frac{\partial^2 F_1}{\partial z^2}-\frac{\partial^2 F_3}{\partial z\partial x}\right)$$

$$= \left(\frac{\partial^2 F_2}{\partial y\partial x}+\frac{\partial^2 F_3}{\partial z\partial x}-\frac{\partial^2 F_1}{\partial y^2}-\frac{\partial^2 F_1}{\partial z^2}\right)i+\left(\frac{\partial^2 F_1}{\partial x\partial y}+\frac{\partial^2 F_3}{\partial z\partial y}-\frac{\partial^2 F_2}{\partial x^2}-\frac{\partial^2 F_2}{\partial z^2}\right)j$$
$$+\left(\frac{\partial^2 F_1}{\partial x\partial z}+\frac{\partial^2 F_2}{\partial y\partial z}-\frac{\partial^2 F_3}{\partial x^2}-\frac{\partial^2 F_3}{\partial y^2}\right)k$$

$$= \left(\frac{\partial^2 F_2}{\partial x\partial y}+\frac{\partial^2 F_3}{\partial x\partial z}+\frac{\partial^2 F_1}{\partial x^2}-\frac{\partial^2 F_1}{\partial x^2}-\frac{\partial^2 F_1}{\partial y^2}-\frac{\partial^2 F_1}{\partial z^2}\right)i$$
$$+\left(\frac{\partial^2 F_1}{\partial y\partial x}+\frac{\partial^2 F_3}{\partial y\partial z}+\frac{\partial^2 F_2}{\partial y^2}-\frac{\partial^2 F_2}{\partial x^2}-\frac{\partial^2 F_2}{\partial y^2}-\frac{\partial^2 F_2}{\partial z^2}\right)j$$
$$+\left(\frac{\partial^2 F_1}{\partial z\partial x}+\frac{\partial^2 F_2}{\partial z\partial y}+\frac{\partial^2 F_3}{\partial z^2}-\frac{\partial^2 F_3}{\partial x^2}-\frac{\partial^2 F_3}{\partial y^2}-\frac{\partial^2 F_3}{\partial z^2}\right)k$$

$$= i\frac{\partial}{\partial x}\left(\frac{\partial F_1}{\partial x}+\frac{\partial F_2}{\partial y}+\frac{\partial F_3}{\partial z}\right)-\left(\frac{\partial^2}{\partial x^2}+\frac{\partial^2}{\partial y^2}+\frac{\partial^2}{\partial z^2}\right)F_1 i$$
$$+ j\frac{\partial}{\partial y}\left(\frac{\partial F_1}{\partial x}+\frac{\partial F_2}{\partial y}+\frac{\partial F_3}{\partial z}\right)-\left(\frac{\partial^2}{\partial x^2}+\frac{\partial^2}{\partial y^2}+\frac{\partial^2}{\partial z^2}\right)F_2 j$$
$$+ k\frac{\partial}{\partial z}\left(\frac{\partial F_1}{\partial x}+\frac{\partial F_2}{\partial y}+\frac{\partial F_3}{\partial z}\right)-\left(\frac{\partial^2}{\partial x^2}+\frac{\partial^2}{\partial y^2}+\frac{\partial^2}{\partial z^2}\right)F_3 k$$

$$= \nabla\left(\frac{\partial F_1}{\partial x}+\frac{\partial F_2}{\partial y}+\frac{\partial F_3}{\partial z}\right)-\left(\frac{\partial^2}{\partial x^2}+\frac{\partial^2}{\partial y^2}+\frac{\partial^2}{\partial z^2}\right)(F_1 i + F_2 j + F_3 k)$$

or $\quad \nabla \times (\nabla \times \vec{F}) = \nabla (\nabla \cdot \vec{F}) - \nabla^2 \vec{F}.$

We could arrive at the same result more conveniently by applying the symbolic method as follows:

$$\text{Curl (Curl F)} = \nabla \times (\nabla \times F)$$

Treating ∇'s as algebraic vectors ∇_1 and ∇_2 and then using the formula of vector triple product, we get

$$\nabla \times (\nabla \times F) = \nabla_1 \times (\nabla_2 \times F) = \nabla_2 (\nabla_1 \cdot F) - (\nabla_1 \times \nabla_2) F$$
$$= \nabla (\nabla \cdot F) - \nabla^2 F \qquad \text{(dropping the suffixes of dels).}$$

EXAMPLE 9.31. If $\vec{a}$ and $\vec{b}$ are constant vectors, show that

$$(i)\ \nabla \left(\frac{\vec{a}\cdot\vec{r}}{r^n}\right) = \frac{\vec{a}}{r^n} - n\frac{(\vec{a}\cdot\vec{r})}{r^{n+2}}\vec{r} \qquad (ii)\quad \nabla \times \{\vec{a} \times (\vec{b} \times \vec{r})\} = \vec{a} \times \vec{b}.$$

SOLUTION: (i) $\nabla \left(\frac{\vec{a}\cdot\vec{r}}{r^n}\right) = \nabla \left(\frac{1}{r^n}\vec{a}\cdot\vec{r}\right) = \frac{1}{r^n}\nabla (\vec{a}\cdot\vec{r}) + (\vec{a}\cdot\vec{r})\, \nabla\left(\frac{1}{r^n}\right)$

$$= \frac{1}{r^n}\vec{a} + (\vec{a}\cdot\vec{r})\,\nabla\left(\frac{1}{r^n}\right) = \frac{\vec{a}}{r^n} + (\vec{a}\cdot\vec{r})(-n)r^{-n-2}\vec{r} \quad (\text{since } \nabla(\vec{a}\cdot\vec{r}) = \vec{a})$$

$$= \frac{1}{r^n}\vec{a} - \frac{n(\vec{a}\cdot\vec{r})}{r^{n+2}}\vec{r}.$$

(ii) $\quad \vec{a} \times (\vec{b} \times \vec{r}) = (\vec{a}\cdot\vec{r})\,\vec{b} - (\vec{a}\cdot\vec{b})\,\vec{r}$

$$\therefore\ \nabla \times [a \times (b \times r)] = \nabla \times \{(\vec{a}\cdot\vec{r})\,\vec{b}\} - \nabla \times \{(\vec{a}\cdot\vec{b})\,\vec{r}\}$$

$$= \nabla(\vec{a}\cdot\vec{r}) \times \vec{b} + (\vec{a}\cdot\vec{r})\,\nabla \times \vec{b} - \{\nabla(\vec{a}\cdot\vec{b}) \times \vec{r} + (\vec{a}\cdot\vec{b})\,\nabla \times \vec{r}\}$$

$$= \vec{a} \times \vec{b} + 0 + 0 + 0, \quad (\text{since } \nabla(\vec{a}\cdot\vec{r}) = \vec{a} \text{ and } \nabla \times \vec{r} = 0) = \vec{a} \times \vec{b}.$$

EXAMPLE 9.32. If $\vec{a}$ is a constant vector function, show that

$$\nabla \times \left(\frac{\vec{a}\times\vec{r}}{r^3}\right) = -\frac{\vec{a}}{r^3} + \frac{3(\vec{a}\cdot\vec{r})}{r^5}\vec{r}.$$

SOLUTION: $\nabla \times \left(\frac{\vec{a}\times\vec{r}}{r^3}\right) = \left(\nabla\cdot\frac{\vec{r}}{r^3}\right)\vec{a} - (\vec{a}\cdot\nabla)\frac{1}{r^3}\vec{r}$ (note the steps)

Now, $\quad \nabla \cdot \frac{\vec{r}}{r^3} = \nabla(r^{-3})\cdot\vec{r} + \frac{1}{r^3}\nabla\cdot\vec{r}$

$$= (-3r^{-5}\vec{r})\cdot\vec{r} + \frac{1}{r^3}(3) = -3r^{-5}r^2 + \frac{3}{r^3} = 0 \quad (\text{as } \vec{r}\cdot\vec{r} = r^2)$$

and $\quad (\vec{a}\cdot\nabla)\,\vec{r}\,r^{-3} = r^{-3}(\vec{a}\cdot\nabla)\,\vec{r} + \vec{r}\,(\vec{a}\cdot\nabla)\,r^{-3} = r^{-3}\vec{a} + [\vec{a}\cdot\nabla r^{-3}]\,\vec{r}$

$$\therefore \quad \nabla\left(\vec{a}\times\frac{\vec{r}}{r^3}\right) = 0 - \frac{\vec{a}}{r^3} - (-3)\frac{\vec{a}\cdot\vec{r}}{r^5}\vec{r} = -\frac{\vec{a}}{r^3} + \frac{3(\vec{a}\cdot\vec{r})}{r^5}\vec{r}. \quad (\text{Since } \nabla r^{-3} = -3\cdot r^{-5}\vec{r})$$

Hence Proved

EXAMPLE 9.33. If $r^2 = x^2 + y^2 + z^2$ show that $\nabla^2 f(r) = f''(r) + \frac{2}{r} f'(r)$ and hence deduce that $\nabla^4 f(r) = f^{iv}(r) + \frac{4}{r} f'''(r)$.

SOLUTION: We know that $\nabla^2\phi = \frac{\partial^2\phi}{\partial x^2} + \frac{\partial^2\phi}{\partial y^2} + \frac{\partial^2\phi}{\partial z^2}$

Let $\phi = f(r)$ then $\frac{\partial\phi}{\partial x} = \frac{d\phi}{dr}\frac{\partial r}{\partial x} = \frac{x}{r} f'(r)$

hence $$\frac{\partial^2\phi}{\partial x^2} = \frac{\partial}{\partial x}\left[\frac{x}{r} f'(r)\right] = \frac{x}{r}\frac{\partial}{\partial x} f'(r) + f'(r)\frac{\partial}{\partial x}\left(\frac{x}{r}\right)$$

$$= \frac{x}{r}\left(\frac{d}{dr} f'(r)\right)\frac{\partial r}{\partial x} + f'(r)\frac{r(1) - x\frac{\partial r}{\partial x}}{r^2} = \frac{x^2}{r^2} f''(r) + f'(r)\left(\frac{1}{r} - \frac{x}{r^2}\cdot\frac{x}{r}\right)$$

$$= \frac{x^2}{r^2} f''(r) + \frac{1}{r} f'(r) - \frac{x^2}{r^3} f'(r)$$

Similarly, $$\frac{\partial^2\phi}{\partial y^2} = \frac{y^2}{r^2} f''(r) + \frac{1}{r} f'(r) - \frac{y^2}{r^3} f'(r)$$

and $$\frac{\partial^2\phi}{\partial z^2} = \frac{z^2}{r^2} f''(r) + \frac{1}{r} f'(r) - \frac{z^2}{r^3} f'(r)$$

$$\therefore \quad \nabla^2 f(r) = \sum \frac{\partial^2}{\partial x^2} f(x) = \frac{f''(r)}{r^2}(x^2 + y^2 + z^2) + \frac{3}{r} f'(r) - \frac{f'(r)}{r^3}(x^2 + y^2 + z^2)$$

or $$\nabla^2 f(r) = f''(r) + \frac{3}{r} f'(r) - \frac{1}{r} f'(r) = f''(r) + \frac{2}{r} f'(r). \quad ...(1)$$

Hence Proved

Further, to derive the expression for $\nabla^4 f(r)$ let us write

$$F(r) = f''(r) + \frac{2}{r} f'(r) \quad \text{then} \quad \nabla^4 f(r) = \nabla^2 F(r) = F''(r) + \frac{2}{r} F'(r) \quad ...(2)$$

Now, $$F'(r) = \frac{d}{dr}[f''(r) + \frac{2}{r} f'(r)] = f'''(r) + \frac{2}{r} f''(r) - \frac{2}{r^2} f'(r) \quad ...(3)$$

and $$F''(r) = f^{(iv)}(r) + \frac{d}{dr}\left[\frac{2}{r} f''(r)\right] - \frac{d}{dr}\left[\frac{2}{r^2} f'(r)\right]$$

$$= f^{(iv)}(r) + \frac{2}{r} f'''(r) - \frac{2}{r^2} f''(r) - \frac{2}{r^2} f''(r) + \frac{4}{r^3} f'(r)$$

$$= f^{(iv)}(r) + \frac{2}{r} f'''(r) - \frac{4}{r^2} f''(r) + \frac{4}{r^3} f'(r) \quad ...(4)$$

Using (3) and (4) in (2), we get

$$\nabla^4 f(r) = f^{(iv)}(r) + \frac{2}{r} f'''(r) - \frac{4}{r^2} f''(r) + \frac{4}{r^3} f'(r) + \frac{2}{r}\left[f'''(r) + \frac{2}{r} f''(r) - \frac{2}{r^2} f'(r)\right]$$

$$= f^{(iv)}(r) + \frac{4}{r} f'''(r).$$ **Hence the result.**

EXAMPLE 9.34. If $\phi(x, y, z) = x^2yz$, $\vec{F} = x^2yi + y^2x^3j - 3x^2z^2k$ and $\vec{G} = 2xz^2i - yzj + x^2y^3k$, find the values of the following expressions at the point (1, 2, 1)

(i) $(\vec{F}\cdot\nabla)\phi$ *(ii)* $\nabla\cdot(\phi\vec{F})$

(iii) $(\vec{F}\cdot\nabla)G$ *(iv)* $(\vec{F}\times\nabla)\times G$.

SOLUTION: Here $\phi = x^2yz$; $F = x^2y\,i + y^2x^3\,j - 3x^2z^2\,k$ and $G = 2xz^2\,i - yz\,j + x^2y^3k$

Therefore $\phi = 2$ at (1, 2, 1) and $\nabla\phi = 2xyzi + x^2z\,j + x^2yk = 4i + j + 2k$ at (1, 2, 1)

$$\vec{F} = x^2yi + y^2x^3j - 3x^2z^2k = 2i + 4j - 3k \quad \text{at} \quad (1, 2, 1)$$

$$\vec{G} = 2xz^2i - yzj + x^2y^3k = 2i - 2j + 8k \quad \text{at} \quad (1, 2, 1)$$

$$\nabla\cdot\vec{F} = 2xy + 2yx^3 - 6x^2z = 2 \quad \text{at} \quad (1, 2, 1)$$

$$\nabla\cdot\vec{G} = 2z^2 - z + 0 = 1 \quad \text{at} \quad (1, 2, 1)$$

$$\nabla\times\vec{F} = 6xz^2j + (3x^2y^2 - x^2)\,k = 6j + 11k \quad \text{at} \quad (1, 2, 1)$$

$$\nabla\times\vec{G} = (3x^2y^2 + y)\,i + (4xz - 2xy^3)\,j = 14i - 12j \quad \text{at} \quad (1, 2, 1)$$

(i) $(\vec{F}\cdot\nabla)\phi = \vec{F}\cdot(\nabla\phi) = (2i + 4j - 3k)\cdot(4i + j + 2k) = 6$ at (1, 2, 1)

(ii) $\nabla\cdot(\phi\vec{F}) = \nabla\phi\cdot\vec{F} + \phi\,\nabla\cdot\vec{F} = (4i + j + 2k)\cdot(2i + 4j - 3k) + 2\,(2) = 10$ at (1, 2, 1)

(iii)
$$(\vec{F}\cdot\nabla)\vec{G} = \left(2\frac{\partial}{\partial x} + 4\frac{\partial}{\partial y} - 3\frac{\partial}{\partial z}\right)(2xz^2i - yzj + x^2y^3k)$$

$$= 2\,(2z^2i + 2xy^3k) + 4\,(-zj + 3x^2y^2k) - 3\,(4xzi - yj)$$

$$= (4z^2 - 12xz)\,i + (-4z + 3y)\,j + (4xy^3 + 12x^2y^2)\,k$$

$$= 8i + 2j + 80k \quad \text{at} \quad (1, 2, 1)$$

(iv) $(\vec{F}\times\nabla)\times\vec{G} = \nabla(\vec{F}\cdot\vec{G}) - (\nabla\cdot\vec{G})\,\vec{F}$ (since $\vec{F}$ is here constant at (1, 2, 1))

$$= \nabla\,[(2i + 4j - 3k)\cdot(2xz^2i - yzj + x^2y^3k)] - 1\,(2i + 4j - 3k)$$

$$= \nabla\,(4xz^2 - 4yz - 3x^2y^3) - (2i + 4j - 3k)$$

$$= [(4z^2 - 6xy^3)\,i + (-4z - 9x^2y^2)\,j + (8xz - 4y)\,k]_{(1,2,1)} - (2i + 4j - 3k)$$

$$= -44i - 40j + 0k - 2i - 4j + 3k = -46i - 44j + 3k.$$ **Ans.**

EXAMPLE 9.35. Show that the field of force given by $\vec{F} = (y^2\cos x + z^3)\,i + (2y\sin x - 4)\,j + (3xz^2 + 2)\,k$ is conservative and find its scalar potential. Also find the work done in moving the particle in the field from a point A (0, 1, – 1) to point B (π/2, – 1, 2).

SOLUTION:
$$\nabla\times\vec{F} = \begin{vmatrix} i & j & k \\ \dfrac{\partial}{\partial x} & \dfrac{\partial}{\partial y} & \dfrac{\partial}{\partial z} \\ y^2\cos x + z^3 & 2y\sin x - 4 & 3xz^2 + 2 \end{vmatrix}$$

$$= i\,(0 - 0) + j\,(0 + 3z^2 - 3z^2) + k\,(2y\cos x - 2y\cos x - 0) = 0.$$

∴ $\vec{F}$ represents a conservative (or irrotational) field. The corresponding scalar potential ϕ is given by $\nabla\phi = \vec{F} = (y^2\cos x + z^3)\,i + (2y\sin x - 4)\,j + (3xz^2 + 2)\,k$

Also we have $d\phi = \frac{\partial\phi}{\partial x}dx + \frac{\partial\phi}{\partial y}dy + \frac{\partial\phi}{\partial z}dz = \nabla\phi \cdot \vec{dr}$

Therefore $d\phi = (y^2 \cos x + z^3)\, dx + (2y \sin x - 4)\, dy + (3xz^2 + 2)\, dz$

$= d\,(y^2 \sin x + xz^3 - 4y + 2z)$

Hence $\phi = y^2 \sin x + xz^3 - 4y + 2z + c$

Next the required work done $= \int \vec{F} \cdot \vec{dr} = \int_A^B \nabla\phi \cdot \vec{dr} = \int_A^B \left(\frac{\partial\phi}{\partial x}dx + \frac{\partial\phi}{\partial y}dy + \frac{\partial\phi}{\partial z}dz\right) = \int_A^B d\phi$

$= [\phi]_{(0,1,-1)}^{(\pi/2,-1,2)} = \left[y^2 \sin x + xz^3 - 4y + 2z\right]_{(0,1,-1)}^{(\pi/2,-1,2)} = 4\pi + 15.$

Hence Proved.

EXAMPLE 9.36. Show that the vector field $\vec{F} = \frac{\vec{r}}{r^3}$ is irrotational as well as solenoidal. Also, find the scalar potential. **(UPTU 2001-02; 2005-06)**

SOLUTION: $\vec{F} = \frac{\vec{r}}{r^3} = \frac{xi + yj + zk}{r^3}$ hence

$$\text{div}\,(\vec{F}) = \nabla \cdot \vec{F} = \Sigma \frac{\partial}{\partial x}\left(\frac{x}{r^3}\right) = \Sigma\left(\frac{1}{r^3} - \frac{3x}{r^4} \cdot \frac{x}{r}\right) = \frac{3}{r^3} - \frac{3}{r^5}(x^2 + y^2 + z^2) = 0$$

Therefore $\vec{F}$ is a solenoidal vector.

Next
$$\text{Curl } \vec{F} = \nabla \times \vec{F} = \begin{vmatrix} i & j & k \\ \partial/\partial x & \partial/\partial y & \partial/\partial z \\ x/r^3 & y/r^3 & z/r^3 \end{vmatrix} = \Sigma i\left[\frac{\partial}{\partial y}\left(\frac{z}{r^3}\right) - \frac{\partial}{\partial z}\left(\frac{y}{r^3}\right)\right]$$

$$= \Sigma i\left[z\frac{\partial}{\partial y}r^{-3} - y\frac{\partial}{\partial z}r^{-3}\right] = \Sigma i\left[z\,(-3)\,r^{-4}\frac{y}{r} - y\,(-3)\,r^{-4}\frac{z}{r}\right] = 0.$$

Therefore $\vec{F}$ is irrotational also. Thus, there exists a scalar potential ϕ such that

$$\vec{F} = \nabla\phi \quad \text{hence} \quad \frac{\partial\phi}{\partial x} = \frac{x}{r^3},\ \frac{\partial\phi}{\partial y} = \frac{y}{r^3},\ \frac{\partial\phi}{\partial z} = \frac{z}{r^3} \qquad \therefore \qquad \vec{F} \cdot d\vec{r} = d\phi.$$

Thus, $d\phi = \frac{x\,dx + y\,dy + z\,dz}{r^3} = \frac{r\,dr}{r^3} = \frac{dr}{r^2} \quad \therefore\ \phi = \int \frac{dr}{r^2} = -\frac{1}{r} + C.$ **Hence Proved.**

EXAMPLE 9.37. Find the directional derivative of the scalar function $\phi = (x^2 + y^2 + z^2)^{-1/2}$ at a point (3, 1, 2) in the direction of the vector (yz, zx, xy).

SOLUTION: $\nabla\phi = \sum i\frac{\partial\phi}{\partial x} = \sum i\frac{\partial}{\partial x}\left(\frac{1}{r}\right) = -\sum \frac{i}{r^2}\frac{\partial r}{\partial x} = -\sum \frac{ix}{r^3}$ where $r^2 = x^2 + y^2 + z^2$

$$\text{Directional derivative} = \nabla\phi \cdot \frac{iyz + jzx + kxy}{\sqrt{y^2z^2 + z^2x^2 + x^2y^2}} = \frac{-3xyz}{(x^2 + y^2 + z^2)^{3/2}\sqrt{y^2z^2 + z^2x^2 + x^2y^2}}$$

$$\text{at } (3, 1, 2)\ \nabla\phi = \frac{-3.3.1.2}{(3^2 + 1^2 + 2^2)^{3/2}\sqrt{3^2.1^2 + 3^2.2^2 + 1^2.2^2}} = \frac{-9}{49\sqrt{14}}.$$ **Ans.**

EXAMPLE 9.38. Show that the vector field $\vec{A} = (x^2 - y^2 + x)\,i - (2xy + y)\,j$ is irrotational and find its scalar potential. **(UPTU 2009-10)**

SOLUTION: $\vec{A} = (x^2 - y^2 + x)\,i - (2xy + y)\,j.$

Therefore
$$\nabla \times \vec{A} = \begin{vmatrix} i & j & k \\ \dfrac{\partial}{\partial x} & \dfrac{\partial}{\partial y} & \dfrac{\partial}{\partial z} \\ x^2 - y^2 + x & -2xy - y & 0 \end{vmatrix} = i(0) + j(0) + k(-2y + 2y) = 0$$

Hence $\vec{A}$ is irrotational. Let its scalar potential be ϕ so that $\vec{A} = \nabla \phi$

$$\Rightarrow \quad \frac{\partial \phi}{\partial x} = x^2 - y^2 + x, \quad \frac{\partial \phi}{\partial y} = -2xy - y$$

On integration *w.r.t.* x, we get

$$\phi = \frac{x^3}{3} - xy^2 + \frac{x^2}{2} + c(y)$$

$$\therefore \quad \frac{\partial \phi}{\partial y} = -2xy + \frac{dc}{dy}. \quad \text{But} \quad \frac{\partial \phi}{\partial y} = -2xy - y.$$

$$\Rightarrow \quad \frac{dc}{dy} = -y. \quad \text{Hence} \quad c = \frac{-y^2}{2}.$$

$$\therefore \quad \phi = \frac{x^3}{3} - xy^2 + \frac{x^2}{2} - \frac{y^2}{2}. \quad \textbf{Ans.}$$

EXAMPLE 9.39. A vector field is given by $\vec{A} = (x^2 + xy^2)i + (y^2 + x^2y)j.$
Show that the field is irrotational and find the scalar potential.

SOLUTION: $\vec{A} = (x^2 + xy^2)i + (y^2 + x^2y)j$

For A to be irrotational $\nabla \times \vec{A} = 0$ where $\nabla \equiv i\dfrac{\partial}{\partial x} + j\dfrac{\partial}{\partial y}.$

$$\therefore \left(i\frac{\partial}{\partial x} + j\frac{\partial}{\partial y}\right) \times [(x^2 + xy^2)i + (y^2 + x^2y)j] = k\left[\frac{\partial}{\partial x}(y^2 + x^2y) - \frac{\partial}{\partial y}(x^2 + xy^2)\right]$$

$$= k\,(2xy - 2xy) = 0. \quad \text{Hence } \vec{A} \text{ is irrotational.}$$

Let the scalar potential of $\vec{A}$ be ϕ then $\vec{A} = \nabla \phi$

$$\Rightarrow \quad \frac{\partial \phi}{\partial x} = x^2 + xy^2 \quad \text{and} \quad \frac{\partial \phi}{\partial y} = y^2 + x^2y.$$

On integration *w.r.t.* x, we get $\phi = \dfrac{x^3}{3} + \dfrac{x^2y^2}{2} + c(y)$

$$\therefore \quad \frac{\partial \phi}{\partial y} = 2\frac{x^2y}{2} + \frac{dc}{dy}. \quad \text{But} \quad \frac{\partial \phi}{\partial y} = y^2 + x^2y \quad \Rightarrow \quad \frac{dc}{dy} = y^2. \quad \text{Hence } c = \frac{y^3}{3}$$

Therefore $\phi = \dfrac{x^3}{3} + \dfrac{x^2y^2}{2} + \dfrac{y^3}{3}.$ **Ans.**

EXAMPLE 9.40. A fluid motion is given by $\vec{V} = (y + z)i + (z + x)j + (x + y)k$. Is this motion irrotational? If so, find the velocity potential. Is the motion possible for incompressible fluid? **(UPTU 2003-04)**

SOLUTION: Velocity vector $\vec{V} = (y + z)i + (z + x)j + (x + y)k.$

$$\therefore \qquad \text{Curl } \vec{V} = \begin{vmatrix} i & j & k \\ \dfrac{\partial}{\partial x} & \dfrac{\partial}{\partial y} & \dfrac{\partial}{\partial z} \\ y+z & z+x & x+y \end{vmatrix} = i(1-1) + j(1-1) + k(1-1) = 0.$$

Hence motion is irrotational. Let the velocity potential be ϕ so that

$$\vec{V} = \nabla \phi = (y + z)i + (z + x)j + (x + y)k = i\frac{\partial \phi}{\partial x} + j\frac{\partial \phi}{\partial y} + k\frac{\partial \phi}{\partial z}$$

$$\Rightarrow \qquad \frac{\partial \phi}{\partial x} = y + z, \quad \frac{\partial \phi}{\partial y} = z + x, \quad \frac{\partial \phi}{\partial z} = x + y.$$

Integrating $\dfrac{\partial \phi}{\partial x} = y + z$ *w.r.t.* x, we get $\phi = xy + xz + c\,(y, z)$.

Differentiating partially *w.r.t.* y, we get

$$\frac{\partial \phi}{\partial y} = x + \frac{\partial c}{\partial y} \quad \text{but} \quad \frac{\partial \phi}{\partial y} = z + x. \text{ Hence } \frac{\partial c}{\partial y} = z \quad \Rightarrow \quad c = yz$$

Thus $\phi = xy + xz + yz$ which satisfies $\dfrac{\partial \phi}{\partial z} = x + y$ also.

Therefore the velocity potential $\phi = xy + yz + zx$.

Also, we have
$$\nabla \cdot \vec{V} = \frac{\partial}{\partial x}(y + z) + \frac{\partial}{\partial y}(z + x) + \frac{\partial}{\partial z}(x + y) = 0$$

Hence, the motion is possible for incompressible fluid since $\vec{V}$ is solenoidal. **Hence Proved.**

EXAMPLE 9.41. In what direction from $(3, 1, -2)$ is the directional derivative of $\phi = x^2y^2z^4$ maximum and what is its magnitude ?

SOLUTION:
$$\nabla\phi = i\frac{\partial \phi}{\partial x} + j\frac{\partial \phi}{\partial y} + k\frac{\partial \phi}{\partial z} = 2xy^2z^4 i + 2yx^2z^4 j + 4x^2y^2z^3 k$$

$$\therefore \qquad [\nabla\phi]_{(3,\,1,\,-2)} = 96i + 288j - 288k$$

we know that directional derivative is maximum in the direction of the normal to the given surface, that is, in the direction of $\nabla\phi$.

$$\therefore \text{ Magnitude Max. } \nabla\phi = \sqrt{96^2 + 288^2 + (-288)^2} = 96\sqrt{1 + 9 + 9} = 417.6 \text{ approx.}$$ **Ans.**

EXAMPLE 9.42. Find the directional derivative of $\phi = x^2yz + 4xz^2$ at the point (1, –2, –1) in the direction of the vector $2i - j - 2k$.

SOLUTION: Here $\nabla\phi = (2xyz + 4z^2)\, i + x^2zj + (x^2y + 8xz)\, k$

and $[\nabla\phi]_{(1,\,-2,\,-1)} = 8i - j - 10k.$

Given vector is $\vec{a} = 2i - j - 2k$

∴ Directional derivative of ϕ in the direction of the vector $\vec{a}$.

$$= \frac{\nabla\phi\cdot\vec{a}}{|\vec{a}|} = \frac{(8i-j-10k)\cdot(2i-j-2k)}{\sqrt{2^2+(-1)^2+(-2)^2}} = \frac{16+1+20}{\sqrt{9}} = \frac{37}{3}. \quad \textbf{Ans.}$$

EXAMPLE 9.43. Find the directional derivative of $\phi = 5x^2y - 5y^2z + \frac{5}{2}z^2x$ at the point (1, 1, 1) in the direction of the line $\frac{x-1}{2} = \frac{y-3}{-2} = \frac{z}{1}$. **(UPTU 2003-04)**

SOLUTION: $\phi = 5x^2y - 5y^2z + \frac{5}{2}z^2x$

Hence $\nabla\phi = \left(10xy + \frac{5}{2}z^2\right)i + (5x^2 - 10yz)j + (-5y^2 + 5zx)k$

∴ $\nabla\phi$ at $(1, 1, 1) = \frac{25}{2}i - 5j + 0k$

Unit vector in the direction of line $\frac{x-1}{2} = \frac{y-3}{-2} = \frac{z}{1}$, is

$$\frac{2i-2j+k}{\sqrt{4+4+1}} = \frac{1}{3}(2i - 2j + k) = \hat{n}.$$

Hence the directional derivative of ϕ along the above line

$$= \nabla\phi\cdot\hat{n} = \left(\frac{25}{2}i - 5j\right)\cdot\left(\frac{2}{3}i - \frac{2}{3}j + \frac{1}{3}k\right) = \frac{25}{2}\left(\frac{2}{3}\right) - 5\left(\frac{-2}{3}\right) + 0$$

$$= \frac{25}{3} + \frac{10}{3} = \frac{35}{3}. \quad \textbf{Ans.}$$

EXAMPLE 9.44. Find the directional derivative of $\frac{1}{r^2}$ in the direction of $\vec{r} = xi + yj + zk$.

(UPTU 2005-06)

SOLUTION: Let $\phi = \frac{1}{r^2}$ ∴ $\nabla\phi = i\frac{\partial\phi}{\partial x} + j\frac{\partial\phi}{\partial y} + k\frac{\partial\phi}{\partial z} = \sum i\frac{\partial}{\partial x}\left(\frac{1}{r^2}\right) = \sum\frac{-2}{r^3}\frac{\partial r}{\partial x}i$

or $\nabla\phi = \frac{-2}{r^3}\sum i\frac{x}{r} = \frac{-2}{r^4}\cdot\vec{r}$

The directional derivative of ϕ along $\vec{r} = \nabla\phi\cdot\frac{\vec{r}}{r} = \frac{-2}{r^4}\frac{1}{r}\vec{r}\cdot\vec{r} = \frac{-2}{r^5}\cdot r^2 = \frac{-2}{r^3}$. **Ans.**

EXAMPLE 9.45. Define the gradient, divergence and curl.

(*i*) If $f(x, y, z) = 3x^2y - y^3z^2$, find gradient f at $(1, -2, -1)$.

(*ii*) If $\vec{F} = xz^3i - 2x^2yzj + 2yz^4k$ find divergence and curl of $\vec{F}$.

(UPTU 2006-07)

SOLUTION: (*i*) Given $f = 3x^2y - y^3z^2$. Gradient of f is given by

$$\nabla f = i\frac{\partial f}{\partial x} + j\frac{\partial f}{\partial y} + k\frac{\partial f}{\partial z} = i(6xy) + j(3x^2 - 3y^2z^2) + k(-2y^3z)$$

$[\nabla f]$ at $(1, -2, -1) = -12i - 9j - 16k$. **Ans.**

(*ii*) Given $\vec{F} = xz^3i - 2x^2yzj + 2yz^4k$. Divergence of $\vec{F}$ is given by

$$\nabla\cdot\vec{F} = \frac{\partial}{\partial x}(xz^3) + \frac{\partial}{\partial y}(-2x^2yz) + \frac{\partial}{\partial z}(2yz^4) = z^3 - 2x^2z + 8yz^3 \quad \textbf{Ans.}$$

and Curl of
$$\vec{F} = \nabla \times \vec{F} = \begin{vmatrix} i & j & k \\ \dfrac{\partial}{\partial x} & \dfrac{\partial}{\partial y} & \dfrac{\partial}{\partial z} \\ xz^3 & -2x^2yz & 2yz^4 \end{vmatrix} = i(2z^4 + 2x^2y) + j(3xz^2 - 0) + k(-4xyz - 0)$$

$$= 2(z^4 + x^2y)\,i + 3xz^2\,j - 4xyzk. \quad \textbf{Ans.}$$

EXAMPLE 9.46. (*a*) Find the directional derivative of $\frac{1}{r}$ in the direction of $\vec{r} = xi + yj + zk$.

(UPTU 2002-03)

(*b*) Find the directional derivative of $f(x, y, z) = 2x^2 + 3y^2 + z^2$ at the point $P(2, 1, 3)$ in the direction of the vector $\vec{a} = i - 2k$. **(UPTU 2008-09)**

SOLUTION: (*a*) $|\vec{r}| = r$,
$$\nabla\left(\frac{1}{r}\right) = i\frac{\partial}{\partial x}\left(\frac{1}{r}\right) + j\frac{\partial}{\partial y}\left(\frac{1}{r}\right) + k\frac{\partial}{\partial z}\left(\frac{1}{r}\right) = -\frac{1}{r^2}\frac{\partial r}{\partial x}i - \frac{1}{r^2}\frac{\partial r}{\partial y}j - \frac{1}{r^2}\frac{\partial r}{\partial z}k$$

$$= -\frac{1}{r^2}\left[\frac{x}{r}i + \frac{y}{r}j + \frac{z}{r}k\right] = -\frac{1}{r^3}\cdot\vec{r}.$$

$\therefore$ Directional derivative of $\frac{1}{r}$ along
$$\vec{r} = \nabla\left(\frac{1}{r}\right)\cdot\hat{r} = \frac{-1}{r^3}\vec{r}\cdot\frac{\vec{r}}{r} = \frac{-1}{r^4}\cdot r^2 = \frac{-1}{r^2} = \frac{-1}{x^2 + y^2 + z^2}.$$

Ans.

(*b*)
$$f(x, y, z) = 2x^2 + 3y^2 + z^2, \text{ hence } \nabla f = i\frac{\partial f}{\partial x} + j\frac{\partial f}{\partial y} + k\frac{\partial f}{\partial z} = 4x\,i + 6y\,j + 2zk$$
$$= 8i + 6j + 6k \text{ at } P(2, 1, 3).$$

The directional derivative of f at P in the direction of $\vec{a} = i - 2k$, is equal to

$$\Delta f\cdot\frac{\vec{a}}{|\vec{a}|} = (8i + 6j + 6k)\cdot\frac{(i - 2k)}{\sqrt{1^2 + (-2)^2}}$$

$$= \frac{|8 - 12|}{\sqrt{5}} = \frac{4}{\sqrt{5}}. \quad \textbf{Ans.}$$

EXAMPLE 9.47. Find the angle between the surfaces $x^2 + y^2 + z^2 = 9$ and $z = x^2 + y^2 - 3$ at the point $(2, -1, 2)$. **(UPTU 2002-03)**

SOLUTION: Normal to the surface $\phi_1 \equiv x^2 + y^2 + z^2 - 9 = 0$ at the point $(2, -1, 2)$

$$= [\nabla \phi_1] \text{ at } (2, -1, 2) = [2xi + 2yj + 2zk]_{(2, -1, 2)} = 4i - 2j + 4k$$

Normal to the surface $\phi_2 = x^2 + y^2 - 3 - z = 0$ at the point $(2, -1, 2)$

$$[\nabla \phi_2]_{(2, -1, 2)} = [2xi + 2yj - k]_{(2, -1, 2)} = 4i - 2j - k.$$

Angle between the two surfaces at $(2, -1, 2)$

$$= \cos^{-1} \frac{(\nabla\phi_1 \cdot \nabla\phi_2)}{|\nabla\phi_1||\nabla\phi_2|} = \cos^{-1}\left\{\frac{(4i - 2j + 4k)\cdot(4i - 2j - k)}{\sqrt{16+4+16}\sqrt{16+4+1}}\right\}$$

$$= \cos^{-1}\left(\frac{16+4-4}{\sqrt{(36)(21)}}\right) = \cos^{-1}\left(\frac{8}{3\sqrt{21}}\right). \quad \textbf{Ans.}$$

EXAMPLE 9.48. **(a)** Prove that div (grad r^n) $= n(n + 1)r^{n-2}$ where $r^2 = x^2 + y^2 + z^2$.

Hence show that $\nabla^2\left(\frac{1}{r}\right) = 0$. **(UPTU 2002-03; 2004-05)**

(b) If $r = (x^2 + y^2 + z^2)^{1/2}$, evaluate $\nabla^2(\log r)$. **[UPTU 2011-12]**

SOLUTION: (a) div (grad r^n) $= \nabla \cdot (\nabla r^n) = \nabla^2 r^n = \left(\frac{\partial^2}{\partial x^2} + \frac{\partial^2}{\partial y^2} + \frac{\partial^2}{\partial z^2}\right)(r^n)$

$$= \sum \frac{\partial^2}{\partial x^2}(r^n) = \sum \frac{\partial}{\partial x}\left(nr^{n-1}\frac{\partial r}{\partial x}\right) = \sum \frac{\partial}{\partial x}\left(nr^{n-1}\frac{x}{r}\right) = \sum \frac{\partial}{\partial x}(nxr^{n-2})$$

$$= \sum\left\{nx(n-2)r^{n-3}\frac{\partial r}{\partial x} + nr^{n-2}\right\} = \sum\left\{n(n-2)xr^{n-3}\cdot\frac{x}{r} + nr^{n-2}\right\}$$

$$= n(n-2)r^{n-4}\Sigma x^2 + 3nr^{n-2} = n(n-2)r^{n-2} + 3nr^{n-2} = n(n+1)r^{n-2}.$$

Next for $n = -1$, $\nabla^2\left(\frac{1}{r}\right) = [n(n+1)r^{n-2}]_{n=-1} = 0.$ **Hence Proved.**

(b) $\nabla^2(\log r) = \nabla \cdot (\nabla \log r) = \left(\frac{\partial^2}{\partial x^2} + \frac{\partial^2}{\partial y^2} + \frac{\partial^2}{\partial z^2}\right)\log r = \sum \frac{\partial^2}{\partial x^2}(\log r)$

Now, $$\frac{\partial}{\partial x}(\log r) = \frac{d}{dr}(\log r)\frac{\partial r}{\partial x} = \frac{1}{r}\cdot\frac{x}{r} = \frac{x}{r^2}$$

and $$\frac{\partial^2}{\partial x^2}(\log r) = \frac{\partial}{\partial x}\left(\frac{x}{r^2}\right) = 1\cdot\frac{1}{r^2} + x\cdot\frac{d}{dx}\left(\frac{1}{r^2}\right)\frac{\partial r}{\partial x}$$

$$= \frac{1}{r^2} + x\left(\frac{-2}{r^3}\right)\cdot\frac{x}{r} = \frac{1}{r^2} - \frac{2x^2}{r^4}$$

$$\Rightarrow \quad \sum \frac{\partial^2}{\partial x^2}(\log r) = \sum \frac{1}{r^2} - \frac{2}{r^4}\Sigma x^2 = \frac{3}{r^2} - \frac{2}{r^4}\cdot r^2 = \frac{1}{r^2} \quad \textbf{Ans.}$$

EXAMPLE 9.49. If $\vec{a}$ is a constant vector, evaluate div $(\vec{r}\times\vec{a})$ and curl $(\vec{r}\times\vec{a})$ where $\vec{r} = xi + yj + zk$ **(UPTU 2002)**

SOLUTION: We know that $\nabla\cdot(\vec{F}\times\vec{G}) = \vec{G}\cdot\nabla\times\vec{F} - \vec{F}\cdot\nabla\times\vec{G}$

Taking $\vec{F} = \vec{r}$ and $\vec{G} = \vec{a}$ where $\vec{a}$ is constant, we get

$$\nabla\cdot(\vec{r}\times\vec{a}) = \vec{a}\cdot\nabla\times\vec{r} - \vec{r}\cdot\nabla\times\vec{a} = 0 - 0 = 0 \qquad \textbf{Ans.}$$

Next, using $\nabla\times(F\times G) = \vec{F}\ (\text{div }\vec{G}) - \vec{G}\ (\text{div }\vec{F}) + (\vec{G}\cdot\nabla)\vec{F} - (\vec{F}\cdot\nabla)\vec{G}$

$$\therefore \quad \nabla\times(\vec{r}\times\vec{a}) = \vec{r}(\nabla\cdot\vec{a}) - a(\nabla\cdot\vec{r}) + (\vec{a}\cdot\nabla)\vec{r} - (\vec{r}\cdot\nabla)\vec{a}$$

$$= 0 - 3a - \left(x\frac{\partial}{\partial x} + y\frac{\partial}{\partial y} + \frac{\partial}{\partial z}\right)\vec{a} + \left(a_1\frac{\partial}{\partial x} + a_2\frac{\partial}{\partial y} + a_3\frac{\partial}{\partial z}\right)\vec{r}$$

$$= -3a - 0 + \vec{a} = -2\vec{a}. \qquad \textbf{Ans.}$$

EXAMPLE 9.50. If $\vec{r} = xi + yj + zk$ and $r = |\vec{r}|$, show that

(*i*) $\text{div}(\vec{r}\,\phi) = 3\phi + \vec{r}\cdot\text{grad}\,\phi$ (*ii*) $\text{div}(\hat{r}) = \dfrac{2}{r}$. **(UPTU 2001)**

SOLUTION: (*i*) We know that $\nabla\cdot(\phi\vec{F}) = \nabla\phi\cdot\vec{F} + \phi\nabla\cdot\vec{F}$

$$\therefore \quad \nabla\cdot(\phi\vec{r}) = \nabla\phi\cdot\vec{r} + \phi\nabla\cdot\vec{r} = \vec{r}\cdot\nabla\phi + 3\phi \qquad \textbf{Hence Proved.}$$

(*ii*) Further $\nabla\cdot(\hat{r}) = \nabla\cdot\left(\dfrac{1}{r}\vec{r}\right) = \nabla\left(\dfrac{1}{r}\right)\cdot\vec{r} + \dfrac{1}{r}\nabla\cdot\vec{r} = -\dfrac{1}{r^2}\Sigma\dfrac{xi}{r}\cdot\vec{r} + \dfrac{1}{r}(3)$

$$= \vec{r}\cdot\left(-\frac{\vec{r}}{r^3}\right) + \frac{3}{r} = \frac{3}{r} - \frac{r^2}{r^3} = \frac{2}{r}. \qquad \textbf{Ans.}$$

EXAMPLE 9.51. Find the directional derivative of $\nabla\,.\,(\nabla f)$ at the point $(1, -2, 1)$ in the direction of the normal to the surface $xy^2z = 3x + z^2$ where $f = 2x^3y^2z^4$.

(UPTU 2008-09)

SOLUTION: $\nabla\,.\,(\nabla f) = \nabla^2 f = \dfrac{\partial^2 f}{\partial x^2} + \dfrac{\partial^2 f}{\partial y^2} + \dfrac{\partial^2 f}{\partial z^2}$ (Here $f = 2x^3y^2z^4$)

$$= 12xy^2z^4 + 4x^3z^4 + 24x^3y^2z^2.$$

$\therefore \quad \nabla(\nabla^2 f) = i(12y^2z^4 + 12x^2z^4 + 72x^2y^2z^2) + j\,(24xyz^4 + 48x^3yz^2) + k\,(48xy^2z^3 + 16x^3z^3 + 48x^3y^2z)$.

At $P(1, -2, 1)$, we have $\nabla(\nabla^2 f) = i\,(48 + 12 + 288) + j\,(-48 - 96) + k(192 + 16 + 192)$

$$= 348\,i - 144\,j + 400\,k.$$

If $\phi = xy^2z - 3x - z^2$ hence

$\nabla\phi = (y^2z - 3)\,i + (2xyz)\,j + (xy^2 - 2z)\,k$ at $P(1, -2, 1)$, $\nabla\phi = i - 4j + 2k$

$\therefore$ Required directional derivative $= \nabla\,(\nabla^2 f))\cdot\dfrac{\nabla\phi}{|\nabla\phi|}$

$$= \frac{348 + 576 + 800}{\sqrt{1+16+4}} = \frac{1724}{\sqrt{21}}. \qquad \textbf{Ans.}$$

EXAMPLE 9.52. Find the values λ and μ so that the surface $\lambda x^2 - \mu yz = (\lambda + z)x$ and $4x^2y + z^3 = 4$ intersect orthogonally at the point $(1, -1, 2)$. **(UPTU 2001)**

SOLUTION: Let $\phi_1 = \lambda x^2 - \mu yz - (\lambda + z)x = 0$ and $\phi_2 = 4x^2y + z^3 - 4 = 0$

then $\nabla\phi_1 = [2\lambda x - (\lambda + z)]\, i + (-\mu z)\, j + (-\mu y - x)k$

and $\nabla\phi_2 = (8xy)i + 4x^2j + 3z^2k$

At $P(1, -1, 2)$ we have $\nabla\phi_1 = (\lambda - 2)i - 2\mu j + (\mu - 1)k$

and $\nabla\phi_2 = -8i + 4j + 12k$

For the surfaces ϕ_1 and ϕ_2 to be orthogonal $\nabla\phi_1 \cdot \nabla\phi_2 = 0$

hence $-8(\lambda - 2) - 8\mu + 12(\mu - 1) = 0$ or $\mu = 2\lambda - 1$

Point $(1, -1, 2)$ lies on $4x^2y + z^3 = 4$ and also on $\lambda x^2 - \mu yz = (\lambda + z)x$

$\therefore$ $\lambda + 2\mu = \lambda + 2 \Rightarrow \mu = 1$ then $\lambda = 1$. **Ans.**

EXERCISE 9A

1. Evaluate $\nabla\phi$ where $\phi = x^3 + y^3 + z^3 - 3xyz$ at the point $(1, -1, 2)$.
2. Obtain the directional derivative of the scalar field $\phi = xy + yz + zx$ at the point $(1, 2, 3)$ in the direction of the vector $3i + 4j + 5k$.
3. Find div $\vec{V}$ and curl $\vec{V}$ if $\vec{V} = \nabla(x^3 + y^3 + z^3 - 3xyz)$.
4. If $\vec{r}$ is the position vector of the $P(x, y, z)$ and $\vec{a}$ and $\vec{b}$ are constant vectors, prove that
$$\vec{a}\cdot\nabla\left\{\vec{b}\cdot\nabla\left(\frac{1}{r}\right)\right\} = \frac{3(\vec{a}\cdot\vec{r})(\vec{b}\cdot\vec{r})}{r^5} - \frac{\vec{a}\cdot\vec{b}}{r^3}.$$
5. If $\phi_1 = x + y + z$, $\phi_2 = x + y$ and $\phi_3 = -(2xz + 2yz + z^2)$ show that
$$\nabla\phi_1\cdot[\nabla\phi_2\times\nabla\phi_3] = 0.$$
6. Given that $\rho\vec{F} = \nabla\phi$ where ρ and ϕ are scalar point functions and $\vec{F}$ is vector point function, find $\vec{F}\cdot\text{Curl}\,\vec{F}$.
7. Find the angle between the two surfaces
$$x^2y + z = 3 \quad\text{and}\quad x\log z - y^2 = 4$$
at their point of inter-section $(-1, 2, 1)$.
8. Determine the values of m and n so that the vector point function
$$\vec{F} = (xyz)^n(x^n i + y^n j + z^n k) \quad\text{is irrotational.}$$
9. Show that $\nabla\times\{\vec{a}\times\nabla(1/r)\} + \nabla\{\vec{a}\cdot\nabla(1/r)\} = 0$ where $\vec{a}$ is constant.
10. If $\vec{F}$ is a solenoidal vector field, show that
$$\text{Curl Curl Curl Curl}\,\vec{F} = \nabla^4\vec{F}.$$
11. If $\vec{a}$ and $\vec{b}$ are constant vectors show that
$$\nabla\{(\vec{r}\times\vec{a})\cdot(\vec{r}\times\vec{b})\} = \vec{b}\times(\vec{r}\times\vec{a}) + \vec{a}\times(\vec{r}\times\vec{b}).$$
12. If the field $\vec{F} = (x - 4y + az)\,i + (bx + 2y + z)\,j + (2x - cy + 3z)\,k$ is conservative, find a, b, c. Also find the work done in moving an object in this field from $(1, 1, 1)$ to the point $(2, 1, 3)$.
13. Show that the vector field $\vec{F} = (x^2 - yz)\,i + (y^2 - zx)\,j + (z^2 - xy)\,k$ is irrotational and find its corresponding scalar field ϕ so that $\vec{F} = \nabla\phi$. Also find the work done by the field if an object gets displaced from the origin to the point $(1, 2, 3)$.
14. Prove that $(y^2 - z^2 + 3yz)\,i + (3xz + 2xy)\,j + (3xy - 2xz + 2)\,k$ is both solenoidal and irrotational.

(UPTU 2008-09)

❑❑❑

CHAPTER

10 Vector Integration

Line Integral, Surface Integrals and Volume Integrals. Green's Theorem, Stoke's Theorem and Gauss Divergence Theorem and Their Applications

Here we shall be introducing the line integrals, surface integrals and volume integrals which have applications in physical problems. Thereafter we shall see transformation of line integral to surface integral through Stoke's theorem and the transformation of volume integral into surface integral through Gauss Divergence Theorem. These theorems are important because of their applications in various problems.

LINE INTEGRALS

We are familiar with the definite integral $I = \int_a^b f(x)\, dx$ carried out along the x-axis. This can be extended to integration along a curve. For this let us consider a curve C defined in terms of s, the arc length, taken as parameter, and let $\vec{F}(s)$ be a vector point function of s. Let the interval $[a, b]$ be divided into n parts $s_{i-1} \le s \le s_i$, $i = 1, 2, \ldots, n$ where $s_0 = s$ at A (when $x = a$) and $s_n = s$ at B (when $x = b$) (See the adjoining figure).

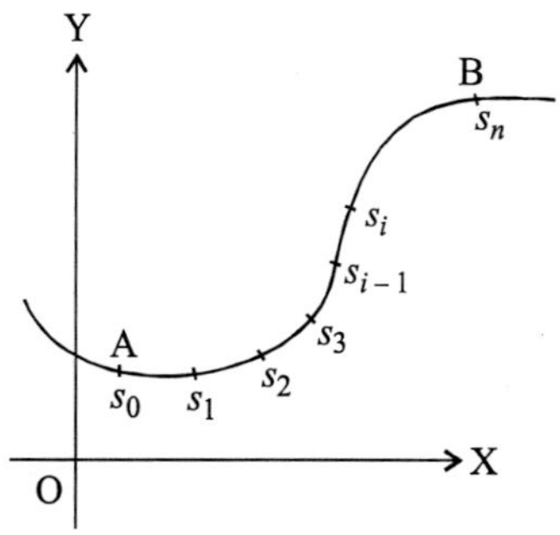

Let $\delta \vec{r}_i = \vec{r}(s_i) - \vec{r}(s_{i-1})$ and let $F_1, F_2, \ldots, F_n$ be the vector components of $\vec{F}$ along the tangent to the curve at the points of division. Next, we form the sum of the scalar products $\vec{F}_1 \cdot \delta\vec{r}_1, \vec{F}_2 \cdot \delta\vec{r}_2, \ldots, \vec{F}_n \cdot \delta\vec{r}_n$, denoted by S_n, that is,

$$S_n = \sum_{i=1}^{n} \vec{F}_i\, \delta\vec{r}_i .$$

Taking the limit of the sum S_n as $n \to \infty$, is defined as the line integral of $\vec{F}$ along the curve C and is written as

$$\int_C \vec{\mathbf{F}}\cdot\vec{dr} \quad \text{where} \quad \vec{dr} = idx + jdy + kdz$$

or as $\int_C \mathrm{F}\cdot\hat{\mathrm{T}}\,ds$ where $\hat{\mathrm{T}}$ is unit vector along the tangent to the curve and ds is an element of arc of the curve C. If the path of integration is a closed curve we use the symbol $\oint \vec{\mathrm{F}}\cdot\vec{dr}$.

Let $\vec{\mathrm{F}} = \mathrm{F}_1 i + \mathrm{F}_2 j + \mathrm{F}_3 k$ and $\vec{dr} = idx + jdy + kdz,$ then we have

$$\int_C \vec{\mathrm{F}}\cdot\vec{dr} = \int_C (\mathrm{F}_1 dx + \mathrm{F}_2 dy + \mathrm{F}_3 dz).$$

In practice, if $\vec{\mathrm{F}} = \vec{v}$ where $\vec{v}$ is the **velocity** of the fluid particle then the line integral $\oint_C \vec{v}\cdot\vec{dr}$ is called the ***CIRCULATION*** of the velocity around a closed curve C. Clearly, **the circulation is zero in a conservative field.**

The line integral $\int_C \vec{\mathrm{F}}\cdot\vec{dr}$ may or may not depend on the path of integration. Let us illustrate this fact by the following problems in two dimensions.

EXAMPLE 10.1. Let the vector field be $\vec{\mathrm{F}} = x^2 i - xyj.$ Find its integral from the origin O to the point P (1, 1)

(*i*) along the straight line OP

(*ii*) along the parabola $y^2 = x$

(*iii*) along the x-axis from O to A (1, 0) and then from A (1, 0) to P (1, 1).

SOLUTION: (*i*) Equation of the line OP is $y = x$ so that $dy = dx$

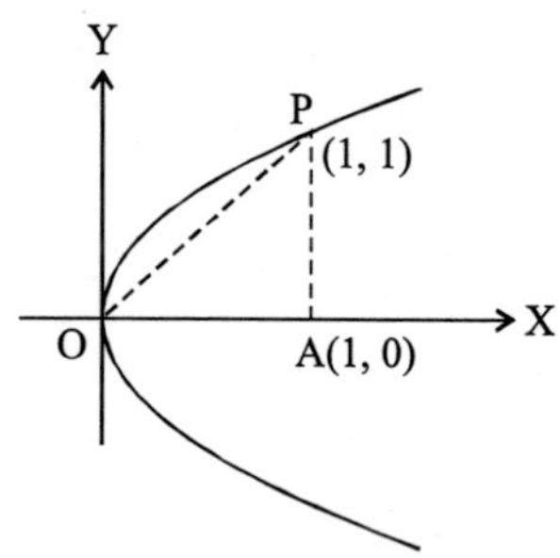

$$\int_{\text{line OP}} \vec{\mathrm{F}}\cdot\vec{dr} = \int_{\text{line OP}} (x^2 i - xyj)\cdot(idx + jdy)$$

$$= \int_{\text{line OP}} x^2 dx - xy\,dy = \int_{\text{line OP}} (x^2 dx - x\times x\,dx) = 0.$$

(*ii*) Equation of the parabola is $y^2 = x$ so that $2ydy = dx$

$$\therefore \quad \int_{\text{curve OP}} \left(x^2 dx - x\frac{1}{2}dx\right) = \int_0^1 \left(x^2 - \frac{x}{2}\right)dx = \left[\frac{x^3}{3} - \frac{x^2}{4}\right]_0^1 = \frac{1}{12}$$

(*iii*) Here we integrate $\vec{\mathrm{F}}$ from O to A along X-axis and from A to P parallel to Y-axis.

$$\therefore \quad \int_{\text{OAP}} \vec{\mathrm{F}}\cdot\vec{dr} = \int_{\text{OAP}} (x^2 dx - xy\,dy) = \int_{\text{OA}} x^2 dx - 0 + \int_{\text{AP}} (0 - 1y\,dy)$$

$$= \int_0^1 x^2 dx - \int_0^1 y\, dy = \left[\frac{x^3}{3}\right]_0^1 - \left[\frac{y^2}{2}\right]_0^1 = \frac{1}{3} - \frac{1}{2} = -\frac{1}{6}.\qquad \textbf{Ans.}$$

This shows from (*i*), (*ii*) and (*iii*) that the line integral of $\vec{F}$ depends upon the path along which it is integrated, hence it is different in (*i*), (*ii*) and (*iii*).

EXAMPLE 10.2. Consider the vector field $\vec{F} = y^2 i + 2xyj$ and integrate it from the origin O to the point P (1, 1) along the paths (*i*), (*ii*) and (*iii*) of the Example 10.1.

SOLUTION: (*i*)

$$\int_{\text{line OP}} \vec{F}\cdot\vec{dr} = \int_{\text{line OP}} (y^2 i + 2xy\, j)\cdot(i\,dx + j\,dy) = \int_{\text{line OP}\,(y=x)} y^2 dx + 2xy\, dy = \int_0^1 x^2 dx + 2x^2 dx$$

$$= \left[\frac{x^3}{3} + \frac{2x^3}{3}\right]_0^1 = \frac{1}{3} + \frac{2}{3} = 1.$$

(*ii*)

$$\int_{\substack{\text{curve}\\ \text{O to P}}} \vec{F}\cdot\vec{dr} = \int y^2 dx + 2xy\, dy = \int_0^1 x\,dx + x\,dx = \left[x^2\right]_0^1 = 1.$$

(*iii*)

$$\int_{\text{OAP}} \vec{F}\cdot\vec{dr} = \int_{\text{OAP}} (y^2 dx + 2xy\, dy) = \int_{\text{OA}} 0 + 0 + \int_{\text{AP}} 0 + 2\cdot 1\cdot y\, dy = 0 + \int_0^1 (0 + 2y)\, dy = \left[y^2\right]_0^1 = 1.$$

Here we note that the line integral of $\vec{F} = y^2 i + 2xyj$ along the paths (*i*), (*ii*), (*iii*) is the same.

Actually the line integral here is independent of the path, since $\vec{F}$ is conservative (i.e., $\nabla \times \vec{F} = 0$).

We can easily show that **if $\vec{F}$ is the gradient of some scalar point function ϕ, the line integral of $\vec{F}$ is independent of the path as follows:**

Suppose

$$\boxed{\vec{F} = \nabla\phi \text{ then } \int \vec{F}\cdot\vec{dr} = \int_A^B \nabla\phi\cdot\vec{dr} = [d\phi]_a^b = [\phi]_a^b = \phi(b) - \phi(a).}\qquad \textbf{Ans.}$$

Thus in this case the line integral of $\vec{F}$ depends only on the end points A and B and not on the path joining A and B.

EXAMPLE 10.3. **(a)** For the vector field $\vec{F} = (3x^2 + 6y)\, i - 14yzj + 20xz^2 k$, calculate $\int_C \vec{F}\cdot\vec{dr}$ from the origin (0, 0, 0) to the point A (1, 1, 1) along the path C

(*i*) $x = t, \quad y = t^2, \quad z = t^3$

(*ii*) the straight line joining the origin O to A of the path C.

(b) If $\vec{A} = (x - y)i + (x + y)j$, evaluate $\oint_C \vec{A}\cdot\vec{dr}$ around the curve C consisting of $y = x^2$ and $y^2 = x$. **[UPTU 2011-12]**

SOLUTION: **(a)** We have $\vec{F} = (3x^2 + 6y)\, i - 14yzj + 20xz^2k$ and $\vec{r} = xi + yj + zk$

$$\therefore \quad \vec{F} \cdot \vec{dr} = (3x^2 + 6y)\, dx - 14\, yz\, dy + 20\, xz^2 dz$$

(*i*) C is the curve $x = t, \quad y = t^2, \quad z = t^3$ so $dx = dt, \; dy = 2t\, dt, \; dz = 3t^2\, dt$ and $t = 0$ at O and $t = 1$ at the point A.

$$\therefore \quad \int_C \vec{F} \cdot \vec{dr} = \int_0^1 (3t^2 + 6t^2)\, dt - 14t^5 \cdot 2t\, dt + 20t^7 \cdot 3t^2\, dt$$

$$= \int_0^1 (9t^2 - 28t^6 + 60t^9)\, dt = \left[\frac{9t^3}{3} - \frac{28t^7}{7} + \frac{60t^{10}}{10}\right]_0^1 = 5. \qquad \textbf{Ans.}$$

(*ii*) The line joining (0, 0, 0) and (1, 1, 1) is given by

$$\frac{x}{1} = \frac{y}{1} = \frac{z}{1} = t, \text{ say} \quad \text{then} \quad x = t, \quad y = t, \quad z = t \text{ and } t \text{ varies from 0 to 1.}$$

$$\therefore \quad \int \vec{F} \cdot \vec{dr} = \int_0^1 (3t^2 + 6t)\, dt - 14t^2 dt + 20t^3 dt = \int_0^1 (6t - 11t^2 + 20t^3)\, dt = \left[3t^2 - \frac{11t^3}{3} + 5t^4\right]_0^1$$

$$= 3 - \frac{11}{3} + 5 = \frac{13}{3}.$$

Let us now find curl $\vec{F}$.

$$\nabla \times \vec{F} = \begin{vmatrix} i & j & k \\ \frac{\partial}{\partial x} & \frac{\partial}{\partial y} & \frac{\partial}{\partial z} \\ 3x^2 + 6y & -14yz & 20xz^2 \end{vmatrix} = (0 + 14\, y)\, i + (0 - 20\, z^2)\, j + (0 - 6)\, k \neq 0.$$

Thus $\vec{F}$, is not conservative and that is why the value of $\int_C \vec{F} \cdot \vec{dr}$ is not the same along the two paths (*i*) and (*ii*) above. **Ans.**

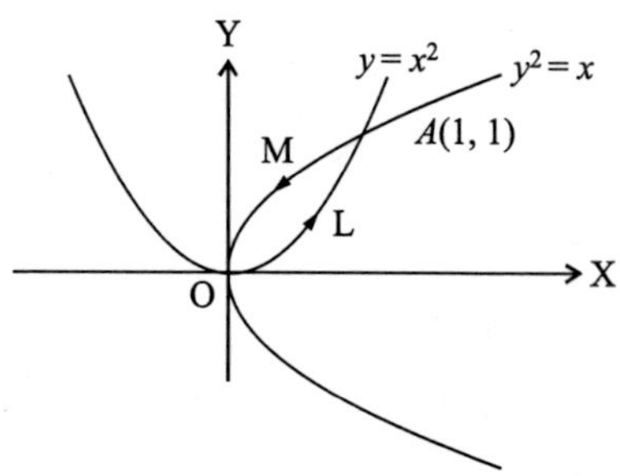

(b) Curve C consists of the portion of the parabola $y = x^2$ from $O(0, 0)$ to $A(1, 1)$ and part of the parabola $y^2 = x$ from A to O.

$$\therefore \quad \oint_C \vec{A} \cdot \vec{dr} = \oint_C [(x - y)i + (x + y)j] \cdot (i\, dx + j\, dy)$$

$$= \oint_C (x - y)dx + (x + y)\, dy$$

$$= \int_{\text{along OLA } (y = x^2)} [(x - y)\, dx + (x + y)\, dy] + \int_{\text{along AMO } (y^2 = x)} (x - y)\, dx + (x + y)\, dy$$

$$= \int_0^1 (x - x^2)\, dx + (x + x^2)\, 2x\, dx + \int_1^0 \left(x - \sqrt{x}\right) dx + (x + \sqrt{x})\frac{dx}{2\sqrt{x}}$$

$$= \int_0^1 (x - x^2 + 2x^2 + 2x^3)\,dx + \int_0^1 \left(x - \sqrt{x} + \frac{\sqrt{x}}{2} + \frac{1}{2} \right) dx$$

$$= \int_0^1 (x + x^2 + 2x^3)\,dx + \int_0^1 \left(-x + \frac{\sqrt{x}}{2} - \frac{1}{2} \right) dx$$

$$= \int_0^1 \left(x + x^2 + 2x^3 - x + \frac{1}{2}\sqrt{x} - \frac{1}{2} \right) dx = \int_0^1 \left(x^2 + 2x^3 + \frac{1}{2}\sqrt{x} - \frac{1}{2} \right) dx$$

$$= \left[\frac{x^3}{3} + \frac{2x^4}{4} + \frac{1}{2} \cdot \frac{2}{3} x\sqrt{x} - \frac{x}{2} \right]_0^1 = \frac{1}{3} + \frac{1}{2} + \frac{1}{3} - \frac{1}{2}$$

$$= \frac{2}{3}$$ **Ans.**

EXAMPLE 10.4. **(*a*)** Evaluate $\int_C \vec{F} \cdot \vec{dr}$ where $\vec{F} = y^2 i + 2xyj$ and

(*i*) C is the straight line path from the point (0, 0) to the point (1, 2)

(*ii*) C is the parabola $y = 2x^2$ from the point (0, 0) to the point (1, 2).

(*b*) Find the work done by the force $\vec{F} = (2y + 3)\, i + xzj + (yz - x)\, k$ when it moves a particle from the point (0, 0, 0) to the point (2, 1, 1) along the curve $x = 2t^2, y = t, z = t^3$. **(UPTU 2010-11)**

SOLUTION: **(*a*)** $\int_C \vec{F} \cdot \vec{dr} = \int_C (y^2 dx + 2xy\, dy)$

(*i*) C is the line $y = 2x$ so $dy = 2dx$,

$$\therefore \quad \int_C \vec{F} \cdot \vec{dr} = \int_C (2x)^2 dx + 2x(2x)\, 2\, dx = \int_C 12x^2\, dx = \left[4x^3\right]_0^1 = 4$$

(*ii*) C is the parabola $y = 2x^2$ and $dy = 4x\, dx$

$$\therefore \quad \int_C \vec{F} \cdot \vec{dr} = \int_0^1 (2x^2)^2 dx + 2x(2x^2) \cdot 4x\, dx = \int_0^1 20x^4 dx = \left[20\frac{x^5}{5} \right]_0^1 = 4$$

Hence, the value of the line integral is independent of the path. Let us check if the field $\vec{F}$ is conservative.

$$\nabla \times \vec{F} = \begin{vmatrix} i & j & k \\ \dfrac{\partial}{\partial x} & \dfrac{\partial}{\partial y} & \dfrac{\partial}{\partial z} \\ y^2 & 2xy & 0 \end{vmatrix} = i\,(0 - 0) + j\,(0 - 0) + k\,(2y - 2y) = 0$$

Thus, the vector field $\vec{F}$ is conservative (or irrotational) and that is why, the value of $\int \vec{F} \cdot \vec{dr}$ is equal along the above two paths (*i*) and (*ii*). **Ans.**

(b) Work done along the curve C is $\int_C \vec{F}\cdot\vec{dr},\ C: x = 2t^2,\ \ y = t,\ \ z = t^3.$

$$\text{Work done} = \int_{O(0,0,0)}^{A(2,1,1)} (2y+3)\,dx + xz\,dy + (yz - x)\,dz \quad \text{as} \quad \vec{r} = xi + yz + zk$$

$$= \int_{t=0}^{1} (2t+3)\,4t\,dt + (2t^2)(t^3)\,dt + (t\cdot t^3 - 2t^2)\,3t^2 dt$$

$$= \int_0^1 (2t^5 + 3t^6 - 6t^4 + 8t^2 + 12t)\,dt = \left[\frac{3t^7}{7} + \frac{2t^6}{6} - \frac{6t^5}{5} + \frac{8t^3}{3} + 6t^2\right]_0^1$$

$$= \frac{3}{7} + \frac{1}{3} - \frac{6}{5} + \frac{8}{3} + 6 = 8.4 \text{ units.} \qquad \textbf{Ans.}$$

GREEN'S LEMMA IN THE PLANE

If two functions $u(x,y)$ and $v(x,y)$ and their partial derivatives are single valued and continuous over a plane region R bounded by a closed curve C, then

$$\int_C u\,dx + v\,dy = \iint_R \left(\frac{\partial v}{\partial x} - \frac{\partial u}{\partial y}\right) dx\,dy$$ **[UPTU 2011-12]**

Proof: Let the closed curve C be such that any straight line parallel to the co-ordinate axes, cuts C in at the most two points. The curve C is divided into parts AEB and BFA where A and B denote the minimum and maximum values of x. Let the equations of the curves AEB and AFB (see the adjoining figure) be $y = y_1(x)$ and $y = y_2(x)$ respectively. If R is the region bounded by the curve C, we have

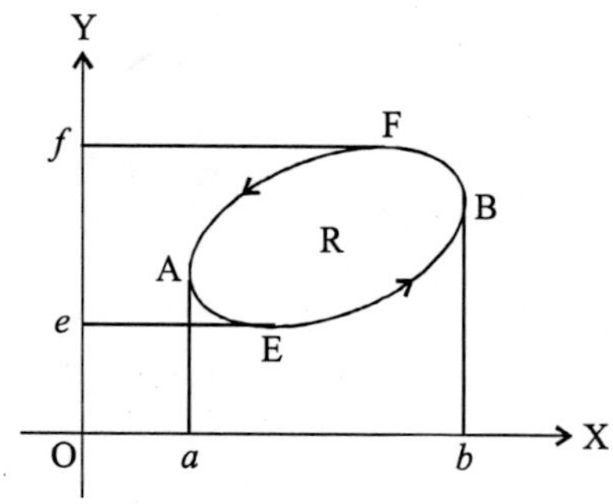

$$\iint_R \frac{\partial u}{\partial y}\,dx\,dy = \int_{x=a}^{b}\left[\int_{y=y_1(x)}^{y_2(x)} \frac{\partial u}{\partial y}\,dy\right] dx =$$

$$\int_{x=a}^{b} \left[u(x,y)\right]_{y=y_1(x)}^{y_2(x)}\,dx = \int_a^b \left[u(x,y_2) - u(x,y_1)\right] dx$$

$$= -\int_a^b u(x,y_1)\,dx - \int_b^a u(x,y_2)\,dx = -\oint_C u\,dx \;\Rightarrow\; \int_C u\,dx = -\iint_R \frac{\partial u}{\partial y}\,dx\,dy \qquad \text{...(1)}$$

Similarly, let the equations of the curve EAF and EBF be $x = x_1(y)$ and $x = x_2(y)$ respectively, then

$$\iint_R \frac{\partial v}{\partial x}\,dx\,dy = \int_{y=e}^{f}\left[\int_{x=x_1(y)}^{x_2(y)} \frac{\partial v}{\partial x}\,dx\right] dy = \int_e^f \left[v(x_2,y) - v(x_1,y)\right] dy$$

$$= \int_f^e v(x_1,y)\,dy + \int_e^f v(x_2,y)\,dy = \oint_C v\,dy \;\Rightarrow\; \oint_C v\,dy = \iint_R \frac{\partial v}{\partial x}\,dx\,dy \qquad \text{...(2)}$$

Adding (1) and (2), we get

$$\oint_C u\,dx + v\,dy = \iint_R \left(\frac{\partial v}{\partial x} - \frac{\partial u}{\partial y}\right) dx\,dy.$$

Now, let us explain the Green's lemma in the vector notation as follows.

We have $u\,dx + v\,dy = (ui + vj) \cdot (i\,dx + j\,dy) = \vec{F} \cdot \vec{dr}$

where $\vec{F} = ui + vj$ and $\vec{r} = xi + yj$ so that $\vec{dr} = idx + jdy$.

Then $\nabla \times \vec{F} = \begin{vmatrix} i & j & k \\ \frac{\partial}{\partial x} & \frac{\partial}{\partial y} & \frac{\partial}{\partial z} \\ u & v & 0 \end{vmatrix} = -\frac{\partial v}{\partial z}\, i + \frac{\partial u}{\partial z}\, j + \left(\frac{\partial v}{\partial x} - \frac{\partial u}{\partial y}\right) k$

Therefore, $(\nabla \times \vec{F}) \cdot k = -\frac{\partial v}{\partial z} i \cdot k + \frac{\partial u}{\partial z} j \cdot k + \left(\frac{\partial v}{\partial x} - \frac{\partial u}{\partial y}\right) k \cdot k = 0 + 0 + \left(\frac{\partial v}{\partial x} - \frac{\partial u}{\partial y}\right)$

Thus, the Green's Lamma in the plane can be written as

$$\oint_C \vec{F} \cdot \vec{dr} = \iint_R (\vec{\nabla} \times \vec{F}) \cdot k\; dx\,dy$$

A generalisation of this to a surface S in space having a curve C as boundary leads quite naturally to Stoke's theorem to be discussed latter.

EXAMPLE 10.5. Verify Green's Lemma in the xy-plane for $\oint_C (xy + y^2)\,dx + x^2 dy$ where C is the closed curve of the region bounded by $y = x$ and $y = x^2$.

SOLUTION: Straight line $y = x$ and the parabola $y = x^2$ intersect at the points (0, 0) and (1, 1). The positive direction in traversing C is as shown in the adjoining figure.

Along $y = x^2$ the given line integral equals $\int u\,dx + v\,dy =$

$$\int_0^1 (x\,(x^2) + x^4)\,dx + x^2\,(2x)\,dx = \int_0^1 (3x^3 + x^4)\,dx = \left[\frac{3x^4}{4} + \frac{x^5}{5}\right]_0^1 = \frac{19}{20}$$

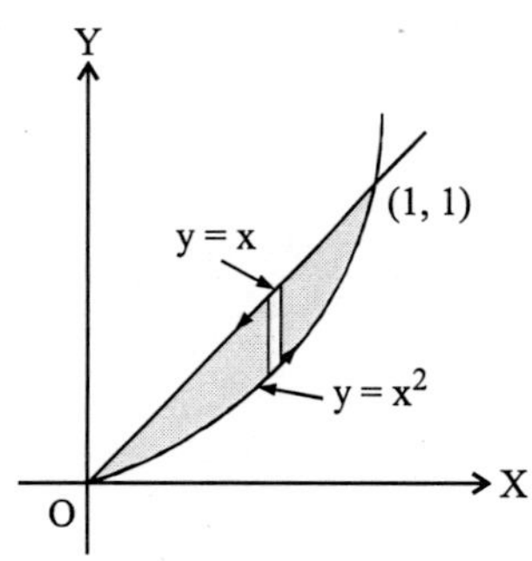

Along $y = x$ from (1, 1) to (0, 0) the line integral equals

$$\int_1^0 (x\,(x) + x^2)\,dx + x^2\,dx = \int_1^0 3x^2\,dx = -1.$$

Therefore, the given line integral $= \frac{19}{20} - 1 = -\frac{1}{20}$.

Next, $\iint_R \left(\frac{\partial v}{\partial x} - \frac{\partial u}{\partial y}\right) dx\,dy = \iint_R \left[\frac{\partial}{\partial x}(x^2) - \frac{\partial}{\partial y}(xy + y^2)\right] dx\,dy$

$$= \iint_R (x - 2y)\,dx\,dy$$

$$= \int_0^1 \int_{x^2}^{x} (x-2y)\, dy\, dx = \int_0^1 \left[xy - y^2\right]_{y=x^2}^{x} dx$$

$$= \int_0^1 \left[x(x) - x^2 - (x^3 - x^4)\right] dx = \left[\frac{x^5}{5} - \frac{x^4}{4}\right] = -\frac{1}{20}.$$

so the Green's lemma is verified.

EXAMPLE 10.6. A vector field $\vec{F}$ is given by $\vec{F} = (\sin y)\, i + x\,(1 + \cos y)\, j$.

Evaluate $\int_C \vec{F} \cdot \vec{dr}$ where C is the ellipse $\dfrac{x^2}{a^2} + \dfrac{y^2}{b^2} = 1, \quad z = 0.$

SOLUTION: $\int_C \vec{F} \cdot \vec{dr} = \int_C [\sin y\, dx + x(1 + \cos y)\, dy]$

It is not easy to evaluate the integral directly whereas using Green's lemma which is applicable here, will facilitate the job.

By Green's lemma, the given line integral is equal to

$$\iint_S \left[\frac{\partial}{\partial x} x(1+\cos y) - \frac{\partial}{\partial y} \sin y\right] dx\, dy \quad \text{where S is the region inside the closed curve } C.$$

$$= \iint_S (1 + \cos y - \cos y)\, dx\, dy = \iint_S dx\, dy = \text{area of S} = \pi\, ab$$

which is the area under the ellipse C : $\dfrac{x^2}{a^2} + \dfrac{y^2}{b^2} = 1, \quad z = 0.$ **Ans.**

EXAMPLE 10.7. Use Green's theorem to find the value of the integral

$$\int_C \vec{F} \cdot \vec{dr} \quad \text{where } \vec{F} = y^3 i - x^3 j \text{ and } C : x^2 + y^2 = a^2, \quad z = 0.$$

SOLUTION : Here $\vec{F} = y^3 i - x^3 j$. $\therefore$ $\int_C \vec{F} \cdot d\vec{r} = \int_C y^3 dx - x^3 dy$

By Green's theorem $\displaystyle\int_C u\, dx + v\, dy = \iint_S \left(\frac{\partial v}{\partial x} - \frac{\partial u}{\partial y}\right) dx\, dy$

where S is the plane area bounded by the closed curve C: $x^2 + y^2 = a^2, \quad z = 0.$

Here $u = y^3$ and $v = -x^3$ hence $\dfrac{\partial v}{\partial x} - \dfrac{\partial u}{\partial y} = -3x^2 - 3y^2$

$$\therefore \quad \iint_S \left(\frac{\partial v}{\partial x} - \frac{\partial u}{\partial y}\right) dx\, dy = \iint_S -3(x^2 + y^2)\, dx\, dy$$

where S is the region inside the circle $x^2 + y^2 = a^2, \quad z = 0$

Changing to polar co-ordinates the above integral becomes

$$-3 \int_0^{2\pi} \int_0^{a} r^2 r\, dr\, d\theta = -3 \int_0^{2\pi} d\theta \int_0^{a} r^3 dr = -3\,(2\pi) \left[\frac{r^4}{4}\right]_0^a = -\frac{3}{2}\,\pi\, a^4. \qquad \textbf{Ans.}$$

EXAMPLE 10.8. Using Green's theorem in the plane, evaluate $\int [\cos y\, i + x(1-\sin y)\, j].\vec{dr}$ for a closed curve given by $x^2 + y^2 = 1,\ \ z = 0.$

SOLUTION: By Green's theorem $\oint_e (udx + vdy) = \iint_R \left(\frac{\partial v}{\partial x} - \frac{\partial u}{\partial y}\right) dxdy$

where C is closed curve enclosing area R

In vector terms, we have $\int_e \vec{F}.\vec{dr} = \iint_R \text{curl}\,\vec{F}\cdot\vec{ds}$

Here $\vec{F} = (\cos y)\, i + x\,(1 - \sin y)\, j$ and $\vec{r} = xi + yj \quad \therefore \quad d\vec{r} = idx + j\,dy.$

Therefore
$$\text{curl}\,\vec{F} = \nabla \times \vec{F} = \begin{vmatrix} i & j & k \\ \frac{\partial}{\partial x} & \frac{\partial}{\partial y} & \frac{\partial}{\partial z} \\ \cos y & x(1-\sin y) & 0 \end{vmatrix}$$

$$= k\left[\frac{\partial}{\partial x}\{x(1-\sin y)\} - \frac{\partial}{\partial y}\cos y\right] = k(1 - \sin y + \sin y) = k$$

$$\Rightarrow \quad \int_c \vec{F}.\vec{dr} = \iint_R k.\vec{ds} = \iint_R k\cdot k\, dx\, dy = \iint_R dx\, dy = \pi \qquad \textbf{Ans.}$$

since, area of the circle $x^2 + y^2 = 1,\ \ z = 0$ is π.

EXAMPLE 10.9. Evaluate by Green's theorem $\int_c e^{-x}(\sin y\, dx + \cos y\, dy)$ where C is the rectangle with vertices $(0, 0), (\pi, 0), (\pi, \pi/2), (0, \pi/2)$.

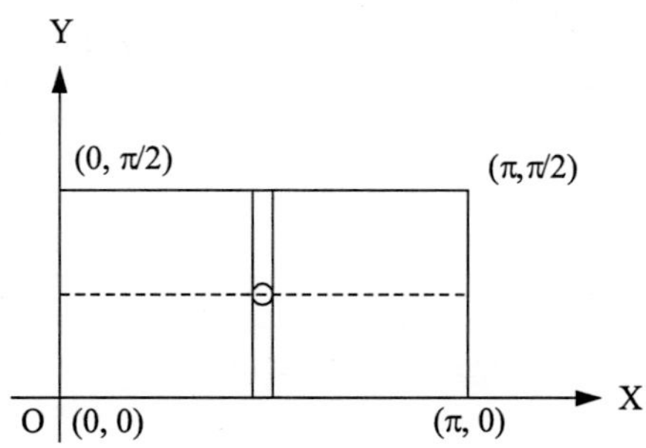

SOLUTION: By Green's theorem $\oint_c f\, dx + g\, dy = \iint_R \left(\frac{\partial g}{\partial x} - \frac{\partial f}{\partial y}\right) dxdy$

where C is a closed curve enclosing the region R. In the given case $f = e^{-x}\sin y,\ g = e^{-x}\cos y.$

$$\int_c e^{-x}\sin y\, dx + e^{-x}\cos y\, dy = \iint_R \left[\frac{\partial}{\partial x}(e^{-x}\cos y) - \frac{\partial}{\partial y}(e^{-x}\sin y)\right] dxdy$$

where R is the region enclosed in the rectangle with vertices $(0, 0), (\pi, 0), (\pi, \pi/2), (0, \pi/2)$

$$= \iint_R (-e^{-x}\cos y - e^{-x}\cos y)\,dydx = -2\iint_R e^{-x}\cos y\,dydx$$

$$= -2\int_0^{\pi} e^{-x}dx \int_0^{\pi/2}\cos y\,dy = -2[-e^{-x}]_0^{\pi}\,[\sin y]_0^{\pi/2}$$

$$= 2(e^{-\pi}-1)\sin \pi/2 = 2\,(e^{-\pi}-1).\quad \textbf{Ans.}$$

EXAMPLE 10.10. Applying Green's theorem, evaluate $\int_C (y-\sin x)\,dx + \cos x\,dy$ where C is the plane triangle enclosed by the lines $y = 0,\ x = \dfrac{\pi}{2},\ y = \dfrac{2x}{\pi}$.

SOLUTION: According to Green's theorem $\int_C u\,dx + v\,dy = \iint_R \left(\dfrac{\partial v}{\partial x} - \dfrac{\partial u}{\partial y}\right)dxdy$

where R is region bounded by the closed curve C.

Here $u = y - \sin x$, $v = \cos x$ and R is the region of the triangle with vertices (0, 0), ($\pi/2$, 0) and ($\pi/2$, 1), which is the curve C.

Here $\dfrac{\partial v}{\partial x} = -\sin x$ and $\dfrac{\partial u}{\partial y} = 1$, therefore by Green's theorem

$$\int_C (y-\sin x)\,dx + \cos x\,dy = \int_0^{\pi/2}\int_0^{2x/\pi} (-\sin x - 1)\,dydx$$

$$= \int_0^{\pi/2} (-\sin x - 1)\,[y]_0^{2x/\pi}\,dx = -\int_0^{\pi/2} \frac{2}{\pi}(1+\sin x)\,x\,dx$$

$$= -\frac{2}{\pi}[x\,(x-\cos x)]_0^{\pi/2} + \frac{2}{\pi}\int_0^{\pi/2} 1\,(x-\cos x)\,dx$$

$$= -\frac{2}{\pi}\left[\frac{\pi^2}{4} - \left(\frac{x^2}{2} - \sin x\right)_0^{\pi/2}\right] = -\frac{2}{\pi}\left[\frac{\pi^2}{4} - \frac{\pi^2}{8} + 1\right]$$

$$= -\left(\frac{\pi}{4} + \frac{2}{\pi}\right).\quad \textbf{Ans.}$$

EXAMPLE 10.11. Apply Green's theorem to evaluate $\oint_C (2y^2dx + 3xdy)$ where C is the boundary of the closed region bounded by $y = x$ and $y = x^2$. **(UPTU 2005-06)**

SOLUTION: By Green's theorem $\oint_C \mathrm{M}dx + \mathrm{N}dy = \iint_R \left(\dfrac{\partial \mathrm{N}}{\partial x} - \dfrac{\partial \mathrm{M}}{\partial y}\right)dxdy$

Here $\mathrm{M} = 2y^2$ and $\mathrm{N} = 3x$ hence $\dfrac{\partial \mathrm{M}}{\partial y} = 4y$ and $\dfrac{\partial \mathrm{N}}{\partial x} = 3$

$$\therefore \quad \int_C 2y^2 dx + 3xdy = \iint (3-4y)dydx = \int_0^1 \int_{y=x^2}^{x} (3-4y)dydx$$

$$= \int_0^1 [3y - 2y^2]_{x^2}^{x} dx = \int_0^1 (3x - 2x^2 - 3x^2 + 2x^4)dx$$

$$= \left[\frac{3x^2}{2} - \frac{5x^3}{3} + \frac{2x^5}{5}\right]_0^1 = \frac{3}{2} - \frac{5}{3} + \frac{2}{5} = \frac{7}{30}. \qquad \textbf{Ans.}$$

EXAMPLE 10.12. Suppose $\vec{F}$ is the force field as $\vec{F} = x^3 i + yj + zk$. Find the work done by $\vec{F}$ along the line from (1, 2, 3) to (3, 5, 7). **(UPTU 2005-06)**

SOLUTION: The line is $\frac{x-1}{3-1} = \frac{y-2}{5-2} = \frac{z-3}{7-3}$ or $\frac{x-1}{2} = \frac{y-2}{3} = \frac{z-3}{4} = \alpha$, say

then $\alpha = 0$ at (1, 2, 3) and $\alpha = 1$ at (3, 5, 7)

$$\text{and the work done} = \int \vec{F}\cdot d\vec{r} = \int x^3 dx + ydy + zdz$$

$$= \int_{\alpha=0}^{1} (2\alpha+1)^3 2d\alpha + (3\alpha+2)3d\alpha + (4\alpha+3)4d\alpha$$

$$= \left[\frac{2(2\alpha+1)^4}{2\cdot 4} + \frac{3(3\alpha+2)^2}{3\cdot 2} + \frac{4(4\alpha+3)^2}{4\cdot 2}\right]_0^1$$

$$= \frac{81}{4} + \frac{25}{2} + \frac{49}{2} - \frac{1}{4} - \frac{4}{2} - \frac{9}{2} = 50.5 \text{ units.} \qquad \textbf{Ans.}$$

EXAMPLE 10.13. Apply Green's theorem to evaluate $\int_C (2x^2 - y^2)dx + (x^2 + y^2)dy$ where C is the boundary of the area enclosed by the X-axis and the upper half of $x^2 + y^2 = a^2$. **(UPTU 2004-05)**

SOLUTION: By Green's theorem $\int_C udx + vdy = \iint_S \left(\frac{\partial v}{\partial x} - \frac{\partial u}{\partial y}\right) dxdy$

Here $u = 2x^2 - y^2$ and $v = x^2 + y^2$ hence

$$\int_C (2x^2 - y^2)dx + (x^2 + y^2)dy = \iint_S (2x + 2y)dxdy = \int_{-a}^{a} \int_0^{\sqrt{a^2 - x^2}} (2x + 2y)dydx$$

$$= \int_{-a}^{a} [2xy + y^2]_0^{\sqrt{a^2-x^2}} dx = \int_{-a}^{a} \left[2x\sqrt{a^2 - x^2} + a^2 - x^2\right] dx$$

$$= 0 + 2\int_0^a (a^2 - x^2)dx = 2\left[a^2 x - \frac{x^3}{3}\right]_0^a = \frac{4a^3}{3}. \qquad \textbf{Ans.}$$

EXAMPLE 10.14. Using Green's theorem evaluate $\int_C x^2 y dx + x^2 dy$ where C is the boundary described counter clockwise of the triangle with vertices (0, 0), (1, 0), (1, 1).

(UPTU 2003-04)

SOLUTION: By Green's theorem $\int_C u dx + v dy = \iint_S \left(\frac{\partial v}{\partial x} - \frac{\partial u}{\partial y} \right) dx dy$

Here $u = x^2 y$ and $v = x^2$ hence

$$\int_C x^2 y dx + x^2 dy = \iint_S (2x - x^2) dx dy = \int_0^1 \int_0^x (2x - x^2) dy dx$$

$$= \int_0^1 [2xy - x^2 y]_0^x \, dx = \int_0^1 (2x^2 - x^3) dx = \left[\frac{2x^3}{3} - \frac{x^4}{4} \right]_0^1 = \frac{5}{12}. \qquad \textbf{Ans.}$$

EXERCISE 10A

1. Evaluate $\oint_C \vec{F}\cdot\vec{dr}$ where $\vec{F} = (\sin z)\, i + (\cos x)\, j + (\sin y)\, k$ and C is the boundary of the rectangle determined by the planes $x = 0, x = \pi, y = 0, y = 1$ and $z = 3$.

2. Determine the value of the line integral $\int_C \vec{F}\cdot\vec{dr}$ where C is the curve along the path $x^2 = 4y,\ \ 3x^3 = 8z$ from $x = 0$ to $x = 2$ and $\vec{F} = 3x^2 i + (2xz - y)\, j + zk.$
[UPTU 2011-12]

3. Find the work done in moving a particle free round the ellipse $\frac{x^2}{16}+\frac{y^2}{9} = 1$ in the plane $z = 0$, if the force field is given by $\vec{F} = (3x - 4y + 2z)\, i + (4x + 2y - 3z^2)\, j + (2xz - 4y^2 + z^3)\, k.$

4. Find the work done when a particle moves along the curve $x = a \cos \theta, y = a \sin \theta, z = b\theta$ from $\theta = \pi/4$ to $\theta = \pi/2$ under the action of the force field
$$\vec{F} = -\,3a \sin^2\theta \cos \theta\; i + a\,(2 \sin \theta - 3 \sin^3 \theta)\, j + (b \sin 2\theta)\, k.$$

5. Find $\int_C \vec{F}\cdot\vec{dr}$ where $\vec{F} = (x^2 + y^2)\, i + (x^2 - y^2)\, j + z^2 k$ along the boundary of the triangle with vertices $(1, 1, 0), (0, 1, 0), (0, 0, 0)$.

6. Evaluate $\int_C \vec{F}\cdot\vec{dr}$ where C is the rectilinear path from the origin (0, 0, 0) to B (1, 0, 0) to C (1, 1, 1) and $\vec{F} = 2yi + xj + z^2 k.$

7. Verify Green's lemma for $\oint_C (3x^2 - 8y^2)\, dx + (4y - 6xy)\, dy$ where C is boundary of the area enclosed by the curves $y = \sqrt{x}$ and $y = x^2$.

8. Use Green's lemma to evaluate $\int (2x\, dy - 3y\, dx)$ around the square with vertices $(0, 2), (2, 0), (-2, 0), (0, -2)$.

9. Verify Green's lemma for the line integral $\int (x^2 + y^2)\, dx + xy\, dy$ taken over $x^2 + y^2 = 4,\ x \ge 0, y \ge 0$.

10. Use Green's lemma to evaluate the line integral $\int (x^2 - y^2)\, dx - 4xy\, dy$ over the region bounded by $y^2 = 4x, y = 0, x = 1$ in the first quadrant.

❑❑❑

SURFACE INTEGRAL

Let $\vec{F}$ be a vector point function and S a given two sided surface. Subdivide the area S into p elements of area δS_i, $i = 1, 2, ..., p$. Let $P_i(x_i, y_i, z_i)$ be any point in δS_i and let $\hat{n}_i$ be unit normal vector drawn outwards to δS_i at the point P_i and let $\vec{F}(x_i, y_i, z_i) = \vec{F}_i$.

Now consider $\sum_{i=1}^{p} (\vec{F}_i \cdot \hat{n}_i)\, \delta S_i$ where $\vec{F}_i \cdot \hat{n}_i$ is the normal component of $\vec{F}_i$ at P_i. Then the limit of this sum as $p \to \infty$ in such a way that the largest dimension of δS_i approaches to zero, if it exists, is called **the surface integral** of the normal component of $\vec{F}$ over S and is denoted by

$$\iint_S \vec{\mathbf{F}} \cdot \hat{\boldsymbol{n}}\, ds$$

Thus we have $\iint_S \vec{F} \cdot \hat{n}\, ds = \lim_{p\to\infty} (\vec{F}_i \cdot \hat{n}_i)\, \delta S_i = \iint_S \vec{F} \cdot \vec{ds}$

where $\vec{ds}$ is the vector $(dy\, dz\, i + dz\, dx\, j + dx\, dy\, k)$.

If $\vec{F} = F_1 i + F_2 j + F_3 k$ then $\iint_S \vec{\mathbf{F}} \cdot \vec{ds} = \iint_S \mathbf{F}_1\, dy\, dz + \mathbf{F}_2\, dz\, dx + \mathbf{F}_3\, dx\, dy$.

EXAMPLE 10.15. **(a)** Evaluate $\iint_S \vec{A} \cdot \vec{ds}$ where $\vec{A} = xi + (z^2 - zx) j - xy\, k$ and S is the surface of the triangle with vertices (2, 0, 0), (0, 2, 0), (0, 0, 4).

(b) Evaluate $\iint_S \vec{F} \cdot \hat{n}\, ds$ where $\vec{F} = 18z\, i - 12j + 3y\, k$ and S is the part of the plane $2x + 3y + 6z = 12$ in the first octant. **[UPTU 2011-12]**

SOLUTION: Here S is the plane surface $\frac{x}{2} + \frac{y}{2} + \frac{z}{4} = 1$ or $2x + 2y + z = 4$ bounded by the lines $x + y = 2,\ z = 0;\ 2y + z = 4,\ x = 0$ and $2x + z = 4,\ y = 0$.

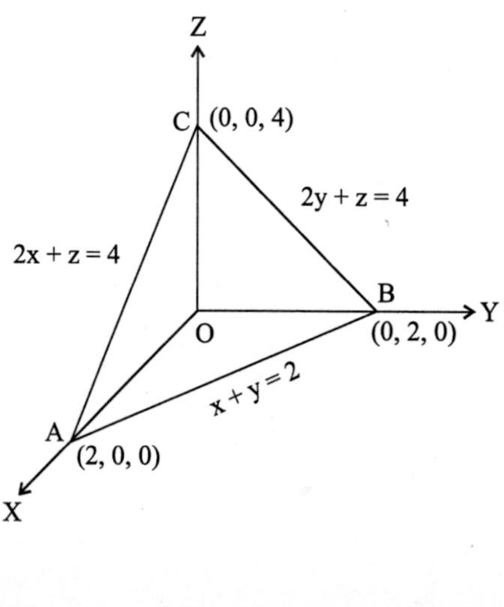

$$\vec{A} \cdot \vec{ds} = [xi + (z^2 - zx)\,j - xyk] \cdot (idydz + jdzdx + kdxdy)$$
$$= xdydz + (z^2 - zx)\,dzdx - xy\,dx\,dy$$

$$\therefore \quad \iint_S \vec{A}\cdot\vec{ds} = \iint_S (x\,dy\,dz + (z^2 - zx)\,dz\,dx - xy\,dx\,dy)$$
$$= \int_{\Delta OBC}\int x\,dy\,dz + \int_{\Delta OAC}\int (z^2 - zx)\,dz\,dx - \int_{\Delta OAB}\int xy\,dx\,dy$$

For points on the given surface, we have $2x = 4 - 2y - z$ or $x = 2 - y - \frac{1}{2}z$

Hence
$$\iint_S \vec{A}\cdot\vec{ds} = \int_0^2\int_0^{4-2y}\left(2 - y - \frac{1}{2}z\right) dz\,dy + \int_0^2\int_0^{4-2x}(z^2 - zx)\,dz\,dx - \int_0^2\int_0^{2-x} xy\,dy\,dx$$
$$= \int_0^2\left[2z - yz - \frac{z^2}{4}\right]_0^{4-2y} dy + \int_0^2\left[\frac{z^3}{3} - \frac{xz^2}{2}\right]_0^{4-2x} dx - \int_0^2\left[x\frac{y^2}{2}\right]_0^{2-x} dx$$
$$= \int_0^2\left[2(4-2y) - y(4-2y) - \frac{1}{4}(4-2y)^2\right] dy$$
$$+ \int_0^2\left[\frac{1}{3}(4-2x)^3 - \frac{x}{2}(4-2x)^2\right] dx + \int_0^2 \frac{x}{2}(2-x)^2\,dx$$
$$= -\frac{22}{3}.$$ **Ans.**

(b) Equation of the plane is $f(x, y, z) \equiv 2x + 3y + 6z = 12$.

$\therefore$ Unit vector along the normal to the plane

$$= \hat{n} = \frac{2i + 3j + 6k}{\sqrt{4 + 9 + 36}} = \frac{2i + 3j + 6k}{7}.$$

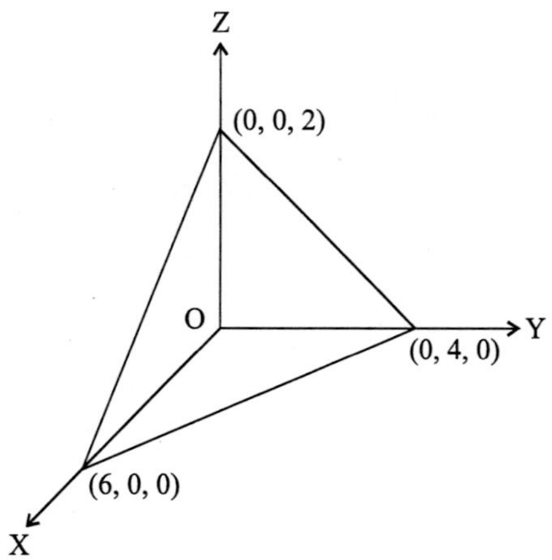

Therefore

$$\iint_S \vec{F}\cdot\hat{n}\,ds = \iint (18z\,i - 12j + 3yk)\cdot\frac{(2i+3j+6k)}{7}\,dx$$

But $$dS = \frac{dx\,dy}{\hat{n}\cdot k} = \frac{dy\,dz}{\hat{n}\cdot i} = \frac{dz\,dx}{\hat{n}\cdot j}$$

Consider the projection of the given plane on $y = 0$, as $2x + 6z = 12 \quad \therefore \quad z = 2 - x/3$.

$$\therefore \quad \iint_S \vec{F}\cdot\hat{n}\,ds = \iint \frac{36z-36}{7}\,ds = \iint \frac{36(z-1)}{7}\frac{dx\,dz}{\hat{n}\cdot j}$$

$$= \iint \frac{36}{7}(z-1)\frac{dz\,dx}{\frac{(2i+3j+6k)}{7}\cdot j} = 12\int_{x=0}^{6}\int_{z=0}^{2-x/3}(z-1)\,dz\,dx$$

$$= 12\int_0^6\left[\frac{z^2}{2} - z\right]_0^{2-x/3} dx = \frac{12}{2}\int_0^6\left[\left(2-\frac{x}{3}\right)^2 - 2\cdot\left(2-\frac{x}{3}\right)\right]dx$$

$$= 6\int_0^6\left(\frac{x^2}{9} - \frac{2x}{3}\right)dx = 6\left[\frac{x^3}{27} - \frac{x^2}{3}\right]_0^6$$

$$= 6\left[\frac{6^3}{27} - \frac{6^2}{3}\right] = -24. \quad \textbf{Ans.}$$

EXAMPLE 10.16. If $\vec{F} = 4xzi - y^2j + yzk$ evaluate $\iint_S \vec{F}\cdot\vec{ds}$ where S is the surface of the cube bounded by $x = 0, \ x = 1, \ y = 0, \ y = 1, \ z = 0, \ z = 1$.

SOLUTION: The given integral has to be calculated over the six faces of the given unit cube. For the face ADEG $(x = 1)$, we have

$$\iint_{ADEG} \vec{F}\cdot\hat{n}\,ds = \int_0^1\int_0^1 (4zi - y^2j + yzk)\cdot i\,dy\,dz = \int_0^1\int_0^1 4z\,dy\,dz = \int_0^1 4z\,dz\int_0^1 dy = 2$$

For the face OBFC $(x = 0)$,

$$\iint_{OBFC} \vec{F}\cdot\hat{n}\,ds = \int_0^1\int_0^1 (4(0)zi - y^2j + yzk)\cdot(-i)\,dy\,dz$$

$$= \int_0^1\int_0^1 0\,dy\,dz = 0$$

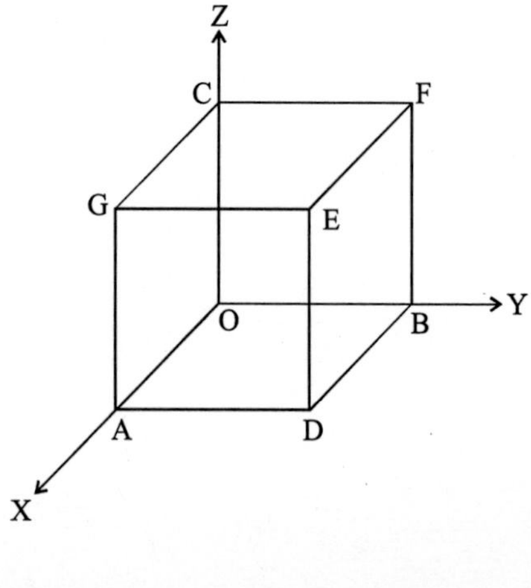

For the face DBFE $(y = 1)$

$$\iint_{DBFE} \vec{F}\cdot\hat{n}\,ds = \int_0^1\int_0^1 (4xzi - 1^2j + (1)zk)\cdot j\,dx\,dz$$

$$= \int_0^1\int_0^1 -dx\,dz = -1$$

For the face OAGC ($y = 0$),

$$\int\limits_{OAGC}\int \vec{F}\cdot\hat{n}\, ds = \int_0^1\int_0^1 (4xzi - 0j + 0k)\cdot(-j)\, dx\, dz = 0$$

For the face GEFC ($z = 1$),

$$\int\limits_{EEFC}\int \vec{F}\cdot\hat{n}\, ds = \int_0^1\int_0^1 (4x)(1)\, i - y^2 j + y(1)k)\cdot k\, dx\, dy = \int_0^1\int_0^1 y\, dx\, dy = \frac{1}{2}$$

and for the face OBDA ($z = 0$),

$$\int\limits_{OBDA}\int \vec{F}\cdot\hat{n}\, ds = \int_0^1\int_0^1 \left[(4x(0)i - y^2 j + y(0)k\right]\cdot(k)\, dx\, dy = 0$$

Adding these yields

$$\iint\limits_{S} \vec{F}\cdot\hat{n}\, ds = 2 + 0 - 1 + 0 + \frac{1}{2} + 0 = \frac{3}{2}. \qquad \textbf{Ans.}$$

EXAMPLE 10.17. Find the value of the surface integral

$$\iint\limits_{S} 2x^2 y\, dy\, dz - y^2 dz\, dx + 4xz^2\, dx\, dy$$

where S is the curved surface of the cylinder $y^2 + z^2 = 9$ bounded by the planes $x = 0,\ x = 2$.

SOLUTION: We know that $\hat{n}\, ds = i\, dy\, dz + j\, dz\, dx + k\, dx\, dy$ in terms of the projection of $\vec{ds}$ on the co-ordinate planes. Taking $\vec{F} = 2x^2yi - y^2j + 4xz^2k$ the given integral can be written as $\iint\limits_{S} \vec{F}\cdot\hat{n}\, ds$.

To find $\hat{n}$, let $\phi = y^2 + z^2 - 9$ then $\nabla\phi = 0i + 2yj + 2zk$

hence $$\hat{n} = \frac{\nabla\phi}{|\nabla\phi|} = \frac{2yj + 2zk}{2\sqrt{y^2 + z^2}} = \frac{yj + zk}{3}$$

$$\therefore \qquad \vec{F}\cdot\hat{n} = 0 - \frac{y^3}{3} + \frac{4}{3}xz^3$$

Thus, the given integral $= \dfrac{1}{3}\iint\limits_{S}(-y^3 + 4xz^3)\, ds$

To evaluate it, let us employ the cylindrical polar co-ordinates $y = r\sin\theta$, $z = r\cos\theta$, $x = x$ so that $ds = r\, d\theta\, dx$

$$\therefore \qquad \text{Above Integral} = \frac{1}{3}\iint\limits_{S}(-r^3\sin^3\theta + 4xr^3\cos^3\theta)\, r\, d\theta\, dx \quad (\text{here } r = 3)$$

$$= \frac{1}{3}\int_0^{2\pi}\int_0^{2}(-81\sin^3\theta + 4(81)x\cos^3\theta)\,dx\,d\theta = 27\int_0^{2\pi}\left[-x\sin^3\theta + 2x^2\cos^3\theta\right]_0^2 d\theta$$

$$= -\,54\int_0^{2\pi}\sin^3\theta\,d\theta + 216\int_0^{2\pi}\cos^3\theta\,d\theta = 0 + 432\int_0^{\pi}\cos^3\theta\,d\theta$$

[using the property $\int_0^{2a} f(x)\,dx = \begin{cases} 2\int_0^a f(x)\,dx & \text{if } f(2a-x) = f(x) \\ 0 & \text{if } f(2a-x) = -f(x) \end{cases}$]

Also, $\int_0^{\pi}\cos^3\theta\,d\theta = 0$ by the same property.

Hence the value of the given integral is zero. **Ans.**

EXAMPLE 10.18. Evaluate the surface integral $\iint_S xdydz + ydzdx + zdxdy$ where S is the portion of the plane $x + 2y + 3z = 6$ lying in the first octanct. **(UPTU 2003-04)**

SOLUTION: The given integral $= I_1 + I_2 + I_3$ where $I_1 = \iint_S xdydz$, $I_2 = \iint_S ydzdx$, $I_3 = \iint_S zdxdy$.

Plane is $x + 2y + 3z = 6$.

For I_1, $x = 6 - 2y - 3z$, for I_2, $y = 3 - \frac{x}{2} - \frac{3}{2}z$ and I_3, $z = 2 - \frac{1}{3}x - \frac{2}{3}y$

$$\therefore \quad I_1 = \int_0^3\int_0^{2-2y/3}(6-2y-3z)dzdy = \int_0^3\left[6z - 2yz - \frac{3z^2}{2}\right]_0^{2-2y/3} dy = 6 \text{ (on simplification)}$$

and
$$I_2 = \int_0^6\int_0^{6-3z}\left(3 - \frac{x}{2} - \frac{3z}{2}\right)dxdz = \int_0^6\left[3x - \frac{x^2}{4} - \frac{3xz}{2}\right]_0^{6-3z} dz = 72 \text{ (on simplification)}$$

and
$$I_3 = \int_0^6\int_0^{3-x/2}\left(2 - \frac{1}{3}x - \frac{2}{3}y\right)dydx = \int_0^6\left[2y - \frac{xy}{3} - \frac{y^2}{3}\right]_0^{3-x/2} dx = 18 \text{ (on simplification)}$$

$\therefore$ Required integral $= 6 + 72 + 18 = 96$. **Ans.**

STOKE'S CIRCULATION THEOREM

> ***If S is an open surface (two sided) bounded by a closed non-intersecting curve C and if a vector field $\vec{F}$ has continuous partial derivatives, then***
>
> $$\oint_C \vec{F}\cdot\vec{dr} = \iint_S (\nabla\times\vec{F})\cdot\hat{n}\,ds = \iint_S (\nabla\times\vec{F})\cdot\vec{ds}$$
>
> ***where C is traversed in positive direction.***

PROOF. Let $\vec{F} = F_1 i + F_2 j + F_3 k$ then

$$\vec{F}\cdot\vec{dr} = F_1 dx + F_2 dy + F_3 dz \qquad ...(1)$$

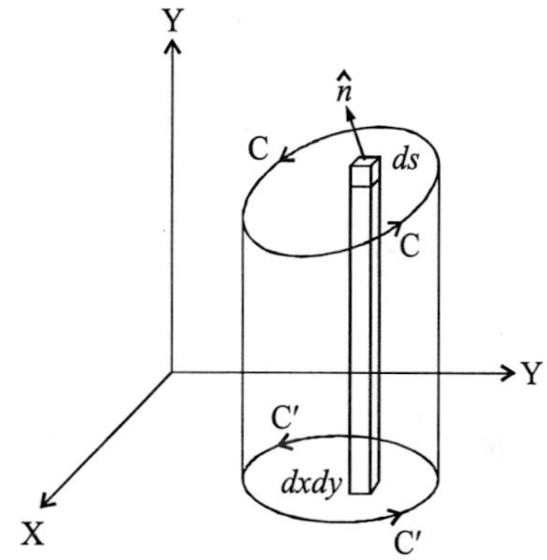

Therefore

$$\iint(\nabla\times\vec{F})\cdot\hat{n}\,ds = \iint_S(\nabla\times F_1 i)\cdot\hat{n}\,ds + \iint_S(\nabla\times F_2 j)\cdot\hat{n}\,ds + \iint_S(\nabla\times F_3 k)\cdot\hat{n}\,ds \qquad ...(2)$$

where $\hat{n}$ is the unit normal vector to the surface S at the point P.

Let us consider the first term on the R.H.S. of (2) as

$$\iint_S(\nabla\times F_1 i)\cdot\hat{n}\,ds = \iint_S\left(i\frac{\partial}{\partial x}+j\frac{\partial}{\partial y}+k\frac{\partial}{\partial z}\right)\times F_1 i\cdot\hat{n}\,ds$$

$$= \iint\left(\frac{\partial F_1}{\partial z}j-\frac{\partial F_1}{\partial y}k\right)\cdot\hat{n}\,ds = \iint_S\left(\frac{\partial F_1}{\partial z}j\cdot\hat{n}-\frac{\partial F_1}{\partial y}k\cdot\hat{n}\right)ds \qquad ...(3)$$

In the above integral ds is the elementary surface whose projection on the xy plane is $dxdy$ (see the above figure). Therefore, $(k\cdot\hat{n})\,ds = dx\,dy$ ...(4)

Now, let the equation of the surface S be $z = f(x, y)$ then

$$F_1(x, y, z) = F_1(x, y, f(x, y)) \quad \text{can be taken as a function of } x, y \text{ only.}$$
$$= \phi(x, y), \text{ say.}$$

We can then write

$$\frac{\partial\phi}{\partial y} = \frac{\partial F_1}{\partial y}\cdot 1+\frac{\partial F_1}{\partial z}\frac{\partial z}{\partial y} \qquad ...(5)$$

Same way we can write $\vec{r} = xi + yj + zk = xi + yj + f(x, y)\,k$ on the surface S. Therefore $\vec{r}$ can be treated as a function x and y only, then we can write

$$\frac{\partial\vec{r}}{\partial y} = \frac{\partial\vec{r}}{\partial y}\cdot 1+\frac{\partial\vec{r}}{\partial z}\cdot\frac{\partial z}{\partial y} = 1\,j + k\,\frac{\partial z}{\partial y} \quad \left(\text{since } \frac{\partial\vec{r}}{\partial y} = j \quad\text{and}\quad \frac{\partial\vec{r}}{\partial z} = k\right).$$

Since $\frac{\partial \vec{r}}{\partial y}$ is tangential to the surface S hence $\frac{\partial \vec{r}}{\partial y}$ is perpendicular to $\hat{n}$, therefore

$$\frac{\partial \vec{r}}{\partial y} \cdot \hat{n} = 0 \quad \Rightarrow \quad \hat{n} \cdot j + \hat{n} \cdot k \frac{\partial z}{\partial y} = 0 \qquad ...(6)$$

Multiplying (6) by $\frac{\partial F_1}{\partial z}$ and using it in (3), we get

$$\iint_S (\nabla \times F_1 i) \cdot \hat{n}\, ds = \iint_S \left(-\hat{n} \cdot k \frac{\partial F_1}{\partial z} \frac{\partial z}{\partial y} - \hat{n} \cdot k \frac{\partial F_1}{\partial y} \right) ds = -\iint_S \left(\frac{\partial F_1}{\partial y} + \frac{\partial F_1}{\partial z} \frac{\partial z}{\partial y} \right) (\hat{n} \cdot k)\, ds$$

$$= -\iint_{S'} \frac{\partial \phi}{\partial y}\, dx\, dy \qquad \text{(On using (5))}$$

where S′ is the projection of S on the *xy*-plane.

Now in the *xy*-plane using Green's lemma, we get

$$\int_{C'} \phi(x, y)\, dx = \iint_{S'} -\frac{\partial \phi}{\partial y}\, dx\, dy \qquad ...(7)$$

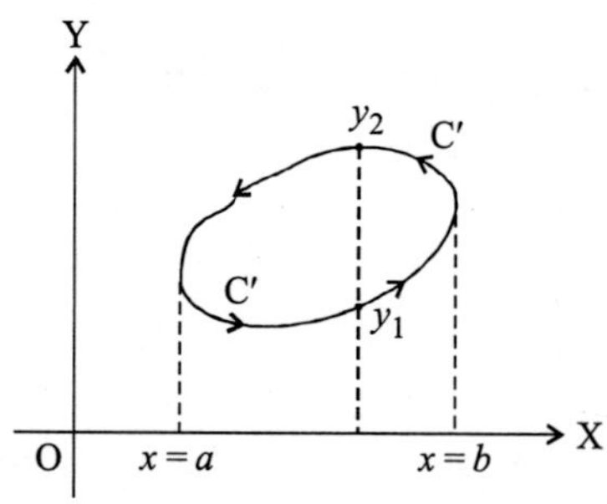

where C′ is the projection of the curve C on the *xy*-plane.

$$\Rightarrow \quad -\iint_{S'} \frac{\partial \phi}{\partial y}\, dx\, dy = -\int_a^b [\phi(x, y_2) - \phi(x, y_1)]\, dx$$

$$= \int_b^a \phi(x, y_2)\, dx + \int_a^b \phi(x, y_1)\, dx$$

where y_1 and y_2 are the lower and upper boundaries for same x respectively.

Hence $$-\iint_{S'} \frac{\partial \phi}{\partial y}\, dx\, dy = \int_{C'} \phi(x, y)\, dx = \int_C F_1\, dx$$

Thus $$\iint_S (\nabla \times i F_1)\, ds = \int_C F_1\, dx$$

Similarly $$\iint_S (\nabla \times j F_2)\, ds = \int_C F_2\, dy \quad \text{and} \quad \iint_S (\nabla \times k F_3)\, ds = \int_C F_3\, dz.$$

Adding these gives $$\iint_S (\nabla \times F) \cdot \vec{ds} = \int_C F_1\, dx + \int_C F_2\, dy + \int_C F_3\, dz = \int_C \vec{F} \cdot \vec{dr},$$

> **Important Corollary:** If two surfaces S_1 and S_2 have same bounding curve C, then
>
> $$\iint_{S_1} (\nabla \times F) \cdot \vec{ds} = \iint_{S_2} (\nabla \times F) \cdot \vec{ds} \quad \text{since each is equal to } \int_C \vec{F} \cdot \vec{dr}.$$

EXAMPLE 10.19. Use Stoke's theorem to evaluate

$$\int_C [(x+2y)dx + (x-z)dy + (y-z)dz]$$

where C is the boundary of the triangle with vertices (2, 0, 0), (0, 3, 0) and (0, 0, 6) oriented in the anti clockwise direction. **(UPTU 2008-09)**

SOLUTION: The given integral $= \int_C \vec{F}\cdot\overrightarrow{dr}$ where $\vec{F} = (x + 2y)i + (x - z)\,j + (y - z)k$ and $\vec{r} = xi + yi + zk.$

By Stoke's theorem $\int_C \vec{F}\cdot\overrightarrow{dr} = \iint_S (\nabla\times\vec{F})\cdot\hat{n}\,ds$

$$\nabla\times\vec{F} = \begin{vmatrix} i & j & k \\ \dfrac{\partial}{\partial x} & \dfrac{\partial}{\partial y} & \dfrac{\partial}{\partial z} \\ x+2y & x-z & y-z \end{vmatrix} = i\,(1 + 1) + j\,(0 - 0) + k\,(1 - 2) = 2i - k$$

and $\hat{n}$ is the unit vector along the normal to the plane $\dfrac{x}{2}+\dfrac{y}{3}+\dfrac{z}{6} = 1$ or $3x + 2y + z = 6.$ which passes through A(2, 0, 0), B(0, 3, 0) and (0, 0, 6).

$$\therefore \qquad \hat{n} = \frac{3i+2j+k}{\sqrt{9+4+1}}$$

$$\therefore \qquad \iint_S (\nabla\times F)\cdot\hat{n}\,ds = \iint_S (2i-k)\cdot\frac{3i+2j+k}{\sqrt{14}}ds$$

$$= \iint_S \frac{(6-1)}{\sqrt{14}}\,ds = \frac{5}{\sqrt{14}} \text{ area of } \Delta ABC.$$

$$= \frac{5}{\sqrt{14}}\frac{1}{2}\left|\overrightarrow{AB}\times\overrightarrow{AC}\right| = \frac{5}{2\sqrt{14}}\,|\,(2i - 3j)\times(2i - 6k)\,|$$

$$= \frac{5}{2\sqrt{14}}|12j + 6k + 18i| = \frac{5}{2\sqrt{14}}\sqrt{12^2+6^2+18^2} = 15. \qquad \textbf{Ans.}$$

EXAMPLE 10.20. Verify Stoke's theorem for $\vec{F} = (2x - y)\,i - yz^2 j - y^2 zk$ where S is the upper half of the surface of the sphere $x^2 + y^2 + z^2 = 1$ and C is its boundary.

(UPTU 2001-02)

SOLUTION: The boundary C of the given open surface S is the circle in the xy-plane with centre at the origin and radius one. Let $x = \cos t,\ y = \sin t,\ z = 0,\ 0 \le t \le 2\pi$ be the parametric equation of C then

$$\oint_C \vec{F}\cdot\overrightarrow{dr} = \oint_C (2x-y)\,dx - yz^2dy - y^2z\,dz = \int_C (2x-y)dx+0+0 \quad \text{as } C: x^2 + y^2 = 1,\ z = 0.$$

$$= \int_0^{2\pi} (2\cos t - \sin t)(-\sin t)\, dt = \int_0^{2\pi} \left[-\sin 2t + \frac{1-\cos 2t}{2}\right] dt$$

$$= \left[\frac{1}{2}\cos 2t + \frac{t}{2} - \frac{1}{4}\sin 2t\right]_0^{2\pi} = \pi.$$

Next $\nabla \times \vec{F} = \begin{vmatrix} i & j & k \\ \dfrac{\partial}{\partial x} & \dfrac{\partial}{\partial y} & \dfrac{\partial}{\partial z} \\ 2x-y & -yz^2 & -y^2 z \end{vmatrix} = i\,(-2yz + 2yz) + j\,(0-0) + k\,(0+1) = k$

$$\therefore \quad \iint_S (\nabla \times \vec{F}) \cdot \hat{n}\, ds = \iint_S k \cdot \hat{n}\, ds = \iint_{S'} dx\, dy \quad (\text{as } \hat{n}\ k\ ds = dx\, dy)$$

where S' is the projection of S on the xy-plane. Therefore, the above integral equals

$$\int_{x=-1}^{1} \int_{y=-\sqrt{1-x^2}}^{\sqrt{1-x^2}} dy\, dx = 2\,(2) \int_0^1 \int_0^{\sqrt{1-x^2}} dy\, dx = \int_0^1 \sqrt{1-x^2}\, dx = \pi \quad \text{area of the circle.}$$

and the Stoke's theorem is verified.

EXAMPLE 10.21. Verify Stoke's theorem for the function $\vec{F} = xyi + yzj + z^2k$ over the cube whose vertices are (0, 0, 0), (*a*, 0, 0), (0, *a*, 0), (0, 0, *a*), (*a*, *a*, 0), (0, *a*, *a*), (*a*, 0, *a*), (*a*, *a*, *a*) if the face of the cube in XOY plane is missing.

SOLUTION: As mentioned earlier, if S_1 and S_2 are two surfaces having the same boundary curve C, we have

$$\iint_{S_1} (\nabla \times \vec{F}) \cdot \vec{ds} = \iint_{S_2} (\nabla \times \vec{F}) \cdot \vec{ds}$$

since both the above integrals are equal to $\int_C \vec{F} \cdot \vec{dr}$. In the present case, in place of considering five faces of the open cube, we consider the single face in the XY-plane.

Here $\nabla \times \vec{F} = \begin{vmatrix} i & j & k \\ \dfrac{\partial}{\partial x} & \dfrac{\partial}{\partial y} & \dfrac{\partial}{\partial z} \\ xy & yz & z^2 \end{vmatrix} = i\,(0-y) + j\,(0-0) + k\,(0-x) = -iy - kx$

Now, for the bounding curve, we have

$$\iint_S (\nabla \times \vec{F}) \cdot \hat{n}\, ds = \iint_S (-iy - kx) \cdot k\, ds = \iint_S -x\, dx\, dy = \int_0^a \int_0^a -x\, dx\, dy = -\int_0^a \left[\frac{x^2}{2}\right]_0^a dy = -\frac{a^3}{2}$$

Here $\vec{F} \cdot \vec{dr} = xy\, dx + yz\, dy + z^2\, dz = xy\, dx$ since $z = 0$ for C

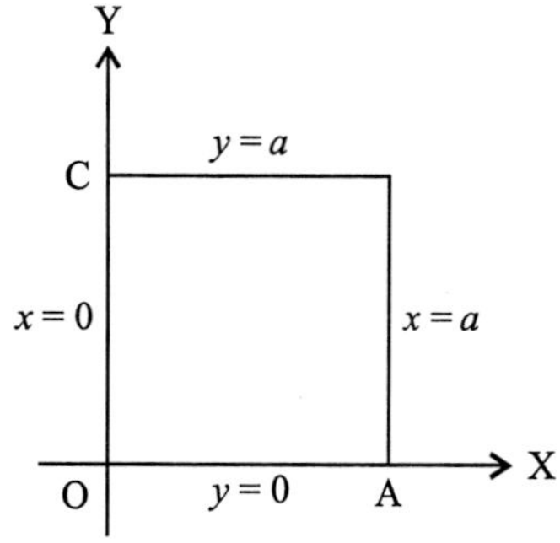

$$\therefore \quad \int_{OA} \vec{F}\cdot \vec{dr} = \int_{OA} xy\,dx = 0 \quad (\text{as } y = 0 \text{ on OA}).$$

$$\int_{AB} \vec{F}\cdot \vec{dr} = \int_{AB} xy\,dx = 0 \quad (\text{as } x = a \text{ along AB so that } dx = 0).$$

$$\int_{BC} \vec{F}\cdot \vec{dr} = \int_{BC} xy\,dx = \int_a^0 ax\,dx = -\frac{a^3}{2} \quad (\text{as } y = a \text{ on BC})$$

$$\text{and} \quad \int_{CO} \vec{F}\cdot \vec{dr} = \int_{CO} xy\,dx = 0 \quad (\text{as } x = 0, \text{ along CO}).$$

Adding the above four integrals, we get

$$\int_C \vec{F}\cdot \vec{dr} = -\frac{a^3}{2}.$$

Thus $\iint_S (\nabla \times \vec{F})\cdot \hat{n}\,ds = \int_C \vec{F}\cdot \vec{dr} = \frac{-a^3}{2}$ and hence the Stoke's theorem is verified.

Hence Proved.

EXAMPLE 10.22. Find the value of the surface integral $\iint_S (\nabla \times \vec{F})\cdot \vec{ds}$ taken over the portion of the surface $x^2 + y^2 - 2ax + az = 0$ and the bounding curve in the plane $z = 0$ where $\vec{F} = (y^2 + z^2 - x^2)\,i + (z^2 + x^2 - y^2)\,j + (x^2 + y^2 - z^2)\,k$.

SOLUTION: The given surface meets the plane $z = 0$ in the circle $x^2 + y^2 - 2ax = 0,\ z = 0$.

The surface integral of $\nabla \times \vec{F}$ over the given surface is the same as the surface integral of $\nabla \times \vec{F}$ over the area of the circle $x^2 + y^2 - 2ax = 0,\ z = 0$.

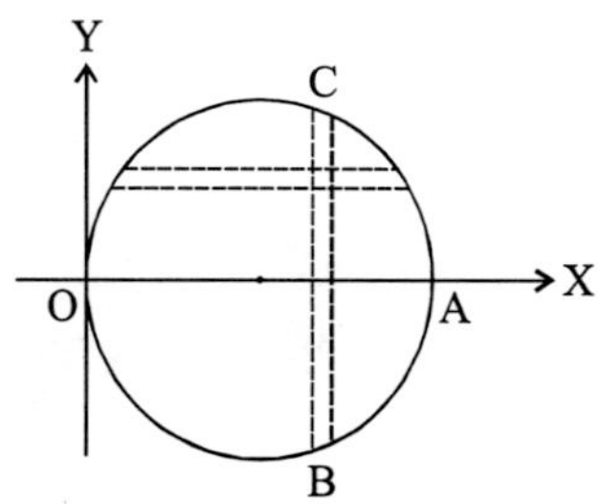

We have $\vec{F} = (y^2 + z^2 - x^2)\,i + (z^2 + x^2 - y^2)\,j + (x^2 + y^2 - z^2)\,k$

$$\therefore \nabla \times \vec{F} = \begin{vmatrix} i & j & k \\ \frac{\partial}{\partial x} & \frac{\partial}{\partial y} & \frac{\partial}{\partial z} \\ y^2 + z^2 - x^2 & z^2 + x^2 - y^2 & x^2 + y^2 - z^2 \end{vmatrix}$$

$$= i\,(2y - 2z) + j\,(2z - 2x) + k\,(2x - 2y)$$

Next, for the area OBAC (where $z = 0$)

$$\nabla \times \vec{F} = 2yi - 2xj + (2x - 2y)\,k \quad \text{and} \quad \vec{ds} = \hat{n}\,ds = k\,dx\,dy$$

$$\therefore \iint_S (\nabla \times \vec{F})\cdot \vec{ds} = \int_0^{2a} \int_{-\sqrt{2ax - x^2}}^{\sqrt{2ax - x^2}} (2x - 2y)\,dy\,dx = \int_0^{2a} \int_{-\sqrt{2ax - x^2}}^{\sqrt{2ax - x^2}} 2x\,dy\,dx - \int_0^{2a} \int_{-\sqrt{2ax - x^2}}^{\sqrt{2ax - x^2}} 2y\,dy\,dx$$

$$= 2\int_0^{2a}\int_0^{\sqrt{2ax-x^2}} 2x\, dy\, dx - 0 \quad \text{(by property of definite integrals)}$$

$$= 4\int_0^{2a} x[y]_0^{\sqrt{2ax-x^2}}\, dx = 4\int_0^{2a} x\sqrt{2ax-x^2}\, dx \quad (\text{putting } x = 2a\sin^2\theta)$$

$= 2\pi a^2$. **Ans.**

EXAMPLE 10.23. Evaluate $\oint_c \vec{F}\cdot d\vec{r}$ by Stoke's theorem where $F = y^2 i + x^2 j - (x+z)y^2 k$ and C is the boundary of the triangle with vertices (0, 0, 0), (1, 0, 0) and (1, 1, 0).

(UPTU 2000-01)

SOLUTION: By Stoke's theorem $\int_c \vec{F}.d\vec{r} = \iint_S curl\ \vec{F}.d\vec{s}$ where $\vec{F} = y^2 i + x^2 i - (x+z)y^2 k$,

C is the triangle OAB and S is the region inside C.

$$\text{Curl } \vec{F} = \begin{vmatrix} i & j & k \\ \dfrac{\partial}{\partial x} & \dfrac{\partial}{\partial y} & \dfrac{\partial}{\partial z} \\ y^2 & x^2 & -(x+z)y^2 \end{vmatrix} = i[-2y(x+z) - O] + j[O + y^2] + k[2x - 2y]$$

$$= -2y(x+z)i + y^2 j + 2(x-y)k$$

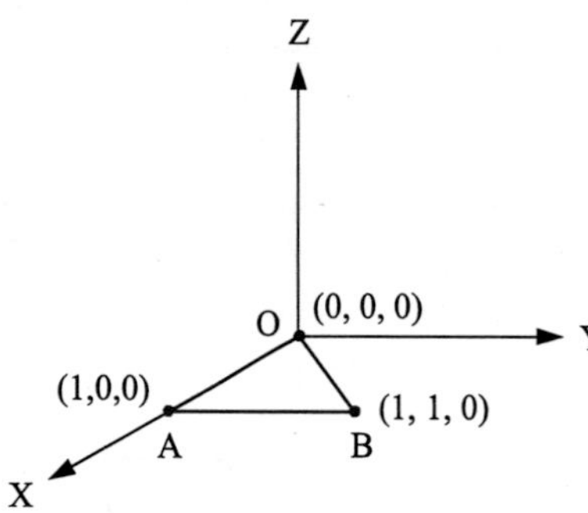

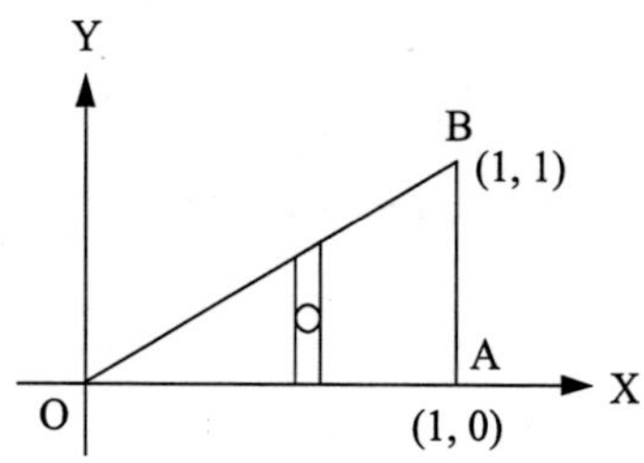

$$\therefore \quad \iint_5 (\nabla\times\vec{F}).k\, dxdy = \iint_s [-2y(x+z)i.k + y^2 j.k + 2(x-y)k\cdot k]dxdy$$

$$= \iint 0 + 0 + 2(x-y)dxdy = 2\int_0^1\int_0^x (x-y)dydx = 2\int_0^1 \left[xy - \frac{y^2}{2}\right]_0^x dx$$

$$= 2\int_0^1 \left(x^2 - \frac{x^2}{2}\right)dx = \int_0^1 x^2 dx = \frac{1}{3}. \quad \textbf{Ans.}$$

EXAMPLE 10.24. Evaluate $\oint_C \vec{F}.d\vec{s}$ where $\vec{F} = xyi - x^2j + (x+z)k$ and S in the region of the plane $2x + 2y + z = 6$ in the first octant.

SOLUTION: $\vec{F} = xyi - x^2j + (x+z)k$ and the surface S is the plane as shown in the adjoining figure. Equation of the plane is $2x + 2y + z = 6$

$\therefore$ Unit vector along the normal to the plane

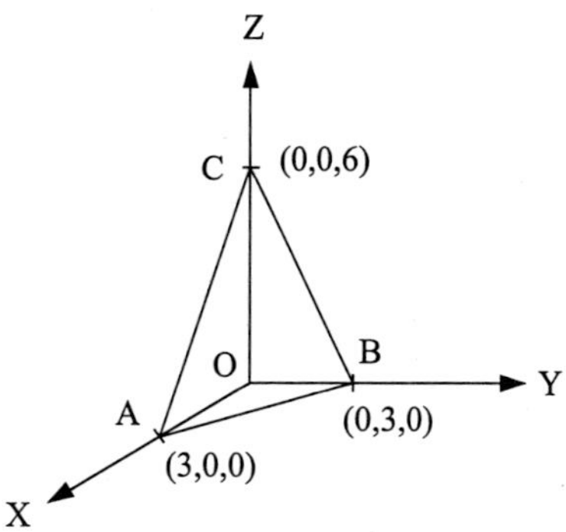

$$= \hat{n} = \frac{2i+2j+k}{\sqrt{4+4+1}} = \frac{1}{3}(2i+2j+k)$$

$\int\int_S \vec{F}.d\vec{s} = \int\int_S \vec{F}.\hat{n}\,ds$. Also we have $dxdy = \hat{n}.k\,ds$

$$\text{or} \quad ds = \frac{dxdy}{\frac{1}{3}(2i+2j+k).k} = 3\,dxdy$$

$$\therefore \quad \int\int \vec{F}.d\vec{s} = \int\int_5 [xyi - x^2j + (x+z)k].\frac{1}{3}(2i+2j+k).3dxdy$$

$$= \int\int_{\nabla OAB} (2xy - 2x^2 + x + z)dxdy = \int\int(2xy - 2x^2 + x + 6 - 2x - 2y)dxdy$$

$$= \int_0^3 \int_0^{3-x} (2xy - 2x^2 - x - 2y + 6)dydx = \int_0^3 [xy^2 - 2x^2y - xy - y^2 + 6y]_0^{3-x} dx$$

$$= \int_0^3 [x(3-x)^2 - 2x^2(3-x) - x(3-x) - (3-x)^2 + 6(3-x)]dx$$

$$= \int_0^3 (3-x)[x(3-x) - 2x^2 - x - (3-x) + 6]dx$$

$$= \int_0^3 (3-x)[-3x^2 + 3x + 3]dx = \int_0^3 (x-3)(3x^2 - 3x - 3)dx$$

$$= 3\int_0^3 (x^3 - 4x^2 + 2x + 3)dx = 3\left[\frac{x^4}{4} - \frac{4x^3}{3} + x^2 + 3x\right]_0^3$$

$$= 3\left[\frac{81}{4} - 4(9) + 9 + 9\right] = 27\left[\frac{9}{4} - 4 + 2\right] = \frac{27}{4}. \qquad \textbf{Ans.}$$

EXAMPLE 10.25. State Stroke's theorem and verify this theorem for $\vec{F} = xzi - yj + x^2yk$ where S is the surface of the region bounded by $x = 0$, $y = 0$, $z = 0$ and $2x + y + 2z = 8$ which is not included on xz plane. **(UPTU 2006-07)**

SOLUTION: The Stoke's theorem is $\oint_C \vec{F}\cdot\vec{dr} = \iint_S (\nabla\times F)\cdot\vec{dS}$ where S is an open surface bounded by the closed curve C. Here $\vec{F} = xzi - yj + x^2yk.$

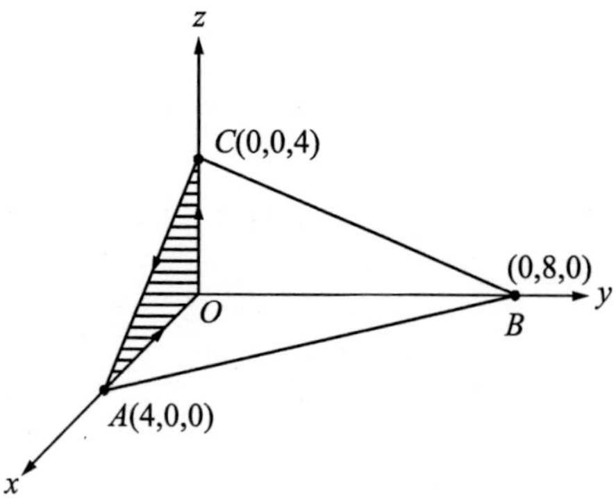

$$\therefore \quad \oint_C \vec{F}\cdot\vec{dr} = \oint_C (xzi - yj + x^2yk)\cdot(idx + jdy + kdz)$$

$$= \oint_C xzdx - ydy + x^2ydz = \int_{AO} + \int_{OC} + \int_{CA}$$

$$= \int_4^0 x(0)dx + \int_0^4 x^2ydz + \int_{x+z=4} xzdx$$

$$= 0 + 0 + \int_0^4 x(4-x)dx = \left[2x^2 - \frac{x^3}{3}\right]_0^4$$

$$= 32 - \frac{64}{3} = \frac{32}{3}.$$

Next, for the surface integral $\nabla\times\vec{F} = \begin{vmatrix} i & j & k \\ \dfrac{\partial}{\partial x} & \dfrac{\partial}{\partial y} & \dfrac{\partial}{\partial z} \\ xz & -y & x^2y \end{vmatrix} = i(x^2 - 0) + j(x - 2xy) + k(0)$

$$= x^2i + x(1 - 2y)j$$

$$\iint_S [\nabla\times\vec{F}]\cdot\vec{dS} = \iint_S [x^2i + x(1-2y)j]\cdot\hat{n}\, dS$$

$\nabla(2x + y + 2z) = 2i + j + 2k$ hence $\hat{n} = \dfrac{2i + j + 2k}{\sqrt{4+1+4}} = \dfrac{1}{3}(2i + j + 2k)$

Thus $\quad \vec{F}\cdot\hat{n} = [x^2i + x(1 - 2y)j]\cdot\dfrac{1}{3}(2i + j - 2k) = \dfrac{1}{3}(2x^2 + x - 2xy).$

$\therefore$ R.H.S. $= \displaystyle\int_0^4\int_0^{4-x} (2x^2 + x - 2xy)dxdz$ where y is obtained from $2x + y + z = 8$.

$$= \int_0^4\int_0^{4-x} [2x^2 + x - 2x(8 - 2x - 2z)dz\,dx = \int_0^4\int_0^{4-x} (6x^2 - 15x + 4xz)dzdx$$

$$= \int_0^4 [(6x^2 - 15x)z + 2xz^2]_0^{4-x}\, dx$$

$$= \int_0^4 (6x^2 - 15x)(4-x) + 2x(4-x)^2 dx = \int_0^4 (-4x^3 + 23x^2 + 28x)dx$$

$$= \left[-x^4 + \frac{23}{3}x^3 - 14x^2\right]_0^4 = 16\left[-16 + \frac{23\times 4}{3} - 14\right] = \frac{32}{3}.$$

Hence Verified.

EXAMPLE 10.26. Verify Stoke's theorem for the vector $\vec{F} = zi + xj + yk$ taken over the half of the sphere $x^2 + y^2 + z^2 = a^2$ lying above the XY-plane. **(UPTU 2002-03)**

SOLUTION: S is the upper half of $x^2 + y^2 + z^2 = a^2$ and C is the boundary of S as the circle $x^2 + y^2 = a^2, z = 0$. By Stoke's theorem $\oint_C \vec{F}\cdot\vec{dr} = \iint_S (\nabla \times \vec{F})\cdot \hat{n}\, ds$

$$\text{L.H.S.} = \oint_C (zi + xj + yk)\cdot(idx + jdy + kdz) = \oint_C zdx + x\,dy + y\,dz$$

$$= \oint_C 0 + x\,dy + 0 = 2\int_{-a}^{a} \sqrt{a^2 - y^2}\,dy = 4\int_0^a \sqrt{1 - y^2}\,dy$$

$$= 4\left[\frac{y}{2}\sqrt{a^2 - y^2} + \frac{a^2}{2}\sin^{-1}\frac{y}{a}\right]_0^a = 4\left[0 + \frac{a^2}{2}\cdot\frac{\pi}{2}\right] = \pi a^2.$$

Next, $$\nabla \times \vec{F} = \begin{vmatrix} i & j & k \\ \frac{\partial}{\partial x} & \frac{\partial}{\partial y} & \frac{\partial}{\partial z} \\ z & x & y \end{vmatrix} = i + j + k$$

and $$\nabla\phi = \left(i\frac{\partial}{\partial x} + j\frac{\partial}{\partial y} + k\frac{\partial}{\partial z}\right)(x^2 + y^2 + z^2 - a^2) = 2xi + 2yj + 2zk$$

$$\therefore \quad \hat{n} = \frac{2xi + 2yj + 2zk}{\sqrt{4x^2 + 4y^2 + 4z^2}} = \frac{1}{a}(xi + yj + 2k)$$

$$\therefore \quad \text{Curl } \vec{F}\cdot\hat{n} = (i + j + k)\cdot\frac{1}{a}(xi + yj + zk) = \frac{1}{a}(x + y + z)$$

and $$\iint_S \nabla\times\vec{F}\cdot\hat{n}ds = \iint_S \frac{1}{a}(x + y + z)dx$$

Converting to spherical polar co-ordinates $x = a\sin\theta\cos\phi$, $y = a\sin\theta\sin\phi$, $z = a\cos\theta$

$$\text{R.H.S.} = \iint(\sin\theta\cos\phi + \sin\theta\sin\phi + \cos\theta)\,ds \quad \text{where } ds = a^2\sin\theta\,d\theta\,d\phi$$

$$= a^2\int_0^{\pi/2}\int_0^{2\pi}\left[\sin^2\theta(\cos\phi + \sin\phi) + \sin\theta\cos\theta\right]d\phi\,d\theta$$

$$= a^2\int_0^{\pi/2}\left[\sin^2\theta(\sin\phi - \cos\phi) + \phi\sin\theta\cos\theta\right]_0^{2\pi} d\theta$$

$$= 0 + \frac{a^2}{2}\int_0^{\pi/2}\cdot 2\pi\sin 2\theta\,d\theta$$

$$= \pi a^2\left[-\frac{1}{2}\cos 2\theta\right]_0^{\pi/2} = \pi a^2 = \text{L.H.S.} \qquad \textbf{Hence verified}$$

VOLUME INTEGRALS

GAUSS DIVERGENCE THEOREM

The surface integral of the normal component of a vector point function $\vec{F}$ taken over a closed surface S enclosing a volume V is equal to the volume integral of the divergence of $\vec{F}$ taken throughout the volume V bounded by S, i.e.,

$$\iint_S \vec{F}\cdot\vec{ds} = \iint_S \vec{F}\cdot\hat{n}\,ds = \iiint_V (\nabla\cdot\vec{F})\,dv$$ **[UPTU 2011-12]**

Let $\vec{F} = F_1 i + F_2 j + F_3 k$ then $\nabla\cdot\vec{F} = \frac{\partial F_1}{\partial x}+\frac{\partial F_2}{\partial y}+\frac{\partial F_3}{\partial z}$

In the rectangular co-ordinate system $dv = dx\,dy\,dz$. Let the unit vector $\hat{n}$ along the normal to the surface S at the point (x, y, z) be given by $\hat{n} = n_1 i + n_2 j + n_3 k$ so that

$$n_1 = \hat{n}\cdot i = \cos\alpha,\quad n_2 = \hat{n}\cdot j = \cos\beta \quad\text{and}\quad n_3 = \hat{n}\cdot k = \cos\gamma$$

Then $\vec{F}\cdot\hat{n} = (F_1 i + F_2 j + F_3 k)\cdot(n_1 i + n_2 j + n_3 k) = F_1\cos\alpha + F_2\cos\beta + F_3\cos\gamma$
therefore the above statement of divergence theorem can be writtten as

$$\iint_S (F_1\cos\alpha+F_2\cos\beta+F_3\cos\gamma)\,ds = \iiint_V \left(\frac{\partial F_1}{\partial x}+\frac{\partial F_2}{\partial y}+\frac{\partial F_3}{\partial z}\right)dx\,dy\,dz.$$

PROOF: Let us prove $\iint_S F_3\cos\gamma\,ds = \iiint_V \frac{\partial F_3}{\partial z}\,dx\,dy\,dz$

Now S is the closed surface such that any line parallel to the co-ordinate axes, cuts S in at most two points. Let the equations of the lower and upper portions S_1 and S_2 of the surface S be $z = f_1(x, y)$ and $z = f_2(x, y)$ respectively. If R denotes the projection of S on the xy plane, then

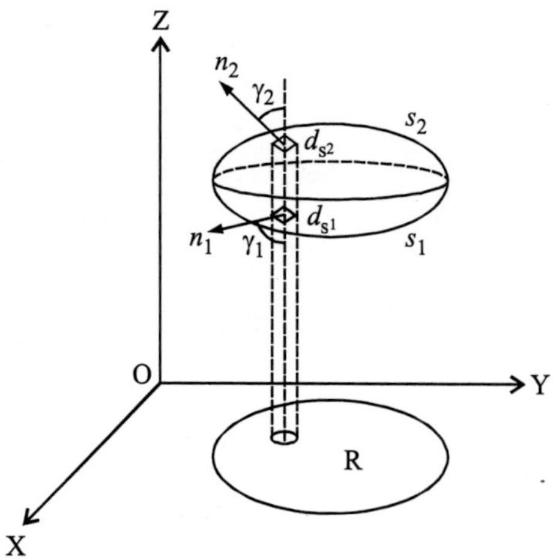

$$\iiint_V \frac{\partial F_3}{\partial z}\,dx\,dy\,dz = \iint_R \left[\int_{f_1(x,y)}^{f_2(x,y)} \frac{\partial F_3}{\partial z}\,dz\right]dy\,dx$$

$$= \iint_R \left[F_3(x, y, z)\right]_{z=f_1(x,y)}^{f_2(x,y)}\,dx\,dy$$

$$= \iint_R \left[F_3(x, y, f_2) - F_3(x, y, f_1)\right]dy\,dx$$

For the upper portion S_2, $dx\,dy = \cos\gamma_2\,ds_2 = k\cdot\hat{n}_2\,ds_2$ since the normal $\hat{n}_2$ to S_2 makes an acute angle γ_2 with k.

For the lower portion S_1, $dy\,dx = -\cos\gamma_1\,ds_1 = -k\cdot\hat{n}_1\,ds_1$ since the normal $\hat{n}_1$ to S_1 makes an obtuse angle γ_1 with k.

Therefore $\iint_R F_3(x, y, f_2)\,dy\,dx = \iint_{S_2} F_3\,k\cdot\hat{n}_2\,ds_2$

and $$\iint_R F_3(x, y, f_1)\, dy\, dx = -\iint_{S_1} F_3\, k \cdot \hat{n}_1\, ds_1$$

$$\therefore \quad \iint_R F_3(x, y, f_2)\, dy\, dx - \iint_R F_3(x, y, f_1)\, dy\, dx = \iint_{S_2} F_3\, k \cdot \hat{n}_2\, ds_2 + \iint_{S_1} F_3\, k \cdot \hat{n}_1\, ds_1$$

$$= \iint_S F_3\, k \cdot \hat{n}\, ds$$

Thus $$\iiint_V \frac{\partial F_3}{\partial z}\, dv = \iint_S F_3\, k \cdot \hat{n}\, ds.$$

Similarly, by projecting S on the other co-ordinates planes, we get

$$\iiint_V \frac{\partial F_1}{\partial x}\, dv = \iint_S F_1\, i \cdot \hat{n}\, ds \quad \text{and} \quad \iiint_V \frac{\partial F_2}{\partial y}\, dv = \iint_S F_2\, j \cdot \hat{n}\, ds$$

Adding these gives

$$\iiint_V \left(\frac{\partial F_1}{\partial x} + \frac{\partial F_2}{\partial y} + \frac{\partial F_3}{\partial z} \right) dv = \iint_S (F_1 i + F_2 j + F_3 k) \cdot \hat{n}\, ds \quad \text{or} \quad \iiint_V \nabla \cdot \vec{F}\, dv = \iint_S \vec{F} \cdot \hat{n}\, ds.$$

EXAMPLE 10.27. Verify the divergence theorem for the vector field $\vec{F} = 4xi - 2y^2j + z^2k$ taken over the region bounded by the cylinder $x^2 + y^2 = 4$, $z = 0$, $z = 3$.

SOLUTION: This being a case of closed surface we have, by divergence theorem

$$\iiint_V \nabla \cdot \vec{F}\, dv = \iint_S \vec{F} \cdot \hat{n}\, ds$$

$$\text{The volume integral} = \iiint_V \nabla \cdot \vec{F}\, dv = \iiint_V \left[\frac{\partial}{\partial x}(4x) + \frac{\partial}{\partial y}(-2y^2) + \frac{\partial}{\partial z}(z^2) \right] dx\, dy\, dz$$

$$= \int_{x=-2}^{2} \int_{y=-\sqrt{4-x^2}}^{\sqrt{4-x^2}} \int_{z=0}^{3} (4 - 4y + 2z)\, dx\, dy\, dz$$

$$= \int_{x=-2}^{2} \int_{y=-\sqrt{4-x^2}}^{\sqrt{4-x^2}} \left[4z - 4yz + z^2 \right]_{z=0}^{3} dx\, dy = \int_{-2}^{2} \int_{-\sqrt{4-x^2}}^{\sqrt{4-x^2}} (21 - 12y)\, dy\, dx$$

$$= 2\int_{-2}^{2} \int_{0}^{\sqrt{4-x^2}} 21\, dy\, dx - 0 \quad \text{(by property of definite integrals)}$$

$$= 84\,\pi \quad \text{(on simplification).}$$

The surface S consists of three surfaces, one being the base S_1 $(z = 0)$, second being the top S_2 $(z = 3)$ and third the curved surface S_3 of the cylinder $x^2 + y^2 = 4$ between $z = 0$ and $z = 3$.

Thus the surface integral $= \iint_S \vec{F}\cdot\hat{n}\, ds = \iint_{S_1} \vec{F}\cdot\hat{n}\, ds_1 + \iint_{S_2} \vec{F}\cdot\hat{n}\, ds_2 + \iint_{S_3} \vec{F}\cdot\hat{n}\, ds_3$

On S_1 $(z = 0)$, $\hat{n} = -k$ and $\vec{F} = 4xi - 2y^2j$ $\therefore$ $\vec{F}\cdot\hat{n} = 0$

therefore $\iint_{S_1} \vec{F}\cdot\hat{n}\, ds_1 = 0.$

On S_2 $(z = 3)$, $\hat{n} = k$ and $\vec{F} = 4xi - 2y^2j + 9k$ $\therefore$ $\vec{F}\cdot\hat{n} = 9$

therefore $\iint_{S_2} \vec{F}\cdot\hat{n}\, ds_2 = 9 \iint_{S_2} ds_2 = 9\,(\pi\, 2^2) = 36\,\pi.$

On S_3, the curved surface of the cylinder $(x^2 + y^2 = 4)$, the normal to $x^2 + y^2 = 4$ is in the direction of $\nabla\,(x^2 + y^2) = 2xi + 2yj$

$\therefore$ The unit vector along the normal to the curved surface is given by

$$\hat{n} = \frac{2xi + 2yj}{\sqrt{4x^2 + 4y^2}} = \frac{xi + yj}{2} \quad (\text{as } x^2 + y^2 = 4)$$

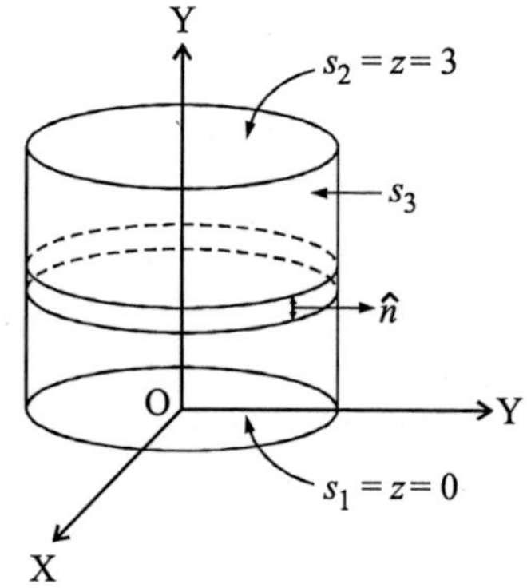

hence $\vec{F}\cdot\hat{n} = (4xi - 2y^2j + z^2k)\cdot\dfrac{xi + yj}{2} = 2x^2 - y^3$

$$\therefore \iint_{S_3} \vec{F}\cdot\hat{n}\, ds_3 = \iint_{S_3} (2x^2 - y^3)\, ds_3$$

Writing $x = 2\cos\theta$, $y = 2\sin\theta$

then $ds_3 = 2\, d\theta\, dz$, and we get

$$\iint_{S_3} \vec{F}\cdot\hat{n}\, ds_3 = \int_{\theta=0}^{2\pi}\int_{z=0}^{3} \left[2(2\cos\theta)^2 - (2\sin\theta)^3\right] 2\, dz\, d\theta$$

$$= 48 \int_0^{2\pi} (\cos^2\theta - \sin^3\theta)\, d\theta = 96 \int_0^{\pi} \cos^2\theta + 0 = 48\,\pi$$

Thus, the given surface integral $= 0 + 36\pi + 48\pi = 84\pi$, which agrees with the value of the volume integral. **Hence Proved.**

EXAMPLE 10.28. If a vector function $\vec{F}$ is defined as $\vec{F} = xy^2i + yz^2j + zx^2k$, evaluate $\iint_S \vec{F}\cdot\vec{ds}$ over the surface given by $x^2 + y^2 + z^2 = 1$.

SOLUTION: Since $S : x^2 + y^2 + z^2 = 1$ is a closed surface enclosing the volume V of the sphere, then by divergence theorem, we have

$$\iint_S \vec{F}\cdot\vec{ds} = \iiint_V \nabla\cdot\vec{F}\, dv$$

Now, $\nabla \cdot \vec{F} = \dfrac{\partial}{\partial x}(xy^2) + \dfrac{\partial}{\partial y}(yz^2) + \dfrac{\partial}{\partial z}(zx^2) = y^2 + z^2 + x^2$

$$\therefore \iiint_V \nabla \cdot \vec{F}\, dv = \iiint_{x^2+y^2+z^2=1} (x^2 + y^2 + z^2)\, dx\, dy\, dz$$

$$= \int_{\theta=0}^{\pi} \int_{\phi=0}^{2\pi} \int_{r=0}^{1} r^2 (r^2 \sin\theta)\, dr\, d\theta\, d\phi = \left[\frac{r^5}{5}\right]_0^1 [\phi]_0^{2\pi} \int_0^{\pi} \sin\theta\, d\theta$$

(on using spherical polar co-ordinates in which $z = r\cos\theta, x = r\sin\theta\cos\phi, y = r\sin\theta\sin\phi$)

$$= \frac{1}{5} \times 2\pi \left[-\cos\theta\right]_0^{\pi} = \frac{4\pi}{5}.$$ **Ans.**

EXAMPLE 10.29. The vector field $\vec{F} = x^2 i + zj + yzk$ is defined over the volume of the cuboid given by $0 \le x \le a,\ 0 \le y \le b,\ 0 \le z \le c$ enclosing the surface S, evaluate the surface integral $\iint_S \vec{F} \cdot \vec{ds}$. **(UPTU 2001-02)**

SOLUTION: The surface integral $\iint_S \vec{F} \cdot \vec{ds}$ has to be evaluated as the sum of six integrals corresponding to the six faces of the cuboid. Since S is a closed surface, the Gauss divergence theorem is applicable, hence

$$\iint_S \vec{F} \cdot \vec{ds} = \iiint_V \nabla \cdot \vec{F}\, dv$$

Therefore, in place of evaluating the surface integral $\iint_S \vec{F} \cdot \vec{ds}$ as the sum of six integrals, we shall evaluate the volume integral $\iiint_V \nabla \cdot \vec{F}\, dv$.

Now, $\text{div}\ \vec{F} = \left(i\dfrac{\partial}{\partial x} + j\dfrac{\partial}{\partial y} + k\dfrac{\partial}{\partial z}\right) \cdot (x^2 i + zj + yzk) = 2x + 0 + y$

$$\therefore \iiint_V \nabla \cdot \vec{F}\, dv = \int_{x=0}^{a} \int_{y=0}^{b} \int_{z=0}^{c} (2x + y)\, dz\, dy\, dx = \int_0^a \int_0^b (2x + y)[z]_0^c\, dy\, dx = c \int_0^a \int_0^b (2x + y)\, dy\, dx$$

$$= c \int_0^a \left[2xy + \frac{y^2}{2}\right]_0^b dx = c \int_0^a \left(2xb + \frac{b^2}{2}\right) dx$$

$$= cb\left[x^2 + \frac{bx}{2}\right]_0^a = bc\left(a^2 + \frac{ab}{2}\right) = abc\left(a + \frac{b}{2}\right).$$ **Ans.**

EXAMPLE 10.30. Use divergence theorem to evaluate $\iint \vec{F}.d\vec{s}$ where $\vec{F} = x^3i + y^3j + z^3k$ and S is the surface of the sphere $x^2 + y^2 + z^2 = a^2$

SOLUTION: Here S is a closed surface, hence by divergence theorem

$$\iint_S \vec{F}.d\vec{s} = \iiint_V \nabla.\vec{F}\,dv = \iiint_V \nabla.(x^3i + y^3j + z^3k)\,dxdydz$$

$$= \iiint (3x^2 + 3y^2 + 3z^2)dx\,dy\,dz$$

Transforming it into spherical polar co-ordinates $x = r \sin\theta \cos\phi$, $y = r \sin\theta \sin\phi$ and $z = r\cos\theta$

$$\text{Given integral} = 3\int_{\theta=0}^{\pi}\int_{\phi=0}^{2\pi}\int_{r=0}^{a} r^2.r^2 \sin\theta\, dr d\phi d\theta$$

$$\left(\text{as } dxdydz = \frac{\partial(x,y,z)}{\partial(r,\theta,\phi)}dr\,d\theta\,d\phi = r^2\sin\theta\,dr\,d\theta\,d\phi\right)$$

$$= 3\int_{\theta=0}^{\pi}\sin\theta d\theta \int_{\phi=0}^{2\pi} d\phi \int_{r=0}^{a} r^4 dr$$

$$= 3.2.2\pi.\frac{a^5}{5} = 12\pi a^5/5. \quad \textbf{Ans.}$$

EXAMPLE 10.31. Verify Gauss divergence theorem for the vector $\vec{F} = yi + xj + z^3k$ taken over the cylindrical region $x^2 + y^2 = 9$; $z = 0$, $z = 6$.

SOLUTION: Here $\vec{F} = yi + xj + z^3k$ and S is the total surface of the circular cylinder which consist of (i) the curved surface (ii), circular plate at $z = 0$ and (iii) the circular plate at $z = 6$.

$$\iint_S \vec{F}\cdot\hat{n}\,ds = \iint_{(i)} (yi + xj + z^3k)\cdot\frac{xi + yj}{3}ds + \iint_{(ii)} (yi + xj + 0^3k)\cdot(-k)ds + \iint_{(iii)} (yi + xj + 6^3k)\cdot k ds$$

$$= \frac{1}{3}\iint_{(i)} (yx + xy)ds + 0 + \iint_{(iii)} 6^3 ds = 0 + 0 + 6^3\,\pi.\,3^2$$

$$\iiint_V \nabla.\vec{F}\,dv = \iiint\left(\frac{\partial}{\partial x}y + \frac{\partial}{\partial y}x + \frac{\partial}{\partial z}z^3\right)dxdydz = \iiint 3z^2 dxdydz$$

$$= 4\int_0^3\int_0^{\sqrt{9-x^2}}\int_0^6 3z^2\,dzdydx = 4\int_0^3\int_0^{\sqrt{9-x^2}} [z^3]_0^6 dydx \quad \text{(using symmetry)}$$

$$= 4\cdot 6^3\cdot\frac{1}{2}\frac{9\pi}{2} = 9\pi\cdot 6^3. \quad \textbf{Hence Verified.}$$

EXAMPLE 10.32. Given $\vec{U} = (x, -y, z)$ and S: $x^2 + y^2 = a^2$, $z = 0$, $z = b$ verify divergence theorem.

SOLUTION: Here $\vec{U} = xi - yj + zk$ and S is a closed surface consisting of (i) surface of cylinder (ii) circular plate $x^2 + y^2 = a^2$ at $z = 0$ and (iii) circular plate $x^2 + y^2 = a^2$ at $z = b$ and at $z = 0$.

$$\iint_S \vec{U}\cdot d\vec{s} = \iint_{(i)} (xi - yj + zk)\cdot\frac{xi + yj}{a}ds + \iint_{(ii)} (xi - yj - 0k)\cdot(-k)ds + \iint_{(iii)} (xi - yj + bk)\cdot kds$$

$$= \iint_{(i)} \frac{x^2 - y^2}{a}ds + 0 + \iint_{(iii)} (xi - yj + bk)\cdot kds = 0 + 0 + b\iint_{(iii)} ds = b\pi a^2$$

$$\iiint_V \nabla.\vec{U} = \iiint\left(i\frac{\partial}{\partial x} + j\frac{\partial}{\partial y} + k\frac{\partial}{\partial z}\right)\cdot(xi - yj + zk)dxdydz = \iiint(1 - 1 + 1)dxdydz$$

$$= \text{Volume of the cylinder } = \pi a^2 b \quad \therefore \quad \iint_S \vec{F}\cdot\vec{ds} = \iiint_V \nabla\cdot\vec{F}dv$$

Hence divergence theorem verified.

EXAMPLE 10.33. Find $\iint_S \vec{F}\cdot\hat{n}\,ds$ where $\vec{F} = (2x + 3z)i - (xz + y)j + (y^2 + 2z)k$ and S is the surface of the sphere having centre at $(3, -1, 2)$ and radius 3. **(UPTU 2000-01; 2005-06)**

SOLUTION: Here S is the surface of the sphere $(x - 3)^2 + (y + 1)^2 + (z - 2)^2 = 9$.

Unit vector $\hat{n}$ along the normal to the above surface

$$= \frac{2(x-3)i + 2(y+1)j + 2(z-2)k}{\sqrt{4(x-3)^2 + 4(y+1)^2 + 4(z-2)^2}} = \frac{1}{3}[(x-3)i + (y+1)j + (z-2)k].$$

Here $\vec{F} = (2x + 3z)\,i - (xz + y)\,j + (y^2 + 2z)\,k$.

Since S is a closed surface, we can use divergence theorem .

Hence, $$\iint_s F.\hat{n}ds = \iiint_V \nabla.\vec{F}\,dv$$

$$= \iiint_V \left[\frac{\partial}{\partial x}(2x + 3z) - \frac{\partial}{\partial y}(xz + y) + \frac{\partial}{\partial z}(y^2 + 2z)\right]dxdydz$$

$$= \iint_V (2 - 1 + 2)dxdydz = 3 \text{ Volume of the sphere}$$

$$= 3.\frac{4}{3}\pi.3^3 = 108\pi. \quad \textbf{Ans.}$$

Example 10.34. Verify Divergence theorem for $\vec{F} = (x^2 - yz)i + (y^2 - zx)j + (z^2 - xy)k$ taken over the rectangular parallelopiped $0 \le x \le a,\ 0 \le y \le b,\ 0 \le z \le c$.

(UPTU 2001 Sp. Exam.)

Solution: Here $\vec{F} = (x^2 - yz)i + (y^2 - zx)j + (z^2 - xy)k$

$\therefore$ $\text{Div } \vec{F} = 2x + 2y + 2z = 2(x + y + z)$

We intend to show that $\iint_S \vec{F}\cdot\hat{n}\, ds = \iiint_V \nabla\cdot\vec{F}\, dv$

$$\text{R.H.S.} = \int_0^c\int_0^b\int_0^a 2(x+y+z)\, dxdydz = \int_0^c\int_0^b \left[x^2 + 2xy + 2xz\right]_0^a dydz$$

$$= \int_0^c\int_0^b (a^2 + 2ay + 2az)\, dydz = \int_0^c \left[a^2y + ay^2 + 2ayz\right]_0^b dz$$

$$= \int_0^c (a^2b + ab^2 + 2abz)\, dz = \left[a^2bz + ab^2z + abz^2\right]_0^c = abc\,(a + b + c).$$

To evaluate the surface integral we divide the closed surface S into 6 parts which are 6 faces of the cuboid.

S_1 = face OABC ($z = 0$), S_2 = face DEFG ($z = c$)

S_3 = face OADG ($y = 0$), S_4 = face BEFC ($y = b$)

S_5 = face ABED ($x = a$), S_6 = OCFG. ($x = 0$),

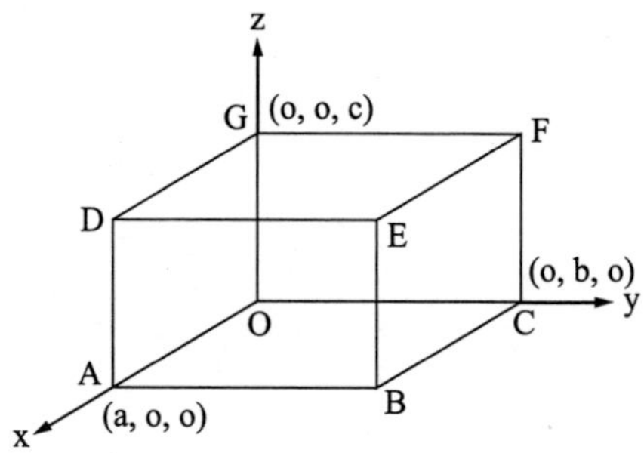

$$\iint_{S_1} \vec{F}\cdot\hat{n}\, ds = \iint_{S_1} (x^2i + y^2j - xyk)\cdot(-k)\, dxdy$$

$$= \int_0^b\int_0^a xy\, dxdy = \int_0^b \left[\frac{x^2}{2}y\right]_0^a dy$$

$$= \int_0^b \frac{a^2y}{2}\, dy = \frac{a^2b^2}{4}.$$

$$\iint_{S_2} \vec{F}\cdot\hat{n}\, ds = \iint_{S_2} [x^2 - cy)i + (y^2 - cx)j + (c^2 - xy)k]\cdot k\, dxdy$$

$$= \int_0^b\int_0^a (c^2 - xy)\, dxdy = \int_0^b \left[c^2x - \frac{x^2y}{2}\right]_0^a dy$$

$$= \int_0^b \left(c^2a - \frac{a^2y}{2}\right) dy = \left[c^2ay - \frac{a^2y^2}{4}\right]_0^b = c^2ab - \frac{a^2b^2}{4}.$$

$$\iint_{S_3} \vec{F}\cdot\hat{n}\,ds = \iint_{S_3}(x^2i - zxj + z^2k)\cdot(-j)\,dxdz = \int_0^c\int_0^a zx\,dxdz$$

$$= \int_0^c\left(\frac{zx^2}{2}\right)_0^a dz = \frac{a^2c^2}{4}.$$

$$\iint_{S_4} \vec{F}\cdot\hat{n}\,ds = \iint_{S_4}[(x^2 - bz)i + (b^2 - zx)j + (z^2 - bx)k]\cdot j\,dxdz$$

$$= \int_0^c\int_0^a (b^2 - zx)\,dxdz = \int_0^c\left[b^2x - \frac{zx^2}{2}\right]_0^a dz = \int_0^c\left(b^2a - \frac{za^2}{2}\right)dz$$

$$= b^2ac - \frac{a^2c^2}{4}.$$

$$\iint_{S_6} \vec{F}\cdot\hat{n}\,ds = \iint_{S_6}(-yzi + y^2j + z^2k)\cdot(-i)\,dydz = \int_0^b\int_0^c yz\,dydz$$

$$= \int_0^b\left[\frac{yz^2}{2}\right]_0^c dy = \int_0^b \frac{yc^2}{2}dy = \frac{b^2c^2}{4}.$$

$$\iint_{S_5} \vec{F}\cdot\hat{n}\,ds = \iint_{S_5}[(a^2 - yz)i + (y^2 - az)j + (z^2 - ay)k)\cdot i\,dydz$$

$$= \int_0^b\int_0^c (a^2 - yz)\,dydz = \int_0^b\left[a^2z - \frac{yz^2}{2}\right]_0^c dy$$

$$= \int_0^b\left(a^2c - \frac{yc^2}{2}\right)dy = \left[a^2cy - \frac{y^2c^2}{4}\right]_0^b = a^2bc - \frac{b^2c^2}{4}.$$

$$\therefore \quad \iint_S \vec{F}\cdot\hat{n}\,ds = \frac{a^2b^2}{4} + \frac{b^2c^2}{4} + \frac{c^2a^2}{4} + c^2ab + b^2ac + a^2bc - \frac{a^2b^2}{4} - \frac{b^2c^2}{4} - \frac{c^2a^2}{4}$$

$$= abc\,(a + b + c) = \text{L.H.S.}$$ **Hence Verified.**

EXAMPLE 10.35. State Gauss divergence theorem and verify this theorem evaluating the following surface integral as triple integral. $\iint_S x^3dydz + x^2ydz\,dx + x^2zdx\,dy$ where S is the closed surface consisting of the cylinder $x^2 + y^2 = a^2$ $(0 \le z \le b)$ and the circular discs at $z = 0$ and $z = b$ $(x^2 + y^2 \le a^2)$. **(UPTU 2006-07)**

SOLUTION: By Gauss divergence theorem $\iint_S \vec{F}\cdot\hat{n}dS = \iiint_V \nabla\cdot\vec{F}\,dv$. Here $\vec{F} = x^3i + x^2yj + x^2zk$, hence $\text{div}\,\vec{F} = \nabla\cdot\vec{F} = 3x^2 + x^2 + x^2 = 5x^2$. The volume integral $\iiint_V \nabla\cdot\vec{F}dv = \iiint 5x^2\,dx\,dy\,dz$ in which z varies from 0 to b, y from $-\sqrt{a^2 - x^2}$ to $\sqrt{a^2 - x^2}$ and x from $-a$ to a.

$$\therefore \quad \text{R.H.S.} = 5\int_{-a}^{a}\int_{-\sqrt{a^2-x^2}}^{\sqrt{a^2-x^2}}\int_{0}^{b} x^2 dz\,dy\,dx = 20b\int_0^a\int_0^{\sqrt{a^2-x^2}} x^2 dy\,dx = 20b\int_0^a x^2\sqrt{a^2-x^2}\,dx$$

$$= \frac{5\pi a^4 b}{4} \quad \text{(on making substitution } x = a\sin\theta\text{)}.$$

For L.H.S. the surface S consists of three parts. S_1 : circular base in the plane $z = 0$, S_2 : the circular top in the plane $z = b$ and S_3 : curved surfaces of the cylinder $x^2 + y^2 = a^2$

$$\iint_{S_1} \vec{F}\cdot\hat{n}\,dS = 0 \qquad \text{(since } z = 0 \text{ and } dz = 0\text{)}$$

$$\iint_{S_2} \vec{F}\cdot\hat{n}\,dS = \int_{-a}^{a}\int_{-\sqrt{a^2-x^2}}^{\sqrt{a^2-x^2}} x^2 b\,dxdy \quad \text{(since } z = b \text{ and } dz = 0\text{)}$$

$$= 4b\int_0^a\int_0^{\sqrt{a^2-x^2}} x^2\,dydx = 4b\int_0^a x^2\sqrt{a^2-x^2}\,dx = \frac{\pi}{4}a^4 b.$$

For S_3 $\nabla(x^2+y^2) = 2xi + 2yj$ hence $\hat{n} = \dfrac{2xi+2yj}{\sqrt{4x^2+4y^2}} = \dfrac{xi+yj}{a}$

$$\therefore \quad \vec{F}\cdot\hat{n} = \frac{1}{a}(x^3 i + x^2 yj + x^2 zk)\cdot(xi+yj) = \frac{1}{a}(x^4 + x^2y^2) = \frac{x^2}{a}(x^2+y^2) = ax^2$$

$$\therefore \quad \iint_{S_3}\vec{F}\cdot\hat{n}\,ds = \iint ax^2\,ds$$

Since $x^2 + y^2 = a^2$ we put $x = a\cos\theta$, $y = a\sin\theta$ hence $ds = a\,d\theta dz$

$$= a\int_0^{2\pi}\int_0^{b} a^2\cos^2\theta\cdot a\,d\theta\,dz = a^4 b\int_0^{2\pi}\cos^2\theta\,d\theta = \frac{a^4 b}{2}\int_0^{2\pi}(1+\cos 2\theta)d\theta$$

$$= \frac{a^4 b}{2}(2\pi + 0) = \pi a^4 b$$

$$\therefore \quad \text{L.H.S.} = 0 + \frac{\pi}{4}a^4 b + \pi a^4 b = \frac{5}{4}\pi a^4 b = \text{R.H.S.} \qquad \textbf{Hence Verified.}$$

EXAMPLE 10.36. Use divergence theorem to show that $\oint_S \nabla r^2\cdot\vec{ds} = 6V$ where S is any closed surface enclosing a volume V. **(UPTU 2003)**

SOLUTION: Since S is a closed surface enclosing a volume V, by Gauss divergence theorem we have

$$\oint_S \nabla r^2\cdot ds = \iiint_V \nabla\cdot\nabla r^2 dv$$

$$= \iiint_V \nabla^2 r^2\,dv = \iiint_V\left(\frac{\partial^2}{\partial x^2}+\frac{\partial^2}{\partial y^2}+\frac{\partial^2}{\partial z^2}\right)(x^2+y^2+z^2)\,dv$$

$$= \iiint (2+2+2)\,dv = 6V. \qquad \textbf{Hence Proved.}$$

EXAMPLE 10.37. Use divergence theorem to evaluate the surface integral $\iint_S x\,dy\,dz + y\,dz\,dx + z\,dx\,dy$ where S is the portion of the plane $x + 2y + 3z = 6$ lying in the first octant and planes $x = 0,\ y = 0,\ z = 0.$ **(UPTU 2004)**

SOLUTION: Here $\vec{F} = xi + yj + zk$ and divergence $\vec{F} = \frac{\partial x}{\partial x} + \frac{\partial y}{\partial y} + \frac{\partial z}{\partial z} = 3$, hence by divergence theorem $\iint_S x\,dy\,dz + y\,dz\,dx + z\,dx\,dy = \iiint_V 3\,dv$

$$= 3\int_0^6 \int_0^{3-x/2} \int_0^{2-x/3-2y/3} dz\,dy\,dx = 3\int_0^6 \int_0^{3-x/2} \left(2 - \frac{x}{3} - \frac{2y}{3}\right) dy\,dx$$

$$= 3\int_0^6 \left[2y - \frac{xy}{3} - \frac{y^2}{3}\right]_0^{\frac{6-x}{2}} dx = 3\int_0^6 \frac{1}{6}\left(\frac{6-x}{2}\right)^2 - \frac{1}{3}\left(\frac{6-x}{2}\right)^2 dx$$

$$= \int_0^6 \frac{1}{4}(6-4)^2\,dx = \frac{-1}{4.3}\left[(6-x)^3\right]_0^6 = \frac{216}{12} = 18. \quad \textbf{Ans.}$$

EXAMPLE 10.38. Evaluate $\iint_S (a^2x^2 + b^2y^2 + c^2z^2)^{-1/2}\,ds$ where S is the surface of the ellipsoid $ax^2 + by^2 + cz^2 = 1.$ **(UPTU 2011-12)**

SOLUTION: $\phi = ax^2 + by^2 + cz^2 - 1 = 0 \quad \therefore \quad \nabla\phi = (2ax)i + 2(yb)j + 2(zc)k$

$$\therefore \quad \hat{n} = \frac{\nabla\phi}{|\nabla\phi|} = \frac{2(axi + byj + czk)}{\sqrt{4a^2x^2 + 4b^2y^2 + 4c^2z^2}} = \frac{axi + byj + czk}{\sqrt{a^2x^2 + b^2y^2 + c^2z^2}}$$

$$\vec{F}\cdot\hat{n} = (a^2x^2 + b^2y^2 + c^2z^2)^{-1/2} \Rightarrow \vec{F} = xi + yj + zk \quad \text{since} \quad ax^2 + bx^2 + cz^2 = 1.$$

Hence by divergence theorem $\iint_S \vec{F}\cdot\overrightarrow{dS} = \iiint_V \nabla\cdot\vec{F}dv$

$$\text{or} \quad \iint_S \vec{F}\cdot\overrightarrow{ds} = \iiint_V 3\,dv = 3V = 3\cdot\frac{4\pi}{3}\frac{1}{\sqrt{a}}\frac{1}{\sqrt{b}}\frac{1}{\sqrt{c}} = \frac{4\pi}{\sqrt{abc}}. \quad \textbf{Ans.}$$

EXERCISE 10B

1. Evaluate the surface integral $\iint_S \vec{F} \cdot \vec{ds}$ where $\vec{F} = xi + yj + zk$ and S is the surface of the sphere $x^2 + y^2 + z^2 = 1$.

2. Evaluate the surface integral $\iint_S (yzi + zxj + xyk) \cdot \vec{ds}$ where S is the surface of the sphere $x^2 + y^2 + z^2 = 1$ in the first octant. **(UPTU 2004-05**

3. Verify Stoke's theorem for the vector field $\vec{F} = (x^2 + y^2)\, i + 2xyj$ in the rectangular region in the XY-plane bounded by the lines $x = 0$, $x = a$, $y = 0$, $y = b$. **(UPTU 2003)**

4. Evaluate the line integral $\oint_C \vec{F} \cdot \vec{dr}$ using Stoke's theorem if C is the boundary of the triangle ABC cut-off the co-ordinate planes by the plane $2x + 2y + z = 4$ and the vector field

$$\vec{F} = 2y\,(1 - x)\, i + (x - x^2 - y^2)\, j + (x^2 + y^2 + z^2)\, k.$$

5. Use Stoke's theorem to calculate the line integral $\int_C ydx + zdy + xdz$ where C is the inter-section of the sphere $x^2 + y^2 + z^2 = a^2$ and the plane $x + z = a$.

6. Evaluate $\int_C (xy\,dx + xy^2\,dy)$ by using Stoke's theorem taking C to be a square in the xy-plane with vertices (1, 0), (– 1, 0), (0, 1) and (0, – 1).

7. Evaluate $\iint_S \nabla \cdot \vec{F}\, \vec{ds}$ where S is the surface bounded by the inter-section of the sphere $x^2 + y^2 + z^2 = a^2$ and the cylinder $x^2 + y^2 = ax$ above the XY-plane and $\vec{F} = -2xyi + x^2j + zk$.

8. Use Stoke's theorem to find $\int_C \vec{F} \cdot \vec{dr}$ if $\vec{F} = -2yi + xj + zk$ and C is the loop of the curve $x^3 + y^3 = 3axy$.

9. Verify Gauss divergence theorem for the function $\vec{F} = x^2i + zj + yzk$ over the unit cube $x = 0$, $x = 1$; $y = 0$, $y = 1$; $z = 0$, $z = 1$. **(UPTU 2002)**

10. Evaluate $\iint_S (y^2z^3i + z^2x^3j + x^2y^3k) \cdot \vec{ds}$ where S is the curved surface of the sphere $x^2 + y^2 + z^2 = a^2$ above the plane $z = 0$.

11. Evaluate by divergence theorem $\iint_S (lx^2 + my^2 + nz^2)\, ds$ over the sphere $(x - a)^2 + (y - b)^2 + (z - c)^2 = \rho^2$, l, m, n being the direction cosines of the outward drawn normal to the sphere.

12. Use Gauss divergence theorem for evaluating the surface integral $\iint_S \vec{F} \cdot \vec{ds}$ where S is the closed surface bounded by the cylinder $x^2 + y^2 = a^2$ and the planes $z = 0, z = h$ and $\vec{F} = i\,yz + j\,zx + kz^2$.

13. Evaluate $\iint_S xz^2 dy\,dz + (x^2 y - z^3)\,dz\,dx + (2xy + y^2 z)\,dx\,dy$ where S is the surface enclosing the region bounded by the hemisphere $x^2 + y^2 + z^2 = 4$ above the XY-plane.

14. Verify the divergence theorem for the function $\vec{F} = 4xzi - y^2 j + yzk$, taken over the cube bounded by the planes $x = 0,\ x = 1;\ y = 0,\ y = 1;\ z = 0,\ z = 1.$ **(UPTU 2009-10)**

❑❑❑

UNIT-V
MATRICES

Elementary Row and Column Transformation
Rank of Matrix
Linear Dependence and Independence of Vectors
Consistency of Linear System of Equations
and their Solution
Characteristic Equation
Caley-Hamilton Theorem
Eigen Values and Eigen Vectors
Diagonalisation of Square Matrix
Complex and Unitry Matrices
Applications to Engineering Problems

CHAPTER

11 Matrices — Elementary Operations and Rank of a Matrix

Definitions and Properties of Matrices, Elementary Row and Column Operations, Linear Transformations, Rank of a Matrix and Normal Form

DEFINITIONS

Matrices have variety of applications, for example, in the solution of linear algebraic system of simultaneous equations, in the solution of ordinary and partial differential equations, etc.

A matrix is a rectangular arrangement of mn elements real or complex in m rows and n columns. Such a matrix A is usually denoted by

$$A = \begin{bmatrix} a_{11} & a_{12} & a_{13} & \cdots & a_{1n} \\ a_{21} & a_{22} & a_{23} & \cdots & a_{2n} \\ a_{31} & a_{32} & a_{33} & \cdots & a_{3n} \\ \cdots & \cdots & \cdots & \cdots & \cdots \\ \cdots & \cdots & \cdots & \cdots & \cdots \\ a_{m1} & a_{m2} & a_{m3} & \cdots & a_{mn} \end{bmatrix} = [a_{ij}] \quad \text{where} \quad i = 1, 2, .., m; \quad j = 1, 2, ..., n.$$

and is said to be of **order** $m \times n$. Matrices as a whole are denoted by capital letters while corresponding small letters with suffices represent its **elements**. The elements $a_{11}, a_{12}, a_{13} \ldots a_{mn}$ may be real or complex. Here we have a **double suffix notation** in which the first suffix indicates the row and the second suffix indicates column in which the element is located. Thus, a_{ij} is an element of ith row and j^{th} column. Unlike determinants, a matrix does not have any value. If $m = n$ the matrix is said to be a **square matrix** of order n. A matrix may also have only one row or only one column.

Thus, $[a_1, \ a_2 \, a_3 \ldots a_n]$ is known as **row matrix or row vector** whereas $\begin{bmatrix} b_1 \\ b_2 \\ \vdots \\ b_n \end{bmatrix}$ is known as **column matrix or column vector.**

In a square matrix A the elements $a_{11}, a_{22}, ..., a_{nn}$ form the **leading diagonal** (or **main diagonal** or **principal diagonal**). The sum of the diagonal elements, that is, $\sum_{i=1}^{n} a_{ii}$ is called the **trace** of the square matrix A.

If all the elements of a matrix are zero it is called **zero matrix or null matrix**. Further, a square matrix in which all the elements except those in the leading diagonal are zero is called a **diagonal matrix**. In other words, if $a_{ij} = 0$ for all $i \neq j$ then A is a diagonal matrix. For example,

$$\begin{bmatrix} 2 & 0 & 0 & 0 \\ 0 & 0 & 0 & 0 \\ 0 & 0 & 1 & 0 \\ 0 & 0 & 0 & 3 \end{bmatrix} \text{ is a diagonal matrix.}$$

In particular, a diagonal matrix in which all the leading diagonal elements are equal, is called a **scalar matrix**. If in a diagonal matrix all the elements in the leading diagonal are 1, *i.e.*, if

$$\begin{aligned} a_{ij} &= 0 \quad \text{for} \quad i \neq j \\ &= 1 \quad \text{for} \quad i = j \end{aligned}$$

it is called a **unit matrix** and is generally denoted by I_n.

For example,

$$I_3 = \begin{bmatrix} 1 & 0 & 0 \\ 0 & 1 & 0 \\ 0 & 0 & 1 \end{bmatrix} \text{ is a unit matrix of order 3.}$$

A square matrix is called an **upper triangular matrix** if all its elements below the leading diagonal are zero, *i.e.*, if $a_{ij} = 0$ for $i > j$. Similarly, a square matrix is called a **lower triangular matrix** if all its elements above the leading diagonal are zero, *i.e.* if $a_{ij} = 0$ for $i < j$.

Thus, the matrix $\begin{bmatrix} 1 & 2 & 5 \\ 0 & 3 & 4 \\ 0 & 0 & 2 \end{bmatrix}$ is an upper triangular matrix.

and $\begin{bmatrix} 2 & 0 & 0 \\ 1 & -1 & 0 \\ -3 & 4 & 0 \end{bmatrix}$ is a lower triangular matrix.

Every square matrix A is associated with a number called its **determinant** written as

$$|A| = \begin{vmatrix} a_{11} & a_{12} & \cdots & \cdots & a_{1n} \\ a_{21} & a_{22} & \cdots & \cdots & a_{2n} \\ \cdots & \cdots & \cdots & \cdots & \cdots \\ \cdots & \cdots & \cdots & \cdots & \cdots \\ a_{n1} & a_{n2} & \cdots & \cdots & a_{nn} \end{vmatrix}$$

If $|A| = 0$ the matrix A is said to be **singular** and if $|A| \neq 0$ it is called **non-singular**.

For example, the matrix

$$A = \begin{bmatrix} 2 & 4 & 3 \\ 1 & 0 & 2 \\ -3 & 5 & 1 \end{bmatrix} \text{ is non-singular since } |A| \neq 0, \text{ whereas the matrix}$$

$$B = \begin{bmatrix} 2 & 1 & -2 \\ 3 & 0 & 5 \\ 4 & 2 & -4 \end{bmatrix} \text{ is singular since } |B| = 0.$$

A square matrix A in which $a_{ij} = a_{ji}$ for all i, j, is called **symmetric** whereas, if $a_{ij} = -a_{ji}$ for all i, j it is said to be **skew-symmetric**. It evidently follows that in a **skew-symmetric matrix** $a_{ii} = 0$.

For example, the matrix

$$\begin{bmatrix} a & h & g \\ h & b & f \\ g & f & c \end{bmatrix} \text{ is symmetric, whereas the matrix}$$

$$\begin{bmatrix} 0 & 2 & -3 \\ -2 & 0 & -5 \\ 3 & 5 & 0 \end{bmatrix} \text{ is skew-symmetric.}$$

Two matrices are said be **equal** if and only if they are of the same order and each-element of one matrix is equal to the corresponding element of the other.

MATRIX OPERATIONS

1. Addition and Subtraction.

Two matrices of the same order can be added (or subtracted) by adding (or subtracting) their corresponding elements.

$$\text{Thus, } \begin{bmatrix} 1 & 3 & -2 \\ 4 & 0 & 2 \\ 0 & 2 & 3 \\ 1 & -1 & 2 \end{bmatrix} + \begin{bmatrix} 2 & 5 & 1 \\ 3 & 2 & -1 \\ 4 & 1 & 2 \\ -3 & 2 & 4 \end{bmatrix} = \begin{bmatrix} 3 & 8 & -1 \\ 7 & 2 & 1 \\ 4 & 3 & 5 \\ -2 & 1 & 6 \end{bmatrix} \text{ and } \begin{bmatrix} 7 & 5 \\ 3 & 4 \end{bmatrix} - \begin{bmatrix} 1 & 2 \\ -4 & 6 \end{bmatrix} = \begin{bmatrix} 6 & 3 \\ 7 & -2 \end{bmatrix}.$$

2. Multiplication of a Matrix by a Scalar (or Scalar Multiplication)

When a matrix is multiplied by a scalar, say k, then its every element gets multiplied by k.

$$\text{Thus, } 3\begin{bmatrix} 1 & 2 & -1 & 1 \\ 4 & -1 & 3 & 6 \\ 0 & 5 & 2 & 4 \end{bmatrix} = \begin{bmatrix} 3 & 6 & -3 & 3 \\ 12 & -3 & 9 & 18 \\ 0 & 15 & 6 & 12 \end{bmatrix}.$$

If A, B, C are any three matrices of same order then the following properties hold good.

(*i*) A + (B + C) = (A + B) + C (Associativity of addition)

(*ii*) A + B = B + A (Commutativity of addition)

(*iii*) k (A + B) = k A + kB, k being scalar (distributivity of addition with respect to a scalar).

3. Matrix Multiplication

Two matrices A and B can be multiplied as A . B only when the number of columns in A is equal to the number of rows in B. This is known as **compatibility condition** *for the product AB* to exist. Thus, the product AB of matrices $A(l \times m)$ and $B(p \times n)$ can exist only when $p = m$. Such matrices are also called **conformable**, and the product matrix C(= AB) would be of order $l \times n$.

We write $$C = (c_{ij})$$

where $$c_{ij} = a_{i1}\, b_{1j} + a_{i2}\, b_{2j} + ... + a_{im}\, b_{mj} = \sum_{k=1}^{m} a_{ik} b_{kj}$$

For example,

$$\begin{bmatrix} a_1 & b_1 & c_1 \\ a_2 & b_2 & c_2 \\ a_3 & b_3 & c_3 \\ a_4 & b_4 & c_4 \end{bmatrix} \begin{bmatrix} p_1 & p_2 \\ q_1 & q_2 \\ r_1 & r_2 \end{bmatrix} = \begin{bmatrix} a_1p_1 + b_1q_1 + c_1r_1 & a_1p_2 + b_1q_2 + c_1r_2 \\ a_2p_1 + b_2q_1 + c_2r_1 & a_2p_2 + b_2q_2 + c_2r_2 \\ a_3p_1 + b_3q_1 + c_3r_1 & a_3p_2 + b_3q_2 + c_3r_2 \\ a_4p_1 + b_4q_1 + c_4r_1 & a_4p_2 + b_4q_2 + c_4r_2 \end{bmatrix}$$

In the product AB, the matrix A is said to be **post - multiplied** by the matrix B and the matrix B is said to be **premultiplied** by A. In general, AB ≠ BA; in other words, *the matrices* AB *and* BA are **not necessarily equal even** *if they are* **conformable for multiplication**.

For instance, if

$$A = \begin{bmatrix} 1 & 2 \\ -1 & 1 \end{bmatrix} \quad \text{and} \quad B = \begin{bmatrix} 3 & 4 \\ 2 & 1 \end{bmatrix}$$

then $$AB = \begin{bmatrix} 7 & 6 \\ -1 & -3 \end{bmatrix} \quad \text{and} \quad BA = \begin{bmatrix} -1 & 10 \\ 1 & 5 \end{bmatrix}$$

hence AB ≠ BA. Thus, the matrix multiplication is not commutative.

Note here that for any square matrix A

AI = IA = A where I is a unit matrix of the same order as that of A and

A0 = 0A = 0 where 0 is null matrix of the same order as that of A.

Also we denote AA by A^2, A^2A by A^3; $(A^2)^2$ by A^4, etc.

4. Transpose of a Matrix

A matrix obtained by interchanging the rows and columns of a matrix A is defined as the **transpose** of A and is denoted by A′ or by A^T. For example,

if $$A = \begin{bmatrix} 2 & 3 & 1 \\ -1 & 5 & 3 \end{bmatrix} \quad \text{then} \quad A' = \begin{bmatrix} 2 & -1 \\ 3 & 5 \\ 1 & 3 \end{bmatrix}.$$

Thus, if A is of order $m \times n$, its transpose A′ will be of order $n \times m$. Further, A is called **symmetric matrix if A′ = A** and is called **skew symmetric if A′ = – A.**

Also, following observations are important to note and verify:

(*i*) $(A')' = A$.

(*ii*) $(A + B)' = A' + B'$ for any two matrices of same order.

(*iii*) Any square matrix A can be expressed as the sum of a symmetric and a skew-symmetric matrix as follows:

$$A = \frac{1}{2}(A + A') + \frac{1}{2}(A - A')$$

Here $(A + A')$ is symmetric as $(A + A')' = A' + A$

and $(A - A')$ is skew-symmetric as $(A - A')' = A' - A = -(A - A')$.

(*iv*) If A and B are square matrices of same order then $(AB)' = B'A'$

and, in general, $(ABCD \ldots)' = \ldots D'C'B'A'$.

Properties of Matrix Multiplication

(*i*) If the matrices A, B and C are compatible for multiplication then A(BC) = (AB)C, that is, matrix multiplication is associative.

(*ii*) A(B + C) = AB + AC, that is, matrices are distributive with respect to multiplication and addition, provided B and C are of the same order and AB and AC exist.

(*iii*) If A and B are compatible for multiplication and k is some scalar, then

$$k(AB) = (kA)B = A(kB).$$

(*iv*) If m and n are positive integers then for any square matrix A, we have

$$A^m A^n = A^{m+n} \text{ and } (A^m)^n = A^{mn}.$$

(*v*) For any two $n \times n$ matrices A and B, we have

$$|AB| = |A|\,|B|.$$

EXAMPLE 11.1. For any two square matrices A and B conformable for multiplication, are the following assertions true ? If not, correct them.

(*i*) $(A + B)^2 = A^2 + 2AB + B^2$ (*ii*) $(A + B)(A - B) = A^2 - B^2$.

SOLUTION: (*i*) **Not true**, because $BA \neq AB$ in general.

The correct form is

$$(A + B)^2 = (A + B)(A + B) = A^2 + AB + BA + B^2.$$

(*ii*) **Not true**, because $BA \neq AB$.

The form should be

$$(A + B)(A - B) = A(A - B) + B(A - B) = A^2 - AB + BA - B^2.$$ **Ans.**

EXAMPLE 11.2. For the matrices

$$A=\begin{bmatrix}3&4&2&1\\1&2&3&1\\0&1&2&3\end{bmatrix} \text{ and } B=\begin{bmatrix}1&2\\3&4\\1&0\\-1&1\end{bmatrix} \text{ verify that } (AB)'=B'A'.$$

SOLUTION: $AB=\begin{bmatrix}3&4&2&1\\1&2&3&1\\0&1&2&3\end{bmatrix}\begin{bmatrix}1&2\\3&4\\1&0\\-1&1\end{bmatrix}=\begin{bmatrix}16&23\\9&11\\2&7\end{bmatrix} \quad \therefore \ (AB)'=\begin{bmatrix}16&9&2\\23&11&7\end{bmatrix}.$

Next $B'A'=\begin{bmatrix}1&3&1&-1\\2&4&0&1\end{bmatrix}\begin{bmatrix}3&1&0\\4&2&1\\2&3&2\\1&1&3\end{bmatrix}$

$$=\begin{bmatrix}3+12+2-1&16+6+3-1&0+3+2-3\\6+16+0+1&2+8+0+1&0+4+0+3\end{bmatrix}=\begin{bmatrix}16&9&2\\23&11&7\end{bmatrix}.$$ **Hence true.**

EXAMPLE 11.3. Express the matrix $\begin{bmatrix}3&1&-2\\2&1&7\\-4&5&3\end{bmatrix}$ as the sum of two matrices, one symmetric and the other skew-symmetric.

SOLUTION: Let $A=\begin{bmatrix}3&1&-2\\2&1&7\\-4&5&3\end{bmatrix}$ then $A'=\begin{bmatrix}3&2&-4\\1&1&5\\-2&7&3\end{bmatrix}$

Now, $\frac{1}{2}(A+A')=\begin{bmatrix}3&3/2&-3\\3/2&1&6\\-3&6&3\end{bmatrix}=C$, say.

and $\frac{1}{2}(A-A')=\begin{bmatrix}0&-1/2&1\\1/2&0&1\\-1&-1&0\end{bmatrix}=D$, say.

Then $C+D=\begin{bmatrix}3&3/2&-3\\3/2&1&6\\-3&6&3\end{bmatrix}+\begin{bmatrix}0&-1/2&1\\1/2&0&1\\-1&-1&0\end{bmatrix}=\begin{bmatrix}3&1&-2\\2&1&7\\-4&5&3\end{bmatrix}=A$

$\therefore$ $A=C+D$ where C is symmetric and D is skew-symmetric. **Ans.**

ADJOINT OF A MATRIX

Let $A = (a_{ij})_{n \times n}$ be a square matrix of order $n \times n$, then its corresponding determinant is

$$|A| = \begin{vmatrix} a_{11} & a_{12} & \cdots & \cdots & a_{1n} \\ a_{21} & a_{22} & \cdots & \cdots & a_{2n} \\ \cdots & \cdots & \cdots & \cdots & \cdots \\ \cdots & \cdots & \cdots & \cdots & \cdots \\ a_{n1} & a_{n2} & \cdots & \cdots & a_{nn} \end{vmatrix}$$

Let $A_{11}, A_{12}, A_{13}, \ldots$ be the co-factors of $a_{11}, a_{12}, a_{13}, \ldots,$ in $|A|$ then the transpose of the cofactor matrix

$$\begin{bmatrix} A_{11} & A_{12} & \cdots & \cdots & A_{1n} \\ A_{21} & A_{22} & \cdots & \cdots & A_{2n} \\ \cdots & \cdots & \cdots & \cdots & \cdots \\ \cdots & \cdots & \cdots & \cdots & \cdots \\ A_{n1} & A_{n2} & \cdots & \cdots & A_{nn} \end{bmatrix} \text{ is } \begin{bmatrix} A_{11} & A_{21} & \cdots & \cdots & A_{n1} \\ A_{12} & A_{22} & \cdots & \cdots & A_{n2} \\ \cdots & \cdots & \cdots & \cdots & \cdots \\ \cdots & \cdots & \cdots & \cdots & \cdots \\ A_{1n} & A_{2n} & \cdots & \cdots & A_{nn} \end{bmatrix}$$

which is called the **adjoint** of A and is denoted by **adj (A)**.

Thus, the adjoint of a square matrix A is the transpose of its co-factor matrix.

For example, if $A = \begin{bmatrix} 2 & 5 & 3 \\ 3 & 1 & 2 \\ 1 & 2 & 1 \end{bmatrix}$

then $A_{11} = \begin{vmatrix} 1 & 2 \\ 2 & 1 \end{vmatrix} = -3,\quad A_{12} = -\begin{vmatrix} 3 & 2 \\ 1 & 1 \end{vmatrix} = -1,\quad A_{13} = \begin{vmatrix} 3 & 1 \\ 1 & 2 \end{vmatrix} = 5,$

$A_{21} = -\begin{vmatrix} 5 & 3 \\ 2 & 1 \end{vmatrix} = 1,\quad A_{22} = \begin{vmatrix} 2 & 3 \\ 1 & 1 \end{vmatrix} = -1,\quad A_{23} = -\begin{vmatrix} 2 & 5 \\ 1 & 2 \end{vmatrix} = 1,$

$A_{31} = \begin{vmatrix} 5 & 3 \\ 1 & 2 \end{vmatrix} = 7,\quad A_{32} = -\begin{vmatrix} 2 & 3 \\ 3 & 2 \end{vmatrix} = 5,\quad A_{33} = \begin{vmatrix} 2 & 5 \\ 3 & 1 \end{vmatrix} = -13.$

$$\therefore \quad \text{adj}(A) = \begin{vmatrix} -3 & -1 & 5 \\ 1 & -1 & 1 \\ 7 & 5 & -13 \end{vmatrix}' = \begin{vmatrix} -3 & 1 & 7 \\ -1 & -1 & 5 \\ 5 & 1 & -13 \end{vmatrix}.$$

INVERSE OF A MATRIX

Let A be a non-singular matrix of order $n \times n$ and B be another matrix of the same order such that

$$BA = I = AB$$

where I is the unit matrix of order n, then B is said to be the inverse of A, denoted by A^{-1} so that $AA^{-1} = I = A^{-1}A$.

Here A^{-1} is unique and $A^{-1} = \dfrac{adj(A)}{|A|}$.

To establish this relation,

let $$A = \begin{bmatrix} a_{11} & a_{12} & \cdots & \cdots & a_{1n} \\ a_{21} & a_{22} & \cdots & \cdots & a_{2n} \\ \cdots & \cdots & \cdots & \cdots & \cdots \\ \cdots & \cdots & \cdots & \cdots & \cdots \\ a_{n1} & a_{n2} & \cdots & \cdots & a_{nn} \end{bmatrix},$$

then consider A. adj (A)

$$= \begin{bmatrix} a_{11} & a_{12} & \cdots & \cdots & a_{1n} \\ a_{21} & a_{22} & \cdots & \cdots & a_{2n} \\ \cdots & \cdots & \cdots & \cdots & \cdots \\ \cdots & \cdots & \cdots & \cdots & \cdots \\ a_{n1} & a_{n2} & \cdots & \cdots & a_{nn} \end{bmatrix} \begin{bmatrix} A_{11} & A_{21} & \cdots & \cdots & A_{n1} \\ A_{12} & A_{22} & \cdots & \cdots & A_{n2} \\ \cdots & \cdots & \cdots & \cdots & \cdots \\ \cdots & \cdots & \cdots & \cdots & \cdots \\ A_{1n} & A_{2n} & \cdots & \cdots & A_{nn} \end{bmatrix}$$

$$= \begin{bmatrix} |A| & 0 & 0 & \cdots & 0 \\ 0 & |A| & 0 & \cdots & 0 \\ \cdots & \cdots & \cdots & \cdots & \cdots \\ \cdots & \cdots & \cdots & \cdots & \cdots \\ 0 & 0 & 0 & \cdots & |A| \end{bmatrix}$$

$$= |A| \begin{bmatrix} 1 & 0 & 0 & \cdots & \cdots & 0 \\ 0 & 1 & 0 & \cdots & \cdots & 0 \\ \cdots & \cdots & \cdots & \cdots & \cdots & \cdots \\ \cdots & \cdots & \cdots & \cdots & \cdots & \cdots \\ 0 & 0 & 0 & \cdots & \cdots & 1 \end{bmatrix} = |A|\ I_n$$

Since $|A| \neq 0$, dividing both sides by $|A|$, we get

$$A\frac{adj(A)}{|A|} = I \qquad \boxed{\Rightarrow \quad \mathbf{A^{-1} = \frac{adj(A)}{|A|}}}.$$

Another important result $(AB)^{-1} = B^{-1}A^{-1}$ can be shown to be true as follows:

$$(AB)(B^{-1}A^{-1}) = A(BB^{-1})A^{-1} = AIA^{-1} = AA^{-1} = I$$

Similarly, $$(B^{-1}A^{-1})(AB) = (B)^{-1}(A^{-1}A)B = B^{-1}IB = B^{-1}B = I$$

Therefore, we have $\boxed{\mathbf{(AB)^{-1} = B^{-1}A^{-1}}}$.

ADDITIONAL USEFUL PROPERTIES

(i) $\boxed{\mathbf{(A')^{-1} = (A^{-1})'}}$

we know that $(AA^{-1}) = I$. Taking transpose on both sides, we get $(AA^{-1})' = I' = I$

$\Rightarrow \quad (A^{-1})'A' = I.$ Hence, $(A')^{-1} = (A^{-1})'$.

(ii) $\boxed{|\mathbf{AB}| = |\mathbf{A}|\ |\mathbf{B}|}$ where A and B are any two square matrices both of order n and $|A| = |(a_{ij})|$ and $|B| = |(b_{ij})|$.

Then $\quad AB = (c_{ij})_{n\times n}$ where $c_{ij} = \sum_{k=1}^{n} a_{ik}b_{kj}$

Now by the rule of multiplication of two determinants of the same order, we have

$$|A|\,|B| = \left|\left(\sum_{k=1}^{n} a_{ik}b_{kj}\right)\right| = |c_{ij}| = |AB|.$$

It can be extended to any finite number of square matrices of same order, as

$$\boxed{|\mathbf{ABCD}\ \ldots\ | = |\mathbf{A}|\ |\mathbf{B}|\ |\mathbf{C}|\ |\mathbf{D}|\ \ldots.}$$

(iii) If A is a matrix of order $n \times n$, then

$$\boxed{|adj\ (\mathbf{A})\ | = |\mathbf{A}|^{n-1}}\ .$$

Earlier we have already established that

$$A\ adj\ (A) = |A|\ I = \begin{bmatrix} |A| & 0 & 0 & \cdots & \cdots & 0 \\ 0 & |A| & 0 & \cdots & \cdots & 0 \\ \cdots & \cdots & \cdots & \cdots & \cdots & \cdots \\ \cdots & \cdots & \cdots & \cdots & \cdots & \cdots \\ 0 & 0 & 0 & \cdots & \cdots & |A| \end{bmatrix}$$

Using the above property (ii), here we get

$$|A.\ adj\ (A)| = |A|\ |adj\ (A)| = \begin{vmatrix} |A| & 0 & 0 & \cdots & \cdots & 0 \\ 0 & |A| & 0 & \cdots & \cdots & 0 \\ \cdots & \cdots & \cdots & \cdots & \cdots & \cdots \\ \cdots & \cdots & \cdots & \cdots & \cdots & \cdots \\ 0 & 0 & 0 & \cdots & \cdots & |A| \end{vmatrix} = |A|^n$$

$\Rightarrow \qquad |adj\ (A)| = |A|^{n-1}.$

(iv) **ORTHOGONAL MATRIX.** An **orthogonal matrix** is one whose transpose is its inverse. Thus, a square matrix A for which $AA' = A'A = I$ is called orthogonal.

For example, let $\quad A = \begin{bmatrix} -2/3 & 1/3 & 2/3 \\ 2/3 & 2/3 & 1/3 \\ 1/3 & -2/3 & 2/3 \end{bmatrix}$

then $\quad A^{-1} = \begin{bmatrix} -2/3 & 2/3 & 1/3 \\ 1/3 & 2/3 & -2/3 \\ 2/3 & 1/3 & 2/3 \end{bmatrix} = A'$ hence A is orthogonal.

EXAMPLE 11.4. **(a)** Find the adjoint and the inverse of the matrix

$$A = \begin{bmatrix} 2 & 4 & 3 \\ 0 & 1 & 1 \\ 2 & 2 & -1 \end{bmatrix}.$$

(b) If $A = \begin{bmatrix} 1 & 2 & 1 \\ a & 0 & 4 \\ 1 & 1 & 1 \end{bmatrix}$ and $adj(adj\ A) = A$, find 'a'. **[UPTU 2011-12]**

SOLUTION: **(a)** The co-factors of the elements of $A\ (= a_{ij})$ are:

$$A_{11} = \begin{vmatrix} 1 & 1 \\ 2 & -1 \end{vmatrix} = -3, \qquad A_{12} = -\begin{vmatrix} 0 & 1 \\ 2 & -1 \end{vmatrix} = 2, \qquad A_{13} = \begin{vmatrix} 0 & 1 \\ 2 & 2 \end{vmatrix} = -2,$$

$$A_{21} = -\begin{vmatrix} 4 & 3 \\ 2 & -1 \end{vmatrix} = 10, \qquad A_{22} = \begin{vmatrix} 2 & 3 \\ 2 & -1 \end{vmatrix} = -8, \qquad A_{23} = -\begin{vmatrix} 2 & 4 \\ 2 & 2 \end{vmatrix} = 4,$$

$$A_{31} = \begin{vmatrix} 4 & 3 \\ 1 & 1 \end{vmatrix} = 1, \qquad A_{32} = -\begin{vmatrix} 2 & 3 \\ 0 & 1 \end{vmatrix} = -2, \qquad A_{33} = \begin{vmatrix} 2 & 4 \\ 0 & 1 \end{vmatrix} = 2$$

$$\therefore \qquad adj\,A = \begin{bmatrix} -3 & 2 & -2 \\ 10 & -8 & 4 \\ 1 & -2 & 2 \end{bmatrix}' = \begin{bmatrix} -3 & 10 & 1 \\ 2 & -8 & -2 \\ -2 & 4 & 2 \end{bmatrix}.$$

Also, $|A| = a_{11} A_{11} + a_{12} A_{12} + a_{13} A_{13} = 2(-3) + 4(2) + 3(-2) = -4$

$$\therefore \qquad A^{-1} = \frac{1}{|A|}\, adj\,A = -\frac{1}{4} \begin{bmatrix} -3 & 10 & 1 \\ 2 & -8 & -2 \\ -2 & 4 & 2 \end{bmatrix}. \qquad \textbf{Ans.}$$

(b) $A = \begin{bmatrix} 1 & 2 & 1 \\ a & 0 & 4 \\ 1 & 1 & 1 \end{bmatrix}$ $\quad \therefore \quad adj(A) = \begin{bmatrix} -4 & -a+4 & a \\ -1 & 0 & 1 \\ 8 & a-4 & -2a \end{bmatrix}' = \begin{bmatrix} -4 & -1 & 8 \\ 4-a & 0 & a-4 \\ a & 1 & -2a \end{bmatrix}$

$$\therefore \qquad adj(adj\ A) = \begin{bmatrix} 4-a & a(4-a) & 4-a \\ 2(4-a) & 0 & 4-a \\ 4-a & 4(4-a) & 4-a \end{bmatrix}' = \begin{bmatrix} 4-a & 2(4-a) & 4-a \\ a(4-a) & 0 & 4(4-a) \\ 4-a & 4-a & 4-a \end{bmatrix}$$

$$= (4-a)\begin{bmatrix} 1 & 2 & 1 \\ a & 0 & 4 \\ 1 & 1 & 1 \end{bmatrix} = \begin{bmatrix} 1 & 2 & 1 \\ a & 0 & 4 \\ 1 & 1 & 1 \end{bmatrix} \quad \text{(given)}$$

Hence $\qquad a = 3 \qquad$ **Ans.**

EXAMPLE 11.5. If $D = \begin{bmatrix} d_1 & 0 & 0 & 0 & \cdots & 0 \\ 0 & d_2 & 0 & 0 & \cdots & 0 \\ \cdots & \cdots & \cdots & \cdots & \cdots & \cdots \\ \cdots & \cdots & \cdots & \cdots & \cdots & \cdots \\ 0 & 0 & 0 & \cdots & \cdots & d_n \end{bmatrix}$ where $d_1, d_2, ..., d_n \neq 0$,

show that $D^{-1} = \begin{bmatrix} 1/d_1 & 0 & 0 & 0 & \cdots & 0 \\ 0 & 1/d_2 & 0 & 0 & \cdots & 0 \\ \cdots & \cdots & \cdots & \cdots & \cdots & \cdots \\ \cdots & \cdots & \cdots & \cdots & \cdots & \cdots \\ 0 & 0 & 0 & \cdots & \cdots & 1/d_n \end{bmatrix}$.

SOLUTION: $adj\ D = \begin{bmatrix} d_2 d_3 \ldots d_n & 0 & 0 & \cdots & \cdots & 0 \\ 0 & d_1 d_3 \cdots d_n & 0 & \cdots & \cdots & 0 \\ \cdots & \cdots & \cdots & \cdots & \cdots & \cdots \\ \cdots & \cdots & \cdots & \cdots & \cdots & \cdots \\ 0 & 0 & 0 & \cdots & \cdots & d_1 d_2 \cdots d_{n-1} \end{bmatrix}$

$$= (d_1 d_2 d_3 \cdots d_n) \begin{bmatrix} \frac{1}{d_1} & 0 & 0 & \cdots & \cdots & 0 \\ 0 & \frac{1}{d_2} & 0 & \cdots & \cdots & 0 \\ \cdots & \cdots & \cdots & \cdots & \cdots & \cdots \\ \cdots & \cdots & \cdots & \cdots & \cdots & \cdots \\ 0 & 0 & 0 & \cdots & \cdots & \frac{1}{d_n} \end{bmatrix}$$

Now, $|D| = d_1 d_2 d_3 \ .... \ d_n$

$$\therefore \quad D^{-1} = \frac{adj\ D}{|D|} = \begin{bmatrix} \frac{1}{d_1} & 0 & 0 & \cdots & \cdots & 0 \\ 0 & \frac{1}{d_2} & 0 & \cdots & \cdots & 0 \\ \cdots & \cdots & \cdots & \cdots & \cdots & \cdots \\ \cdots & \cdots & \cdots & \cdots & \cdots & \cdots \\ 0 & 0 & 0 & \cdots & \cdots & \frac{1}{d_n} \end{bmatrix}.$$ **Hence Proved.**

EXAMPLE 11.6. **(a)** Show that the matrix $\begin{bmatrix} \cos\theta & 0 & \sin\theta \\ 0 & 1 & 0 \\ -\sin\theta & 0 & \cos\theta \end{bmatrix}$ is orthogonal.

(b) The matrix $A = \begin{bmatrix} 2 & 0 & -1 \\ 5 & 1 & 0 \\ 0 & 1 & 3 \end{bmatrix}$ satisfies the matrix equation

$A^3 - 6A^2 + 11A - I = 0$, where I is the identity matrix of order 3. Find A^{-1}.

[UPTU 2011-12]

SOLUTION: (a) Let $A = \begin{bmatrix} \cos\theta & 0 & \sin\theta \\ 0 & 1 & 0 \\ -\sin\theta & 0 & \cos\theta \end{bmatrix}$ then $A' = \begin{bmatrix} \cos\theta & 0 & -\sin\theta \\ 0 & 1 & 0 \\ \sin\theta & 0 & \cos\theta \end{bmatrix}$

Now $$AA' = \begin{bmatrix} \cos^2\theta + \sin^2\theta & 0 & 0 \\ 0 & 1 & 0 \\ 0 & 0 & \sin^2\theta + \cos^2\theta \end{bmatrix} = \begin{bmatrix} 1 & 0 & 0 \\ 0 & 1 & 0 \\ 0 & 0 & 1 \end{bmatrix} = I$$

Therefore A is an orthogonal matrix. **Hence Proved.**

(b) Multiplying both sides of $A^3 - 6A^2 + 11A - I = 0$ by A^{-1} we get $A^2 - 6A + 11I - A^{-1} = 0$

or $$A^{-1} = A^2 - 6A + 11I$$

Here $$A^2 = \begin{bmatrix} 2 & 0 & -1 \\ 5 & 1 & 0 \\ 0 & 1 & 3 \end{bmatrix}\begin{bmatrix} 2 & 0 & -1 \\ 5 & 1 & 0 \\ 0 & 1 & 3 \end{bmatrix} = \begin{bmatrix} 4 & -1 & -5 \\ 15 & 1 & -5 \\ 5 & 4 & 9 \end{bmatrix}$$

$\therefore$ $$A^{-1} = \begin{bmatrix} 4 & -1 & -5 \\ 15 & 1 & -5 \\ 5 & 4 & 9 \end{bmatrix} - 6\begin{bmatrix} 2 & 0 & -1 \\ 5 & 1 & 0 \\ 0 & 1 & 3 \end{bmatrix} + \begin{bmatrix} 11 & 0 & 0 \\ 0 & 11 & 0 \\ 0 & 0 & 11 \end{bmatrix}$$

$$= \begin{bmatrix} 3 & -1 & 1 \\ -15 & 6 & -5 \\ 4 & -2 & 2 \end{bmatrix}$$ **Ans.**

Solving the Simultaneous Equations by Matrix Inversion

Consider the following n equations with n unknowns $x_1, x_2 \ldots, x_n$

$$a_{11}x_1 + a_{12}x_2 + \ldots + a_{1n}x_n = b_1$$

$$a_{21}x_1 + a_{22}x_2 + \ldots + a_{2n}x_n = b_2$$

$$\ldots\ldots\ldots\ldots\ldots\ldots = \ldots$$

$$\ldots\ldots\ldots\ldots\ldots\ldots = \ldots$$

$$a_{n1}x_1 + a_{n2}x_2 + \ldots + a_{nn}x_n = b_n$$

We can write the above equations in matrix equation form as AX = B ...(1)

where $A = \begin{bmatrix} a_{11} & a_{12} & \cdots & \cdots & a_{1n} \\ a_{21} & a_{22} & \cdots & \cdots & a_{2n} \\ \cdots & \cdots & \cdots & \cdots & \cdots \\ \cdots & \cdots & \cdots & \cdots & \cdots \\ a_{n1} & a_{n2} & \cdots & \cdots & a_{nn} \end{bmatrix}$, $X = \begin{bmatrix} x_1 \\ x_2 \\ \vdots \\ x_n \end{bmatrix}$, $B = \begin{bmatrix} b_1 \\ b_2 \\ \vdots \\ b_n \end{bmatrix}$.

Premultiplying by A^{-1} on both sides of (1), gives $A^{-1}(AX) = A^{-1}B$.

Since matrices are associative w.r. t. multiplication, we have

$$(A^{-1}A)X = A^{-1}B \quad \text{or} \quad IX = A^{-1}B \qquad \text{or} \qquad X = A^{-1}B$$

or
$$\begin{bmatrix} x_1 \\ x_2 \\ \vdots \\ x_n \end{bmatrix} = \frac{1}{|A|}\begin{bmatrix} A_{11} & A_{21} & \cdots & \cdots & A_{n1} \\ A_{12} & A_{22} & \cdots & \cdots & A_{n2} \\ \cdots & \cdots & \cdots & \cdots & \cdots \\ \cdots & \cdots & \cdots & \cdots & \cdots \\ A_{1n} & A_{2n} & \cdots & \cdots & A_{nn} \end{bmatrix}\begin{bmatrix} b_1 \\ b_2 \\ \vdots \\ b_n \end{bmatrix} \qquad ...(2)$$

The matrix $A^{-1}B$ is also a column matrix. Equating the values of $x_1, x_2, ..., x_n$ to the corresponding elements in the product on the R.H.S. of (2), we get the desired solution.

EXAMPLE 11.7. Write the following equations in the matrix form $AX = B$ and solve

$$3x + 2y + 4z = 7$$
$$2x + y + z = 4$$
$$x + 3y + 5z = 2.$$

SOLUTION: $AX = B$

or
$$\begin{bmatrix} 3 & 2 & 4 \\ 2 & 1 & 1 \\ 1 & 3 & 5 \end{bmatrix}\begin{bmatrix} x \\ y \\ z \end{bmatrix} = \begin{bmatrix} 7 \\ 4 \\ 2 \end{bmatrix}.$$

Here
$$A = \begin{bmatrix} 3 & 2 & 4 \\ 2 & 1 & 1 \\ 1 & 3 & 5 \end{bmatrix}, \quad X = \begin{bmatrix} x \\ y \\ z \end{bmatrix}, \quad B = \begin{bmatrix} 7 \\ 4 \\ 2 \end{bmatrix}.$$

Hence
$$adj\, A = \begin{bmatrix} 2 & -9 & 5 \\ 2 & 11 & -7 \\ -2 & 5 & -1 \end{bmatrix}' = \begin{bmatrix} 2 & 2 & -2 \\ -9 & 11 & 5 \\ 5 & -7 & -1 \end{bmatrix}$$

and $|A| = 3(2) + 2(-9) + 4(5) = 6 - 18 + 20 = 8$

$\therefore$
$$A^{-1} = \frac{adj\, A}{|A|} = \frac{1}{8}\begin{bmatrix} 2 & 2 & -2 \\ -9 & 11 & 5 \\ 5 & -7 & -1 \end{bmatrix}$$

Now,
$$X = A^{-1}B = \frac{1}{8}\begin{bmatrix} 2 & 2 & -2 \\ -9 & 11 & 5 \\ 5 & -7 & -1 \end{bmatrix}\begin{bmatrix} 7 \\ 4 \\ 2 \end{bmatrix} = \frac{1}{8}\begin{bmatrix} 14+8-4 \\ -63+44+10 \\ 35-28-2 \end{bmatrix} = \frac{1}{8}\begin{bmatrix} 18 \\ -9 \\ 5 \end{bmatrix}$$

Hence $x = 18/8$, $y = -9/8$, $z = 5/8$. **Ans.**

EXERCISE 11A

1. If $A = \begin{bmatrix} 1 & 2 \\ 3 & 4 \end{bmatrix}$ show that $A^2 - 5A = 2I$ where I is the unit matrix of order 2, and hence find A^4.

2. If $A = \begin{bmatrix} 1 & 1 & 1 \\ 1 & 1 & 1 \\ 1 & 1 & 1 \end{bmatrix}$ and $B = \begin{bmatrix} 1 & 2 & 3 \\ 4 & 5 & 6 \\ 7 & 8 & 9 \end{bmatrix}$ verify that $(AB)' = B'A'$.

3. Let $A = \begin{bmatrix} -1 & 2 & 3 \\ 1 & 3 & -1 \\ 4 & 5 & 1 \end{bmatrix}$ show that $(A^2)' = (A')^2$.

4. Express the matrix $\begin{bmatrix} 2 & 1 & 3 \\ -1 & 4 & 1 \\ 0 & 2 & -2 \end{bmatrix}$ as the sum of a symmetric and a skew-symmetric matrix.

5. Let $A = \begin{bmatrix} a_1 & b_1 \\ a_2 & b_2 \end{bmatrix}$ show that $A^{-1} = \dfrac{1}{a_1b_2 - a_2b_1}\begin{bmatrix} b_2 & -b_1 \\ -a_2 & a_1 \end{bmatrix}$ and hence find the inverse of the matrix $B = \begin{bmatrix} \cos\theta & -\sin\theta \\ \sin\theta & \cos\theta \end{bmatrix}$. Is B orthogonal ?

6. Determine the inverse of the matrix $\begin{bmatrix} 5 & -2 & 4 \\ -2 & 1 & 1 \\ 4 & 1 & 0 \end{bmatrix}$.

7. Given that $A = \begin{bmatrix} 0 & 1 & -1 \\ 2 & -3 & 4 \\ 3 & -2 & 3 \end{bmatrix}$ and $B = \begin{bmatrix} 1 & -1 & 1 \\ -3 & 2 & -1 \\ -2 & 1 & -1 \end{bmatrix}$, verify

(*i*) $(A^2)' = (A')^2$ (*ii*) $(AB)^{-1} = B^{-1}A^{-1}$.

8. If $A = \begin{bmatrix} -1 & -2 & -2 \\ 2 & 1 & -2 \\ 2 & -2 & 1 \end{bmatrix}$ show that *adj* A = 3 A′ and find *adj* (*adj* A).

9. Find $\begin{bmatrix} 1 & -\tan\theta \\ \tan\theta & 1 \end{bmatrix}\begin{bmatrix} 1 & \tan\theta \\ -\tan\theta & 1 \end{bmatrix}^{-1}$.

10. Solve by matrix inversion method the following set of equations

$$x + 3y + 3z = 1$$
$$x + 4y + 3z = 0$$
$$x + 3y + 4z = 2.$$

❑❑❑

LINEAR TRANSFORMATIONS

Consider three co-ordinate systems in a plane denoted by x_1x_2 – system, y_1y_2 – system and w_1w_2– system. Let us assume that these are related by the transformations

$$\left.\begin{aligned} y_1 &= a_{11}x_1 + a_{12}x_2 \\ y_2 &= a_{21}x_1 + a_{22}x_2 \end{aligned}\right\} \quad ...(1)$$

and

$$\left.\begin{aligned} x_1 &= b_{11}w_1 + b_{12}w_2 \\ x_2 &= b_{21}w_1 + b_{22}w_2 \end{aligned}\right\} \quad ...(2)$$

which are particular linear transformations. On putting (2) in (1) we see that the y_1y_2 – system can be obtained directly from the w_1w_2 – system by a single linear transformation

$$\left.\begin{aligned} y_1 &= c_{11}w_1 + c_{12}w_2 \\ y_2 &= c_{21}w_1 + c_{22}w_2 \end{aligned}\right\} \quad ...(3)$$

This yields that

$$\left.\begin{aligned} y_1 &= a_{11}(b_{11}w_1 + b_{12}w_2) + a_{12}(b_{21}w_1 + b_{22}w_2) \\ y_2 &= a_{21}(b_{11}w_1 + b_{12}w_2) + a_{22}(b_{21}w_1 + b_{22}w_2) \end{aligned}\right\} \quad ...(4)$$

Comparing (4) with (3), gives

$$\left.\begin{aligned} c_{11} &= a_{11}b_{11} + a_{12}b_{21}, \qquad c_{12} = a_{11}b_{12} + a_{12}b_{22} \\ c_{21} &= a_{21}b_{11} + a_{22}b_{21}, \qquad c_{22} = a_{21}b_{12} + a_{22}b_{22}) \end{aligned}\right\} \quad ...(5)$$

Our above calculation provides motivation for matrix multiplication. The transformation (1) can be written in compact form as $y = \text{A}x$ ...(6)

where $y = \begin{bmatrix} y_1 \\ y_2 \end{bmatrix}$, $\text{A} = \begin{bmatrix} a_{11} & a_{12} \\ a_{21} & a_{22} \end{bmatrix}$, $x = \begin{bmatrix} x_1 \\ x_2 \end{bmatrix}$

and (2) can be written as $x = \text{B}w$ ...(7)

where $x = \begin{bmatrix} x_1 \\ x_2 \end{bmatrix}$, $\text{B} = \begin{bmatrix} b_{11} & b_{12} \\ b_{21} & b_{22} \end{bmatrix}$, $w = \begin{bmatrix} w_1 \\ w_2 \end{bmatrix}$.

Thus, substituting the linear transformation (7) in (6) we get the co-efficient matrix C of the composite transformation as

$$y = \text{A}x = \text{A}(\text{B}w) = \text{AB}w = \text{C}w \text{ where } \boxed{\text{C} = \text{AB}.}$$

The idea can be easily extended to m variables $y_1, y_2, ..., y_m$ and n variables $x_1, x_2, ..., x_n$ and p variables $w_1, w_2, ..., w_p$. The matrix A is of order $m \times n$, B of order $n \times p$ therefore C is of order $m \times p$.

For example, let (x, y) be the co-ordinates of a point P referred to a set of rectangular axes OX, OY. Then its co-ordinates (x', y') referred to another set of rectangular axes OX′, OY′ obtained by rotating the former axes OX, OY through an angle θ, are given by

$$x' = x\cos\theta + y\sin\theta, \quad y' = -x\sin\theta + y\cos\theta$$

Similarly, the relations of the form

$$\begin{aligned} x' &= l_1x + m_1y + n_1z \\ y' &= l_2x + m_2y + n_2z \\ z' &= l_3x + m_3y + n_3z \end{aligned}$$

define a linear transformation from (x, y, z) to (x', y', z') in 3D.

In general, the relation $Y = AX$

$$\text{where}\quad Y=\begin{bmatrix} y_1 \\ y_2 \\ \vdots \\ y_n \end{bmatrix},\ A=\begin{bmatrix} a_{11} & a_{12} & \cdots & \cdots & a_{1n} \\ a_{21} & a_{22} & \cdots & \cdots & a_{2n} \\ \cdots & \cdots & \cdots & \cdots & \cdots \\ \cdots & \cdots & \cdots & \cdots & \cdots \\ a_{n1} & a_{n2} & \cdots & \cdots & a_{nn} \end{bmatrix},\ X=\begin{bmatrix} x_1 \\ x_2 \\ \vdots \\ x_n \end{bmatrix}$$

gives an example of a linear transformation from n variables $x_1, x_2, ..., x_n$ to n variables $y_1, y_2, ..., y_n$, *i.e.*, the transformation of the vector X to the vector Y. This transformation is called **linear** because the linear relations $A(x_1 + x_2) = Ax_1 + Ax_2$ and $A(bx) = bA(x)$ hold good.

If the determinant of the co-efficient matrix is zero, *i.e.*, $|A| = 0$, the transformation is called **singular** otherwise it is called **non-singular**. For a non-singular transformation $Y = AX$ we can invert it to get the inverse transformation $X = A^{-1}Y$. A non-singular transformation is sometimes referred to as **regular transformation**.

The linear transformation $Y = AX$ is called **orthogonal** if it transforms $x_1^2 + x_2^2 + ... + x_n^2$ into $y_1^2 + y_2^2 + ... + y_n^2$. Let us show it.

If $\quad X = [x_1, x_2, ..., x_n]$ then $X' = \begin{bmatrix} x_1 \\ x_2 \\ \vdots \\ x_n \end{bmatrix}$ hence $XX' = x_1^2 + x_2^2 + ... + x_n^2$

Similarly, let $Y = [y_1, y_2, ..., y_n]$ then $Y' = \begin{bmatrix} y_1 \\ y_2 \\ \vdots \\ y_n \end{bmatrix}$ hence $YY' = y_1^2 + y_2^2 + ... + y_n^2$

Thus, if $Y = AX$ is an orthogonal transformation, then

$$X'X = Y'Y = (AX)'(AX) = X'A'AX$$

which is possible only when $A'A = I$ so that $\boxed{A' = A^{-1} \text{ for orthogonal transformation}}$.

ELEMENTARY OPERATIONS (OR TRANSFORMATIONS)

The following operations are known as **elementary operations or elementary transformations**:

(*i*) **Interchanging any two rows (columns):** This is denoted by R_{ij} for interchanging i^{th} row and j^{th} row, and by C_{ij} for interchanging i^{th} column and j^{th} column.

(*ii*) **Multiplying any row (column) by a non-zero quantity:** This is denoted by kR_i for multiplying i^{th} row by k and by kC_i for multiplying ith column by k $(k \neq 0)$.

(*iii*) **Adding a constant times the elements of a row (column) to the corresponding elements of another row (column):** It is denoted by $R_m + kR_n$, where k times the elements of nth row are added to the elements of mth row. Similarly, $C_m + kC_n$ for columns.

ELEMENTARY MATRICES

When an elementary transformation is applied to a unit matrix the resulting matrix is called **elementary matrix**.

For instance, consider

$$I_3 = \begin{bmatrix} 1 & 0 & 0 \\ 0 & 1 & 0 \\ 0 & 0 & 1 \end{bmatrix}$$

Applying the operation $R_2 \leftrightarrow R_3$, we get

$$I_3 \sim \begin{bmatrix} 1 & 0 & 0 \\ 0 & 0 & 1 \\ 0 & 1 & 0 \end{bmatrix} \quad \text{which is an elementry matrix.}$$

Similarly, applying $R_1 \to R_1 + kR_3$ on I_3, we get

$$I_3 \sim \begin{bmatrix} 1 & 0 & k \\ 0 & 1 & 0 \\ 0 & 0 & 1 \end{bmatrix} \quad \text{which is also an elementary matrix.}$$

AN IMPORTANT DEDUCTION

Every row (column) transformation on a matrix can be affected by pre (post) multiplication by corresponding elementary matrix.

Consider the matrix A given by

$$A = \begin{bmatrix} 2 & 3 & 4 \\ 5 & 6 & 7 \\ 3 & 5 & 9 \end{bmatrix}$$

On applying row transformton $R_3 \to R_3 - 4R_1$ on A, we get a matrix B given by

$$B = \begin{bmatrix} 2 & 3 & 4 \\ 5 & 6 & 7 \\ -5 & -7 & -7 \end{bmatrix}$$

Now we shall show that premultiplication of A by corresponding elementary matrix will give us B.

$$I = \begin{bmatrix} 1 & 0 & 0 \\ 0 & 1 & 0 \\ 0 & 0 & 1 \end{bmatrix} \text{ hence elementary matrix } (R_3 - 4R_1) = \begin{bmatrix} 1 & 0 & 0 \\ 0 & 1 & 0 \\ -4 & 0 & 0 \end{bmatrix}$$

$$\therefore \quad (R_3 - 4R_1) \times A = \begin{bmatrix} 1 & 0 & 0 \\ 0 & 1 & 0 \\ -4 & 0 & 1 \end{bmatrix} \begin{bmatrix} 2 & 3 & 4 \\ 5 & 6 & 7 \\ 3 & 5 & 9 \end{bmatrix} = \begin{bmatrix} 2 & 3 & 4 \\ 5 & 6 & 7 \\ -5 & -7 & -7 \end{bmatrix} = B$$

INVERSE OF A MATRIX BY GAUSS-JORDAN METHOD

The elementary row operations which reduce a given square matrix A $(n \times n)$ to a unit matrix I $(n \times n)$, when applied to the unit matrix I, give the inverse of A. Suppose the successive row operations which reduce the given square matrix A to I result from the pre-multiplication of A by the elementary matrices $R_1, R_2, ..., R_n$ so that $R_n R_{n-1} R_{n-2} \cdots R_2 R_1 A = I$.

Post-multiplying both sides by A^{-1}, we get

$$(R_n R_{n-1} R_{n-2} \cdots R_2 . R_1 A) A^{-1} = IA^{-1}$$

or $\quad R_n R_{n-1} R_{n-2} \cdots R_1 (AA^{-1}) = A^{-1}$

or $\quad \mathbf{R_n R_{n-1} R_{n-2} \cdots R_1 I = A^{-1}}$.

However, in practice, to find A^{-1} we write A and I side-by-side and the same row operations are performed on both and as soon as A is reduced to I the other matrix will be nothing but A^{-1}.

EXAMPLE 11.8. Transform the following matrix to upper triangular form **(Echelon Form).**

$$A = \begin{bmatrix} 1 & 3 & 3 \\ 2 & 4 & 10 \\ 3 & 8 & 4 \end{bmatrix}$$

(UPTU 2001)

SOLUTION: Applying $R_2 \to R_2 - 2R_1$ and $R_3 \to R_3 - 3R_1$, we get

$$A = \begin{bmatrix} 1 & 3 & 3 \\ 2 & 4 & 10 \\ 3 & 8 & 4 \end{bmatrix} \sim \begin{bmatrix} 1 & 3 & 3 \\ 0 & -2 & 4 \\ 0 & -1 & -5 \end{bmatrix}$$

Again applying $R_3 \to R_3 - 1/2R_2$, we get

$$A \sim \begin{bmatrix} 1 & 3 & 3 \\ 0 & -2 & -1 \\ 0 & 0 & -13 \end{bmatrix}$$ which is in the upper triangular form. **Ans.**

EXAMPLE 11.9. Find the inverse of the matrix A given by

$$A = \begin{bmatrix} 3 & -3 & 4 \\ 2 & -3 & 4 \\ 0 & -1 & -1 \end{bmatrix}$$

by applying elementary row transformations using Gauss-Jordan method.

(UPTU I Sem. 2002)

SOLUTION: We write $A = I_3 A$,

that is, $$\begin{bmatrix} 3 & -3 & 4 \\ 2 & -3 & 4 \\ 0 & -1 & 1 \end{bmatrix} = \begin{bmatrix} 1 & 0 & 0 \\ 0 & 1 & 0 \\ 0 & 0 & 1 \end{bmatrix} \begin{bmatrix} 3 & -3 & 4 \\ 2 & -3 & 4 \\ 0 & -1 & 1 \end{bmatrix}$$

Applying $R_1 \to R_1 - R_2$, we get

$$\begin{bmatrix} 1 & 0 & 0 \\ 2 & -3 & 4 \\ 0 & -1 & 1 \end{bmatrix} = \begin{bmatrix} 1 & -1 & 0 \\ 0 & 1 & 0 \\ 0 & 0 & 1 \end{bmatrix} A$$

Next, applying $R_2 \to R_2 - 4R_3$, we get

$$\begin{bmatrix} 1 & 0 & 0 \\ 0 & -3 & 4 \\ 0 & -1 & 1 \end{bmatrix} = \begin{bmatrix} 1 & -1 & 0 \\ -2 & 3 & 0 \\ 0 & 0 & 1 \end{bmatrix} A$$

Again, applying $R_2 \to R_2 - 4R_2$, gives

$$\begin{bmatrix} 1 & 0 & 0 \\ 0 & 1 & 0 \\ 0 & -1 & 1 \end{bmatrix} = \begin{bmatrix} 1 & -1 & 0 \\ -2 & 3 & -4 \\ 0 & 0 & 1 \end{bmatrix} A$$

then, applying $R_3 \to R_3 + R_2$, we get

$$\begin{bmatrix} 1 & 0 & 0 \\ 0 & 1 & 0 \\ 0 & 0 & 1 \end{bmatrix} = \begin{bmatrix} 1 & -1 & 0 \\ -2 & 3 & -4 \\ -2 & 3 & -3 \end{bmatrix} A$$

Hence $$A^{-1} = \begin{bmatrix} 1 & -1 & 0 \\ -2 & 3 & -4 \\ -2 & 3 & -3 \end{bmatrix}.$$ **Ans.**

EXAMPLE 11.10. Using row operations compute the inverse of the matrix

$$A = \begin{bmatrix} 1 & 0 & 4 \\ 2 & -2 & 1 \\ -1 & 1 & -1 \end{bmatrix}.$$

SOLUTION: $A = \begin{bmatrix} 1 & 0 & 4 \\ 2 & -2 & 1 \\ -1 & 1 & -1 \end{bmatrix}$, $I = \begin{bmatrix} 1 & 0 & 0 \\ 0 & 1 & 0 \\ 0 & 0 & 1 \end{bmatrix}$

$$A \sim \begin{bmatrix} 1 & 0 & 4 \\ 0 & 0 & -1 \\ -1 & 1 & -1 \end{bmatrix}, \qquad I \sim \begin{bmatrix} 1 & 0 & 0 \\ 0 & 1 & 2 \\ 0 & 0 & 1 \end{bmatrix} \text{ (on applying } R_2 \to R_2 + 2R_3)$$

$$A \sim \begin{bmatrix} 1 & 0 & 4 \\ 0 & 0 & -1 \\ 0 & 1 & 3 \end{bmatrix}, \qquad I \sim \begin{bmatrix} 1 & 0 & 0 \\ 0 & 1 & 2 \\ 1 & 0 & 1 \end{bmatrix} \text{ (on applying } R_3 \to R_3 + R_1)$$

$$A \sim \begin{bmatrix} 1 & 0 & 4 \\ 0 & 1 & 3 \\ 0 & 0 & -1 \end{bmatrix}, \qquad I \sim \begin{bmatrix} 1 & 0 & 0 \\ 1 & 0 & 1 \\ 0 & 1 & 2 \end{bmatrix} \text{ (on applying } R_2 \leftrightarrow R_3)$$

$$A \sim \begin{bmatrix} 1 & 0 & 4 \\ 0 & 1 & 0 \\ 0 & 0 & -1 \end{bmatrix}, \qquad I \sim \begin{bmatrix} 1 & 4 & 8 \\ 1 & 3 & 7 \\ 0 & 1 & 2 \end{bmatrix}$$

(on applying $R_1 \to R_1 + 4R_3$ and $R_2 \to R_2 + 3\,R_3$)

$$A \sim \begin{bmatrix} 1 & 0 & 0 \\ 0 & 1 & 0 \\ 0 & 0 & 1 \end{bmatrix}, \qquad I \sim \begin{bmatrix} 1 & 4 & 8 \\ 1 & 3 & 7 \\ 0 & -1 & -2 \end{bmatrix} \text{ (on applying } R_3 \to -R_3)$$

Therefore $A^{-1} = \begin{bmatrix} 1 & 4 & 8 \\ 1 & 3 & 7 \\ 0 & -1 & -2 \end{bmatrix}$. **Ans.**

EXAMPLE 11.11. **(a)** Using the method of elementary row transformations to compute the inverse of

$$\begin{bmatrix} 1 & 2 & 5 \\ 2 & 3 & 1 \\ -1 & 1 & 1 \end{bmatrix}.$$

(b) Using elementary row transformations, find the inverse of the matrix

$$A = \begin{bmatrix} 2 & 3 & 4 \\ 4 & 3 & 1 \\ 1 & 2 & 4 \end{bmatrix}$$

[UPTU 2011-12]

SOLUTION: (a) Let $A = \begin{bmatrix} 1 & 2 & 5 \\ 2 & 3 & 1 \\ -1 & 1 & 1 \end{bmatrix}$ and $I_3 = \begin{bmatrix} 1 & 0 & 0 \\ 0 & 1 & 0 \\ 0 & 0 & 1 \end{bmatrix}$

We apply row transformations on A as well as I simultaneously, till A becomes I then the fate of I will be A^{-1}.

Applying $R_2 \to R_2 + 2R_1$ and $R_3 \to R_3 + R_1$ we get

$$A \sim \begin{bmatrix} 1 & 2 & 5 \\ 0 & -1 & -9 \\ 0 & 3 & 6 \end{bmatrix} \quad \text{and} \quad I \sim \begin{bmatrix} 1 & 0 & 0 \\ -2 & 1 & 0 \\ 1 & 0 & 1 \end{bmatrix} \text{ Now applying } R_3 \to R_3 + 3R_2$$

$$A \sim \begin{bmatrix} 1 & 2 & 5 \\ 0 & -1 & -9 \\ 0 & 0 & -21 \end{bmatrix} \quad \text{and} \quad I \sim \begin{bmatrix} 1 & 0 & 0 \\ -2 & 1 & 0 \\ -5 & 3 & 1 \end{bmatrix} \text{ Now applying } R_2 \to -R_2,\ R_3 \to -\frac{1}{21} R_3$$

$$A \sim \begin{bmatrix} 1 & 2 & 5 \\ 0 & 1 & 9 \\ 0 & 0 & 1 \end{bmatrix} \qquad \text{and} \qquad I \sim \begin{bmatrix} 1 & 0 & 0 \\ 2 & -1 & 0 \\ \frac{5}{21} & \frac{-3}{21} & \frac{-1}{21} \end{bmatrix}.$$

Next applying $R_2 \to R_2 - 9R_3$ and $R_1 \to R_1 - 5R_3$ we get

$$A \sim \begin{bmatrix} 1 & 2 & 0 \\ 0 & 1 & 0 \\ 0 & 0 & 1 \end{bmatrix}, \qquad \text{and} \qquad I \sim \begin{bmatrix} \frac{-4}{21} & \frac{15}{21} & \frac{5}{21} \\ \frac{-3}{21} & \frac{6}{21} & \frac{9}{21} \\ \frac{5}{21} & \frac{-3}{21} & \frac{-1}{21} \end{bmatrix} = \frac{1}{21}\begin{bmatrix} -4 & 15 & 5 \\ -3 & 6 & 9 \\ 5 & -3 & -1 \end{bmatrix}$$

Finally, applying $R_1 \to R_1 - 2R_2$, we get

$$A \sim \begin{bmatrix} 1 & 0 & 0 \\ 0 & 1 & 0 \\ 0 & 0 & 1 \end{bmatrix}, \qquad \text{and} \qquad I \sim \frac{1}{21}\begin{bmatrix} 2 & 3 & -13 \\ -3 & 6 & 9 \\ 5 & -3 & -1 \end{bmatrix}$$

Thus $\qquad A^{-1} = \dfrac{1}{21}\begin{bmatrix} 2 & 3 & -13 \\ -3 & 6 & 9 \\ 5 & -3 & -1 \end{bmatrix}$. **Ans.**

(*b*) We have $A = IA$ where I is unit matrix of order 3×3.

or
$$\begin{bmatrix} 2 & 3 & 4 \\ 4 & 3 & 1 \\ 1 & 2 & 4 \end{bmatrix} = \begin{bmatrix} 1 & 0 & 0 \\ 0 & 1 & 0 \\ 0 & 0 & 1 \end{bmatrix}\begin{bmatrix} 2 & 3 & 4 \\ 4 & 3 & 1 \\ 1 & 2 & 4 \end{bmatrix}$$

Applying $R_1 \leftrightarrow R_3$ then $R_2 \to R_2 - 4R_1$ and $R_3 \to R_3 - 2R_1$, we get

$$\begin{bmatrix} 1 & 2 & 4 \\ 0 & -5 & -15 \\ 0 & -1 & -4 \end{bmatrix} = \begin{bmatrix} 0 & 0 & 1 \\ 0 & 1 & -4 \\ 1 & 0 & -2 \end{bmatrix} A$$

Now applying $R_2 \to -\frac{1}{5}R_2$, we get

$$\begin{bmatrix} 1 & 2 & 4 \\ 0 & 1 & 3 \\ 0 & -1 & -4 \end{bmatrix} = \begin{bmatrix} 0 & 0 & 1 \\ 0 & -\frac{1}{5} & -\frac{4}{15} \\ 1 & 0 & -2 \end{bmatrix} A$$

Now operating $R_3 \to R_3 + R_2$ and $R_3 \to -R_3$ given

$$\begin{bmatrix} 1 & 2 & 4 \\ 0 & 1 & 3 \\ 0 & 0 & 1 \end{bmatrix} = \begin{bmatrix} 0 & 0 & 1 \\ 0 & +\frac{1}{5} & -\frac{1}{15} \\ -1 & \frac{1}{5} & \frac{6}{5} \end{bmatrix} A$$

Next $R_2 \to R_2 - 3R_3$, $R_1 \to R_1 - 2R_2$ and $R_1 \to R_1 - 4R_3$, we get

$$\begin{bmatrix} 1 & 0 & 0 \\ 0 & 1 & 0 \\ 0 & 0 & 1 \end{bmatrix} = \begin{bmatrix} -2 & \frac{4}{5} & \frac{9}{5} \\ 3 & -\frac{4}{5} & -\frac{14}{5} \\ -1 & \frac{1}{5} & \frac{6}{5} \end{bmatrix}$$

$\therefore$ $$A^{-1} = \frac{1}{5}\begin{bmatrix} -10 & 4 & 9 \\ 15 & -4 & -14 \\ -5 & 1 & 6 \end{bmatrix}$$ **Ans.**

EXAMPLE 11.12. Find the inverse of the matrix $\begin{bmatrix} 0 & 1 & 2 \\ 1 & 2 & 3 \\ 3 & 1 & 1 \end{bmatrix}$ by E – row operations.

(UPTU 2000-01; 2003-04)

SOLUTION: Let $A = \begin{bmatrix} 0 & 1 & 2 \\ 1 & 2 & 3 \\ 3 & 1 & 1 \end{bmatrix}$ and $I = \begin{bmatrix} 1 & 0 & 0 \\ 0 & 1 & 0 \\ 0 & 0 & 1 \end{bmatrix}$

These we write as $\left[\begin{array}{ccc|ccc} 0 & 1 & 2 & 1 & 0 & 0 \\ 1 & 2 & 3 & 0 & 1 & 0 \\ 3 & 1 & 1 & 0 & 0 & 1 \end{array}\right]$ (Applying $R_1 \leftrightarrow R_2$)

$$= \left[\begin{array}{ccc|ccc} 1 & 2 & 3 & 0 & 1 & 0 \\ 0 & 1 & 2 & 1 & 0 & 0 \\ 3 & 1 & 1 & 0 & 0 & 1 \end{array}\right] \quad \text{(Next applying } R_3 \to R_3 - 3R_1)$$

$$= \left[\begin{array}{ccc|ccc} 1 & 2 & 3 & 0 & 1 & 0 \\ 0 & 1 & 2 & 1 & 0 & 0 \\ 0 & -5 & -8 & 0 & -3 & 1 \end{array}\right] \quad \text{(Next } R_3 \to R_3 + 5R_2)$$

$$= \left[\begin{array}{ccc|ccc} 1 & 2 & 3 & 0 & 1 & 0 \\ 0 & 1 & 2 & 1 & 0 & 0 \\ 0 & 0 & 2 & 5 & -3 & 1 \end{array}\right] \quad \text{(Next } R_2 \to R_2 - R_3)$$

$$= \left[\begin{array}{ccc|ccc} 1 & 2 & 3 & 0 & 1 & 0 \\ 0 & 1 & 0 & -4 & 3 & 1 \\ 0 & 0 & 2 & 5 & -3 & 1 \end{array}\right] \quad \text{(Next } R_1 \to R_1 - 2R_2)$$

$$= \left[\begin{array}{ccc|ccc} 1 & 0 & 3 & 8 & -5 & 2 \\ 0 & 1 & 0 & -4 & 3 & 1 \\ 0 & 0 & 2 & 5 & -3 & 1 \end{array}\right] \quad \text{(Next } R_3 \to \tfrac{1}{2} R_3)$$

$$= \left[\begin{array}{ccc|ccc} 1 & 0 & 3 & 8 & -5 & 1/2 \\ 0 & 1 & 0 & -4 & 3 & 1 \\ 0 & 0 & 1 & 5/2 & -3/2 & 1/2 \end{array}\right] \quad (\text{Next } R_1 \to R_1 - 3R_3)$$

$$= \left[\begin{array}{ccc|ccc} 1 & 0 & 0 & 1/2 & -1/2 & 1/2 \\ 0 & 1 & 0 & -4 & 3 & 1 \\ 0 & 0 & 1 & 5/2 & -3/2 & 1/2 \end{array}\right]$$

Then $$A^{-1} = \begin{bmatrix} \frac{1}{2} & \frac{-1}{2} & \frac{1}{2} \\ -4 & 3 & 1 \\ \frac{5}{2} & \frac{-3}{2} & \frac{1}{2} \end{bmatrix} = \frac{1}{2}\begin{bmatrix} 1 & -1 & 1 \\ -8 & 6 & 2 \\ 5 & -3 & 1 \end{bmatrix}.$$ **Ans.**

EXAMPLE 11.13. Find the inverse of the matrix $A = \begin{bmatrix} 1 & 2 & 6 \\ 2 & 5 & 15 \\ 6 & 15 & 46 \end{bmatrix}$ by using elementary row operations.

SOLUTION: $$A = \begin{bmatrix} 1 & 2 & 6 \\ 2 & 5 & 15 \\ 6 & 15 & 46 \end{bmatrix} = \begin{bmatrix} 1 & 2 & 6 \\ 2 & 5 & 15 \\ 6 & 15 & 46 \end{bmatrix}\begin{bmatrix} 1 & 0 & 0 \\ 0 & 1 & 0 \\ 0 & 0 & 1 \end{bmatrix} = AI$$

Applying $R_3 \to R_3 - 3R_2$ on A and I both $= \begin{bmatrix} 1 & 2 & 6 \\ 2 & 5 & 15 \\ 0 & 0 & 1 \end{bmatrix}\begin{bmatrix} 1 & 0 & 0 \\ 0 & 1 & 0 \\ 0 & -3 & 1 \end{bmatrix}$

Next applying $R_2 \to R_2 - 2R_1$ on A and I, we get

$$\begin{bmatrix} 1 & 2 & 6 \\ 0 & 1 & 3 \\ 0 & 0 & 1 \end{bmatrix}\begin{bmatrix} 1 & 0 & 0 \\ -2 & 1 & 0 \\ 0 & -3 & 1 \end{bmatrix}$$

Next applying $R_1 \to R_1 - 2R_2$ and then $R_2 \to R_2 - 3R_3$, we get

$$\begin{bmatrix} 1 & 0 & 0 \\ 0 & 1 & 0 \\ 0 & 0 & 1 \end{bmatrix}\begin{bmatrix} 5 & -2 & 0 \\ -2 & 10 & -3 \\ 0 & -3 & 1 \end{bmatrix}$$

$\therefore$ $$A^{-1} = \begin{bmatrix} 5 & -2 & 0 \\ -2 & 10 & -3 \\ 0 & -3 & 1 \end{bmatrix}.$$ **Ans.**

EXAMPLE 11.14. Compute the inverse of the following matrix by using elementary transformations

$$\begin{bmatrix} 2 & -6 & -2 & -3 \\ 5 & -13 & -4 & -7 \\ -1 & 4 & 1 & 2 \\ 0 & 1 & 0 & 1 \end{bmatrix}.$$

SOLUTION: We find the inverse applying row oparations only. Write A = IA

or
$$\begin{bmatrix} 2 & -6 & -2 & -3 \\ 5 & -13 & -4 & -7 \\ -1 & 4 & 1 & 2 \\ 0 & 1 & 0 & 1 \end{bmatrix} = \begin{bmatrix} 1 & 0 & 0 & 0 \\ 0 & 1 & 0 & 0 \\ 0 & 0 & 1 & 0 \\ 0 & 0 & 0 & 1 \end{bmatrix} \begin{bmatrix} 2 & -6 & -2 & -3 \\ 5 & -13 & -4 & -7 \\ -1 & 4 & 1 & 2 \\ 0 & 1 & 0 & 1 \end{bmatrix}$$

Applying $R_1 \leftrightarrow R_3$ and $R_2 \leftrightarrow R_3$ on both sides, we get

$$\begin{bmatrix} -1 & 4 & 1 & 2 \\ 2 & -6 & -2 & -3 \\ 5 & -13 & -4 & -7 \\ 0 & 1 & 0 & 1 \end{bmatrix} = \begin{bmatrix} 0 & 0 & 1 & 0 \\ 1 & 0 & 0 & 0 \\ 0 & 1 & 0 & 0 \\ 0 & 0 & 0 & 1 \end{bmatrix} A$$

Then applying $R_2 \to R_2 + 2R_1$ and $R_3 \to R_3 + 5R_1$ then $R_1 \to -R_1$, we get

$$\begin{bmatrix} 1 & -4 & -1 & -2 \\ 0 & 2 & 0 & 1 \\ 0 & 7 & 1 & 3 \\ 0 & 1 & 0 & 1 \end{bmatrix} = \begin{bmatrix} 0 & 0 & -1 & 0 \\ 1 & 0 & 2 & 0 \\ 0 & 1 & 5 & 0 \\ 0 & 0 & 0 & 1 \end{bmatrix} A$$

Next, applying $R_1 \to R_1 + 2R_2$ and $R_2 \to \frac{1}{2} R_2$, we get

$$\begin{bmatrix} 1 & 0 & -1 & 0 \\ 0 & 1 & 0 & 1/2 \\ 0 & 7 & 1 & 3 \\ 0 & 1 & 0 & 1 \end{bmatrix} = \begin{bmatrix} 2 & 0 & 3 & 0 \\ 1/2 & 0 & 1 & 0 \\ 0 & 1 & 5 & 0 \\ 0 & 0 & 0 & 1 \end{bmatrix} A$$

Next, applying $R_3 \to R_3 - 7R_2$ and $R_4 \to R_4 - R_2$, we get

$$\begin{bmatrix} 1 & 0 & -1 & 0 \\ 0 & 1 & 0 & 1/2 \\ 0 & 0 & 1 & -1/2 \\ 0 & 0 & 0 & 1/2 \end{bmatrix} = \begin{bmatrix} 2 & 0 & 3 & 0 \\ 1/2 & 0 & 1 & 0 \\ -7/2 & 1 & -2 & 0 \\ -1/2 & 0 & -1 & 1 \end{bmatrix} A$$

Next, applying $R_3 \to R_3 + R_4$ and $R_2 \to R_2 - R_4$, we get

$$\begin{bmatrix} 1 & 0 & -1 & 0 \\ 0 & 1 & 0 & 0 \\ 0 & 0 & 1 & 0 \\ 0 & 0 & 0 & 1/2 \end{bmatrix} = \begin{bmatrix} 2 & 0 & 3 & 0 \\ 1 & 0 & 2 & -1 \\ -4 & 1 & -3 & 1 \\ -1/2 & 0 & -1 & 1 \end{bmatrix} A$$

Finally applying $R_1 \to R_1 + R_3$ and $R_4 \to 2R_4$, we get

$$\begin{bmatrix} 1 & 0 & 0 & 0 \\ 0 & 1 & 0 & 0 \\ 0 & 0 & 1 & 0 \\ 0 & 0 & 0 & 1 \end{bmatrix} = \begin{bmatrix} -2 & 1 & 0 & 1 \\ 1 & 0 & 2 & -1 \\ -4 & 1 & -3 & 1 \\ -1 & 0 & -2 & 2 \end{bmatrix} A$$

Therefore $$A^{-1} = \begin{bmatrix} -2 & 1 & 0 & 1 \\ 1 & 0 & 2 & -1 \\ -4 & 1 & -3 & 1 \\ -1 & 0 & -2 & 2 \end{bmatrix}.$$ **Ans.**

EXAMPLE 11.15. Find the inverse of the following matrix M by applying elementary transformations

$$M = \begin{bmatrix} 0 & 2 & 1 & 3 \\ 1 & 1 & -1 & -2 \\ 1 & 2 & 0 & 1 \\ -1 & 1 & 2 & 6 \end{bmatrix}.$$ **(UPTU 2003)**

SOLUTION: We write $M = I_4M$

or $$\begin{bmatrix} 0 & 2 & 1 & 3 \\ 1 & 1 & -1 & -2 \\ 1 & 2 & 0 & 1 \\ -1 & 1 & 2 & 6 \end{bmatrix} = \begin{bmatrix} 1 & 0 & 0 & 0 \\ 0 & 1 & 0 & 0 \\ 0 & 0 & 1 & 0 \\ 0 & 0 & 0 & 1 \end{bmatrix} M \quad \text{operating } R_1 \leftrightarrow R_2$$

$$\begin{bmatrix} 1 & 1 & -1 & -2 \\ 0 & 2 & 1 & 3 \\ 1 & 2 & 0 & 1 \\ -1 & 1 & 2 & 6 \end{bmatrix} = \begin{bmatrix} 0 & 1 & 0 & 0 \\ 1 & 0 & 0 & 0 \\ 0 & -1 & 1 & 0 \\ 0 & 1 & 0 & 1 \end{bmatrix} M \quad \text{operating } R_3 - R_1 \text{ and } R_4 + R_1$$

$$\begin{bmatrix} 1 & 1 & -1 & -2 \\ 0 & 2 & 1 & 3 \\ 0 & 1 & 1 & 3 \\ 0 & 2 & 1 & 4 \end{bmatrix} = \begin{bmatrix} 0 & 1 & 0 & 0 \\ 1 & 0 & 0 & 0 \\ 0 & -1 & 1 & 0 \\ 0 & 1 & 0 & 1 \end{bmatrix} M \quad \text{operating } R_4 - R_2 \text{ and } R_1 + R_3$$

$$\begin{bmatrix} 1 & 2 & 0 & 1 \\ 0 & 2 & 1 & 3 \\ 0 & 1 & 1 & 3 \\ 0 & 0 & 0 & 1 \end{bmatrix} = \begin{bmatrix} 0 & 0 & 1 & 0 \\ 1 & 0 & 0 & 0 \\ 0 & -1 & 1 & 0 \\ -1 & 1 & 0 & 1 \end{bmatrix} M \quad \text{next applying } R_2 - R_3$$

$$\begin{bmatrix} 1 & 2 & 0 & 1 \\ 0 & 1 & 0 & 0 \\ 0 & 1 & 1 & 3 \\ 0 & 0 & 0 & 1 \end{bmatrix} = \begin{bmatrix} 0 & 0 & 1 & 0 \\ 1 & 0 & 0 & 0 \\ 0 & -1 & 1 & 0 \\ -1 & 1 & 0 & 1 \end{bmatrix} M \quad \text{next applying } R_1 - 2R_2 \text{ and } R_3 - R_2$$

$$\begin{bmatrix} 1 & 0 & 0 & 0 \\ 0 & 1 & 0 & 0 \\ 0 & 0 & 1 & 0 \\ 0 & 0 & 0 & 1 \end{bmatrix} = \begin{bmatrix} -2 & -2 & 3 & 0 \\ 1 & 1 & -1 & 0 \\ -1 & -2 & 2 & 0 \\ -1 & 1 & 0 & 1 \end{bmatrix} M \quad \text{next applying } R_1 - R_4 \text{ and } R_3 - 3R_1$$

$$\begin{bmatrix} 1 & 0 & 0 & 0 \\ 0 & 1 & 0 & 0 \\ 0 & 0 & 1 & 0 \\ 0 & 0 & 0 & 1 \end{bmatrix} = \begin{bmatrix} -1 & -3 & 3 & -1 \\ 1 & 1 & -1 & 0 \\ 2 & -5 & 2 & -3 \\ -1 & 1 & 0 & 1 \end{bmatrix} M \quad \text{or} \quad I = M^{-1}M$$

Hence $M^{-1} = \begin{bmatrix} -1 & -3 & 3 & -1 \\ 1 & 1 & -1 & 0 \\ 2 & -5 & 2 & -3 \\ -1 & 1 & 0 & 1 \end{bmatrix}$. **Ans.**

RANK OF A MATRIX

(UPTU 2003)

A matrix is said to be of ***rank r*** if

(*i*) it has at least one non-zero minor of order r, and

(*ii*) every minor of order greater than r vanishes.

In other words, the rank of a matrix A is equal to the order of the highest ordered non-zero minor in A and is denoted by $r(A)$ or by $\rho(A)$.

Thus, **the rank of a non-singular square matrix of order *n* is *n*.** For example,

let $A = \begin{bmatrix} 1 & 2 \\ 3 & 4 \end{bmatrix}$ then $\rho(A) = 2$.

If $A = \begin{bmatrix} 1 & 2 \\ 2 & 4 \end{bmatrix}$ then $\rho(A) = 1$

Next, let $A = \begin{bmatrix} 1 & 4 & 3 \\ 2 & 5 & 2 \\ 3 & 6 & 1 \end{bmatrix}$. Here $|A| = 0$ and $\begin{vmatrix} 1 & 4 \\ 2 & 5 \end{vmatrix} \neq 0$ hence $\rho(A) = 2$.

Two matrices A and B are said to be **equivalent** if they are of the same order and of same rank and we write $A \sim B$. It is important to note here that when a matrix is subjected to one or more elementary operations, its rank does not change.

EXAMPLE 11.16. Find the rank of the matrix

$$A = \begin{bmatrix} 6 & 1 & 3 & 8 \\ 4 & 2 & 6 & -1 \\ 10 & 3 & 9 & 7 \\ 16 & 4 & 12 & 15 \end{bmatrix}.$$

SOLUTION: $A \sim \begin{bmatrix} 1 & 6 & 3 & 8 \\ 2 & 4 & 6 & -1 \\ 3 & 10 & 9 & 7 \\ 4 & 16 & 12 & 15 \end{bmatrix}$ (on operating C_{12} on A)

$\sim \begin{bmatrix} 1 & 6 & 3 & 8 \\ 0 & -8 & 0 & -17 \\ 0 & -8 & 0 & -17 \\ 0 & -8 & 0 & -17 \end{bmatrix}$ (on operating $R_2 \to R_2 - 2R_1$, $R_3 \to R_3 - 3R_1$, $R_4 \to R_4 - 4R_1$)

$\sim \begin{bmatrix} 1 & 0 & 0 & 0 \\ 0 & -8 & 0 & -17 \\ 0 & -8 & 0 & -17 \\ 0 & -8 & 0 & -17 \end{bmatrix}$ (on operating $C_2 \to C_2 - 6C_1$, $C_3 \to C_3 - 3C_1$, $C_4 \to C_4 - 8C_1$)

$\sim \begin{bmatrix} 1 & 0 & 0 & 0 \\ 0 & 1 & 0 & 0 \\ 0 & 1 & 0 & 0 \\ 0 & 1 & 0 & 0 \end{bmatrix}$ (on operating $C_2 \to -\frac{1}{8}C_2$ and then $C_4 \to C_4 - 17C_2$)

$\sim \begin{bmatrix} 1 & 0 & 0 & 0 \\ 0 & 1 & 0 & 0 \\ 0 & 0 & 0 & 0 \\ 0 & 0 & 0 & 0 \end{bmatrix}$ (on operating $R_3 \to R_3 - R_2$ and $R_4 \to R_4 - R_2$)

which clearly establishes that rank (A) = 2. **Ans.**

EXAMPLE 11.17. Find the values of a and b such that the rank of matrix $A = \begin{bmatrix} 1 & -2 & 3 & 1 \\ 2 & 1 & -1 & 2 \\ 6 & -2 & a & b \end{bmatrix}$ is 2.

SOLUTION: $A \sim \begin{bmatrix} 1 & -2 & 3 & 1 \\ 0 & 5 & -7 & 0 \\ 0 & -5 & a+3 & b-6 \end{bmatrix}$ (On applying $R_3 \to R_3 - 3R_2$ and $R_2 \to R_2 - 2R_1$)

$$A \sim \begin{bmatrix} 1 & -2 & 3 & 1 \\ 0 & 5 & -7 & 0 \\ 0 & 0 & a-4 & b-6 \end{bmatrix} \text{ On applying } R_3 \to R_3 + R_2$$

Since rank of A is given to be 2 we have $a = 4, b = 6$. **Ans.**

EXAMPLE 11.18. Find the rank of the matrix $\begin{bmatrix} 0 & 1 & 2 & -2 \\ 4 & 0 & 2 & 6 \\ 2 & 1 & 3 & 1 \end{bmatrix}$.

SOLUTION:

$$A = \begin{bmatrix} 0 & 1 & 2 & -2 \\ 4 & 0 & 2 & 6 \\ 2 & 1 & 3 & 1 \end{bmatrix}$$

$$\sim \begin{bmatrix} 2 & 1 & 3 & 1 \\ 4 & 0 & 2 & 6 \\ 0 & 1 & 2 & -2 \end{bmatrix} \text{ (using } R_1 \leftrightarrow R_3)$$

$$\sim \begin{bmatrix} 2 & 1 & 3 & 1 \\ 0 & -2 & -4 & 4 \\ 0 & 1 & 2 & -2 \end{bmatrix} \text{ (using } R_2 \to R_2 - 2R_1)$$

$$\sim \begin{bmatrix} 2 & 1 & 3 & 1 \\ 0 & 0 & 0 & 0 \\ 0 & 1 & 2 & -2 \end{bmatrix} \text{ (using } R_2 \to R_2 + 2R_1)$$

$\therefore$ $\rho(A) = 2$ (since one minor of order two $= \begin{vmatrix} 2 & 1 \\ 0 & 1 \end{vmatrix} = 2 \neq 0$). **Ans.**

EXAMPLE 11.19. **(a)** The rank of matrix $\begin{bmatrix} \mu & -1 & 0 \\ 0 & \mu & -1 \\ -1 & 0 & \mu \end{bmatrix}$ is 2, for μ equal to

(i) any row number **(ii)** 3 **(iii)** 1 **(iv)** 2.

(*b*) Find the value of P for which the matrix

$$A = \begin{bmatrix} 3 & P & P \\ P & 3 & P \\ P & P & 3 \end{bmatrix} \text{ is to be of rank 1.}$$

[MTU 2011-12]

SOLUTION: (a) let $A = \begin{bmatrix} \mu & -1 & 0 \\ 0 & \mu & -1 \\ -1 & 0 & \mu \end{bmatrix}$

Since row or column operations on a matrix do not change its rank.

$\therefore$ Operating $R_1 \leftrightarrow R_3$ $\qquad A \sim \begin{bmatrix} -1 & 0 & \mu \\ 0 & \mu & -1 \\ \mu & -1 & 0 \end{bmatrix}$

Then applying $R_3 \to R_3 + \mu R_1$, gives $A \sim \begin{bmatrix} -1 & 0 & \mu \\ 0 & \mu & -1 \\ 0 & -1 & \mu^2 \end{bmatrix}$

and applying $R_3 \to R_3 + \frac{1}{\mu} R_2$, we get $A \sim \begin{bmatrix} -1 & 0 & \mu \\ 0 & \mu & -1 \\ 0 & 0 & \mu^2 - \frac{1}{\mu} \end{bmatrix}$

Since rank of A is 2 we must have $\mu^2 - \frac{1}{\mu} = 0$ or $\mu^3 = 1$ $\therefore$ $\mu = 1$. **Ans (iii)**

(*b*) For $\rho(A)$ to be less than 3.

$$|A| = 0 \Rightarrow \begin{vmatrix} 3 & P & P \\ P & 3 & P \\ P & P & 3 \end{vmatrix} = 0$$

or $\quad 3[9 - P^2] - P[3P - P^2] + P[P^2 - 3P] = 0$

or $\quad 2P^3 - 9P^2 + 27 = 0 \Rightarrow P = 3$

Thus, for $P = 3$, $A = \begin{bmatrix} 3 & 3 & 3 \\ 3 & 3 & 3 \\ 3 & 3 & 3 \end{bmatrix}$

whose rank is obviously 1. **Ans.**

EXAMPLE 11.20. Find the rank of the matrix $\begin{bmatrix} 1 & 1 & 1 & 1 \\ 1 & 3 & -2 & 1 \\ 2 & 0 & -3 & 2 \\ 3 & 3 & 0 & 3 \end{bmatrix}$.

SOLUTION:

$$A = \begin{bmatrix} 1 & 1 & 1 & 1 \\ 1 & 3 & -2 & 1 \\ 2 & 0 & -3 & 2 \\ 3 & 3 & 0 & 3 \end{bmatrix} \sim \begin{bmatrix} 1 & 1 & 1 & 1 \\ 0 & 2 & -3 & 0 \\ 0 & -2 & -5 & 0 \\ 0 & 0 & -3 & 0 \end{bmatrix}$$

Applying $R_2 \to R_2 - R_1$, $R_3 \to R_3 - 2R_1$, $R_4 \to R_4 - 3R_1$

$$A \sim \begin{bmatrix} 1 & 1 & 1 & 1 \\ 0 & 2 & -3 & 0 \\ 0 & 0 & -8 & 0 \\ 0 & 0 & -3 & 0 \end{bmatrix} \quad \text{(on applying } R_3 \to R_3 + R_2\text{)}$$

$$A \sim \begin{bmatrix} 1 & 1 & 1 & 1 \\ 0 & 2 & 3 & 0 \\ 0 & 0 & -8 & 0 \\ 0 & 0 & 0 & 0 \end{bmatrix} \quad \text{(On applying } R_4 \to R_4 - \frac{3}{8}R_3\text{)}$$

$\therefore \quad \rho(A) = 3.$ **Ans.**

EXAMPLE 11.21. Find the rank of the matrix $\begin{bmatrix} 2 & 3 & 4 & -1 \\ 5 & 2 & 0 & -1 \\ -4 & 5 & 12 & -1 \end{bmatrix}$.

SOLUTION: Let

$$A = \begin{bmatrix} 2 & 3 & 4 & -1 \\ 5 & 2 & 0 & -1 \\ -4 & 5 & 12 & -1 \end{bmatrix}$$

Appling elementary row operations $R_2 \to R_2 - R_1$ and $R_3 \to R_3 - R_1$, we get

$$A \sim \begin{bmatrix} 2 & 3 & 4 & -1 \\ 3 & -1 & -4 & 0 \\ -6 & 2 & 8 & 0 \end{bmatrix}$$

Next, applying $R_3 \to \mathrm{R}_3 + 2R_2$, we get

$$A \sim \begin{bmatrix} 2 & 3 & 4 & -1 \\ 3 & -1 & -4 & 0 \\ 0 & 0 & 0 & 0 \end{bmatrix}$$

Therefore, the rank of $A = 2$. **Ans.**

EXAMPLE 11.22. Find the rank of the matrix $\begin{bmatrix} 5 & 6 & 7 & 8 \\ 6 & 7 & 8 & 9 \\ 11 & 12 & 13 & 14 \\ 16 & 17 & 18 & 19 \end{bmatrix}$.

SOLUTION: Let

$$A = \begin{bmatrix} 5 & 6 & 7 & 8 \\ 6 & 7 & 8 & 9 \\ 11 & 12 & 13 & 14 \\ 16 & 17 & 18 & 19 \end{bmatrix}$$

Operating $C_4 \to C_4 - C_3$, $C_3 \to C_3 - C_2$, $C_2 \to C_2 - C_1$, we get

$$A \sim \begin{bmatrix} 5 & 1 & 1 & 1 \\ 6 & 1 & 1 & 1 \\ 11 & 1 & 1 & 1 \\ 16 & 1 & 1 & 1 \end{bmatrix}$$

Next operating $R_4 \to R_4 - R_3, R_3 \to R_3 - R_2$ and $R_2 \to R_1$, we get

$$A \sim \begin{bmatrix} 5 & 1 & 1 & 1 \\ 1 & 0 & 0 & 0 \\ 5 & 0 & 0 & 0 \\ 5 & 0 & 0 & 0 \end{bmatrix}$$

Now operating $C_3 \to C_3 - C_2$, and $C_4 \to C_4 - C_2$, we get

$$A \sim \begin{bmatrix} 5 & 1 & 0 & 0 \\ 1 & 0 & 0 & 0 \\ 5 & 0 & 0 & 0 \\ 5 & 0 & 0 & 0 \end{bmatrix}$$

Finaly subtracting $5R_2$ from R_1, R_3 and R_4 gives

$$A \sim \begin{bmatrix} 0 & 1 & 0 & 0 \\ 1 & 0 & 0 & 0 \\ 0 & 0 & 0 & 0 \\ 0 & 0 & 0 & 0 \end{bmatrix}$$

Clearly rank of $A = 2$. **Ans.**

EXAMPLE 11.23. Use elementary transformations to reduce the following matrix A to triangular form and hence find the rank of A.

$$A = \begin{bmatrix} 2 & 3 & -1 & -1 \\ 1 & -1 & -2 & -4 \\ 3 & 1 & 3 & -2 \\ 6 & 3 & 0 & -7 \end{bmatrix}.$$ **(UPTU 2005-06)**

SOLUTION: $$A \sim \begin{bmatrix} 1 & -1 & -2 & -4 \\ 0 & 5 & 3 & 7 \\ 0 & 4 & 9 & 10 \\ 0 & 9 & 12 & 17 \end{bmatrix}$$

(On applying $R_1 \leftrightarrow R_2$ then $R_2 \to R_2 - 2R_1$, $R_3 \to R_3 - 3R_1$ and $R_4 \to R_4 - 6R_1$)

$$A \sim \begin{bmatrix} 1 & -1 & -2 & -4 \\ 0 & 1 & -6 & -3 \\ 0 & 4 & 9 & 10 \\ 0 & 0 & 0 & 0 \end{bmatrix}$$ (On applying $R_4 \to R_4 - R_2 - R_3$ and $R_2 \to R_2 - R_3$)

$$A \sim \begin{bmatrix} 1 & -1 & -2 & -4 \\ 0 & 1 & -6 & -3 \\ 0 & 0 & 33 & 22 \\ 0 & 0 & 0 & 0 \end{bmatrix}$$ (On applying $R_3 \to R_3 - 4R_2$)

which is in triangular form and hence rank of $A = 3$. **Ans.**

EXAMPLE 11.24. Find the rank of the matrix $\begin{bmatrix} 2 & 3 & -2 & 4 \\ 3 & -2 & 1 & 2 \\ 3 & 2 & 3 & 4 \\ -2 & 4 & 0 & 5 \end{bmatrix}$. **(UPTU 2006-07)**

SOLUTION: $A = \begin{bmatrix} 2 & 3 & -2 & 4 \\ 3 & -2 & 1 & 2 \\ 3 & 2 & 3 & 4 \\ -2 & 4 & 0 & 5 \end{bmatrix} \sim \begin{bmatrix} -1 & 5 & -3 & 2 \\ 3 & -2 & 1 & 2 \\ 0 & 4 & 2 & 2 \\ 0 & 7 & -2 & 9 \end{bmatrix}$

(On applying $R_4 \to R_4 + R_1,\ R_3 \to R_3 - R_2,\ R_1 \to R_1 - R_2$)

$\sim \begin{bmatrix} 1 & -5 & 3 & -2 \\ 0 & 13 & -8 & 8 \\ 0 & 4 & 2 & 2 \\ 0 & 7 & -2 & 9 \end{bmatrix}$ (On applying $R_2 \to R_2 + 3R_1$ and $R_1 \to -R_1$)

$\sim \begin{bmatrix} 1 & -5 & 3 & -2 \\ 0 & 1 & -14 & 2 \\ 0 & 4 & 2 & 2 \\ 0 & 7 & -2 & 9 \end{bmatrix}$ (On applying $R_2 \to R_2 - 3R_3$)

$\sim \begin{bmatrix} 1 & -5 & 3 & -2 \\ 0 & 1 & -14 & 2 \\ 0 & 0 & 58 & -6 \\ 0 & 0 & 96 & -5 \end{bmatrix}$ (On applying $R_3 \to R_3 - 4R_2$ and $R_4 \to R_4 - 7R_1$)

$\sim \begin{bmatrix} 1 & -5 & 3 & -2 \\ 0 & 1 & -14 & 2 \\ 0 & 0 & 1 & -3/29 \\ 0 & 0 & 96 & -5 \end{bmatrix}$ (On applying $R_3 \to \frac{1}{58}R_3$)

$\sim \begin{bmatrix} 1 & -5 & 3 & -2 \\ 0 & 1 & -14 & 2 \\ 0 & 0 & 1 & -3/29 \\ 0 & 0 & 0 & 143/29 \end{bmatrix}$ (On applying $R_4 \to R_4 - 96R_3$)

Hence the rank of A is 4. **Ans.**

NORMAL FORM OF A MATRIX

Every non-zero matrix A of rank r can be reduced by a sequence of elementary transformations, to the form $\begin{bmatrix} I_r & 0 \\ 0 & 0 \end{bmatrix}$ which is called the **normal form** of A.

In other words, we can **perform any number of row or column transformations such that the top left element becomes 1 and the other elements on the leading diagonal should be zero or one. Count the numbers 1 on the leading diagonal and that is the rank of the matrix.**

Because each elementary transformation can be affected by pre-multiplication or post-multiplication with a suitable elementary matrix and each elementary matrix is non-singular, therefore, we have an important result as follows:

Corresponding to every matrix A of rank r there exist non-singular matrices P and Q such that

$$\mathbf{PAQ} = \begin{bmatrix} \mathbf{I}_r & \mathbf{0} \\ \mathbf{0} & \mathbf{0} \end{bmatrix}.$$

Here, if A is a matrix of order $m \times n$ then P and Q are square matrices of order $m \times m$ and $n \times n$ respectively.

EXAMPLE 11.25. Find two non-singular matrices P and Q such that PAQ is in normal form where $A = \begin{bmatrix} 1 & 1 & 2 \\ 1 & 2 & 3 \\ 0 & -1 & -1 \end{bmatrix}$.

SOLUTION: We write

$$I\,A\,I = \begin{bmatrix} 1 & 0 & 0 \\ 0 & 1 & 0 \\ 0 & 0 & 1 \end{bmatrix} \begin{bmatrix} 1 & 1 & 2 \\ 1 & 2 & 3 \\ 0 & -1 & -1 \end{bmatrix} \begin{bmatrix} 1 & 0 & 0 \\ 0 & 1 & 0 \\ 0 & 0 & 1 \end{bmatrix}$$

$$(\text{Applying } R_2 \to R_2 - R_1) = \begin{bmatrix} 1 & 0 & 0 \\ -1 & 1 & 0 \\ 0 & 0 & 1 \end{bmatrix} \begin{bmatrix} 1 & 1 & 2 \\ 0 & 1 & 1 \\ 0 & -1 & -1 \end{bmatrix} \begin{bmatrix} 1 & 0 & 0 \\ 0 & 1 & 0 \\ 0 & 0 & 1 \end{bmatrix}$$

$$(\text{Applying } R_3 \to R_3 + R_2) = \begin{bmatrix} 1 & 0 & 0 \\ -1 & 1 & 0 \\ -1 & 1 & 1 \end{bmatrix} \begin{bmatrix} 1 & 1 & 2 \\ 0 & 1 & 1 \\ 0 & 0 & 0 \end{bmatrix} \begin{bmatrix} 1 & 0 & 0 \\ 0 & 1 & 0 \\ 0 & 0 & 1 \end{bmatrix}$$

$$(\text{Applying } C_3 \to C_3 - C_2) = \begin{bmatrix} 1 & 0 & 0 \\ -1 & 1 & 0 \\ -1 & 1 & 1 \end{bmatrix} \begin{bmatrix} 1 & 1 & 1 \\ 0 & 1 & 0 \\ 0 & 0 & 0 \end{bmatrix} \begin{bmatrix} 1 & 0 & 0 \\ 0 & 1 & -1 \\ 0 & 0 & 1 \end{bmatrix}$$

$$(\text{Applying } C_2 \to C_2 - C_1 \text{ and } C_3 \to C_3 - C_1) = \begin{bmatrix} 1 & 0 & 0 \\ -1 & 1 & 0 \\ -1 & 1 & 1 \end{bmatrix} \begin{bmatrix} 1 & 0 & 0 \\ 0 & 1 & 0 \\ 0 & 0 & 0 \end{bmatrix} \begin{bmatrix} 1 & -1 & -1 \\ 0 & 1 & -1 \\ 0 & 0 & 1 \end{bmatrix}$$

Therefore $P = \begin{bmatrix} 1 & 0 & 0 \\ -1 & 1 & 0 \\ -1 & 1 & 1 \end{bmatrix}$ and $Q = \begin{bmatrix} 1 & -1 & -1 \\ 0 & 1 & -1 \\ 0 & 0 & 1 \end{bmatrix}$

and the normal form is $\begin{bmatrix} 1 & 0 & 0 \\ 0 & 1 & 0 \\ 0 & 0 & 0 \end{bmatrix}$. **Ans.**

EXAMPLE 11.26. For the matrix $A = \begin{bmatrix} 2 & 1 & -3 & -6 \\ 3 & -3 & 1 & 2 \\ 1 & 1 & 1 & 2 \end{bmatrix}$ find its rank and the non-singular matrices P and Q such that PAQ is in normal form. **(UPTU 2003)**

SOLUTION: We have

$$A = \begin{bmatrix} 2 & 1 & -3 & -6 \\ 3 & -3 & 1 & 2 \\ 1 & 1 & 1 & 2 \end{bmatrix} = \begin{bmatrix} 1 & 0 & 0 \\ 0 & 1 & 0 \\ 0 & 0 & 1 \end{bmatrix} A \begin{bmatrix} 1 & 0 & 0 & 0 \\ 0 & 1 & 0 & 0 \\ 0 & 0 & 1 & 0 \\ 0 & 0 & 0 & 1 \end{bmatrix}$$

Applying $R_1 \leftrightarrow R_3$, we get

$$\begin{bmatrix} 1 & 1 & 1 & 2 \\ 3 & -3 & 1 & 2 \\ 2 & 1 & -3 & -6 \end{bmatrix} = \begin{bmatrix} 0 & 0 & 1 \\ 0 & 1 & 0 \\ 1 & 0 & 0 \end{bmatrix} A \begin{bmatrix} 1 & 0 & 0 & 0 \\ 0 & 1 & 0 & 0 \\ 0 & 0 & 1 & 0 \\ 0 & 0 & 0 & 1 \end{bmatrix}$$

Now, performing $R_2 \to R_2 - 3\,R_1$ and $R_3 \to R_3 - 2R_1$, we get

$$\begin{bmatrix} 1 & 1 & 1 & 2 \\ 0 & -6 & -2 & -4 \\ 0 & -1 & -5 & -10 \end{bmatrix} = \begin{bmatrix} 0 & 0 & 1 \\ 0 & 1 & -3 \\ 1 & 0 & -2 \end{bmatrix} A \begin{bmatrix} 1 & 0 & 0 & 0 \\ 0 & 1 & 0 & 0 \\ 0 & 0 & 1 & 0 \\ 0 & 0 & 0 & 1 \end{bmatrix}$$

Next, applying $C_2 \to C_2 - C_1$, $C_3 \to C_3 - C_1$ and $C_4 \to C_4 - 2C_1$, gives

$$\begin{bmatrix} 1 & 0 & 0 & 0 \\ 0 & -6 & -2 & -4 \\ 0 & -1 & -5 & -10 \end{bmatrix} = \begin{bmatrix} 0 & 0 & 1 \\ 0 & 1 & -3 \\ 1 & 0 & -2 \end{bmatrix} A \begin{bmatrix} 1 & -1 & -1 & -2 \\ 0 & 1 & 0 & 0 \\ 0 & 0 & 1 & 0 \\ 0 & 0 & 0 & 1 \end{bmatrix}$$

Again, applying $R_2 \leftrightarrow R_3$ and $R_3 \to R_3 - 6R_2$ and $R_2 \to -R_2$, gives

$$\begin{bmatrix} 1 & 0 & 0 & 0 \\ 0 & 1 & 5 & 10 \\ 0 & 0 & 28 & 56 \end{bmatrix} = \begin{bmatrix} 0 & 0 & 1 \\ -1 & 0 & 2 \\ -6 & 1 & 9 \end{bmatrix} A \begin{bmatrix} 1 & -1 & -1 & -2 \\ 0 & 1 & 0 & 0 \\ 0 & 0 & 1 & 0 \\ 0 & 0 & 0 & 1 \end{bmatrix}$$

Now, operating $C_4 \to C_4 - 2C_3$ and $C_3 \to C_3 - 5C_2$, gives

$$\begin{bmatrix} 1 & 0 & 0 & 0 \\ 0 & 1 & 0 & 0 \\ 0 & 0 & 28 & 0 \end{bmatrix} = \begin{bmatrix} 0 & 0 & 1 \\ -1 & 0 & 2 \\ -6 & 1 & 9 \end{bmatrix} A \begin{bmatrix} 1 & -1 & 4 & 0 \\ 0 & 1 & -5 & 0 \\ 0 & 0 & 1 & -2 \\ 0 & 0 & 0 & 1 \end{bmatrix}$$

Finally, operating $R_4 \to \frac{1}{28} R_4$, we get

$$\begin{bmatrix} 1 & 0 & 0 & 0 \\ 0 & 1 & 0 & 0 \\ 0 & 0 & 1 & 0 \end{bmatrix} = \begin{bmatrix} 0 & 0 & 1 \\ -1 & 0 & 2 \\ -\frac{3}{14} & \frac{1}{28} & \frac{9}{28} \end{bmatrix} A \begin{bmatrix} 1 & -1 & 4 & 0 \\ 0 & 1 & -5 & 0 \\ 0 & 0 & 1 & -2 \\ 0 & 0 & 0 & 1 \end{bmatrix} = PAQ$$

where $P = \begin{bmatrix} 0 & 0 & 1 \\ -1 & 0 & 2 \\ -\frac{3}{14} & \frac{1}{28} & \frac{9}{28} \end{bmatrix}$ and $Q = \begin{bmatrix} 1 & -1 & 4 & 0 \\ 0 & 1 & -5 & 0 \\ 0 & 0 & 1 & -2 \\ 0 & 0 & 0 & 1 \end{bmatrix}$.

Obviously the rank of $A = 3$. **Ans.**

EXAMPLE 11.27. For the matrix $A = \begin{bmatrix} 1 & 2 & 3 & -2 \\ 2 & -2 & 1 & 3 \\ 3 & 0 & 4 & 1 \end{bmatrix}$, find non- singular matrixes P and Q such that PAQ is in normal form.

SOLUTION: $A_{3 \times 4} = I_3 A I_4$, that is,

$$\begin{bmatrix} 1 & 2 & 3 & -2 \\ 2 & -2 & 1 & 3 \\ 3 & 0 & 4 & 1 \end{bmatrix} = \begin{bmatrix} 1 & 0 & 0 \\ 0 & 1 & 0 \\ 0 & 0 & 1 \end{bmatrix} A \begin{bmatrix} 1 & 0 & 0 & 0 \\ 0 & 1 & 0 & 0 \\ 0 & 0 & 1 & 0 \\ 0 & 0 & 0 & 1 \end{bmatrix}$$

When we apply row operation on A, it is to applied on I_3 also and when column operation on A is applied on A it is to be applied on I_4 also. Now $R_3 \to R_3 - 3R_1$ and $R_2 \to R_2 - 2R_1$, gives

$$\begin{bmatrix} 1 & 2 & 3 & -2 \\ 0 & -6 & -5 & 7 \\ 0 & -6 & -5 & 7 \end{bmatrix} = \begin{bmatrix} 1 & 0 & 0 \\ -2 & 1 & 0 \\ -3 & 0 & 1 \end{bmatrix} A \begin{bmatrix} 1 & 0 & 0 & 0 \\ 0 & 1 & 0 & 0 \\ 0 & 0 & 1 & 0 \\ 0 & 0 & 0 & 1 \end{bmatrix}$$

Now applying $R_3 \to R_3 - R_2$ and $R_2 \to - R_2$, we get

$$\begin{bmatrix} 1 & 2 & 3 & -2 \\ 0 & 6 & 5 & -7 \\ 0 & 0 & 0 & 0 \end{bmatrix} = \begin{bmatrix} 1 & 0 & 0 \\ 2 & -1 & 0 \\ -1 & -1 & 1 \end{bmatrix} A \begin{bmatrix} 1 & 0 & 0 & 0 \\ 0 & 1 & 0 & 0 \\ 0 & 0 & 1 & 0 \\ 0 & 0 & 0 & 1 \end{bmatrix}$$

Next, operating $C_3 \to C_3 - 3C_1$ and $C_4 \to C_4 + C_2$, we get

$$\begin{bmatrix} 1 & 2 & 0 & 0 \\ 0 & 6 & 5 & -1 \\ 0 & 0 & 0 & 0 \end{bmatrix} = \begin{bmatrix} 1 & 0 & 0 \\ 2 & -1 & 0 \\ -1 & -1 & 1 \end{bmatrix} A \begin{bmatrix} 1 & 0 & -3 & 0 \\ 0 & 1 & 0 & 1 \\ 0 & 0 & 1 & 0 \\ 0 & 0 & 0 & 1 \end{bmatrix}$$

Now applying $C_2 \to \frac{1}{2}C_2$ and $C_4 \to 5C_4 + C_3$, we get

$$\begin{bmatrix} 1 & 1 & 0 & 0 \\ 0 & 3 & 5 & 0 \\ 0 & 0 & 0 & 0 \end{bmatrix} = \begin{bmatrix} 1 & 0 & 0 \\ 2 & -1 & 0 \\ -1 & -1 & 1 \end{bmatrix} A \begin{bmatrix} 1 & 0 & -3 & -3 \\ 0 & \frac{1}{2} & 0 & 5 \\ 0 & 0 & 1 & 0 \\ 0 & 0 & 0 & 5 \end{bmatrix}$$

Then applying $C_2 \to C_2 - C_1$, we get

$$\begin{bmatrix} 1 & 0 & 0 & 0 \\ 0 & 3 & 5 & 0 \\ 0 & 0 & 0 & 0 \end{bmatrix} = \begin{bmatrix} 1 & 0 & 0 \\ 2 & -1 & 0 \\ -1 & -1 & 1 \end{bmatrix} A \begin{bmatrix} 1 & -1 & -3 & -3 \\ 0 & \frac{1}{2} & 0 & 5 \\ 0 & 0 & 1 & 1 \\ 0 & 0 & 0 & 5 \end{bmatrix}$$

Then oprate $C_3 \to 5C_2 - 3C_3$ to get

$$\begin{bmatrix} 1 & 0 & 0 & 0 \\ 0 & 3 & 0 & 0 \\ 0 & 0 & 0 & 0 \end{bmatrix} = \begin{bmatrix} 1 & 0 & 0 \\ 2 & -1 & 0 \\ -1 & -1 & 1 \end{bmatrix} A \begin{bmatrix} 1 & -1 & 4 & -3 \\ 0 & \frac{1}{2} & \frac{5}{2} & 5 \\ 0 & 0 & -3 & 1 \\ 0 & 0 & 0 & 5 \end{bmatrix}$$

Finally, oprating $R_2 \to \frac{1}{3}R_2$ gives

$$\begin{bmatrix} 1 & 0 & 0 & 0 \\ 0 & 1 & 0 & 0 \\ 0 & 0 & 0 & 0 \end{bmatrix} = \begin{bmatrix} 1 & 0 & 0 \\ \frac{2}{3} & \frac{-1}{3} & 0 \\ -1 & -1 & 1 \end{bmatrix} A \begin{bmatrix} 1 & -1 & 4 & -3 \\ 0 & \frac{1}{2} & \frac{5}{2} & 5 \\ 0 & 0 & -3 & 1 \\ 0 & 0 & 0 & 5 \end{bmatrix} = PAQ. \qquad \textbf{Ans.}$$

EXAMPLE 11.28. Reduce the following matrix into its normal form and hence find its rank:

$$A = \begin{bmatrix} 2 & 3 & -1 & -1 \\ 1 & -1 & -2 & -4 \\ 3 & 1 & 3 & -2 \\ 6 & 3 & 0 & -7 \end{bmatrix}.$$ **(UPTU 2005)**

SOLUTION: Applying $R_1 \leftrightarrow R_2$, we get

$$A \sim \begin{bmatrix} 1 & -1 & -2 & -4 \\ 2 & 3 & -1 & -1 \\ 3 & 1 & 3 & -2 \\ 6 & 3 & 0 & -7 \end{bmatrix} \quad \text{(by } R_2 \to R_2 - 2R_1,\ R_3 \to R_3 - 3R_1,\ R_4 \to R_4 - 6R_1\text{)}$$

$$\sim \begin{bmatrix} 1 & -1 & -2 & -4 \\ 0 & 5 & 3 & 7 \\ 0 & 4 & 9 & 10 \\ 0 & 9 & 12 & 17 \end{bmatrix} \quad \text{(by } C_2 \to C_2 + C_1,\ C_3 \to C_3 + 2C_1,\ C_4 \to C_4 + 4C_1\text{)}$$

$$\sim \begin{bmatrix} 1 & 0 & 0 & 0 \\ 0 & 5 & 3 & 7 \\ 0 & 4 & 9 & 10 \\ 0 & 9 & 12 & 17 \end{bmatrix} \quad \text{(by } R_4 \to R_4 - R_2 - R_3\text{)}$$

$$\sim \begin{bmatrix} 1 & 0 & 0 & 0 \\ 0 & 5 & 3 & 7 \\ 0 & 4 & 9 & 10 \\ 0 & 0 & 0 & 0 \end{bmatrix} \quad \text{(by } R_2 \to R_2 - R_3\text{)}$$

$$\sim \begin{bmatrix} 1 & 0 & 0 & 0 \\ 0 & 1 & -6 & -3 \\ 0 & 4 & 9 & 10 \\ 0 & 0 & 0 & 0 \end{bmatrix} \quad \text{(by } R_3 \to R_3 - 4R_2\text{)}$$

$$\sim \begin{bmatrix} 1 & 0 & 0 & 0 \\ 0 & 1 & -6 & -3 \\ 0 & 0 & 33 & 22 \\ 0 & 0 & 0 & 0 \end{bmatrix} \quad \text{(by } C_3 \to C_3 + 6C_2, C_4 \to C_4 + 3C_2\text{)}$$

$$\sim \begin{bmatrix} 1 & 0 & 0 & 0 \\ 0 & 1 & 0 & 0 \\ 0 & 0 & 33 & 22 \\ 0 & 0 & 0 & 0 \end{bmatrix} \quad \text{(by } C_3 \to \frac{1}{33} C_1 \text{ and } C_4 \to C_4 - 22\,C_3\text{)}$$

$$\sim \begin{bmatrix} 1 & 0 & 0 & 0 \\ 0 & 1 & 0 & 0 \\ 0 & 0 & 1 & 0 \\ 0 & 0 & 0 & 0 \end{bmatrix}$$

$$\sim \begin{bmatrix} I_3 & 0 \\ 0 & 0 \end{bmatrix}$$

Hence the rank $(A) = 3$. **Ans.**

EXAMPLE 11.29. Find two non-singular matrix P and Q such the PAQ is in the normal form

where $A = \begin{bmatrix} 1 & -1 & -1 \\ 1 & 1 & 1 \\ 3 & 1 & 1 \end{bmatrix}$. **(UPTU 2001)**

SOLUTION: We write $A = I\,AI$, that is

$$\begin{bmatrix} 1 & -1 & -1 \\ 1 & 1 & 1 \\ 3 & 1 & 1 \end{bmatrix} = \begin{bmatrix} 1 & 0 & 0 \\ 0 & 1 & 0 \\ 0 & 0 & 1 \end{bmatrix} A \begin{bmatrix} 1 & 0 & 0 \\ 0 & 1 & 0 \\ 0 & 0 & 1 \end{bmatrix}$$

Applying $C_2 \to C_2 + C_1$ and $C_3 \to C_3 + C_1$, we get

$$\begin{bmatrix} 1 & 0 & 0 \\ 1 & 2 & 2 \\ 3 & 4 & 4 \end{bmatrix} = \begin{bmatrix} 1 & 0 & 0 \\ 0 & 1 & 0 \\ 0 & 0 & 1 \end{bmatrix} A \begin{bmatrix} 1 & 1 & 1 \\ 0 & 1 & 0 \\ 0 & 0 & 1 \end{bmatrix}$$

Now appling $R_2 \to R_2 - R_1$ and $R_3 \to R_3 - 3R_1$, we get

$$\begin{bmatrix} 1 & 0 & 0 \\ 0 & 2 & 2 \\ 0 & 4 & 4 \end{bmatrix} = \begin{bmatrix} 1 & 0 & 0 \\ -1 & 1 & 0 \\ -3 & 0 & 1 \end{bmatrix} A \begin{bmatrix} 1 & 1 & 1 \\ 0 & 1 & 0 \\ 0 & 0 & 1 \end{bmatrix}$$

Next applying $C_3 \to C_3 - C_2$ and $C_2 \to \frac{1}{2}C_2$, we get

$$\begin{bmatrix} 1 & 0 & 0 \\ 0 & 1 & 0 \\ 0 & 2 & 0 \end{bmatrix} = \begin{bmatrix} 1 & 0 & 0 \\ -1 & 1 & 0 \\ -3 & 0 & 1 \end{bmatrix} A \begin{bmatrix} 1 & 1/2 & 0 \\ 0 & 1/2 & -1 \\ 0 & 0 & 1 \end{bmatrix}$$

Finally applying $R_3 \to R_3 - 2R_2$, we get

$$\begin{bmatrix} 1 & 0 & 0 \\ 0 & 1 & 0 \\ 0 & 0 & 0 \end{bmatrix} = \begin{bmatrix} 1 & 0 & 0 \\ -1 & 1 & 0 \\ -1 & -2 & 1 \end{bmatrix} A \begin{bmatrix} 1 & 1/2 & 0 \\ 0 & 1/2 & -1 \\ 0 & 0 & 1 \end{bmatrix} = PAQ$$

$$\begin{bmatrix} I_2 & 0 \\ 0 & 0 \end{bmatrix} = PAQ, \quad \text{where } P = \begin{bmatrix} 1 & 0 & 0 \\ -1 & 1 & 0 \\ -1 & -2 & 0 \end{bmatrix} \text{ and } Q = \begin{bmatrix} 1 & 1/2 & 0 \\ 0 & 1/2 & -1 \\ 0 & 0 & 1 \end{bmatrix}. \quad \textbf{Ans.}$$

EXAMPLE 11.30. Find the rank of the matrix $\begin{bmatrix} 3 & -1 & 2 \\ -6 & 2 & 4 \\ -3 & 1 & 2 \end{bmatrix}$ by reducing it to its normal form.

SOLUTION: Let $A = \begin{bmatrix} 3 & -1 & 2 \\ -6 & 2 & 4 \\ -3 & 1 & 2 \end{bmatrix}$

Apply the elementary row operations $C_1 \leftrightarrow C_2$ and $C_2 \to C_2 + 3C_1$ and $C_1 \to -C_1$, we get

$$A \sim \begin{bmatrix} 1 & 0 & 2 \\ -2 & 0 & 4 \\ -1 & 0 & 2 \end{bmatrix}$$

Next apply $R_2 \to R_2 + 2R_1$ and $R_3 \to R_3 + R_1$ and get

$$A \sim \begin{bmatrix} 1 & 0 & 2 \\ 0 & 0 & 8 \\ 0 & 0 & 4 \end{bmatrix}$$

Then apply $C_2 \leftrightarrow C_3$, $R_1 \to R_1 - \frac{1}{4}R_2$, $R_3 \to R_3 - \frac{1}{2}R_2$ and $R_2 \to \frac{1}{8}R_2$ and get

$$A \sim \begin{bmatrix} 1 & 0 & 0 \\ 0 & 1 & 0 \\ 0 & 0 & 0 \end{bmatrix} = \begin{bmatrix} I_2 & 0 \\ 0 & 0 \end{bmatrix}.$$ Hence rank of A is 2. **Ans.**

EXAMPLE 11.31. Reduce the matrix A to its normal form when $A = \begin{bmatrix} 1 & 2 & -1 & 4 \\ 2 & 4 & 3 & 4 \\ 1 & 2 & 3 & 4 \\ -1 & -2 & 6 & -7 \end{bmatrix}$

hence find the rank of A. **(UPTU 2004-05)**

SOLUTION: $A \sim \begin{bmatrix} 1 & 2 & -1 & 4 \\ 0 & 0 & 5 & -4 \\ 0 & 0 & 4 & 0 \\ 0 & 0 & 5 & -3 \end{bmatrix}$ (On applying $R_2 \to R_2 - 2R_1$, $R_3 \to R_3 - R_1$, $R_4 \to R_4 + R_1$)

$\sim \begin{bmatrix} 1 & 4 & -1 & 2 \\ 0 & -4 & 5 & 0 \\ 0 & 0 & 4 & 0 \\ 0 & -3 & 5 & 0 \end{bmatrix}$ (On applying $C_2 \leftrightarrow C_4$)

$\sim \begin{bmatrix} 1 & 0 & 0 & 0 \\ 0 & -4 & 5 & 0 \\ 0 & 0 & 4 & 0 \\ 0 & -3 & 5 & 0 \end{bmatrix}$ (On applying $C_2 \to C_2 - 4C_1$, $C_3 \to C_3 + C_1$, $C_4 \to C_4 - 2C_1$)

$$\sim \begin{bmatrix} 1 & 0 & 0 & 0 \\ 0 & 1 & 5 & 0 \\ 0 & 4 & 4 & 0 \\ 0 & 2 & 5 & 0 \end{bmatrix} \quad \text{(On applying } C_2 \to C_2 + C_3\text{)}$$

$$\sim \begin{bmatrix} 1 & 0 & 0 & 0 \\ 0 & 1 & 5 & 0 \\ 0 & 0 & -16 & 0 \\ 0 & 0 & -5 & 0 \end{bmatrix} \quad \text{(On applying } R_3 \to R_3 - 4R_2 \text{ and } R_4 \to R_4 - 2R_2\text{)}$$

$$\sim \begin{bmatrix} 1 & 0 & 0 & 0 \\ 0 & 1 & 0 & 0 \\ 0 & 0 & 1 & 0 \\ 0 & 0 & 0 & 0 \end{bmatrix}$$

(On applying $R_2 \to R_2 + R_4$ and $R_3 \to \frac{-1}{16}R_3$ and then $R_4 \to R_4 - 5R_3$)

$$= \begin{bmatrix} I_3 & 0 \\ 0 & 0 \end{bmatrix}$$ which is the required normal form and hence rank of A is 3. **Ans.**

EXAMPLE 11.32. Find the rank of the following matrix using the elementary transformations

$$\begin{bmatrix} 1 & -3 & 1 & 2 \\ 0 & 1 & 2 & 3 \\ 3 & 4 & 1 & -2 \end{bmatrix}.$$ **(UPTU 2009-10)**

SOLUTION: $$A = \begin{bmatrix} 1 & -3 & 1 & 2 \\ 0 & 1 & 2 & 3 \\ 3 & 4 & 1 & -2 \end{bmatrix} \quad \text{Applying } R_3 \to R_3 - 3R_1$$

$$\sim \begin{bmatrix} 1 & -3 & 1 & 2 \\ 0 & 1 & 2 & 3 \\ 0 & 13 & -2 & -8 \end{bmatrix} \quad \text{now operating } R_3 \to R_3 - 13R_2$$

$$\sim \begin{bmatrix} 1 & -3 & 1 & 2 \\ 0 & 1 & 2 & 3 \\ 0 & 0 & -28 & -47 \end{bmatrix} \quad \text{next applying } R_1 \to R_1 + 3R_2$$

$$\sim \begin{bmatrix} 1 & 0 & 7 & 11 \\ 0 & 1 & 2 & 3 \\ 0 & 0 & -28 & -47 \end{bmatrix}$$

Its rank = 3 as $\begin{vmatrix} 0 & 7 & 11 \\ 1 & 2 & 3 \\ 0 & -28 & -47 \end{vmatrix} \neq 0$. Thus, $\rho(A) = 3$. **Ans.**

EXAMPLE 11.33. Find all the values of μ for which the rank of the matrix

$$A = \begin{bmatrix} \mu & -1 & 0 & 0 \\ 0 & \mu & -1 & 0 \\ 0 & 0 & \mu & -1 \\ -6 & 11 & -6 & 1 \end{bmatrix} \text{ is equal to 3.}$$ **(UPTU 2008-09)**

SOLUTION: $A = \begin{bmatrix} \mu & -1 & 0 & 0 \\ 0 & \mu & -1 & 0 \\ 0 & 0 & \mu & -1 \\ -6 & 11 & -6 & 1 \end{bmatrix}$. Applying $R_3 \to R_3 + R_4$

$$\sim \begin{bmatrix} \mu & -1 & 0 & 0 \\ 0 & \mu & -1 & 0 \\ -6 & 11 & \mu-6 & 0 \\ -6 & 11 & -6 & 1 \end{bmatrix}$$

$\rho(A) = 4$ if $\mu \neq 6$ since $\begin{vmatrix} \mu & -1 & 0 \\ 0 & \mu & -1 \\ -6 & 11 & \mu-6 \end{vmatrix} \neq 0$ when $\mu \neq 6$.

and if $\mu = 6$ we have

$$A \sim \begin{bmatrix} 6 & -1 & 0 & 0 \\ 0 & 6 & -1 & 0 \\ -6 & 11 & 0 & 0 \\ -6 & 11 & -6 & 1 \end{bmatrix}$$

and then $\rho(A) = 3$ since $\begin{vmatrix} 6 & -1 & 0 \\ 0 & 6 & -1 \\ -6 & 11 & 0 \end{vmatrix} \neq 0.$ **Ans.**

EXERCISE 11B

1. Find the inverse of the matrix $A = \begin{bmatrix} 2 & 4 & 3 & 2 \\ 3 & 6 & 5 & 2 \\ 2 & 5 & 2 & -3 \\ 4 & 5 & 14 & 14 \end{bmatrix}$.

2. Reduce the matrix A to diagonal form $A = \begin{bmatrix} -1 & 2 & -2 \\ 1 & 2 & 1 \\ -1 & -1 & 0 \end{bmatrix}$. **(UPTU 2003)**

3. Find the rank of the following matrix by reducing it to normal form

$$A = \begin{bmatrix} 1 & 2 & -1 & 3 \\ 4 & 1 & 2 & 1 \\ 3 & -1 & 1 & 2 \\ 1 & 2 & 0 & 1 \end{bmatrix}.$$ **(UPTU 2001) (Sp. Ex. 2001)**

4. If $A = \begin{bmatrix} 3 & -3 & 4 \\ 2 & -3 & 4 \\ 0 & -1 & 1 \end{bmatrix}$. Find two non-singular matrices P and Q such that $PAQ = I$ and hence find A^{-1}.

5. Find the rank of the matrix $\begin{bmatrix} 1 & 2 & 3 & 2 \\ 2 & 3 & 5 & 1 \\ 1 & 3 & 4 & 5 \end{bmatrix}$. **(UPTU 2000, 2001, 2003, 2004)**

6. Find the rank of the matrix $\begin{bmatrix} 2 & -4 & 3 & 1 & 0 \\ 1 & -2 & 1 & -4 & 2 \\ 0 & 1 & -1 & 3 & 1 \\ 4 & -7 & 4 & -4 & 5 \end{bmatrix}$ by reducing it to normal form.

7. For the matrix $A = \begin{bmatrix} 1 & -1 & 2 & -3 \\ 4 & 1 & 0 & 2 \\ 0 & 3 & 1 & 4 \\ 0 & 1 & 0 & 2 \end{bmatrix}$ find non-singular matrices P and Q such that PAQ is in the normal form. Hence find the rank of A.

8. Find the value of b such that the rank of the matrix $\begin{bmatrix} 1 & 5 & 4 \\ 0 & 3 & 2 \\ b & 13 & 10 \end{bmatrix}$ is 2. **(UPTU 2008-09)**

9. Reduce the following matrix to normal form and hence find its rank $\begin{bmatrix} 5 & 3 & 14 & 4 \\ 0 & 1 & 2 & 1 \\ 1 & -1 & 2 & - \end{bmatrix}$.

(UPTU 2010-11)

CHAPTER

12 Linear Dependence of Vectors and Consistency of Equations

Linear Dependence and Independence of Vectors. Consistence of Linear System of Simultaneous Equations and their Solution.

LINEAR DEPENDENCE AND INDEPENDENCE OF VECTORS

We are familiar with the vectors like $2i - 3j + 5k$ having components 2, –3, 5 along the co-ordinate axes. In general, any quantity having n components is *called* **a vector of order n**. As such the co-efficients of unknowns in a linear equation form a vector. Elements of a row matrix or a column matrix also form vectors.

The set of vectors $X_1, X_2, \ldots X_n$ is said to be **linearly dependent** if there exist scalars $c_1, c_2, c_3, \ldots, c_n$, not all zero, such that

$$c_1X_1 + c_2X_2 + \ldots + c_nX_n = 0 \quad \ldots(1)$$

and, if no such scalars other than zero exist, the vectors $X_1, X_2, \ldots, X_n$ are **called linearly independent**.

In the above definition, out of $c_1, c_2, \ldots, c_n$ if at least one of them is non-zero, say $c_1 \neq 0$ then the relation (1) can be written as

$$X_1 = -\frac{c_2}{c_1}X_2 - \frac{c_3}{c_1}X_3 \ldots - \frac{c_n}{c_1}X_n$$

and we say that vector X_1 is a linear combination of the vectors $X_2, X_3, \ldots, X_n$. In other words, if atleast one vector can be expressed as a linear combination of the other vectors then the vectors $X_1, X_2, .. X_n$ are said to be **linearly dependent**; and if not a single vector can be expressed as a linear combination of the others then the vectors are *called* **linearly independent**.

EXAMPLE 12.1. Investigate if the following set of vectors are linearly independent or dependent:

$$X_1 = [1, 2, 3]', \ X_2 = [3, -2, 1]', \ X_3 = [1, -6, -5]'.$$

SOLUTION: Let us assume that the given vectors are linearly dependent then there must exist scalars c_1, c_2, c_3 such that $c_1X_1 + c_2X_2 + c_3X_3 = 0 \quad \ldots (1)$

or $$c_1\begin{bmatrix} 1 \\ 2 \\ 3 \end{bmatrix} + c_2\begin{bmatrix} 3 \\ -2 \\ 1 \end{bmatrix} + c_3\begin{bmatrix} 1 \\ -6 \\ -5 \end{bmatrix} = \begin{bmatrix} 0 \\ 0 \\ 0 \end{bmatrix}$$

$$\Rightarrow \quad \begin{aligned} c_1+3c_2+c_3 &= 0 \\ 2c_1-2c_2-6c_3 &= 0 \\ 3c_1+c_2-5c_3 &= 0 \end{aligned}$$

which are homogeneous equations having trivial solution $c_1 = c_2 = c_3 = 0$. However, for non-trivial solution let us find the rank of the co-efficient matrix.

$$\begin{bmatrix} 1 & 3 & 1 \\ 2 & -2 & -6 \\ 3 & 1 & -5 \end{bmatrix} \sim \begin{bmatrix} 1 & 3 & 1 \\ 0 & -8 & -8 \\ 0 & -8 & -8 \end{bmatrix} \sim \begin{bmatrix} 1 & 3 & 1 \\ 0 & -8 & -8 \\ 0 & 0 & 0 \end{bmatrix}$$

The rank of this matrix is 2 which is less than the number of unknowns. Therefore, non-trivial solution exists. Thus, we have two independent equations as $c_1 + 3c_2 + c_3 = 0$ and $c_2 + c_3 = 0$. Third one can be derived from these two.

Since, the number of arbitrary solutions is equal to the number of unknowns minus the rank of the co-efficient matrix, we take $c_3 = \lambda$ then $c_2 = -\lambda$ and hence $c_1 = 2\lambda$.

Putting these values of c_1, c_2, c_3 in (1), we have

$$2\lambda X_1 - \lambda X_2 + \lambda X_3 = 0 \quad \text{or} \quad 2X_1 - X_2 + X_3 = 0.$$

Therefore, the vectors are dependent. **Ans.**

EXAMPLE 12.2. Are the vectors $X_1 = (1, 3, 4, 2)$, $X_2 = (3, -5, 2, 2)$, $X_3 = (2, -1, 3, 2)$ linearly dependent? If so, express one of these as a linear combination of others.

SOLUTION: If X_1, X_2, X_3 are linearly dependent then $\lambda_1 X_1 + \lambda_2 X_2 + \lambda_3 X_3 = 0$ where not all $\lambda_1, \lambda_2, \lambda_3$ are zero.

Hence we have $\lambda_1 + 3\lambda_2 + 2\lambda_3 = 0,\quad 3\lambda_1 - 5\lambda_2 - \lambda_3 = 0,$

$4\lambda_1 + 2\lambda_2 + 3\lambda_3 = 0,\quad 2\lambda_1 + 2\lambda_2 + 2\lambda_3 = 0$ or $\lambda_1 + \lambda_2 + \lambda_3 = 0$

Solving two equations, say first and third of the above four, we get

$\lambda_1 = 1,\ \lambda_2 = 1,\ \lambda_3 = -2$ which do satisfy the other two equations:

Hence X_1, X_2, X_3 are linearly dependent and the relation between them is $X_1 + X_2 - 2X_3 = 0$. **Ans.**

EXAMPLE 12.3. Examine the following system of vectors for linear dependence. If dependent, find the relation between them. $X_1 = (1, 1, -1, 1)$, $X_2 = (1, -1, 2, -1)$, $X_3 = (3, 1, 0, 1)$.

SOLUTION: X_1, X_2, X_3 will be dependent if there exist $\lambda_1, \lambda_2, \lambda_3$, not all zero, such that

$$\lambda_1 X_1 + \lambda_2 X_2 + \lambda_3 X_3 = 0.$$

$$\Rightarrow \quad \lambda_1 + \lambda_2 + 3\lambda_3 = 0, \quad \lambda_1 - \lambda_2 + \lambda_3 = 0,$$

$$-\lambda_1 + 2\lambda_2 + 0\lambda_3 = 0, \quad \text{and} \quad \lambda_1 - \lambda_2 + \lambda_3 = 0.$$

$$\Rightarrow \quad \lambda_1 = 2\lambda_2,\ \lambda_2 = -\lambda_3 \quad \text{or} \quad \frac{\lambda_1}{2} = \frac{\lambda_2}{1} = \frac{\lambda_3}{-1}$$ which are not all zero.

Hence X_1, X_2, X_3 are dependent and the relation between them is $2X_1 + X_2 - X_3 = 0$. **Ans.**

EXAMPLE 12.4. Find the value of λ for which the vectors $(1, -2, \lambda)$, $(2, -1, 5)$ and $(3, -5, 7\lambda)$ are linearly dependent. **(UPTU 2006-07)**

SOLUTION: Vectors are $x_1 = (1, -2, \lambda)$, $x_2 = (2, -1, 5)$ and $x_3 = (3, -5, 7\lambda)$.

Consider $\lambda_1 x_1 + \lambda_2 x_2 + \lambda_3 x_3 = 0$ which gives the following equations

$$\lambda_1 + 2\lambda_2 + 3\lambda_3 = 0, \quad -2\lambda_1 - \lambda_2 - 5\lambda_3 = 0 \quad \text{and} \quad \lambda\lambda_1 + 5\lambda_2 + 7\lambda_3 = 0.$$

For x_1, x_2, x_3 to be linearly dependent the above system must have a non-zero solution.

Hence $$\begin{vmatrix} 1 & 2 & 3 \\ -2 & -1 & -5 \\ \lambda & 5 & 7\lambda \end{vmatrix} = 0$$

which on solving gives $\lambda = 5/14$ which is the required value of λ. **Ans.**

EXAMPLE 12.5. Show that the vectors $x_1 = (1, 2, 4)$, $x_2 = (2, -1, 3)$, $x_3 = (0, 1, 2)$ and $x_4 = (-3, 7, 2)$ are linearly dependent and find the relation between them. **(UPTU 2003)**

SOLUTION: Let $a_1x_1 + a_2x_2 + a_3x_3 + a_4x_4 = 0$ where a_1, a_2, a_3, a_4 are scalars, not all zero

that is, $a_1(1, 2, 4) + a_2(2, -1, 3) + a_3(0, 1, 2) + a_4(-3, 7, 2) = 0$.

$$\Rightarrow \quad a_1 + 2a_2 - 3a_4 = 0 \qquad ...(1)$$

$$2a_1 - a_2 + a_3 + 7a_4 = 0 \qquad ...(2)$$

$$4a_1 + 3a_2 + 2a_3 + 2a_4 = 0 \qquad ...(3)$$

Subtracting 2 times (2) from (3), gives $5a_2 - 12a_4 = 0$ ...(4)

From (1) and (4) we get $\dfrac{a_1}{-9} = \dfrac{a_2}{12} = \dfrac{a_4}{5} = \lambda$, say.

Putting these values of a_1, a_2, a_4, in (3) gives $a_3 = -5\lambda$

It implies that $a_1 = -9$, $a_2 = 12$, $a_3 = -5$ and $a_4 = 5$ are the scalars, not all zero. Hence vectors are linearly dependent. The relation between the vectors is $9x_1 - 12x_2 + 5x_3 - 5x_4 = 0$. **Ans.**

CONSISTENCY OF SIMULTANEOUS EQUATIONS

Consider the following system of m linear simultaneous equations in n unknowns:

$$a_{11}x_1 + a_{12}x_2 + ... + a_{1n}x_n = b_1$$
$$a_{21}x_1 + a_{22}x_2 + ... + a_{2n}x_n = b_2$$
$$\cdots\cdots\cdots\cdots\cdots\cdots$$
$$\cdots\cdots\cdots\cdots\cdots\cdots$$
$$a_{m1}x_1 + a_{m2}x_2 + ... + a_{mn}x_n = b_m$$

Solving this system is finding all possible solutions. Of course, the above system can be written in matrix form as $AX = B$ ...(1)

where the coefficient matrix A, the column matrix X and the column matrix B, are given by

$$A = \begin{bmatrix} a_{11} & a_{12} & \cdots & \cdots & a_{1n} \\ a_{21} & a_{22} & \cdots & \cdots & a_{2n} \\ \cdots & \cdots & \cdots & \cdots & \cdots \\ \cdots & \cdots & \cdots & \cdots & \cdots \\ a_{m1} & a_{m2} & \cdots & \cdots & a_{mn} \end{bmatrix}, \quad X = \begin{bmatrix} x_1 \\ x_2 \\ \vdots \\ x_n \end{bmatrix}, \quad B = \begin{bmatrix} b_1 \\ b_2 \\ \vdots \\ b_m \end{bmatrix}$$

If B = 0, *i.e,.* if $b_1 = b_2 = ... = b_m = 0$ then (1) is called the **system of homogeneous equations.**

When (1) has a solution it is said to be ***consistent,*** otherwise the system is *called* **inconsistent.** A consistent system of equations may have only one solution called ***unique solution,*** or **infinitely many solutions.**

$$\text{Consider the matrix} = \begin{bmatrix} a_{11} & a_{12} & \cdots & \cdots & a_{1n} & b_1 \\ a_{21} & a_{22} & \cdots & \cdots & a_{2n} & b_2 \\ \cdots & \cdots & \cdots & \cdots & \cdots & \cdots \\ \cdots & \cdots & \cdots & \cdots & \cdots & \cdots \\ a_{m1} & a_{m2} & \cdots & \cdots & a_{mn} & b_m \end{bmatrix} \quad \text{where B is appended to A.}$$

It is called ***augmented matrix*** written as [A :B].

We find here the ranks of the co-efficinent matrix A and of the augmented matrix [A:B]. The golden rule for the consistency of the given system of equations, is as follows:

> (*i*) rank (A) ≠ rank (A : B) the system is **inconsistent** hence no solution.
>
> (*ii*) rank (A) = rank (A : B) the system is **consistent** and following two cases arise.
>
> (*a*) if rank (A) = rank (A : B) = n the number of unknowns, we have a **unique solution.**
>
> (*b*) if rank (A) = rank (A : B) < n there are **infinitely many solutions.**

In the case of linear homogeneous system of simultaneous equations AX = 0 we find the rank of the co-efficient matrix A as r, say. Now, if $r = n$ the above homogeneous system has trivial solution $x_1 = x_2 = ... = x_n = 0$. But if $r < n$, the above system gives non-trivial solutions. Actually, in this case there would exist r solutions where r unknowns are expressed as a linear combination of the remaining $(n - r)$ unknowns.

EXAMPLE 12.6. **(a)** Discuss the consistency of the following system of equations

$$\begin{aligned} x+y+2z+w &= 5 \\ 2x+3y-z-2w &= 2 \\ 4x+5y+3z &= 7. \end{aligned}$$

(b) Show that the system of equations:

$$\begin{aligned} 3x + 4y + 5z &= A \\ 4x + 5y + 6z &= B \\ 5x + 6y + 7z &= C \end{aligned}$$

are consistent only when A, B, C are in A.P. **[UPTU 2011-12]**

SOLUTION: **(a)** Matrix equation is AX = B where

$$A = \begin{bmatrix} 1 & 1 & 2 & 1 \\ 2 & 3 & -1 & -2 \\ 4 & 5 & 3 & 0 \end{bmatrix}, \quad X = \begin{bmatrix} x \\ y \\ z \\ w \end{bmatrix}, \quad B = \begin{bmatrix} 5 \\ 2 \\ 7 \end{bmatrix}.$$

The augmented matrix $= [A : B] = \begin{bmatrix} 1 & 1 & 2 & 1 & \vdots & 5 \\ 2 & 3 & -1 & -2 & \vdots & 2 \\ 4 & 5 & 3 & 0 & \vdots & 7 \end{bmatrix}$.

Operating $R_3 \to R_3 - 2R_2$ and $R_2 \to R_2 - 2R_1$, gives

$$[A : B] \sim \begin{bmatrix} 1 & 1 & 2 & 1 & \vdots & 5 \\ 0 & 1 & -5 & -4 & \vdots & -8 \\ 0 & -1 & 5 & 4 & \vdots & 3 \end{bmatrix}$$

Now applying $R_3 \to R_3 + R_2$, gives

$$[A : B] \sim \begin{bmatrix} 1 & 1 & 2 & 1 & \vdots & 5 \\ 0 & 1 & -5 & -4 & \vdots & -8 \\ 0 & 0 & 0 & 0 & \vdots & -5 \end{bmatrix}$$

Rank $[A : B] = 3$ since $\begin{vmatrix} 2 & 1 & 5 \\ -5 & -4 & -8 \\ 0 & 0 & -5 \end{vmatrix} \neq 0$

and Rank $(A) =$ rank of $\begin{bmatrix} 1 & 1 & 2 & 1 \\ 0 & 1 & -5 & -4 \\ 0 & 0 & 0 & 0 \end{bmatrix} = 2.$

Since $\rho\,[A : B] \neq \rho\,(A)$ the given system of equations is inconsistent. **Ans.**

(b) The given system of equations of can be written as

$$\begin{bmatrix} 3 & 4 & 5 \\ 4 & 5 & 6 \\ 5 & 6 & 7 \end{bmatrix} \begin{bmatrix} x \\ y \\ z \end{bmatrix} = \begin{bmatrix} A \\ B \\ C \end{bmatrix} \quad \text{or} \quad MX = N.$$

where $M = \begin{bmatrix} 3 & 4 & 5 \\ 4 & 5 & 6 \\ 5 & 6 & 7 \end{bmatrix}$, $X = \begin{bmatrix} x \\ y \\ z \end{bmatrix}$ and $N = \begin{bmatrix} A \\ B \\ C \end{bmatrix}$.

Now, the augmented matrix $Z = [M : N]$, that is,

$$Z = \begin{bmatrix} 3 & 4 & 5 & A \\ 4 & 5 & 6 & B \\ 5 & 6 & 7 & C \end{bmatrix}$$

Applying $R_3 \to R_3 - R_2$ and $R_2 \to R_2 - R_1$, we get

$$Z \sim \begin{bmatrix} 3 & 4 & 5 & A \\ 1 & 1 & 1 & B - A \\ 1 & 1 & 1 & C - B \end{bmatrix}$$

Again applying $R_3 \to R_3 - R_2$, gives

$$Z \sim \begin{bmatrix} 3 & 4 & 5 & A \\ 1 & 1 & 1 & B-A \\ 0 & 0 & 0 & C-B-B+A \end{bmatrix}$$

Here $P(M) = 2$ and for $\rho(M) = \rho(Z)$, we have

$C - 2B + A = 0$ hence A, B, C are in A.P. **Hence Proved.**

EXAMPLE 12.7. If possible, solve the following system of equations

$$2x+3y+4z=11$$

$$x+5y+7z=15$$

$$3x+11y+13z=25.$$

SOLUTION: The given system can be written in matrix equation form, as $AX = B$

where $A = \begin{bmatrix} 2 & 3 & 4 \\ 1 & 5 & 7 \\ 3 & 11 & 13 \end{bmatrix}$, $X = \begin{bmatrix} x \\ y \\ z \end{bmatrix}$, $B = \begin{bmatrix} 11 \\ 15 \\ 25 \end{bmatrix}$.

$$[A:B] = \begin{bmatrix} 2 & 3 & 4 & \vdots & 11 \\ 1 & 5 & 7 & \vdots & 15 \\ 3 & 11 & 13 & \vdots & 25 \end{bmatrix} \sim \begin{bmatrix} 1 & 5 & 7 & \vdots & 15 \\ 0 & -7 & -10 & \vdots & -19 \\ 0 & -4 & -8 & \vdots & -20 \end{bmatrix}$$

(on applying $R_1 \leftrightarrow R_2, R_2 \to R_2 - 2R_1$ and $R_3 \to R_3 - 3R_1$)

$$\sim \begin{bmatrix} 1 & 5 & 7 & \vdots & 15 \\ 0 & 1 & 2 & \vdots & 5 \\ 0 & -7 & -10 & \vdots & -19 \end{bmatrix} \text{ (on applying } R_3 \to -\frac{1}{4}R_3 \text{ and } R_2 \leftrightarrow R_3)$$

$$\sim \begin{bmatrix} 1 & 5 & 7 & \vdots & 15 \\ 0 & 1 & 2 & \vdots & 5 \\ 0 & 0 & 4 & \vdots & 16 \end{bmatrix} \text{ (on operating } R_3 \to R_3 + 7R_2)$$

$$\sim \begin{bmatrix} 1 & 5 & 7 & \vdots & 15 \\ 0 & 1 & 0 & \vdots & -3 \\ 0 & 0 & 1 & \vdots & 4 \end{bmatrix} \text{ (on applying } R_3 \to \frac{1}{4}R_3 \text{ and } R_2 \to R_2 - 2R_3)$$

$$\sim \begin{bmatrix} 1 & 0 & 0 & \vdots & 2 \\ 0 & 1 & 0 & \vdots & -3 \\ 0 & 0 & 1 & \vdots & 4 \end{bmatrix} \text{ (on applying } R_1 \to R_1 - 5R_2 - 7R_3)$$

$\therefore$ $\rho\,[A : B] = 3$ and $\rho\,(A) = 3$

Since $\rho\,(A : B) = \rho\,(A) = 3$ = number of unknowns, hence unique solution as $x = 2$, $y = -3$, $z = 4$.

Ans.

EXAMPLE 12.8. Discuss the consistency of the following system of equations for various values of λ:

$$2x_1 - 3x_2 + 6x_3 - 5x_4 = 3$$
$$x_2 - 4x_3 + x_4 = 1$$
$$4x_1 - 5x_2 + 8x_3 - 9x_4 = \lambda.$$

And, if consistent, solve it.

SOLUTION: The matrix equation is $AX = B$. The augmented matrix is

$$[A:B] = \begin{bmatrix} 2 & -3 & 6 & -5 & \vdots & 3 \\ 0 & 1 & -4 & 1 & \vdots & 1 \\ 4 & -5 & 8 & -9 & \vdots & \lambda \end{bmatrix} \sim \begin{bmatrix} 2 & -3 & 6 & -5 & \vdots & 3 \\ 0 & 1 & -4 & 1 & \vdots & 1 \\ 0 & 1 & -4 & 1 & \vdots & \lambda - 6 \end{bmatrix} \text{(on operating } R_3 \to R_3 - 2R_1)$$

$$\sim \begin{bmatrix} 2 & -3 & 6 & -5 & \vdots & 3 \\ 0 & 1 & -4 & 1 & \vdots & 1 \\ 0 & 0 & 0 & 0 & \vdots & \lambda - 7 \end{bmatrix} \text{(on operating } R_3 \to R_3 - R_2)$$

Now, when $\lambda \neq 7$, rank $(A : B) = 3$ and rank $(A) = 2$ hence the system is inconsistent.

And when $\lambda = 7$, rank $(A : B) = 2$ and rank $(A) = 2$ hence consistent. Also, rank $(A : B) =$ rank $(A) < 4$, the number of unknowns, therefore we shall have infinitely many solutions.

To obtain many solutions when $\lambda = 7$, we have two independent equations as given below:

$$x_2 - 4x_3 + x_4 = 1 \quad ...(1)$$

$$2x_1 - 3x_2 + 6x_3 - 5x_4 = 3 \quad ...(2)$$

Here the number of arbitrary solutions = number of unknowns – rank of A

Therefore, if we take $x_3 = t_1$ and $x_4 = t_2$ then from (1), we get $x_2 = 4t_1 - t_2 + 1$

and from (2) $\quad 2x_1 = 3\,(4t_1 - t_2 + 1) - 6t_1 + 5t_2 + 3$

or $\quad x_1 = 3t_1 + t_2 + 3$

Giving arbitrary values to t_1 and t_2 we get infinitely many values of x_1 and x_2. **Ans.**

EXAMPLE 12.9. Test for consistency of the following set of equations

$$2x - 3y + 5z = 1$$
$$3x + y - z = 2$$
$$x + 4y - 6z = 1$$

and, if consistent, solve the system.

SOLUTION: Given system of equations can be written as $AX = B$

where $A = \begin{bmatrix} 2 & -3 & 5 \\ 3 & 1 & -1 \\ 1 & 4 & -6 \end{bmatrix}, \quad X = \begin{bmatrix} x \\ y \\ z \end{bmatrix}, \quad B = \begin{bmatrix} 1 \\ 2 \\ 1 \end{bmatrix}.$

The Augmented matrix is given by

$$(A:B) = \begin{bmatrix} 2 & -3 & 5 & : & 1 \\ 3 & 1 & -1 & : & 2 \\ 1 & 4 & -6 & : & 1 \end{bmatrix}$$

$$\sim \begin{bmatrix} 1 & 4 & -6 & : & 1 \\ 0 & -11 & 17 & : & -1 \\ 0 & -11 & 17 & : & -1 \end{bmatrix} \quad \text{(on applying } R_1 \leftrightarrow R_3 \text{ and then } R_2 - 3R_1 \text{ and } R_3 - 2R_1)$$

$$\sim \begin{bmatrix} 1 & 4 & -6 & : & 1 \\ 0 & -11 & 17 & : & -1 \\ 0 & 0 & 0 & : & 0 \end{bmatrix} \quad \text{(on operating } R_3 \to R_3 - R_2) \qquad \text{...(1)}$$

Obviously here rank (A : B) = 2, rank (A) = 2 hence the system is consistent. Also since the common value of the ranks is less than the number of unknowns, we can have infinitely many solutions.

From (1) it follows that we have two independent equations,

$$x + 4y - 6z = 1 \qquad \text{...(2)}$$

and $$-11y + 17z = -1 \qquad \text{...(3)}$$

Here number of arbitrary solutions = number of unknowns – rank (A)

Therefore, we set $z = t$, then from (3) we get $y = \dfrac{(1+17t)}{11}$.

and from (2) we get $x = 1 - 4y + 6z = 1 - \dfrac{4}{11}(1 + 17t) + 6t = \dfrac{(7-2t)}{11}$.

Thus, the infinitely many solutions are given by

$$x = \frac{7-2t}{11}, \quad y = \frac{1+17t}{11}, \quad z = t \text{ where } t \text{ is an arbitrary quantity.}$$ **Ans.**

EXAMPLE 12.10. Test for consistency of the following system of equations and, if yes, solve them:

$$x_1 + 2x_2 - x_3 = 3$$
$$3x_1 - x_2 + 2x_3 = 1$$
$$2x_1 - 2x_2 + 3x_3 = 2$$
$$x_1 - x_2 + x_3 = -1.$$

(UPTU 2002-03)

SOLUTION: The given system in the matrix equation form is AX = B;

where $A = \begin{bmatrix} 1 & 2 & -1 \\ 3 & -1 & 2 \\ 2 & -2 & 3 \\ 1 & -1 & 1 \end{bmatrix}$, $X = \begin{bmatrix} x_1 \\ x_2 \\ x_3 \end{bmatrix}$, $B = \begin{bmatrix} 3 \\ 1 \\ 2 \\ -1 \end{bmatrix}$

$$\therefore \quad (A:B) = \begin{bmatrix} 1 & 2 & -1 & : & 3 \\ 3 & -1 & 2 & : & 1 \\ 2 & -2 & 3 & : & 2 \\ 1 & -1 & 1 & : & -1 \end{bmatrix} \sim \begin{bmatrix} 1 & 2 & -1 & : & 3 \\ 0 & -7 & 5 & : & -8 \\ 0 & -6 & 5 & : & -4 \\ 0 & -3 & 2 & : & -4 \end{bmatrix}$$

(on operating $R_2 \to R_2 - 3R_1$, $R_3 \to R_3 - 2R_1$ and $R_4 \to R_4 - R_1$).

$$\sim \begin{bmatrix} 1 & 2 & -1 & \vdots & 3 \\ 0 & -1 & 0 & \vdots & -4 \\ 0 & 0 & 1 & \vdots & 4 \\ 0 & -3 & 2 & \vdots & -4 \end{bmatrix} \text{(on performing } R_2 \to R_2 - R_3 \text{ and } R_3 \to R_3 - 2R_4\text{)}$$

$$\sim \begin{bmatrix} 1 & 2 & -1 & \vdots & 3 \\ 0 & 1 & 0 & \vdots & 4 \\ 0 & 0 & 1 & \vdots & 4 \\ 0 & 0 & 2 & \vdots & 8 \end{bmatrix} \text{(on performing } R_4 \to R_4 + 3R_2 \text{ and } R_2 \to -R_2\text{)}$$

$$\sim \begin{bmatrix} 1 & 2 & -1 & \vdots & 3 \\ 0 & 1 & 0 & \vdots & 4 \\ 0 & 0 & 1 & \vdots & 4 \\ 0 & 0 & 0 & \vdots & 0 \end{bmatrix} \text{(on performing } R_4 \to R_4 - 2R_3\text{)}$$

Therefore, rank (A : B) = rank (A) = 3 = number of unknowns, hence unique solution. To obtain this unique solution, we have

$$(A:B) \sim \begin{bmatrix} 1 & 0 & 0 & \vdots & -1 \\ 0 & 1 & 0 & \vdots & 4 \\ 0 & 0 & 1 & \vdots & 4 \\ 0 & 0 & 0 & \vdots & 0 \end{bmatrix}, \text{(on operating } R_1 \to R_1 + R_3 \text{ and } R_1 \to R_1 - 2R_2\text{)}.$$

Therefore, the unique solution is $x = -1$, $y = 4$, $z = 4$. **Ans.**

EXAMPLE 12.11. Determine the value(s) of λ for which the following set of equations possess a non-trivial solution and find the solutions for real λ:

$$3x + y - \lambda z = 0$$
$$4x - 2y - 3z = 0$$
$$2\lambda x + 4y + \lambda z = 0.$$

SOLUTION: Given equations are homogeneous and these will have non-trivial solution when the determinant of the co-efficient matrix is equal to 0, *i.e*; when

$$\begin{vmatrix} 3 & 1 & -\lambda \\ 4 & -2 & -3 \\ 2\lambda & 4 & \lambda \end{vmatrix} = 0$$

or $3(-2\lambda + 12) - (4\lambda + 6\lambda) - \lambda(16 + 4\lambda) = 0$

or $\lambda^2 + 8\lambda - 9 = 0 \Rightarrow \lambda = -9, 1.$

For $\lambda = -9$, the set of equations are:

$$3x + y + 9z = 0$$
$$4x - 2y - 3z = 0$$
$$-18x + 4y - 9z = 0$$

Taking any two of the above equations, say first two, the solution is:

$$\frac{x}{15}=\frac{y}{45}=\frac{z}{-10} \quad \text{or} \quad \frac{x}{3}=\frac{y}{9}=\frac{z}{-2}.$$

$\Rightarrow$ $x=\dfrac{-3k}{2}$, $y=\dfrac{-9k}{2}$, $z=k$ where k is an arbitrary quantity, is the solution for $\lambda=-9$.

Similarly, for $\lambda=1$ the set of equations are:

$$\begin{aligned} 3x+y-z &= 0\\ 4x-2y-3z &= 0\\ 2x+4y+z &= 0 \end{aligned}$$

Taking any two of the above equations, say first two, the solution is $\dfrac{x}{-5}=\dfrac{y}{5}=\dfrac{z}{-10}$

$\Rightarrow$ $x=\dfrac{k'}{2}$, $y=-\dfrac{k'}{2}$, $z=k'$ where k' is another arbitrary quantity, is the solution for $\lambda=1$. **Ans.**

EXAMPLE 12.12. Find, for what values of a and b, the equations

$$\begin{aligned} x+2y+3z &= 6\\ x+3y+5z &= 9\\ 2x+5y+az &= b \end{aligned}$$

have (i) no solution (ii) a unique solution (iii) more than one solution.

SOLUTION: The given system of equations is $AX=B$ where

$$A=\begin{bmatrix}1&2&3\\1&3&5\\2&5&a\end{bmatrix}, X=\begin{bmatrix}x\\y\\z\end{bmatrix}, \quad B=\begin{bmatrix}6\\9\\b\end{bmatrix}$$

The augmented matrix $D=\begin{bmatrix}1&2&3&:&6\\1&3&5&:&9\\2&5&a&:&b\end{bmatrix}$

$$D\sim\begin{bmatrix}1&2&3&:&6\\0&1&2&:&3\\0&1&a-6&:&b-12\end{bmatrix} \quad \text{(On applying } R_2\to R_2-R_1 \text{ and } R_3\to R_3-2R_1)$$

$$\sim\begin{bmatrix}1&2&3&:&6\\0&1&2&:&3\\0&0&a-8&:&b-15\end{bmatrix} \quad \text{(On applying } R_3\to R_3-R_2)$$

If $a=8, b\neq 15$ we have $\rho(A)=2$, $\rho(D)=3$ so $\rho(A)\neq\rho(D)$ hence system has no solution.

Similarly, if $a\neq 8$ we have $\rho(A)=3$, $\rho(D)=3$

$\therefore$ $\rho(A)=\rho(D)=$ number of unknowns hence the system will have unique solution.

Further, if $a=8$ and $b=15$, we have $\rho(A)=2$, $\rho(D)=2$

$\therefore$ $\rho(A)=\rho(D)<3$ hence more than one solution. **Ans.**

EXAMPLE 12.13. **(*a*)** For what values of η the equations

$$x + y + z = 1$$
$$x + 2y + 4z = \eta$$
$$x + 4y + 10z = \eta^2.$$

have a solution and solve them completely in each case. **(UPTU 2001 Q.B.)**

(*b*) Test for consistency and solve the following system of equations

$$2x - y + 3z = 8$$
$$-x + 2y + z = 4$$
$$3x + y - 4z = 0.$$

SOLUTION: **(*a*)** The given equation can be written as matrix equation as

$$AX = B \text{ where } A = \begin{bmatrix} 1 & 1 & 1 \\ 1 & 2 & 4 \\ 1 & 4 & 10 \end{bmatrix}, X = \begin{bmatrix} x \\ y \\ z \end{bmatrix}, B = \begin{bmatrix} 1 \\ \eta \\ \eta^2 \end{bmatrix}.$$

Augmented matrix
$$D = \begin{bmatrix} 1 & 1 & 1 & : & 1 \\ 1 & 2 & 4 & : & \eta \\ 1 & 4 & 10 & : & \eta^2 \end{bmatrix}$$

Applying $R_3 \to R_3 - R_2$ and $R_2 \to R_2 - R_1$, we get

$$D \sim \begin{bmatrix} 1 & 1 & 1 & : & 1 \\ 0 & 1 & 3 & : & \eta - 1 \\ 0 & 2 & 6 & : & \eta^2 - \eta \end{bmatrix}$$

Applying $R_3 \to R_3 - 2R_2$,

$$D \sim \begin{bmatrix} 1 & 1 & 1 & : & 1 \\ 0 & 1 & 3 & : & \eta - 1 \\ 0 & 0 & 0 & : & \eta^2 - 3\eta + 2 \end{bmatrix}$$

Here $\quad \eta^2 - 3\eta + 2 = (\eta - 1)(\eta - 2).$

For $\eta = 1, 2,$ $\rho(A) = 2$ and $\rho(D) = 2$ hence consistent.

Hence for $\eta = 1$ and $\eta = 2$ we have many solutions as $\rho(A) = \rho(D) <$ the number of unknowns

For $\eta = 1$, if we take $z = t$ then $x = 2t + 1$ and $y = -3t$ given the solution where t is arbitrary

For $\eta = 2$, if we take $z = t$, then $x = 2t$ and $y = 1 - 3t$ is the solution where t is arbitrary.

Ans.

(*b*) The given system of equations is $Ax = B$

where $$A = \begin{bmatrix} 2 & -1 & 3 \\ -1 & 2 & 1 \\ 3 & 1 & -4 \end{bmatrix}, \quad X = \begin{bmatrix} x \\ y \\ z \end{bmatrix}, \quad B = \begin{bmatrix} 8 \\ 4 \\ 0 \end{bmatrix}$$

The augmented matrix $$D = \left[\begin{array}{ccc|c} 2 & -1 & 3 & 8 \\ -1 & 2 & 1 & 4 \\ 3 & 1 & -4 & 0 \end{array}\right]$$

Applying $R_1 \to R_1 + R_2$ and $R_3 \to R_3 + 3R_2$, we get

$$D \sim \left[\begin{array}{ccc|c} 1 & 1 & 4 & 12 \\ -1 & 2 & 1 & 4 \\ 0 & 7 & -1 & 12 \end{array}\right]$$

Now operating $R_2 \to R_2 + R_1$, gives

$$D \sim \left[\begin{array}{ccc|c} 1 & 1 & 4 & 12 \\ 0 & 3 & 5 & 16 \\ 0 & 7 & -1 & 12 \end{array}\right]$$

Next, $R_1 \to R_1 + 4R_3$ and $R_2 \to R_2 + 5R_3$, gives

$$D \sim \begin{bmatrix} 1 & 29 & 0 & 60 \\ 0 & 38 & 0 & 76 \\ 0 & 7 & -1 & 12 \end{bmatrix}$$

$$\sim \left[\begin{array}{ccc|c} 1 & 29 & 0 & 60 \\ 0 & 1 & 0 & 2 \\ 0 & 7 & -1 & 12 \end{array}\right] \quad \left(R_2 \to \frac{1}{38} R_2\right)$$

$$\sim \left[\begin{array}{ccc|c} 1 & 0 & 0 & 2 \\ 0 & 1 & 0 & 2 \\ 0 & 0 & -1 & -2 \end{array}\right] \quad \text{Using } R_1 \to R_1 - 29R_2 \text{ and } R_3 \to R_3 - 7R_2$$

$$\sim \left[\begin{array}{ccc|c} 1 & 0 & 0 & 2 \\ 0 & 1 & 0 & 2 \\ 0 & 0 & 1 & 2 \end{array}\right] \quad \text{Using } R_3 \to (-1)\, R_3$$

We have $\rho(A) = 3$ and $\rho(D) = 3$ and number of unknowns = 3.

$\therefore$ as $\rho(A) = \rho(D) = 3$ = no. of unknowns,

we have unique solution as $x = y = z = 2$. **Ans.**

EXAMPLE 12.14. Consider the following system of equations in which p is a real parameter.

$$x_1 - 2x_2 + 3x_3 = -1$$
$$2x_1 - 3x_2 + 4x_3 = -1$$
$$px_1 + x_2 + (p^2 - p - 3)\, x_3 = 2p + 2$$

By using row operations on the associated matrix obtain a row equivalent triangular form. Also

(i) find the values of p (if any) for which the system is inconsistent.

(ii) find the values of p (if any) for which the system has an infinite number of solutions.

(iii) solve the system in the case when $p = 2$, explaining carefully how you obtain the solution from the triangular form.

SOLUTION: The given equations can be written as $AX = B$

where $A = \begin{bmatrix} 1 & -2 & 3 \\ 2 & -3 & 4 \\ p & 1 & p^2-p-3 \end{bmatrix}, X = \begin{bmatrix} x_1 \\ x_2 \\ x_3 \end{bmatrix}, \quad B = \begin{bmatrix} -1 \\ -1 \\ 2p+2 \end{bmatrix}.$

$$\text{The augmented matrix} = D = [A : B] = \left[\begin{array}{ccc|c} 1 & -2 & 3 & -1 \\ 2 & -3 & 4 & -1 \\ p & 1 & p^2-p-3 & 2p+2 \end{array}\right]$$

To convert it into triangular form, we apply $R_2 \to R_2 - 2R_1$ and $R_3 \to R_3 - pR_1$ and get:

$$D \sim \left[\begin{array}{ccc|c} 1 & -2 & 3 & -1 \\ 0 & 1 & -2 & 1 \\ 0 & 1+2p & p^2-4p-3 & 3p+2 \end{array}\right]$$

Next, applying $R_3 \to R_3 - (1 + 2p)\, R_2$, we get

$$D \sim \left[\begin{array}{ccc|c} 1 & -2 & 3 & -1 \\ 0 & 1 & -2 & 1 \\ 0 & 0 & p^2-1 & p+1 \end{array}\right] = D', \text{ say} \qquad \textbf{Ans.}$$

(i) Now if $p = 1$ then $\rho(A) = 2$ and $\rho(D) = 3$. Since $\rho(A) \neq \rho(D)$ then system is inconsistent.

(ii) And if $p = -1$ then $\rho(A) = 2$ and $\rho(D) = 2$. Since $\rho(A) = \rho(D) < 3$, the number of unknowns, hence we get many solutions. **Ans.**

(iii) If $p = 2$, $D' = \left[\begin{array}{ccc|c} 1 & -2 & 3 & -1 \\ 0 & 1 & -2 & 1 \\ 0 & 0 & 3 & 3 \end{array}\right]$. On applying $R_3 \to \frac{1}{3} R_3$ and $R_2 \to R_2 + 2R_3$, we get

$D' \sim \left[\begin{array}{ccc|c} 1 & -2 & 3 & -1 \\ 0 & 1 & 0 & 3 \\ 0 & 0 & 1 & 1 \end{array}\right]$. Lastly applying $R_1 \to R_1 - 3R_3$ and $R_1 \to R_1 + 2R_2$, we get

$D' \sim \left[\begin{array}{ccc|c} 1 & 0 & 0 & 2 \\ 0 & 1 & 0 & 3 \\ 0 & 0 & 1 & 1 \end{array}\right]$. Hence the solution for $p = 2$ is $x_1 = 2,\ x_2 = 3,\ x_3 = 1$ **Ans.**

EXAMPLE 12.15. Find the value of λ for which the equations

$$(\lambda - 1)x + (3\lambda + 1)y + 2\lambda z = 0$$
$$(\lambda - 1)x + (4\lambda - 2)y + (\lambda + 3)z = 0$$
$$2x + (3\lambda + 1)y + 3(\lambda - 1)z = 0$$

are consistent and find the ratios of $x : y : z$ when λ has the smallest of these values. What happens when λ has the greatest of these values.

SOLUTION: The given equations are homogeneous equations in x, y, z and will be consistent, if

$$\begin{vmatrix} \lambda-1 & 3\lambda+1 & 2\lambda \\ \lambda-1 & 4\lambda-2 & \lambda+3 \\ 2 & 3\lambda+1 & 3(\lambda-1) \end{vmatrix} = 0. \quad \text{(Applying } R_2 \to R_2 - R_1\text{)}$$

or
$$\begin{vmatrix} \lambda-1 & 3\lambda+1 & 2\lambda \\ 0 & \lambda-3 & 3-\lambda \\ 2 & 3\lambda+1 & 3(\lambda-1) \end{vmatrix} = 0 \quad \text{(Now applying } C_3 \to C_3 + C_2\text{)}$$

or
$$\begin{vmatrix} \lambda-1 & 3\lambda+1 & 5\lambda+1 \\ 0 & \lambda-3 & 0 \\ 2 & 3\lambda+1 & 6\lambda-2 \end{vmatrix} = 0$$

Expanding w.r.t. the elements of second row, we get

$$(\lambda-3)\begin{vmatrix} \lambda-1 & 5\lambda+1 \\ 2 & 6\lambda-2 \end{vmatrix} = 0 \quad \text{or} \quad (\lambda-3)\begin{vmatrix} \lambda-3 & -\lambda+3 \\ 2 & 2(3\lambda-1) \end{vmatrix} = 0$$

or $(\lambda - 3)^2 [2(3\lambda - 1) + 2] = 0$ or $6\lambda(\lambda - 3)^2 = 0.$

Thus the given equations are consistent when $\lambda = 0, 3$. When λ is least, equal to 0, then equations are $-x + y = 0$, $-x - 2y + 3z = 0$ and $2x + y - 3z = 0$ which gives $x = y = z$.

Further when λ is greatest, equal to 3, the given equations are identical as

$2x + 10y + 6z = 0$ or $x + 5y + 3z = 0.$ **Ans.**

EXAMPLE 12.16. For what values of the parameters λ and μ do the system of equations $x + y + z = 6$, $x + 2y + 3z = 10$, $x + 2y + \lambda z = \mu$, have (i) no solution (ii) unique solutions (iii) more than one solution ? **(UPTU 2002)**

SOLUTION: Given equations can be written as $AX = B$

where $A = \begin{bmatrix} 1 & 1 & 1 \\ 1 & 2 & 3 \\ 1 & 2 & \lambda \end{bmatrix}$, $X = \begin{bmatrix} x \\ y \\ z \end{bmatrix}$ and $B = \begin{bmatrix} 6 \\ 10 \\ \mu \end{bmatrix}$

Augmented matrix $[A : B] = \left[\begin{array}{ccc|c} 1 & 1 & 1 & 6 \\ 1 & 2 & 3 & 10 \\ 1 & 2 & \lambda & \mu \end{array}\right]$

Appling $R_3 \to R_3 - R_2$ and $R_2 \to R_2 - R_1$, we get

$$D = [A \mid B] \sim \left[\begin{array}{ccc|c} 1 & 1 & 1 & 6 \\ 0 & 1 & 2 & 4 \\ 0 & 0 & \lambda-3 & \mu-10 \end{array}\right]$$

Case I: $\lambda = 3$, $\mu \neq 10$, we have $\rho(A) = 2$, $\rho(D) = 3$ and since $\rho(A) \neq \rho(D)$, hence no solution.

Case II: $\lambda \neq 3$, μ has any value, then $\rho(A) = 3$, $\rho(D) = 3$ since $\rho(A) = \rho(D) = 3$ = number of unknowns, hence unique solution.

Case III: $\lambda = 3$ and $\mu = 10$ then $\rho(A) = 2$, $\rho(D) = 2$ since $\rho(A) = \rho(D) <$ number of unknowns hence many solutions. **Ans.**

EXAMPLE 12.17. Solve the system of equations

$$2x_1 + 3x_2 + x_3 = 9$$
$$x_1 + 2x_2 + 3x_3 = 6$$
$$3x_1 + x_2 + 2x_3 = 8$$

by Gauss elimination method. **(UPTU 2006-07)**

SOLUTION: The augmented matrix is

$$[A : B] = \left[\begin{array}{ccc|c} 2 & 3 & 1 & 9 \\ 1 & 2 & 3 & 6 \\ 3 & 1 & 2 & 8 \end{array}\right] \sim \left[\begin{array}{ccc|c} 1 & 2 & 3 & 6 \\ 2 & 3 & 1 & 9 \\ 3 & 1 & 2 & 8 \end{array}\right] \quad (R_1 \leftrightarrow R_2)$$

$$\sim \left[\begin{array}{ccc|c} 1 & 2 & 3 & 6 \\ 0 & -1 & -5 & -3 \\ 0 & -5 & -7 & -10 \end{array}\right] \quad (\text{using } R_2 \to R_2 - 2R_1 \text{ and } R_3 \to R_3 - 3R_1)$$

$$\sim \left[\begin{array}{ccc|c} 1 & 0 & -7 & 0 \\ 0 & +1 & +5 & +3 \\ 0 & -5 & -7 & -10 \end{array}\right] \quad \begin{array}{l}(\text{using } R_1 \to R_1 + 2R_2 \text{ and } R_2 \to -R_2) \\ (\text{and using } R_3 \to R_3 + 5R_2)\end{array}$$

$$\sim \left[\begin{array}{ccc|c} 1 & 0 & -7 & 0 \\ 0 & 1 & 5 & 3 \\ 0 & 0 & 18 & 5 \end{array}\right]$$

$$\sim \left[\begin{array}{ccc|c} 1 & 0 & -7 & 0 \\ 0 & 1 & 5 & 3 \\ 0 & 0 & 1 & 5/18 \end{array}\right] \quad (\text{using } R_3 \to \frac{1}{18}R_3)$$

$$\sim \left[\begin{array}{ccc|c} 1 & 0 & 0 & 35/18 \\ 0 & 1 & 0 & 29/18 \\ 0 & 0 & 1 & 5/18 \end{array}\right] \quad (\text{using } R_1 \to R_1 + 7R_3 \text{ and } R_2 \to R_2 - 5R_3)$$

Thus we have the unique solution as

$$x_1 = \frac{35}{18}, \quad x_2 = \frac{29}{18}, \quad x_3 = \frac{5}{18}.$$ **Ans.**

EXAMPLE 12.18. Test the consistency of the following system of equations and hence find the solution.

$$4x_1 - x_2 = 12$$
$$-x_1 + 5x_2 - 2x_3 = 0$$
$$-2x_2 + 4x_3 = -8.$$

(UPTU 2005-06)

SOLUTION: The augmented matrix [A : B] is equal to

$$\left[\begin{array}{ccc|c} 4 & -1 & 0 & 12 \\ -1 & 5 & -2 & 0 \\ 0 & -2 & 4 & -8 \end{array}\right] \quad (\text{apply } R_1 \leftrightarrow R_2 \text{ and } R_2 \to R_2 + 4R_1 \text{ and } R_1 \to -R_1)$$

$$\sim \left[\begin{array}{ccc|c} 1 & -5 & 2 & 0 \\ 0 & 19 & -8 & 12 \\ 0 & -2 & 4 & -8 \end{array}\right] \quad \left(\text{apply } R_2 \leftrightarrow R_3 \text{ and } R_2 \to -\frac{1}{2}R_2\right)$$

$$\sim \left[\begin{array}{ccc|c} 1 & -5 & 2 & 0 \\ 0 & 1 & -2 & 4 \\ 0 & 19 & -8 & 12 \end{array}\right] \quad (\text{apply } R_3 \to R_3 - 19\,R_2)$$

$$\sim \left[\begin{array}{ccc|c} 1 & -5 & 2 & 0 \\ 0 & 1 & -2 & 4 \\ 0 & 0 & 30 & -64 \end{array}\right] \quad (\text{which is in the echelon form.})$$

Clearly, rank (A) = Rank [A : B] = 3 = Number of unknowns and hence we get unique solution. Then the equations become

$$30x_3 = -64, \quad x_2 - 2x_3 = 4 \quad \text{and} \quad x_1 - 5x_2 + 2x_3 = 0$$

$$\therefore \quad x_3 = -\frac{32}{15}, \quad x_2 = -\frac{4}{15}, \quad x_1 = \frac{44}{15}. \qquad \textbf{Ans.}$$

EXAMPLE 12.19. Find the value of λ such that the following equations have unique solution:

$$\lambda x + 2y - 2z - 1 = 0$$
$$4x + 2\lambda y - z - 2 = 0$$
$$6x + 6y + \lambda z - 3 = 0$$

and, by matrix method, solve these when $\lambda = 2$. **(UPTU 2002)**

SOLUTION: The given system of equations will have unique solution when $\begin{vmatrix} \lambda & 2 & -2 \\ 4 & 2\lambda & -1 \\ 6 & 6 & \lambda \end{vmatrix} \neq 0,$

that is $(\lambda - 2)(\lambda^2 + 2\lambda + 15) \neq 0 \Rightarrow$ when $\lambda \neq 2$ we may have unique solution since $\lambda^2 + 2\lambda + 15 = 0$ gives complex values of λ.

Next, when $\lambda = 2$, the augmented matrix D is given by

$$D = \left[\begin{array}{ccc|c} 2 & 2 & -2 & 1 \\ 4 & 4 & -1 & 2 \\ 6 & 6 & 2 & 3 \end{array}\right] \sim \left[\begin{array}{ccc|c} 2 & 2 & -2 & 1 \\ 0 & 0 & 3 & 0 \\ 0 & 0 & 8 & 0 \end{array}\right]$$

Obviously, rank (D) = rank (A) = 2 which is less than 3, the number of unknowns.
Hence many solutions exist as follows.

Let us take $y = k_1$, $z = k_2$ then $x = k_1 - k_2 + \dfrac{1}{2}$ where k_1 and k_2 are arbitrary constants. **Ans.**

Example 12.20. Show that the row vectors of the matrix $\begin{bmatrix} 1 & 2 & -2 \\ -1 & 3 & 0 \\ 0 & -2 & 1 \end{bmatrix}$ are linearly independent.

(UPTU 2009-10)

Solution: We know that for a square matrix A if $|A| = 0$, that is if the matrix is singular then its row vectors are linearly dependent.

Here $|A| = \begin{vmatrix} 1 & 2 & -2 \\ -1 & 3 & 0 \\ 0 & -2 & 1 \end{vmatrix} = 1(3) - 2(-1) - 2(2) \neq 0$, thus A is non-singular,

hence the row vectors of A are linearly independent. **Hence Proved.**

Example 12.21. Determine b such that the system of homogeneous equations

$$\begin{aligned} 2x + y + 2z &= 0 \\ x + y + 3z &= 0 \\ 4x + 3y + bz &= 0 \end{aligned}$$

has (*i*) trivial solution (*ii*) non-trivial solution.
Also find the non-trivial solution using matrix method.

Solution: Let $A = \begin{bmatrix} 2 & 1 & 2 \\ 1 & 1 & 3 \\ 4 & 3 & b \end{bmatrix}$ and system of equations is $AX = 0$.

$$|A| = \begin{vmatrix} 2 & 1 & 2 \\ 1 & 1 & 3 \\ 4 & 3 & b \end{vmatrix} = 2(b-9) - (b-12) + 2(3-4) = b - 8$$

Hence if $b \neq 8$ then $|A| \neq 0$ hence trivial solution exists, that is, $x = y = z = 0$.
Further, if $b = 8$, $|A| = 0$ hence non-trivial solution exists.
For the non-trivial solution, we have

$$2x + y + 2z = 0$$
$$x + y + 3z = 0$$

$$\Rightarrow \quad \frac{x}{3-2} = \frac{y}{2-6} = \frac{z}{2-1} = K, \text{ say}$$

$$\Rightarrow \quad x = K, \; y = -4K, \; z = K \text{ where } K \text{ is arbitrary.}$$ **Ans.**

EXERCISE 12A

1. Using the rank concept investigate the consistency of the system of equations

$$\begin{aligned} 4x-2y+6z &= 8 \\ x+y-3z &= -1 \\ 15x-3y+9z &= 21. \end{aligned}$$

and, if consistent, solve it.

2. Show that the system of equations

$$\begin{aligned} x+y+3z &= 6 \\ 2x+3y+z &= 8 \\ x+5y+7z &= 20 \\ x+z &= 10 \end{aligned}$$

is consistent and has unique solution.

3. Investigate whether the set of equations

$$\begin{aligned} 2x-y-z &= 2 \\ x+2y+z &= 2 \\ 4x-7y-5z &= 2. \end{aligned}$$

is consistent or not and, if consistent, solve it.

4. Investigate if the following system of equations is consistent or not.

$$\begin{aligned} x_1+x_2+x_3 &= 3 \\ 2x_1-x_2+3x_3 &= 1 \\ 4x_1+x_2+5x_3 &= 3 \\ 3x_1-2x_2+x_3 &= 4. \end{aligned}$$

5. Determine the value(s) of λ for which the following equations

$$\begin{aligned} 2x-3y+z &= \lambda x \\ 2x-3y+2z &= \lambda y \\ -x+2y+0z &= \lambda z \end{aligned}$$

can possess a non-trivial solution.

6. Determine the values of λ and μ so that the equations

$$\begin{aligned} 2x+3y+5z &= 9 \\ 7x+3y-2z &= 8 \\ 2x+3y+\lambda z &= \mu \end{aligned}$$

have (*i*) no solution, (*ii*) a unique solution and (*iii*) infinitely many solutions.

7. Solve by matrix method, the system of equations

$$\begin{aligned} x+3y-z-\omega &= 0 \\ 3x+y+2z-2\omega &= 4 \\ 4x-3y-4z-3\omega &= -13. \end{aligned}$$

8. Test for consistency the following system of equations

$$\begin{aligned} x+y+z &= 3 \\ x+2y+3z &= 4 \\ x+4y+9z &= 4 \\ 2x+3y+9z &= 6. \end{aligned}$$

9. Solve the following system of equations using elementary transformations

$$\begin{aligned} 2x+y-z+3\omega &= 11 \\ x-2y+z+\omega &= 8 \\ 4x+7y+2z+\omega &= 0 \\ 3x+5y+4z+4\omega &= 17. \end{aligned}$$

10. Find the values of a and b such that the system of equations

$$\begin{aligned} 3x-2y+z &= 6 \\ 5x-8y+9z &= 3 \\ 2x+y+az &= -b \end{aligned}$$

may have unique solution.

11. Test whether the vectors (1, 1, 1, 3), (1, 2, 3, 4) and (2, 3, 4, 7) are linearly dependent or not. If dependent, find the relation between them.

12. Show that the square matrix

$$\begin{bmatrix} \cos\phi\cos\theta & \sin\phi & \cos\phi\sin\theta \\ -\sin\phi\cos\theta & \cos\phi & -\sin\phi\sin\theta \\ -\sin\theta & 0 & \cos\theta \end{bmatrix} \text{ is orthogonal.}$$

13. Consider the linear transformations

$$\begin{aligned} x_1 &= y_1 - 2y_2 + y_3 \\ x_2 &= 2y_1 + y_2 - 3y_3, \end{aligned}$$

and

$$\begin{aligned} y_1 &= z_1 + 2z_2 \\ y_2 &= 2z_1 - z_2 \\ y_3 &= 2z_1 + 3z_2. \end{aligned}$$

Using matrix method find the linear transformation from (z_1, z_2) to (x_1, x_2).

14. Given the linear transformation $Y = AX$ where

$$Y = [y_1, y_2, y_3]', \quad X = [x_1, x_2, x_3]' \quad \text{and} \quad A = \begin{bmatrix} 1 & 1 & 2 \\ 1 & 2 & 5 \\ 1 & 3 & 3 \end{bmatrix},$$

find the co-ordinates (x_1, x_2, x_3) corresponding to (2, 0, 5) in Y.

15. Show that the transformation

$$\begin{aligned} y_1 &= x_1 - x_2 + x_3 \\ y_2 &= 3x_1 - x_2 + 2x_3 \\ y_3 &= 2x_1 - 2x_2 + 3x_3 \end{aligned}$$

is non-singular. Find the inverse transformation.

❑❑❑

CHAPTER

13 Eigen Values, Eigen Vectors, Diagonalising a Matrix and Complex Matrix

Characteristic Equation, Eigen Values and Eigen Vectors, Caley-Hamilton Theorem. Diagonalising a Matrix, complex and Unitary Matrices.

CHARACTERISTIC EQUATION, EIGEN VALUES AND EIGEN VECTORS

Consider the linear transformation $Y = AX$ which transforms the column vector $X = \begin{bmatrix} x_1 \\ x_2 \\ \vdots \\ x_n \end{bmatrix}$ into the column vector $\begin{bmatrix} y_1 \\ y_2 \\ \vdots \\ y_n \end{bmatrix}$ where A, the transformation matrix, is given by

$$A = \begin{bmatrix} a_{11} & a_{12} & a_{13} & \cdots & \cdots & a_{1n} \\ a_{21} & a_{22} & a_{23} & \cdots & \cdots & a_{2n} \\ \cdots & \cdots & \cdots & \cdots & \cdots & \cdots \\ \cdots & \cdots & \cdots & \cdots & \cdots & \cdots \\ a_{n1} & a_{n2} & a_{n3} & \cdots & \cdots & a_{nn} \end{bmatrix}.$$

It may be interesting to consider another transformation $Y = \lambda X$ (where λ is a parameter) under which a vector X is transformed into a vector Y having same direction as that of X but having different magnitude. This implies that

$$\boxed{\lambda X = AX \quad \text{or} \quad (A - \lambda I)X = 0} \qquad \dots(1)$$

which represents the following system of homogeneous equations in $x_1, x_2, \dots, x_n$

$$\left.\begin{array}{l}(a_{11}-\lambda)\,x_1+a_{12}x_2+\cdots+a_{1n}x_n=0\\ a_{21}x_1+(a_{22}-\lambda)x_2+\cdots+a_{2n}x_n=0\\ \cdots\quad \cdots\cdots\quad\cdots\quad\cdots\quad\cdots\cdots\cdots\\ \cdots\quad \cdots\cdots\quad\cdots\quad\cdots\quad\cdots\cdots\cdots\\ a_{n1}x_1+a_{n2}x_2+\cdots+(a_{nn}-\lambda)x_n=0\end{array}\right\} \quad ...(2)$$

This system of equations has trivial solution $X = 0$ for any λ, but we are interested only in the non-trivial solutions of (2) which is possible only when the matrix $(A - \lambda I)$ is singular, *i.e.* when $|A - \lambda I| = 0$

or
$$\begin{vmatrix} a_{11}-\lambda & a_{12} & a_{13} & \cdots & a_{1n} \\ a_{21} & a_{22}-\lambda & a_{23} & \cdots & a_{2n} \\ \cdots & \cdots & \cdots & \cdots & \cdots \\ \cdots & \cdots & \cdots & \cdots & \cdots \\ \cdots & \cdots & \cdots & \cdots & \cdots \\ a_{n1} & a_{n2} & a_{n3} & \cdots & a_{nn}-\lambda \end{vmatrix} = 0 \quad ...(3)$$

This is called the **characteristic equation** of the square matrix A.

In (3) when the determinant is expanded we get nth degree polynomial equation in λ which will have n roots, say $\lambda_1, \lambda_2, ..., \lambda_n,$ called **eigen values or characteristic values or characteristic roots or latent roots** *of* the matrix A.

Corresponding to n eigen values we can get n values of the column vector X by solving the equations (1), that is, $AX = \lambda X$. ...(4)

These column vectors are called **eigen vectors or characteristic vectors** of the matrix A.

Let us take up the frequently occurring case when the matrix A is of the order 3×3.

Let
$$A = \begin{bmatrix} a_{11} & a_{12} & a_{13} \\ a_{21} & a_{22} & a_{23} \\ a_{31} & a_{32} & a_{33} \end{bmatrix}$$

The characteristic equation $|A - \lambda I| = 0$ becomes

$$\begin{vmatrix} a_{11}-\lambda & a_{12} & a_{13} \\ a_{21} & a_{22}-\lambda & a_{23} \\ a_{31} & a_{32} & a_{33}-\lambda \end{vmatrix} = 0$$

which, on simplification, gives

$$\lambda^3-(a_{11}+a_{22}+a_{33})\,\lambda^2+\left\{\begin{vmatrix} a_{11} & a_{12} \\ a_{21} & a_{22} \end{vmatrix}+\begin{vmatrix} a_{22} & a_{23} \\ a_{32} & a_{33} \end{vmatrix}+\begin{vmatrix} a_{11} & a_{13} \\ a_{31} & a_{33} \end{vmatrix}\right\}\lambda-\begin{vmatrix} a_{11} & a_{12} & a_{13} \\ a_{21} & a_{22} & a_{23} \\ a_{31} & a_{32} & a_{33} \end{vmatrix}=0 \quad ...(5)$$

An important *property* to note here, is "**if A is a symmetric matrix, the eigen vectors of the equation (A – λI) X = 0 are pairwise orthogonal**" Thus, if $\begin{bmatrix} x_1 \\ x_2 \\ \vdots \\ x_n \end{bmatrix}$ and $\begin{bmatrix} x_1' \\ x_2' \\ \vdots \\ x_n' \end{bmatrix}$ are two eigen vectors of a symmetric matrix A then $x_1x_1' + x_2x_2' + x_3x_3' + ... + x_nx_n' = 0$.

Some Useful Deductions

If λ is a characteristic value of a square matrix A, then

(*i*) $\lambda + k$ is a characteristic value of $A + kI$

(*ii*) $k\lambda$ is a characteristic value of kA

(*iii*) $\frac{1}{\lambda}$ is a characteristic value of A^{-1}. In other words $\frac{|A|}{\lambda}$ is an eigen value of the matrix adj (A). **(UPTU 2001)**

(*iv*) λ^2 is a characteristic values of A^2.

Let us deduce the above results.

Since λ is a eigen value of A we have $(A - \lambda I)X = 0$ ... (1)

(*i*) We can write (1) as $(A + kI - kI - \lambda I)X = 0$ or $(A + kI - (\lambda + k)X = 0$

which establishes that $A + kI$ has eigen value $\lambda + k$.

(*ii*) Multiplying (1) by k, we get $k(A - \lambda I)X = 0$

or $[kA - (\lambda\, k)\, I]\, X = 0$

which establishes that the matrix kA has eigen value as $\lambda\, k$.

(*iii*) Premultiplying (1) by A^{-1}, we get

$$\{A^{-1}(A - \lambda I)\}X = 0 \quad \text{or} \quad (I - \lambda A^{-1})X = 0$$

Multiplying throughout by λ^{-1}, we get

$$(\lambda^{-1} I - A^{-1})X = 0 \quad \text{or} \quad (A^{-1} - \lambda^{-1} I)X = 0$$

which establishes that $\frac{1}{\lambda}$ is eigen value of A^{-1}.

(*iv*) Premultiplying (1) by A, we get

$$\{A(A - \lambda I)\}X = 0 \quad \text{or} \quad (A^2 - \lambda A)X = 0$$

or $(A^2 - \lambda(A - \lambda I) - \lambda^2 I)X = 0$ or $(A^2 - \lambda^2 I)X - \lambda(A - \lambda I)X = 0$

But $(A - \lambda I)X = 0$ hence $(A^2 - \lambda^2 I)X = 0$

which shows that λ^2 is the eigen value of the matrix A^2.

Note here, that the deduction (iv) can be extended to the fact that λ^n will be the eigen value of the square matrix A^n if λ is the eigen value of the square matrix A.

EXAMPLE 13.1. Show that, for any square matrix A

(i) the sum of all its eigen values is equal to the sum of the elements of its principal diagonal. **(UPTU 2002-03)**

(ii) the product of all the eigen values of A is equal to | A |. **(UPTU 2003-04)**

SOLUTION: Characteristic equation of A is $|A - \lambda I| = 0$

or $$\begin{vmatrix} a_{11}-\lambda & a_{12} & a_{13} & \cdots & a_{1n} \\ a_{21} & a_{22}-\lambda & a_{23} & \cdots & a_{2n} \\ \cdots & \cdots & \cdots & \cdots & \cdots \\ a_{n1} & a_{n2} & a_{n3} & \cdots & a_{nn}-\lambda \end{vmatrix} = 0 \quad \text{... (1)}$$

or $(-1)^n[\lambda^n + c_1\lambda^{n-1} + c_2\lambda^{n-2} + ... + c_{n-1}\lambda + c_n] = (-1)^n(\lambda - \lambda_1)(\lambda - \lambda_2) ... (\lambda - \lambda_n) = 0$

where $\lambda_1, \lambda_2, ... \lambda_n$ are eigen values of A.

(i) In the expansion of (1) co-efficient of λ^{n-1} is $a_{11} + a_{22} + a_{33} + ... + a_{nn}$ which is the sum of the eigen values $= \lambda_1 + \lambda_2 + ... + \lambda_n$.

(ii) Putting $\lambda = 0$ in (1) we get $c_n = \lambda_1, \lambda_2 ... \lambda_n = |A|$. **Hence Proved.**

EXAMPLE 13.2. Find the sum of the characteristic roots of the matrix A^2, given that $A = \begin{bmatrix} 3 & 0 & 0 \\ 8 & 4 & 0 \\ 6 & 2 & 5 \end{bmatrix}$.

(UPTU 2003)

SOLUTION: The characteristic equation for the matrix $A = \begin{bmatrix} 3 & 0 & 0 \\ 8 & 4 & 0 \\ 6 & 2 & 5 \end{bmatrix}$ is $|A - \lambda I| = 0$

or $\begin{vmatrix} 3-\lambda & 0 & 0 \\ 8 & 4-\lambda & 0 \\ 6 & 2 & 5-\lambda \end{vmatrix} = 0$ whose roots as characteristic roots are $\lambda = 3, 4, 5$.

We know that the characteristic roots of the matrix A^2 are λ^2 as 9, 16, 25, whose sum = 9 + 16 + 25 = 50. **Ans.**

EXAMPLE 13.3. If matrix A is defined as $A = \begin{bmatrix} -1 & 0 & 0 \\ 2 & -3 & 0 \\ 1 & 4 & 2 \end{bmatrix}$ then the eigen values of A^3 are

(i) –1, –27, 8 (ii) 1, 8, 27 (iii) (–1, 8, –27) (iv) (–1, –8, –27).

SOLUTION: The characteristic equation of A is $|A - \lambda I| = 0$

or $\begin{vmatrix} -1-\lambda & 0 & 0 \\ 2 & -3-\lambda & 0 \\ 1 & 4 & 2-\lambda \end{vmatrix} = 0$ or $(-1-\lambda)(-3-\lambda)(2-\lambda) = 0$

Hence $\lambda = -1$, $\lambda = -3$ and $\lambda = 2$ are eigen values of A. We know that the eigen values of A^3 are the cubes of the eigen values of A, that is (–1, –27, 8). Hence (iii) **Ans.**

EXAMPLE 13.4. (*a*) Find the eigen values and eigen vectors of the matrix $A = \begin{bmatrix} 4 & -5 \\ 1 & -2 \end{bmatrix}$.

(*b*) Find the eigen values and corresponding eigen vectors of the matrix $A = \begin{bmatrix} -5 & 2 \\ 2 & -2 \end{bmatrix}$.

(UPTU 2008-09)

SOLUTION: (*a*) The characteristic equation of the given matrix A, is

$$|A - \lambda I| = 0 \quad \text{or} \quad \begin{vmatrix} 4-\lambda & -5 \\ 1 & -2-\lambda \end{vmatrix} = 0$$

or $\lambda^2 - 2\lambda - 3 = 0$ which gives $\lambda = -1$ and 3.

Therefore, the eigen values of A are $\lambda = -1$, $\lambda = 3$.

Now, when $\lambda = -1$ the equation $(A - \lambda I)\, X = 0$ is equivalent to the system of linear equations

$$5x_1 - 5x_2 = 0 \quad \text{and} \quad x_1 - x_2 = 0.$$

The obvious solution of this system is the vector $\begin{bmatrix} 1 \\ 1 \end{bmatrix}$.

Of course, this vector multiplied by any scalar is also a solution. Thus, for any number $c_1 \neq 0$, $X = c_1 \begin{bmatrix} 1 \\ 1 \end{bmatrix}$ is an eigen vector corresponding to $\lambda = -1$.

Next, when $\lambda = 3$ the equation $(A - \lambda I)X = 0$ is equivalent to one equation only $x_1 - 5x_2 = 0$.

which has $\begin{bmatrix} 5 \\ 1 \end{bmatrix}$ as obvious solution. In general for any non-zero number c_2, $X = c_2 \begin{bmatrix} 5 \\ 1 \end{bmatrix}$ is the other eigen vector of A corresponding to $\lambda = 3$. **Ans.**

(*b*) Characterstic equation of A is $\begin{vmatrix} -5-\lambda & 2 \\ 2 & -2-\lambda \end{vmatrix} = 0$

or $\lambda^2 + 7\lambda + 6 = 0$ whose roots are $\lambda = -1, -6$ as eigen values of A.

For eigen vectors we have $(A - \lambda I)X = 0$ as $(-5 - \lambda)\,x + 2y = 0$ and $2x + (-2 - \lambda)y = 0$

When $\lambda = -1$ we have $-4x + 2y = 0$ and $2x - y = 0$ which are same.

$\therefore$ $y = 2x$ and the eigen vector is $\begin{bmatrix} 1 \\ 2 \end{bmatrix}$.

When $\lambda = -6$, we have $x + 2y = 0$ and $2x + 4y = 0$ which are same

$\therefore$ $x = -2y$ and the eigen vector is $\begin{bmatrix} -2 \\ 1 \end{bmatrix}$.

Thus, for $\lambda = -1$ the eigen vector is $\begin{bmatrix} 1 \\ 2 \end{bmatrix}$ and for $\lambda = -6$ the eigen vector is $\begin{bmatrix} -2 \\ 1 \end{bmatrix}$. **Ans.**

EXAMPLE 13.5. Find the eigen values and eigen vectors of the matrix

(a) $A=\begin{bmatrix}2 & 1 & 1\\ 2 & 3 & 2\\ 3 & 3 & 4\end{bmatrix}$. (b) $A=\begin{bmatrix}1 & 1 & 1\\ 1 & 1 & 1\\ 1 & 1 & 1\end{bmatrix}$ **[UPTU 2011-12]**

SOLUTION: **(a)** The characteristic equation $|A - \lambda I| = 0$ is $\begin{vmatrix}2-\lambda & 1 & 1\\ 2 & 3-\lambda & 2\\ 3 & 3 & 4-\lambda\end{vmatrix} = 0$

which on expansion, gives $\lambda^3 - 9\lambda^2 + 15\lambda - 7 = 0$ or $(\lambda - 1)(\lambda - 1)(\lambda - 7) = 0$.

$\therefore$ The eigen values of A are $\lambda_1 = 1$, $\lambda_2 = 1$, $\lambda_3 = 7$, here $\lambda_1 = \lambda_2$.

Now, the eigen vectors are given by $(A - \lambda I)X = 0$. ... (1)

Let us first take-up the non-repeated eigen value $\lambda_3 = 7$.

Equation (1) gives

$$-5x_1 + x_2 + x_3 = 0$$
$$2x_1 - 4x_2 + 2x_3 = 0$$
$$3x_1 + 3x_2 - 3x_3 = 0$$

From the last two equations, we get $\frac{x_1}{1} = \frac{x_2}{2} = \frac{x_3}{3}$.

Therefore, the eigen vector of A for $\lambda = 7$, is $X_1 = k_1\begin{bmatrix}1\\2\\3\end{bmatrix}$ (where k_1 is a non-zero constant)

Next, for $\lambda_1 = \lambda_2 = 1$ the equation (1), gives only one independent equation as

$$x_1 + x_2 + x_3 = 0$$

Since, the matrix A is not symmetric we can get two linearly independent vectors X_2 and X_3 by taking $x_1 = 0$ and $x_2 = 0$ respectively. When $x_1 = 0$ we have $x_2 + x_3 = 0$

On setting $x_2 = k_2$ we have $x_3 = -k_2$ and hence one eigen vector of A, for $\lambda = 1$ is

$$X_2 = k_2\begin{bmatrix}0\\1\\-1\end{bmatrix}$$ where k_2 is a non-zero arbitrary constant.

When we take $x_2 = 0$ we have $x_1 + x_3 = 0$ and on setting $x_3 = -k_3$ we have $x_1 = k_3$ hence the other eigen vector of A for $\lambda = 1$, is $X_3 = k_3\begin{bmatrix}1\\0\\-1\end{bmatrix}$ where k_3 is any non-zero constant.

Thus, the eigen vectors of the given matrix A are:

$$X_1 = \begin{bmatrix}1\\2\\3\end{bmatrix},\quad X_2 = \begin{bmatrix}0\\1\\-1\end{bmatrix},\quad X_3 = \begin{bmatrix}1\\0\\-1\end{bmatrix}.$$

It can be easily shown that X_1, X_2, X_3 are linearly independent. **Ans.**

(b) The characteristic equation is $|A - \lambda I| = 0$

or $$\begin{vmatrix} 1-\lambda & 1 & 1 \\ 1 & 1-\lambda & 1 \\ 1 & 1 & 1-\lambda \end{vmatrix} = 0$$

or $(1 - \lambda)[(1 - \lambda)^2 - 1] - 1[1 - \lambda - 1] + 1(1 - (1 - \lambda)] = 0$

or $(1 - \lambda)^3 - (1 - \lambda) + \lambda + \lambda = 0$ or $1 - \lambda^3 - 3\lambda + 3\lambda^2 - 1 + 3\lambda = 0$

or $\lambda^3 - 3\lambda^2 = 0 \Rightarrow \lambda = 0, 0, 3.$

$\therefore$ Eigne Values are 0, 0 and 3. **Ans.**

For eigen vectors X we have $(A - \lambda I)X = 0$ which gives

$(1 - \lambda)x + y + z = 0,\quad x + (1 - \lambda)y + z = 0$ and $x + y + (1 - \lambda)z = 0$

Now for $\lambda = 3$ we have $-2x + y + z = 0,\quad x - 2y + z = 0,\quad x + y - 2z = 0$

which gives the eigen vector as $\frac{x}{3} = \frac{y}{3} = \frac{z}{3}$ or vector (1, 1, 1).

For $\lambda = 0$ we have one equation only as $x + y + z = 0$

At $z = 1$, we have $x + y = -1 \quad \therefore$ vector is (1, –2, 1)

and at $z = -1$, we have $x + y = 1 \quad \therefore$ vector is (1, 0, –1). **Ans.**

EXAMPLE 13.6. Determine the characteristic roots and the corresponding characteristic vectors of the matrix

$$A = \begin{bmatrix} 2 & 1 & 0 \\ 0 & 2 & 1 \\ 0 & 0 & 2 \end{bmatrix}.$$

SOLUTION: The characteristic equation is $|A - \lambda I| = 0$

or $$\begin{vmatrix} 2-\lambda & 1 & 0 \\ 0 & 2-\lambda & 1 \\ 0 & 0 & 2-\lambda \end{vmatrix} = 0 \quad \text{or} \quad (2 - \lambda)^3 = 0$$

Hence the characteristic roots are $\lambda = 2, 2, 2.$

For characteristic vectors we write $(A - \lambda I)\, X = 0$

$$\Rightarrow \quad \begin{aligned} (2 - \lambda)\, x + y &= 0 \\ (2 - \lambda)\, y + z &= 0 \\ (2 - \lambda)\, z &= 0 \end{aligned}$$

On taking $\lambda = 2$ we have $y = 0,\ z = 0$

Thus the eigen vector can be taken as $x = t, y = 0, z = 0$ where t is arbitrary or (1, 0, 0).

That is $\begin{bmatrix} t \\ 0 \\ 0 \end{bmatrix}$. Thus three characteristic vectors can be obtained by taking three different values of t. **Ans.**

EXAMPLE 13.7. Obtain the eigen values and eigen vectors for the matrix

$$A = \begin{bmatrix} 2 & 0 & 1 \\ 0 & 2 & 0 \\ 1 & 0 & 2 \end{bmatrix}$$ and verify that the eigen vectors are orthogonal.

SOLUTION: The characteristic equation for the matrix A is $|A - \lambda I| = 0$

or $$\begin{vmatrix} 2-\lambda & 0 & 1 \\ 0 & 2-\lambda & 0 \\ 1 & 0 & 2-\lambda \end{vmatrix} = 0 \quad \text{or} \quad (2-\lambda)^3 - (2-\lambda) = 0$$

or $$(2-\lambda)(2-\lambda+1)(2-\lambda-1) = 0 \quad \text{or} \quad (1-\lambda)(2-\lambda)(3-\lambda) = 0$$

Hence the eigen values of A are $\lambda = 1, 2, 3$.

For eigen vectors X of A we have $(A - \lambda I)X = 0$ where $X = \begin{bmatrix} x \\ y \\ z \end{bmatrix}$

or $$\begin{bmatrix} 2-\lambda & 0 & 1 \\ 0 & 2-\lambda & 0 \\ 1 & 0 & 2-\lambda \end{bmatrix} \begin{bmatrix} x \\ y \\ z \end{bmatrix} = 0$$

or $$(2-\lambda)x + z = 0, \quad (2-\lambda)y = 0, \quad x + (2-\lambda)z = 0$$

Now, when $\lambda = 1$ we have $x + z = 0$, $y = 0$, that is if $x = 1$, $z = -1$, $y = 0$

and the eigen vector for $\lambda = 1$ is $X_1 = \begin{bmatrix} 1 \\ 0 \\ -1 \end{bmatrix}$

Next, when $\lambda = 2$ we have $z = 0, x = 0, y = y$

and the eigen vector for $\lambda = 2$ is $X_2 = \begin{bmatrix} 0 \\ 1 \\ 0 \end{bmatrix}$

Next, when $\lambda = 3$ we have $-x + z = 0$, $y = 0$ or $z = x$, $y = 0$

Thus if we take $x = 1$ then $z = 1, y = 0$, and the eigen vector for $\lambda = 3$ is $X_3 = \begin{bmatrix} 1 \\ 0 \\ 1 \end{bmatrix}$. **Ans.**

Now, for orthogonality of eigen vectors

$$X_1 \cdot X_2 = 1.0 + 0.1 + (-1)\,0 = 0$$

$$X_1 \cdot X_3 = 1.1 + 0.0 + (-1)\,1 = 0$$

and $$X_2 \cdot X_3 = 0.1 + 1.0 + 0.1 = 0$$

Therefore X_1, X_2, X_3 are orthogonal to each other. **Hence Verified.**

EXAMPLE 13.8. Obtain the characteristic values and characteristic vectors of the matrix

$$A=\begin{bmatrix}-2 & 2 & 1\\ 2 & 1 & 2\\ 1 & 2 & 6\end{bmatrix}.$$

and verify that these vectors of A are pairwise orthogonal.

SOLUTION: The characteristic equation of A is $|A-\lambda I|=0$.

i.e., $$\begin{vmatrix}-2-\lambda & 2 & 1\\ 2 & 1-\lambda & 2\\ 1 & 2 & 6-\lambda\end{vmatrix}=0 \quad \text{or} \quad \lambda^3-5\lambda^2-17\lambda+21=0$$

or $(\lambda-1)(\lambda+3)(\lambda-7)=0$ Thus, the eigen values of A are $\lambda=1,-3,7$.

When $\lambda=1$ the equation $(A-\lambda I)X=0$ gives the following equations

$$\begin{aligned}-3x_1+2x_2+x_3 &= 0\\ 2x_1+2x_3 &= 0\\ x_1+2x_2+5x_3 &= 0\end{aligned}$$

which is equivalent to two equations only $x_1+2x_2+5x_3=0$ and $x_2+2x_3=0$

Setting here $x_3=-k_1$ we get $x_2=2k_1$ and in turn $x_1=k_1$. Therefore, all eigen vectors corresponding to $\lambda=1$ are given by $X_1=c_1\begin{bmatrix}1\\2\\-1\end{bmatrix}$ where c_1 may be any non-zero number.

Next, when $\lambda=-3$ the equation $(A+3I)X=0$ gives the following equations

$$\begin{aligned}x_1+2x_2+x_3 &= 0\\ 2x_1+4x_2+2x_3 &= 0\\ x_1+2x_2+9x_3 &= 0\end{aligned}$$

which is equivalent to $x_3=0$ and $x_1+2x_2+x_3=0$ or $x_1+2x_2=0$

Thus, setting $x_2=k_2$ we get $x_1=-2k_2$, Therefore, all eigen vectors of A corresponding to $\lambda=-3$ are of the form $X_2=c_2\begin{bmatrix}2\\-1\\0\end{bmatrix}$ $(c_2\neq 0)$.

Finally, when $\lambda=7$ the equation $(A-\lambda I)X=0$ gives the equations

$$\begin{aligned}-9x_1+2x_2+x_3 &= 0\\ 2x_1-6x_2+2x_3 &= 0\\ x_1+2x_2-x_3 &= 0\end{aligned}$$

which is equivalent to two equations $x_1+2x_2-x_3=0$ and $5x_2-2x_3=0$.

Setting $x_3=5k_3$ we get $x_2=2k_3$ and $x_1=k_3$. Therefore, the eigen vector of A for $\lambda=7$, is

$$X_3=c_3\begin{bmatrix}1\\2\\5\end{bmatrix}, \quad (c_3\neq 0).$$

Further, X_1, X_2, X_3 are linearly independent eigen vectors of A. In the present case, since the matrix A is symmetric we expect the eigen vectors to be pairwise orthogonal.

$$X_1 . X_2 = (1)(2) + 2(-1) + (-1)(0) = 0, \quad X_2 . X_3 = (2)(1) + (-1)(2) + (0)(5) = 0$$

and $X_3 . X_1 = (1)(1) + (2)(2) + 5(-1) = 0$ **hence verified that the eigen vectors are pairwise orthogonal.**

EXAMPLE 13.9. Find the eigen values and eigen vectors of the matrix

$$A = \begin{bmatrix} 8 & -6 & 2 \\ -6 & 7 & -4 \\ 2 & -4 & 3 \end{bmatrix}$$

and test for the orthogonality of the vectors. **(UPTU 2005, 2010-11)**

SOLUTION: The characteristic equation of A is $|A - \lambda I| = 0$

or $$\begin{vmatrix} 8-\lambda & -6 & 2 \\ -6 & 7-\lambda & -4 \\ 2 & -4 & 3-\lambda \end{vmatrix} = 0 \quad \text{or} \quad \lambda^3 - 18\lambda^2 + 45\lambda = 0.$$

or $\lambda(\lambda-3)(\lambda-15) = 0$ Thus, the eigen values of A are $\lambda_1 = 0, \lambda_2 = 3, \lambda_3 = 15$.

The eigen vector X_1 of A corresponding to $\lambda_1 = 0$ is the solution of the system of equations $(A - \lambda_1 I)X = 0$, that is,

$$8x_1 - 6x_2 + 2x_3 = 0$$
$$-6x_1 + 7x_2 - 4x_3 = 0$$
$$2x_1 - 4x_2 + 3x_3 = 0$$

Eliminating x_1 from the last two equations, gives $-5x_2 + 5x_3 = 0$ or $x_2 = x_3$.

Setting $x_3 = k_1$, we get $x_2 = k_1$ hence $x_1 = \frac{k_1}{2}$.

∴ The eigen vector of A corresponding to $\lambda = 0$, is $X_1 = k_1\begin{bmatrix} 1/2 \\ 1 \\ 1 \end{bmatrix} = k_1\begin{bmatrix} 1 \\ 2 \\ 2 \end{bmatrix}$ $(k_1 \neq 0)$

Similarly, for $\lambda = 3$ we get $X_2 = k_2\begin{bmatrix} 2 \\ 1 \\ -2 \end{bmatrix}$ $(k_2 \neq 0)$ and for $\lambda = 15$ we get $X_3 = k_3\begin{bmatrix} 2 \\ -2 \\ 1 \end{bmatrix}$ $(k_3 \neq 0)$.

Further, since A is symmetric matrix, the eigen vectors X_1, X_2, X_3 should be mutually orthogonal.

$$X_1.X_2 = (1)(2) + (2)(1) + (2)(-2) = 0$$
$$X_2.X_3 = (2)(2) + (1)(-2) + (-2)(1) = 0$$
$$X_3.X_1 = (2)(1) + (-2)(2) + (1)(2) = 0.$$ **Hence verified.**

EXAMPLE 13.10. Find the eigen values and eigen vectors for the following matrix

$$A = \begin{bmatrix} -9 & 2 & 6 \\ 5 & 0 & -3 \\ -16 & 4 & 11 \end{bmatrix}.$$

SOLUTION: The characteristic equation is $|A - \lambda I| = 0$

or $$\begin{vmatrix} -9-\lambda & 2 & 6 \\ 5 & -\lambda & -3 \\ -16 & 4 & -\lambda+11 \end{vmatrix} = 0$$

or $\lambda^3 - 2\lambda^2 - \lambda + 2 = 0$ or $(\lambda^2 - 1)(\lambda - 2) = 0$

whose roots are $\lambda = 1, -1, 2$ which are eigen values of A.

For eigen vectors, we write $(A - \lambda I)X = 0$

or $$\begin{bmatrix} -9-\lambda & 2 & 6 \\ 5 & -\lambda & -3 \\ -16 & 4 & -\lambda+11 \end{bmatrix} \begin{bmatrix} x \\ y \\ z \end{bmatrix} = \begin{bmatrix} 0 \\ 0 \\ 0 \end{bmatrix}.$$

Now for $\lambda = 1$, we have

$$-10x + 2y + 6z = 0, \quad 5x - y - 3z = 0, \quad -16x + 4y + 10z = 0.$$

Here first two equations are same and solving the last two, we get

$$\frac{x}{1} = \frac{y}{-1} = \frac{z}{2}$$

Hence the eigen vector corresponding to $\lambda = 1$ is $(1, -1, 2)$.

Similarly, for $\lambda = -1$, we have the equations for eigen vectors as

$$-8x + 2y + 6z = 0, \quad 5x + y - 3z = 0, \quad -16x + 4y + 12z = 0.$$

Here first and third equations are same and solving the first two, we get

$$\frac{x}{2} = \frac{-y}{-1} = \frac{z}{3}$$

Hence the eigen vector corresponding to $\lambda = -1$ is $(2, -1, 3)$.

And, for $\lambda = 2$, we have

$$-11x + 2y + 6z = 0, \quad 5x - 2y - 3z = 0, \quad -16x + 4y + 9z = 0$$

which on solving, give $$\frac{x}{2} = \frac{y}{-1} = \frac{z}{4}$$

Hence the eigen vector corresponding to $\lambda = 2$ is $(2, -1, 4)$. Thus the eigen values are $1, -1, 2$ and the corresponding eigen vectors are $(1, -1, 2)$, $(2, -1, 3)$ and $(2, -1, 4)$. **Ans.**

EXAMPLE 13.11. Find the eigen values and eigen vectors of the matrix

$$A = \begin{bmatrix} -2 & 2 & -3 \\ 2 & 1 & -6 \\ -1 & -2 & 0 \end{bmatrix}.$$

SOLUTION: The characteristic equation of A is $|A - \lambda I| = 0$

or $$\begin{vmatrix} -2-\lambda & 2 & -3 \\ 2 & 1-\lambda & -6 \\ -1 & -2 & -\lambda \end{vmatrix} = 0 \quad \text{or} \quad \lambda^3 + \lambda^2 - 21\lambda - 45 = 0$$

or $$(\lambda + 3)(\lambda^2 - 2\lambda - 15) = 0 \quad \text{or} \quad (\lambda + 3)^2 (\lambda - 5) = 0$$

Therefore, the eigen values of A are $\lambda = -3, -3, 5$.

For eigen vectors, we have $(A - \lambda I)X = 0$ where $X = \begin{bmatrix} x \\ y \\ z \end{bmatrix}$

For $\lambda = 5$ we have
$$-7x + 2y - 3z = 0$$
$$2x - 4y - 3z = 0$$
$$-x - 2y - 5z = 0$$

$\Rightarrow$ $\dfrac{x}{8} = \dfrac{y}{16} = \dfrac{z}{-8}$ or $\dfrac{x}{1} = \dfrac{y}{2} = \dfrac{z}{-1}$ hence one eigen vector is $(1, 2, -1)$.

For $\lambda = -3$ we have
$$x + 2y - 3z = 0$$
$$2x + 4y - 6z = 0$$
$$-x - 2y + 3z = 0$$

which are one equation only as $x + 2y - 3z = 0$

Therefore one eigen vector is $(1, 1, 1)$ and the second eigen vector can be taken as $(2, -1, 0)$. Thus, for $\lambda = 5$ the eigen vector is $(1, 2, -1)$ whereas for $\lambda = -3$ we have eigen vectors as $(1, 1, 1)$ and $(2, -1, 0)$. **Ans.**

EXAMPLE 13.12. Find the eigen values and eigen vectors of the matrix

$$\text{A} = \begin{bmatrix} 3 & 1 & 4 \\ 0 & 2 & 6 \\ 0 & 0 & 5 \end{bmatrix}.$$ **(UPTU 2001 Sp. Exam.)**

SOLUTION: The characteristic equation of A is $|\text{A} - \lambda \text{I}| = 0$

or $$\begin{vmatrix} 3-\lambda & 1 & 4 \\ 0 & 2-\lambda & 6 \\ 0 & 0 & 5-\lambda \end{vmatrix} = 0 \quad \text{or} \quad (3-\lambda)(2-\lambda)(5-\lambda) = 0.$$ Hence $\lambda = 2, 3, 5$.

Thus the eigen values of A are 2, 3 and 5.

For eigen vectors, we have $(A - \lambda I)X = 0$ where $x = (x_1, x_2, x_3)$.

or
$$(3 - \lambda)x_1 + x_2 + 4x_3 = 0$$
$$(2 - \lambda)x_2 + 6x_3 = 0$$
$$(5 - \lambda)x_3 = 0$$

For $\lambda = 2$, we have $x_1 + x_2 + 4x_3 = 0$, $6x_3 = 0$ and $3x_3 = 0$, $\therefore$ $x_1 + x_2 = 0$

Hence eigen vector for $\lambda = 2$ is $(1, -1, 0)$.

For $\lambda = 3$, we have $x_2 + 4x_3 = 0$, $-x_2 + 6x_3 = 0$ and $2x_3 = 0$ hence $x_2 = x_3 = 0$ and thus the eigen vector for $\lambda = 3$ is $(1, 0, 0)$

For $\lambda = 5$ we have $-2x_1 + x_2 + 4x_3 = 0$, $-3x_2 + 6x_3 = 0$

Hence $x_2 = 2x_3$ and $x_1 = 3x_3$

hence eigen vector for $\lambda = 5$ is $(3, 2, 1)$. **Ans.**

EXAMPLE 13.13. Find the characteristic equation of the matrix $\begin{bmatrix} 1 & 2 & 2 \\ 0 & 2 & 1 \\ -1 & 2 & 2 \end{bmatrix}$. Also find the eigen values and eigen vectors of the matrix. **(UPTU 2006-07)**

SOLUTION: $A = \begin{bmatrix} 1 & 2 & 2 \\ 0 & 2 & 1 \\ -1 & 2 & 2 \end{bmatrix}$. Its characteristic equation is $|A - \lambda I| = 0$

or
$$\begin{vmatrix} 1-\lambda & 2 & 2 \\ 0 & 2-\lambda & 1 \\ -1 & 2 & 2-\lambda \end{vmatrix} \quad \text{or} \quad (1 - \lambda)[(2 - \lambda)^2 - 2] + (-1)(2 - 4 + 2\lambda) = 0$$

or
$$(1 - \lambda)[\lambda^2 - 4\lambda + 2 + 2] = 0 \quad \text{or} \quad (1 - \lambda)(2 - \lambda)^2 = 0.$$

Hence the eigen values of A are $\lambda = 1, 2, 2$.

For eigen vectors we have $(A - \lambda I)X = 0$

or
$$(1 - \lambda)x_1 + 2x_2 + 2x_3 = 0$$
$$0x_1 + (2 - \lambda)x_2 + x_3 = 0$$
$$-1x_1 + 2x_2 + (2 - \lambda)x_3 = 0$$

When $\lambda = 1$, we have $x_2 + x_3 = 0$ and $-x_1 + 2x_2 + x_3 = 0$

$$\Rightarrow \quad \frac{x_1}{1} = \frac{x_2}{1} = \frac{x_3}{-1} \quad \therefore \text{ eigen vector is } (1, 1, -1)$$

When $\lambda = 2$, we have $-x_1 + 2x_2 + 2x_3 = 0$, $x_3 = 0$ and $-x_1 + 2x_2 = 0$

$\Rightarrow$ $x_3 = 0$ and $x_1 = 2x_2$

The eigen vector for $\lambda = 2$ is $(2, 1, 0)$ only. **Ans.**

CALEY HAMILTON THEOREM

Every square matrix satisfies its own characteristic equation. **(UPTU 2001 Sp. Exam.)**

PROOF: For a square matrix A of order n the characteristic equation is $|A - \lambda I| = 0$

or $(-1)^n [\lambda^n + a_1\lambda^{n-1} + a_2\lambda^{n-2} + ... + a_n] = 0$ ...(1)

Replacing λ by A, gives $A^n + a_1A^{n-1} + a_2 A^{n-2} + ... + a_n I = 0$ which is to be proved.

We know that $[A - \lambda I]$ *adj* $[A - \lambda I] = |A - \lambda I|\, I$ ...(2)

Since adj $(A - \lambda I)$ has elements as cofactors of the elements of $|A - \lambda I|$ and the elements of $[A - \lambda I]$ are at the most of first degree in λ. The elements of adj $(A - \lambda I)$ are polynomials in λ of degree $n - 1$ or less.

Therefore, adj $(A - \lambda I)$ can be written as a matrix polynomial in λ, as

$$\text{adj}(A - \lambda I) = B_0\lambda^{n-1} + B_1\lambda^{n-2} + ... + B_{n-1}. \quad ...(3)$$

Using (1) and (3), equation (2) becomes

$$(-1)^n [\lambda^n + a_1\lambda^{n-1} + a_2\lambda^{n-2} + ... + a_n] I = (A - \lambda I) [B_0\lambda^{n-1} + B_1\lambda^{n-2} + ... + B^{n-1}]$$

Comparing the co-efficients of terms of same powers of λ on both sides, we get

$$(-1)^n I = -B_0 I$$
$$(-1)^n a_1 I = AB_0 - B_1 I$$
$$(-1)^n a_2 I = AB_1 - B_2 I$$
$$.............. \quad ....$$
$$(-1)^n a_n I = AB_{n-1}$$

Premultiplying the above equations by $A^n, A^{n-1}, A^{n-2}, ..., I$ respectively, and then adding them, we get

$$(-1)^n[A^n + a_1A^{n-1} + a_2A^{n-2} + ... + a_n I] = -B_0A^n + A^nB_0 - B_1A^{n-1} + A^{n-1}B_1 - B_2A^{n-2} + ... + AB_{n-1} = 0$$

$\Rightarrow$ $A^n + a_1A^{n-1} + a_2A^{n-2} + ... + a_n I = 0$, hence the theorem.

The theorem can be used to find A^{-1} as follows:

Premultiplying the above equation by A^{-1}, gives

$$A^{n-1} + a_1A^{n-2} + a_2A^{n-3} + ... + a_{n-1}I + a_nA^{-1} = 0$$

$$\therefore \quad A^{-1} = \frac{-1}{a_n}[A^{n-1} + a_1A^{n-2} + a_2A^{n-3} + ... + a_{n-1} I].$$

EXAMPLE 13.14. (*a*) Verify Caley-Hamilton theorem for the matrix

$A = \begin{bmatrix} 1 & 2 \\ 2 & -1 \end{bmatrix}$ and hence find A^{-1}. **(UPTU 2008-09)**

(*b*) Use Caley-Hamilton theorem to express

$A^5 - 4A^4 - 7A^3 + 11 A^2 - A - 10 I$ in terms of A where $A = \begin{bmatrix} 1 & 4 \\ 2 & 3 \end{bmatrix}$.

SOLUTION: (*a*) Characterstic equation is $|A - \lambda I| = 0$

or $\begin{vmatrix} 1-\lambda & 2 \\ 2 & -1-\lambda \end{vmatrix} = 0$ or $\lambda^2 - 5 = 0.$

By Caley-Hamilton theorem $A^2 - 5I = 0$

$$A^2 - 5I = \begin{bmatrix} 1 & 2 \\ 2 & -1 \end{bmatrix}\begin{bmatrix} 1 & 2 \\ 2 & -1 \end{bmatrix} - \begin{bmatrix} 5 & 0 \\ 0 & 5 \end{bmatrix} = \begin{bmatrix} 5 & 0 \\ 0 & 5 \end{bmatrix} - \begin{bmatrix} 5 & 0 \\ 0 & 5 \end{bmatrix} = \begin{bmatrix} 0 & 0 \\ 0 & 0 \end{bmatrix}$$ **Hence the theorem.**

Multiplying $A^2 - 5I = 0$ by A^{-1}, we get $A - 5A^{-1} = 0$

or $$A^{-1} = \frac{1}{5}A = \begin{bmatrix} \frac{1}{5} & \frac{2}{5} \\ \frac{2}{5} & \frac{-1}{5} \end{bmatrix}.$$ **Ans.**

(*b*) Characteristic equation of A is $|A - \lambda I| = 0$

or $\begin{vmatrix} 1-\lambda & 4 \\ 2 & 3-\lambda \end{vmatrix} = 0$ or $\lambda^2 - 4\lambda - 5 = 0$ ... (1)

By Caley-Hamilton theorem, A satisfies (1) hence $A^2 - 4A - 5I = 0.$ ... (2)

Therefore the given expression

$$A^5 - 4A^4 - 7A^3 + 11A^2 - A - 10I = A^3(A^2 - 4A - 5I) - 2A^3 + 11A^2 - A - 10I$$

$$= A^3(O) - 2A^3 + 11A^2 - A - 10I = -2A(A^2 - 4A - 5I) + 3A^2 - 11A - 10I$$

$$= -2A(O) + 3A^2 - 11A - 10I = 3(A^2 - 4A - 5I) + A + 5I = O + A + 5I$$

$$= \begin{bmatrix} 6 & 4 \\ 2 & 8 \end{bmatrix}.$$ **Ans.**

EXAMPLE 13.15. Verify Caley-Hamilton theorem for the matrix

$$A = \begin{bmatrix} 1 & 2 & -2 \\ -1 & 3 & 0 \\ 0 & -2 & 1 \end{bmatrix}$$

and use it to find A^{-1} and A^4.

SOLUTION: The characteristic equation of A is $|A - \lambda I| = 0$

or $$\begin{vmatrix} 1-\lambda & 2 & -2 \\ -1 & 3-\lambda & 0 \\ 0 & -2 & 1-\lambda \end{vmatrix} = 0$$

which, on expansion, gives $\lambda^3 - 5\lambda^2 + 9\lambda - 1 = 0.$ This means, we are to verify that

$A^3 - 5A^2 + 9A - I = 0.$ ...(1)

Now $$A^2 = \begin{bmatrix} 1 & 2 & -2 \\ -1 & 3 & 0 \\ 0 & -2 & 1 \end{bmatrix}\begin{bmatrix} 1 & 2 & -2 \\ -1 & 3 & 0 \\ 0 & -2 & 1 \end{bmatrix} = \begin{bmatrix} -1 & 12 & -4 \\ -4 & 7 & 2 \\ 2 & -8 & 1 \end{bmatrix}$$

and $$A^3 = A^2A = \begin{bmatrix} -1 & 12 & -4 \\ -4 & 7 & 2 \\ 2 & -8 & 1 \end{bmatrix}\begin{bmatrix} 1 & 2 & -2 \\ -1 & 3 & 0 \\ 0 & -2 & 1 \end{bmatrix} = \begin{bmatrix} -13 & 42 & -2 \\ -11 & 9 & 10 \\ 10 & -22 & -3 \end{bmatrix}$$

$$\therefore \quad A^3 - 5A^2 + 9A - I = \begin{bmatrix} -13+5+9-1 & 42-60+18-0 & -2+20-18-0 \\ -11+20-9-0 & 9-35+27-1 & 10-10+0-0 \\ 10-10+0-0 & -22+40-18-0 & -3-5+9-1 \end{bmatrix} = O$$

Hence the theorem is verified.

Now, multiplying (1) throughout by, A^{-1}, we get $A^2 - 5A + 9I - A^{-1} = 0$

$$\therefore \quad A^{-1} = A^2 - 5A + 9I = \begin{bmatrix} -1-5+9 & 12-10+0 & -4+10+0 \\ -4+5+0 & 7-15+9 & 2+0+0 \\ 2+0+0 & -8+10+0 & 1-5+9 \end{bmatrix} = \begin{bmatrix} 3 & 2 & 6 \\ 1 & 1 & 2 \\ 2 & 2 & 5 \end{bmatrix}$$

Next, to find A^4 we multiply (1) by A and get $A^4 - 5A^3 + 9A^2 - A = 0$

$$\therefore \quad A^4 = 5A^3 - 9A^2 + A = \begin{bmatrix} -65+9+1 & 210-108+2 & -10+36-2 \\ -55+36-1 & 45-63+3 & 50-18+0 \\ 50-18+0 & -110+72-2 & -15-9+1 \end{bmatrix} = \begin{bmatrix} -55 & 104 & 24 \\ -20 & -15 & 32 \\ 32 & -40 & -23 \end{bmatrix}. \quad \textbf{Ans.}$$

EXAMPLE 13.16. Verify Caley-Hamilton theorem for the matrix

$$A = \begin{bmatrix} 2 & 1 & 1 \\ 0 & 1 & 0 \\ 1 & 1 & 2 \end{bmatrix}$$

Also express $A^8 - 5A^7 + 7A^6 - 3A^5 + A^4 - 5A^3 + 8A^2 - 2A + I$ as quadratic polynomial in A. **(UPTU 2003)**

SOLUTION: The characteristic equation of A is $|A - \lambda I| = 0$

or $$\begin{vmatrix} 2-\lambda & 1 & 1 \\ 0 & 1-\lambda & 0 \\ 1 & 1 & 2-\lambda \end{vmatrix} = 0$$

or $(2 - \lambda)(1 - \lambda)(2 - \lambda) - (1 - \lambda) = 0$ or $\lambda^3 - 5\lambda^2 + 7\lambda - 3 = 0$

By Caley-Hamilton theorem we have $A^3 - 5A^2 + 7A - 3I = 0$.

Now $A^2 = \begin{bmatrix} 2 & 1 & 1 \\ 0 & 1 & 0 \\ 1 & 1 & 2 \end{bmatrix} \begin{bmatrix} 2 & 1 & 1 \\ 0 & 1 & 0 \\ 1 & 1 & 2 \end{bmatrix} = \begin{bmatrix} 5 & 4 & 4 \\ 0 & 1 & 0 \\ 4 & 4 & 5 \end{bmatrix}$

and $A^3 = \begin{bmatrix} 5 & 4 & 4 \\ 0 & 1 & 0 \\ 4 & 4 & 5 \end{bmatrix} \begin{bmatrix} 2 & 1 & 1 \\ 0 & 1 & 0 \\ 1 & 1 & 2 \end{bmatrix} = \begin{bmatrix} 14 & 13 & 13 \\ 0 & 1 & 0 \\ 13 & 13 & 14 \end{bmatrix}$

$$\therefore \quad A^3 - 5A^2 + 7A - 3I = \begin{bmatrix} 14 & 13 & 13 \\ 0 & 1 & 0 \\ 13 & 13 & 14 \end{bmatrix} - \begin{bmatrix} 25 & 20 & 20 \\ 0 & 5 & 0 \\ 20 & 20 & 25 \end{bmatrix} + \begin{bmatrix} 14 & 7 & 7 \\ 0 & 7 & 0 \\ 7 & 7 & 14 \end{bmatrix} - \begin{bmatrix} 3 & 0 & 0 \\ 0 & 3 & 0 \\ 0 & 0 & 3 \end{bmatrix}$$

$$= \begin{bmatrix} 0 & 0 & 0 \\ 0 & 0 & 0 \\ 0 & 0 & 0 \end{bmatrix}.$$ Hence the theorem is verified.

Next, $A^8 - 5A^7 + 7A^6 - 3A^5 + A^4 - 5A^3 + 8A^2 - 2A + I$

$$= A^5 (A^3 - 5A^2 + 7A - 3I) + A (A^3 - 5A^2 + 8A - 2I) + I$$
$$= 0 + A (A^3 - 5A^2 + 7A - 3I) + A^2 + A + I$$
$$= 0 + 0 + A^2 + A + I = A^2 + A + I.$$ **Ans.**

EXAMPLE 13.17. Verify Caley-Hamiltion theorem for the matrix

$$A = \begin{bmatrix} 0 & 0 & 1 \\ 3 & 1 & 0 \\ 2 & 1 & 4 \end{bmatrix}.$$ Hence compute A^{-1}.

SOLUTION: The characteristic equation of A is $|A - \lambda I| = 0$

or $\begin{vmatrix} -\lambda & 0 & 1 \\ 3 & 1-\lambda & 0 \\ 2 & 1 & 4-\lambda \end{vmatrix} = 0$ or $\lambda^3 - 5\lambda^2 + 2\lambda - 1 = 0.$ ...(1)

By Caley-Hamilton theorem, matrix A should satisfy (1), that is,

$A^3 - 5A^2 + 2A - I = 0$ should be true. ...(2)

Now $A^2 = \begin{bmatrix} 0 & 0 & 1 \\ 3 & 1 & 0 \\ 2 & 1 & 4 \end{bmatrix} \begin{bmatrix} 0 & 0 & 1 \\ 3 & 1 & 0 \\ 2 & 1 & 4 \end{bmatrix} = \begin{bmatrix} 2 & 1 & 4 \\ 3 & 1 & 3 \\ 11 & 5 & 18 \end{bmatrix}$

and $A^3 = A \,.\, A^2 = \begin{bmatrix} 0 & 0 & 1 \\ 3 & 1 & 0 \\ 2 & 1 & 4 \end{bmatrix} \begin{bmatrix} 2 & 1 & 4 \\ 3 & 1 & 3 \\ 11 & 5 & 18 \end{bmatrix} = \begin{bmatrix} 11 & 5 & 18 \\ 9 & 4 & 15 \\ 51 & 23 & 83 \end{bmatrix}$

$$\therefore\ A^3 - 5A^2 + 2A - I = \begin{bmatrix} 11 & 5 & 18 \\ 9 & 4 & 15 \\ 51 & 23 & 83 \end{bmatrix} + \begin{bmatrix} -10 & -5 & -20 \\ -15 & -5 & -15 \\ -55 & -25 & -90 \end{bmatrix} + \begin{bmatrix} 0 & 0 & 2 \\ 6 & 2 & 0 \\ 4 & 2 & 8 \end{bmatrix} + \begin{bmatrix} -1 & 0 & 0 \\ 0 & -1 & 0 \\ 0 & 0 & -1 \end{bmatrix}$$

$$= \begin{bmatrix} 0 & 0 & 0 \\ 0 & 0 & 0 \\ 0 & 0 & 0 \end{bmatrix}$$ Hence Caley-Hamilton theorem is verified.

Premultiplying (2) by A^{-1}, we get $A^2 - 5A + 2I - A^{-1} = 0$ or $A^{-1} = A^2 - 5A + 2I$

or $$A^{-1} = \begin{bmatrix} 2 & 1 & 4 \\ 3 & 1 & 3 \\ 11 & 5 & 18 \end{bmatrix} + \begin{bmatrix} 0 & 0 & -5 \\ -15 & -5 & 0 \\ -10 & -5 & -20 \end{bmatrix} + \begin{bmatrix} 2 & 0 & 0 \\ 0 & 2 & 0 \\ 0 & 0 & 2 \end{bmatrix} = \begin{bmatrix} 4 & 1 & -1 \\ -12 & -2 & 3 \\ 1 & 0 & 0 \end{bmatrix}.$$ **Ans.**

EXAMPLE 13.18. Find the characteristic equation of the matrix

$$A = \begin{bmatrix} 1 & -1 & 1 \\ 4 & 1 & 0 \\ 8 & 1 & 1 \end{bmatrix}.$$ Hence find A^{-1}.

SOLUTION: The characteristic eqaution of A is $|A - \lambda I| = 0$

or $$\begin{vmatrix} 1-\lambda & -1 & 1 \\ 4 & 1-\lambda & 0 \\ 8 & 1 & 1-\lambda \end{vmatrix} = 0$$ or $(1-\lambda)^3 + 4 - 4\lambda + 4 - 8 + 8\lambda = 0$

or $\lambda^3 - 3\lambda^2 - \lambda - 1 = 0$ is the required characterstic equation. ...(1)

Now by Caley-Hamilton theorem A satisfies (1) hence we have $A^3 - 3A^2 - A - I = 0$

Multiplying throughout by A^{-1}, we get $A^2 - 3A - I = A^{-1}$. ...(2)

Here $$A^2 = \begin{bmatrix} 1 & -1 & 1 \\ 4 & 1 & 0 \\ 8 & 1 & 1 \end{bmatrix}\begin{bmatrix} 1 & -1 & 1 \\ 4 & 1 & 0 \\ 8 & 1 & 1 \end{bmatrix} = \begin{bmatrix} 5 & -1 & 2 \\ 8 & -3 & 4 \\ 20 & -6 & 9 \end{bmatrix}$$

Hence from (2), we get

$$A^{-1} = \begin{bmatrix} 5 & -1 & 2 \\ 8 & -3 & 4 \\ 20 & -6 & 9 \end{bmatrix} + \begin{bmatrix} -3 & 3 & -3 \\ -12 & -3 & 0 \\ -24 & -3 & -3 \end{bmatrix} + \begin{bmatrix} -1 & 0 & 0 \\ 0 & -1 & 0 \\ 0 & 0 & -1 \end{bmatrix}$$

$$= \begin{bmatrix} 1 & 2 & -1 \\ -4 & -7 & 4 \\ -4 & -9 & 5 \end{bmatrix}.$$ **Ans.**

EXAMPLE 13.19. Find the inverse of the matrix $A = \begin{bmatrix} 7 & -1 & 3 \\ 6 & 1 & 4 \\ 2 & 4 & 8 \end{bmatrix}$ using Caley Hamilton theorem.

SOLUTION: The characteristic equation of the matrix A is $|A - \lambda I| = 0$

or $$\begin{vmatrix} 7-\lambda & -1 & 3 \\ 6 & 1-\lambda & 4 \\ 2 & 4 & 8-\lambda \end{vmatrix} = 0$$

or $(7 - \lambda)[(1 - \lambda)(8 - \lambda) - 16] + 1[48 - 6\lambda - 8] + 3[24 - 2 + 2\lambda] = 0$

or $(7 - \lambda)(\lambda^2 - 9\lambda - 8) - 6\lambda + 40 + 6\lambda + 66 = 0$ or $\lambda^3 - 16\lambda^2 + 55\lambda - 50 = 0$

By Caley Hamilton theorem $A^3 - 16A^2 + 55A - 50I = 0$

$\Rightarrow$ $A^2 - 16A + 55I - 50A^{-1} = 0$

$$\therefore \quad A^{-1} = \frac{1}{50}[A^2 - 16A + 55I]$$

Here, $$A^2 = \begin{bmatrix} 7 & -1 & 3 \\ 6 & 1 & 4 \\ 2 & 4 & 8 \end{bmatrix}\begin{bmatrix} 7 & -1 & 3 \\ 6 & 1 & 4 \\ 2 & 4 & 8 \end{bmatrix} = \begin{bmatrix} 49 & 4 & 41 \\ 56 & 11 & 54 \\ 54 & 34 & 86 \end{bmatrix}$$

$$\therefore \quad A^{-1} = \frac{1}{50}\begin{bmatrix} 49 & 4 & 41 \\ 56 & 11 & 54 \\ 54 & 34 & 86 \end{bmatrix} - \frac{16}{50}\begin{bmatrix} 7 & -1 & 3 \\ 6 & 1 & 4 \\ 2 & 4 & 8 \end{bmatrix} + \frac{1}{50}\begin{bmatrix} 55 & 0 & 0 \\ 0 & 55 & 0 \\ 0 & 0 & 55 \end{bmatrix}$$

$$= \frac{1}{50}\left[\begin{array}{c|c|c} 49 - 112 + 55 & 4 + 16 \quad 0 & 41 - 48 + 0 \\ 56 - 96 + 0 & 11 - 16 + 55 & 54 - 64 + 0 \\ 54 - 32 + 0 & 34 - 64 + 0 & 86 - 128 + 55 \end{array}\right]$$

$$= \frac{1}{50}\begin{bmatrix} -8 & 20 & -7 \\ -40 & 50 & -10 \\ 22 & -30 & 13 \end{bmatrix}$$

Thus, $$A^{-1} = \begin{bmatrix} -4/25 & 2/5 & -7/50 \\ -4/5 & 1 & -1/5 \\ 11/25 & -3/5 & 13/50 \end{bmatrix}.$$ **Ans.**

EXAMPLE 13.20. Verity Caley-Hamilton theorem and find the inverse of the matrix $\begin{bmatrix} 7 & 2 & -2 \\ -6 & -1 & 2 \\ 6 & 2 & -1 \end{bmatrix}$.

SOLUTION: Characteristic equation is $|A - \lambda I| = 0$ or $\begin{vmatrix} 7-\lambda & 2 & -2 \\ -6 & -1-\lambda & 2 \\ 6 & 2 & -1-\lambda \end{vmatrix} = 0$

which gives $\lambda^3 - 5\lambda^2 + 7\lambda - 3 = 0$. By Caley Hamilton theorem, A must satisfy it.

Hence $A^3 - 5A^2 + 7A - 3I = 0$ should be true.

Now $$A^2 = \begin{bmatrix} 7 & 2 & -2 \\ -6 & -1 & 2 \\ 6 & 2 & -1 \end{bmatrix}\begin{bmatrix} 7 & 2 & -2 \\ -6 & -1 & 2 \\ 6 & 2 & -1 \end{bmatrix} = \begin{bmatrix} 25 & 8 & -8 \\ -24 & -7 & 8 \\ 24 & 8 & -7 \end{bmatrix}$$

and $$A^3 = \begin{bmatrix} 25 & 8 & -8 \\ -24 & -7 & 8 \\ 24 & 8 & -7 \end{bmatrix}\begin{bmatrix} 7 & 2 & -2 \\ -6 & -1 & 2 \\ 6 & 2 & -1 \end{bmatrix} = \begin{bmatrix} 79 & 26 & -26 \\ -78 & -25 & 26 \\ 78 & 26 & -25 \end{bmatrix}$$

Hence $$A^3 - 5A^2 + 7A - 3I = \begin{bmatrix} 79 & 26 & -26 \\ -78 & 25 & 26 \\ 78 & 26 & -25 \end{bmatrix} - \begin{bmatrix} 125 & 40 & -40 \\ -120 & -35 & 40 \\ 120 & 40 & -35 \end{bmatrix} + \begin{bmatrix} 49 & 14 & -14 \\ -42 & -7 & 14 \\ 42 & 14 & -7 \end{bmatrix} - \begin{bmatrix} 3 & 0 & 0 \\ 0 & 3 & 0 \\ 0 & 0 & 3 \end{bmatrix} = 0$$

Hence the theorem is verified.

To find A^{-1} we premultiply the above equation by A^{-1} and get

$$A^2 - 5A + 7I - 3A^{-1} = 0 \quad \therefore \quad A^{-1} = \frac{1}{3}[A^2 - 5A + 7I]$$

or $$A^{-1} = \frac{1}{3}\left\{\begin{bmatrix} 25 & 8 & -8 \\ -24 & -7 & 8 \\ 24 & 8 & -7 \end{bmatrix} - 5\begin{bmatrix} 7 & 2 & -2 \\ -6 & -1 & 2 \\ 6 & 2 & -1 \end{bmatrix} + 7\begin{bmatrix} 1 & 0 & 0 \\ 0 & 1 & 0 \\ 0 & 0 & 1 \end{bmatrix}\right\} = \frac{1}{3}\begin{bmatrix} -3 & -2 & 2 \\ 6 & 5 & -2 \\ -6 & -2 & 5 \end{bmatrix}.$$ **Ans.**

EXAMPLE 13.21. State Caley-Hamilton theorem. Using this theorem find the inverse of the matrix A given by $A = \begin{bmatrix} 2 & -1 & 1 \\ -1 & 2 & -1 \\ 1 & -1 & 2 \end{bmatrix}$. **(UPTU 2003-04; 2004-05; 2005-06; 2011-12)**

SOLUTION: Characteristic equation of A is $|A - \lambda I| = 0$

or $$\begin{vmatrix} 2-\lambda & -1 & 1 \\ -1 & 2-\lambda & -1 \\ 1 & -1 & 2-\lambda \end{vmatrix} = 0 \quad \text{or} \quad \lambda^3 - 6\lambda^2 + 9\lambda - 4 = 0.$$

By Caley-Hamilton theorem $A^3 - 6A^2 + 9A - 4I = 0$

Multiplying both sides by A^{-1}, we get $A^2 - 6A + 9I - 4A^{-1} = 0$.

Now, $$A^2 = \begin{bmatrix} 2 & -1 & 1 \\ -1 & 2 & -1 \\ 1 & -1 & 2 \end{bmatrix}\begin{bmatrix} 2 & -1 & 1 \\ -1 & 2 & -1 \\ 1 & -1 & 2 \end{bmatrix} = \begin{bmatrix} 6 & -5 & 5 \\ -5 & 6 & -5 \\ 5 & -5 & 6 \end{bmatrix}$$

$$\therefore \quad A^2 - 6A + 9I = \begin{bmatrix} 6 & -5 & 5 \\ -5 & 6 & -5 \\ 5 & -5 & 6 \end{bmatrix} + \begin{bmatrix} -12 & 6 & -6 \\ 6 & -12 & 6 \\ -6 & 6 & -12 \end{bmatrix} + \begin{bmatrix} 9 & 0 & 0 \\ 0 & 9 & 0 \\ 0 & 0 & 9 \end{bmatrix}$$

$$= \begin{bmatrix} 3 & 1 & -1 \\ 1 & 3 & 1 \\ -1 & 1 & 3 \end{bmatrix} \quad \therefore \quad A^{-1} = \frac{1}{4}\begin{bmatrix} 3 & 1 & -1 \\ 1 & 3 & 1 \\ -1 & 1 & 3 \end{bmatrix}.$$ **Ans.**

EXAMPLE 13.22. (*a*) Verify Caley-Hamilton theorem for the matrix $\begin{bmatrix} 1 & 2 & 3 \\ 2 & 4 & 5 \\ 3 & 5 & 6 \end{bmatrix}$. Also, find its inverse using this theorem. **(UPTU 2006-07)**

(*b*) If $A = \begin{bmatrix} 1 & 0 & 0 \\ 1 & 0 & 1 \\ 0 & 1 & 0 \end{bmatrix}$ then show that

$$A^n = A^{n-2} + A^2 - I \quad \text{for} \quad n \geq 3.$$ **(UPTU 2008-09)**

SOLUTION: (*a*) Let $A = \begin{bmatrix} 1 & 2 & 3 \\ 2 & 4 & 5 \\ 3 & 5 & 6 \end{bmatrix}$. Its characteristic equation is $|A - \lambda I| = 0$

or $$\begin{vmatrix} 1-\lambda & 2 & 3 \\ 2 & 4-\lambda & 5 \\ 3 & 5 & 6-\lambda \end{vmatrix} = 0 \quad \text{or} \quad \lambda^3 - 11\lambda^2 - 4\lambda + 1 = 0$$

By Caley-Hamilton theorem we have $A^3 - 11A^2 - 4A + I = 0$.

Here, $$A^2 = \begin{bmatrix} 1 & 2 & 3 \\ 2 & 4 & 5 \\ 3 & 5 & 6 \end{bmatrix}\begin{bmatrix} 1 & 2 & 3 \\ 2 & 4 & 5 \\ 3 & 5 & 6 \end{bmatrix} = \begin{bmatrix} 14 & 25 & 31 \\ 25 & 45 & 56 \\ 31 & 56 & 70 \end{bmatrix}$$

and $$A^3 = \begin{bmatrix} 1 & 2 & 3 \\ 2 & 4 & 5 \\ 3 & 5 & 6 \end{bmatrix}\begin{bmatrix} 14 & 25 & 31 \\ 25 & 45 & 56 \\ 31 & 56 & 70 \end{bmatrix} = \begin{bmatrix} 157 & 283 & 353 \\ 283 & 510 & 636 \\ 353 & 636 & 793 \end{bmatrix}$$

$\therefore \quad A^3 - 11A^2 - 4A + I = 0$ (on simplification).

Next pre-multiplying both sides by A^{-1}, we get

$$A^2 - 11A - 4I + A^{-1} = 0$$

$$\therefore \quad A^{-1} = -A^2 + 11A + 4I$$

$$= \begin{bmatrix} -14 & -25 & -31 \\ -25 & -45 & -56 \\ -31 & -56 & -70 \end{bmatrix} + \begin{bmatrix} +11 & +22 & +33 \\ +22 & +44 & +55 \\ +33 & +55 & +66 \end{bmatrix} + \begin{bmatrix} +4 & 0 & 0 \\ 0 & +4 & 0 \\ 0 & 0 & +4 \end{bmatrix}$$

$$= \begin{bmatrix} 1 & -3 & 2 \\ -3 & 3 & -1 \\ 2 & -1 & 0 \end{bmatrix}.$$ **Ans.**

(*b*) The characterstic equation of A is $|A-\lambda I|=0$

or $$\begin{vmatrix} 1-\lambda & 0 & 0 \\ 1 & -\lambda & 1 \\ 0 & 1 & -\lambda \end{vmatrix}=0 \quad \text{or} \quad (1-\lambda)(\lambda^2-1)=0 \quad \text{or} \quad \lambda^3-\lambda^2-\lambda+1=0$$

By Caley-Hamilton theorem we have $A^3-A^2-A+1=0$, hence $A^3=A+A^2-I \Rightarrow A^n=A^{n-2}+A^2-I$ is true for $n=3$.

$\Rightarrow$ $A^4=A^3+A^2-A=A^2+A-I+A^2-A=2A^2-I \Rightarrow A^n=A^{n-2}+A^2-I$ is true for $n=4$

Next, $A^5=A^4+A^3-A^2=2A^2-I+A^3-A^2=A^3+A^2-I \Rightarrow A^n=A^{n-2}+A^2-I$ for $n=5$

Similarly we can show that $A^n=A^{n-2}+A^2-I$ is true for $n=6, 7, \ldots$

Thus $A^n=A^{n-2}+A^2-I$ is true for $n \geq 3$. **Hence Proved.**

DIAGONALISING A SQUARE MATRIX

THEOREM: For any square matrix A of order n having linearly independent eigen vectors there exists a matrix P such that $P^{-1}AP$ is a diagonal matrix.

PROOF: For convenience, we choose A as a 3×3 square matrix. Let its eigen values be $\lambda_1, \lambda_2, \lambda_3$ and corresponding eigen vectors be:

$$X_1=\begin{bmatrix} x_1 \\ y_1 \\ z_1 \end{bmatrix}, \quad X_2=\begin{bmatrix} x_2 \\ y_2 \\ z_2 \end{bmatrix}, \quad X_3=\begin{bmatrix} x_3 \\ y_3 \\ z_3 \end{bmatrix}, \quad \text{so that } AX_1=\lambda_1X_1,\ AX_2=\lambda_2X_2,\ AX_3=\lambda_3X_3.$$

Now consider the square matrix

$$[X_1\ X_2\ X_3]=\begin{bmatrix} x_1 & x_2 & x_3 \\ y_1 & y_2 & y_3 \\ z_1 & z_2 & z_3 \end{bmatrix}=P \text{ say, then we have}$$

$$A.P=A[X_1 X_2 X_3]=[AX_1, AX_2, AX_3]=[\lambda_1X_1, \lambda_2X_2, \lambda_3X_3]$$

$$=\begin{bmatrix} \lambda_1x_1 & \lambda_2x_2 & \lambda_3x_3 \\ \lambda_1y_1 & \lambda_2y_2 & \lambda_3y_3 \\ \lambda_1z_1 & \lambda_2z_2 & \lambda_3z_3 \end{bmatrix}=\begin{bmatrix} x_1 & x_2 & x_3 \\ y_1 & y_2 & y_3 \\ z_1 & z_2 & z_3 \end{bmatrix}\begin{bmatrix} \lambda_1 & 0 & 0 \\ 0 & \lambda_2 & 0 \\ 0 & 0 & \lambda_3 \end{bmatrix}$$

$$=PD \quad \text{where D is the diagonal matrix} \begin{bmatrix} \lambda_1 & 0 & 0 \\ 0 & \lambda_2 & 0 \\ 0 & 0 & \lambda_3 \end{bmatrix}$$

$$\Rightarrow \quad P^{-1}(AP)=P^{-1}(PD)=(P^{-1}P)D=ID=D.$$

The matrix P which diagonalizes the given matrix A, is called the **MODAL MATRIX** and the resulting diagonal matrix D is called the **SPECTRAL MATRIX** of A. The modal matrix P is found by grouping the eigen vectors of A into square matrix and diagonal matrix has the eigen values of A as its elements.

A transformation of the form $M^{-1}AM$ on a square matrix A is called **similarity transformation.**

EXAMPLE 13.23. (*a*) Let $A = \begin{bmatrix} 6 & -2 & 2 \\ -2 & 3 & -1 \\ 2 & -1 & 3 \end{bmatrix}$. Find the modal matrix M such that $M^{-1}AM$ is a diagonal matrix.

(*b*) Find a matrix P which diagonalizes the matrix $A = \begin{bmatrix} 4 & 1 \\ 2 & 3 \end{bmatrix}$ and verify $P^{-1}AP = D$ where D is a diagonal matrix. **(UPTU 2008-09)**

SOLUTION: (*a*) The characteristic equation of matrix A is $|A - \lambda I| = 0$

or $$\begin{vmatrix} 6-\lambda & -2 & 2 \\ -2 & 3-\lambda & -1 \\ 2 & -1 & 3-\lambda \end{vmatrix} = 0$$

or $(6 - \lambda)[(3 - \lambda)^2 - 1] + 2(-6 + 2\lambda + 2) + 2(2 - 6 + 2\lambda) = 0$

or $-\lambda^3 + 12\lambda^2 - 36\lambda + 32 = 0$ or $(\lambda - 2)^2(\lambda - 8) = 0$

Hence, the eigen values are 2, 2 and 8.

For eigen vectors we have $(A - \lambda I)X = 0$

or $$(6 - \lambda)x - 2y + 2z = 0,$$
$$-2x + (3 - \lambda)y - z = 0,$$
$$2x - y + (3 - \lambda)z = 0.$$

For $\lambda = 8$, we have $-2x - 2y + 2z = 0$

$$-2x - 5y - z = 0$$
$$2x - y - 5z = 0$$

which gives $x = 2$, $y = -1$, $z = 1$, hence the eigen vector for $\lambda = 8$, is $\begin{bmatrix} 2 \\ -1 \\ 1 \end{bmatrix}$.

Next, for $\lambda = 2$, we have $$4x - 2y + 2z = 0$$
$$-2x + y - z = 0$$
$$2x - y + z = 0$$

but actually we have here only one equation as $2x - y + z = 0$

Here, let us take $x = t$ then we have $y - z = 2t$

Now for $z = 0$, $y = 2t$, and for $y = 0$, $z = -2t$.

Thus the eigen vectors for $\lambda = 2$ are $\begin{bmatrix} t \\ 2t \\ 0 \end{bmatrix}$ and $\begin{bmatrix} t \\ 0 \\ -2t \end{bmatrix}$

or two eigen vectors for $\lambda = 2$ are $\begin{bmatrix} 1 \\ 2 \\ 0 \end{bmatrix}$ and $\begin{bmatrix} -1 \\ 0 \\ 2 \end{bmatrix}$

Therefore the modal matrix $M = \begin{bmatrix} 2 & 1 & -1 \\ -1 & 2 & 0 \\ 1 & 0 & 2 \end{bmatrix}$

Now we can calculate M^{-1}, then it is easy to verify that

$$M^{-1}AM = \begin{bmatrix} 2 & 0 & 0 \\ 0 & 2 & 0 \\ 0 & 0 & 8 \end{bmatrix} \text{ the diagonal matrix.} \qquad \textbf{Ans.}$$

(*b*) $A = \begin{bmatrix} 4 & 1 \\ 2 & 3 \end{bmatrix}$. The characterstic equation is $|A - \lambda I| = 0$.

or $\begin{vmatrix} 4-\lambda & 1 \\ 2 & 3-\lambda \end{vmatrix} = 0$ or $\lambda^2 - 7\lambda + 10 = 0 \quad \therefore \quad \lambda = 2,\ 5$

Thus, eigen values of A are 2 and 5.

For eigen vectors, we have $(A - \lambda I)X = 0$ where $X = \begin{bmatrix} x \\ y \end{bmatrix}$

or $(4-\lambda)x + y = 0$ and $2x + (3-\lambda)y = 0$

When $\lambda = 2$ we have $2x + y = 0 \quad \therefore \quad x = 1$ then $y = -2$

$\therefore$ $\begin{bmatrix} 1 \\ -2 \end{bmatrix}$ is the eigen vector for $\lambda = 2$.

When $\lambda = 5$ we have $-x + y = 0 \quad \therefore \quad x = 1$ then $y = 1$

$\therefore$ $\begin{bmatrix} 1 \\ 1 \end{bmatrix}$ is the eigen vector for $\lambda = 5$

Therefore modal matrix $P = \begin{bmatrix} 1 & 1 \\ -2 & 1 \end{bmatrix}$ then $P^{-1} = \frac{1}{3}\begin{bmatrix} 1 & -1 \\ 2 & 1 \end{bmatrix}$

Thus, $$P^{-1}AP = \frac{1}{3}\begin{bmatrix} 1 & -1 \\ 2 & 1 \end{bmatrix}\begin{bmatrix} 4 & 1 \\ 2 & 3 \end{bmatrix}\begin{bmatrix} 1 & 1 \\ -2 & 1 \end{bmatrix}$$

$$= \frac{1}{3}\begin{bmatrix} 2 & -2 \\ 10 & 5 \end{bmatrix}\begin{bmatrix} 1 & 1 \\ -2 & 1 \end{bmatrix} = \frac{1}{3}\begin{bmatrix} 6 & 0 \\ 0 & 15 \end{bmatrix} = \begin{bmatrix} 2 & 0 \\ 0 & 5 \end{bmatrix} = D. \qquad \textbf{Hence verified.}$$

EXAMPLE 13.24. Find the matrix P such that $P^{-1}AP$ is a diagonal matrix where $A = \begin{bmatrix} 2 & -2 & 3 \\ 1 & 1 & 1 \\ 1 & 3 & -1 \end{bmatrix}$.

SOLUTION: The characteristic equation of A is $|A - \lambda I| = 0$

or $\begin{vmatrix} 2-\lambda & -2 & 3 \\ 1 & 1-\lambda & 1 \\ 1 & 3 & -1-\lambda \end{vmatrix} = 0$ or $\lambda^3 - 2\lambda^2 - 5\lambda + 6 = 0$

or $(\lambda + 2)(\lambda - 1)(\lambda - 3) = 0.$

Therefore, the eigen values of A are $\lambda_1 = -2$, $\lambda_2 = 1$, $\lambda_3 = 3$.

For characteristic vectors X, we have $(A - \lambda I)X = 0$

$$\text{or} \quad \begin{bmatrix} 2-\lambda & -2 & 3 \\ 1 & 1-\lambda & 1 \\ 1 & 3 & -1-\lambda \end{bmatrix} \begin{bmatrix} x_1 \\ x_2 \\ x_3 \end{bmatrix} = \begin{bmatrix} 0 \\ 0 \\ 0 \end{bmatrix}$$

For $\lambda = \lambda_1 = -2$, we have the equations as

$$4x_1 - 2x_2 + 3x_3 = 0 \quad ...(1)$$

$$x_1 + 3x_2 + x_3 = 0 \quad ...(2)$$

$$x_1 + 3x_2 + x_3 = 0 \quad ...(3)$$

Since above equations (2) and (3) are same, solving (1) and (2), gives

$$\frac{x_1}{-11} = \frac{x_2}{-1} = \frac{x_3}{14} \quad \text{or} \quad \frac{x_1}{11} = \frac{x_2}{1} = \frac{x_3}{-14} \quad ...(4)$$

And for $\lambda = \lambda_2 = 1$ the eigen vector is given by the equations

$$x_1 - 2x_2 + 3x_3 = 0 \quad ...(5)$$

$$x_1 + 0x_2 + x_3 = 0 \quad ...(6)$$

$$x_1 + 3x_2 - 2x_3 = 0 \quad ...(7)$$

Solving any two of these equations, say (5) and (6), we get

$$\frac{x_1}{-2} = \frac{x_2}{2} = \frac{x_3}{2} \quad \text{or} \quad \frac{x_1}{1} = \frac{x_2}{-1} = \frac{x_3}{1}. \quad ...(8)$$

Finally, for $\lambda = \lambda_3 = 3$ the eigen vector is given by the equations

$$-x_1 - 2x_2 + 3x_3 = 0 \quad ...(9)$$

$$x_1 - 2x_2 + x_3 = 0 \quad ...(10)$$

$$x_1 + 3x_2 - 4x_3 = 0 \quad ...(11)$$

Solving any two of the above equations, say (9) and (10), we get

$$\frac{x_1}{4} = \frac{x_2}{4} = \frac{x_3}{4} \quad \text{or} \quad \frac{x_1}{1} = \frac{x_2}{1} = \frac{x_3}{1} \quad ...(12)$$

Thus, we have the eigen vectors

$$X_1 = \begin{bmatrix} 11 \\ 1 \\ -14 \end{bmatrix}, \quad X_2 = \begin{bmatrix} 1 \\ -1 \\ -1 \end{bmatrix}, \quad X_3 = \begin{bmatrix} 1 \\ 1 \\ 1 \end{bmatrix}$$

Therefore, the modal matrix P is given by

$$P = \begin{bmatrix} 11 & 1 & 1 \\ 1 & -1 & 1 \\ -14 & -1 & 1 \end{bmatrix} \quad \therefore \quad P^{-1} = \frac{-1}{30} \begin{bmatrix} 0 & -2 & 2 \\ -15 & 25 & -10 \\ -15 & -3 & -12 \end{bmatrix}$$

$$\text{Hence} \quad P^{-1}AP = \frac{-1}{30} \begin{bmatrix} 0 & -2 & 2 \\ -15 & 25 & -10 \\ -15 & -3 & -12 \end{bmatrix} \begin{bmatrix} 2 & -2 & 3 \\ 1 & 1 & 1 \\ 1 & 3 & -1 \end{bmatrix} \begin{bmatrix} 11 & 1 & 1 \\ 1 & -1 & 1 \\ -14 & -1 & 1 \end{bmatrix} = \begin{bmatrix} -2 & 0 & 0 \\ 0 & 1 & 0 \\ 0 & 0 & 3 \end{bmatrix}. \quad \textbf{Ans.}$$

EXAMPLE 13.25. Diagonalise the matrix $A = \begin{bmatrix} -1 & 1 & 2 \\ 0 & -2 & 1 \\ 0 & 0 & -3 \end{bmatrix}$.

SOLUTION: The characteristic equation is $|A - \lambda I| = 0$

or $\begin{vmatrix} -1-\lambda & 1 & 2 \\ 0 & -2-\lambda & 1 \\ 0 & 0 & -3-\lambda \end{vmatrix} = 0$ or $(-1-\lambda)(-2-\lambda)(-3-\lambda) = 0$

Hence eigen values are $\lambda = -1, -2, -3$.

For $\lambda = -1$ the eigen vector is given by $(A - \lambda I)X = 0$

or
$$(-1+1)x + y + 2z = 0$$
$$-y + z = 0$$
$$-2z = 0$$

Hence eigen vector is $[1 \quad 0 \quad 0]'$.

For $\lambda = -2$ we have the equations as

$$x + y + 2z = 0, \quad z = 0, \quad z = 0 \quad \text{or} \quad y = -x \quad \text{and} \quad z = 0$$

Hence eigen vector is $[1 \quad -1 \quad 0]'$.

For $\lambda = -3$ we have the equations as

$$2x + y + 2z = 0$$
$$y + z = 0$$
$$0z = 0$$

Taking $z = 1$ then $y = -1$ and $x = -\frac{1}{2}$

Hence eigen vector is $[1 \quad 2, -2]'$.

Therefore, the modal matrix $P = \begin{bmatrix} 1 & 1 & 1 \\ 0 & -1 & 2 \\ 0 & 0 & -2 \end{bmatrix}$

Here $|P| = 1(-1)(-2) = 2$

$$\therefore \quad \frac{1}{2}\text{adj}(P) = \frac{1}{2}\begin{bmatrix} 2 & 2 & 3 \\ 0 & -2 & -2 \\ 0 & 0 & -1 \end{bmatrix} = P^{-1}$$

$$\therefore \quad D = P^{-1}AP = \frac{1}{2}\begin{bmatrix} 2 & 2 & 3 \\ 0 & -2 & -2 \\ 0 & 0 & -1 \end{bmatrix}\begin{bmatrix} -1 & 1 & 2 \\ 0 & -2 & 1 \\ 0 & 0 & -3 \end{bmatrix}\begin{bmatrix} 1 & 1 & 1 \\ 0 & -1 & 2 \\ 0 & 0 & -2 \end{bmatrix}$$

$$= \frac{1}{2}\begin{bmatrix} -2 & -2 & -3 \\ 0 & 4 & 4 \\ 0 & 0 & 3 \end{bmatrix}\begin{bmatrix} 1 & 1 & 1 \\ 0 & -1 & 2 \\ 0 & 0 & -2 \end{bmatrix} = \begin{bmatrix} -1 & 0 & 0 \\ 0 & -2 & 0 \\ 0 & 0 & -3 \end{bmatrix}. \quad \textbf{Ans.}$$

EXAMPLE 13.26. Diagonalise the matrix $\begin{bmatrix} 1 & 6 & 1 \\ 1 & 2 & 0 \\ 0 & 0 & 3 \end{bmatrix}$. **(UPTU 2006-07)**

SOLUTION: Let $A = \begin{bmatrix} 1 & 6 & 1 \\ 1 & 2 & 0 \\ 0 & 0 & 3 \end{bmatrix}$. Its characteristic equation is $|A - \lambda I| = 0$

or $\begin{vmatrix} 1-\lambda & 6 & 1 \\ 1 & 2-\lambda & 0 \\ 0 & 0 & 3-\lambda \end{vmatrix} = 0$ or $(3 - \lambda)[(1 - \lambda)(2 - \lambda) - 6] = 0$

or $(3 - \lambda)(\lambda^2 - 3\lambda - 4) = 0$ hence $\lambda = -1, 3, 4$ are the eigen values of A.

Next, for eigen vectors of A we have $(A - \lambda I)X = 0$

or
$$(1 - \lambda)x + 6y + z = 0$$
$$x + (2 - \lambda)y + 0z = 0$$
$$0 + 0 + (3 - \lambda)z = 0$$

Now for $\lambda = -1$, we have $z = 0$, $2x + 6y = 0$, $x + 3y = 0$.

$\therefore$ eigen vector for $\lambda = -1$ is $[-3 \quad 1 \quad 0]'$.

For $\lambda = 3$, we have $-2x + 6y + z = 0$, $x - y = 0$

$\Rightarrow$ eigen vector is $[1 \quad 1 \quad -4]'$.

For $\lambda = 4$, we have $-3x + 6y + z = 0$, $x - 2y = 0$, $z = 0$

Hence the eigen vector is $[2 \quad 1 \quad 0]'$.

$\therefore$ Modal matrix $P = \begin{bmatrix} -3 & 1 & 2 \\ 1 & 1 & 1 \\ 0 & -4 & 0 \end{bmatrix}$

$\therefore$ $P^{-1} = \dfrac{1}{20}\begin{bmatrix} -4 & 8 & 1 \\ 0 & 0 & -5 \\ 4 & 12 & 4 \end{bmatrix}$

$\therefore$ Required diagonalised matrix $= P^{-1}AP$

$$= \frac{1}{20}\begin{bmatrix} -4 & 8 & 1 \\ 0 & 0 & -5 \\ 4 & 12 & 4 \end{bmatrix}\begin{bmatrix} 1 & 6 & 1 \\ 1 & 2 & 0 \\ 0 & 0 & 3 \end{bmatrix}\begin{bmatrix} -3 & 1 & 2 \\ 1 & 1 & 1 \\ 0 & -4 & 0 \end{bmatrix}$$

$$= \frac{1}{20}\begin{bmatrix} 4 & -8 & -1 \\ 0 & 0 & -15 \\ 16 & 48 & 16 \end{bmatrix}\begin{bmatrix} -3 & 1 & 2 \\ 1 & 1 & 1 \\ 0 & -4 & 0 \end{bmatrix}$$

$$= \begin{bmatrix} -1 & 0 & 0 \\ 0 & 3 & 0 \\ 0 & 0 & 4 \end{bmatrix}.$$ **Ans.**

EXAMPLE 13.27. Reduce the matrix A to diagonal form where $A = \begin{bmatrix} -1 & 2 & -2 \\ 1 & 2 & 1 \\ -1 & -1 & 0 \end{bmatrix}$.

(UPTU 2001, 2004, 2005-06)

SOLUTION: We first find the eigen values of A. The characteristic equation is $|A - \lambda I| = 0$

or $\begin{vmatrix} -1-\lambda & 2 & -2 \\ 1 & 2-\lambda & 1 \\ -1 & -1 & -\lambda \end{vmatrix} = 0$ or $(\lambda - 1)(5 - \lambda^2) = 0.$

Hence eigen values are $1, \sqrt{5}, -\sqrt{5}$.

For eigen vectors $(A - \lambda I)X = 0$ where $X' = [x_1, x_2, x_3]$ or $X = \begin{bmatrix} x_1 \\ x_2 \\ x_3 \end{bmatrix}$

Now for $\lambda = 1$, we have
$$-2x_1 + 2x_2 - 2x_3 = 0$$
$$x_1 + x_2 + x_3 = 0$$
$$-x_1 - x_2 - x_3 = 0$$

$\Rightarrow \quad \dfrac{x_1}{1} = \dfrac{x_2}{0} = \dfrac{x_3}{-1}$. Hence, the eigen vector for $\lambda = 1$, is $\begin{bmatrix} 1 \\ 0 \\ -1 \end{bmatrix} = [1 \quad 0 \quad -1]'$.

For $\lambda = \sqrt{5}$ we have $(-1-\sqrt{5})x_1 + 2x_2 - 2x_3 = 0$
$$x_1 + (2-\sqrt{5})x_2 + x_3 = 0$$
$$-x_1 - x_2 - \sqrt{5}x_3 = 0$$

$$\Rightarrow \quad \frac{x_1}{(\sqrt{5}-1)^2} = \frac{x_2}{\sqrt{5}-1} = \frac{x_3}{1-\sqrt{5}}$$

$$\Rightarrow \quad \frac{x_1}{\sqrt{5}-1} = \frac{x_2}{1} = \frac{x_3}{-1}.$$

Hence, the eigen vector for $\lambda = \sqrt{5}$, is $\begin{bmatrix} \sqrt{5}-1 \\ 1 \\ -1 \end{bmatrix} = \left[\sqrt{5}-1 \quad 1 \quad -1\right]'$.

Similarly, for $\lambda = -\sqrt{5}$ the eigen vector is $\begin{bmatrix} \sqrt{5}+1 \\ -1 \\ 1 \end{bmatrix} = \left[\sqrt{5}+1 \quad -1 \quad 1\right]'$.

$\therefore$ The modal matrix $P = \begin{bmatrix} 1 & \sqrt{5}-1 & \sqrt{5}+1 \\ 0 & 1 & -1 \\ -1 & -1 & 1 \end{bmatrix}$

Required diagonal matrix $D = P^{-1}AP$.

Now $$P^{-1} = \frac{1}{2\sqrt{5}}\begin{bmatrix} 0 & -2\sqrt{5} & -2\sqrt{5} \\ 1 & 2+\sqrt{5} & 1 \\ 1 & 2-\sqrt{5} & 1 \end{bmatrix}$$

$\therefore$ $$D = \frac{1}{2\sqrt{5}}\begin{bmatrix} 0 & -2\sqrt{5} & -2\sqrt{5} \\ 1 & 2+\sqrt{5} & 1 \\ +1 & 2-\sqrt{5} & 1 \end{bmatrix}\begin{bmatrix} -1 & 2 & -2 \\ 1 & 2 & 1 \\ -1 & -1 & 0 \end{bmatrix}\begin{bmatrix} 1 & \sqrt{5}-1 & \sqrt{5}+1 \\ 0 & 1 & -1 \\ -1 & -1 & 1 \end{bmatrix}$$

$$= \frac{1}{2\sqrt{5}}\begin{bmatrix} 0 & -2\sqrt{5} & -2\sqrt{5} \\ 1 & 2+\sqrt{5} & 1 \\ 1 & 2-\sqrt{5} & 1 \end{bmatrix}\begin{bmatrix} 1 & -\sqrt{5}+5 & -\sqrt{5}-5 \\ 0 & \sqrt{5} & \sqrt{5} \\ -1 & -\sqrt{5} & -\sqrt{5} \end{bmatrix}$$

$$= \frac{1}{2\sqrt{5}}\begin{bmatrix} 2\sqrt{5} & 0 & 0 \\ 0 & 10 & 0 \\ 0 & 0 & -10 \end{bmatrix} = \begin{bmatrix} 1 & 0 & 0 \\ 0 & \sqrt{5} & 0 \\ 0 & 0 & -\sqrt{5} \end{bmatrix}. \quad \textbf{Ans.}$$

EXAMPLE 13.28. The eigen vectors of a matrix A corresponding to the eigen values 1, 1, 3 are $[1, 0, -1]^1$, $[0, 1, -1]$, and $[1, 1, 0]'$ respectively. Find A.

SOLUTION: The modal matrix of A is $P = \begin{bmatrix} 1 & 0 & 1 \\ 0 & 1 & 1 \\ -1 & -1 & 0 \end{bmatrix}$.

We know that $P^{-1}AP = D = \begin{bmatrix} 1 & 0 & 0 \\ 0 & 1 & 0 \\ 0 & 0 & 3 \end{bmatrix}$ then we are to find A.

Pre multiplying both sides by P we get $AP = PD$

and then post multiplying both sides by P^{-1} we get $A = PDP^{-1}$

$$PD = \begin{bmatrix} 1 & 0 & 1 \\ 0 & 1 & 1 \\ -1 & -1 & 0 \end{bmatrix}\begin{bmatrix} 1 & 0 & 0 \\ 0 & 1 & 0 \\ 0 & 0 & 3 \end{bmatrix} = \begin{bmatrix} 1 & 0 & 3 \\ 0 & 1 & 3 \\ -1 & -1 & 0 \end{bmatrix}$$

$\therefore$ $$A = \begin{bmatrix} 1 & 0 & 3 \\ 0 & 1 & 3 \\ -1 & -1 & 0 \end{bmatrix}\frac{1}{2}\begin{bmatrix} 1 & -1 & -1 \\ -1 & 1 & -1 \\ 1 & 1 & 1 \end{bmatrix} = \frac{1}{2}\begin{bmatrix} 4 & 2 & 2 \\ 2 & 4 & 2 \\ 0 & 0 & 2 \end{bmatrix} = \begin{bmatrix} 2 & 1 & 1 \\ 1 & 2 & 1 \\ 0 & 0 & 1 \end{bmatrix}. \quad \textbf{Ans.}$$

COMPLEX MATRIX

HERMITIAN AND SKEW-HERMITIAN MATRICES

(UP TU. 2001-02)

Hermitian matrix is a generalisation of symmetric matrix whereas skew-Hermitian matrix is that of skew-symmetrix matrix when the elements of the matrix are complex quantities.

Consider the matrix Z of complex numbers z, given by

$$Z = (z_{mn}) \quad \text{where} \quad z_{mn} = a_{mn} + i\, b_{mn}, \quad i = \sqrt{-1}$$

The matrix $\overline{Z}$, conjugate of the matrix Z, is given by

$$\overline{Z} = (\overline{z}_{mn}) \quad \text{where} \quad \overline{z}_{mn} = a_{mn} - i\, b_{mn}.$$

Let us define a matrix Z^* as transpose of $\overline{Z}$, that is, $Z^* = (\overline{Z})'$.

The matrix Z is called **Hermitian** if $\boxed{Z^* = Z \quad \text{then} \quad \overline{z}_{nm} = z_{mn}}$ for all m, n.

$$\Rightarrow \quad a_{nm} - i\, b_{nm} = a_{mn} + i\, b_{mn} \quad \text{hence} \quad a_{nm} = a_{mn} \quad \text{and} \quad b_{mn} = -\, b_{nm} \quad \text{for all } m, n.$$

When $m = n$ we have $b_{nn} = -\, b_{nn}$, i.e., $b_{nn} = 0$ for all n.

Thus, in a Hermitian matrix the leading diagonal elements are all real.

Next, the matrix Z is called **skew-Hermitian** if $\boxed{Z^* = -\,Z}$ which means that

$$\overline{z}_{nm} = -\, z_{mn}, \quad \text{i.e.,} \quad a_{nm} = -\, a_{mn} \quad \text{and} \quad b_{nm} = b_{mn} \quad \text{for all } m, n.$$

When $m = n$ we have $a_{nn} = -\, a_{nn}$ hence $a_{nn} = 0$ for all n.

Thus, in a skew-Hermitian matrix all the leading diagonal elements are purely imaginary.

For example, the matrix

$$\begin{bmatrix} 2 & 1-i & -2-3i \\ 1+i & -3 & 3+2i \\ -2+3i & 3-2i & 5 \end{bmatrix} \quad \text{is a third order Hermitian matrix.}$$

And the matrix

$$\begin{bmatrix} 2i & 3+i & 1-i \\ -3+i & 3i & 2+3i \\ -1-i & -2+3i & 4i \end{bmatrix} \quad \text{is a third order skew-Hermitian matrix.}$$

UNITARY MATRIX

A square matrix is said to be **unitary** if $AA^* = I = A^*A$ where $A^* = (\bar{A})'$, that is, A^* is the transpose of the conjugate of A.

EXAMPLE 13.29. If A and B are any two complex matrices, show that

(*i*) $(A^*)^* = A$ (*ii*) $(A+B)^* = A^* + B^*$ (*iii*) $(AB)^* = B^*A^*$.

SOLUTION: (*i*) $A^* = (\bar{A})'$ hence $(A^*)^* = (\bar{\bar{A}}')' = A$

(*ii*) $(A+B)^* = (\overline{A+B})' = (\bar{A}+\bar{B})' = \bar{A}'+\bar{B}' = A^*+B^*$

(*iii*) $(AB)^* = (\overline{AB})' = \overline{(AB)'} = (\overline{B'A'}) = \bar{B}'\bar{A}' = B^*A^*$. **Hence Proved.**

EXAMPLE 13.30. Show that every complex square matrix A can be expressd as a sum $R + iS$ uniquely where R and S are both Hermitian.

SOLUTION: We can write $A = \left(\dfrac{A+A^*}{2}\right) + i\left(\dfrac{A-A^*}{2i}\right) = R + iS$

where $R = \dfrac{1}{2}(A+A^*)$ and $S = \dfrac{1}{2i}(A-A^*)$.

Now we are to show that R and S are Hermitian.

$$R^* = \frac{1}{2}(A+A^*)^* = \frac{1}{2}[A^* + (A^*)^*A] = \frac{1}{2}[A^*+A] = R$$

hence R is Hermitian.

Next $S^* = \left[\dfrac{1}{2i}(A-A^*)\right]^* = -\dfrac{1}{2i}(A-A^*)^* = -\dfrac{1}{2i}[A^*-(A^*)^*] = -\dfrac{1}{2i}[A^*-A] = \dfrac{1}{2i}(A-A^*) = S.$

Hence S is also Hermitian. **Hence Proved.**

EXAMPLE 13.31. Express the matrix $A = \begin{bmatrix} 1+i & 2 & 5-5i \\ 2i & 2+i & 4+2i \\ -1+i & -4 & 7 \end{bmatrix}$

as the sum of a Hermitian and a skew Hermitian matrix.

SOLUTION: $A = \begin{bmatrix} 1+i & 2 & 5-5i \\ 2i & 2+i & 4+2i \\ -1+i & -4 & 7 \end{bmatrix}$ hence $\bar{A} = \begin{bmatrix} 1-i & 2 & 5+5i \\ -2i & 2-i & 4-2i \\ -1-i & -4 & 7 \end{bmatrix}$

$$A^* = (\bar{A})' = \begin{bmatrix} 1-i & -2i & -1-i \\ 2 & 2-i & -4 \\ 5+5i & 4-2i & 7 \end{bmatrix}$$

Now $$\frac{A+A^*}{2} = \begin{bmatrix} 1 & 1-i & 2-3i \\ 1+i & 2 & i \\ 2+3i & -i & 7 \end{bmatrix} = P, \text{ say.}$$

and $$\frac{A-A^*}{2} = \begin{bmatrix} i & 1+i & 3-2i \\ -1+i & i & 4+i \\ -3-2i & -4+i & 0 \end{bmatrix} = Q, \text{ say.}$$

$$A = \frac{A+A^*}{2}+\frac{A-A^*}{2} = P+Q$$

where P is Hermitian and Q is skew-Hermitian **Hence Proved.**

EXAMPLE 13.32. (*i*) Define unitary matrix and show that the matrix $\begin{bmatrix} \alpha+i\gamma & -\beta+i\delta \\ \beta+i\delta & \alpha+i\gamma \end{bmatrix}$ is unitary if $\alpha^2+\beta^2+\gamma^2+\delta^2 = 1$.

(UPTU 2010-11)

(*ii*) Show that the matrix $\frac{1}{\sqrt{3}}\begin{bmatrix} 1 & 1+i \\ 1-i & -1 \end{bmatrix}$ is unitary.

(UPTU 2001-02, 2002-03, 2005-06)

SOLUTION: A square matrix A is called unitary if $AA^\theta = I = A^\theta A$ where A^θ denotes the transpose of the conjugate of A.

(*i*) Here $$A = \begin{bmatrix} \alpha+i\gamma & -\beta+i\delta \\ \beta+i\delta & \alpha-i\gamma \end{bmatrix} \text{ then } A^\theta = (\bar{A})' = \begin{bmatrix} \alpha-i\gamma & -\beta-i\delta \\ \beta-i\delta & \alpha+i\gamma \end{bmatrix}' = \begin{bmatrix} \alpha-i\gamma & \beta-i\delta \\ -\beta-i\delta & \alpha+i\gamma \end{bmatrix}$$

Hence $$AA^\theta = \begin{bmatrix} \alpha+i\gamma & -\beta+i\delta \\ \beta+i\delta & \alpha-i\gamma \end{bmatrix}\begin{bmatrix} \alpha-i\gamma & \beta-i\delta \\ -\beta-i\delta & \alpha+i\gamma \end{bmatrix} = \begin{bmatrix} \alpha^2+\gamma^2+\beta^2+\delta^2 & \alpha\beta+\gamma\delta-\alpha\beta-\gamma\delta \\ \alpha\beta+\gamma\delta-\alpha\beta-\gamma\delta & \beta^2+\delta^2+\alpha^2+\gamma^2 \end{bmatrix}$$

$$= (\alpha^2+\beta^2+\gamma^2+\delta^2)\begin{bmatrix} 1 & 0 \\ 0 & 1 \end{bmatrix}$$

Therefore, for A to be unitary, $\alpha^2+\beta^2+\gamma^2+\delta^2 = 1$.

(*ii*) Here $$A = \frac{1}{\sqrt{3}}\begin{bmatrix} 1 & 1+i \\ 1-i & -1 \end{bmatrix} \text{ hence } A^\theta = (\bar{A})' = \frac{1}{\sqrt{3}}\begin{bmatrix} 1 & 1-i \\ 1+i & -1 \end{bmatrix}' = \frac{1}{\sqrt{3}}\begin{bmatrix} 1 & 1+i \\ 1-i & -1 \end{bmatrix}$$

$$\therefore\ AA^\theta = \frac{1}{3}\begin{bmatrix} 1 & 1+i \\ 1-i & -1 \end{bmatrix}\begin{bmatrix} 1 & 1+i \\ 1-i & -1 \end{bmatrix} = \frac{1}{3}\begin{bmatrix} 3 & 0 \\ 0 & 3 \end{bmatrix} = \begin{bmatrix} 1 & 0 \\ 0 & 1 \end{bmatrix}$$

Therefore, A is a unitary matrix. **Hence Proved.**

EXAMPLE 13.33. **(a)** If $A = \begin{bmatrix} 2+i & 3 & -1+3i \\ -5 & i & 4-2i \end{bmatrix}$ verify that $A^* A$ is Hermitian matrix where A^* is the conjugate transpose of A. **(UPTU 2002-03)**

(*b*) Express the Hermitian matrix $A = \begin{bmatrix} 1 & -i & 1+i \\ i & 0 & 2-3i \\ 1-i & 2+3i & 2 \end{bmatrix}$ as $P + iQ$ where P is a real symmetric matrix and Q is a real skew symmetric matrix. **[UPTU 2011-12]**

SOLUTION: (a) $A' = \begin{bmatrix} 2+i & -5 \\ 3 & i \\ -1+3i & 4-2i \end{bmatrix}$. Hence $A^* = (\bar{A}') = \begin{bmatrix} 2-i & -5 \\ 3 & -i \\ -1-3i & 4+2i \end{bmatrix}$

$$\therefore \quad A^* A = \begin{bmatrix} 2-i & -5 \\ 3 & -i \\ -1-3i & 4+2i \end{bmatrix} \begin{bmatrix} 2+i & 3 & -1+3i \\ -5 & i & 4-2i \end{bmatrix}$$

$$= \begin{bmatrix} 30 & 6-8i & -19+17i \\ 6+8i & 10 & -5+5i \\ -19-17i & -5-5i & 30 \end{bmatrix} \quad \text{which is Hermitian.}$$

Hence verified.

(*b*) Here $A = \begin{bmatrix} 1 & -i & 1+i \\ i & 0 & 2-3i \\ 1-i & 2+3i & 2 \end{bmatrix}$ is a Hermitian matrix since $(\bar{A})' = A$.

Next,
$$A = \begin{bmatrix} 1 & 0 & 1 \\ 0 & 0 & 2 \\ 1 & 2 & 2 \end{bmatrix} + i \begin{bmatrix} 0 & -1 & 1 \\ 1 & 0 & -3 \\ -1 & 3 & 0 \end{bmatrix} = P + iQ$$

where
$$P = \begin{bmatrix} 1 & 0 & 1 \\ 0 & 0 & 2 \\ 1 & 2 & 2 \end{bmatrix} \text{ and } Q = \begin{bmatrix} 0 & -1 & 1 \\ 1 & 0 & -3 \\ -1 & 3 & 0 \end{bmatrix}$$

Clearly P is a symmetric matrix and Q is a skew-symmetric matrix. **Ans.**

EXAMPLE 13.34. **(a)** If $A = \begin{bmatrix} 0 & 1+2i \\ -1+2i & 0 \end{bmatrix}$ obtain the matrix $(I - A)(I + A)^{-1}$ and show that it is unitary. **(UPTU 2001)**

(*b*) Show that

$$A = \begin{bmatrix} i & 0 & 0 \\ 0 & 0 & i \\ 0 & i & 0 \end{bmatrix}$$

is skew Hermitian and also unitary. **[UPTU 2011-12]**

SOLUTION: $I - A = \begin{bmatrix} 1 & -1-2i \\ 1-2i & 1 \end{bmatrix}$ and $I + A = \begin{bmatrix} 1 & 1+2i \\ -1+2i & 1 \end{bmatrix}$

$$\therefore \quad (I+A)^{-1} = \frac{1}{|I+A|} \text{ adj } (I+A) = \frac{1}{6}\begin{bmatrix} 1 & 1-2i \\ -1-2i & 1 \end{bmatrix}'$$

$$\text{Therefore} \quad (I - A)(I + A)^{-1} = \begin{bmatrix} 1 & -1-2i \\ 1-2i & 1 \end{bmatrix}\left(\frac{1}{6}\right)\begin{bmatrix} 1 & -1-2i \\ 1-2i & 1 \end{bmatrix}$$

$$= \frac{1}{6}\begin{bmatrix} -4 & -2-4i \\ 2-4i & -4 \end{bmatrix} = B, \text{ say.}$$

$$\therefore \quad B' = \frac{1}{6}\begin{bmatrix} -4 & 2-4i \\ -2-4i & -4 \end{bmatrix} \text{ and } \overline{B}' = \frac{1}{6}\begin{bmatrix} -4 & 2+4i \\ -2+4i & -4 \end{bmatrix} = B^*$$

$$\text{Now} \quad B^* B = \frac{1}{6}\begin{bmatrix} -4 & 2+4i \\ -2+4i & -4 \end{bmatrix}\frac{1}{6}\begin{bmatrix} -4 & -2-4i \\ 2-4i & -4 \end{bmatrix} = \frac{1}{36}\begin{bmatrix} 36 & 0 \\ 0 & 36 \end{bmatrix} = I$$

Therefore, $B = (I - A)(I + A)^{-1}$ is a unitary matrix. **Hence Proved**

(*b*) For matrix A to be skew Hermitian

$$A^{\theta} = -A \quad \text{i.e.} \quad \overline{(A')} = -A$$

$$\text{Now} \quad A' = \begin{bmatrix} i & 0 & 0 \\ 0 & 0 & i \\ 0 & i & 0 \end{bmatrix} \text{ and } (\overline{A}') = \begin{bmatrix} -i & 0 & 0 \\ 0 & 0 & -i \\ 0 & -i & 0 \end{bmatrix} = A^{\theta}$$

$$\therefore \quad \overline{(A')} = -A = A^{\theta}$$

Hence the matrix A is skew Hermitian. Next, for a matrix A to be unitary it must satisfy $AA^{\theta} = I$

$$AA^{\theta} = \begin{bmatrix} i & 0 & 0 \\ 0 & 0 & i \\ 0 & i & 0 \end{bmatrix}\begin{bmatrix} -i & 0 & 0 \\ 0 & 0 & -i \\ 0 & -i & 0 \end{bmatrix} = \begin{bmatrix} -i^2+0+0 & 0+0+0 & 0+0+0 \\ 0+0+0 & 0+0-i^2 & 0+0+0 \\ 0+0+0 & 0+0+0 & 0+0-i^2 \end{bmatrix}$$

$$= \begin{bmatrix} 1 & 0 & 0 \\ 0 & 1 & 0 \\ 0 & 0 & 1 \end{bmatrix}$$

Hence matrix A is unifary. **Hence Proved.**

EXAMPLE 13.35. Show that the eigen values of a unitary matrix are unit modulus.

(UPTU 2001 Sp. Exam.)

SOLUTION: We know that a square matrix A is a unitary matrix if $A^* A = I = A A^*$. If λ is an eigen value of A and X its eigen vector, then

$$AX = \lambda X \Rightarrow (AX)^* = (\lambda X)^* \Rightarrow X^*A^* = \overline{\lambda} X^*$$

We therefore can write $X^*A^*(AX) = \left(\overline{\lambda} X^*\right)(\lambda X)$

$\Rightarrow$ $X^*(A^*A)X = \lambda\bar{\lambda}(X^*X)$ or $X^*(I)X = \lambda\bar{\lambda}(X^*X)$ since $A^*A = I$.

or $X^*X = \lambda\bar{\lambda}(X^*X)$ $\Rightarrow$ $(1-\lambda\bar{\lambda})\,X^*X = 0$

Since X is an eigen vector, $X^*X \neq 0$

$\Rightarrow$ $1-\lambda\bar{\lambda}=0$ $\Rightarrow$ $|\lambda|^2 = 1$ $\therefore$ $|\lambda| = 1$. **Hence proved.**

EXAMPLE 13.36. Prove that the eigen values of a Hemitian matrix are all real.

(UPTU 2003)

SOLUTION: If λ is an eigen value of a Hemitian matrix A then there exists an eigen vector X such that $AX = \lambda X$. Premultiplying both sides by X^*, gives

$$X^*AX = X^*\lambda X \qquad ...(1)$$

Taking transpose and conjugate on both sides of the above equation

$$(X^*AX)^* = (\lambda X^*X)^* \quad \Rightarrow \quad X^*A^*(X^*)^* = X^*(X^*)^*\lambda^*$$

or $X^*A^*X = X^*X\bar{\lambda}$. Since A is Hemitian $A^* = A$, we get

$$X^*AX = \bar{\lambda}X^*X \qquad ...(2)$$

From (1) and (2), we get $X^*(\lambda X) = \bar{\lambda}X^*X$

$\Rightarrow$ $(\lambda - \bar{\lambda})X^*X = 0$. But $X^*X \neq 0$ hence

$\lambda = \bar{\lambda}$ which means λ is real. **Hence Proved.**

EXAMPLE 13.37. Show that the eigen values of a skew-Hemiltain matrix is either zero or purely imaginary. **(UPTU 2003)**

SOLUTION: If A is a skew-Hemitan matrix, iA is Hemiltian. If λ is an eigen value of A then $AX = \lambda X$ $\Rightarrow$ $(iA)X = (i\lambda)A$ hence $i\lambda$ is eigen value of iA.

But iA is Hemiltian matrix hence $i\lambda$ must be real.

Which implies that λ is either zero or purely imaginary. **Hence Proved.**

EXAMPLE 13.38. Prove that the eigen values of an idempotent matrix are either zero or unity.

SOLUTION: We know that square matrix A is an **idempotent matrix** if $A^2 = A$.

If X is an eigen vector of A corresponding to eigen value λ, then $AX = \lambda X$. Premultiplying on both sides by A, gives $A(AX) = A(\lambda X) = \lambda(AX)$

or $(AA)X = \lambda(\lambda X) = \lambda^2 X$ or $A^2X = \lambda^2 X$

$\Rightarrow$ $AX = \lambda^2 X$ (as A is idempotent)

or $\lambda X = \lambda^2 X$ $\Rightarrow$ $\lambda^2 = \lambda$ or $\lambda(\lambda - 1) = 0$

Hence $\lambda = 0$ or 1. **Hence Proved.**

EXERCISE 13A

1. Verify Caley-Hamilton theorem for the matrix

$$A = \begin{bmatrix} 8 & -8 & -2 \\ 4 & -3 & -2 \\ 3 & -4 & 1 \end{bmatrix}$$

and hence find A^{-1}.

2. Verify Caley-Hamilton theorem for the matrix $\begin{bmatrix} 1 & 4 \\ 2 & 3 \end{bmatrix}$ and use it to simplify the expression

$$A^5 + 5A^4 - 6A^3 + 2A^2 - 4A + 7I.$$

3. Find the eigen values and eigen vectors for matrix

$$\begin{bmatrix} 1 & -6 & -4 \\ 0 & 4 & 2 \\ 0 & -6 & -3 \end{bmatrix}.$$

4. Determine the eigen values and eigen vectors of the matrix

$$\begin{bmatrix} 4 & 2 & -2 \\ -5 & 3 & 2 \\ -2 & 4 & 1 \end{bmatrix}.$$

5. Show that any square matrix A and its transpose A' have same characteristic equation.

6. If λ is an eigen value of an orthogonal matrix then show that $1/\lambda$ is also its eigen value.

7. Show that the matrix $A = \begin{bmatrix} -9 & 4 & 4 \\ -8 & 3 & 4 \\ -16 & 8 & 7 \end{bmatrix}$ can be reduced to the diagonal form with diagonal elements as eigen values of A.

8. Find the eigen values and eigen vectors of the matrix A given by

$$A = \begin{bmatrix} 1 & 1 & 3 \\ 1 & 5 & 1 \\ 3 & 1 & 1 \end{bmatrix}$$

Also, diagonalise A and then find A^4.

9. If $A = \begin{bmatrix} 3 & -3 & 4 \\ 2 & -3 & 4 \\ 0 & -1 & 1 \end{bmatrix}$, find two non-singular matrices

P and Q such that $PAQ = I$ and hence find A^{-1}.

[**Hint:** A = I A I. Operating $R_1 \to R_1 - R_2$ on both sides

$$\begin{bmatrix} 1 & 0 & 0 \\ 2 & -3 & 4 \\ 0 & -1 & 1 \end{bmatrix} = \begin{bmatrix} 1 & -1 & 0 \\ 0 & 1 & 0 \\ 0 & 0 & 1 \end{bmatrix} A \begin{bmatrix} 1 & 0 & 0 \\ 0 & 1 & 0 \\ 0 & 0 & 1 \end{bmatrix} \text{ then apply } R_2 \to R_2 - 2R_1$$

$$\begin{bmatrix} 1 & 0 & 0 \\ 0 & -3 & 4 \\ 0 & -1 & 1 \end{bmatrix} = \begin{bmatrix} 1 & -1 & 0 \\ -2 & 3 & 0 \\ 0 & 0 & 1 \end{bmatrix} A \begin{bmatrix} 1 & 0 & 0 \\ 0 & 1 & 0 \\ 0 & 0 & 1 \end{bmatrix} \text{ then apply } C_2 \to -C_2$$

$$\begin{bmatrix} 1 & 0 & 0 \\ 0 & 3 & 4 \\ 0 & 1 & 1 \end{bmatrix} = \begin{bmatrix} 1 & -1 & 0 \\ -2 & 3 & 0 \\ 0 & 0 & 1 \end{bmatrix} A \begin{bmatrix} 1 & 0 & 0 \\ 0 & -1 & 0 \\ 0 & 0 & 1 \end{bmatrix} \text{ then apply } R_2 \leftrightarrow R_3$$

and then $R_3 \to R_3 - 3R_2$, then $C_3 \to C_3 - C_2$ to finally get

$$\begin{bmatrix} 1 & 0 & 0 \\ 0 & 1 & 0 \\ 0 & 0 & 1 \end{bmatrix} = \begin{bmatrix} 1 & -1 & 0 \\ - & 0 & 1 \\ -2 & 3 & -3 \end{bmatrix} A \begin{bmatrix} 1 & 0 & 0 \\ 0 & -1 & 0 \\ 0 & 0 & 1 \end{bmatrix} \text{ that is } I = PAQ.$$

Next $A^{-1} = QP$]

10. Verify Caley-Hamilton theorem for the matrix $\begin{bmatrix} -1 & 2 & 0 \\ -1 & 1 & 2 \\ 1 & 2 & 1 \end{bmatrix}$. Also obtain A^{-1} and A^3.

11. Show that the matrix $A = \begin{bmatrix} 3 & 1 & -1 \\ -2 & 1 & 2 \\ 0 & 1 & 2 \end{bmatrix}$ is diagonalizable. Hence find P such that $P^{-1}AP$ is diagonal matrix. **(UPTU 2008-09) (MTU 2011-12)**

12. Prove that every Hamiltian matrix can be written as A + *i*B where A is real and symmetric and B is real and skew symmetric.

13. Express the matrix $A = \begin{bmatrix} i & 2-3i & 4+5i \\ 6+i & 0 & 4-5i \\ -i & 2-i & 2+i \end{bmatrix}$ as the sum of a Hermitian and a skew Hermitian matrix.

(UPTU 2009-10)

❑❑❑

Short Answer Type Questions

1. $D^{n-2}\,\{x^n\}$ is

   (a) 0 (b) $n!$ (c) $(n-2)!$ (d) $n!\,x^2/2$.

2. If $y = \sin x$ then y_{10} is equal

   (a) $\sin x$ (b) $-\sin x$ (c) $\cos x$ (d) $-\cos x$.

3. If $y = 1 + x + \dfrac{x^2}{2!} + \dfrac{x^3}{3!} + \ldots$ then y_n is equal to

   (a) 0 (b) 1 (c) y (d) none of these.

4. If $y = xe^x$ then y_n is equal to

   (a) $(x + n)\,e^x$ (b) xe^x (c) e^x (d) $(x + 1)e^x$.

5. If $y = e^{-x} \sin x$ then y_4 is

   (a) $4y$ (b) $-4y$ (c) $\pm y$ (d) $2y$.

6. If $y = (\sin^{-1} x)^2$ then $(1 - x^2)y'' - xy'$ is

   (a) 0 (b) 4 (c) 2 (d) -2

7. If $y^2 = P(x)$ is a polynomial of degree 3 then $2\dfrac{d}{dx}\left\{y^3 \dfrac{d^2y}{dx^2}\right\}$ is equal to

   (a) $P'''(x) + P'(x)$ (b) $P''(x)\,P'''(x)$ (c) $P(x)\,P'''(x)$ (d) a constant.

8. If $y^3 + x^3 = 3axy$ then $\dfrac{d^2y}{dx^2} = \ldots$

9. (a) If $y = \cos^4 x$ then $y_n = \ldots$

   (b) If $y = \dfrac{ax + b}{cx + d}$ then $2y_1y_2 = \ldots$

   (c) If $y = \cos x \cdot \cos 2x \cdot \cos 3x$ then $y_n = \ldots$

   (d) If $y = \tan^{-1}\left(\dfrac{1+x}{1-x}\right)$ then $y_n = \ldots$

   (e) If $y = \sinh x$ then $y_{2n+1} = \ldots$

10. If f is a homogeneous function of degree n in x and y then $x^2\dfrac{\partial^2 f}{\partial x^2}+2xy\dfrac{\partial^2 f}{\partial x^2 y}+y^2\dfrac{\partial^2 f}{\partial y^2}$ is equal to

(*a*) $n(n-1)f$ (*b*) nf (*c*) n^2f (*d*) 0.

11. If $f(x, y) = 0$ then $\dfrac{dy}{dx} = \ldots$

(*a*) $\dfrac{\partial f}{\partial x}$ (*b*) $\dfrac{-\partial f/\partial x}{\partial f/\partial y}$ (*c*) $\dfrac{\partial f/\partial y}{\partial f/\partial x}$ (*d*) $\dfrac{-\partial f/\partial y}{\partial f/\partial x}$.

12. If $z = f(x, y)$ where $x = \phi(t)$ and $y = \phi(-t)$ then $\dfrac{dz}{dt}$ is equal to

(*a*) $\dfrac{\partial z/\partial x}{\partial z/\partial y}$ (*b*) $\dfrac{\partial z}{\partial x}\dfrac{dx}{dt}+\dfrac{\partial z}{\partial y}\dfrac{dy}{dt}$ (*c*) $\dfrac{\dfrac{-\partial z}{\partial x}}{\dfrac{\partial z}{\partial t}}$ (*d*) none of these.

13. If $u = \sin^{-1}\left(\dfrac{x}{y}\right)+\tan^{-1}\left(\dfrac{y}{x}\right)$ then the value of $x\dfrac{\partial u}{\partial x}+y\dfrac{\partial u}{\partial y}$ is equal to

(*a*) 2 (*b*) 1 (*c*) 0 (*d*) none of these.

14. Fill in the blanks:

(*a*) If $x = r\cos\theta$, $y = r\sin\theta$ then $\left(\dfrac{\partial y}{\partial r}\right)_\theta = \ldots$

(*b*) If $u = f\left(\dfrac{y}{x}\right)$ then $x\dfrac{\partial u}{\partial x}+y\dfrac{\partial u}{\partial y} = \ldots$

(*c*) If $u = \log_e\left(\dfrac{x^4+y^4}{x^2-y^2}\right)$ then $x\dfrac{\partial u}{\partial x}+y\dfrac{\partial u}{\partial y} = \ldots$

(*d*) If $u = \cos^{-1}\left(\dfrac{x+y}{\sqrt{x}+\sqrt{y}}\right)$ then $x\dfrac{\partial u}{\partial x}+y\dfrac{\partial u}{\partial y} = \ldots$

(*e*) If $f(x, y) = x^3y - xy^3$ then $\left(\dfrac{1}{\partial f/\partial x}+\dfrac{1}{\partial f/\partial y}\right)_{(1,\,2)} = \ldots$

15. Number of asymptotes of the curve $y = \dfrac{x^2+1}{x^2-1}$ parallel to y-axis are

(*a*) 1 (*b*) 2 (*c*) 3 (*d*) 4.

16. Curve $x^3 + y^3 = 3axy$ is symmetrical about

(*a*) x-axis (*b*) y-axis (*c*) $x = y$ (*d*) $x = -y$.

17. Asymptotes of the curve $r\theta = a$

(*a*) $r\sin\theta = a$ (*b*) $r\cos\theta = a$ (*c*) $r = a\cos\theta$ (*d*) none of these.

18. If $f(x, y) = x^4 + y^4 + x^2y^2$ then $\left(\dfrac{\partial^2 f}{\partial x^2}\right)_{(1,1)}$

(*a*) 12 (*b*) 14 (*c*) 10 (*d*) none of these.

19. If $x = r\cos\theta$, $y = r\sin\theta$ then $\dfrac{\partial(x, y)}{\partial(r, \theta)}$ is

(*a*) $\dfrac{1}{r}$ (*b*) r (*c*) r^2 (*d*) $\dfrac{1}{r^2}$.

20. If $x = r\cos\theta$, $y = r\sin\theta$ then $\dfrac{\partial(r, \theta)}{\partial(x, y)}$ is

(*a*) r (*b*) $\dfrac{1}{r}$ (*c*) r^2 (*d*) $\dfrac{1}{r^2}$.

21. The value of $\dfrac{\partial(u, v, w)}{\partial(x, y, z)} \cdot \dfrac{\partial(x, y, z)}{\partial(u, v, w)}$ is

(*a*) 1 (*b*) 2 (*c*) 0 (*d*) none of these.

22. The function (u, v, w) which are the function of three independent variables x, y, z are dependent if $\dfrac{\partial(u, v, w)}{\partial(x, y, z)}$ is

(*a*) finite (*b*) unity (*c*) zero (*d*) none of these.

23. What error in the common logarithm of a number will be produced by an error of 1% in the number

(*a*) 0.01 (*b*) 0.02 (*c*) 0 (*d*) none of these.

24. The period T of a simple pendulum is $T = 2\pi\sqrt{\dfrac{l}{g}}$. The maximum error in T due to possible errors upto 1% in l and 2.5% in g will be

(*a*) 2.75% (*b*) 1.75% (*c*) 2.5% (*d*) none of these.

25. $F(x, y)$ will have minimum value if

(*a*) $rt\text{-}s^2 = 0$ (*b*) $rt\text{-}s^2 < 0$ (*c*) $rt\text{-}s^2 > 0$ and $r > 0$

(*d*) $rt\text{-}s^2 > 0$ and $r < 0$.

26. True/False

(*a*) $f_x = 0$ and $f_y = 0$ yields critical or stationary points of f.

(*b*) Extrema occur only at stationary points.

27. Fill in the blanks

(*a*) Triangle of maximum area inscribed in a circle is ______.

(*b*) If perimeter of a triangle is constant, its area is maximum when triangle is ______.

(*c*) Rectangular solid of maximum volume that can be inscribed in a sphere is ______.

(*d*) $\dfrac{\partial(u, v)}{\partial(r, s)} \cdot \dfrac{\partial(r, s)}{\partial(x, y)} =$ _______.

(*e*) If $x + y + z = a$ when the maximum value of $xyz =$ _______.

28. If A is matrix of order $m \times n$ such that AB and BA are defined then matrix B is an

(*a*) $m \times n$ matrix (*c*) $n \times m$ matrix

(*c*) $n \times n$ matrix (*d*) $m \times m$ matrix.

29. A square matrix A is said to be symmetric if

(*a*) $A^t = -A$ (*b*) $A^t \neq A$ (*c*) $A^t = A$ (*d*) none of these.

30. A square matrix A is said to be unitary matrix if

(*a*) $A = A^\theta$ (*b*) $AA^\theta = I$ (*c*) $A = \overline{A}$ (*d*) $A^t = A$.

31. If $A = \begin{bmatrix} 5 & 0 & 0 \\ 0 & 3 & 0 \\ 0 & 0 & 2 \end{bmatrix}$ then A is

(*a*) unit matrix (*b*) diagonal matrix (*c*) complex matrix (*d*) none of these.

32. If $A = \begin{bmatrix} 0 & 5 & -7 \\ -5 & 0 & 9 \\ 7 & -9 & 0 \end{bmatrix}$ then A is

(*a*) symmetric (*b*) diagonal (*c*) upper triangular (*d*) skew symmetric.

33. If $A = \begin{bmatrix} 4 & x+2 \\ 2x-3 & x+1 \end{bmatrix}$ is symmetric then $x =$

(*a*) 3 (*b*) 2 (*c*) 5 (*d*) 1.

34. If A, B are symmetric matrix of same order then $AB - BA$ is

(*a*) symmetric (*b*) skew-symmetric (*c*) null matrix (*d*) unit matrix.

35. If A, B, C are invertible matrices then $(ABC)^{-1}$ is equal to

(*a*) $B^{-1}C^{-1}A^{-1}$ (*b*) $C^{-1}A^{-1}B^{-1}$ (*c*) $C^{-1}B^{-1}A^{-1}$ (*d*) $A^{-1}B^{-1}C^{-1}$.

36. If $A = \begin{bmatrix} 1 & 3 & -5 \\ 5 & 2 & 7 \\ 0 & 3 & 7 \end{bmatrix}$ then trace of matrix is

(*a*) 10 (*b*) –2 (*c*) 14 (*d*) 10.

37. A and B are two non zero square matrices such that $A \cdot B = \mathbf{0}$ then

(*a*) both A and B are singular

(*b*) either of them is singular

(*c*) neither matrix is singular

(*d*) none of these.

38. Let A is a square matrix which of the following are symmetric matrices

(*a*) $A + A^t$ (*b*) $A \cdot A^t$ (*c*) $A - A^t$ (*d*) $A^t \cdot A$.

39. If I_3 is identity matrix of order 3 then $(I_3)^{-1}$ is

(*a*) 0 (*b*) $3I_3$ (*c*) I_3 (*d*) none of these.

40. If A is symmetric as well as skew symmetric then A is

(*a*) diagonal matrix (*b*) null matrix (*c*) triangular matrix (*d*) unit matrix.

41. If $A = [a_{ij}]_{m\times n}$ is a matrix of s.t. $a_{ij} = 1 \ \forall \ i, j$ then

(*a*) rank $(A) > 1$ (*b*) rank $(A) = 1$ (*c*) rank $(A) = m$ (*d*) rank $(A) = n$.

42. If A and B are two matrices such that rank$(A) = m$, rank$(B) = n$ then

(*a*) rank $(AB) = mn$

(*b*) rank $(AB) \geq$ rank of (A)

(*c*) rank $(AB) \geq$ rank of B

(*d*) rank $AB \leq$ min (rank (A), rank (B)).

43. The rank of unit matrix I_n is

(*a*) less than n (*b*) greater than n (*c*) n (*d*) 1

44. Rank of $\begin{bmatrix} 0 & 1 & 2 \\ 0 & 1 & 3 \\ 0 & 1 & 4 \end{bmatrix}$ is

(*a*) 0 (*b*) 1 (*c*) 2 (*d*) 3.

45. Rank of $\begin{bmatrix} 0 & 1 & 2 \\ 0 & 2 & 4 \\ 0 & 3 & 6 \end{bmatrix}$ is

(*a*) 0 (*b*) 1 (*c*) 2 (*d*) 3.

46. Rank of the matrix $\begin{bmatrix} 1 & 0 & 0 & 0 \\ 0 & 1 & 0 & 0 \\ 0 & 0 & 0 & 1 \end{bmatrix}$ is

(*a*) 0 (*b*) 1 (*c*) 2 (*d*) 3.

47. If the system of equations $x + 2y + 3z = 1$, $2x + y + 3z = 2$, $5x + 5y + 9z = 4$ is consistent then there is

(*a*) infinite solutions (*b*) unique solution (*c*) none of these (*d*) can't say

48. If $A = \begin{bmatrix} a & b \\ c & d \end{bmatrix}$ then A^{-1} is

(*a*) $\dfrac{1}{ad-bc}\begin{bmatrix} a & b \\ c & d \end{bmatrix}$

(*b*) $\dfrac{1}{ad-bc}\begin{bmatrix} d & -c \\ -b & a \end{bmatrix}$

(*c*) $\dfrac{1}{|A|}\begin{bmatrix} a & d \\ c & b \end{bmatrix}$

(*d*) $\dfrac{1}{|A|}\begin{bmatrix} d & a \\ c & b \end{bmatrix}$.

49. The column vectors of the matrix $\begin{bmatrix} 1 & 2 & 1 \\ 2 & 6 & 0 \\ 3 & 4 & 1 \end{bmatrix}$ are

(*a*) linearly independent

(*b*) linearly dependent

(*c*) none of these

(*d*) can't say.

50. If the system of equations $x + 2y - 3 = 1$, $(p + 2)\, z = 3$, $(2p + 1)\, y + z = 2$ is inconsistent then the value of p is

(*a*) -2 (*b*) $-\dfrac{1}{2}$ (*c*) 0 (*d*) 2.

51. If $\lambda_1, \lambda_2 \ldots \lambda_n$ are latent roots of matrix A then A^m has the latent roots

(*a*) $\lambda_1^n, \lambda_2^n, \ldots \lambda_n^n$ (*b*) $\lambda_1^m, \lambda_2^m, \ldots \lambda_n^m$ (*c*) $\frac{1}{\lambda_1}, \frac{1}{\lambda_2} \cdots \frac{1}{\lambda_n}$ (*d*) none of these.

52. True/False

(*a*) Every symmetric matrix can be expressed as a sum of its transpose and matrix itself.

(*b*) If A is unitary, A^{-1} is also unitary.

(*c*) Every square matrix satisfies its own characteristic equation.

(*d*) Then are infinitely many eigen vectors corresponding to single eigen value.

(*e*) Any superset of family of linearly dependent set of vectors is linearly dependent.

53. Fill in the blanks:

(*a*) Characteristic roots of Hermitian matrix are all _______

(*b*) Characteristic roots of skew-Hermitian matrix are either _______ of _______

(*c*) The sum of all eigen values of A is equal to _______

(*d*) Inverse of a square matrix exists if it is a _______ matrix.

(*e*) Characteristic roots of skew-symmetric matrix are either _______ or _______

54. If $\vec{r} = xi + yj + zk$ then div $\vec{r}$ is

(*a*) 1 (*b*) 0 (*c*) 3 (*d*) –3.

55. If $\vec{r} = xi + yj + zk$ then curl $\vec{r}$ is

(*a*) 1 (*b*) 0 (*c*) i (*d*) none of these.

56. If $\vec{r} = xi + yj + zk$ then grad (r^n) is

(*a*) 0 (*b*) $n\,r^{n-2}\vec{r}$ (*c*) $\frac{\vec{r}\,r^{n-2}}{n!}$ (*d*) none of these.

57. Vector $\vec{F}$ is a solenoidal if

(*a*) $\nabla \times \vec{F} = 0$ (*b*) $\nabla \cdot \vec{F} = 0$ (*c*) $\nabla \cdot (\nabla \vec{F}) = 0$ (*d*) none of these.

58. If $\vec{F} = xy^2 i - 2y^2 z^3 j + xyz^2 k$ then $div\,\vec{F}$ at (1, –1, 1) is

(*a*) 0 (*b*) 1 (*c*) 2 (*d*) 3.

59. Fill in the blanks

(*a*) If $f = x^2y + 2xyz + z^2$ then curl grad f is _______

(*b*) Curl of constant vector is _______

(*c*) div $\vec{r}$ is _______

(*d*) curl $\vec{r}$ is _______

(*e*) div curl $\vec{A}$ = _______

60. True/False

(*a*) The Gauss divergence theorem relates certain surface integrals to line integrals.

(*b*) The vector field $\vec{F} = xi - yj$ is divergence free and irrotational.

SEMESTER I

MATHEMATICS-I

Answers and Hints To Exercises

EXERCISE 1A

1. (*i*) $\dfrac{4(-1)^{n+1}n!}{(x+2)^{n+1}}+\dfrac{9(-1)^n\,n!\,2^n}{2(2x+3)^{n+1}}.$ (*ii*) $(-1)^n\,n!\left\{\dfrac{16}{(x-2)^{n+1}}-\dfrac{1}{(x-1)^{n+1}}\right\}.$

2. (*i*) $2^{n-1}\left\{2^{n-2}\cos\left(4x+\dfrac{n\pi}{2}\right)-\cos\left(2x+\dfrac{n\pi}{2}\right)\right\}$

(*ii*) $(-1)^n\,n!\left\{\dfrac{1}{(x-1)^{n+1}}-\dfrac{1}{(x+1)^{n+1}}+\dfrac{2(n+1)}{(x-1)^{n+2}}\right\}.$

3. $\dfrac{1}{4a}\sec^4\theta/2.$

7. $(-1)^n(n-1)!\,\sin^n\theta\,\sin n\theta$ where $\theta=\cot^{-1}x.$

10. $\dfrac{1}{2}e^x\left[(101)^{n/2}\sin(10x+n\tan^{-1}0)-5^{n/2}\sin(2x+n\tan^{-1}2)\right]$

11. $(-1)^n\,n!\left[\dfrac{1}{\left(x-\frac{1}{2}\right)^{n+1}}-\dfrac{1}{\left(x-\frac{1}{3}\right)^{n+1}}\right]$

EXERCISE 1B

1. (*i*) $(x^2-n^2+n)\sin\left(x+\dfrac{n\pi}{2}\right)-2nx\cos\left(x+\dfrac{n\pi}{2}\right).$

(*ii*) $e^x\left[\log x+\dfrac{{}^nc_1}{x}-\dfrac{{}^nc_2}{x^2}+\dfrac{{}^nc_3\cdot 2!}{x^3}-\cdots+(-1)^{n-1}\dfrac{(n-1)!}{x^n}\right].$

2. $\left[\textbf{Hint: } y_1=\dfrac{x}{x+1}+\log(x+1)=1-\dfrac{1}{x+1}+\log(x+1),\text{ then differentiate } (n-1)\text{ times.}\right]$

5. $\left[\textbf{Hint: } \text{Putting } x=a\tan\theta \text{ gives } y=\tan^{-1}\left(\dfrac{1+\tan\theta}{1-\tan\theta}\right)=\tan^{-1}\left(\tan\left(\dfrac{\pi}{4}+\theta\right)\right)=\dfrac{\pi}{4}+\theta=\dfrac{\pi}{4}+\tan^{-1}\dfrac{x}{a}.\right]$

8. [**Hint:** $\sin x\,f(x)=\cos x$ then differentiate n times.]

9. $\left[\textbf{Hint: } y_1=\dfrac{dy}{dt}\Big/\dfrac{dx}{dt}=-\dfrac{a\sin at}{\cos t}\ or\ y_1^2=\dfrac{a^2(1-y^2)}{1-x^2}.\right]$

10. $y_n(0)=0$ when n is even
$=(-1)^n\,m(m-1)(m-2)\ldots(m-2n)$ when n is odd.

12. $-n(n-1)\sin(n\pi/2).$

13. $(a^2-n^2)[a^2-(n-2)^2][a^2-(n-4)^2]\ldots\begin{cases}0 & \text{if } n \text{ is even}\\(a^2-1)\,a & \text{if } n \text{ is odd.}\end{cases}$

EXERCISE 1C

2. [**Hint :** Put $x+2x^2=t$ and use Binomial expansion of $(1+t)^{1/2}$]

3. $1-\dfrac{\left(x-\frac{\pi}{2}\right)^2}{2!}+\dfrac{\left(x-\frac{\pi}{2}\right)^4}{4!}-\ldots$

7. 0.7%.

8. (*i*) -5.024 cm^3 (*ii*) -5.024 cm^2.

10. [**Hint :** $f(mx)=f[x+(m-1)x]=f(x+h)$

11. [**Hint :** $x=\log(1+y)$]

12. $\left[\textbf{Hint:} \text{ Take } y=\log\sec x \text{ then } \dfrac{dy}{dx}=\tan x = x+\dfrac{x^3}{3}+\dfrac{2}{5}x^5+\ldots \text{ then integrate.}\right]$

13. 0.85 approx.

14. 0.515078

16. 2.61205.

EXERCISE 2A

1. $y=-\dfrac{x}{2},\ y=-x+1,\ y=+x+1.$

2. $y=x,\ 2y=-x+1,\ 2y=-x-1.$

3. $y=0,\ y=-x+1,\ y=-x-1.$

4. $ay=\pm bx.$

5. $y=b,\ x+a=0.$

6. $y=0,\ \ y=x,\ \ y=x\pm 1.$

7. $y=x,\ y=\dfrac{x}{2},\ y=\dfrac{(x+1)}{2}.$

8. $x=2a,\ y=x+a,\ y=-x-a$

9. $x=0,\ y=0,\ x=1,\ y=1$

10. $y=-x,\ y=x,\ y=x+1$

EXERCISE 2B

1. (*i*) $\dfrac{y^2}{c}$ (*ii*) $a \sec \dfrac{x}{a}.$

2. (*i*) $3\,(axy)^{1/3}$ [**Hint :** Convert into parametric form $x=a\cos^3\theta,\ y=a\sin^3\theta$.]

   (*ii*) $\dfrac{2}{\sqrt{a}}(x+y)^{3/2}.$ [**Hint :** Convert into parametric form $x=a\cos^4\theta,\ y=a\sin^4\theta$.]

3. $a \cot t.$

5. $\dfrac{3}{2}.$

6. $3a/2$ [**Hint :** Here the x-axis and y-axis are tangents to the curve at the origin hence either of the two formulae can be applied. Dividing throughout by xy gives $\dfrac{x^2}{2y}+\dfrac{y^2}{2x}=\dfrac{3a}{2}.$ Since

$$\lim_{x,y\to 0}\frac{x^2}{2y}=\rho.\ \lim_{x,y\to 0}\frac{y^2}{2x}=\lim_{x,y\to 0}\left(\frac{2y}{x^2}\cdot\frac{xy}{4}\right)=\lim_{x,y\to 0}\frac{2y}{x^2}\cdot\lim_{x,y\to 0}\frac{xy}{4}=\frac{1}{\rho}\cdot 0=0$$

$\therefore\ \rho + 0 = 3\,a/2$

8. $\dfrac{1}{4a}$

9. $a\sqrt{2}$.

13. $\rho_0 = 1$.

17. $(ax)^{2/3} - (by)^{2/3} = (a^2 + b^2)^{2/3}$.

$$\text{Centre of curvature} = \left(\frac{a^2+b^2}{a\cos^3\theta}, \frac{-\sin^3\theta\,(a^2+b^2)}{\cos^3\theta}\right)$$

[**Hint :** Write $x = a\sec\theta$, $y = b\tan\theta$.]

18. $\dfrac{5\sqrt{5}}{4}$, $\left(\dfrac{\pi-10}{4}, \dfrac{9}{4}\right)$

EXERCISE 2C

1.

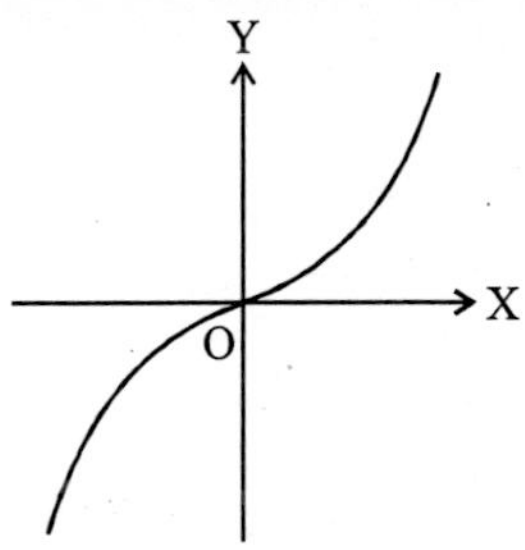

2.

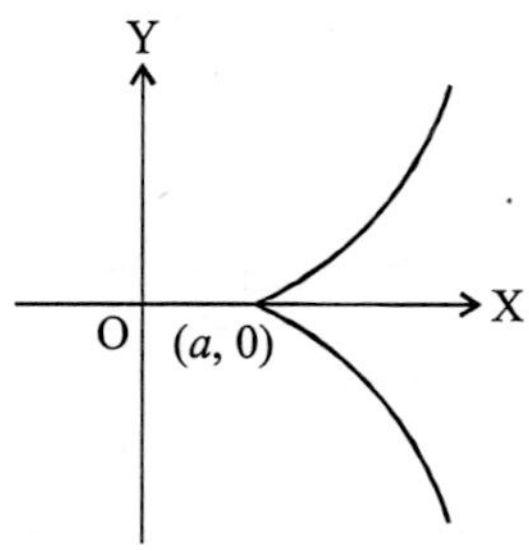

3.

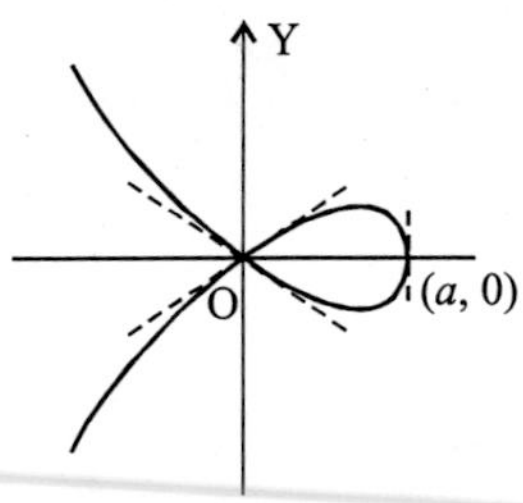

4.

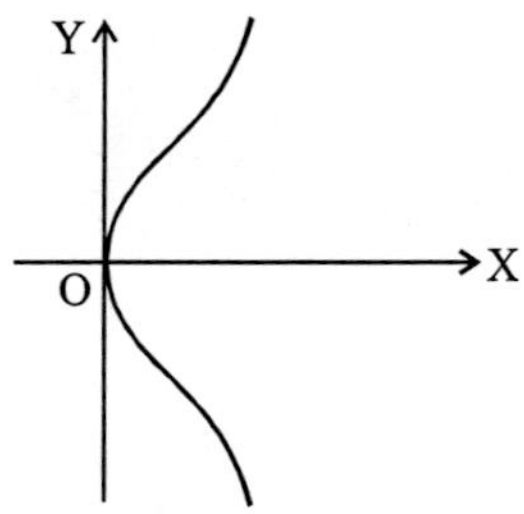

5.

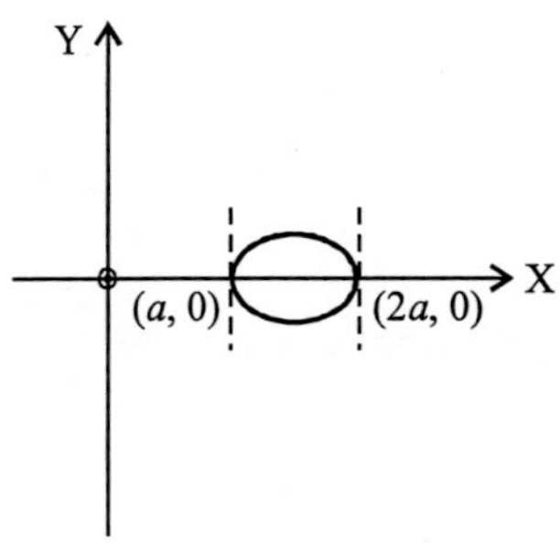

6.

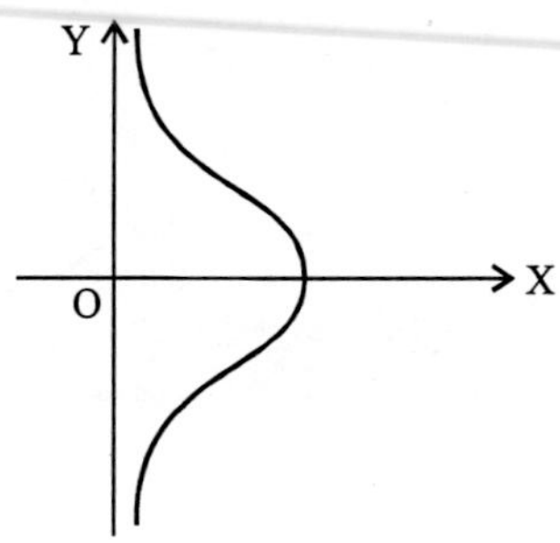

7.

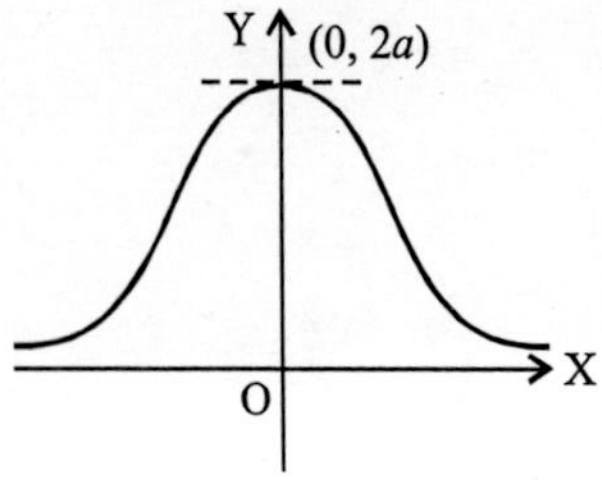

8.

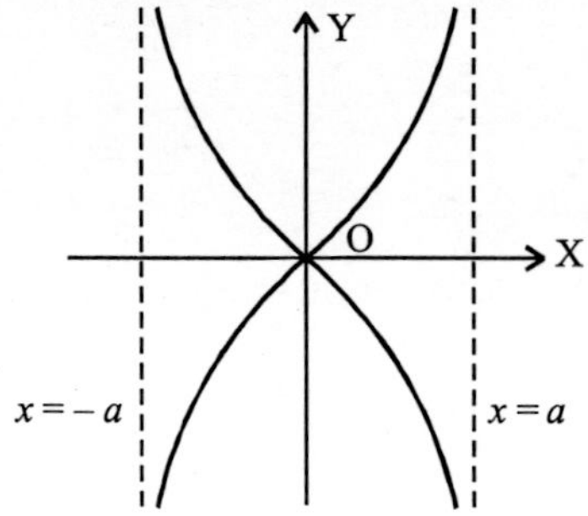

9.

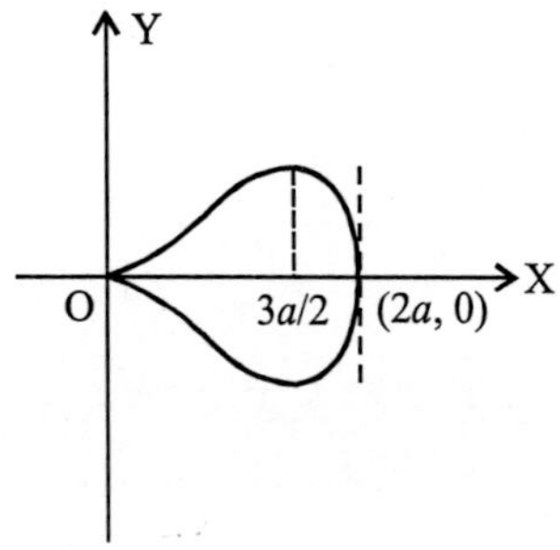

10.

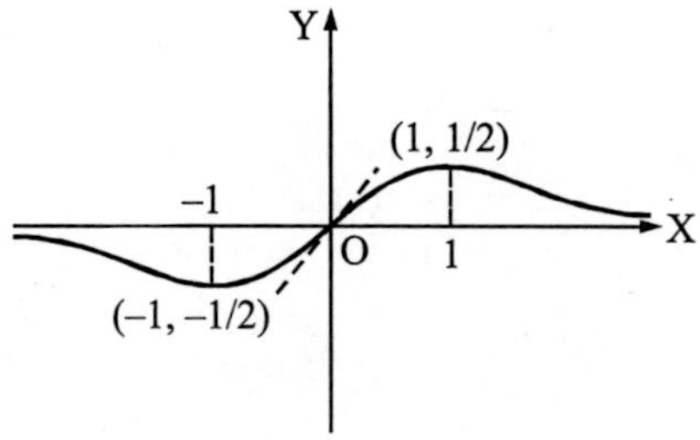

11.

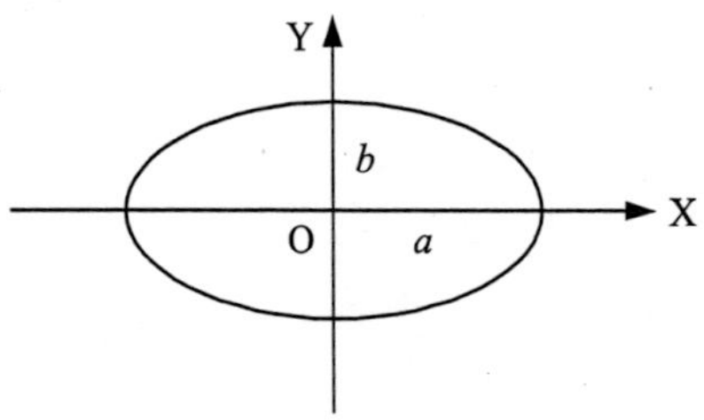

12.

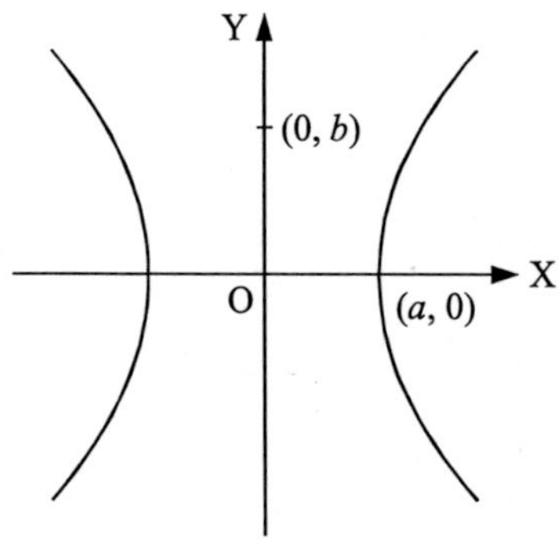

13.

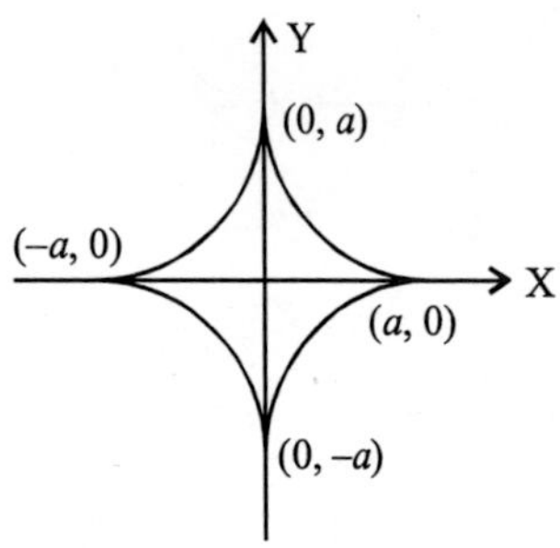

14.

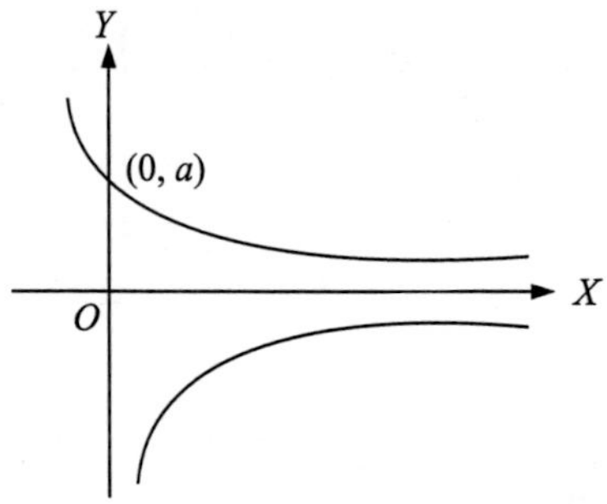

EXERCISE 2D

1. (*i*)

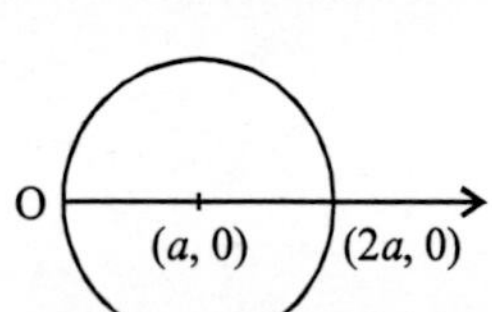

(*ii*)

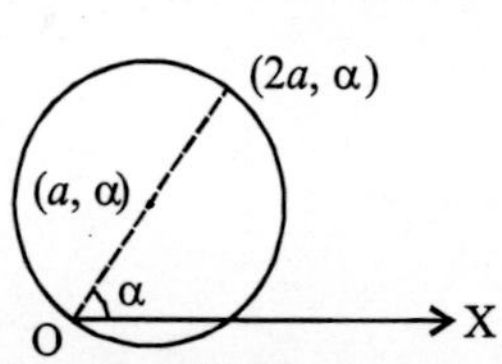

2.

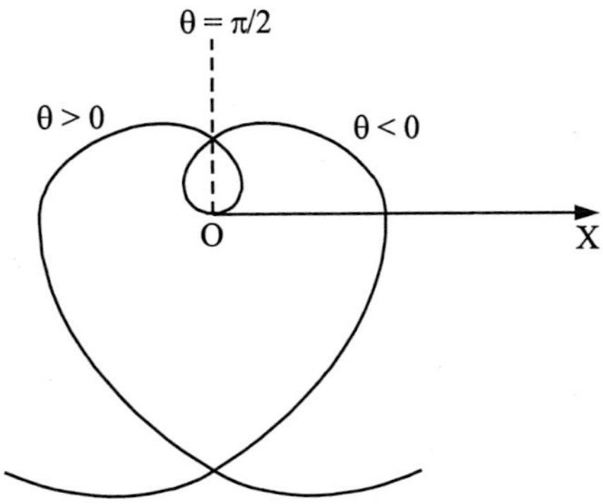
θ = π/2
θ > 0
θ < 0
O
X

3.

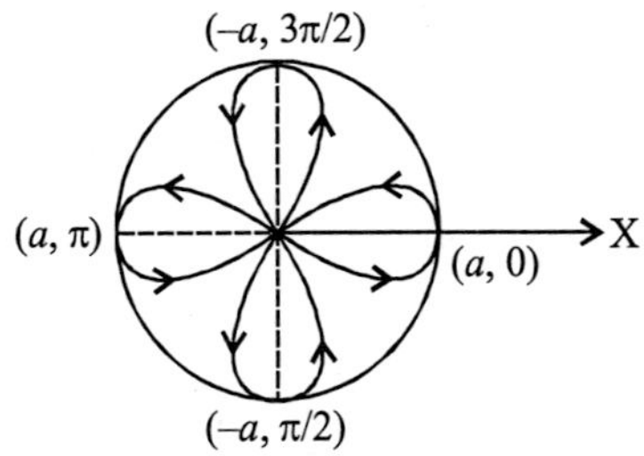
(–a, 3π/2)
(a, π)
X
(a, 0)
(–a, π/2)

4.

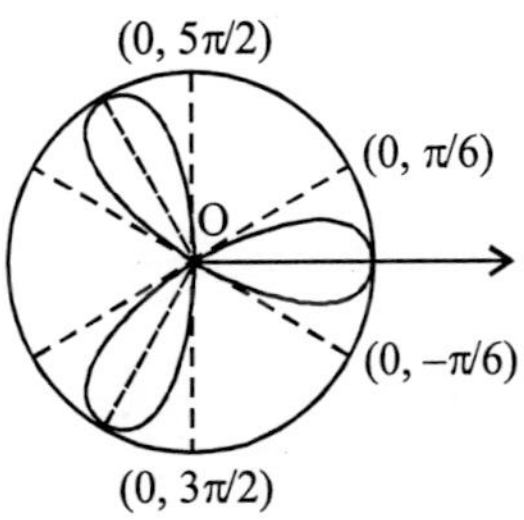
(0, 5π/2)
(0, π/6)
O
X
(0, –π/6)
(0, 3π/2)

5.

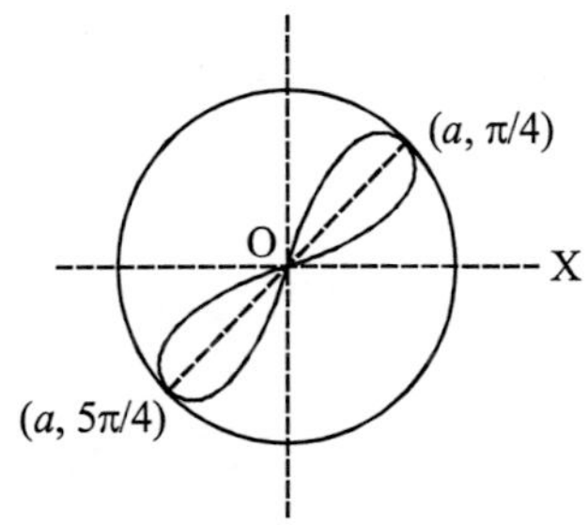
(a, π/4)
O
X
(a, 5π/4)

6.

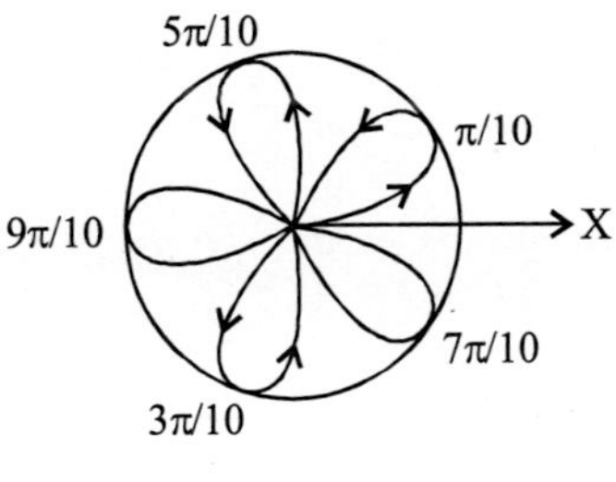
5π/10
π/10
9π/10
X
7π/10
3π/10

7.

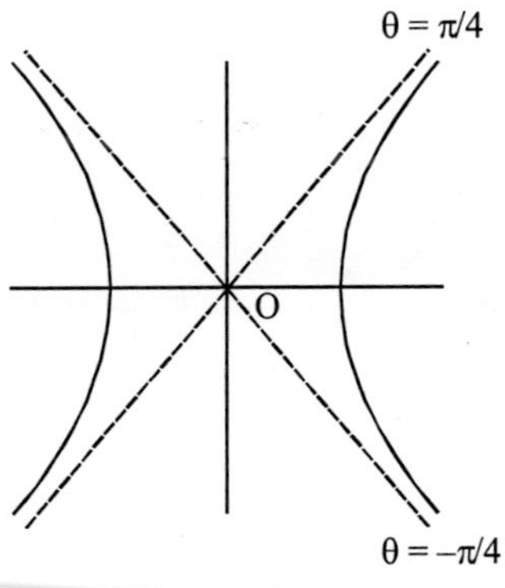
θ = π/4
O
θ = –π/4

8.

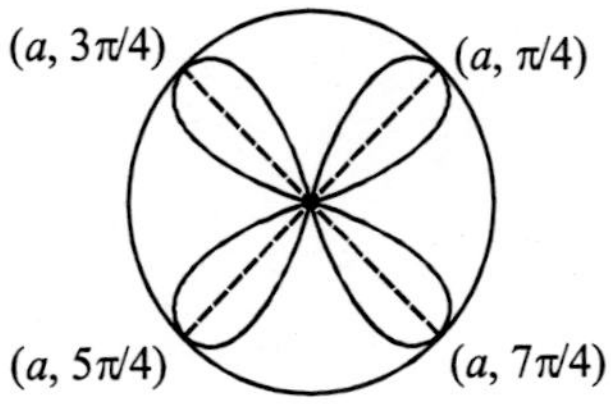
(a, 3π/4)
(a, π/4)
(a, 5π/4)
(a, 7π/4)

9.

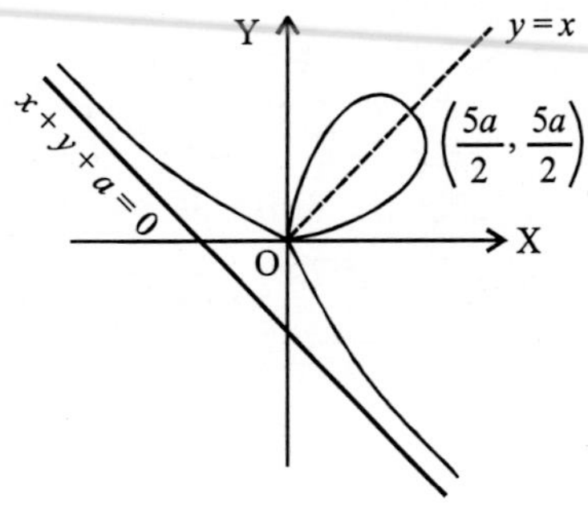
Y
y = x
x + y + a = 0
(5a/2, 5a/2)
O
X

EXERCISE 3A

2. 2. **3.** $\dfrac{x^2 - y^2}{x^2 + y^2}$. **4.** $\dfrac{2}{u}$.

8. $\dfrac{1}{x+r}$. **10.** 2.

EXERCISE 3B

1. $\dfrac{81}{8} + \dfrac{49\pi}{2}$.

2. $\left[\textbf{Hint: } \text{We are actually to prove that } x\dfrac{\partial u}{\partial x} + y\dfrac{\partial u}{\partial y} = 0\right]$

4. $\dfrac{1}{144}\tan u\,(\tan^2 u - 1)$.

5. 0.

[**Hint :** $y = x \sin u$ means $u = \sin^{-1}(y/x)$ as such u is a homogeneous function of degree 0].

7. $\dfrac{6}{5}\tan u\left(\dfrac{6}{5}\sec^2 u - 1\right)$. **8.** $\dfrac{-2\cos u\,(1+\sin^2 u)}{\sin^3 u}$.

9. $\dfrac{4x^2y^2z^2}{x^2+y^2+z^2}$. **11.** $-u$.

12. -1 [**Hint :** $u = \dfrac{1}{2}\log(x^2+y^2)$. Let $z = x^2 + y^2$ then $z = e^{2u} = f(u)$].

13. 0.

EXERCISE 3C

2. $\dfrac{d^2y}{dx^2} = \dfrac{\Delta}{(hx+by+f)^3}$ where $\Delta = \begin{vmatrix} a & h & g \\ h & b & f \\ g & f & c \end{vmatrix}$. **1.** $-\dfrac{x^2+2xy+2y^2}{x^2+4xy+y^2}$.

4. 0.

[**Hint :** From $du = \dfrac{\partial u}{\partial x}dx + \dfrac{\partial u}{\partial y}dy$ we can write $\dfrac{du}{dx} = \dfrac{\partial u}{\partial x}\cdot 1 + \dfrac{\partial u}{\partial y}\dfrac{dy}{dx}$.

Let $x^3 + y^3 + 3xy - 1 = 0$ be denoted by $f(x, y) = 0$ from which we get

$\dfrac{dy}{dx} = -\dfrac{\partial f}{\partial x}\Big/\dfrac{\partial f}{\partial y}$, etc.]

5. [**Hint:** $df = \dfrac{\partial f}{\partial x}dx + \dfrac{\partial f}{\partial y}dy + \dfrac{\partial f}{\partial z}dz = 0$. Dividing throughout by dx gives

$$\frac{\partial f}{\partial x} + \frac{\partial f}{\partial y}\frac{dy}{dx} + \frac{\partial f}{\partial z}\frac{dz}{dx} = 0 \qquad \text{...(1)}$$

Let $\phi(x, y, z)\ xyz = \text{constt}$, hence $d\phi = \dfrac{\partial \phi}{\partial x}dx + \dfrac{\partial \phi}{\partial y}dy + \dfrac{\partial \phi}{\partial z}dz = 0.$

Dividing throughout by dx, gives $\dfrac{\partial \phi}{\partial x} + \dfrac{\partial \phi}{\partial y}\dfrac{dy}{dx} + \dfrac{\partial \phi}{\partial z}\dfrac{dz}{dx} = 0$ ... (2).

Then eliminate $\dfrac{dz}{dx}$ between (1) and (2).]

6. 2.

7. [**Hint:** Let $\dfrac{x}{y} = X, \dfrac{y}{z} = Y, \dfrac{z}{x} = Z$ where X, Y, Z are themselves functions of x, y, z. Then

$$\frac{\partial u}{\partial x} = \frac{\partial u}{\partial X}\frac{\partial X}{\partial x} + \frac{\partial u}{\partial Y}\frac{\partial Y}{\partial x} + \frac{\partial u}{\partial Z}\frac{\partial Z}{\partial x} = \frac{1}{y}\frac{\partial u}{\partial X} + 0 - \frac{z}{x^2}\frac{\partial u}{\partial Z}$$

or $x\dfrac{\partial u}{\partial x} = \dfrac{x}{y}\dfrac{\partial u}{\partial X} - \dfrac{z}{x}\dfrac{\partial u}{\partial Z} = X\dfrac{\partial u}{\partial X} - Z\dfrac{\partial u}{\partial Z}$. Similarly write for $y\dfrac{\partial u}{\partial y}$ and $z\dfrac{\partial u}{\partial y}$ and add.]

8. $a = 1, b = \dfrac{2}{3}$ or $a = \dfrac{2}{3}, b = 1$.

[**Hint:** $z = z(u, v)$ where $u = x + ay, \quad v = x + by$.

$$\frac{\partial z}{\partial x} = \frac{\partial z}{\partial u}\frac{\partial u}{\partial x} + \frac{\partial z}{\partial v}\frac{\partial v}{\partial x} = \frac{\partial z}{\partial u}\cdot 1 + \frac{\partial z}{\partial v}\cdot 1 \quad \Rightarrow \quad \frac{\partial}{\partial x} \equiv \frac{\partial}{\partial u} + \frac{\partial}{\partial v}$$

$$\frac{\partial z}{\partial y} = \frac{\partial z}{\partial u}\frac{\partial u}{\partial y} + \frac{\partial z}{\partial v}\frac{\partial v}{\partial y} = \frac{\partial z}{\partial u}\cdot a + \frac{\partial z}{\partial v}\cdot b \quad \Rightarrow \quad \frac{\partial}{\partial y} \equiv a\frac{\partial}{\partial u} + b\frac{\partial}{\partial v}$$

then $\dfrac{\partial^2 z}{\partial x^2} = \dfrac{\partial}{\partial x}\left(\dfrac{\partial z}{\partial x}\right) = \left(\dfrac{\partial}{\partial u} + \dfrac{\partial}{\partial v}\right)\left(\dfrac{\partial z}{\partial u} + \dfrac{\partial z}{\partial v}\right),$

$$\frac{\partial^2 z}{\partial y^2} = \frac{\partial}{\partial y}\left(\frac{\partial z}{\partial y}\right) = \left(a\frac{\partial}{\partial u} + b\frac{\partial}{\partial v}\right)\left(a\frac{\partial z}{\partial u} + b\frac{\partial z}{\partial v}\right)$$

and $\dfrac{\partial z}{\partial x \partial y} = \dfrac{\partial}{\partial x}\left(\dfrac{\partial z}{\partial y}\right) = \left(\dfrac{\partial}{\partial u} + \dfrac{\partial}{\partial v}\right)\left(a\dfrac{\partial z}{\partial u} + b\dfrac{\partial z}{\partial v}\right)$, etc.]

14. [**Hint:** Write $ax^2 + 2hxy + by^2 = z$ then $u = f(z), \ v = \phi(z)$ then proceed.]

15. 0. [**Hint:** Put $2x - 3y = x', \ 3y - 4z = y'$ and $4z - 2x = z'$].

EXERCISE 4A

1. $-\sin^3 x \sin^2 y \sin z$.

2. [**Hint :** $x = r \sin \theta \cos \phi, \ y = r \sin \theta \sin \phi, \ z = r \cos \theta$.]

3. [**Hint :** $\dfrac{\partial(u, v)}{\partial(x, y)} = \dfrac{1}{\dfrac{\partial(x, y)}{\partial(u, v)}}$].

5. [**Hint :** Let $f_1 = x^2 + y^2 + u^2 - v^2 = 0$ and $f_2 = uv + xy = 0$

then $\dfrac{\partial(u, v)}{\partial(x, y)} = (-1)^2 \dfrac{\partial(f_1, f_2)}{\partial(x, y)} \Big/ \dfrac{\partial(f_1, f_2)}{\partial(u, v)}$, etc.]

6. $\dfrac{(x-y)(y-z)(z-x)}{(u-v)(v-w)(w-x)}$.

7. $\dfrac{1}{(x-y)(x-z)}$. $\left[\textbf{Hint:} \text{ Take } f_1 = u - xyz = 0, f_2 = v - x^2 - y^2 - z^2 = 0, f_3 = w - x - y - z = 0\right.$

then $\left.\dfrac{\partial x}{\partial u} = -\dfrac{\partial(f_1, f_2, f_3)}{\partial(u, y, z)} \Big/ \dfrac{\partial(f_1, f_2, f_3)}{\partial(x, y, z)}.\right]$

8. $\left(\dfrac{\partial v}{\partial y}\right)_{(x,u)} = \dfrac{z-y}{v-z}$ and $\left(\dfrac{\partial y}{\partial v}\right)_{(x,z)} = \dfrac{v-a}{u-y}$

$\left[\textbf{Hint:} \text{ Take } f_1 \equiv x + y + z + u + v - a, f_2 \equiv x^2 + y^2 + z^2 + u^2 + v^2 - b\right.$

Here $\dfrac{\partial v}{\partial y} = -\dfrac{\partial(f_1, f_2)}{\partial(z, y)} \Big/ \dfrac{\partial(f_1, f_2)}{\partial(z, u)}$ and $\left.\dfrac{\partial y}{\partial v} = -\dfrac{\partial(f_1, f_2)}{\partial(v, u)} \Big/ \dfrac{\partial(f_1, f_2)}{\partial(y, u)}.\right]$

9. Functionally dependent, $u^2 = v + 2w$.

10. Yes, functionally dependent, relation is $v = \sin u$.

11. Functionally dependent because $\dfrac{\partial(u,v)}{\partial(x,y)} = \dfrac{\partial(u,v)}{\partial(y,z)} = \dfrac{\partial(u,v)}{\partial(z,x)} = 0$. The relation is $v(1-u) = 1$.

13. $\dfrac{vw}{(u-v)(u-w)}$ 14. $v = uw$.

15. $-e^{2v} \sec u$ 16. $uv = 1 + w$.

EXERCISE 5A

1. 2 %. 2. 1.75 % 3. 4 %.

4. Rs. 1238.40. 5. 2.96. 6. $3\left(1+\dfrac{\pi}{4}\right)$ %.

7. 2 %. 8. $-\dfrac{1}{6}$%. 9. – 0.5 %.

10. (3.186, 3.814) 11. 0.284 ft. 13. 31.45

EXERCISE 5B

1. u has neither maxima nor minima at (0, 0) and u is minimum at $\left(\sqrt{2}, -\sqrt{2}\right)$ and $\left(-\sqrt{2}, \sqrt{2}\right)$.

3. Max. value $= \dfrac{3\sqrt{3}}{8}$ at $\left(m\pi + \dfrac{\pi}{3}, n\pi + \dfrac{\pi}{3}\right)$, $m, n \in I$.

Min. value $= \dfrac{-3\sqrt{3}}{8}$ at $\left(m\pi + \dfrac{2\pi}{3}, n\pi + \dfrac{2\pi}{3}\right)$, $m, n \in I$.

5. $\dfrac{28}{5}$ is the max. value and 3 is the min. value. 6. $\dfrac{c^2}{a^2+b^2}$.

7. $\left(\frac{8}{5}, \frac{3}{5}\right)$ and $\left(\frac{-8}{5}, \frac{-3}{5}\right)$.

8. $\frac{1}{3}$.

9. $x = 2, y = 3, z = 1$.

10. 40, 40, 40.

11. $x = \frac{2a\Delta}{a^2 + b^2 + c^2}, y = \frac{2b\Delta}{a^2 + b^2 + c^2}, z = \frac{2c\Delta}{a^2 + b^2 + c^2}$.

[**Hint:** Let ABC be the given triangle with sides a, b, c and area Δ. Let P be any point inside the triangle whose distances from the sides are denoted by x, y, z from BC, CA and AB respectively. Let $u = x^2 + y^2 + z^2$, then the relation between x, y, z is $\frac{1}{2}ax + \frac{by}{2} + \frac{cz}{2} = ar$ (Δ PBC) + ar (Δ PCA) + ar (Δ PAB) = Δ or $ax + by + cz = 2\Delta$, etc.]

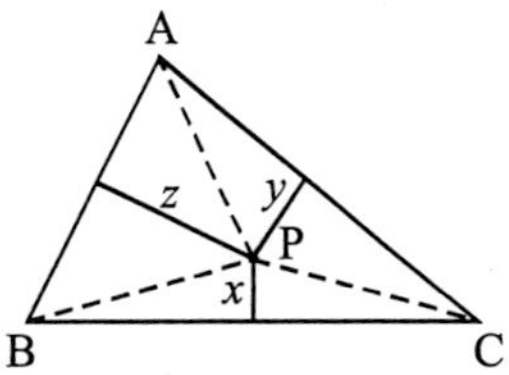

13. $\left(\frac{l}{a}\right)^l \left(\frac{m}{b}\right)^m \left(\frac{n}{c}\right)^n$

14. $\tan^{-1}\sqrt{2}$

16. $\frac{1}{8}$ [**Hint:** $\cos A \cos B \cos C = \cos A \cos B \cos (A + B) = f(A, B)$, etc.]

EXERCISE 6A

1. 1/2.

2. $a\pi/4$. [**Hint:** $I = \int_0^{a\sqrt{3}} x\, dx \int_0^{\sqrt{x^2+a^2}} \frac{dy}{\left(\sqrt{x^2+a^2}\right)^2 + y^2}$, etc.]

3. $\frac{\pi}{4}$. $\left[\textbf{Hint: } I = \int_1^{\infty}\int_{x^2}^{\infty} \frac{dx\, dy}{x^4 + y^2}.\right]$

4. $\frac{1}{ab}(\sin 1 - \cos 1)$.

5. $\frac{a\pi}{2}$. $\left[\textbf{Hint: } I = \int_0^a \frac{-dx}{a-x} \int_x^{\sqrt{ax}} \frac{y\, dy}{\sqrt{ax - y^2}}.\right]$

6. $\frac{3}{8}a^4$. [**Hint:** $I = I_1 + I_2$

where $I_1 = \int_a^{2a} \int_0^{2a-y} xy\, dxdy$ and $I_2 = \int_0^a \int_0^{\sqrt{ay}} xy\, dx\, dy$.]

7. $\int_0^a \int_{-\sqrt{a^2-x^2}}^{\sqrt{a^2-x^2}} f(x, y)\, dy\, dx$.

8. $3e^4 - 7$.

9. (i) $1 - \frac{1}{\sqrt{2}}$. (ii) $\log\left(\frac{1+e}{2}\right)$

10. 1. [**Hint:** Change the order of integration.]

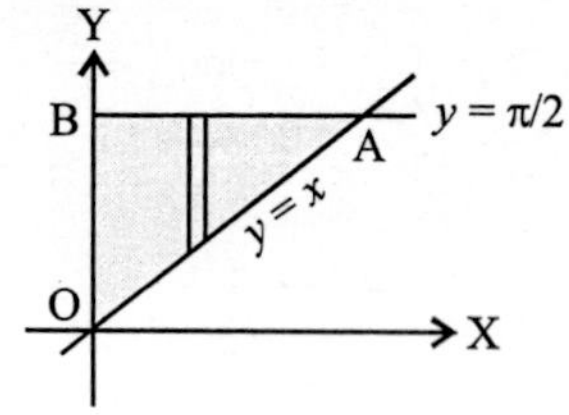

11. $\frac{1}{e}$. [**Hint:** Change the order of integration.] **12.** $\frac{e-1}{e}$.

13. $\frac{1}{3a^2}\left[(1-a^2)^{3/2}-1\right]$.

[**Hint:** Evaluate by changing the order of integration, see the figure.]

14. $\pi\,[f(0) - f(a)]$

[**Hint:** Change the order of integration then $I = \int_0^a f'(y)\,dy \int_y^a \frac{dx}{\sqrt{(a-x)(x-y)}}$. In the inner integral put $x = a\cos^2\theta + y\sin^2\theta$, etc.]

EXERCISE 6B

1. $\frac{a^3}{2}\left(\frac{a}{\pi}\sin\frac{\pi}{a} - \cos\frac{\pi}{a}\right)$. **2.** 2π.

3. $\frac{\pi}{16}(b^4 - a^4)$.

4. $8a^2 = \left(\frac{\pi}{2} - \frac{5}{3}\right)$. [**Hint:** convert to polar co-cordinates]

5. $\frac{3}{2}a^{-2/3}\left(\frac{3\sqrt{\pi}\,\Gamma\!\left(\frac{2}{3}\right)}{\Gamma\!\left(\frac{1}{6}\right)} - 1\right)$.

6. π. [**Hint:** As shown in the figure area of integration is divided as $(y-1)^2 = 2x - x^2$.]

7. $\frac{a^4}{8}$. **8.** $\frac{3}{4}\pi a^4$. **9.** $\frac{\pi}{4}$.

10. $1 - \frac{1}{\sqrt{2}}$. **11.** $\frac{\pi a^2}{6}$. **12.** $\frac{1}{2}$.

EXERCISE 7A

1. $\frac{1}{48}$. **2.** 0.

3. $\frac{1}{4}(e^2 - 8e + 13)$. **4.** $\frac{4\pi}{3}abc$.

5. $\frac{4\pi}{3}(a^2 - b^2)^{3/2}$. [**Hint:** Volume of the upper half of the hole = $\iint_R z\,dx\,dy$ where z is obtained from $x^2 + y^2 + z^2 = a^2$ and R is the radius of the circle in the XOY plane, *i.e.* $x^2 + y^2 = b^2$.

$\therefore\ V_1 = 2\iint_R \sqrt{a^2 - x^2 - y^2}\,dx\,dy = 2\int_0^{2\pi}\int_0^{b} \sqrt{a^2 - r^2}\cdot r\,d\theta\,dr$ etc. Required Volume = $\frac{4}{3}\pi a^3 - V_1$.]

6. $\frac{1}{8a^2}\left[\frac{\pi}{4}-\frac{1}{2}\tan^{-1}\frac{1}{\sqrt{2}}\right]$.

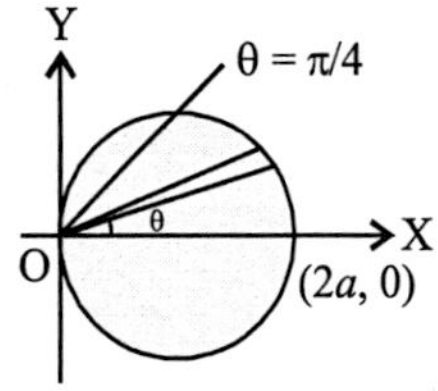

[**Hint:** $x = a + \sqrt{a^2 - y^2} \Rightarrow (x-a)^2 = a^2 - y^2 = x^2 + y^2 - 2ax = 0$

or $r = 2a\cos\theta$ (see figure.)]

7. $\frac{\pi h a^2}{2}$.

8. $\frac{\Gamma l\,\Gamma m\,\Gamma n}{(l+m+n)\,\Gamma(l+m+n)}$.

[**Hint:** $I = \int_0^1 x^{l-1}\,dx \int_0^{1-x} y^{m-1}\,dy \int_0^{1-x-y} z^{n-1}\,dz$

$= \frac{1}{n}\int_0^1 x^{l-1}\,dx \int_0^{1-x} y^{m-1}(1-x-y)\,dy$. Now putting $y = (1-x)\,t$, we get

$I = \frac{1}{n}\int_0^1 x^{l-1}(1-x)^{m+n}\,dx \int_0^1 t^{m-1}(1-t)^n\,dt$ or $I = \frac{1}{n}\,\beta(l, m+n+1)\,\beta(m, n+1)$ etc.]

9. $\frac{2\pi a^5}{15}$.

[**Hint:** Limits for z are to be taken from the sphere and those for x and y from $x^2+y^2=ax$.]

10. $\frac{\pi a^5}{10}$. [**Hint:** Convert to spherical polar co-ordinates then $I = \int_0^{\pi/2}\int_0^{\pi/2}\int_0^{a} r^4 \sin\theta\,d\theta\,d\phi\,dr$.]

11. $\frac{4}{15}\pi a^3 bc$

[**Hint:** Convert to ellipsoidal polar co-ordinates $x = ar\sin\theta\cos\phi$, $y = b\,r\sin\theta\sin\phi$, $z = c\,r\cos\theta$.]

12. $\frac{8}{3}$. [**Hint:** Put $x+y=u$, $x-y=v$ hence $dx\,dy = \left|\frac{\partial(x,y)}{\partial(u,v)}\right| du\,dv = \frac{1}{2}\,du\,dv$.

$\therefore$ Given integral $= \int_0^2\int_0^2\left[\left(\frac{u+v}{2}\right)^2+\left(\frac{u-v}{2}\right)^2\right]\frac{1}{2}\,du\,dv$.]

13. $\frac{4}{27}(27-8)(8-1)(343-125)$ [**Hint:** Put $xy=u$, $yz=v$, $zx=w$ then $\frac{\partial(u,v,w)}{\partial(x,y,z)} = 2xyz$ and

$dx\,dy\,dz = \frac{1}{2xyz}\,du\,dv\,dw = \frac{1}{2\sqrt{uvw}}\,du\,dv\,dw$, etc.]

EXERCISE 8A

1. $\dfrac{a^2-b^2}{(a^2+b^2)^2}$. **[Hint:** Consider $\int_0^\infty x\,e^{-(a-ib)x}\,dx$. Integrate it by parts and then its real part gives the value of the integral.]

2. $\dfrac{24}{(\log 4)^5}$ **[Hint:** Put $4^x = e^{-t}$, etc.]

3. **[Hint:** Put $x = e^{-t}$.]

4. $\dfrac{(-1)^n \lceil n+1}{(m+1)^{n+1}}$

6. $\dfrac{1}{5005}$ $\left[\textbf{Hint:} \text{ Given integral} = \int_0^\infty \dfrac{x^{5-1}dx}{(1+x)^{5+10}} + \int_0^\infty \dfrac{x^{10-1}\,dx}{(1+x)^{10+5}} = \beta(5, 10) + \beta(10, 5)\right]$

7. $\left[\textbf{Hint:} \text{ Substitute } \dfrac{x}{a+x} = \dfrac{t}{a+1}.\right]$

8. $\dfrac{\beta(m, n)}{2^m}$.

9. **[Hint:** Put $x = 1/t$ and proceed.]

10. $\left[\textbf{Hint:} \text{ The given integral} = \int_0^\infty \dfrac{x^{n-1}dx}{(1+x)^{n+(1-n)}} = \beta(n, 1-n).\right]$

12. $\left[\textbf{Hint:} \beta(n, n) = \int_0^1 x^{n-1}(1-x)^{n-1}dx = \int_0^{1/2} x^{n-1}(1-x)^{n-1}dx + \int_{1/2}^1 x^{n-1}(1-x)^{n-1}dx.\right.$

$\left.\text{In the second Integral on the right put } x = 1-t, \text{ etc.}\right]$

14. $2n\beta(m, n)$

15. $\dfrac{abc}{16}$.

EXERCISE 9A

1. $9i - 3j + 15k$.

2. $\dfrac{46}{5\sqrt{2}}$.

3. $6(x+y+z)$, 0.

4. **[Hint:** Use $\nabla(\vec{b}\cdot\vec{r}) = \vec{b}$].

5. 0.

6. 0.

7. $\cos^{-1}\dfrac{5}{\sqrt{18\times 17}}$.

8. $m = 0, n = 1$.

10. **[Hint:** curl curl $\vec{F} = \nabla\times(\nabla\times\vec{F}) = \nabla(\nabla\cdot\vec{F}) - \nabla^2\vec{F} = 0 - \nabla^2\vec{F}$, repeat the process]

11. [**Hint:** First prove that $(\vec{r} \times \vec{a}) \times (\vec{r} \cdot \vec{b}) = (\vec{r} \cdot \vec{r})(\vec{a} \cdot \vec{b}) - (\vec{r} \cdot \vec{a})(\vec{r} \cdot \vec{b})$ then apply ∇ to it.]

12. $a = 2, b = -4, c = -1$, work done $= \frac{43}{2}$ units.

13. $\phi = \frac{1}{3}(x^3 + y^3 + z^3) - xyz + C$ and work done = 6 units.

14. [**Hint:** $\nabla \cdot \vec{F} = 0$ and $\nabla \times \vec{F} = 0$].

EXERCISE 10A

1. -2

2. 16. [**Hint:** $\vec{F} \cdot \vec{dr} = 3x^2\, dx + (2xz - y)dy + z\, dz$

Since $y = \frac{x^2}{4}$ and $z = \frac{3x^3}{8}$ we have $dy = \frac{2x\, dx}{4}$ and $dz = \frac{9x^2\, dx}{8}$

Hence $\vec{F} \cdot \vec{dr} = 3x^2\, dx + \left\{2x\left(\frac{3}{8}\right)x^3 - \frac{x^2}{4}\right\}\frac{2x}{4}dx + \frac{3x^3}{8}\left(\frac{9x^2}{8}dx\right)$]

3. 96π. **4.** $\frac{1}{2}(a^2 + b^2)$. **5.** 1/3.

6. $\frac{4}{3}$. **7.** $\frac{3}{2}$. **8.** 40.

9. $-\frac{8}{3}$ in each case. **10.** -2.

EXERCISE 10B

1. 4π. **2.** $\frac{3}{8}$. **3.** $2ab^2$, in each integral.

4. -2 [**Hint:** $\nabla \times \vec{F} = 2yi - 2xj - k$, $\hat{n} = \frac{\nabla\phi}{|\nabla\phi|} = \frac{1}{3}(2i + 2j + k)$

$$\therefore \iint_s \nabla \times \vec{F} \cdot \hat{n}\, ds = \int_0^2 \int_0^{2-x} (4y - 4x - 1)\, dy\, dx].$$

5. $\frac{-\pi a^2}{\sqrt{2}}$. [**Hint:** $\hat{n} = \frac{i+k}{\sqrt{2}}$, $\nabla \times \vec{F} \cdot \hat{n}\, ds = -\sqrt{2}\, ds$

$\therefore \iint_S \nabla \times \vec{F} \cdot \hat{n}\, ds = -\sqrt{2}\,\pi r^2$ where r is the radius of the circle cut-off by the plane in the given sphere.]

6. $\frac{5}{3}$.

7. $\frac{1}{4}\pi a^3$.

8. $\frac{9a^2}{r}$ [**Hint:** $\iint \nabla \times \vec{F} \cdot \hat{n}\, ds = \iint 3\, k \cdot k\, dx\, dy = 3\iint_S dx\, dy$. Then convert into polars].

9. $\frac{3}{2}$ in each case.

10. $\frac{\pi a^8}{64}$.

11. $\frac{8}{3}\pi\,(a + b + c)\,\rho^3$.

12. $\pi a^2 h^2$ [**Hint:** By Gauss Theorem $\iint_S \vec{F} \cdot \vec{ds} = \iiint_V \nabla \cdot \vec{F}\, dv = \iiint_V 2z\, dx\, dy\, dz$. Here z varies from 0 to h and x, y vary according to $x^2 + y^2 = a^2, z = 0$.]

13. $\frac{64}{5}\pi$.

14. $\iint_S \vec{F} \cdot \vec{ds} = \iiint_V \nabla \cdot \vec{F}\, dx\, dy dz = -\frac{3}{2}$.

EXERCISE 11A

1. $\begin{bmatrix} 199 & 290 \\ 435 & 635 \end{bmatrix}$. [**Hint :** $A^2 = 5A + 2I$ hence $A^4 = (5A + 2I)^2$

or $A^4 = 25A^2 + 20AI + 4I = 25\,(5A + 2I) + 20\,A + 4I = 145A + 54\,I$.]

4. $\frac{1}{2}(A + A') + \frac{1}{2}(A - A') = \begin{bmatrix} 2 & 0 & 3/2 \\ 0 & 4 & 3/2 \\ 3/2 & 3/2 & -2 \end{bmatrix} + \begin{bmatrix} 0 & 1 & 3/2 \\ -1 & 0 & -1/2 \\ -3/2 & 1/2 & 0 \end{bmatrix}$.

5. $B^{-1} = \begin{bmatrix} \cos\theta & \sin\theta \\ -\sin\theta & \cos\theta \end{bmatrix}$. Yes, orthogonal.

6. $-\frac{1}{37}\begin{bmatrix} -1 & 4 & -6 \\ 4 & -16 & -13 \\ -6 & -13 & 1 \end{bmatrix}$.

8. $|A|\, A$

9. $\begin{bmatrix} \cos 2\theta & -\sin 2\theta \\ \sin 2\theta & \cos 2\theta \end{bmatrix}$.

10. $x = 1, y = -1, z = 1$.

EXERCISE 11B

1. $\begin{bmatrix} -23 & 29 & -64/5 & -18/5 \\ 10 & -12 & 26/5 & 7/5 \\ 1 & -2 & 6/5 & 2/5 \\ 2 & -2 & 3/5 & 1/5 \end{bmatrix}$

2. $\begin{bmatrix} -1 & 0 & 0 \\ 0 & 4 & 0 \\ 0 & 0 & 5 \end{bmatrix}$

3. rank (A) = 3

4. $P = \begin{bmatrix} 1 & -1 & 0 \\ 0 & 0 & 1 \\ -2 & 3 & -3 \end{bmatrix}$ and $Q = \begin{bmatrix} 1 & 0 & 0 \\ 0 & -1 & 1 \\ 0 & 0 & 1 \end{bmatrix}$ $A^{-1} = \begin{bmatrix} 1 & -1 & 0 \\ -2 & 3 & -4 \\ -2 & 3 & -3 \end{bmatrix}$

5. Rank = 2

6. Normal form = $\begin{bmatrix} I_3 & 0 \\ 0 & 0 \end{bmatrix}$ $\therefore$ Rank = 3

7. $P = \begin{bmatrix} 1 & 0 & 0 & 0 \\ 0 & 0 & 0 & 1 \\ 0 & 0 & 1 & -3 \\ 1/3 & -1/12 & -2/3 & 29/12 \end{bmatrix}$, $Q = \begin{bmatrix} 1 & 1 & -2 & -3 \\ 0 & 1 & 0 & -2 \\ 0 & 0 & 1 & 2 \\ 0 & 0 & 0 & 1 \end{bmatrix}$ Rank of $A = 4$.

8. 2

9. 3

EXERCISE 12A

1. Consistent, $\rho(A) = \rho(A : B) = 2$. $x = 1, y = -2 + 3k, z = k$ where k is arbitrary.

2. Inconsistent.

3. Consistent $x = \dfrac{6+K}{5}, y = \dfrac{2-3K}{5}, z = K$ where $K \in R$.

4. Inconsistent.

5. $\lambda = 1, -3$.

6. (*i*) unique solution for $\lambda \neq 5$.

(*ii*) no solution for $\lambda = 5$ and $\mu \neq 9$.

(*iii*) many solution for $\lambda = 5, \mu = 9$.

7. Many solutions, $x = \dfrac{35K-31}{75}, y = \dfrac{5K-65}{75}, z = \dfrac{-7K+164}{75}$, $w = K$ where K is arbitrary.

8. Inconsistent.

9. $x = 2, y = -1, z = 1, \omega = 3$.

10. $a \neq 3$ and b may have any value.

11. Vectors are linearly dependent. If the given vectors are X, Y, Z then Z = X + Y.

13. $x_1 = -z_1 + 7z_2, x_2 = -2z_1 - 6z_2$.

14. $\left(\dfrac{13}{5}, \dfrac{11}{5}, \dfrac{7}{5}\right)$

15. $\begin{bmatrix} 1/2 & 1/2 & -1/2 \\ -5/2 & 1/2 & 1/2 \\ -2 & 0 & 1 \end{bmatrix}$

EXERCISE 13A

1. $\dfrac{1}{6}\begin{bmatrix} -11 & 16 & 10 \\ -10 & 14 & 8 \\ -7 & 8 & 8 \end{bmatrix}$.

2. $\begin{bmatrix} 1861 & 3676 \\ 1838 & 3699 \end{bmatrix}$.

3. Eigen values are 1, 1, 3 and corresponding eigen vectors are $\begin{bmatrix} 1 \\ -2 \\ 3 \end{bmatrix}, \begin{bmatrix} 2 \\ -2 \\ 3 \end{bmatrix}, \begin{bmatrix} 2 \\ 1 \\ -2 \end{bmatrix}$.

4. Eigen values are 1, 2, 5 and corresponding eigen vectors are $\begin{bmatrix} 2 \\ 1 \\ 4 \end{bmatrix}, \begin{bmatrix} 1 \\ 1 \\ 2 \end{bmatrix}, \begin{bmatrix} 0 \\ 1 \\ 1 \end{bmatrix}$.

8. $-2, 3, 6$ Eigen vectors $\begin{bmatrix} 1 \\ 0 \\ -1 \end{bmatrix}, \begin{bmatrix} 1 \\ -1 \\ 1 \end{bmatrix}, \begin{bmatrix} 1 \\ 2 \\ 1 \end{bmatrix}$, $P = \begin{bmatrix} 1 & 1 & 1 \\ 0 & -1 & 2 \\ -1 & 1 & 1 \end{bmatrix}$, $D = P^{-1}AP$ = diagonal matrix of A.

9. $Q = \begin{bmatrix} 1 & 0 & 0 \\ 0 & -1 & 1 \\ 0 & 0 & 1 \end{bmatrix}$, $P = \begin{bmatrix} 1 & -1 & 0 \\ 0 & 0 & 1 \\ -2 & 3 & -3 \end{bmatrix}$, $A^{-1} = \begin{bmatrix} 1 & -1 & 0 \\ -2 & 3 & -4 \\ -2 & 3 & -3 \end{bmatrix}$.

10. $A^{-1} = \begin{bmatrix} -2/3 & -2 & 0 \\ 1 & -2/3 & -2 \\ -1 & -2 & -2/3 \end{bmatrix}$ and $A^3 = \begin{bmatrix} -1 & 10 & 12 \\ 1 & 11 & 10 \\ -1 & 16 & 17 \end{bmatrix}$.

11. Eigen values are 1, 2, 3. Eigen vectors are

$\begin{bmatrix} 1 \\ -1 \\ 1 \end{bmatrix}, \begin{bmatrix} 1 \\ 0 \\ 1 \end{bmatrix}$ and $\begin{bmatrix} 0 \\ 1 \\ 1 \end{bmatrix}$, $P = \begin{bmatrix} 1 & 1 & 0 \\ -1 & 0 & 1 \\ 1 & 1 & 1 \end{bmatrix}$, $P^{-1}AP = \begin{bmatrix} 1 & 0 & 0 \\ 0 & 2 & 0 \\ 0 & 0 & 3 \end{bmatrix}$.

13. $$A = \frac{A + A^*}{2} + \frac{A - A^*}{2} = \begin{bmatrix} i & 4 - zi & 2 + 3i \\ 4 + 2i & 0 & 3 - 2i \\ 2 - 3i & 3 + 2i & 2 + i \end{bmatrix} + \begin{bmatrix} 0 & -2 - i & 2 + 2i \\ 2 - i & 0 & 1 - 3i \\ -2 + 2i & -1 + 3i & 0 \end{bmatrix}$$

ANSWERS TO SHORT ANSWER TYPE QUESTIONS

1. (*d*) **2.** (*b*) **3.** (*c*) **4.** (*a*) **5.** (*b*)

6. (*c*) **7.** (*c*) **8.** $y_2 = \dfrac{-2a^3xy}{(y^2 - ax)^3}$

9. (*a*) $y_n = \dfrac{1}{2} 2^n \cos\left(2x + \dfrac{n\pi}{2}\right) + \dfrac{1}{8} \cdot 4^n \cos\left(4x + \dfrac{n\pi}{2}\right)$

(*b*) $2y_1y_2 = \dfrac{-4c\,(ad - bx)}{(cx + d)^3}$

(*c*) $y_n = \dfrac{1}{4}\left[6^n \cos\left(6x + \dfrac{n\pi}{2}\right) + 4^n \cos\left(4x + \dfrac{n\pi}{2}\right) + 2^n \cos\left(2x + \dfrac{n\pi}{2}\right)\right]$

(*d*) $y_n = (-1)^{n-1}\,(n-1)!\,\sin^n\theta\,\sin n\theta,\ \theta = \cot^{-1} x$

(*e*) $y_{2n+1} \cosh x$

10. (*a*) **11.** (*b*) **12.** (*b*) **13.** (*c*)

14. (*a*) sin θ (*b*) 0 (*c*) 2 (*d*) $-\frac{1}{2}\tan u$ (*e*) $-\frac{13}{22}$

15. (*b*) **16.** (*c*) **17.** (*a*) **18.** (*b*) **19.** (*b*)

20. (*b*) **21.** (*a*) **22.** (*c*) **23.** (*a*) **24.** (*b*)

25. (*c*) **26.** (*a*) True (*b*) True

27. (*a*) $\frac{a^2}{4}\sqrt{3}$ (it is equilateral Δ with side *a*) (*b*) an equilateral triangle

(*c*) cube (*d*) $\frac{\partial(u, v)}{\partial(x, y)}$ (*e*) $\left(\frac{a}{3}\right)^3$

28. (*b*) **29.** (*c*) **30.** (*b*) **31.** (*b*) **32.** (*d*)

33. (*c*) **34.** (*b*) **35.** (*c*) **36.** (*a*) **37.** (*d*)

38. (*a*) **39.** (*c*) **40.** (*b*) **41.** (*b*) **42.** (*a*)

43. (*c*) **44.** (*c*) **45.** (*b*) **46.** (*d*) **47.** (*b*)

48. (*b*) **49.** (*a*) **50.** (*a*) **51.** (*a*)

52. (*a*) False (*b*) True (*c*) True (*d*) True (*e*) True

53. (*a*) real (*b*) zero or purely imaginary (*c*) trace of the matrix

(*d*) non-singular (*e*) zero or purely imaginary

54. (*c*) **55.** (*b*) **56.** (*b*) **57.** (*b*) **58.** (*d*)

59. (*a*) zero (*b*) zero (*c*) 3 (*d*) zero (*e*) zero

60. (*a*) False (*b*) True.

B.TECH.

(Sem I) Theory Examination 2011-12

MATHEMATICS-I

EAS 103

Time : 3 Hours *Total Marks : 100*

Note: Attempt questions from each Section as indicated. The symbols have their usual meaning.

SECTION-A

Attempt all parts of this question. Each part carries 2 marks. **2 × 10 = 20**

1. (*a*) Find y_n if $y = \dfrac{x^n - 1}{x - 1}$.

(*b*) What is the asymptote of the curve $y^2(2a - x) = x^3$?

(*c*) If $u = x^2yz - 4y^2z^2 + 2xz^3$, find the value of

$$x\frac{\partial u}{\partial x} + y\frac{\partial u}{\partial y} + z\frac{\partial u}{\partial z}$$

(*d*) Calculate: $\dfrac{\partial(u, v)}{\partial(x, y)}$

for $x = e^u \cos v$ and $y = e^u \sin v$.

(*e*) Find the value of: $\left|\left(-\dfrac{5}{2}\right)\right.$

(*f*) Find the value of the integral $\iint_R xy\, dx\, dy$, where R is the region bounded by the x-axis, the line $y = 2x$ and the parabola $x^2 = 4ay$.

(*g*) Show that the vector: $\vec{V} = 3y^4z^2\hat{i} + 4x^3z^2\hat{j} + 3x^2y^2\hat{k}$ is solenoidal.

(*h*) State Green's theorem for a plane region.

(*i*) The matrix: $A = \begin{bmatrix} 2 & 0 & -1 \\ 5 & 1 & 0 \\ 0 & 1 & 3 \end{bmatrix}$

satisfies the matrix equation $A^3 - 6A^2 + 11A - I = 0$ where I is an identity matrix of order 3. Find A^{-1}.

(*j*) Show that the matrix: $A = \begin{bmatrix} 3 & 1 & -1 \\ -2 & 1 & 2 \\ 0 & 1 & 3 \end{bmatrix}$ is diagonalizable.

SECTION-B

Attempt any *three* parts of the questions. **10 × 3 = 30**

2. (*a*) If $y(x) = \sin px + \cos px$, prove that $y_n(x) = p^n[1 + (-1)^n \sin 2px)]^{1/2}$

Hence, show that $y_8(\pi) = \left(\frac{1}{2}\right)^{31/2}$ when $p = \frac{1}{4}$.

(*b*) Evaluate: $[(3.82)^2 + 2(2.1)^3]^{1/5}$ using theory of approximation.

(*c*) Evaluate: $\displaystyle\int_0^1 \int_0^{\sqrt{1-x^2}} \frac{e^y}{(e^y+1)\sqrt{1-x^2-y}}\,dx\,dy$.

(*d*) Evaluate: $\iint_S \vec{F}\cdot\hat{n}\,dS$ where $\vec{F} = 18z\hat{i} - 12\hat{j} + 3y\hat{k}$ and S is the part of the plane $2x + 3y + 6z = 12$ in the first octant.

(*e*) If: $A = \begin{bmatrix} 1 & 2 & 1 \\ a & 0 & 4 \\ 1 & 1 & 1 \end{bmatrix}$ and $\text{adj}(\text{adj } A) = A$, find a.

SECTION-C

All questions of this Section are compulsory. Attempt any *two* parts from each question. **10 × 5 = 50**

2. (*a*) If $y = \left(\frac{1+x}{1-x}\right)^{1/2}$, prove that

$$(1-x^2)y_n - [2(n-1)x + 1]y_{n-1} - (n-1)(n-2)y_{n-2} = 0.$$

(*b*) Obtain the series for $\log_e(1 + x)$ and then find the series for $\log_e\left(\frac{1+x}{1-x}\right)$ and hence, determine the value of $\log_e\left(\frac{11}{9}\right)$ up to five places of decimal.

(*c*) Find the asymptotes of: $y^3 - x^2y - 2xy^2 + 2x^3 - 7xy + 3y^2 + 2x^2 + 2x + 2y + 1 = 0$

4. (*a*) If: $u = \sin^{-1}\left(\frac{x^{1/3} - y^{1/3}}{x^{1/2} - y^{1/2}}\right)^{1/2}$, show that $x\frac{\partial u}{\partial x} + y\frac{\partial u}{\partial y} = -\frac{1}{12}\tan u$.

(*b*) Are the functions: $u = \frac{x-y}{x+z}$, $v = \frac{x+z}{y+z}$

functionally dependent? If so, find the relation between them.

(*c*) Using the Lagrange's method, find the maximum and minimum distances from the origin to the curve $3x^2 + 4xy + 6y^2 = 140$.

5. (*a*) Prove that: $B(m, n) = \int_0^1 \frac{x^{m-1} + x^{n-1}}{(1+x)^{m+n}} dx.$

(*b*) Find the volume of the tetrahedron bounded by the plane $\frac{x}{a}+\frac{y}{b}+\frac{z}{c} = 1$ and the coordinate planes.

(*c*) Find the volume of the solid which is bounded by the surfaces $2z = x^2 + y^2$ and $z = x$.

6. (*a*) If $\vec{F} = (\vec{a}\cdot\vec{r})\vec{r}$, where $\vec{a}$ is a constant vector, find curl $\vec{F}$ and proved that it is perpendicular to $\vec{a}$.

(*b*) Find the work done in moving a particle in the force field: $\vec{F} = 3x^2\hat{i} + (2xz - y)\hat{j} + z\hat{k}$ along the curve $x^2 = 4y$ and $3x^3 = 8z$ from $x = 0$ to $x = 2$.

(*c*) Prove that:

$$\iint_S \frac{1}{\sqrt{a^2x^2 + b^2y^2 + c^2z^2}} dS = \frac{4\pi}{\sqrt{abc}}$$

where S is the ellipsoid $ax^2 + by^2 + cz^2 = 1$.

7. (*a*) Find the value of P for which the matrix:

$$A = \begin{bmatrix} 3 & P & P \\ P & 3 & P \\ P & P & 3 \end{bmatrix} \text{ is of rank } 1.$$

(*b*) Show that the system of equation:

$$3x + 4y + 5z = A$$
$$4x + 5y + 6z = B$$
$$5x + 6y + 7z = C$$

are consistent only if A, B and C are in arithmetic progression (A.P.)

(*c*) Show that:

$$A = \begin{bmatrix} i & 0 & 0 \\ 0 & 0 & i \\ 0 & i & 0 \end{bmatrix} \text{ is Skew-Hermitian and also unitary.}$$

B.TECH.

(Sem I) Theory Examination 2011-12

MATHEMATICS-I

EAS 103

Time : 3 Hours *Total Marks : 100*

SECTION-A

1. All parts of this question are compulsory: **(2 × 10 = 20)**

(*a*) Find the nth derivative of $x^{n-1} \log x$.

(*b*) Find the Taylor's series expansion of: $f(xy) = x^3 + xy$ about point (2, 1)

(*c*) If $u = e^x \sin y$ and $v = e^x \cos y$, evaluate: $\frac{\partial(u, v)}{\partial(x, y)}$.

(*d*) Find the minimum value of $x^2 + y^2 + 6x + 12$.

(e) Find the eigen values of the matrix $\begin{bmatrix} 1 & 1 & 1 \\ 1 & 1 & 1 \\ 1 & 1 & 1 \end{bmatrix}$.

(*f*) Calculate the inverse of the matrix. $\begin{bmatrix} 1 & 2 \\ 5 & 7 \end{bmatrix}$

(*g*) Evaluate $\int_0^1 \int_1^2 \int_2^3 xyz \, dx \, dy \, dz$

(*h*) Evaluate the area enclosed between the parabola $y = x^2$ and the straight line $y = x$.

(*i*) Find the magnitude of the gradient of the function $f = xyz$ at (1, 0, 2).

(*j*) Write the statement of divergence theorem for a given vector field $\vec{F}$.

SECTION-B

2. Attempt any *three* parts of the following: **(2 × 10 = 20)**

(*a*) Find the eigen values and eigen vectors of the following matrix:

$$\begin{bmatrix} 2 & -1 & 1 \\ -1 & 2 & -1 \\ 1 & -2 & 2 \end{bmatrix}$$

(*b*) If $y = a \cos(\log x) + b \sin(\log x)$. find $y_n(0)$.

(*c*) The angles of triangle are calculated from the sides a, b, c. If small changes δa, δb and δc are made in the sides, find δA, δB and δC where Δ is the area of the triangle and A, B, C are angles opposite to sides, a, b, c respectively. Also show that $\delta A + \delta B + \delta C = 0$.

(*d*) Find the volume bounded by the elliptic paraboloids $z = x^2 + 9y^2$ and $z = 18 - x^2 - 9y^2$.

(*e*) If $\vec{A} = (x-y)\hat{i} + (x+y)j$, evaluate $\oint_C \vec{A} \cdot d\vec{r}$ around the curve C consisting of $y = x^2$ and $y^2 = x$.

SECTION-C

2. Attempt any *two* parts from each question. All questions are compulsory. **(5 × 2 × 5 = 50)**

(*a*) If $y = \tan^{-1}\left(\dfrac{a+x}{a-x}\right)$, prove that:

$(a^2 + x^2)y_{n+2} + 2(n+1)xy_{n+1} + n(n+1)y_n = 0$

(b) If $u(x, y, z) = \log(\tan x + \tan y + \tan z)$, prove that

$$\sin 2x \frac{\partial u}{\partial x} + \sin 2y \frac{\partial u}{\partial y} + \sin 2z \frac{\partial u}{\partial z} = 2$$

(*c*) Trace the curve: $r^2 = a^2 \cos 2\theta$.

4. (*a*) If $y_1 = \dfrac{x_2\, x_3}{x_1}$, $y_2 = \dfrac{x_3\, x_1}{x_2}$ and $y_3 = \dfrac{x_1\, x_2}{x_3}$, find the value of $\dfrac{\partial(y_1, y_2, y_3)}{\partial(x_1, x_2, x_3)}$.

(*b*) Find the extreme values of: $f(x, y) = x^3 + y^3 - 3\,axy$

(*c*) If the base radius and height of a cone are measured as 4 cm and 8 cm, with a possible error of 0.04 and 0.08 cm respectively, calculate the percentage (%) error in calculating volume of the cone.

5. (*a*) Define curl of a vector. Prove the following vector identity:

$$\text{Div}(\vec{u} \times \vec{v}) = \text{Curl}\,\vec{u} \cdot \vec{v} - \text{Curl}\,\vec{v} \cdot \vec{u}.$$

(*b*) If $r = (x^2 + y^2 + z^3)^{1/2}$; evaluate $\nabla^2(\log r)$.

(*c*) Find the surface area of the plane $x + 2y + 2z = 12$ cut off by $x = 0, y = 0$ and $x^2 + y^2 = 16$.

6. (*a*) Express the Hermitian Matrix:

$$A = \begin{bmatrix} 1 & -i & 1+i \\ i & 0 & 2-3i \\ 1-i & 2+3i & 2 \end{bmatrix}$$

as $P + iQ$ where P is a real symmetric and Q is a real skew symmetric matrix.

(*b*) Using elementary row transformation, find the inverse of the following matrix.

$$A = \begin{bmatrix} 2 & 3 & 4 \\ 4 & 3 & 1 \\ 1 & 2 & 4 \end{bmatrix}$$

(c) State and verify Cayley-Hamilton theorem for the following matrix:

$$A = \begin{bmatrix} 2 & -1 & 1 \\ -1 & 2 & -1 \\ 1 & -1 & 2 \end{bmatrix}$$

7. (*a*) Find the mass of a plate which is formed by the coordinate planes and the plane $\frac{x}{a} + \frac{y}{b} + \frac{z}{c} = 1$, the density is given by $\rho = k\,xyz$.

(*b*) Using Beta and Gamma functions, evaluate $\int_0^\infty \frac{dx}{1+x^4}$.

(*c*) Evaluate the integral $\int_0^a \int_0^{\sqrt{a^2-y^2}} (x^2 + y^2)\,dx\,dy$ by changing into polar coordinates.

B.TECH.

(Sem I) ODD Semester Theory Examination 2010-11

MATHEMATICS-I

EAS 103

Time : 3 Hours *Total Marks : 100*

SECTION-A

All parts of this question are compulsory: **2 × 10 = 20**

1. (*a*) If $u = f\left(\frac{y}{x}\right)$ then

$$x\frac{\partial u}{\partial x} + y\frac{\partial u}{\partial y} = \ldots$$

(*b*) The curve $x^{2/3} + y^{2/3} = a^{2/3}$

is symmetrical about . . .

Indicate True or False of the following statements:

(*c*) (*i*) Two functions u and v are functionally dependent if their Jacobian with respect to x and y is zero. **(True/False)**

(*ii*) If $(x, y) = 1 - x^2y^2$, then stationary point is (0, 0). **(True/False)**

(*d*) (*i*) The minimum value of $f(x, y) = x^2 + y^2$ is zero. **(True/False)**

(*ii*) If u, v are functions of r, s which are themselves function of x, y then

$$\frac{\partial(u, v)}{\partial(x, y)} = \frac{\partial(u, v)}{\partial(r, s)} \times \frac{\partial(r, s)}{\partial(x, y)}.$$ **(True/False)**

Pick the correct answer of the choices given below:

(*e*) The eigen values of $\begin{bmatrix} 1 & 1 & 1 \\ 1 & 1 & 1 \\ 1 & 1 & 1 \end{bmatrix}$ are

(*a*) 0, 0, 0 (*b*) 0, 0, 1 (*c*) 0, 0, 3 (*d*) 1, 1, 1

(*f*) The rank of matrix $\begin{bmatrix} 1 & 1 & 1 \\ 1 & -1 & 0 \\ 1 & 1 & 1 \end{bmatrix}$ is

(*a*) 0 (*b*) 1 (*c*) 2 (*d*) 3

(*g*) $\frac{\beta(m+1, n)}{\beta(m, n)}$ is equal to

(*a*) $\frac{m}{n}$ (*b*) $\frac{m+1}{n}$ (*c*) $\frac{m-1}{n}$ (*d*) $\frac{m}{m+n}$.

(*h*) The value of the integral $\int_0^\infty e^{-x^2} dx$ is

(*a*) $\frac{2}{\sqrt{\pi}}$ (*b*) $\frac{\sqrt{\pi}}{2}$ (*c*) $\frac{\pi}{2}$ (*d*) $\frac{2}{\pi}$.

Fill up the blanks with the correct answer:

(*i*) The Gauss divergence theorem relates certain surface integrals to _________.

(volume integrals/line integrals)

(*j*) The vector field $\vec{F} = xi - yj$ is divergence free _________.

(but not irrotational/and irrotational)

SECTION-B

Attempt any **three** parts of the following: **10 × 3 = 30**

2. (*a*) If $y = \sin(a \sin^{-1} x)$. Find $(y_n)_0$.

(*b*) If u, v, w are the roots of the equation

$(x - a)^3 + (x - b)^3 + (x - c)^3 = 0$, then find $\frac{\partial(u, v, w)}{\partial(a, b, c)}$.

(*c*) Find the eigen values and eigen vectors of the matrix

$$A = \begin{bmatrix} 8 & -6 & 2 \\ -6 & 7 & -4 \\ 2 & -4 & 3 \end{bmatrix}.$$

(*d*) Change the order of integration in

$I = \int_0^2 \int_{x^2/4}^{3-x} xy\, dy\, dx$ and hence evaluate it.

(*e*) Find the volume enclosed between the two surfaces $Z = 8 - x^2 - y^2$ and $Z = x^2 + 3y^2$.

SECTION-C

Attempt any **two** parts from each question. All questions are compulsory. **5 × 2 × 5 = 50**

3. (*a*) Trace the curve $y^2(a - x) = x^3$.

(*b*) If $Z = f(x + ct) + \varphi(x - ct)$ show that $\frac{\partial^2 z}{\partial t^2} = c^2 \frac{\partial^2 z}{\partial x^2}$.

(*c*) Expand e^{ax} sin in the powers of x and y as far as terms of third degree.

4. (*a*) A rectangular box, open at the top, is to have a volume of 32 cubic feet. Determine the dimensions of the box requiring least material for its construction.

(*b*) If $u_1 = \frac{x_2 x_3}{x_1}$, $u_2 = \frac{x_3\, x_1}{x_2}$ and $u_3 = \frac{x_1\, x_2}{x_3}$ find the value of $\frac{\partial(u_1, u_2, u_3)}{\partial(x_1, x_2, x_3)}$.

(*c*) Find the percentage of error in calculating the area of an ellipse $\frac{x^2}{a^2} + \frac{y^2}{b^2} = 1$, when error of +1% is made in measuring the major and minor axes.

5. (*a*) Test for consistency and solve the following system of equations

$$2x - y + 3z = 8$$
$$-x + 2y + z = 4$$
$$3x + y - 4z = 0.$$

(*b*) Reduce the following matrix to normal form and hence find its rank:

$$\begin{bmatrix} 5 & 3 & 14 & 4 \\ 0 & 1 & 2 & 1 \\ 1 & -1 & 2 & 0 \end{bmatrix}.$$

(*c*) Show that the matrix

$$\begin{bmatrix} a+ic & -b+id \\ b+id & a-ic \end{bmatrix}$$ is unitary if and only if $a^2 + b^2 + c^2 + d^2 = 1$.

6. (*a*) Prove that

$$\int_0^1 \frac{dx}{\sqrt{1+x^4}} = \frac{1}{4\sqrt{2}} \beta\left(\frac{1}{4}, \frac{1}{2}\right).$$

(*b*) Evaluate

$$\iiint x^{l-1} y^{m-1} z^{n-1} \, dx \, dy \, dz, \quad \text{where } x > 0, y > 0, z > 0$$

under the condition $\left(\frac{x}{a}\right)^p + \left(\frac{y}{b}\right)^q + \left(\frac{z}{c}\right)^t \le 1.$

(*c*) Find the area of one loop of the lemniscate $r^2 = a^2 \cos 2\theta$.

7. (*a*) Find the directional derivative of $\varphi(x, y, z) = xy^2 + yz^3$ at the point $(2, -1, 1)$ in the direction of the normal to the surface $x \log z - y^2 + 4 = 0$ at $(2, -1, 1)$.

(*b*) If all second order derivatives of φ and $\vec{v}$ are continuous, then show that

(*i*) curl (grad ϕ) $= \vec{0}$

(*ii*) div(curl $\vec{v}$) $= 0$

(*c*) Find the work done by the force $\vec{f} = (2y+3)\hat{i} + xz\,\hat{j} + (yz - x)\hat{k}$ when it moves a particle from the point (0, 0, 0) to the point (2, 1, 1) along the curve $x = 2t^2$, $y = t$ and $z = t^3$.

B.TECH.

Semester I ODD Semester Theory Examination, 2010-11

MATHEMATICS-I

TAS-104

Time : 3 Hours *Total Marks : 100*

SECTION-A

Note: (1) Attempt all questions.

(2) All questions carry equal marks.

1. Attempt any **two** parts of the following: **(10 × 2 = 20)**

(*a*) (*i*) If $N = \begin{bmatrix} 0 & 1+2 \\ -1+2i & 0 \end{bmatrix}$ is a matrix, then show that $(1 - N)(1 + N)^{-1}$ is unitary matrix.

(*ii*) Find the rank of the following matrix by reducing it into Echelon form

$$\begin{bmatrix} 2 & 3 & -1 & -1 \\ 1 & -1 & -2 & -4 \\ 3 & 1 & 3 & -2 \\ 6 & 3 & 0 & -7 \end{bmatrix}.$$

(*b*) For what values of λ, the equations

$x + y + z = 1;$ $x + 2y + 4z = \lambda,$ $x + 4y + 10z = \lambda^2$

have a solution and solve them completely in each case.

(*c*) Find the eigen values and the corresponding eigen vectors of the following matrix and hence diagonalize it

$$\begin{bmatrix} 2 & 0 & 1 \\ 0 & 3 & 0 \\ 1 & 0 & 2 \end{bmatrix}$$

2. Attempt any **two** parts of the following: **(10 × 2 = 20)**

(*a*) State Leibnitz theorem.

Find $(y_n)_0$ when $\sin\sqrt{y} = x$.

(*b*) If $u = (x^2 + y^2 + z^2)^{-1/2}$, show that:

(*i*) $x\dfrac{\partial u}{\partial x} + y\dfrac{\partial u}{\partial y} + z\dfrac{\partial u}{\partial z} = -u$

(*ii*) $\dfrac{\partial^2 u}{\partial x^2} + \dfrac{\partial^2 u}{\partial y^2} + \dfrac{\partial^2 u}{\partial z^2} = 0.$

(*c*) Expand $f(x, y) = e^x \cos y$ about $\left(1, \dfrac{\pi}{4}\right)$ in Taylor's series upto second degree terms.

3. Attempt any **two** parts of the following: **(10 × 2 = 20)**

(*a*) If $u = 2y + z$, $v = x - 2y + 3z$ and $w = 2xy - xz + 4yz - 2z^2$, show that they are not independent. Find the relation between u, v and w.

(*b*) Find the dimensions of a rectangular box of maximum capacity whose surface is given when the box is open at the top.

(*c*) (*i*) The diameter and height of a right circular cylinder are found by measurement to be 8.0 cm and 12.5 cm respectively, with possible errors of 0.05 in each measurement. Find the maximum possible approximate error in the computed volume.

(*ii*) Trace the curve $r = a \sin 3\theta$.

4. Attempt any **four** parts of the following: **(5 × 4 = 20)**

(*a*) Prove that $\int_0^\infty \sqrt{y}\, e^{-y^2} dy \times \int_a^\infty y^{-1/2} e^{-y^2}\, dy = \dfrac{\pi}{2\sqrt{2}}$.

(*b*) Change the order of integration in the double integral

$$\int_0^a \int_{\sqrt{ax-x^2}}^{\sqrt{ax}} f(x, y)\, dx\, dy .$$

(*c*) Using the transformation $x + y = u$, $y = uv$, evaluate $\iint xy(1-x-y)^{1/2}\, dx\, dy$, integration being extended over the area of the triangle bounded by the lines $x = 0$, $y = 0$ and $x + y = 1$.

(*d*) Show that the area bounded by the curve $x^n + y^n = a^n$ and the coordinate axis in the first quadrant, is

$$\frac{a^2 \{\Gamma(1/n)\}^2}{2n\,\Gamma(2/n)} .$$

(*e*) Evaluate $\iiint_S x\, dV$, where S is the region bounded by the surfaces $y = x^2$, $y = x + 2$, $4z = x^2 + y^2$ and $z = x + 3$.

(*f*) Find the mass of a plate which is formed by the coordinate planes and the plane $\dfrac{x}{a} + \dfrac{y}{b} + \dfrac{z}{c} = 1$, the variable density being $Kxyz$.

5. Attempt any **two** parts of the following: **(10 × 2 = 20)**

(*a*) If $r = |\vec{r}|$ where $\vec{r} = x\hat{i} + y\hat{j} + z\hat{k}$, prove that:

(*i*) $\nabla f(r) = f'(r)\nabla r$,

(*ii*) grad $r = \dfrac{\vec{r}}{r}$,

(*iii*) $\nabla f(r) \times \vec{r} = 0$

(*iv*) grad$\left(\dfrac{1}{r}\right) = -\dfrac{\vec{r}}{r^3}$

(*v*) $\nabla \log r = \dfrac{\vec{r}}{r^2}$

(*vi*) grad $r^n = n\, r^{n-2}\,\vec{r}$.

(*b*) State Gauss-Divergence theorem. Verify this theorem for $\vec{F} = (2x - z)\hat{i} + x^2 y\hat{j} - xz^2\hat{k}$ taken over the region bounded by $x = 0,\ x = 1,\ y = 0,\ y = 1$ and $z = 0,\ z = 1$.

(*c*) Verify Green's theorem for

$$\int_C [(2x^2 - y^2)\, dx + (x^2 + y^2)\, dy],$$

where C is the boundary of the area enclosed by the x-axis and the upper half of the circle $x^2 + y^2 = a^2$.

B.TECH.

(Sem I) ODD Semester Theory Examination 2009-10

ENGINEERING MATHEMATICS-I

EAS 103

Time : 3 Hours *Total Marks : 100*

SECTION-A

All parts of this question are compulsory **2 × 10 = 20**

1. (*a*) If $f(x) = f(0) + kf_1(0) + \frac{k^2}{L^2} f_2(\theta\, k),\ 0 < \theta < 1$ then the value of θ when $k = 1$ and $f(x) = (1 - x)^{5/2}$ is given as _____.

(*b*) The shortest distance from the point $(1, 2, -1)$ to the sphere $x^2 + y^2 + z^2 = 24$ shall be _____.

(*c*) The Jacobian $J\left(\frac{u, v}{x, y}\right)$ for $u = e^x \sin y,\ v = x \log \sin y$ shall be _____.

(*d*) For the curve $ay^2 = x^2(a - x)$, which of the following statement(s) is/are Incorrect?

(*i*) Curve passes through origin

(*ii*) Curve is symmetrical about y axis

(*iii*) Curve has two branches

(*iv*) Curve has no tangents at origin

(*e*) If P and Q are non-singular matrices, then for Matrix 'M', which of the following statements are correct ?

(*i*) Rank $(PMQ) >$ Rank M

(*ii*) Rank $(PMQ) =$ Rank M

(*iii*) Rank $(PMQ) <$ Rank M

(*iv*) Rank $(PMQ) =$ Rank $M +$ Rank (PQ)

(*f*) If λ is an eigen value of the matrix 'M' then for the matrix $(M - \lambda I)$, which of the following statement(s) is/are correct?

(*i*) Skew symmetric

(*ii*) Non singular

(*iii*) Singular

(*iv*) None of these.

Indicate True/False for the following statements:

(*g*) For $\int_0^{\infty}\int_x^{\infty} f(x\,y)\,dx\,dy$, the change of order of integration makes it to

(*i*) $\int_0^{\infty}\int_x^{\infty} f(x\,y)\,dx\,dy$ True/False

(*ii*) $\int_x^{\infty}\int_0^{\infty} f(x\,y)\,dx\,dy$ True/False

(*iii*) $\int_0^{\infty}\int_0^{y} f(x\,y)\,dx\,dy$ True/False

(*iv*) $\int_0^{\infty}\int_0^{x} f(x\,y)\,dx\,dy$ True/False

(*h*) The value of $\left|\overline{-\frac{1}{2}}\right.$ is given by

(*i*) $\sqrt{\pi}$ True/False

(*ii*) $2\sqrt{\pi}$ True/False

(*iii*) $-\sqrt{\pi}$ True/False

(*iv*) $-2\sqrt{\pi}$ True/False

Pick up the correct option from the following:

(*i*) If $\vec{F}$ is the velocity of a fluid particle then $\int_C \vec{F}\cdot d\bar{r}$ represents

(*i*) Work done

(*ii*) Circulation

(*iii*) Flux

(*iv*) Conservative field.

(*j*) The value of $\iint_S \vec{F}\cdot\hat{n}\,ds$ where

$\vec{F} = ax\hat{i} + by\,\hat{j} + cz\hat{k}$; a, b, c being constants, is given by

(*i*) $\frac{\pi}{3}(a+b+c)$

(*ii*) $\frac{4\pi}{3}(a+b+c)$

(*iii*) $2\pi(a+b+c)$

(*iv*) $\pi(a+b+c)$

SECTION-B

Attempt any three parts of the following: **10 × 3 = 30**

1. (*a*) Determine the values of '*a*' and '*b*' for which the following system of equations

$3x + 5y - az = 7,$

$x - by + 4z = -3,$

$ax + 4y - 5z = 4$

has (*i*) No solution

(*ii*) A unique solution

(*iii*) Infinite no. of solutions.

(*b*) Find the value of $D^n\{x^{n-1} \log x\}$, where $D^n \equiv \dfrac{d^n}{dx^n}$.

(*c*) If $u = \dfrac{(x+y)}{z}$, $v = \dfrac{(y+z)}{x}$, $w = \dfrac{y(x+y+z)}{(xz)}$, then show that u, v, w are not independent and find the relation between them.

(*d*) A rigid body is rotating with constant angular velocity $\vec{w}$ about a fixed axis. If $\vec{v}$ is the linear velocity of any point of the body then prove that curl $\vec{v} = 2\vec{w}$.

(*e*) Assuming $\Gamma n \Gamma 1-n$ = p cosec *n*p, $0 < n < 1$, show that $\int_0^\infty \dfrac{x^{p-1}}{1+x} dx = \left(\dfrac{\pi}{\sin p\pi}\right)$; $0 < p < 1$.

SECTION-C

All questions of this section are compulsory. Attempt any two parts from each question. **10 × 5 = 50**

1. (*a*) If $x = \sin\left(\dfrac{\log y}{a}\right)$, then evaluate the value $(1 - x^2)y_{n+2} - (2n + 1)x\, y_{n+1} - (n^2 + a^2)y_n = 0$ with usual symbols.

(*b*) If $u = \cos^{-1}\left(\dfrac{x+y}{\sqrt{x}+\sqrt{y}}\right)$, show that

$$x\frac{\partial u}{\partial x} + y\frac{\partial u}{\partial y} + \frac{1}{2}\cot u = 0$$

(*c*) Verify Euler's theorem for $z = \dfrac{x^{1/3} + y^{1/3}}{x^{1/2} + y^{1/2}}$.

4. (*a*) In a plane ΔABC, find the maximum value of $\cos A \cos B \cos C$.

(*b*) If $x = e^v \sec u$, $y = e^v \tan u$ then evaluate $\dfrac{\partial(x, y)}{\partial(u, v)}$.

(*c*) The power '*P*' required to propel a steamer of length '*l*' at a speed '*u*' is given by $P = \lambda u^3 l^3$ where λ is constant. If *u* is increased by 3% and *l* is decreased by 1%, find the corresponding increase in '*P*'.

5. (*a*) Show that row vectors of the matrix $\begin{bmatrix} 1 & 2 & -2 \\ -1 & 3 & 0 \\ 0 & -2 & 1 \end{bmatrix}$ are linearly independent.

(*b*) Find the rank of the following matrix using the elementary transformations $\begin{bmatrix} 1 & -3 & 1 & 2 \\ 0 & 1 & 2 & 3 \\ 3 & 4 & 1 & -2 \end{bmatrix}$.

(*c*) Express the matrix $A = \begin{bmatrix} i & 2-3i & 4+5i \\ 6+i & 0 & 4-5i \\ -i & 2-i & 2+i \end{bmatrix}$ as a sum of Hermitian and Skew Hermitian matrix.

6. Interpret the physical meaning of $curl\,\vec{F}$ and $div\,\vec{F}$.

(*b*) Verify the divergence theorem for the function

$\vec{F} = 4xz\,i - y^2\,j + yzk$; taken over the cube bounded by planes $x = 0$; $x = 1$; $y = 0, y = 1$; $z = 0, z = 1$.

(*c*) If a vector field is given by

$\vec{F} = (x^2 - y^2 + x)\,i - (2xy + y)\,j$. Is this field irrotational? If so, find its scalar potential.

7. (*a*) Evaluate $\iint_R \left(1 - \frac{x^2}{a^2} + \frac{y^2}{b^2}\right) dx\,dy$ over the first quadrant of the ellipse $\frac{x^2}{a^2} + \frac{y^2}{b^2} = 1$.

(*b*) Find the mass of the region bounded by ellipsoid $\frac{x^2}{a^2} + \frac{y^2}{b^2} + \frac{z^2}{c^2} = 1$; if the density varies as the square of the distance from the centre.

(*c*) A triangular prism is formed by the planes whose equations are $ay = bx$, $y = 0$ and $x = a$. Find the volume of this prism between the plane $z = 0$ and the surface $z = c + xy$.

B.TECH.

Semester I Examination, 2008-09

MATHEMATICS-I

TAS-104

Time : 3 Hours *Total Marks : 100*

SECTION-A

All parts of this question are compulsory **2 × 10 = 20**

1. Attempt any two parts of the following:

(*a*) Find all values of μ for which rank of the matrix

$$A = \begin{bmatrix} \mu & -1 & 0 & 0 \\ 0 & \mu & -1 & 0 \\ 0 & 0 & \mu & -1 \\ -6 & 11 & -6 & 1 \end{bmatrix}$$

is equal to 3.

(*b*) If $A = \begin{bmatrix} 1 & 0 & 0 \\ 1 & 0 & 1 \\ 0 & 1 & 0 \end{bmatrix}$, then show that

$A^n = A^{n-2} + A^2 - I$ for $n \geq 3$.

(*c*) Show that the matrix $A = \begin{bmatrix} 3 & 1 & -1 \\ -2 & 1 & 2 \\ 0 & 1 & 2 \end{bmatrix}$ is diagonalizable. Hence, find P such that $P^{-1}AP$ is a diagonal matrix.

2. Attempt any two parts of the following:

(*a*) Find $(y_n)_0$ when $y = \sin (a \sin^{-1} x)$.

(*b*) If $u = e^{xyz}$, show that

$$\frac{\partial^3 u}{\partial x \partial y \partial z} = (1 + 3xyz + x^2y^2z^2)\, e^{xyz}.$$

(*c*) Trace the curve $y^2(a + x) = x^2(3a - x)$.

3. Attempt any two parts of the following:

(*a*) Show that the functions $u = x^2 + y^2 + z^2$, $v = x + y + z$, $w = yz + zx + xy$ are not independent of oneanother.

(*b*) The height h and the semi-vertical angle α of a cone are measured, and from them A, the total surface area of the cone, including the base, is calculated. If h and α are in error by small quantities δh and $\delta\alpha$ respectively, find corresponding error in the area. Show further that, if $\alpha = \frac{\pi}{6}$, an error of +1 per cent in h will be approximately compensated by an error of $-0^\circ.33$ in α.

(*c*) Determine the points where the function $x^3 + y^3 - 3axy$ has a maximum or minimum.

(*d*) Find the point upon the plane $ax + by + cz = p$ at which the function $f = x^2 + y^2 + z^2$ has a minimum value and find this minimum f.

4. Attempt any two parts of the following:

(*a*) Evaluate

$$\iint_R xy\,dx\,dy$$

where R is the quadrant of the circle $x^2 + y^2 = a^2$ where $x \geq 0$ and $y \geq 0$.

(*b*) Find the volume common to the cylinders

$x^2 + y^2 = a^2$ and $x^2 + z^2 = a^2$.

(*c*) Evaluate

$$\iiint_R (x - 2y + z)\,dx\,dy\,dz$$

where $R: 0 \leq x \leq 1,\ 0 \leq y \leq x^2,\ 0 \leq z \leq x + y$.

5. Attempt any two parts of the following:

(*a*) Find the directional derivative of $f(x, y, z) = 2x^2 + 3y^2 + z^2$ at the point $P(2, 1, 3)$ in the direction of the vector $\vec{a} = \hat{i} - 2\hat{k}$.

(*b*) Show that

$\iint_S \vec{F} \cdot \hat{n}\,dS = \frac{3}{2}$, where $\vec{F} = 4xzi - y^2 j + yzk$ and S is the surface of the cube bounded by the planes $x = 0,\ x = 1,\ y = 0,\ y = 1,\ z = 0,\ z = 1$.

(*c*) Use the Stoke's theorem to evaluate

$$\int_C [(x + 2y)\,dx + (x - z)\,dy + (y - z)dz]$$

where C is the boundary of the triangle with vertices (2, 0, 0), (0, 3, 0) and (0, 0, 6) oriented in the anti-clockwise direction.

B.TECH.

(Only for the candidates admitted/Readmitted in the Session 2008-09)

Semester I Examination, 2008-09

ELEMENTS OF ENGINEERING MATHEMATICS-I

EAS-103

Time : 3 Hours *Total Marks : 100*

SECTION-A

All parts of this question are compulsory. **2 × 10 = 20**

1. (*a*) For which value of '*b*' the rank of the matrix

$$A = \begin{bmatrix} 1 & 5 & 4 \\ 0 & 3 & 2 \\ b & 13 & 10 \end{bmatrix} \text{ is } 2,\ b = \text{\_\_\_\_\_\_}.$$

(*b*) Determine the constants a and b such that the curl of vector

$\overline{A} = (2xy + 3yz)i + (x^2 + axz - 4z^2)\,j + (3xy + byz)k$ is zero, $a =$ ______, $b =$ ______.

(*c*) The nth derivative (y_n) of the function $y = x^2 \sin x$ at $x = 0$ is ______.

(*d*) With usual notations, match the items on right hand side with those on left hand side for properties of Max^{m} and minimum.

(*i*) Max^{m}	(*p*) $rt - s^2 = 0$
(*ii*) Min^{m}	(*q*) $rt - s^2 < 0$
(*iii*) Saddle point	(*r*) $rt - s^2 > 0,\ r > 0$
(*iv*) Failure case	(*s*) $rt - s^2 > 0$ and $r < 0$

(*e*) Match the items on the right hand side with those of left hand side for the following special functions: (Full marks is awarded if all matchings are correct)

(*i*) $\beta(p, q)$	(*p*) $\Gamma\left(\frac{1}{2}\right)$
(*ii*) $\dfrac{\Gamma p\,\Gamma q}{\Gamma(p+q)}$	(*q*) $\displaystyle\int_0^\infty \frac{y^{p-1}}{(1+y)(p+q)}\,dy$
(*iii*) $\sqrt{\pi}$	(*r*) $\beta(p, q)$
(*iv*) $\dfrac{\pi}{\sin p\pi}$	(*s*) $\Gamma p\,\Gamma(1-p)$

Indicate True or False for the following statements:

(*f*) (*i*) If $|A| = 0$, then at least one eigen value is zero. (True/False)

(*ii*) A^{-1} exists iff 0 is an eigen value of A. (True/False)

(*iii*) If $|A| \neq 0$, then A is known as singular matrix. (True/False)

(*iv*) Two vectors X and Y is said to be orthogonal Y, $X^T Y = Y^T X \neq 0$. (True/False)

(*g*) (*i*) The curve $y^2 = 4ax$ is symmetric about x-axis. (True/False)

(*ii*) The curve $x^3 + y^3 = 3axy$ is symmetric about the line $y = -x$. (True/False)

(*iii*) The curve $x^2 + y^2 = a^2$ is symmetric about both the axis x and y. (True/False)

(*iv*) The curve $x^3 - y^3 = 3axy$ is symmetric about the line $y = x$. (True/False)

Pick the correct answer of the choices given below:

(*h*) If $\overline{r} = xi + yj + zk$ is position vector, then value of $\nabla(\log r)$ is

(*i*) $\dfrac{\overline{r}}{r}$ (*ii*) $\dfrac{\overline{r}}{r^2}$ (*iii*) $-\dfrac{\overline{r}}{r^3}$ (*iv*) none of the above.

(*i*) The Jacobian $\dfrac{\partial(u, v)}{\partial(x, y)}$ for the function $u = e^x \sin y$, $v = (x + \log \sin y)$ is

(*i*) 1 (*ii*) $\sin x \sin y - xy \cos x \cos y$ (*iii*) 0 (*iv*) $\dfrac{e^x}{x}$.

(*j*) The volume of the solid under the surface $az = x^2 + y^2$ and whose base R is the circle $x^2 + y^2 = a^2$ is given as

(*i*) $\dfrac{\pi}{2a}$ (*ii*) $\dfrac{\pi a^3}{2}$ (*iii*) $\dfrac{4}{3}\pi a^3$ (*iv*) None of the above.

SECTION-B

Attempt any three parts of the following: **10 × 3 = 30**

2. (*a*) If $y = (\sin^{-1} x)^2$ prove that $y_n(0) = 0$ for n odd and $y_n(0) = 2 \cdot 2^2 \cdot 4^2 \cdot 6^2 \ldots (n-2)^2$, $n \neq 2$ for n is even.

(*b*) Find the dimension of rectangular box of maximum capacity whose surface area is given when (a) box is open at the top (b) box is closed.

(*c*) Find a matrix P which diagonalizes the matrix $A = \begin{bmatrix} 4 & 1 \\ 2 & 3 \end{bmatrix}$, verify $P^{-1}AP = D$ where D is the diagonal matrix.

(*d*) Find the area and the mass contained in the first quadrant enclosed by the curve $\left(\dfrac{x}{a}\right)^{\alpha} + \left(\dfrac{y}{b}\right)^{\beta} = 1$ where $\alpha > 0$, $\beta > 0$ given that density at any point $P(x, y)$ is $k\sqrt{xy}$.

(*e*) Using the divergence theorem, evaluate the surface integral $\iint_S (yz\, dy\, dz + sz\, dz\, dx + xy\, dy\, dx)$ where $S : x^2 + y^2 + z^2 = 4$.

SECTION-C

Attempt any two parts from each question. All questions are compulsory.

3. (*a*) Trace the curve $r^2 = a^2 \cos 2\theta$.

(*b*) If $u = \log\left(\dfrac{(x^2+y^2)}{(x+y}\right)$, prove that $x\dfrac{\partial u}{\partial x} + y\dfrac{\partial u}{\partial y} = 1$.

(*c*) If $V = f(2x - 3y, 3y - 4z, 4z - 2x)$, compute the value of $6V_x + 4V_y + 3V_z$.

4. (*a*) The temperature 'T' at any point (x, y, z) in space is $T(xyz) = K\,xyz^2$ where K is constant. Find the highest temperature on the surface of the sphere $x^2 + y^2 + z^2 = a^2$.

(*b*) Verify the chain rule for Jacobians if $x = u$, $y = u \tan v$, $z = w$.

(*c*) The time 'T' of a complete oscillation of a simple pendulum of length 'L' is governed by the equation $T = 2\pi\sqrt{\dfrac{L}{g}}$, g is constant, find the approximate error in the calculated value of T corresponding to an error of 2% in the value of L.

5. (*a*) Determine 'b' such that the system of homogeneous equation $2x + y + 2z = 0$; $x + y + 3z = 0$; $4x + 3y + bz = 0$ has (*i*) Trivial solution (*ii*) Non-Trivial solution. Find the Non-Trivial solution using matrix method.

(*b*) Verify Cayley-Hamilton theorem for the matrix $A = \begin{pmatrix} -1 & 2 \\ 2 & -1 \end{pmatrix}$.

(*c*) Find the eigen value and corresponding eigen vectors of the matrix $I = \begin{pmatrix} -5 & 2 \\ 2 & -2 \end{pmatrix}$.

6. (*a*) Find the directional derivative of $\nabla \cdot (\nabla f)$ at the point $(1, -2, 1)$ in the direction of the normal to the surface $xy^2z = 3x + z^2$ where $f = 2x^3y^2z^4$.

(*b*) Find the area of the region in the first quadrant bounded by the curves

$$y = x,\ y = \frac{1}{x},\ y = \frac{x}{4}.$$

(*c*) Prove that $(y^2 - z^2 + 3yz)i + (3xz + 2xy)j + (3xy + 2xz + 2)k$ is both solenoidal and irrotational.

7. (*a*) Changing the order of integration of $\int_0^\infty \int_0^\infty e^{-xy} \sin nx \, dx \, dy$,

show that $\displaystyle\int_0^\infty \left(\frac{\sin nx}{x}\right) dx = \frac{\pi}{2}$.

(*b*) Determine the area bounded by the curves $xy = 2$, $4y = x^2$ and $y = 4$.

(*c*) For a β function, show that $\beta(p, q) = \beta(p + 1, q) + \beta(p, q + 1)$.

Index

L

M

N

P

R

S

T

U

❑❑❑

Win Prizes!

ATTENTION READERS

We request you for your frank assessment regarding some of the aspects of the book given as under:

Elements of Engineering Mathematics–I

By

Dr. V.P. Mishra

Dr. Jyoti Sinha

Please fill up the given space in neat capital letters.

(i) What topics of your syllabus that are important from you examination point of view are not covered in the book ?

....

....

....

....

(ii) What are the chapters wherein the treatment of the subject matter is not systematic or organised or updated ?

....

....

....

....

(iii) Have you come across misprints/mistakes/factual inaccuracies in the book? Please specify the chapters and the page numbers.

....

....

....

....

(iv) What other books on the same subject have you found/heard better that the present book? Please specify in terms of price and quality.

....

....

....

....

(v) Further suggestions and comments for the improvement of the book:

....

....

....

....

(PS: If need be please attach a separate sheet to write your views.)

Other Details :

(i) Who recommended you the book ? (Please tick in the box near the option relevant to you.)

☐ Teachers ☐ Friends ☐ Bookseller

(ii) Name of the recommending teacher, his designation and address.

....
....
....
....

(iii) Name and address of the bookseller you purchased the book form:

....
....
....
....

(iv) Name and address of your institution (Please mention the University or Board, as the case may be)

....
....
....
....

(v) Your name and complete postal address:

....
....
....
....

(vi) Course you are enrolled in:

....
....
....
....

The best assessment will be awarded each month. The award will be in the form of our publications as decided by the Editorial Board.

Please mail the filled up coupon at your earliest to:

S.K. Kataria & Sons®

4885/109, Prakash Mahal, Dr. Subhash Bhargav Lane,
Opposite Delhi Medical Association, Daryaganj, New Delhi-110002 (INDIA)
Phone: +91-11-23243489, 23269324, 43551243 Telefax: +91-11-23243489
e-mail: katariabooks@yahoo.com; katariabook@gmail.com
Website: www.skkatariaandsons.com; *For e-books:* www.katsonbooks.com